AF619256

ALLE ZEIT WACH
1842

Ivan Kolář
Peter W. Michor
Jan Slovák

Natural Operations in Differential Geometry

Springer-Verlag Berlin Heidelberg GmbH

Ivan Kolář and Jan Slovák
Department of Algebra and Geometry
Faculty of Science
Masaryk University
Janáčkovo nám 2a
CS-662 95 Brno, Czechoslovakia

Peter W. Michor
Institut für Mathematik
der Universität Wien
Strudlhofgasse 4
A-1090 Wien, Austria

Mathematics Subject Classification (1991): 53-02, 53-01, 58-02, 58-01, 53A55, 53C05, 58A20

DOI 10.1007/978-3-662-02950-3

Library of Congress Cataloging-in-Publication Data available.

Originally published by Springer-Verlag Berlin Heidelberg New York in 1993
MyCopy version of the original edition 1993

Typesetting: Camera ready by the authors using Springer TEX macropackage
41/3140 - 5 4 3 2 1 0 - Printed on acid-free paper
www.springer.com/mycopy

TABLE OF CONTENTS

PREFACE

The aim of this work is threefold:

First it should be a monographical work on natural bundles and natural operators in differential geometry. This is a field which every differential geometer has met several times, but which is not treated in detail in one place. Let us explain a little, what we mean by naturality.

Exterior derivative commutes with the pullback of differential forms. In the background of this statement are the following general concepts. The vector bundle $\Lambda^k T^*M$ is in fact the value of a functor, which associates a bundle over M to each manifold M and a vector bundle homomorphism over f to each local diffeomorphism f between manifolds of the same dimension. This is a simple example of the concept of a natural bundle. The fact that exterior derivative d transforms sections of $\Lambda^k T^*M$ into sections of $\Lambda^{k+1}T^*M$ for every manifold M can be expressed by saying that d is an operator from $\Lambda^k T^*M$ into $\Lambda^{k+1}T^*M$. That exterior derivative d commutes with local diffeomorphisms now means, that d is a natural operator from the functor $\Lambda^k T^*$ into functor $\Lambda^{k+1}T^*$. If $k > 0$, one can show that d is the unique natural operator between these two natural bundles up to a constant. So even linearity is a consequence of naturality. This result is archetypical for the field we are discussing here. A systematic treatment of naturality in differential geometry requires to describe all natural bundles, and this is also one of the undertakings of this book.

Second this book tries to be a rather comprehensive textbook on all basic structures from the theory of jets which appears in different branches of differential geometry. Even though Ehresmann in his original papers from 1951 underlined the conceptual meaning of the notion of an r-jet for differential geometry, jets have been mostly used as a purely technical tool in certain problems in the theory of systems of partial differential equations, in singularity theory, in variational calculus and in higher order mechanics. But the theory of natural bundles and natural operators clarifies once again that jets are one of the fundamental concepts in differential geometry, so that a thorough treatment of their basic properties plays an important role in this book. We also demonstrate that the central concepts from the theory of connections can very conveniently be formulated in terms of jets, and that this formulation gives a very clear and geometric picture of their properties.

This book also intends to serve as a self-contained introduction to the theory of Weil bundles. These were introduced under the name 'les espaces des points proches' by A. Weil in 1953 and the interest in them has been renewed by the recent description of all product preserving functors on manifolds in terms of products of Weil bundles. And it seems that this technique can lead to further interesting results as well.

Third in the beginning of this book we try to give an introduction to the fundamentals of differential geometry (manifolds, flows, Lie groups, differential forms, bundles and connections) which stresses naturality and functoriality from the beginning and is as coordinate free as possible. Here we present the Frölicher-Nijenhuis bracket (a natural extension of the Lie bracket from vector fields to vector valued differential forms) as one of the basic structures of differential geometry, and we base nearly all treatment of curvature and Bianchi identities on it. This allows us to present the concept of a connection first on general fiber bundles (without structure group), with curvature, parallel transport and Bianchi identity, and only then add G-equivariance as a further property for principal fiber bundles. We think, that in this way the underlying geometric ideas are more easily understood by the novice than in the traditional approach, where too much structure at the same time is rather confusing. This approach was tested in lecture courses in Brno and Vienna with success.

A specific feature of the book is that the authors are interested in general points of view to different structures in differential geometry. The modern development of the global differential geometry clarified that differential geometric objects form fiber bundles over manifolds as a rule. Nijenhuis revisited the classical theory of geometric objects from this point of view. Each type of geometric objects can be interpreted as a rule F transforming every m-dimensional manifold M into a fibered manifold $FM \to M$ over M and every local diffeomorphism $f : M \to N$ into a fibered manifold morphism $Ff : FM \to FN$ over f. The geometric character of F is then expressed by the functoriality condition $F(g \circ f) = Fg \circ Ff$. Hence the classical bundles of geometric objects are now studied in the form of the so called lifting functors or (which is the same) natural bundles on the category $\mathcal{M}f_m$ of all m-dimensional manifolds and their local diffeomorphisms. An important result by Palais and Terng, completed by Epstein and Thurston, reads that every lifting functor has finite order. This gives a full description of all natural bundles as the fiber bundles associated with the r-th order frame bundles, which is useful in many problems. However in several cases it is not sufficient to study the bundle functors defined on the category $\mathcal{M}f_m$. For example, if we have a Lie group G, its multiplication is a smooth map $\mu : G \times G \to G$. To construct an induced map $F\mu : F(G \times G) \to FG$, we need a functor F defined on the whole category $\mathcal{M}f$ of all manifolds and all smooth maps. In particular, if F preserves products, then it is easy to see that $F\mu$ endows FG with the structure of a Lie group. A fundamental result in the theory of the bundle functors on $\mathcal{M}f$ is the complete description of all product preserving functors in terms of the Weil bundles. This was deduced by Kainz and Michor, and independently by Eck and Luciano, and it is presented in chapter VIII of this book. At several other places we then compare and contrast the properties of the product preserving bundle functors and the non-product-preserving ones, which leads us to interesting geometric results. Further, some functors of modern differential geometry are defined on the category of fibered manifolds and their local isomorphisms, the general connections being the simplest example. Last but not least we remark that Eck has recently introduced the general concepts of gauge natural bundles and gauge natural operators. Taking into account the present role of the gauge theories in theoretical physics and mathematics, we devote the last chapter of the book to this subject.

If we interpret geometric objects as bundle functors defined on a suitable category over manifolds, then some geometric constructions have the role of natural transformations. Several other ones represent natural operators, i.e. they map sections of certain fiber bundles to sections of other ones and commute with the action of local isomorphisms. Hence geometric means natural in such situations. That is why we develop a rather general theory of bundle functors and natural operators in this book. The principal advantage of interpreting geometric as natural is that we obtain a well-defined concept. Then we can pose, and sometimes even solve, the problem of determining all natural operators of a prescribed type. This gives us the complete list of all possible geometric constructions of the type in question. In some cases we even discover new geometric operators in this way.

Our practical experience taught us that the most effective way how to treat natural operators is to reduce the question to a finite order problem, in which the corresponding jet spaces are finite dimensional. Since the finite order natural operators are in a simple bijection with the equivariant maps between the corresponding standard fibers, we can apply then several powerful tools from classical algebra and analysis, which can lead rather quickly to a complete solution of the problem. Such a passing to a finite order situation has been of great profit in other branches of mathematics as well. Historically, the starting point for the reduction to the jet spaces is the famous Peetre theorem saying that every linear support non-increasing operator has locally finite order. We develop an essential generalization of this technique and we present a unified approach to the finite order results for both natural bundles and natural operators in chapter V.

The primary purpose of chapter VI is to explain some general procedures, which can help us in finding all the equivariant maps, i.e. all natural operators of a given type. Nevertheless, the greater part of the geometric results is original. Chapter VII is devoted to some further examples and applications, including Gilkey's theorem that all differential forms depending naturally on Riemannian metrics and satisfying certain homogeneity conditions are in fact Pontryagin forms. This is essential in the recent heat kernel proofs of Atiyah Singer Index theorem. We also characterize the Chern forms as the only natural forms on linear symmetric connections. In a special section we comment on the results of Kirillov and his colleagues who investigated multilinear natural operators with the help of representation theory of infinite dimensional Lie algebras of vector fields. In chapter X we study systematically the natural operators on vector fields and connections. Chapter XI is devoted to a general theory of Lie derivatives, in which the geometric approach clarifies, among others, the relations to natural operators.

The material for chapters VI, X and sections 12, 30–32, 47, 49, 50, 52–54 was prepared by the first author (I.K.), for chapters I, II, III, VIII by the second author (P.M.) and for chapters V, IX and sections 13–17, 33, 34, 48, 51 by the third author (J.S.). The authors acknowledge A. Cap, M. Doupovec, and J. Janyška, for reading the manuscript and for several critical remarks and comments and A. A. Kirillov for commenting section 34.

The joint work of the authors on the book has originated in the seminar of the first two authors and has been based on the common cultural heritage of Middle Europe. The authors will be pleased if the reader realizes a reflection of those traditions in the book.

CHAPTER I.
MANIFOLDS AND LIE GROUPS

In this chapter we present an introduction to the basic structures of differential geometry which stresses global structures and categorical thinking. The material presented is standard - but some parts are not so easily found in text books: we treat initial submanifolds and the Frobenius theorem for distributions of non constant rank, and we give a very quick proof of the Campbell - Baker - Hausdorff formula for Lie groups. We also prove that closed subgroups of Lie groups are Lie subgroups.

1. Differentiable manifolds

1.1. A *topological manifold* is a separable Hausdorff space M which is locally homeomorphic to $\mathbb{R}^n$. So for any $x \in M$ there is some homeomorphism $u : U \to u(U) \subseteq \mathbb{R}^n$, where U is an open neighborhood of x in M and $u(U)$ is an open subset in $\mathbb{R}^n$. The pair (U, u) is called a *chart* on M.

From topology it follows that the number n is locally constant on M; if n is constant, M is sometimes called a *pure manifold*. We will only consider pure manifolds and consequently we will omit the prefix pure.

A family $(U_\alpha, u_\alpha)_{\alpha \in A}$ of charts on M such that the U_α form a cover of M is called an *atlas*. The mappings $u_{\alpha\beta} := u_\alpha \circ u_\beta^{-1} : u_\beta(U_{\alpha\beta}) \to u_\alpha(U_{\alpha\beta})$ are called the chart changings for the atlas (U_α), where $U_{\alpha\beta} := U_\alpha \cap U_\beta$.

An atlas $(U_\alpha, u_\alpha)_{\alpha \in A}$ for a manifold M is said to be a C^k*-atlas*, if all chart changings $u_{\alpha\beta} : u_\beta(U_{\alpha\beta}) \to u_\alpha(U_{\alpha\beta})$ are differentiable of class C^k. Two C^k-atlases are called C^k*-equivalent*, if their union is again a C^k-atlas for M. An equivalence class of C^k-atlases is called a C^k*-structure* on M. From differential topology we know that if M has a C^1-structure, then it also has a C^1-equivalent C^∞-structure and even a C^1-equivalent C^ω-structure, where C^ω is shorthand for real analytic. By a C^k-manifold M we mean a topological manifold together with a C^k-structure and a chart on M will be a chart belonging to some atlas of the C^k-structure.

But there are topological manifolds which do not admit differentiable structures. For example, every 4-dimensional manifold is smooth off some point, but there are such which are not smooth, see [Quinn, 82], [Freedman, 82]. There are also topological manifolds which admit several inequivalent smooth structures. The spheres from dimension 7 on have finitely many, see [Milnor, 56]. But the most surprising result is that on $\mathbb{R}^4$ there are uncountably many pairwise inequivalent (exotic) differentiable structures. This follows from the results of [Donaldson, 83] and [Freedman, 82], see [Gompf, 83] or [Freedman-Feng Luo, 89] for an overview.

Note that for a Hausdorff C^∞-manifold in a more general sense the following properties are equivalent:

(1) It is paracompact.
(2) It is metrizable.
(3) It admits a Riemannian metric.
(4) Each connected component is separable.

In this book a manifold will usually mean a C^∞-manifold, and smooth is used synonymously for C^∞, it will be Hausdorff, separable, finite dimensional, to state it precisely.

Note finally that any manifold M admits a finite atlas consisting of $\dim M + 1$ (not connected) charts. This is a consequence of topological dimension theory [Nagata, 65], a proof for manifolds may be found in [Greub-Halperin-Vanstone, Vol. I, 72].

1.2. A mapping $f : M \to N$ between manifolds is said to be C^k if for each $x \in M$ and each chart (V, v) on N with $f(x) \in V$ there is a chart (U, u) on M with $x \in U$, $f(U) \subseteq V$, and $v \circ f \circ u^{-1}$ is C^k. We will denote by $C^k(M, N)$ the space of all C^k-mappings from M to N.

A C^k-mapping $f : M \to N$ is called a *C^k-diffeomorphism* if $f^{-1} : N \to M$ exists and is also C^k. Two manifolds are called *diffeomorphic* if there exists a diffeomorphism between them. From differential topology we know that if there is a C^1-diffeomorphism between M and N, then there is also a C^∞-diffeomorphism. All smooth manifolds together with the C^∞-mappings form a category, which will be denoted by $\mathcal{M}f$. One can admit non pure manifolds even in $\mathcal{M}f$, but we will not stress this point of view.

A mapping $f : M \to N$ between manifolds of the same dimension is called a *local diffeomorphism*, if each $x \in M$ has an open neighborhood U such that $f|U : U \to f(U) \subset N$ is a diffeomorphism. Note that a local diffeomorphism need not be surjective or injective.

1.3. The set of smooth real valued functions on a manifold M will be denoted by $C^\infty(M, \mathbb{R})$, in order to distinguish it clearly from spaces of sections which will appear later. $C^\infty(M, \mathbb{R})$ is a real commutative algebra.

The *support* of a smooth function f is the closure of the set, where it does not vanish, $supp(f) = \overline{\{x \in M : f(x) \neq 0\}}$. The *zero set* of f is the set where f vanishes, $Z(f) = \{x \in M : f(x) = 0\}$.

Any manifold admits *smooth partitions of unity*: Let $(U_\alpha)_{\alpha \in A}$ be an open cover of M. Then there is a family $(\varphi_\alpha)_{\alpha \in A}$ of smooth functions on M, such that $supp(\varphi_\alpha) \subset U_\alpha$, $(supp(\varphi_\alpha))$ is a locally finite family, and $\sum_\alpha \varphi_\alpha = 1$ (locally this is a finite sum).

1.4. Germs. Let M and N be manifolds and $x \in M$. We consider all smooth mappings $f : U_f \to N$, where U_f is some open neighborhood of x in M, and we put $f \underset{x}{\sim} g$ if there is some open neighborhood V of x with $f|V = g|V$. This is an equivalence relation on the set of mappings considered. The equivalence class of a mapping f is called the *germ of f at x*, sometimes denoted by $\text{germ}_x f$. The space of all germs at x of mappings $M \to N$ will be denoted by $C_x^\infty(M,N)$. This construction works also for other types of mappings like real analytic or holomorphic ones, if M and N have real analytic or complex structures.

If $N = \mathbb{R}$ we may add and multiply germs, so we get the real commutative algebra $C_x^\infty(M,\mathbb{R})$ of germs of smooth functions at x.

Using smooth partitions of unity (see 1.3) it is easily seen that each germ of a smooth function has a representative which is defined on the whole of M. For germs of real analytic or holomorphic functions this is not true. So $C_x^\infty(M,\mathbb{R})$ is the quotient of the algebra $C^\infty(M,\mathbb{R})$ by the ideal of all smooth functions $f : M \to \mathbb{R}$ which vanish on some neighborhood (depending on f) of x.

1.5. The tangent space of $\mathbb{R}^n$. Let $a \in \mathbb{R}^n$. A *tangent vector* with foot point a is simply a pair (a, X) with $X \in \mathbb{R}^n$, also denoted by X_a. It induces a *derivation* $X_a : C^\infty(\mathbb{R}^n,\mathbb{R}) \to \mathbb{R}$ by $X_a(f) = df(a)(X_a)$. The value depends only on the germ of f at a and we have $X_a(f \cdot g) = X_a(f) \cdot g(a) + f(a) \cdot X_a(g)$ (the derivation property).

If conversely $D : C^\infty(\mathbb{R}^n,\mathbb{R}) \to \mathbb{R}$ is linear and satisfies $D(f \cdot g) = D(f) \cdot g(a) + f(a) \cdot D(g)$ (a derivation at a), then D is given by the action of a tangent vector with foot point a. This can be seen as follows. For $f \in C^\infty(\mathbb{R}^n,\mathbb{R})$ we have

$$\begin{aligned} f(x) &= f(a) + \int_0^1 \tfrac{d}{dt} f(a + t(x-a))dt \\ &= f(a) + \sum_{i=1}^n \int_0^1 \tfrac{\partial f}{\partial x^i}(a + t(x-a))dt\,(x^i - a^i) \\ &= f(a) + \sum_{i=1}^n h_i(x)(x^i - a^i). \end{aligned}$$

$D(1) = D(1 \cdot 1) = 2D(1)$, so $D(\text{constant}) = 0$. Thus

$$\begin{aligned} D(f) &= D(f(a) + \sum_{i=1}^n h_i(x)(x^i - a^i)) \\ &= 0 + \sum_{i=1}^n D(h_i)(a^i - a^i) + \sum_{i=1}^n h_i(a)(D(x^i) - 0) \\ &= \sum_{i=1}^n \tfrac{\partial f}{\partial x^i}(a) D(x^i), \end{aligned}$$

where x^i is the i-th coordinate function on $\mathbb{R}^n$. So we have the expression

$$D(f) = \sum_{i=1}^n D(x^i) \tfrac{\partial}{\partial x^i}|_a(f), \qquad D = \sum_{i=1}^n D(x^i) \tfrac{\partial}{\partial x^i}|_a.$$

Thus D is induced by the tangent vector $(a, \sum_{i=1}^n D(x^i)e_i)$, where (e_i) is the standard basis of $\mathbb{R}^n$.

1.6. The tangent space of a manifold. Let M be a manifold and let $x \in M$ and $\dim M = n$. Let T_xM be the vector space of all derivations at x of $C^\infty_x(M,\mathbb{R})$, the algebra of germs of smooth functions on M at x. (Using 1.3 it may easily be seen that a derivation of $C^\infty(M,\mathbb{R})$ at x factors to a derivation of $C^\infty_x(M,\mathbb{R})$.)

So T_xM consists of all linear mappings $X_x : C^\infty(M,\mathbb{R}) \to \mathbb{R}$ satisfying $X_x(f \cdot g) = X_x(f) \cdot g(x) + f(x) \cdot X_x(g)$. The space T_xM is called the *tangent space of M at x*.

If (U,u) is a chart on M with $x \in U$, then $u^* : f \mapsto f \circ u$ induces an isomorphism of algebras $C^\infty_{u(x)}(\mathbb{R}^n,\mathbb{R}) \cong C^\infty_x(M,\mathbb{R})$, and thus also an isomorphism $T_xu : T_xM \to T_{u(x)}\mathbb{R}^n$, given by $(T_xu.X_x)(f) = X_x(f \circ u)$. So T_xM is an n-dimensional vector space. The dot in $T_xu.X_x$ means that we apply the linear mapping T_xu to the vector X_x — a dot will frequently denote an application of a linear or fiber linear mapping.

We will use the following notation: $u = (u^1,\dots,u^n)$, so u^i denotes the i-th coordinate function on U, and

$$\tfrac{\partial}{\partial u^i}|_x := (T_xu)^{-1}(\tfrac{\partial}{\partial x^i}|_{u(x)}) = (T_xu)^{-1}(u(x),e_i).$$

So $\frac{\partial}{\partial u^i}|_x \in T_xM$ is the derivation given by

$$\tfrac{\partial}{\partial u^i}|_x(f) = \frac{\partial(f \circ u^{-1})}{\partial x^i}(u(x)).$$

From 1.5 we have now

$$\begin{aligned} T_xu.X_x &= \sum_{i=1}^n (T_xu.X_x)(x^i)\tfrac{\partial}{\partial x^i}|_{u(x)} = \\ &= \sum_{i=1}^n X_x(x^i \circ u)\tfrac{\partial}{\partial x^i}|_{u(x)} = \sum_{i=1}^n X_x(u^i)\tfrac{\partial}{\partial x^i}|_{u(x)}. \end{aligned}$$

1.7. The tangent bundle. For a manifold M of dimension n we put $TM := \bigsqcup_{x\in M} T_xM$, the disjoint union of all tangent spaces. This is a family of vector spaces parameterized by M, with projection $\pi_M : TM \to M$ given by $\pi_M(T_xM) = x$.

For any chart (U_α,u_α) of M consider the chart $(\pi_M^{-1}(U_\alpha), Tu_\alpha)$ on TM, where $Tu_\alpha : \pi_M^{-1}(U_\alpha) \to u_\alpha(U_\alpha) \times \mathbb{R}^n$ is given by the formula $Tu_\alpha.X = (u_\alpha(\pi_M(X)), T_{\pi_M(X)}u_\alpha.X)$. Then the chart changings look as follows:

$$\begin{aligned} Tu_\beta \circ (Tu_\alpha)^{-1} : Tu_\alpha(\pi_M^{-1}(U_{\alpha\beta})) = u_\alpha(U_{\alpha\beta}) \times \mathbb{R}^n \to \\ \to u_\beta(U_{\alpha\beta}) \times \mathbb{R}^n = Tu_\beta(\pi_M^{-1}(U_{\alpha\beta})), \end{aligned}$$

$$\begin{aligned} ((Tu_\beta \circ (Tu_\alpha)^{-1})(y,Y))(f) &= ((Tu_\alpha)^{-1}(y,Y))(f \circ u_\beta) \\ &= (y,Y)(f \circ u_\beta \circ u_\alpha^{-1}) = d(f \circ u_\beta \circ u_\alpha^{-1})(y).Y \\ &= df(u_\beta \circ u_\alpha^{-1}(y)).d(u_\beta \circ u_\alpha^{-1})(y).Y \\ &= (u_\beta \circ u_\alpha^{-1}(y), d(u_\beta \circ u_\alpha^{-1})(y).Y)(f). \end{aligned}$$

So the chart changings are smooth. We choose the topology on TM in such a way that all Tu_α become homeomorphisms. This is a Hausdorff topology, since $X, Y \in TM$ may be separated in M if $\pi(X) \neq \pi(Y)$, and in one chart if $\pi(X) = \pi(Y)$. So TM is again a smooth manifold in a canonical way; the triple (TM, π_M, M) is called the *tangent bundle* of M.

1.8. Kinematic definition of the tangent space. Consider $C_0^\infty(\mathbb{R}, M)$, the space of germs at 0 of smooth curves $\mathbb{R} \to M$. We put the following equivalence relation on $C_0^\infty(\mathbb{R}, M)$: the germ of c is equivalent to the germ of e if and only if $c(0) = e(0)$ and in one (equivalently each) chart (U, u) with $c(0) = e(0) \in U$ we have $\frac{d}{dt}|_0(u \circ c)(t) = \frac{d}{dt}|_0(u \circ e)(t)$. The equivalence classes are called velocity vectors of curves in M. We have the following mappings

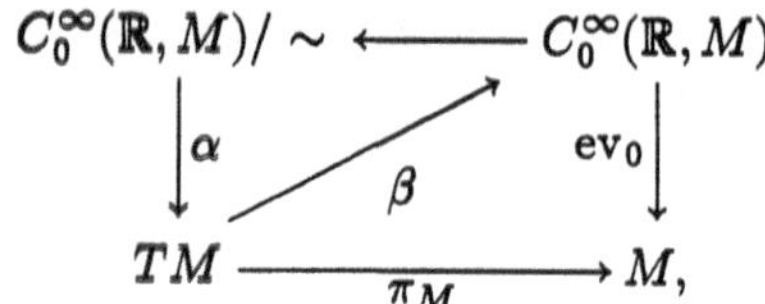

where $\alpha(c)(\mathrm{germ}_{c(0)} f) = \frac{d}{dt}|_0 f(c(t))$ and $\beta : TM \to C_0^\infty(\mathbb{R}, M)$ is given by: $\beta((Tu)^{-1}(y, Y))$ is the germ at 0 of $t \mapsto u^{-1}(y + tY)$. So TM is canonically identified with the set of all possible velocity vectors of curves in M.

1.9. Let $f : M \to N$ be a smooth mapping between manifolds. Then f induces a linear mapping $T_x f : T_x M \to T_{f(x)} N$ for each $x \in M$ by $(T_x f.X_x)(h) = X_x(h \circ f)$ for $h \in C_{f(x)}^\infty(N, \mathbb{R})$. This mapping is linear since $f^* : C_{f(x)}^\infty(N, \mathbb{R}) \to C_x^\infty(M, \mathbb{R})$, given by $h \mapsto h \circ f$, is linear, and $T_x f$ is its adjoint, restricted to the subspace of derivations.

If (U, u) is a chart around x and (V, v) is one around $f(x)$, then

$$\begin{aligned}
(T_x f.\tfrac{\partial}{\partial u^i}|_x)(v^j) &= \tfrac{\partial}{\partial u^i}|_x(v^j \circ f) = \tfrac{\partial}{\partial x^i}(v^j \circ f \circ u^{-1}), \\
T_x f.\tfrac{\partial}{\partial u^i}|_x &= \textstyle\sum_j (T_x f.\tfrac{\partial}{\partial u^i}|_x)(v^j) \tfrac{\partial}{\partial v^j}|_{f(x)} \qquad \text{by 1.7} \\
&= \textstyle\sum_j \tfrac{\partial(v^j \circ f \circ u^{-1})}{\partial x^i}(u(x)) \tfrac{\partial}{\partial v^j}|_{f(x)}.
\end{aligned}$$

So the matrix of $T_x f : T_x M \to T_{f(x)} N$ in the bases $(\frac{\partial}{\partial u^i}|_x)$ and $(\frac{\partial}{\partial v^j}|_{f(x)})$ is just the Jacobi matrix $d(v \circ f \circ u^{-1})(u(x))$ of the mapping $v \circ f \circ u^{-1}$ at $u(x)$, so $T_{f(x)} v \circ T_x f \circ (T_x u)^{-1} = d(v \circ f \circ u^{-1})(u(x))$.

Let us denote by $Tf : TM \to TN$ the total mapping, given by $Tf|T_x M := T_x f$. Then the composition $Tv \circ Tf \circ (Tu)^{-1} : u(U) \times \mathbb{R}^m \to v(V) \times \mathbb{R}^n$ is given by $(y, Y) \mapsto ((v \circ f \circ u^{-1})(y), d(v \circ f \circ u^{-1})(y)Y)$, and thus $Tf : TM \to TN$ is again smooth.

If $f : M \to N$ and $g : N \to P$ are smooth mappings, then we have $T(g \circ f) = Tg \circ Tf$. This is a direct consequence of $(g \circ f)^* = f^* \circ g^*$, and it is the global version of the chain rule. Furthermore we have $T(\mathrm{Id}_M) = \mathrm{Id}_{TM}$.

If $f \in C^\infty(M, \mathbb{R})$, then $Tf : TM \to T\mathbb{R} = \mathbb{R} \times \mathbb{R}$. We then define the *differential* of f by $df := pr_2 \circ Tf : TM \to \mathbb{R}$. Let t denote the identity function on $\mathbb{R}$, then $(Tf.X_x)(t) = X_x(t \circ f) = X_x(f)$, so we have $df(X_x) = X_x(f)$.

1.10. Submanifolds. A subset N of a manifold M is called a *submanifold*, if for each $x \in N$ there is a chart (U, u) of M such that $u(U \cap N) = u(U) \cap (\mathbb{R}^k \times 0)$, where $\mathbb{R}^k \times 0 \hookrightarrow \mathbb{R}^k \times \mathbb{R}^{n-k} = \mathbb{R}^n$. Then clearly N is itself a manifold with $(U \cap N, u|U \cap N)$ as charts, where (U, u) runs through all *submanifold charts* as above and the injection $i : N \hookrightarrow M$ is an embedding in the following sense:

An *embedding* $f : N \to M$ from a manifold N into another one is an injective smooth mapping such that $f(N)$ is a submanifold of M and the (co)restricted mapping $N \to f(N)$ is a diffeomorphism.

If $f : \mathbb{R}^n \to \mathbb{R}^q$ is smooth and the rank of f (more exactly: the rank of its derivative) is q at each point of $f^{-1}(0)$, say, then $f^{-1}(0)$ is a submanifold of $\mathbb{R}^n$ of dimension $n - q$ or empty. This is an immediate consequence of the implicit function theorem.

The following theorem needs three applications of the implicit function theorem for its proof, which can be found in [Dieudonné, I, 60, 10.3.1].

Theorem. *Let $f : W \to \mathbb{R}^q$ be a smooth mapping, where W is an open subset of $\mathbb{R}^n$. If the derivative $df(x)$ has constant rank k for each $x \in W$, then for each $a \in W$ there are charts (U, u) of W centered at a and (V, v) of $\mathbb{R}^q$ centered at $f(a)$ such that $v \circ f \circ u^{-1} : u(U) \to v(V)$ has the following form:*

$$(x_1, \dots, x_n) \mapsto (x_1, \dots, x_k, 0, \dots, 0).$$

So $f^{-1}(b)$ is a submanifold of W of dimension $n - k$ for each $b \in f(W)$. □

1.11. Example: Spheres. We consider the space $\mathbb{R}^{n+1}$, equipped with the standard inner product $\langle x, y \rangle = \sum x^i y^i$. The *n-sphere* S^n is then the subset $\{x \in \mathbb{R}^{n+1} : \langle x, x \rangle = 1\}$. Since $f(x) = \langle x, x \rangle$, $f : \mathbb{R}^{n+1} \to \mathbb{R}$, satisfies $df(x)y = 2\langle x, y \rangle$, it is of rank 1 off 0 and by 1.10 the sphere S^n is a submanifold of $\mathbb{R}^{n+1}$.

In order to get some feeling for the sphere we will describe an explicit atlas for S^n, the *stereographic atlas*. Choose $a \in S^n$ ('south pole'). Let

$$U_+ := S^n \setminus \{a\}, \qquad u_+ : U_+ \to \{a\}^\perp, \qquad u_+(x) = \tfrac{x - \langle x, a \rangle a}{1 - \langle x, a \rangle},$$
$$U_- := S^n \setminus \{-a\}, \qquad u_- : U_- \to \{a\}^\perp, \qquad u_-(x) = \tfrac{x - \langle x, a \rangle a}{1 + \langle x, a \rangle}.$$

From an obvious drawing in the 2-plane through 0, x, and a it is easily seen that u_+ is the usual stereographic projection. We also get

$$u_+^{-1}(y) = \tfrac{|y|^2 - 1}{|y|^2 + 1} a + \tfrac{2}{|y|^2 + 1} y \qquad \text{for } y \in \{a\}^\perp$$

and $(u_- \circ u_+^{-1})(y) = \frac{y}{|y|^2}$. The latter equation can directly be seen from a drawing.

1.12. Products. Let M and N be smooth manifolds described by smooth atlases $(U_\alpha, u_\alpha)_{\alpha \in A}$ and $(V_\beta, v_\beta)_{\beta \in B}$, respectively. Then the family $(U_\alpha \times V_\beta, u_\alpha \times v_\beta : U_\alpha \times V_\beta \to \mathbb{R}^m \times \mathbb{R}^n)_{(\alpha,\beta) \in A \times B}$ is a smooth atlas for the cartesian product $M \times N$. Clearly the projections

$$M \xleftarrow{pr_1} M \times N \xrightarrow{pr_2} N$$

are also smooth. The *product* $(M \times N, pr_1, pr_2)$ has the following universal property:

For any smooth manifold P and smooth mappings $f : P \to M$ and $g : P \to N$ the mapping $(f,g) : P \to M \times N$, $(f,g)(x) = (f(x), g(x))$, is the unique smooth mapping with $pr_1 \circ (f,g) = f$, $pr_2 \circ (f,g) = g$.

From the construction of the tangent bundle in 1.7 it is immediately clear that

$$TM \xleftarrow{T(pr_1)} T(M \times N) \xrightarrow{T(pr_2)} TN$$

is again a product, so that $T(M \times N) = TM \times TN$ in a canonical way.

Clearly we can form products of finitely many manifolds.

1.13. Theorem. *Let M be a connected manifold and suppose that $f : M \to M$ is smooth with $f \circ f = f$. Then the image $f(M)$ of f is a submanifold of M.*

This result can also be expressed as: 'smooth retracts' of manifolds are manifolds. If we do not suppose that M is connected, then $f(M)$ will not be a pure manifold in general, it will have different dimension in different connected components.

Proof. We claim that there is an open neighborhood U of $f(M)$ in M such that the rank of $T_y f$ is constant for $y \in U$. Then by theorem 1.10 the result follows.

For $x \in f(M)$ we have $T_x f \circ T_x f = T_x f$, thus $\operatorname{im} T_x f = \ker(\mathrm{Id} - T_x f)$ and $\operatorname{rank} T_x f + \operatorname{rank}(\mathrm{Id} - T_x f) = \dim M$. Since $\operatorname{rank} T_x f$ and $\operatorname{rank}(\mathrm{Id} - T_x f)$ cannot fall locally, $\operatorname{rank} T_x f$ is locally constant for $x \in f(M)$, and since $f(M)$ is connected, $\operatorname{rank} T_x f = r$ for all $x \in f(M)$.

But then for each $x \in f(M)$ there is an open neighborhood U_x in M with $\operatorname{rank} T_y f \geq r$ for all $y \in U_x$. On the other hand $\operatorname{rank} T_y f = \operatorname{rank} T_y(f \circ f) = \operatorname{rank} T_{f(y)} f \circ T_y f \leq \operatorname{rank} T_{f(y)} f = r$. So the neighborhood we need is given by $U = \bigcup_{x \in f(M)} U_x$. $\square$

1.14. Corollary. *1. The (separable) connected smooth manifolds are exactly the smooth retracts of connected open subsets of $\mathbb{R}^n$'s.*

2. $f : M \to N$ is an embedding of a submanifold if and only if there is an open neighborhood U of $f(M)$ in N and a smooth mapping $r : U \to M$ with $r \circ f = \mathrm{Id}_M$.

Proof. Any manifold M may be embedded into some $\mathbb{R}^n$, see 1.15 below. Then there exists a tubular neighborhood of M in R^n (see [Hirsch, 76, pp. 109–118]), and M is clearly a retract of such a tubular neighborhood. The converse follows from 1.13.

For the second assertion repeat the argument for N instead of $\mathbb{R}^n$. $\square$

1.15. Embeddings into $\mathbb{R}^n$'s. Let M be a smooth manifold of dimension m. Then M can be embedded into $\mathbb{R}^n$, if

(1) $n = 2m + 1$ (see [Hirsch, 76, p 55] or [Bröcker-Jänich, 73, p 73]),
(2) $n = 2m$ (see [Whitney, 44]).
(3) Conjecture (still unproved): The minimal n is $n = 2m - \alpha(m) + 1$, where $\alpha(m)$ is the number of 1's in the dyadic expansion of m.

There exists an immersion (see section 2) $M \to \mathbb{R}^n$, if

(1) $n = 2m$ (see [Hirsch, 76]),
(2) $n = 2m - \alpha(m)$ (see [Cohen, 82]).

2. Submersions and immersions

2.1. Definition. A mapping $f : M \to N$ between manifolds is called a *submersion* at $x \in M$, if the rank of $T_x f : T_x M \to T_{f(x)} N$ equals $\dim N$. Since the rank cannot fall locally (the determinant of a submatrix of the Jacobi matrix is not 0), f is then a submersion in a whole neighborhood of x. The mapping f is said to be a *submersion*, if it is a submersion at each $x \in M$.

2.2. Lemma. *If $f : M \to N$ is a submersion at $x \in M$, then for any chart (V, v) centered at $f(x)$ on N there is chart (U, u) centered at x on M such that $v \circ f \circ u^{-1}$ looks as follows:*

$$(y^1, \dots, y^n, y^{n+1}, \dots, y^m) \mapsto (y^1, \dots, y^n)$$

Proof. Use the inverse function theorem. □

2.3. Corollary. *Any submersion $f : M \to N$ is open: for each open $U \subset M$ the set $f(U)$ is open in N.* □

2.4. Definition. A triple (M, p, N), where $p : M \to N$ is a surjective submersion, is called a *fibered manifold*. M is called the *total space*, N is called the *base*.

A fibered manifold admits local sections: For each $x \in M$ there is an open neighborhood U of $p(x)$ in N and a smooth mapping $s : U \to M$ with $p \circ s = \mathrm{Id}_U$ and $s(p(x)) = x$.

The existence of local sections in turn implies the following universal property:

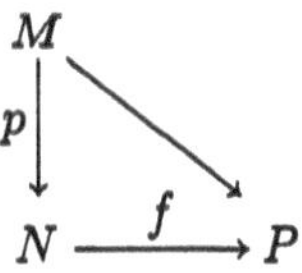

If (M, p, N) is a fibered manifold and $f : N \to P$ is a mapping into some further manifold, such that $f \circ p : M \to P$ is smooth, then f is smooth.

2.5. Definition. A smooth mapping $f : M \to N$ is called an *immersion* at $x \in M$ if the rank of $T_x f : T_x M \to T_{f(x)} N$ equals $\dim M$. Since the rank is maximal at x and cannot fall locally, f is an immersion on a whole neighborhood of x. f is called an immersion if it is so at every $x \in M$.

2.6. Lemma. *If $f : M \to N$ is an immersion, then for any chart (U, u) centered at $x \in M$ there is a chart (V, v) centered at $f(x)$ on N such that $v \circ f \circ u^{-1}$ has the form:*

$$(y^1, \dots, y^m) \mapsto (y^1, \dots, y^m, 0, \dots, 0)$$

Proof. Use the inverse function theorem. □

2.7 Corollary. *If $f : M \to N$ is an immersion, then for any $x \in M$ there is an open neighborhood U of $x \in M$ such that $f(U)$ is a submanifold of N and $f|U : U \to f(U)$ is a diffeomorphism.* □

2.8. Definition. If $i : M \to N$ is an injective immersion, then (M, i) is called an *immersed submanifold* of N.

A submanifold is an immersed submanifold, but the converse is wrong in general. The structure of an immersed submanifold (M, i) is in general not determined by the subset $i(M) \subset N$. All this is illustrated by the following example. Consider the curve $\gamma(t) = (\sin^3 t, \sin t.\cos t)$ in $\mathbb{R}^2$. Then $((-\pi, \pi), \gamma|(-\pi, \pi))$ and $((0, 2\pi), \gamma|(0, 2\pi))$ are two different immersed submanifolds, but the image of the embedding is in both cases just the figure eight.

2.9. Let M be a submanifold of N. Then the embedding $i : M \to N$ is an injective immersion with the following property:

(1) *For any manifold Z a mapping $f : Z \to M$ is smooth if and only if $i \circ f : Z \to N$ is smooth.*

The example in 2.8 shows that there are injective immersions without property (1).

2.10. We want to determine all injective immersions $i : M \to N$ with property 2.9.1. To require that i is a homeomorphism onto its image is too strong as 2.11 and 2.12 below show. To look for all smooth mappings $i : M \to N$ with property 2.9.1 (initial mappings in categorical terms) is too difficult as remark 2.13 below shows.

2.11. Lemma. *If an injective immersion $i : M \to N$ is a homeomorphism onto its image, then $i(M)$ is a submanifold of N.*

Proof. Use 2.7. □

2.12. Example. We consider the 2-dimensional torus $\mathbb{T}^2 = \mathbb{R}^2/\mathbb{Z}^2$. Then the quotient mapping $\pi : \mathbb{R}^2 \to \mathbb{T}^2$ is a covering map, so locally a diffeomorphism. Let us also consider the mapping $f : \mathbb{R} \to \mathbb{R}^2$, $f(t) = (t, \alpha.t)$, where α is irrational. Then $\pi \circ f : \mathbb{R} \to \mathbb{T}^2$ is an injective immersion with dense image, and it is obviously not a homeomorphism onto its image. But $\pi \circ f$ has property 2.9.1, which follows from the fact that π is a covering map.

2.13. Remark. If $f : \mathbb{R} \to \mathbb{R}$ is a function such that f^p and f^q are smooth for some p, q which are relatively prime in $\mathbb{N}$, then f itself turns out to be smooth, see [Joris, 82]. So the mapping $i : t \mapsto \binom{t^p}{t^q}$, $\mathbb{R} \to \mathbb{R}^2$, has property 2.9.1, but i is not an immersion at 0.

2.14. Definition. For an arbitrary subset A of a manifold N and $x_0 \in A$ let $C_{x_0}(A)$ denote the set of all $x \in A$ which can be joined to x_0 by a smooth curve in N lying in A.

A subset M in a manifold N is called *initial submanifold* of dimension m, if the following property is true:

(1) *For each $x \in M$ there exists a chart (U, u) centered at x on N such that $u(C_x(U \cap M)) = u(U) \cap (\mathbb{R}^m \times 0)$.*

The following three lemmas explain the name initial submanifold.

2.15. Lemma. *Let $f : M \to N$ be an injective immersion between manifolds with property 2.9.1. Then $f(M)$ is an initial submanifold of N.*

Proof. Let $x \in M$. By 2.6 we may choose a chart (V, v) centered at $f(x)$ on N and another chart (W, w) centered at x on M such that $(v \circ f \circ w^{-1})(y^1, \dots, y^m) = (y^1, \dots, y^m, 0, \dots, 0)$. Let $r > 0$ be so small that $\{y \in \mathbb{R}^m : |y| < r\} \subset w(W)$ and $\{z \in \mathbb{R}^n : |z| < 2r\} \subset v(V)$. Put

$$U := v^{-1}(\{z \in \mathbb{R}^n : |z| < r\}) \subset N,$$
$$W_1 := w^{-1}(\{y \in \mathbb{R}^m : |y| < r\}) \subset M.$$

We claim that $(U, u = v|U)$ satisfies the condition of 2.14.1.

$$u^{-1}(u(U) \cap (\mathbb{R}^m \times 0)) = u^{-1}(\{(y^1, \dots, y^m, 0 \dots, 0) : |y| < r\}) =$$
$$= f \circ w^{-1} \circ (u \circ f \circ w^{-1})^{-1}(\{(y^1, \dots, y^m, 0 \dots, 0) : |y| < r\}) =$$
$$= f \circ w^{-1}(\{y \in \mathbb{R}^m : |y| < r\}) = f(W_1) \subseteq C_{f(x)}(U \cap f(M)),$$

since $f(W_1) \subseteq U \cap f(M)$ and $f(W_1)$ is C^∞-contractible.

Now let conversely $z \in C_{f(x)}(U \cap f(M))$. Then by definition there is a smooth curve $c : [0,1] \to N$ with $c(0) = f(x)$, $c(1) = z$, and $c([0,1]) \subseteq U \cap f(M)$. By property 2.9.1 the unique curve $\bar{c} : [0,1] \to M$ with $f \circ \bar{c} = c$, is smooth.

We claim that $\bar{c}([0,1]) \subseteq W_1$. If not then there is some $t \in [0,1]$ with $\bar{c}(t) \in w^{-1}(\{y \in \mathbb{R}^m : r \le |y| < 2r\})$ since $\bar{c}$ is smooth and thus continuous. But then we have

$$(v \circ f)(\bar{c}(t)) \in (v \circ f \circ w^{-1})(\{y \in \mathbb{R}^m : r \le |y| < 2r\}) =$$
$$= \{(y, 0) \in \mathbb{R}^m \times 0 : r \le |y| < 2r\} \subseteq \{z \in \mathbb{R}^n : r \le |z| < 2r\}.$$

This means $(v \circ f \circ \bar{c})(t) = (v \circ c)(t) \in \{z \in \mathbb{R}^n : r \le |z| < 2r\}$, so $c(t) \notin U$, a contradiction.

So $\bar{c}([0,1]) \subseteq W_1$, thus $\bar{c}(1) = f^{-1}(z) \in W_1$ and $z \in f(W_1)$. Consequently we have $C_{f(x)}(U \cap f(M)) = f(W_1)$ and finally $f(W_1) = u^{-1}(u(U) \cap (\mathbb{R}^m \times 0))$ by the first part of the proof. □

2.16. Lemma. *Let M be an initial submanifold of a manifold N. Then there is a unique C^∞-manifold structure on M such that the injection $i : M \to N$ is an injective immersion. The connected components of M are separable (but there may be uncountably many of them).*

Proof. We use the sets $C_x(U_x \cap M)$ as charts for M, where $x \in M$ and (U_x, u_x) is a chart for N centered at x with the property required in 2.14.1. Then the chart changings are smooth since they are just restrictions of the chart changings on N. But the sets $C_x(U_x \cap M)$ are not open in the induced topology on M in general. So the identification topology with respect to the charts $(C_x(U_x \cap M), u_x)_{x \in M}$ yields a topology on M which is finer than the induced topology, so it is Hausdorff. Clearly $i : M \to N$ is then an injective immersion. Uniqueness of the smooth structure follows from the universal property of lemma 2.17 below. Finally note that N admits a Riemannian metric since it is separable, which can be induced on M, so each connected component of M is separable. □

2.17. Lemma. *Any initial submanifold M of a manifold N with injective immersion $i : M \to N$ has the universal property 2.9.1:*

For any manifold Z a mapping $f : Z \to M$ is smooth if and only if $i \circ f : Z \to N$ is smooth.

Proof. We have to prove only one direction and we will suppress the embedding i. For $z \in Z$ we choose a chart (U, u) on N, centered at $f(z)$, such that $u(C_{f(z)}(U \cap M)) = u(U) \cap (\mathbb{R}^m \times 0)$. Then $f^{-1}(U)$ is open in Z and contains a chart (V, v) centered at z on Z with $v(V)$ a ball. Then $f(V)$ is C^∞-contractible in $U \cap M$, so $f(V) \subseteq C_{f(z)}(U \cap M)$, and $(u|C_{f(z)}(U \cap M)) \circ f \circ v^{-1} = u \circ f \circ v^{-1}$ is smooth. □

2.18. Transversal mappings. Let M_1, M_2, and N be manifolds and let $f_i : M_i \to N$ be smooth mappings for $i = 1, 2$. We say that f_1 and f_2 are *transversal* at $y \in N$, if

$$\operatorname{im} T_{x_1} f_1 + \operatorname{im} T_{x_2} f_2 = T_y N \quad \text{whenever} \quad f_1(x_1) = f_2(x_2) = y.$$

Note that they are transversal at any y which is not in $f_1(M_1)$ or not in $f_2(M_2)$. The mappings f_1 and f_2 are simply said to be *transversal*, if they are transversal at every $y \in N$.

If P is an initial submanifold of N with injective immersion $i : P \to N$, then $f : M \to N$ is said to be transversal to P, if i and f are transversal.

Lemma. *In this case $f^{-1}(P)$ is an initial submanifold of M with the same codimension in M as P has in N, or the empty set. If P is a submanifold, then also $f^{-1}(P)$ is a submanifold.*

Proof. Let $x \in f^{-1}(P)$ and let (U, u) be an initial submanifold chart for P centered at $f(x)$ on N, i.e. $u(C_x(U \cap P)) = u(U) \cap (\mathbb{R}^p \times 0)$. Then the mapping

$$M \supseteq f^{-1}(U) \xrightarrow{f} U \xrightarrow{u} u(U) \subseteq \mathbb{R}^p \times \mathbb{R}^{n-p} \xrightarrow{pr_2} \mathbb{R}^{n-p}$$

is a submersion at x since f is transversal to P. So by lemma 2.2 there is a chart (V, v) on M centered at x such that we have

$$(pr_2 \circ u \circ f \circ v^{-1})(y^1, \dots, y^{n-p}, \dots, y^m) = (y^1, \dots, y^{n-p}).$$

But then $z \in C_x(f^{-1}(P) \cap V)$ if and only if $v(z) \in v(V) \cap (0 \times \mathbb{R}^{m-n+p})$, so $v(C_x(f^{-1}(P) \cap V)) = v(V) \cap (0 \times \mathbb{R}^{m-n+p})$. □

2.19. Corollary. *If $f_1 : M_1 \to N$ and $f_2 : M_2 \to N$ are smooth and transversal, then the topological pullback*

$$M_1 \underset{(f_1, N, f_2)}{\times} M_2 = M_1 \times_N M_2 := \{(x_1, x_2) \in M_1 \times M_2 : f_1(x_1) = f_2(x_2)\}$$

is a submanifold of $M_1 \times M_2$, and it has the following universal property.

For any smooth mappings $g_1 : P \to M_1$ and $g_2 : P \to M_2$ with $f_1 \circ g_1 = f_2 \circ g_2$ there is a unique smooth mapping $(g_1, g_2) : P \to M_1 \times_N M_2$ with $pr_1 \circ (g_1, g_2) = g_1$ and $pr_2 \circ (g_1, g_2) = g_2$.

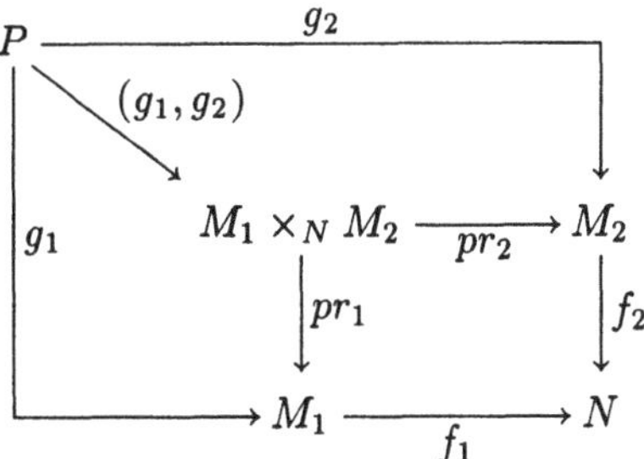

This is also called the pullback property in the category $\mathcal{M}f$ of smooth manifolds and smooth mappings. So one may say, that transversal pullbacks exist in the category $\mathcal{M}f$.

Proof. $M_1 \times_N M_2 = (f_1 \times f_2)^{-1}(\Delta)$, where $f_1 \times f_2 : M_1 \times M_2 \to N \times N$ and where Δ is the diagonal of $N \times N$, and $f_1 \times f_2$ is transversal to Δ if and only if f_1 and f_2 are transversal. □

2.20. The category of fibered manifolds. Consider a fibered manifold (M, p, N) from 2.4 and a point $x \in N$. Since p is a surjective submersion, the injection $i_x : x \to N$ of x into N and $p : M \to N$ are transversal. By 2.19, $p^{-1}(x)$ is a submanifold of M, which is called the *fiber over* $x \in N$.

Given another fibered manifold $(\bar{M}, \bar{p}, \bar{N})$, a *morphism* $(M, p, N) \to (\bar{M}, \bar{p}, \bar{N})$ means a smooth map $f : M \to N$ transforming each fiber of M into a fiber of $\bar{M}$. The relation $f(M_x) \subset \bar{M}_{\bar{x}}$ defines a map $\underline{f} : N \to \bar{N}$, which is characterized by the property $\bar{p} \circ f = \underline{f} \circ p$. Since $\bar{p} \circ f$ is a smooth map, $\underline{f}$ is also smooth by 2.4. Clearly, *all fibered manifolds and their morphisms form a category*, which will be denoted by $\mathcal{FM}$. Transforming every fibered manifold (M, p, N) into its base N and every fibered manifold morphism $f : (M, p, N) \to (\bar{M}, \bar{p}, \bar{N})$ into the induced map $\underline{f} : N \to \bar{N}$ defines the *base functor* $B : \mathcal{FM} \to \mathcal{M}f$.

If (M, p, N) and $(\bar{M}, \bar{p}, N)$ are two fibered manifolds over the same base N, then the pullback $M \times_{(p,N,\bar{p})} \bar{M} = M \times_N \bar{M}$ is called the *fibered product* of M and $\bar{M}$. If p, $\bar{p}$ and N are clear from the context, then $M \times_N \bar{M}$ is also denoted by $M \oplus \bar{M}$. Moreover, if $f_1 : (M_1, p_1, N) \to (\bar{M}_1, \bar{p}_1, \bar{N})$ and $f_2 : (M_2, p_2, N) \to (\bar{M}_2, \bar{p}_2, \bar{N})$ are two $\mathcal{FM}$-morphisms over the same base map $f_0 : N \to \bar{N}$, then the values of the restriction $f_1 \times f_2 | M_1 \times_N M_2$ lie in $\bar{M}_1 \times_{\bar{N}} \bar{M}_2$. The restricted map will be denoted by $f_1 \times_{f_0} f_2 : M_1 \times_N M_2 \to \bar{M}_1 \times_{\bar{N}} \bar{M}_2$ or $f_1 \oplus f_2 : M_1 \oplus M_2 \to \bar{M}_1 \oplus \bar{M}_2$ and will be called the fibered product of f_1 and f_2.

3. Vector fields and flows

3.1. Definition. A *vector field* X on a manifold M is a smooth section of the tangent bundle; so $X : M \to TM$ is smooth and $\pi_M \circ X = \mathrm{Id}_M$. A *local vector field* is a smooth section, which is defined on an open subset only. We denote the set of all vector fields by $\mathfrak{X}(M)$. With point wise addition and scalar multiplication $\mathfrak{X}(M)$ becomes a vector space.

Example. Let (U, u) be a chart on M. Then the $\frac{\partial}{\partial u^i} : U \to TM|U$, $x \mapsto \frac{\partial}{\partial u^i}|_x$, described in 1.6, are local vector fields defined on U.

Lemma. *If X is a vector field on M and (U, u) is a chart on M and $x \in U$, then we have $X(x) = \sum_{i=1}^m X(x)(u^i)\frac{\partial}{\partial u^i}|_x$. We write $X|U = \sum_{i=1}^m X(u^i)\frac{\partial}{\partial u^i}$.* □

3.2. The vector fields $(\frac{\partial}{\partial u^i})_{i=1}^m$ on U, where (U, u) is a chart on M, form a *holonomic frame field.* By a *frame field* on some open set $V \subset M$ we mean $m = \dim M$ vector fields $s_i \in \mathfrak{X}(V)$ such that $s_1(x), \dots, s_m(x)$ is a linear basis of T_xM for each $x \in V$. In general, a frame field on V is said to be *holonomic*, if V can be covered by an atlas $(U_\alpha, u_\alpha)_{\alpha \in A}$ such that $s_i|U_\alpha = \frac{\partial}{\partial u^i_\alpha}$ for all $\alpha \in A$. In the opposite case, the frame field is called *anholonomic*.

With the help of partitions of unity and holonomic frame fields one may construct 'many' vector fields on M. In particular the values of a vector field can be arbitrarily preassigned on a discrete set $\{x_i\} \subset M$.

3.3. Lemma. *The space $\mathfrak{X}(M)$ of vector fields on M coincides canonically with the space of all derivations of the algebra $C^\infty(M, \mathbb{R})$ of smooth functions, i.e. those $\mathbb{R}$-linear operators $D : C^\infty(M, \mathbb{R}) \to C^\infty(M, \mathbb{R})$ with $D(fg) = D(f)g + fD(g)$.*

Proof. Clearly each vector field $X \in \mathfrak{X}(M)$ defines a derivation (again called X, later sometimes called $\mathcal{L}_X$) of the algebra $C^\infty(M, \mathbb{R})$ by the prescription $X(f)(x) := X(x)(f) = df(X(x))$.

If conversely a derivation D of $C^\infty(M, \mathbb{R})$ is given, for any $x \in M$ we consider $D_x : C^\infty(M, \mathbb{R}) \to \mathbb{R}$, $D_x(f) = D(f)(x)$. Then D_x is a derivation at x of $C^\infty(M, \mathbb{R})$ in the sense of 1.5, so $D_x = X_x$ for some $X_x \in T_xM$. In this way we get a section $X : M \to TM$. If (U, u) is a chart on M, we have $D_x = \sum_{i=1}^m X(x)(u^i)\frac{\partial}{\partial u^i}|_x$ by 1.6. Choose V open in M, $V \subset \overline{V} \subset U$, and $\varphi \in C^\infty(M, \mathbb{R})$ such that $\mathrm{supp}(\varphi) \subset U$ and $\varphi|V = 1$. Then $\varphi \cdot u^i \in C^\infty(M, \mathbb{R})$ and $(\varphi u^i)|V = u^i|V$. So $D(\varphi u^i)(x) = X(x)(\varphi u^i) = X(x)(u^i)$ and $X|V = \sum_{i=1}^m D(\varphi u^i)|V \cdot \frac{\partial}{\partial u^i}|V$ is smooth. □

3.4. The Lie bracket. By lemma 3.3 we can identify $\mathfrak{X}(M)$ with the vector space of all derivations of the algebra $C^\infty(M, \mathbb{R})$, which we will do without any notational change in the following.

If X, Y are two vector fields on M, then the mapping $f \mapsto X(Y(f)) - Y(X(f))$ is again a derivation of $C^\infty(M, \mathbb{R})$, as a simple computation shows. Thus there is a unique vector field $[X, Y] \in \mathfrak{X}(M)$ such that $[X, Y](f) = X(Y(f)) - Y(X(f))$ holds for all $f \in C^\infty(M, \mathbb{R})$.

In a local chart (U, u) on M one immediately verifies that for $X|U = \sum X^i \frac{\partial}{\partial u^i}$ and $Y|U = \sum Y^i \frac{\partial}{\partial u^i}$ we have

$$\Big[\sum_i X^i \tfrac{\partial}{\partial u^i}, \sum_j Y^j \tfrac{\partial}{\partial u^j}\Big] = \sum_{i,j} \big(X^i(\tfrac{\partial}{\partial u^i} Y^j) - Y^i(\tfrac{\partial}{\partial u^i} X^j)\big) \tfrac{\partial}{\partial u^j},$$

since second partial derivatives commute. The $\mathbb{R}$-bilinear mapping

$$[\quad,\quad] : \mathfrak{X}(M) \times \mathfrak{X}(M) \to \mathfrak{X}(M)$$

is called the *Lie bracket*. Note also that $\mathfrak{X}(M)$ is a module over the algebra $C^\infty(M, \mathbb{R})$ by point wise multiplication $(f, X) \mapsto fX$.

Theorem. *The Lie bracket* $[\quad,\quad] : \mathfrak{X}(M) \times \mathfrak{X}(M) \to \mathfrak{X}(M)$ *has the following properties:*

$$[X, Y] = -[Y, X],$$
$$[X, [Y, Z]] = [[X, Y], Z] + [Y, [X, Z]], \quad \textit{the Jacobi identity,}$$
$$[fX, Y] = f[X, Y] - (Yf)X,$$
$$[X, fY] = f[X, Y] + (Xf)Y.$$

The form of the Jacobi identity we have chosen says that $\operatorname{ad}(X) = [X, \quad]$ is a derivation for the Lie algebra $(\mathfrak{X}(M), [\quad,\quad])$.

The pair $(\mathfrak{X}(M), [\quad,\quad])$ is the prototype of a *Lie algebra*. The concept of a Lie algebra is one of the most important notions of modern mathematics.

Proof. All these properties can be checked easily for the commutator $[X, Y] = X \circ Y - Y \circ X$ in the space of derivations of the algebra $C^\infty(M, \mathbb{R})$. □

3.5. Integral curves. Let $c : J \to M$ be a smooth curve in a manifold M defined on an interval J. We will use the following notations: $c'(t) = \dot{c}(t) = \frac{d}{dt}c(t) := T_t c.1$. Clearly $c' : J \to TM$ is smooth. We call c' a vector field along c since we have $\pi_M \circ c' = c$.

A smooth curve $c : J \to M$ will be called an *integral curve* or *flow line* of a vector field $X \in \mathfrak{X}(M)$ if $c'(t) = X(c(t))$ holds for all $t \in J$.

3.6. Lemma. *Let X be a vector field on M. Then for any $x \in M$ there is an open interval J_x containing 0 and an integral curve $c_x : J_x \to M$ for X (i.e. $c_x' = X \circ c_x$) with $c_x(0) = x$. If J_x is maximal, then c_x is unique.*

Proof. In a chart (U, u) on M with $x \in U$ the equation $c'(t) = X(c(t))$ is an ordinary differential equation with initial condition $c(0) = x$. Since X is smooth there is a unique local solution by the theorem of Picard-Lindelöf, which even depends smoothly on the initial values, [Dieudonné I, 69, 10.7.4]. So on M there are always local integral curves. If $J_x = (a, b)$ and $\lim_{t \to b-} c_x(t) =: c_x(b)$ exists in M, there is a unique local solution c_1 defined in an open interval containing b with $c_1(b) = c_x(b)$. By uniqueness of the solution on the intersection of the two intervals, c_1 prolongs c_x to a larger interval. This may be repeated (also on the left hand side of J_x) as long as the limit exists. So if we suppose J_x to be maximal, J_x either equals $\mathbb{R}$ or the integral curve leaves the manifold in finite (parameter-) time in the past or future or both. □

3.7. The flow of a vector field. Let $X \in \mathfrak{X}(M)$ be a vector field. Let us write $\operatorname{Fl}_t^X(x) = \operatorname{Fl}^X(t,x) := c_x(t)$, where $c_x : J_x \to M$ is the maximally defined integral curve of X with $c_x(0) = x$, constructed in lemma 3.6. The mapping Fl^X is called the *flow* of the vector field X.

Theorem. *For each vector field X on M, the mapping $\operatorname{Fl}^X : \mathcal{D}(X) \to M$ is smooth, where $\mathcal{D}(X) = \bigcup_{x\in M} J_x \times \{x\}$ is an open neighborhood of $0 \times M$ in $\mathbb{R} \times M$. We have*

$$\operatorname{Fl}^X(t+s,x) = \operatorname{Fl}^X(t, \operatorname{Fl}^X(s,x))$$

in the following sense. If the right hand side exists, then the left hand side exists and we have equality. If both t, $s \geq 0$ or both are ≤ 0, and if the left hand side exists, then also the right hand side exists and we have equality.

Proof. As mentioned in the proof of 3.6, $\operatorname{Fl}^X(t,x)$ is smooth in (t,x) for small t, and if it is defined for (t,x), then it is also defined for (s,y) nearby. These are local properties which follow from the theory of ordinary differential equations.

Now let us treat the equation $\operatorname{Fl}^X(t+s,x) = \operatorname{Fl}^X(t, \operatorname{Fl}^X(s,x))$. If the right hand side exists, then we consider the equation

$$\begin{cases} \frac{d}{dt} \operatorname{Fl}^X(t+s,x) = \frac{d}{du} \operatorname{Fl}^X(u,x)|_{u=t+s} = X(\operatorname{Fl}^X(t+s,x)), \\ \operatorname{Fl}^X(t+s,x)|_{t=0} = \operatorname{Fl}^X(s,x). \end{cases}$$

But the unique solution of this is $\operatorname{Fl}^X(t, \operatorname{Fl}^X(s,x))$. So the left hand side exists and equals the right hand side.

If the left hand side exists, let us suppose that $t, s \geq 0$. We put

$$c_x(u) = \begin{cases} \operatorname{Fl}^X(u,x) & \text{if } u \leq s \\ \operatorname{Fl}^X(u-s, \operatorname{Fl}^X(s,x)) & \text{if } u \geq s. \end{cases}$$

$$\frac{d}{du} c_x(u) = \left.\begin{cases} \frac{d}{du} \operatorname{Fl}^X(u,x) = X(\operatorname{Fl}^X(u,x)) \quad \text{for } u \leq s \\ \frac{d}{du} \operatorname{Fl}^X(u-s, \operatorname{Fl}^X(s,x)) = X(\operatorname{Fl}^X(u-s, \operatorname{Fl}^X(s,x))) \end{cases}\right\} =$$
$$= X(c_x(u)) \quad \text{for } 0 \leq u \leq t+s.$$

Also $c_x(0) = x$ and on the overlap both definitions coincide by the first part of the proof, thus we conclude that $c_x(u) = \operatorname{Fl}^X(u,x)$ for $0 \leq u \leq t+s$ and we have $\operatorname{Fl}^X(t, \operatorname{Fl}^X(s,x)) = c_x(t+s) = \operatorname{Fl}^X(t+s,x)$.

Now we show that $\mathcal{D}(X)$ is open and Fl^X is smooth on $\mathcal{D}(X)$. We know already that $\mathcal{D}(X)$ is a neighborhood of $0 \times M$ in $\mathbb{R} \times M$ and that Fl^X is smooth near $0 \times M$.

For $x \in M$ let J'_x be the set of all $t \in \mathbb{R}$ such that Fl^X is defined and smooth on an open neighborhood of $[0,t] \times \{x\}$ (respectively on $[t,0] \times \{x\}$ for $t < 0$) in $\mathbb{R} \times M$. We claim that $J'_x = J_x$, which finishes the proof. It suffices to show that J'_x is not empty, open and closed in J_x. It is open by construction, and not empty, since $0 \in J'_x$. If J'_x is not closed in J_x, let $t_0 \in J_x \cap (\overline{J'_x} \setminus J'_x)$ and suppose that $t_0 > 0$, say. By the local existence and smoothness Fl^X exists and is smooth near $[-\varepsilon, \varepsilon] \times \{y := \mathrm{Fl}^X(t_0, x)\}$ for some $\varepsilon > 0$, and by construction Fl^X exists and is smooth near $[0, t_0 - \varepsilon] \times \{x\}$. Since $\mathrm{Fl}^X(-\varepsilon, y) = \mathrm{Fl}^X(t_0 - \varepsilon, x)$ we conclude for t near $[0, t_0 - \varepsilon]$, x' near x, and t' near $[-\varepsilon, \varepsilon]$, that $\mathrm{Fl}^X(t + t', x') = \mathrm{Fl}^X(t', \mathrm{Fl}^X(t, x'))$ exists and is smooth. So $t_0 \in J'_x$, a contradiction. □

3.8. Let $X \in \mathfrak{X}(M)$ be a vector field. Its flow Fl^X is called *global* or *complete*, if its domain of definition $\mathcal{D}(X)$ equals $\mathbb{R} \times M$. Then the vector field X itself will be called a *complete vector field*. In this case Fl_t^X is also sometimes called $\exp tX$; it is a diffeomorphism of M.

The *support* $\mathrm{supp}(X)$ of a vector field X is the closure of the set $\{x \in M : X(x) \neq 0\}$.

Lemma. *Every vector field with compact support on M is complete.*

Proof. Let $K = \mathrm{supp}(X)$ be compact. Then the compact set $0 \times K$ has positive distance to the disjoint closed set $(\mathbb{R} \times M) \setminus \mathcal{D}(X)$ (if it is not empty), so $[-\varepsilon, \varepsilon] \times K \subset \mathcal{D}(X)$ for some $\varepsilon > 0$. If $x \notin K$ then $X(x) = 0$, so $\mathrm{Fl}^X(t, x) = x$ for all t and $\mathbb{R} \times \{x\} \subset \mathcal{D}(X)$. So we have $[-\varepsilon, \varepsilon] \times M \subset \mathcal{D}(X)$. Since $\mathrm{Fl}^X(t + \varepsilon, x) = \mathrm{Fl}^X(t, \mathrm{Fl}^X(\varepsilon, x))$ exists for $|t| \leq \varepsilon$ by theorem 3.7, we have $[-2\varepsilon, 2\varepsilon] \times M \subset \mathcal{D}(X)$ and by repeating this argument we get $\mathbb{R} \times M = \mathcal{D}(X)$. □

So on a compact manifold M each vector field is complete. If M is not compact and of dimension ≥ 2, then in general the set of complete vector fields on M is neither a vector space nor is it closed under the Lie bracket, as the following example on $\mathbb{R}^2$ shows: $X = y\frac{\partial}{\partial x}$ and $Y = \frac{x^2}{2}\frac{\partial}{\partial y}$ are complete, but neither $X + Y$ nor $[X, Y]$ is complete.

3.9. f-related vector fields. If $f : M \to M$ is a diffeomorphism, then for any vector field $X \in \mathfrak{X}(M)$ the mapping $Tf^{-1} \circ X \circ f$ is also a vector field, which we will denote f^*X. Analogously we put $f_*X := Tf \circ X \circ f^{-1} = (f^{-1})^*X$.

But if $f : M \to N$ is a smooth mapping and $Y \in \mathfrak{X}(N)$ is a vector field there may or may not exist a vector field $X \in \mathfrak{X}(M)$ such that the following diagram commutes:

$$\begin{array}{ccc} TM & \xrightarrow{Tf} & TN \\ X\uparrow & & \uparrow Y \\ M & \xrightarrow{f} & N. \end{array} \tag{1}$$

Definition. Let $f : M \to N$ be a smooth mapping. Two vector fields $X \in \mathfrak{X}(M)$ and $Y \in \mathfrak{X}(N)$ are called *f-related*, if $Tf \circ X = Y \circ f$ holds, i.e. if diagram (1) commutes.

Example. If $X \in \mathfrak{X}(M)$ and $Y \in \mathfrak{X}(N)$ and $X \times Y \in \mathfrak{X}(M \times N)$ is given by $(X \times Y)(x, y) = (X(x), Y(y))$, then we have:

(2) $X \times Y$ and X are pr_1-related.
(3) $X \times Y$ and Y are pr_2-related.
(4) X and $X \times Y$ are $\mathrm{ins}(y)$-related if and only if $Y(y) = 0$, where $\mathrm{ins}(y)(x) = (x, y)$, $\mathrm{ins}(y) : M \to M \times N$.

3.10. Lemma. *Consider vector fields $X_i \in \mathfrak{X}(M)$ and $Y_i \in \mathfrak{X}(N)$ for $i = 1, 2$, and a smooth mapping $f : M \to N$. If X_i and Y_i are f-related for $i = 1, 2$, then also $\lambda_1 X_1 + \lambda_2 X_2$ and $\lambda_1 Y_1 + \lambda_2 Y_2$ are f-related, and also $[X_1, X_2]$ and $[Y_1, Y_2]$ are f-related.*

Proof. The first assertion is immediate. To show the second let $h \in C^\infty(N, \mathbb{R})$. Then by assumption we have $Tf \circ X_i = Y_i \circ f$, thus:

$$\begin{aligned}(X_i(h \circ f))(x) &= X_i(x)(h \circ f) = (T_x f.X_i(x))(h) = \\ &= (Tf \circ X_i)(x)(h) = (Y_i \circ f)(x)(h) = Y_i(f(x))(h) = (Y_i(h))(f(x)),\end{aligned}$$

so $X_i(h \circ f) = (Y_i(h)) \circ f$, and we may continue:

$$\begin{aligned}[X_1, X_2](h \circ f) &= X_1(X_2(h \circ f)) - X_2(X_1(h \circ f)) = \\ &= X_1(Y_2(h) \circ f) - X_2(Y_1(h) \circ f) = \\ &= Y_1(Y_2(h)) \circ f - Y_2(Y_1(h)) \circ f = [Y_1, Y_2](h) \circ f.\end{aligned}$$

But this means $Tf \circ [X_1, X_2] = [Y_1, Y_2] \circ f$. □

3.11. Corollary. *If $f : M \to N$ is a local diffeomorphism (so $(T_x f)^{-1}$ makes sense for each $x \in M$), then for $Y \in \mathfrak{X}(N)$ a vector field $f^*Y \in \mathfrak{X}(M)$ is defined by $(f^*Y)(x) = (T_x f)^{-1}.Y(f(x))$. The linear mapping $f^* : \mathfrak{X}(N) \to \mathfrak{X}(M)$ is then a Lie algebra homomorphism, i.e. $f^*[Y_1, Y_2] = [f^*Y_1, f^*Y_2]$.*

3.12. The Lie derivative of functions. For a vector field $X \in \mathfrak{X}(M)$ and $f \in C^\infty(M, \mathbb{R})$ we define $\mathcal{L}_X f \in C^\infty(M, \mathbb{R})$ by

$$\begin{aligned}\mathcal{L}_X f(x) &:= \tfrac{d}{dt}|_0 f(\mathrm{Fl}^X(t, x)) \quad \text{or} \\ \mathcal{L}_X f &:= \tfrac{d}{dt}|_0 (\mathrm{Fl}^X_t)^* f = \tfrac{d}{dt}|_0 (f \circ \mathrm{Fl}^X_t).\end{aligned}$$

Since $\mathrm{Fl}^X(t, x)$ is defined for small t, for any $x \in M$, the expressions above make sense.

Lemma. *$\frac{d}{dt}(\mathrm{Fl}^X_t)^* f = (\mathrm{Fl}^X_t)^* X(f)$, in particular for $t = 0$ we have $\mathcal{L}_X f = X(f) = df(X)$.* □

3.13. The Lie derivative for vector fields. For $X, Y \in \mathfrak{X}(M)$ we define $\mathcal{L}_X Y \in \mathfrak{X}(M)$ by

$$\mathcal{L}_X Y := \tfrac{d}{dt}|_0 (\mathrm{Fl}^X_t)^* Y = \tfrac{d}{dt}|_0 (T(\mathrm{Fl}^X_{-t}) \circ Y \circ \mathrm{Fl}^X_t),$$

and call it the *Lie derivative* of Y along X.

Lemma. *$\mathcal{L}_X Y = [X, Y]$ and $\frac{d}{dt}(\mathrm{Fl}^X_t)^* Y = (\mathrm{Fl}^X_t)^* \mathcal{L}_X Y = (\mathrm{Fl}^X_t)^*[X, Y]$.*

Proof. Let $f \in C^\infty(M, \mathbb{R})$ be a function and consider the mapping $\alpha(t, s) := Y(\mathrm{Fl}^X(t, x))(f \circ \mathrm{Fl}^X_s)$, which is locally defined near 0. It satisfies

$$\begin{aligned}&\alpha(t, 0) = Y(\mathrm{Fl}^X(t, x))(f), \\ &\alpha(0, s) = Y(x)(f \circ \mathrm{Fl}^X_s), \\ &\tfrac{\partial}{\partial t}\alpha(0, 0) = \tfrac{\partial}{\partial t}\big|_0 Y(\mathrm{Fl}^X(t, x))(f) = \tfrac{\partial}{\partial t}\big|_0 (Yf)(\mathrm{Fl}^X(t, x)) = X(x)(Yf), \\ &\tfrac{\partial}{\partial s}\alpha(0, 0) = \tfrac{\partial}{\partial s}|_0 Y(x)(f \circ \mathrm{Fl}^X_s) = Y(x)\tfrac{\partial}{\partial s}|_0(f \circ \mathrm{Fl}^X_s) = Y(x)(Xf).\end{aligned}$$

But on the other hand we have

$$\begin{aligned}\tfrac{\partial}{\partial u}|_0\alpha(u,-u) &= \tfrac{\partial}{\partial u}|_0 Y(\mathrm{Fl}^X(u,x))(f\circ \mathrm{Fl}^X_{-u}) =\\ &= \tfrac{\partial}{\partial u}|_0 \Big(T(\mathrm{Fl}^X_{-u})\circ Y\circ \mathrm{Fl}^X_u\Big)_x(f) = (\mathcal{L}_X Y)_x(f),\end{aligned}$$

so the first assertion follows. For the second claim we compute as follows:

$$\begin{aligned}\tfrac{\partial}{\partial t}(\mathrm{Fl}^X_t)^*Y &= \tfrac{\partial}{\partial s}|_0 \Big(T(\mathrm{Fl}^X_{-t})\circ T(\mathrm{Fl}^X_{-s})\circ Y\circ \mathrm{Fl}^X_s\circ \mathrm{Fl}^X_t\Big)\\ &= T(\mathrm{Fl}^X_{-t})\circ \tfrac{\partial}{\partial s}|_0 \Big(T(\mathrm{Fl}^X_{-s})\circ Y\circ \mathrm{Fl}^X_s\Big)\circ \mathrm{Fl}^X_t\\ &= T(\mathrm{Fl}^X_{-t})\circ [X,Y]\circ \mathrm{Fl}^X_t = (\mathrm{Fl}^X_t)^*[X,Y]. \quad\square\end{aligned}$$

3.14. Lemma. *Let $X\in\mathfrak{X}(M)$ and $Y\in\mathfrak{X}(N)$ be f-related vector fields for a smooth mapping $f:M\to N$. Then we have $f\circ \mathrm{Fl}^X_t = \mathrm{Fl}^Y_t\circ f$, whenever both sides are defined. In particular, if f is a diffeomorphism we have $\mathrm{Fl}^{f^*Y}_t = f^{-1}\circ \mathrm{Fl}^Y_t\circ f$.*

Proof. We have $\frac{d}{dt}(f\circ \mathrm{Fl}^X_t) = Tf\circ \frac{d}{dt}\mathrm{Fl}^X_t = Tf\circ X\circ \mathrm{Fl}^X_t = Y\circ f\circ Fl^X_t$ and $f(\mathrm{Fl}^X(0,x)) = f(x)$. So $t\mapsto f(\mathrm{Fl}^X(t,x))$ is an integral curve of the vector field Y on N with initial value $f(x)$, so we have $f(\mathrm{Fl}^X(t,x)) = \mathrm{Fl}^Y(t,f(x))$ or $f\circ \mathrm{Fl}^X_t = \mathrm{Fl}^Y_t\circ f$. $\square$

3.15. Corollary. *Let $X,Y\in\mathfrak{X}(M)$. Then the following assertions are equivalent*

(1) $\mathcal{L}_X Y = [X,Y] = 0$.
(2) $(\mathrm{Fl}^X_t)^*Y = Y$, *wherever defined.*
(3) $\mathrm{Fl}^X_t\circ \mathrm{Fl}^Y_s = \mathrm{Fl}^Y_s\circ \mathrm{Fl}^X_t$, *wherever defined.*

Proof. (1) $\Leftrightarrow$ (2) is immediate from lemma 3.13. To see (2) $\Leftrightarrow$ (3) we note that $\mathrm{Fl}^X_t\circ \mathrm{Fl}^Y_s = \mathrm{Fl}^Y_s\circ \mathrm{Fl}^X_t$ if and only if $\mathrm{Fl}^Y_s = \mathrm{Fl}^X_{-t}\circ \mathrm{Fl}^Y_s\circ \mathrm{Fl}^X_t = \mathrm{Fl}^{(\mathrm{Fl}^X_t)^*Y}_s$ by lemma 3.14; and this in turn is equivalent to $Y = (\mathrm{Fl}^X_t)^*Y$. $\square$

3.16. Theorem. *Let M be a manifold, let $\varphi^i:\mathbb{R}\times M\supset U_{\varphi^i}\to M$ be smooth mappings for $i=1,\dots,k$ where each U_{φ^i} is an open neighborhood of $\{0\}\times M$ in $\mathbb{R}\times M$, such that each φ^i_t is a diffeomorphism on its domain, $\varphi^i_0 = \mathrm{Id}_M$, and $\frac{\partial}{\partial t}\big|_0 \varphi^i_t = X_i\in\mathfrak{X}(M)$. We put $[\varphi^i,\varphi^j]_t = [\varphi^i_t,\varphi^j_t] := (\varphi^j_t)^{-1}\circ(\varphi^i_t)^{-1}\circ\varphi^j_t\circ\varphi^i_t$. Then for each formal bracket expression P of lenght k we have*

$$\begin{aligned}0 &= \tfrac{\partial^\ell}{\partial t^\ell}|_0 P(\varphi^1_t,\dots,\varphi^k_t) \quad \text{for } 1\le \ell < k,\\ P(X_1,\dots,X_k) &= \tfrac{1}{k!}\tfrac{\partial^k}{\partial t^k}|_0 P(\varphi^1_t,\dots,\varphi^k_t)\in\mathfrak{X}(M)\end{aligned}$$

in the sense explained in step 2 of the proof. In particular we have for vector fields $X,Y\in\mathfrak{X}(M)$

$$\begin{aligned}0 &= \tfrac{\partial}{\partial t}\big|_0 (\mathrm{Fl}^Y_{-t}\circ \mathrm{Fl}^X_{-t}\circ \mathrm{Fl}^Y_t\circ \mathrm{Fl}^X_t),\\ [X,Y] &= \tfrac{1}{2}\tfrac{\partial^2}{\partial t^2}|_0(\mathrm{Fl}^Y_{-t}\circ \mathrm{Fl}^X_{-t}\circ \mathrm{Fl}^Y_t\circ \mathrm{Fl}^X_t).\end{aligned}$$

Proof. Step 1. Let $c : \mathbb{R} \to M$ be a smooth curve. If $c(0) = x \in M$, $c'(0) = 0, \dots, c^{(k-1)}(0) = 0$, then $c^{(k)}(0)$ is a well defined tangent vector in T_xM which is given by the derivation $f \mapsto (f \circ c)^{(k)}(0)$ at x.

For we have

$$((f.g) \circ c)^{(k)}(0) = ((f \circ c).(g \circ c))^{(k)}(0) = \sum_{j=0}^{k} \binom{k}{j} (f \circ c)^{(j)}(0)(g \circ c)^{(k-j)}(0)$$
$$= (f \circ c)^{(k)}(0)g(x) + f(x)(g \circ c)^{(k)}(0),$$

since all other summands vanish: $(f \circ c)^{(j)}(0) = 0$ for $1 \leq j < k$.

Step 2. Let $\varphi : \mathbb{R} \times M \supset U_\varphi \to M$ be a smooth mapping where U_φ is an open neighborhood of $\{0\} \times M$ in $\mathbb{R} \times M$, such that each φ_t is a diffeomorphism on its domain and $\varphi_0 = \mathrm{Id}_M$. We say that φ_t is a *curve of local diffeomorphisms* though Id_M.

From step 1 we see that if $\frac{\partial^j}{\partial t^j}|_0 \varphi_t = 0$ for all $1 \leq j < k$, then $X := \frac{1}{k!}\frac{\partial^k}{\partial t^k}|_0 \varphi_t$ is a well defined vector field on M. We say that X is the *first non-vanishing derivative* at 0 of the curve φ_t of local diffeomorphisms. We may paraphrase this as $(\partial_t^k|_0 \varphi_t^*)f = k!\mathcal{L}_X f$.

Claim 3. Let φ_t, ψ_t be curves of local diffeomorphisms through Id_M and let $f \in C^\infty(M, \mathbb{R})$. Then we have

$$\partial_t^k|_0 (\varphi_t \circ \psi_t)^* f = \partial_t^k|_0 (\psi_t^* \circ \varphi_t^*) f = \sum_{j=0}^{k} \binom{k}{j} (\partial_t^j|_0 \psi_t^*)(\partial_t^{k-j}|_0 \varphi_t^*) f.$$

Also the multinomial version of this formula holds:

$$\partial_t^k|_0 (\varphi_t^1 \circ \dots \circ \varphi_t^\ell)^* f = \sum_{j_1+\dots+j_\ell = k} \frac{k!}{j_1! \dots j_\ell!} (\partial_t^{j_1}|_0 (\varphi_t^\ell)^*) \dots (\partial_t^{j_1}|_0 (\varphi_t^1)^*) f.$$

We only show the binomial version. For a function $h(t,s)$ of two variables we have

$$\partial_t^k h(t,t) = \sum_{j=0}^{k} \binom{k}{j} \partial_t^j \partial_s^{k-j} h(t,s)|_{s=t},$$

since for $h(t,s) = f(t)g(s)$ this is just a consequence of the Leibnitz rule, and linear combinations of such decomposable tensors are dense in the space of all functions of two variables in the compact C^∞-topology, so that by continuity the formula holds for all functions. In the following form it implies the claim:

$$\partial_t^k|_0 f(\varphi(t, \psi(t,x))) = \sum_{j=0}^{k} \binom{k}{j} \partial_t^j \partial_s^{k-j} f(\varphi(t, \psi(s,x)))|_{t=s=0}.$$

Claim 4. Let φ_t be a curve of local diffeomorphisms through Id_M with first non-vanishing derivative $k!X = \partial_t^k|_0 \varphi_t$. Then the inverse curve of local diffeomorphisms φ_t^{-1} has first non-vanishing derivative $-k!X = \partial_t^k|_0 \varphi_t^{-1}$.

For we have $\varphi_t^{-1} \circ \varphi_t = \mathrm{Id}$, so by claim 3 we get for $1 \le j \le k$

$$0 = \partial_t^j|_0(\varphi_t^{-1} \circ \varphi_t)^* f = \sum_{i=0}^{j} \tbinom{j}{i}(\partial_t^i|_0\varphi_t^*)(\partial_t^{j-i}(\varphi_t^{-1})^*)f =$$

$$= \partial_t^j|_0\varphi_t^*(\varphi_0^{-1})^*f + \varphi_0^*\partial_t^j|_0(\varphi_t^{-1})^*f,$$

i.e. $\partial_t^j|_0\varphi_t^* f = -\partial_t^j|_0(\varphi_t^{-1})^* f$ as required.

Claim 5. Let φ_t be a curve of local diffeomorphisms through Id_M with first non-vanishing derivative $m!X = \partial_t^m|_0\varphi_t$, and let ψ_t be a curve of local diffeomorphisms through Id_M with first non-vanishing derivative $n!Y = \partial_t^n|_0\psi_t$.

Then the curve of local diffeomorphisms $[\varphi_t, \psi_t] = \psi_t^{-1} \circ \varphi_t^{-1} \circ \psi_t \circ \varphi_t$ has first non-vanishing derivative

$$(m+n)![X,Y] = \partial_t^{m+n}|_0[\varphi_t, \psi_t].$$

From this claim the theorem follows.

By the multinomial version of claim 3 we have

$$\begin{aligned} A_N f :&= \partial_t^N|_0(\psi_t^{-1} \circ \varphi_t^{-1} \circ \psi_t \circ \varphi_t)^* f \\ &= \sum_{i+j+k+\ell=N} \frac{N!}{i!j!k!\ell!}(\partial_t^i|_0\varphi_t^*)(\partial_t^j|_0\psi_t^*)(\partial_t^k|_0(\varphi_t^{-1})^*)(\partial_t^\ell|_0(\psi_t^{-1})^*)f. \end{aligned}$$

Let us suppose that $1 \le n \le m$, the case $m \le n$ is similar. If $N < n$ all summands are 0. If $N = n$ we have by claim 4

$$A_N f = (\partial_t^n|_0\varphi_t^*)f + (\partial_t^n|_0\psi_t^*)f + (\partial_t^n|_0(\varphi_t^{-1})^*)f + (\partial_t^n|_0(\psi_t^{-1})^*)f = 0.$$

If $n < N \le m$ we have, using again claim 4:

$$\begin{aligned} A_N f &= \sum_{j+\ell=N} \frac{N!}{j!\ell!}(\partial_t^j|_0\psi_t^*)(\partial_t^\ell|_0(\psi_t^{-1})^*)f + \delta_N^m\left((\partial_t^m|_0\varphi_t^*)f + (\partial_t^m|_0(\varphi_t^{-1})^*)f\right) \\ &= (\partial_t^N|_0(\psi_t^{-1} \circ \psi_t)^*)f + 0 = 0. \end{aligned}$$

Now we come to the difficult case $m, n < N \le m+n$.

$$\begin{aligned} A_N f &= \partial_t^N|_0(\psi_t^{-1} \circ \varphi_t^{-1} \circ \psi_t)^* f + \tbinom{N}{m}(\partial_t^m|_0\varphi_t^*)(\partial_t^{N-m}|_0(\psi_t^{-1} \circ \varphi_t^{-1} \circ \psi_t)^*)f \\ &\quad + (\partial_t^N|_0\varphi_t^*)f, \qquad (1) \end{aligned}$$

by claim 3, since all other terms vanish, see (3) below. By claim 3 again we get:

$$\begin{aligned} \partial_t^N|_0(\psi_t^{-1} \circ \varphi_t^{-1} \circ \psi_t)^* f &= \sum_{j+k+\ell=N} \frac{N!}{j!k!\ell!}(\partial_t^j|_0\psi_t^*)(\partial_t^k|_0(\varphi_t^{-1})^*)(\partial_t^\ell|_0(\psi_t^{-1})^*)f \\ &= \sum_{j+\ell=N} \tbinom{N}{j}(\partial_t^j|_0\psi_t^*)(\partial_t^\ell|_0(\psi_t^{-1})^*)f + \tbinom{N}{m}(\partial_t^{N-m}|_0\psi_t^*)(\partial_t^m|_0(\varphi_t^{-1})^*)f \qquad (2) \\ &\quad + \tbinom{N}{m}(\partial_t^m|_0(\varphi_t^{-1})^*)(\partial_t^{N-m}|_0(\psi_t^{-1})^*)f + \partial_t^N|_0(\varphi_t^{-1})^* f \\ &= 0 + \tbinom{N}{m}(\partial_t^{N-m}|_0\psi_t^*)m!\mathcal{L}_{-X}f + \tbinom{N}{m}m!\mathcal{L}_{-X}(\partial_t^{N-m}|_0(\psi_t^{-1})^*)f \\ &\quad + \partial_t^N|_0(\varphi_t^{-1})^* f \\ &= \delta_{m+n}^N(m+n)!(\mathcal{L}_X\mathcal{L}_Y - \mathcal{L}_Y\mathcal{L}_X)f + \partial_t^N|_0(\varphi_t^{-1})^* f \\ &= \delta_{m+n}^N(m+n)!\mathcal{L}_{[X,Y]}f + \partial_t^N|_0(\varphi_t^{-1})^* f \end{aligned}$$

From the second expression in (2) one can also read off that

$$\partial_t^{N-m}|_0(\psi_t^{-1}\circ\varphi_t^{-1}\circ\psi_t)^*f = \partial_t^{N-m}|_0(\varphi_t^{-1})^*f. \tag{3}$$

If we put (2) and (3) into (1) we get, using claims 3 and 4 again, the final result which proves claim 5 and the theorem:

$$\begin{aligned} A_N f &= \delta^N_{m+n}(m+n)!\mathcal{L}_{[X,Y]}f + \partial_t^N|_0(\varphi_t^{-1})^*f \\ &\quad + \tbinom{N}{m}(\partial_t^m|_0\varphi_t^*)(\partial_t^{N-m}|_0(\varphi_t^{-1})^*)f + (\partial_t^N|_0\varphi_t^*)f \\ &= \delta^N_{m+n}(m+n)!\mathcal{L}_{[X,Y]}f + \partial_t^N|_0(\varphi_t^{-1}\circ\varphi_t)^*f \\ &= \delta^N_{m+n}(m+n)!\mathcal{L}_{[X,Y]}f + 0. \quad\square \end{aligned}$$

3.17. Theorem. *Let $X_1,\dots,X_m$ be vector fields on M defined in a neighborhood of a point $x\in M$ such that $X_1(x),\dots,X_m(x)$ are a basis for T_xM and $[X_i,X_j]=0$ for all i,j.*

Then there is a chart (U,u) of M centered at x such that $X_i|U = \frac{\partial}{\partial u^i}$.

Proof. For small $t=(t^1,\dots,t^m)\in\mathbb{R}^m$ we put

$$f(t^1,\dots,t^m) = (\mathrm{Fl}^{X_1}_{t^1}\circ\dots\circ\mathrm{Fl}^{X_m}_{t^m})(x).$$

By 3.15 we may interchange the order of the flows arbitrarily. Therefore

$$\tfrac{\partial}{\partial t^i}f(t^1,\dots,t^m) = \tfrac{\partial}{\partial t^i}(\mathrm{Fl}^{X_i}_{t^i}\circ\mathrm{Fl}^{X_1}_{t^1}\circ\dots)(x) = X_i((\mathrm{Fl}^{x_1}_{t^1}\circ\dots)(x)).$$

So T_0f is invertible, f is a local diffeomorphism, and its inverse gives a chart with the desired properties. $\square$

3.18. Distributions. Let M be a manifold. Suppose that for each $x\in M$ we are given a sub vector space E_x of T_xM. The disjoint union $E=\bigsqcup_{x\in M}E_x$ is called a *distribution* on M. We do not suppose, that the dimension of E_x is locally constant in x.

Let $\mathfrak{X}_{loc}(M)$ denote the set of all locally defined smooth vector fields on M, i.e. $\mathfrak{X}_{loc}(M)=\bigcup\mathfrak{X}(U)$, where U runs through all open sets in M. Furthermore let $\mathfrak{X}_E$ denote the set of all local vector fields $X\in\mathfrak{X}_{loc}(M)$ with $X(x)\in E_x$ whenever defined. We say that a subset $\mathcal{V}\subset\mathfrak{X}_E$ *spans* E, if for each $x\in M$ the vector space E_x is the linear span of the set $\{X(x):X\in\mathcal{V}\}$. We say that E is a *smooth distribution* if $\mathfrak{X}_E$ spans E. Note that every subset $\mathcal{W}\subset\mathfrak{X}_{loc}(M)$ spans a distribution denoted by $E(\mathcal{W})$, which is obviously smooth (the linear span of the empty set is the vector space 0). From now on we will consider only smooth distributions.

An *integral manifold* of a smooth distribution E is a connected immersed submanifold (N,i) (see 2.8) such that $T_xi(T_xN)=E_{i(x)}$ for all $x\in N$. We will see in theorem 3.22 below that any integral manifold is in fact an initial submanifold of M (see 2.14), so that we need not specify the injective immersion i. An integral manifold of E is called *maximal* if it is not contained in any strictly larger integral manifold of E.

3.19. Lemma. *Let E be a smooth distribution on M. Then we have:*

1. If (N, i) is an integral manifold of E and $X \in \mathfrak{X}_E$, then i^*X makes sense and is an element of $\mathfrak{X}_{loc}(N)$, which is $i|i^{-1}(U_X)$-related to X, where $U_X \subset M$ is the open domain of X.

2. If (N_j, i_j) are integral manifolds of E for $j = 1,2$, then $i_1^{-1}(i_1(N_1) \cap i_2(N_2))$ and $i_2^{-1}(i_1(N_1) \cap i_2(N_2))$ are open subsets in N_1 and N_2, respectively; furthermore $i_2^{-1} \circ i_1$ is a diffeomorphism between them.

3. If $x \in M$ is contained in some integral submanifold of E, then it is contained in a unique maximal one.

Proof. 1. Let U_X be the open domain of $X \in \mathfrak{X}_E$. If $i(x) \in U_X$ for $x \in N$, we have $X(i(x)) \in E_{i(x)} = T_x i(T_x N)$, so $i^*X(x) := ((T_x i)^{-1} \circ X \circ i)(x)$ makes sense. It is clearly defined on an open subset of N and is smooth in x.

2. Let $X \in \mathfrak{X}_E$. Then $i_j^*X \in \mathfrak{X}_{loc}(N_j)$ and is i_j-related to X. So by lemma 3.14 for $j = 1,2$ we have

$$i_j \circ \mathrm{Fl}_t^{i_j^*X} = Fl_t^X \circ i_j.$$

Now choose $x_j \in N_j$ such that $i_1(x_1) = i_2(x_2) = x_0 \in M$ and choose vector fields $X_1, \dots, X_n \in \mathfrak{X}_E$ such that $(X_1(x_0), \dots, X_n(x_0))$ is a basis of E_{x_0}. Then

$$f_j(t^1, \dots, t^n) := (\mathrm{Fl}_{t^1}^{i_j^*X_1} \circ \dots \circ \mathrm{Fl}_{t^n}^{i_j^*X_n})(x_j)$$

is a smooth mapping defined near zero $\mathbb{R}^n \to N_j$. Since obviously $\frac{\partial}{\partial t^k}|_0 f_j = i_j^* X_k(x_j)$ for $j = 1,2$, we see that f_j is a diffeomorphism near 0. Finally we have

$$\begin{aligned}(i_2^{-1} \circ i_1 \circ f_1)(t^1, \dots, t^n) &= (i_2^{-1} \circ i_1 \circ \mathrm{Fl}_{t^1}^{i_1^*X_1} \circ \dots \circ \mathrm{Fl}_{t^n}^{i_1^*X_n})(x_1) \\ &= (i_2^{-1} \circ \mathrm{Fl}_{t^1}^{X_1} \circ \dots \circ \mathrm{Fl}_{t^n}^{X_n} \circ i_1)(x_1) \\ &= (\mathrm{Fl}_{t^1}^{i_2^*X_1} \circ \dots \circ \mathrm{Fl}_{t^n}^{i_2^*X_n} \circ i_2^{-1} \circ i_1)(x_1) \\ &= f_2(t^1, \dots, t^n).\end{aligned}$$

So $i_2^{-1} \circ i_1$ is a diffeomorphism, as required.

3. Let N be the union of all integral manifolds containing x. Choose the union of all the atlases of these integral manifolds as atlas for N, which is a smooth atlas for N by 2. Note that a connected immersed submanifold of a separable manifold is automatically separable (since it carries a Riemannian metric). □

3.20. Integrable distributions and foliations.

A smooth distribution E on a manifold M is called *integrable*, if each point of M is contained in some integral manifold of E. By 3.19.3 each point is then contained in a unique maximal integral manifold, so the maximal integral manifolds form a partition of M. This partition is called the *foliation* of M induced by the integrable distribution E, and each maximal integral manifold is called a *leaf* of this foliation. If $X \in \mathfrak{X}_E$ then by 3.19.1 the integral curve $t \mapsto \mathrm{Fl}^X(t, x)$ of X through $x \in M$ stays in the leaf through x.

Note, however, that usually a foliation is supposed to have constant dimensions of the leafs, so our notion here is sometimes called a *singular foliation.*

Let us now consider an arbitrary subset $\mathcal{V} \subset \mathfrak{X}_{loc}(M)$. We say that $\mathcal{V}$ is *stable* if for all $X, Y \in \mathcal{V}$ and for all t for which it is defined the local vector field $(\mathrm{Fl}_t^X)^*Y$ is again an element of $\mathcal{V}$.

If $\mathcal{W} \subset \mathfrak{X}_{loc}(M)$ is an arbitrary subset, we call $\mathcal{S}(\mathcal{W})$ the set of all local vector fields of the form $(\mathrm{Fl}_{t_1}^{X_1} \circ \dots \circ \mathrm{Fl}_{t_k}^{X_k})^*Y$ for $X_i, Y \in \mathcal{W}$. By lemma 3.14 the flow of this vector field is

$$\mathrm{Fl}((\mathrm{Fl}_{t_1}^{X_1} \circ \dots \circ \mathrm{Fl}_{t_k}^{X_k})^*Y, t) = \mathrm{Fl}_{-t_k}^{X_k} \circ \dots \circ \mathrm{Fl}_{-t_1}^{X_1} \circ \mathrm{Fl}_t^Y \circ \mathrm{Fl}_{t_1}^{X_1} \circ \dots \circ \mathrm{Fl}_{t_k}^{X_k},$$

so $\mathcal{S}(\mathcal{W})$ is the minimal stable set of local vector fields which contains $\mathcal{W}$.

Now let F be an arbitrary distribution. A local vector field $X \in \mathfrak{X}_{loc}(M)$ is called an *infinitesimal automorphism* of F, if $T_x(\mathrm{Fl}_t^X)(F_x) \subset F_{\mathrm{Fl}^X(t,x)}$ whenever defined. We denote by $aut(F)$ the set of all infinitesimal automorphisms of F. By arguments given just above, $aut(F)$ is stable.

3.21. Lemma. *Let E be a smooth distribution on a manifold M. Then the following conditions are equivalent:*

(1) *E is integrable.*
(2) *$\mathfrak{X}_E$ is stable.*
(3) *There exists a subset $\mathcal{W} \subset \mathfrak{X}_{loc}(M)$ such that $\mathcal{S}(\mathcal{W})$ spans E.*
(4) *$aut(E) \cap \mathfrak{X}_E$ spans E.*

Proof. (1) $\Longrightarrow$ (2). Let $X \in \mathfrak{X}_E$ and let L be the leaf through $x \in M$, with $i : L \to M$ the inclusion. Then $\mathrm{Fl}_{-t}^X \circ i = i \circ \mathrm{Fl}_{-t}^{i^*X}$ by lemma 3.14, so we have

$$\begin{aligned} T_x(\mathrm{Fl}_{-t}^X)(E_x) &= T(\mathrm{Fl}_{-t}^X).T_x i.T_x L = T(\mathrm{Fl}_{-t}^X \circ i).T_x L \\ &= Ti.T_x(\mathrm{Fl}_{-t}^{i^*X}).T_x L \\ &= Ti.T_{Fl^{i^*X}(-t,x)}L = E_{Fl^X(-t,x)}. \end{aligned}$$

This implies that $(\mathrm{Fl}_t^X)^*Y \in \mathfrak{X}_E$ for any $Y \in \mathfrak{X}_E$.

(2) $\Longrightarrow$ (4). In fact (2) says that $\mathfrak{X}_E \subset aut(E)$.

(4) $\Longrightarrow$ (3). We can choose $\mathcal{W} = aut(E) \cap \mathfrak{X}_E$: for $X, Y \in \mathcal{W}$ we have $(\mathrm{Fl}_t^X)^*Y \in \mathfrak{X}_E$; so $\mathcal{W} \subset \mathcal{S}(\mathcal{W}) \subset \mathfrak{X}_E$ and E is spanned by $\mathcal{W}$.

(3) $\Longrightarrow$ (1). We have to show that each point $x \in M$ is contained in some integral submanifold for the distribution E. Since $\mathcal{S}(\mathcal{W})$ spans E and is stable we have

$$T(\mathrm{Fl}_t^X).E_x = E_{\mathrm{Fl}^X(t,x)} \tag{5}$$

for each $X \in \mathcal{S}(\mathcal{W})$. Let $\dim E_x = n$. There are $X_1, \dots, X_n \in \mathcal{S}(\mathcal{W})$ such that $X_1(x), \dots, X_n(x)$ is a basis of E_x, since E is smooth. As in the proof of 3.19.2 we consider the mapping

$$f(t^1, \dots, t^n) := (\mathrm{Fl}_{t^1}^{X_1} \circ \dots \circ \mathrm{Fl}_{t^n}^{X_n})(x),$$

defined and smooth near 0 in $\mathbb{R}^n$. Since the rank of f at 0 is n, the image under f of a small open neighborhood of 0 is a submanifold N of M. We claim that N is an integral manifold of E. The tangent space $T_{f(t^1,\dots,t^n)}N$ is linearly generated by

$$\begin{aligned}\tfrac{\partial}{\partial t^k}(\mathrm{Fl}^{X_1}_{t^1}\circ\dots\circ\mathrm{Fl}^{X_n}_{t^n})(x) &= T(\mathrm{Fl}^{X_1}_{t^1}\circ\dots\circ\mathrm{Fl}^{X_{k-1}}_{t^{k-1}})X_k((\mathrm{Fl}^{X_k}_{t^k}\circ\dots\circ\mathrm{Fl}^{X_n}_{t^n})(x))\\ &= ((\mathrm{Fl}^{X_1}_{-t^1})^*\cdots(\mathrm{Fl}^{X_{k-1}}_{-t^{k-1}})^*X_k)(f(t^1,\dots,t^n)).\end{aligned}$$

Since $\mathcal{S}(\mathcal{W})$ is stable, these vectors lie in $E_{f(t)}$. From the form of f and from (5) we see that $\dim E_{f(t)} = \dim E_x$, so these vectors even span $E_{f(t)}$ and we have $T_{f(t)}N = E_{f(t)}$ as required. □

3.22. Theorem (local structure of foliations). *Let E be an integrable distribution of a manifold M. Then for each $x \in M$ there exists a chart (U,u) with $u(U) = \{y \in \mathbb{R}^m : |y^i| < \varepsilon$ for all $i\}$ for some $\varepsilon > 0$, and an at most countable subset $A \subset \mathbb{R}^{m-n}$, such that for the leaf L through x we have*

$$u(U \cap L) = \{y \in u(U) : (y^{n+1},\dots,y^m) \in A\}.$$

Each leaf is an initial submanifold.

If furthermore the distribution E has locally constant rank, this property holds for each leaf meeting U with the same n.

This chart (U,u) is called a *distinguished chart* for the distribution or the foliation. A connected component of $U \cap L$ is called a *plaque*.

Proof. Let L be the leaf through x, $\dim L = n$. Let $X_1,\dots,X_n \in \mathfrak{X}_E$ be local vector fields such that $X_1(x),\dots,X_n(x)$ is a basis of E_x. We choose a chart (V,v) centered at x on M such that the vectors

$$X_1(x),\dots,X_n(x), \tfrac{\partial}{\partial v^{n+1}}|_x,\dots,\tfrac{\partial}{\partial v^m}|_x$$

form a basis of T_xM. Then

$$f(t^1,\dots,t^m) = (\mathrm{Fl}^{X_1}_{t^1}\circ\dots\circ\mathrm{Fl}^{X_n}_{t^n})(v^{-1}(0,\dots,0,t^{n+1},\dots,t^m))$$

is a diffeomorphism from a neighborhood of 0 in $\mathbb{R}^m$ onto a neighborhood of x in M. Let (U,u) be the chart given by f^{-1}, suitably restricted. We have

$$y \in L \iff (\mathrm{Fl}^{X_1}_{t^1}\circ\dots\circ\mathrm{Fl}^{X_n}_{t^n})(y) \in L$$

for all y and all $t^1,\dots,t^n$ for which both expressions make sense. So we have

$$f(t^1,\dots,t^m) \in L \iff f(0,\dots,0,t^{n+1},\dots,t^m) \in L,$$

and consequently $L \cap U$ is the disjoint union of connected sets of the form $\{y \in U : (u^{n+1}(y),\dots,u^m(y)) = \text{constant}\}$. Since L is a connected immersed submanifold of M, it is second countable and only a countable set of constants can appear in the description of $u(L\cap U)$ given above. From this description it is clear that L is an initial submanifold (2.14) since $u(C_x(L\cap U)) = u(U)\cap(\mathbb{R}^n\times 0)$.

The argument given above is valid for any leaf of dimension n meeting U, so also the assertion for an integrable distribution of constant rank follows. □

3.23. Involutive distributions. A subset $\mathcal{V} \subset \mathfrak{X}_{loc}(M)$ is called *involutive* if $[X,Y] \in \mathcal{V}$ for all $X, Y \in \mathcal{V}$. Here $[X,Y]$ is defined on the intersection of the domains of X and Y.

A smooth distribution E on M is called *involutive* if there exists an involutive subset $\mathcal{V} \subset \mathfrak{X}_{loc}(M)$ spanning E.

For an arbitrary subset $\mathcal{W} \subset \mathfrak{X}_{loc}(M)$ let $\mathcal{L}(\mathcal{W})$ be the set consisting of all local vector fields on M which can be written as finite expressions using Lie brackets and starting from elements of $\mathcal{W}$. Clearly $\mathcal{L}(\mathcal{W})$ is the smallest involutive subset of $\mathfrak{X}_{loc}(M)$ which contains $\mathcal{W}$.

3.24. Lemma. *For each subset $\mathcal{W} \subset \mathfrak{X}_{loc}(M)$ we have*

$$E(\mathcal{W}) \subset E(\mathcal{L}(\mathcal{W})) \subset E(\mathcal{S}(\mathcal{W})).$$

In particular we have $E(\mathcal{S}(\mathcal{W})) = E(\mathcal{L}(\mathcal{S}(\mathcal{W})))$.

Proof. We will show that for $X, Y \in \mathcal{W}$ we have $[X,Y] \in \mathfrak{X}_{E(\mathcal{S}(\mathcal{W}))}$, for then by induction we get $\mathcal{L}(\mathcal{W}) \subset \mathfrak{X}_{E(\mathcal{S}(\mathcal{W}))}$ and $E(\mathcal{L}(\mathcal{W})) \subset E(\mathcal{S}(\mathcal{W}))$.

Let $x \in M$; since by 3.21 $E(\mathcal{S}(\mathcal{W}))$ is integrable, we can choose the leaf L through x, with the inclusion i. Then i^*X is i-related to X, i^*Y is i-related to Y, thus by 3.10 the local vector field $[i^*X, i^*Y] \in \mathfrak{X}_{loc}(L)$ is i-related to $[X,Y]$, and $[X,Y](x) \in E(\mathcal{S}(\mathcal{W}))_x$, as required. □

3.25. Theorem. *Let $\mathcal{V} \subset \mathfrak{X}_{loc}(M)$ be an involutive subset. Then the distribution $E(\mathcal{V})$ spanned by $\mathcal{V}$ is integrable under each of the following conditions.*

1. *M is real analytic and $\mathcal{V}$ consists of real analytic vector fields.*
2. *The dimension of $E(\mathcal{V})$ is constant along all flow lines of vector fields in $\mathcal{V}$.*

Proof. (1) For $X, Y \in \mathcal{V}$ we have $\frac{d}{dt}(\mathrm{Fl}_t^X)^*Y = (\mathrm{Fl}_t^X)^*\mathcal{L}_X Y$, consequently $\frac{d^k}{dt^k}(\mathrm{Fl}_t^X)^*Y = (\mathrm{Fl}_t^X)^*(\mathcal{L}_X)^k Y$, and since everything is real analytic we get for $x \in M$ and small t

$$(\mathrm{Fl}_t^X)^*Y(x) = \sum_{k\geq 0} \frac{t^k}{k!}\frac{d^k}{dt^k}|_0(\mathrm{Fl}_t^X)^*Y(x) = \sum_{k\geq 0}\frac{t^k}{k!}(\mathcal{L}_X)^k Y(x).$$

Since $\mathcal{V}$ is involutive, all $(\mathcal{L}_X)^k Y \in \mathcal{V}$. Therefore we get $(\mathrm{Fl}_t^X)^*Y(x) \in E(\mathcal{V})_x$ for small t. By the flow property of Fl^X the set of all t satisfying $(\mathrm{Fl}_t^X)^*Y(x) \in E(\mathcal{V})_x$ is open and closed, so it follows that 3.21.2 is satisfied and thus $E(\mathcal{V})$ is integrable.

(2) We choose $X_1, \dots, X_n \in \mathcal{V}$ such that $X_1(x), \dots, X_n(x)$ is a basis of $E(\mathcal{V})_x$. For $X \in \mathcal{V}$, by hypothesis, $E(\mathcal{V})_{\mathrm{Fl}^X(t,x)}$ has also dimension n and admits $X_1(\mathrm{Fl}^X(t,x)), \dots, X_n(\mathrm{Fl}^X(t,x))$ as basis for small t. So there are smooth

functions $f_{ij}(t)$ such that

$$[X, X_i](\mathrm{Fl}^X(t,x)) = \sum_{j=1}^{n} f_{ij}(t) X_j(\mathrm{Fl}^X(t,x)).$$

$$\begin{aligned}\tfrac{d}{dt} T(\mathrm{Fl}^X_{-t}).X_i(\mathrm{Fl}^X(t,x)) &= T(\mathrm{Fl}^X_{-t}).[X, X_i](\mathrm{Fl}^X(t,x)) = \\ &= \sum_{j=1}^{n} f_{ij}(t) T(\mathrm{Fl}^X_{-t}).X_j(\mathrm{Fl}^X(t,x)).\end{aligned}$$

So the T_xM-valued functions $g_i(t) = T(\mathrm{Fl}^X_{-t}).X_i(\mathrm{Fl}^X(t,x))$ satisfy the linear ordinary differential equation $\frac{d}{dt} g_i(t) = \sum_{j=1}^{n} f_{ij}(t) g_j(t)$ and have initial values in the linear subspace $E(\mathcal{V})_x$, so they have values in it for all small t. Therefore $T(\mathrm{Fl}^X_{-t})E(\mathcal{V})_{\mathrm{Fl}^X(t,x)} \subset E(\mathcal{V})_x$ for small t. Using compact time intervals and the flow property one sees that condition 3.21.2 is satisfied and $E(\mathcal{V})$ is integrable. □

Example. The distribution spanned by $\mathcal{W} \subset \mathfrak{X}_{loc}(\mathbb{R}^2)$ is involutive, but not integrable, where $\mathcal{W}$ consists of all global vector fields with support in $\mathbb{R}^2 \setminus \{0\}$ and the field $\frac{\partial}{\partial x^1}$; the leaf through 0 should have dimension 1 at 0 and dimension 2 elsewhere.

3.26. By a *time dependent vector field* on a manifold M we mean a smooth mapping $X : J \times M \to TM$ with $\pi_M \circ X = pr_2$, where J is an open interval. An integral curve of X is a smooth curve $c : I \to M$ with $\dot{c}(t) = X(t, c(t))$ for all $t \in I$, where I is a subinterval of J.

There is an associated vector field $\bar{X} \in X(J \times M)$, given by $\bar{X}(t,x) = (1_t, X(t,x)) \in T_t\mathbb{R} \times T_xM$.

By the *evolution operator* of X we mean the mapping $\Phi^X : J \times J \times M \to M$, defined in a maximal open neighborhood of the diagonal in $M \times M$ and satisfying the differential equation

$$\begin{cases} \frac{d}{dt}\Phi^X(t,s,x) = X(t, \Phi^X(t,s,x)) \\ \Phi^X(s,s,x) = x. \end{cases}$$

It is easily seen that $(t, \Phi^X(t,s,x)) = \mathrm{Fl}^{\bar{X}}(t-s, (s,x))$, so the maximally defined evolution operator exists and is unique, and it satisfies

$$\Phi^X_{t,s} = \Phi^X_{t,r} \circ \Phi^X_{r,s}$$

whenever one side makes sense (with the restrictions of 3.7), where $\Phi^X_{t,s}(x) = \Phi(t,s,x)$.

4. Lie groups

4.1. Definition. A *Lie group* G is a smooth manifold and a group such that the multiplication $\mu : G \times G \to G$ is smooth. We shall see in a moment, that then also the inversion $\nu : G \to G$ turns out to be smooth.

We shall use the following notation:
$\mu : G \times G \to G$, multiplication, $\mu(x, y) = x.y$.
$\lambda_a : G \to G$, left translation, $\lambda_a(x) = a.x$.
$\rho_a : G \to G$, right translation, $\rho_a(x) = x.a$.
$\nu : G \to G$, inversion, $\nu(x) = x^{-1}$.
$e \in G$, the unit element.
Then we have $\lambda_a \circ \lambda_b = \lambda_{a.b}$, $\rho_a \circ \rho_b = \rho_{b.a}$, $\lambda_a^{-1} = \lambda_{a^{-1}}$, $\rho_a^{-1} = \rho_{a^{-1}}$, $\rho_a \circ \lambda_b = \lambda_b \circ \rho_a$. If $\varphi : G \to H$ is a smooth homomorphism between Lie groups, then we also have $\varphi \circ \lambda_a = \lambda_{\varphi(a)} \circ \varphi$, $\varphi \circ \rho_a = \rho_{\varphi(a)} \circ \varphi$, thus also $T\varphi.T\lambda_a = T\lambda_{\varphi(a)}.T\varphi$, etc. So $T_e\varphi$ is injective (surjective) if and only if $T_a\varphi$ is injective (surjective) for all $a \in G$.

4.2. Lemma. *$T_{(a,b)}\mu : T_aG \times T_bG \to T_{ab}G$ is given by*

$$T_{(a,b)}\mu.(X_a, Y_b) = T_a(\rho_b).X_a + T_b(\lambda_a).Y_b.$$

Proof. Let $ri_a : G \to G \times G$, $ri_a(x) = (a, x)$ be the right insertion and let $li_b : G \to G \times G$, $li_b(x) = (x, b)$ be the left insertion. Then we have

$$\begin{aligned} T_{(a,b)}\mu.(X_a, Y_b) &= T_{(a,b)}\mu.(T_a(li_b).X_a + T_b(ri_a).Y_b) = \\ &= T_a(\mu \circ li_b).X_a + T_b(\mu \circ ri_a).Y_b = T_a(\rho_b).X_a + T_b(\lambda_a).Y_b. \quad \square \end{aligned}$$

4.3. Corollary. *The inversion $\nu : G \to G$ is smooth and*

$$T_a\nu = -T_e(\rho_{a^{-1}}).T_a(\lambda_{a^{-1}}) = -T_e(\lambda_{a^{-1}}).T_a(\rho_{a^{-1}}).$$

Proof. The equation $\mu(x, \nu(x)) = e$ determines ν implicitly. Since we have $T_e(\mu(e, \ \ )) = T_e(\lambda_e) = \mathrm{Id}$, the mapping ν is smooth in a neighborhood of e by the implicit function theorem. From $(\nu \circ \lambda_a)(x) = x^{-1}.a^{-1} = (\rho_{a^{-1}} \circ \nu)(x)$ we may conclude that ν is everywhere smooth. Now we differentiate the equation $\mu(a, \nu(a)) = e$; this gives in turn

$$\begin{aligned} 0_e &= T_{(a,a^{-1})}\mu.(X_a, T_a\nu.X_a) = T_a(\rho_{a^{-1}}).X_a + T_{a^{-1}}(\lambda_a).T_a\nu.X_a, \\ T_a\nu.X_a &= -T_e(\lambda_a)^{-1}.T_a(\rho_{a^{-1}}).X_a = -T_e(\lambda_{a^{-1}}).T_a(\rho_{a^{-1}}).X_a. \quad \square \end{aligned}$$

4.4. Example. The *general linear group* $GL(n, \mathbb{R})$ is the group of all invertible real $n \times n$-matrices. It is an open subset of $L(\mathbb{R}^n, \mathbb{R}^n)$, given by $\det \neq 0$ and a Lie group.

Similarly $GL(n, \mathbb{C})$, the group of invertible complex $n \times n$-matrices, is a Lie group; also $GL(n, \mathbb{H})$, the group of all invertible quaternionic $n \times n$-matrices, is a Lie group, but the quaternionic determinant is a more subtle instrument here.

4.5. Example. The *orthogonal group* $O(n,\mathbb{R})$ is the group of all linear isometries of $(\mathbb{R}^n, \langle\ ,\ \rangle)$, where $\langle\ ,\ \rangle$ is the standard positive definite inner product on $\mathbb{R}^n$. The *special orthogonal group* $SO(n,\mathbb{R}) := \{A \in O(n,\mathbb{R}) : \det A = 1\}$ is open in $O(n,\mathbb{R})$, since

$$O(n,\mathbb{R}) = SO(n,\mathbb{R}) \sqcup \begin{pmatrix} -1 & 0 \\ 0 & \mathbb{I}_{n-1} \end{pmatrix} SO(n,\mathbb{R}),$$

where $\mathbb{I}_k$ is short for the identity matrix $\mathrm{Id}_{\mathbb{R}^k}$. We claim that $O(n,\mathbb{R})$ and $SO(n,\mathbb{R})$ are submanifolds of $L(\mathbb{R}^n,\mathbb{R}^n)$. For that we consider the mapping $f : L(\mathbb{R}^n,\mathbb{R}^n) \to L(\mathbb{R}^n,\mathbb{R}^n)$, given by $f(A) = A.A^t$. Then $O(n,\mathbb{R}) = f^{-1}(\mathbb{I}_n)$; so $O(n,\mathbb{R})$ is closed. Since it is also bounded, $O(n,\mathbb{R})$ is compact. We have $df(A).X = X.A^t + A.X^t$, so $\ker df(\mathbb{I}_n) = \{X : X + X^t = 0\}$ is the space $\mathfrak{o}(n,\mathbb{R})$ of all skew symmetric $n \times n$-matrices. Note that $\dim \mathfrak{o}(n,\mathbb{R}) = \frac{1}{2}(n-1)n$. If A is invertible, we get $\ker df(A) = \{Y : Y.A^t + A.Y^t = 0\} = \{Y : Y.A^t \in \mathfrak{o}(n,\mathbb{R})\} = \mathfrak{o}(n,\mathbb{R}).(A^{-1})^t$. The mapping f takes values in $L_{sym}(\mathbb{R}^n,\mathbb{R}^n)$, the space of all symmetric $n\times n$-matrices, and $\dim\ker df(A) + \dim L_{sym}(\mathbb{R}^n,\mathbb{R}^n) = \frac{1}{2}(n-1)n + \frac{1}{2}n(n+1) = n^2 = \dim L(\mathbb{R}^n,\mathbb{R}^n)$, so $f : GL(n,\mathbb{R}) \to L_{sym}(\mathbb{R}^n,\mathbb{R}^n)$ is a submersion. Since obviously $f^{-1}(\mathbb{I}_n) \subset GL(n,\mathbb{R})$, we conclude from 1.10 that $O(n,\mathbb{R})$ is a submanifold of $GL(n,\mathbb{R})$. It is also a Lie group, since the group operations are obviously smooth.

4.6. Example. The *special linear group* $SL(n,\mathbb{R})$ is the group of all $n \times n$-matrices of determinant 1. The function $\det : L(\mathbb{R}^n,\mathbb{R}^n) \to \mathbb{R}$ is smooth and $d\det(A)X = \mathrm{trace}(C(A).X)$, where $C(A)^i_j$, the cofactor of A^j_i, is the determinant of the matrix, which results from putting 1 instead of A^j_i into A and 0 in the rest of the j-th row and the i-th column of A. We recall Cramer's rule $C(A).A = A.C(A) = \det(A).\mathbb{I}_n$. So if $C(A) \neq 0$ (i.e. $\mathrm{rank}(A) \geq n-1$) then the linear functional $df(A)$ is non zero. So $\det : GL(n,\mathbb{R}) \to \mathbb{R}$ is a submersion and $SL(n,\mathbb{R}) = (\det)^{-1}(1)$ is a manifold and a Lie group of dimension $n^2 - 1$. Note finally that $T_{\mathbb{I}_n}SL(n,\mathbb{R}) = \ker d\det(\mathbb{I}_n) = \{X : \mathrm{trace}(X) = 0\}$. This space of traceless matrices is usually called $\mathfrak{sl}(n,\mathbb{R})$.

4.7. Example. The *symplectic group* $Sp(n,\mathbb{R})$ is the group of all $2n \times 2n$-matrices A such that $\omega(Ax,Ay) = \omega(x,y)$ for all $x,y \in \mathbb{R}^{2n}$, where ω is the standard non degenerate skew symmetric bilinear form on $\mathbb{R}^{2n}$.

Such a form exists on a vector space if and only if the dimension is even, and on $\mathbb{R}^n \times (\mathbb{R}^n)^*$ the standard form is given by $\omega((x,x^*),(y,y^*)) = \langle x,y^*\rangle - \langle y,x^*\rangle$, i.e. in coordinates $\omega((x^i)_{i=1}^{2n},(y^j)_{j=1}^{2n}) = \sum_{i=1}^n (x^i y^{n+i} - x^{n+i}y^i)$. Any symplectic form on $\mathbb{R}^{2n}$ looks like that after choosing a suitable basis. Let $(e_i)_{i=1}^{2n}$ be the standard basis in $\mathbb{R}^{2n}$. Then we have

$$(\omega(e_i,e_j)^i_j) = \begin{pmatrix} 0 & \mathbb{I}_n \\ -\mathbb{I}_n & 0 \end{pmatrix} =: J,$$

and the matrix J satisfies $J^t = -J$, $J^2 = -\mathbb{I}_{2n}$, $J\binom{x}{y} = \binom{y}{-x}$ in $\mathbb{R}^n \times \mathbb{R}^n$, and $\omega(x,y) = \langle x, Jy\rangle$ in terms of the standard inner product on $\mathbb{R}^{2n}$.

For $A \in L(\mathbb{R}^{2n}, \mathbb{R}^{2n})$ we have $\omega(Ax, Ay) = \langle Ax, JAy \rangle = \langle x, A^t JAy \rangle$. Thus $A \in Sp(n, \mathbb{R})$ if and only if $A^t JA = J$.

We consider now the mapping $f : L(\mathbb{R}^{2n}, \mathbb{R}^{2n}) \to L(\mathbb{R}^{2n}, \mathbb{R}^{2n})$ given by $f(A) = A^t JA$. Then $f(A)^t = (A^t JA)^t = -A^t JA = -f(A)$, so f takes values in the space $\mathfrak{o}(2n, \mathbb{R})$ of skew symmetric matrices. We have $df(A)X = X^t JA + A^t JX$, and therefore

$$\begin{aligned} \ker df(\mathbb{I}_{2n}) &= \{X \in L(\mathbb{R}^{2n}, \mathbb{R}^{2n}) : X^t J + JX = 0\} \\ &= \{X : JX \text{ is symmetric}\} =: \mathfrak{sp}(n, \mathbb{R}). \end{aligned}$$

We see that $\dim \mathfrak{sp}(n, \mathbb{R}) = \frac{2n(2n+1)}{2} = \binom{2n+1}{2}$. Furthermore we have $\ker df(A) = \{X : X^t JA + A^t JX = 0\}$ and $X \mapsto A^t JX$ is an isomorphism $\ker df(A) \to L_{sym}(\mathbb{R}^{2n}, \mathbb{R}^{2n})$, if A is invertible. Thus $\dim \ker df(A) = \binom{2n+1}{2}$ for all $A \in GL(2n, \mathbb{R})$. If $f(A) = J$, then $A^t JA = J$, so A has rank $2n$ and is invertible, and $\dim \ker df(A) + \dim \mathfrak{o}(2n, \mathbb{R}) = \binom{2n+1}{2} + \frac{2n(2n-1)}{2} = 4n^2 = \dim L(\mathbb{R}^{2n}, \mathbb{R}^{2n})$. So $f : GL(2n, \mathbb{R}) \to \mathfrak{o}(2n, \mathbb{R})$ is a submersion and $f^{-1}(J) = Sp(n, \mathbb{R})$ is a manifold and a Lie group. It is the symmetry group of 'classical mechanics'.

4.8. Example. The complex general linear group $GL(n, \mathbb{C})$ of all invertible complex $n \times n$-matrices is open in $L_{\mathbb{C}}(\mathbb{C}^n, \mathbb{C}^n)$, so it is a real Lie group of real dimension $2n^2$; it is also a complex Lie group of complex dimension n^2. The complex special linear group $SL(n, \mathbb{C})$ of all matrices of determinant 1 is a submanifold of $GL(n, \mathbb{C})$ of complex codimension 1 (or real codimension 2).

The complex orthogonal group $O(n, \mathbb{C})$ is the set

$$\{A \in L(\mathbb{C}^n, \mathbb{C}^n) : g(Az, Aw) = g(z, w) \text{ for all } z, w\},$$

where $g(z, w) = \sum_{i=1}^n z^i w^i$. This is a complex Lie group of complex dimension $\frac{(n-1)n}{2}$, and it is *not* compact. Since $O(n, \mathbb{C}) = \{A : A^t A = \mathbb{I}_n\}$, we have $1 = \det_{\mathbb{C}}(\mathbb{I}_n) = \det_{\mathbb{C}}(A^t A) = \det_{\mathbb{C}}(A)^2$, so $\det_{\mathbb{C}}(A) = \pm 1$. Thus $SO(n, \mathbb{C}) := \{A \in O(n, \mathbb{C}) : \det_{\mathbb{C}}(A) = 1\}$ is an open subgroup of index 2 in $O(n, \mathbb{C})$.

The group $Sp(n, \mathbb{C}) = \{A \in L_{\mathbb{C}}(\mathbb{C}^{2n}, \mathbb{C}^{2n}) : A^t JA = J\}$ is also a complex Lie group of complex dimension $n(2n+1)$.

These groups here are the *classical complex Lie groups*. The groups $SL(n, \mathbb{C})$ for $n \geq 2$, $SO(n, \mathbb{C})$ for $n \geq 3$, $Sp(n, \mathbb{C})$ for $n \geq 4$, and five more exceptional groups exhaust all simple complex Lie groups up to coverings.

4.9. Example. Let $\mathbb{C}^n$ be equipped with the standard hermitian inner product $(z, w) = \sum_{i=1}^n z^i \overline{w}^i$. The *unitary* group $U(n)$ consists of all complex $n \times n$-matrices A such that $(Az, Aw) = (z, w)$ for all z, w holds, or equivalently $U(n) = \{A : A^* A = \mathbb{I}_n\}$, where $A^* = \overline{A}^t$.

We consider the mapping $f : L_{\mathbb{C}}(\mathbb{C}^n, \mathbb{C}^n) \to L_{\mathbb{C}}(\mathbb{C}^n, \mathbb{C}^n)$, given by $f(A) = A^* A$. Then f is smooth but not holomorphic. Its derivative is $df(A)X = X^* A + A^* X$, so $\ker df(\mathbb{I}_n) = \{X : X^* + X = 0\} =: \mathfrak{u}(n)$, the space of all skew hermitian matrices. We have $\dim_{\mathbb{R}} \mathfrak{u}(n) = n^2$. As above we may check that $f : GL(n, \mathbb{C}) \to L_{herm}(\mathbb{C}^n, \mathbb{C}^n)$ is a submersion, so $U(n) = f^{-1}(\mathbb{I}_n)$ is a compact real Lie group of dimension n^2.

The *special unitary* group is $SU(n) = U(n) \cap SL(n, \mathbb{C})$. For $A \in U(n)$ we have $|\det_{\mathbb{C}}(A)| = 1$, thus $\dim_{\mathbb{R}} SU(n) = n^2 - 1$.

4.10. Example. The group $Sp(n)$. Let $\mathbb{H}$ be the division algebra of quaternions. Then $Sp(1) := S^3 \subset \mathbb{H} \cong \mathbb{R}^4$ is the group of unit quaternions, obviously a Lie group.

Now let V be a right vector space over $\mathbb{H}$. Since $\mathbb{H}$ is not commutative, we have to distinguish between left and right vector spaces and we choose right ones as basic, so that matrices can multiply from the left. By choosing a basis we get $V = \mathbb{R}^n \otimes_{\mathbb{R}} \mathbb{H} = \mathbb{H}^n$. For $u = (u^i)$, $v = (v^i) \in \mathbb{H}^n$ we put $\langle u, v\rangle := \sum_{i=1}^n \overline{u}^i v^i$. Then $\langle\ ,\ \rangle$ is $\mathbb{R}$-bilinear and $\langle ua, vb\rangle = \overline{a}\langle u, v\rangle b$ for $a, b \in \mathbb{H}$.

An $\mathbb{R}$ linear mapping $A : V \to V$ is called *$\mathbb{H}$-linear* or *quaternionically linear* if $A(ua) = A(u)a$ holds. The space of all such mappings shall be denoted by $L_{\mathbb{H}}(V, V)$. It is real isomorphic to the space of all quaternionic $n \times n$-matrices with the usual multiplication, since for the standard basis $(e_i)_{i=1}^n$ in $V = \mathbb{H}^n$ we have $A(u) = A(\sum_i e_i u^i) = \sum_i A(e_i) u^i = \sum_{i,j} e_j A^j_i u^i$. Note that $L_{\mathbb{H}}(V, V)$ is only a real vector space, if V is a right quaternionic vector space - any further structure must come from a second (left) quaternionic vector space structure on V.

$GL(n, \mathbb{H})$, the group of invertible $\mathbb{H}$-linear mappings of $\mathbb{H}^n$, is a Lie group, because it is $GL(4n, \mathbb{R}) \cap L_{\mathbb{H}}(\mathbb{H}^n, \mathbb{H}^n)$, open in $L_{\mathbb{H}}(\mathbb{H}^n, \mathbb{H}^n)$.

A quaternionically linear mapping A is called isometric or *quaternionically unitary*, if $\langle A(u), A(v)\rangle = \langle u, v\rangle$ for all $u, v \in \mathbb{H}^n$. We denote by $Sp(n)$ the group of all quaternionic isometries of $\mathbb{H}^n$, the *quaternionic unitary group*. The reason for its name is that $Sp(n) = Sp(2n, \mathbb{C}) \cap U(2n)$, since we can decompose the quaternionic hermitian form $\langle\ ,\ \rangle$ into a complex hermitian one and a complex symplectic one. Also we have $Sp(n) \subset O(4n, \mathbb{R})$, since the real part of $\langle\ ,\ \rangle$ is a positive definite real inner product. For $A \in L_{\mathbb{H}}(\mathbb{H}^n, \mathbb{H}^n)$ we put $A^* := \overline{A}^t$. Then we have $\langle u, A(v)\rangle = \langle A^*(u), v\rangle$, so $\langle A(u), A(v)\rangle = \langle A^*A(u), v\rangle$. Thus $A \in Sp(n)$ if and only if $A^*A = \mathrm{Id}$.

Again $f : L_{\mathbb{H}}(\mathbb{H}^n, \mathbb{H}^n) \to L_{\mathbb{H},herm}(\mathbb{H}^n, \mathbb{H}^n) = \{A : A^* = A\}$, given by $f(A) = A^*A$, is a smooth mapping with $df(A)X = X^*A + A^*X$. So we have $\ker df(\mathrm{Id}) = \{X : X^* = -X\} =: \mathfrak{sp}(n)$, the space of quaternionic skew hermitian matrices. The usual proof shows that f has maximal rank on $GL(n, \mathbb{H})$, so $Sp(n) = f^{-1}(\mathrm{Id})$ is a compact real Lie group of dimension $2n(n-1) + 3n$.

The groups $SO(n, \mathbb{R})$ for $n \geq 3$, $SU(n)$ for $n \geq 2$, $Sp(n)$ for $n \geq 2$ and real forms of the exceptional complex Lie groups exhaust all simple compact Lie groups up to coverings.

4.11. Invariant vector fields and Lie algebras. Let G be a (real) Lie group. A vector field ξ on G is called *left invariant*, if $\lambda_a^* \xi = \xi$ for all $a \in G$, where $\lambda_a^* \xi = T(\lambda_{a^{-1}}) \circ \xi \circ \lambda_a$ as in section 3. Since by 3.11 we have $\lambda_a^*[\xi, \eta] = [\lambda_a^* \xi, \lambda_a^* \eta]$, the space $\mathfrak{X}_L(G)$ of all left invariant vector fields on G is closed under the Lie bracket, so it is a sub Lie algebra of $\mathfrak{X}(G)$. Any left invariant vector field ξ is uniquely determined by $\xi(e) \in T_eG$, since $\xi(a) = T_e(\lambda_a).\xi(e)$. Thus the Lie algebra $\mathfrak{X}_L(G)$ of left invariant vector fields is linearly isomorphic to T_eG, and on T_eG the Lie bracket on $\mathfrak{X}_L(G)$ induces a Lie algebra structure, whose bracket is again denoted by $[\ ,\ ]$. This Lie algebra will be denoted as usual by $\mathfrak{g}$, sometimes by $Lie(G)$.

We will also give a name to the isomorphism with the space of left invariant vector fields: $L : \mathfrak{g} \to \mathfrak{X}_L(G)$, $X \mapsto L_X$, where $L_X(a) = T_e\lambda_a.X$. Thus $[X,Y] = [L_X, L_Y](e)$.

A vector field η on G is called *right invariant*, if $\rho_a^*\eta = \eta$ for all $a \in G$. If ξ is left invariant, then $\nu^*\xi$ is right invariant, since $\nu \circ \rho_a = \lambda_{a^{-1}} \circ \nu$ implies that $\rho_a^*\nu^*\xi = (\nu \circ \rho_a)^*\xi = (\lambda_{a^{-1}} \circ \nu)^*\xi = \nu^*(\lambda_{a^{-1}})^*\xi = \nu^*\xi$. The right invariant vector fields form a sub Lie algebra $\mathfrak{X}_R(G)$ of $\mathfrak{X}(G)$, which is again linearly isomorphic to T_eG and induces also a Lie algebra structure on T_eG. Since $\nu^* : \mathfrak{X}_L(G) \to \mathfrak{X}_R(G)$ is an isomorphism of Lie algebras by 3.11, $T_e\nu = -\mathrm{Id} : T_eG \to T_eG$ is an isomorphism between the two Lie algebra structures. We will denote by $R : \mathfrak{g} = T_eG \to \mathfrak{X}_R(G)$ the isomorphism discussed, which is given by $R_X(a) = T_e(\rho_a).X$.

4.12. Lemma. *If L_X is a left invariant vector field and R_Y is a right invariant one, then $[L_X, R_Y] = 0$. Thus the flows of L_X and R_Y commute.*

Proof. We consider $0 \times L_X \in \mathfrak{X}(G \times G)$, given by $(0 \times L_X)(a,b) = (0_a, L_X(b))$. Then $T_{(a,b)}\mu.(0_a, L_X(b)) = T_a\rho_b.0_a + T_b\lambda_a.L_X(b) = L_X(ab)$, so $0 \times L_X$ is μ-related to L_X. Likewise $R_Y \times 0$ is μ-related to R_Y. But then $0 = [0 \times L_X, R_Y \times 0]$ is μ-related to $[L_X, R_Y]$ by 3.10. Since μ is surjective, $[L_X, R_Y] = 0$ follows. □

4.13. Let $\varphi : G \to H$ be a homomorphism of Lie groups, so for the time being we require φ to be smooth.

Lemma. *Then $\varphi' := T_e\varphi : \mathfrak{g} = T_eG \to \mathfrak{h} = T_eH$ is a Lie algebra homomorphism.*

Proof. For $X \in \mathfrak{g}$ and $x \in G$ we have

$$T_x\varphi.L_X(x) = T_x\varphi.T_e\lambda_x.X = T_e(\varphi \circ \lambda_x).X =$$
$$= T_e(\lambda_{\varphi(x)} \circ \varphi).X = T_e(\lambda_{\varphi(x)}).T_e\varphi.X = L_{\varphi'(X)}(\varphi(x)).$$

So L_X is φ-related to $L_{\varphi'(X)}$. By 3.10 the field $[L_X, L_Y] = L_{[X,Y]}$ is φ-related to $[L_{\varphi'(X)}, L_{\varphi'(Y)}] = L_{[\varphi'(X),\varphi'(Y)]}$. So we have $T\varphi \circ L_{[X,Y]} = L_{[\varphi'(X),\varphi'(Y)]} \circ \varphi$. If we evaluate this at e the result follows. □

Now we will determine the Lie algebras of all the examples given above.

4.14. For the Lie group $GL(n,\mathbb{R})$ we have $T_eGL(n,\mathbb{R}) = L(\mathbb{R}^n,\mathbb{R}^n) =: \mathfrak{gl}(n,\mathbb{R})$ and $T\,GL(n,\mathbb{R}) = GL(n,\mathbb{R}) \times L(\mathbb{R}^n,\mathbb{R}^n)$ by the affine structure of the surrounding vector space. For $A \in GL(n,\mathbb{R})$ we have $\lambda_A(B) = A.B$, so λ_A extends to a linear isomorphism of $L(\mathbb{R}^n,\mathbb{R}^n)$, and for $(B,X) \in T\,GL(n,\mathbb{R})$ we get $T_B(\lambda_A).(B,X) = (A.B, A.X)$. So the left invariant vector field $L_X \in \mathfrak{X}_L(GL(n,\mathbb{R}))$ is given by $L_X(A) = T_e(\lambda_A).X = (A, A.X)$.

Let $f : GL(n,\mathbb{R}) \to \mathbb{R}$ be the restriction of a linear functional on $L(\mathbb{R}^n,\mathbb{R}^n)$. Then we have $L_X(f)(A) = df(A)(L_X(A)) = df(A)(A.X) = f(A.X)$, which we may write as $L_X(f) = f(\quad .X)$. Therefore

$$L_{[X,Y]}(f) = [L_X, L_Y](f) = L_X(L_Y(f)) - L_Y(L_X(f)) =$$
$$= L_X(f(\quad .Y)) - L_Y(f(\quad .X)) =$$
$$= f(\quad .X.Y) - f(\quad .Y.X) = L_{XY-YX}(f).$$

So the Lie bracket on $\mathfrak{gl}(n,\mathbb{R}) = L(\mathbb{R}^n,\mathbb{R}^n)$ is given by $[X,Y] = XY - YX$, the usual commutator.

4.15. Example. Let V be a vector space. Then $(V,+)$ is a Lie group, $T_0V = V$ is its Lie algebra, $TV = V\times V$, left translation is $\lambda_v(w) = v+w$, $T_w(\lambda_v).(w,X) = (v+w,X)$. So $L_X(v) = (v,X)$, a constant vector field. Thus the Lie bracket is 0.

4.16. Example. The special linear group is $SL(n,\mathbb{R}) = \det^{-1}(1)$ and its Lie algebra is given by $T_eSL(n,\mathbb{R}) = \ker d\det(\mathbb{I}) = \{X \in L(\mathbb{R}^n,\mathbb{R}^n) : \operatorname{trace} X = 0\} = \mathfrak{sl}(n,\mathbb{R})$ by 4.6. The injection $i : SL(n,\mathbb{R}) \to GL(n,\mathbb{R})$ is a smooth homomorphism of Lie groups, so $T_ei = i' : \mathfrak{sl}(n,\mathbb{R}) \to \mathfrak{gl}(n,\mathbb{R})$ is an injective homomorphism of Lie algebras. Thus the Lie bracket is given by $[X,Y] = XY - YX$.

The same argument gives the commutator as the Lie bracket in all other examples we have treated. We have already determined the Lie algebras as T_eG.

4.17. One parameter subgroups. Let G be a Lie group with Lie algebra $\mathfrak{g}$. A *one parameter subgroup* of G is a Lie group homomorphism $\alpha : (\mathbb{R},+) \to G$, i.e. a smooth curve α in G with $\alpha(0) = e$ and $\alpha(s+t) = \alpha(s).\alpha(t)$.

Lemma. *Let $\alpha : \mathbb{R} \to G$ be a smooth curve with $\alpha(0) = e$. Let $X = \dot\alpha(0) \in \mathfrak{g}$. Then the following assertions are equivalent.*

(1) *α is a one parameter subgroup.*
(2) *$\alpha(t) = \operatorname{Fl}^{L_X}(t,e)$ for all t.*
(3) *$\alpha(t) = \operatorname{Fl}^{R_X}(t,e)$ for all t.*
(4) *$x.\alpha(t) = \operatorname{Fl}^{L_X}(t,x)$, or $\operatorname{Fl}^{L_X}_t = \rho_{\alpha(t)}$, for all t.*
(5) *$\alpha(t).x = \operatorname{Fl}^{R_X}(t,x)$, or $\operatorname{Fl}^{R_X}_t = \lambda_{\alpha(t)}$, for all t.*

Proof. (1) $\Longrightarrow$ (4).

$$\begin{aligned}\tfrac{d}{dt}x.\alpha(t) &= \tfrac{d}{ds}|_0 x.\alpha(t+s) = \tfrac{d}{ds}|_0 x.\alpha(t).\alpha(s) = \tfrac{d}{ds}|_0 \lambda_{x.\alpha(t)}\alpha(s)\\ &= T_e(\lambda_{x.\alpha(t)}).\tfrac{d}{ds}|_0\alpha(s) = L_X(x.\alpha(t)).\end{aligned}$$

By uniqueness of solutions we get $x.\alpha(t) = \operatorname{Fl}^{L_X}(t,x)$.

(4) $\Longrightarrow$ (2). This is clear.

(2) $\Longrightarrow$ (1). We have $\frac{d}{ds}\alpha(t)\alpha(s) = \frac{d}{ds}(\lambda_{\alpha(t)}\alpha(s)) = T(\lambda_{\alpha(t)})\frac{d}{ds}\alpha(s) = T(\lambda_{\alpha(t)})L_X(\alpha(s)) = L_X(\alpha(t)\alpha(s))$ and $\alpha(t)\alpha(0) = \alpha(t)$. So we get $\alpha(t)\alpha(s) = \operatorname{Fl}^{L_X}(s,\alpha(t)) = \operatorname{Fl}^{L_X}_s \operatorname{Fl}^{L_X}_t(e) = \operatorname{Fl}^{L_X}(t+s,e) = \alpha(t+s)$.

(4) $\Longleftrightarrow$ (5). We have $\operatorname{Fl}^{\nu^*\xi}_t = \nu^{-1}\circ\operatorname{Fl}^{\xi}_t\circ\nu$ by 3.14. Therefore we have by 4.11

$$\begin{aligned}(\operatorname{Fl}^{R_X}_t(x^{-1}))^{-1} &= (\nu\circ\operatorname{Fl}^{R_X}_t\circ\nu)(x) = \operatorname{Fl}^{\nu^*R_X}_t(x)\\ &= \operatorname{Fl}^{L_X}_{-t}(x) = x.\alpha(-t).\end{aligned}$$

So $\operatorname{Fl}^{R_X}_t(x^{-1}) = \alpha(t).x^{-1}$, and $\operatorname{Fl}^{R_X}_t(y) = \alpha(t).y$.

(5) $\Longrightarrow$ (3) $\Longrightarrow$ (1) can be shown in a similar way. $\square$

An immediate consequence of the foregoing lemma is that left invariant and the right invariant vector fields on a Lie group are always complete, so they have global flows, because a locally defined one parameter group can always be extended to a globally defined one by multiplying it up.

4.18. Definition. The *exponential mapping* $\exp : \mathfrak{g} \to G$ of a Lie group is defined by

$$\exp X = \mathrm{Fl}^{L_X}(1,e) = \mathrm{Fl}^{R_X}(1,e) = \alpha_X(1),$$

where α_X is the one parameter subgroup of G with $\dot\alpha_X(0) = X$.

Theorem.

(1) $\exp : \mathfrak{g} \to G$ *is smooth.*
(2) $\exp(tX) = \mathrm{Fl}^{L_X}(t,e)$.
(3) $\mathrm{Fl}^{L_X}(t,x) = x.\exp(tX)$.
(4) $\mathrm{Fl}^{R_X}(t,x) = \exp(tX).x$.
(5) $\exp(0) = e$ *and* $T_0 \exp = \mathrm{Id} : T_0\mathfrak{g} = \mathfrak{g} \to T_eG = \mathfrak{g}$, *thus* $\exp$ *is a diffeomorphism from a neighborhood of 0 in* $\mathfrak{g}$ *onto a neighborhood of* e *in* G.

Proof. (1) Let $0 \times L \in \mathfrak{X}(\mathfrak{g} \times G)$ be given by $(0 \times L)(X,x) = (0_X, L_X(x))$. Then $pr_2 \mathrm{Fl}^{0\times L}(t,(X,e)) = \alpha_X(t)$ is smooth in (t,X).

(2) $\exp(tX) = \mathrm{Fl}^{t.L_X}(1,e) = \mathrm{Fl}^{L_X}(t,e) = \alpha_X(t)$.

(3) and (4) follow from lemma 4.17.

(5) $T_0 \exp .X = \frac{d}{dt}|_0 \exp(0 + t.X) = \frac{d}{dt}|_0 \mathrm{Fl}^{L_X}(t,e) = X$. □

4.19. Remark. If G is connected and $U \subset \mathfrak{g}$ is open with $0 \in U$, then the group generated by $\exp(U)$ equals G.

For this group is a subgroup of G containing some open neighborhood of e, so it is open. The complement in G is also open (as union of the other cosets), so this subgroup is open and closed. Since G is connected, it coincides with G.

If G is not connected, then the subgroup generated by $\exp(U)$ is the connected component of e in G.

4.20. Remark. Let $\varphi : G \to H$ be a smooth homomorphism of Lie groups. Then the diagram

$$\begin{array}{ccc} \mathfrak{g} & \xrightarrow{\varphi'} & \mathfrak{h} \\ \downarrow{\scriptstyle \exp^G} & & \downarrow{\scriptstyle \exp^H} \\ G & \xrightarrow{\varphi} & H \end{array}$$

commutes, since $t \mapsto \varphi(\exp^G(tX))$ is a one parameter subgroup of H and $\frac{d}{dt}|_0 \varphi(\exp^G tX) = \varphi'(X)$, so $\varphi(\exp^G tX) = \exp^H(t\varphi'(X))$.

If G is connected and $\varphi, \psi : G \to H$ are homomorphisms of Lie groups with $\varphi' = \psi' : \mathfrak{g} \to \mathfrak{h}$, then $\varphi = \psi$. For $\varphi = \psi$ on the subgroup generated by $\exp^G \mathfrak{g}$ which equals G by 4.19.

4.21. Theorem. *A continuous homomorphism $\varphi : G \to H$ between Lie groups is smooth. In particular a topological group can carry at most one compatible Lie group structure.*

Proof. Let first $\varphi = \alpha : (\mathbb{R}, +) \to G$ be a continuous one parameter subgroup. Then $\alpha(-\varepsilon, \varepsilon) \subset \exp(U)$, where U is an absolutely convex open neighborhood of 0 in $\mathfrak{g}$ such that $\exp|2U$ is a diffeomorphism, for some $\varepsilon > 0$. Put $\beta := (\exp|2U)^{-1} \circ \alpha : (-\varepsilon, \varepsilon) \to \mathfrak{g}$. Then for $|t| < \frac{1}{\varepsilon}$ we have $\exp(2\beta(t)) = \exp(\beta(t))^2 = \alpha(t)^2 = \alpha(2t) = \exp(\beta(2t))$, so $2\beta(t) = \beta(2t)$; thus $\beta(\frac{s}{2}) = \frac{1}{2}\beta(s)$ for $|s| < \varepsilon$. So we have $\alpha(\frac{s}{2}) = \exp(\beta(\frac{s}{2})) = \exp(\frac{1}{2}\beta(s))$ for all $|s| < \varepsilon$ and by recursion we get $\alpha(\frac{s}{2^n}) = \exp(\frac{1}{2^n}\beta(s))$ for $n \in \mathbb{N}$ and in turn $\alpha(\frac{k}{2^n}s) = \alpha(\frac{s}{2^n})^k = \exp(\frac{1}{2^n}\beta(s))^k = \exp(\frac{k}{2^n}\beta(s))$ for $k \in \mathbb{Z}$. Since the $\frac{k}{2^n}$ for $k \in \mathbb{Z}$ and $n \in \mathbb{N}$ are dense in R and since α is continuous we get $\alpha(ts) = \exp(t\beta(s))$ for all $t \in \mathbb{R}$. So α is smooth.

Now let $\varphi : G \to H$ be a continuous homomorphism. Let $X_1, \dots, X_n$ be a linear basis of $\mathfrak{g}$. We define $\psi : \mathbb{R}^n \to G$ as $\psi(t^1, \dots, t^n) = \exp(t^1 X_1) \cdots \exp(t^n X_n)$. Then $T_0\psi$ is invertible, so ψ is a diffeomorphism near 0. Sometimes ψ^{-1} is called a coordinate system of the second kind. $t \mapsto \varphi(\exp^G tX_i)$ is a continuous one parameter subgroup of H, so it is smooth by the first part of the proof. We have $(\varphi \circ \psi)(t^1, \dots, t^n) = (\varphi \exp(t^1 X_1)) \cdots (\varphi \exp(t^n X_n))$, so $\varphi \circ \psi$ is smooth. Thus φ is smooth near $e \in G$ and consequently everywhere on G. □

4.22. Theorem. *Let G and H be Lie groups (G separable is essential here), and let $\varphi : G \to H$ be a continuous bijective homomorphism. Then φ is a diffeomorphism.*

Proof. Our first aim is to show that φ is a homeomorphism. Let V be an open e-neighborhood in G, and let K be a compact e-neighborhood in G such that $K.K^{-1} \subset V$. Since G is separable there is a sequence $(a_i)_{i\in\mathbb{N}}$ in G such that $G = \bigcup_{i=1}^{\infty} a_i.K$. Since H is locally compact, it is a Baire space (V_i open and dense implies $\bigcap V_i$ dense). The set $\varphi(a_i)\varphi(K)$ is compact, thus closed. Since $H = \bigcup_i \varphi(a_i).\varphi(K)$, there is some i such that $\varphi(a_i)\varphi(K)$ has non empty interior, so $\varphi(K)$ has non empty interior. Choose $b \in G$ such that $\varphi(b)$ is an interior point of $\varphi(K)$ in H. Then $e_H = \varphi(b)\varphi(b^{-1})$ is an interior point of $\varphi(K)\varphi(K^{-1}) \subset \varphi(V)$. So if U is open in G and $a \in U$, then e_H is an interior point of $\varphi(a^{-1}U)$, so $\varphi(a)$ is in the interior of $\varphi(U)$. Thus $\varphi(U)$ is open in H, and φ is a homeomorphism.

Now by 4.21 φ and φ^{-1} are smooth. □

4.23. Examples. The exponential mapping on $GL(n,\mathbb{R})$. Let $X \in \mathfrak{gl}(n,\mathbb{R}) = L(\mathbb{R}^n, \mathbb{R}^n)$, then the left invariant vector field is given by $L_X(A) = (A, A.X) \in GL(n,\mathbb{R}) \times \mathfrak{gl}(n,\mathbb{R})$ and the one parameter group $\alpha_X(t) = \mathrm{Fl}^{L_X}(t, \mathbb{I})$ is given by the differential equation $\frac{d}{dt}\alpha_X(t) = L_X(\alpha_X(t)) = \alpha_X(t).X$, with initial condition $\alpha_X(0) = \mathbb{I}$. But the unique solution of this equation is $\alpha_X(t) = e^{tX} = \sum_{k=0}^{\infty} \frac{t^k}{k!} X^k$. So

$$\exp^{GL(n,\mathbb{R})}(X) = e^X = \sum_{k=0}^{\infty} \frac{1}{k!} X^k.$$

If $n = 1$ we get the usual exponential mapping of one real variable. For all Lie subgroups of $GL(n,\mathbb{R})$, the exponential mapping is given by the same formula $\exp(X) = e^X$; this follows from 4.20.

4.24. The adjoint representation. A *representation* of a Lie group G on a finite dimensional vector space V (real or complex) is a homomorphism $\rho : G \to GL(V)$ of Lie groups. Then by 4.13 $\rho' : \mathfrak{g} \to \mathfrak{gl}(V) = L(V,V)$ is a Lie algebra homomorphism.

For $a \in G$ we define $\operatorname{conj}_a : G \to G$ by $\operatorname{conj}_a(x) = axa^{-1}$. It is called the *conjugation* or the *inner automorphism* by $a \in G$. We have $\operatorname{conj}_a(xy) = \operatorname{conj}_a(x)\operatorname{conj}_a(y)$, $\operatorname{conj}_{ab} = \operatorname{conj}_a \circ \operatorname{conj}_b$, and conj is smooth in all variables.

Next we define for $a \in G$ the mapping $\operatorname{Ad}(a) = (\operatorname{conj}_a)' = T_e(\operatorname{conj}_a) : \mathfrak{g} \to \mathfrak{g}$. By 4.13 $\operatorname{Ad}(a)$ is a Lie algebra homomorphism, so we have $\operatorname{Ad}(a)[X,Y] = [\operatorname{Ad}(a)X, \operatorname{Ad}(a)Y]$. Furthermore $\operatorname{Ad} : G \to GL(\mathfrak{g})$ is a representation, called the *adjoint representation* of G, since $\operatorname{Ad}(ab) = T_e(\operatorname{conj}_{ab}) = T_e(\operatorname{conj}_a \circ \operatorname{conj}_b) = T_e(\operatorname{conj}_a) \circ T_e(\operatorname{conj}_b) = \operatorname{Ad}(a) \circ \operatorname{Ad}(b)$. We will use the relations $\operatorname{Ad}(a) = T_e(\operatorname{conj}_a) = T_a(\rho_{a^{-1}}).T_e(\lambda_a) = T_{a^{-1}}(\lambda_a).T_e(\rho_{a^{-1}})$.

Finally we define the (lower case) *adjoint representation* of the Lie algebra $\mathfrak{g}$, $\operatorname{ad} : \mathfrak{g} \to \mathfrak{gl}(\mathfrak{g}) = L(\mathfrak{g},\mathfrak{g})$, by $\operatorname{ad} := \operatorname{Ad}' = T_e \operatorname{Ad}$.

Lemma. *(1) $L_X(a) = R_{\operatorname{Ad}(a)X}(a)$ for $X \in \mathfrak{g}$ and $a \in G$.*
(2) $\operatorname{ad}(X)Y = [X,Y]$ for $X, Y \in \mathfrak{g}$.

Proof. (1) $L_X(a) = T_e(\lambda_a).X = T_e(\rho_a).T_e(\rho_{a^{-1}} \circ \lambda_a).X = R_{\operatorname{Ad}(a)X}(a)$.

(2) Let $X_1, \dots, X_n$ be a linear basis of $\mathfrak{g}$ and fix $X \in \mathfrak{g}$. Then $\operatorname{Ad}(x)X = \sum_{i=1}^n f_i(x).X_i$ for $f_i \in C^\infty(G,\mathbb{R})$ and we have in turn

$$\begin{aligned}
\operatorname{Ad}'(Y)X &= T_e(\operatorname{Ad}(\quad)X)Y = d(\operatorname{Ad}(\quad)X)|_e Y = d(\textstyle\sum f_i X_i)|_e Y = \\
&= \textstyle\sum df_i|_e(Y)X_i = \sum L_Y(f_i)(e).X_i. \\
L_X(x) &= R_{\operatorname{Ad}(x)X}(x) = R(\textstyle\sum f_i(x)X_i)(x) = \sum f_i(x).R_{X_i}(x) \text{ by (1).} \\
[L_Y, L_X] &= [L_Y, \textstyle\sum f_i.R_{X_i}] = 0 + \sum L_Y(f_i).R_{X_i} \text{ by 3.4 and 4.12.} \\
[Y,X] &= [L_Y, L_X](e) = \textstyle\sum L_Y(f_i)(e).R_{X_i}(e) = \operatorname{Ad}'(Y)X = \operatorname{ad}(Y)X. \qquad \square
\end{aligned}$$

4.25. Corollary. *From 4.20 and 4.23 we have*

$$\begin{aligned}
\operatorname{Ad} \circ exp^G &= exp^{GL(\mathfrak{g})} \circ \operatorname{ad} \\
\operatorname{Ad}(exp^G X)Y &= \sum_{k=0}^{\infty} \tfrac{1}{k!} (\operatorname{ad} X)^k Y = e^{\operatorname{ad} X} Y \\
&= Y + [X,Y] + \tfrac{1}{2!}[X,[X,Y]] + \tfrac{1}{3!}[X,[X,[X,Y]]] + \cdots \qquad \square
\end{aligned}$$

4.26. The right logarithmic derivative. Let M be a manifold and let $f : M \to G$ be a smooth mapping into a Lie group G with Lie algebra $\mathfrak{g}$. We define the mapping $\delta f : TM \to \mathfrak{g}$ by the formula $\delta f(\xi_x) := T_{f(x)}(\rho_{f(x)^{-1}}).T_x f.\xi_x$. Then δf is a $\mathfrak{g}$-valued 1-form on M, $\delta f \in \Omega^1(M,\mathfrak{g})$, as we will write later. We call δf the *right logarithmic derivative* of f, since for $f : \mathbb{R} \to (\mathbb{R}^+, \cdot)$ we have $\delta f(x).1 = \frac{f'(x)}{f(x)} = (\log \circ f)'(x)$.

Lemma. *Let $f, g : M \to G$ be smooth. Then we have*

$$\delta(f.g)(x) = \delta f(x) + \mathrm{Ad}(f(x)).\delta g(x).$$

Proof. We just compute:

$$\begin{aligned}\delta(f.g)(x) &= T(\rho_{g(x)^{-1}.f(x)^{-1}}).T_x(f.g) = \\ &= T(\rho_{f(x)^{-1}}).T(\rho_{g(x)^{-1}}).T_{(f(x),g(x))}\mu.(T_x f, T_x g) = \\ &= T(\rho_{f(x)^{-1}}).T(\rho_{g(x)^{-1}}).\left(T(\rho_{g(x)}).T_x f + T(\lambda_{f(x)}).T_x g\right) = \\ &= \delta f(x) + \mathrm{Ad}(f(x)).\delta g(x). \quad \square\end{aligned}$$

Remark. The *left logarithmic derivative* $\delta^{\mathrm{left}} f \in \Omega^1(M, \mathfrak{g})$ of a smooth mapping $f : M \to G$ is given by $\delta^{\mathrm{left}} f.\xi_x = T_{f(x)}(\lambda_{f(x)^{-1}}).T_x f.\xi_x$. The corresponding Leibnitz rule for it is uglier than that for the right logarithmic derivative:

$$\delta^{\mathrm{left}}(fg)(x) = \delta^{\mathrm{left}} g(x) + \mathrm{Ad}(g(x)^{-1})\delta^{\mathrm{left}} f(x).$$

The form $\delta^{\mathrm{left}}(\mathrm{Id}_G) \in \Omega^1(G; \mathfrak{g})$ is also called the *Maurer-Cartan form* of the Lie group G.

4.27. Lemma. *For $\exp : \mathfrak{g} \to G$ and for $g(z) := \dfrac{e^z - 1}{z}$ we have*

$$\delta(\exp)(X) = T(\rho_{\exp(-X)}).T_X \exp = \sum_{p=0}^{\infty} \tfrac{1}{(p+1)!} (\mathrm{ad}\, X)^p = g(\mathrm{ad}\, X).$$

Proof. Wc put $M(X) - \delta(\exp)(X) : \mathfrak{g} \to \mathfrak{g}$. Then

$$\begin{aligned}(s+t)M((s+t)X) &= (s+t)\delta(\exp)((s+t)X) \\ &= \delta(\exp((s+t)\quad))X \qquad \text{by the chain rule,} \\ &= \delta(\exp(s\quad).\exp(t\quad)).X \\ &= \delta(\exp(s\quad)).X + \mathrm{Ad}(\exp(sX)).\delta(\exp(t\quad)).X \qquad \text{by 4.26,} \\ &= s.\delta(\exp)(sX) + \mathrm{Ad}(\exp(sX)).t.\delta(\exp)(tX) \\ &= s.M(sX) + \mathrm{Ad}(\exp(sX)).t.M(tX).\end{aligned}$$

Now we put $N(t) := t.M(tX) \in L(\mathfrak{g}, \mathfrak{g})$, then the above equation gives $N(s+t) = N(s) + \mathrm{Ad}(\exp(sX)).N(t)$. We fix t, apply $\frac{d}{ds}|_0$, and get $N'(t) = N'(0) + \mathrm{ad}(X).N(t)$, where $N'(0) = M(0) + 0 = \delta(\exp)(0) = \mathrm{Id}_{\mathfrak{g}}$. So we have the differential equation $N'(t) = \mathrm{Id}_{\mathfrak{g}} + \mathrm{ad}(X).N(t)$ in $L(\mathfrak{g}, \mathfrak{g})$ with initial condition $N(0) = 0$. The unique solution is

$$N(s) = \sum_{p=0}^{\infty} \tfrac{1}{(p+1)!}\ \mathrm{ad}(X)^p.s^{p+1}, \quad \text{and so}$$

$$\delta(\exp)(X) = M(X) = N(1) = \sum_{p=0}^{\infty} \tfrac{1}{(p+1)!}\ \mathrm{ad}(X)^p. \quad \square$$

4.28. Corollary. *$T_X \exp$ is bijective if and only if no eigenvalue of $\operatorname{ad}(X) : \mathfrak{g} \to \mathfrak{g}$ is of the form $\sqrt{-1}\,2k\pi$ for $k \in \mathbb{Z} \setminus \{0\}$.*

Proof. The zeros of $g(z) = \frac{e^z - 1}{z}$ are exactly $z = \sqrt{-1}\,2k\pi$ for $k \in \mathbb{Z} \setminus \{0\}$. The linear mapping $T_X \exp$ is bijective if and only if no eigenvalue of $g(\operatorname{ad}(X)) = T(\rho_{\exp(-X)}).T_X \exp$ is 0. But the eigenvalues of $g(\operatorname{ad}(X))$ are the images under g of the eigenvalues of $\operatorname{ad}(X)$. $\square$

4.29. Theorem. The Baker-Campbell-Hausdorff formula.
Let G be a Lie group with Lie algebra $\mathfrak{g}$. For complex z near 1 we consider the function $f(z) := \frac{\log(z)}{z-1} = \sum_{n\geq 0} \frac{(-1)^n}{n+1}(z-1)^n$.

Then for X, Y near 0 in $\mathfrak{g}$ we have $\exp X . \exp Y = \exp C(X,Y)$, where

$$\begin{aligned}
C(X,Y) &= Y + \int_0^1 f(e^{t.\operatorname{ad} X}.e^{\operatorname{ad} Y}).X\,dt \\
&= X + Y + \sum_{n\geq 1} \frac{(-1)^n}{n+1} \int_0^1 \Big(\sum_{\substack{k,\ell\geq 0\\ k+\ell\geq 1}} \frac{t^k}{k!\,\ell!} (\operatorname{ad} X)^k (\operatorname{ad} Y)^\ell \Big)^n X\,dt \\
&= X + Y + \sum_{n\geq 1} \frac{(-1)^n}{n+1} \sum_{\substack{k_1,\dots,k_n\geq 0\\ \ell_1,\dots,\ell_n\geq 0\\ k_i+\ell_i\geq 1}} \frac{(\operatorname{ad} X)^{k_1}(\operatorname{ad} Y)^{\ell_1}\dots(\operatorname{ad} X)^{k_n}(\operatorname{ad} Y)^{\ell_n}}{(k_1+\dots+k_n+1)k_1!\dots k_n!\ell_1!\dots\ell_n!} X \\
&= X + Y + \tfrac{1}{2}[X,Y] + \tfrac{1}{12}([X,[X,Y]] - [Y,[Y,X]]) + \cdots
\end{aligned}$$

Proof. Let $C(X,Y) := \exp^{-1}(\exp X . \exp Y)$ for X, Y near 0 in $\mathfrak{g}$, and let $C(t) := C(tX, Y)$. Then

$$\begin{aligned}
T(\rho_{\exp(-C(t))}) \tfrac{d}{dt}(\exp C(t)) &= \delta \exp(C(t)).\dot{C}(t) \\
&= \textstyle\sum_{k\geq 0} \frac{1}{(k+1)!} (\operatorname{ad} C(t))^k \dot{C}(t) = g(\operatorname{ad} C(t)).\dot{C}(t),
\end{aligned}$$

where $g(z) := \frac{e^z-1}{z} = \sum_{k\geq 0} \frac{z^k}{(k+1)!}$. We have $\exp C(t) = \exp(tX)\exp Y$ and $\exp(-C(t)) = \exp(C(t))^{-1} = \exp(-Y)\exp(-tX)$, therefore

$$\begin{aligned}
T(\rho_{\exp(-C(t))}) \tfrac{d}{dt}(\exp C(t)) &= T(\rho_{\exp(-Y)\exp(-tX)}) \tfrac{d}{dt}(\exp(tX)\exp Y) \\
&= T(\rho_{\exp(-tX)}) T(\rho_{\exp(-Y)}) T(\rho_{\exp Y}) \tfrac{d}{dt} \exp(tX) \\
&= T(\rho_{\exp(-tX)}).R_X(\exp(tX)) = X, \quad \text{by 4.18.4 and 4.11.}
\end{aligned}$$

$$X = g(\operatorname{ad} C(t)).\dot{C}(t).$$

$$\begin{aligned}
e^{\operatorname{ad} C(t)} &= \operatorname{Ad}(\exp C(t)) \qquad \text{by 4.25} \\
&= \operatorname{Ad}(\exp(tX)\exp Y) = \operatorname{Ad}(\exp(tX)).\operatorname{Ad}(\exp Y) \\
&= e^{\operatorname{ad}(tX)}.e^{\operatorname{ad} Y} = e^{t.\operatorname{ad} X}.e^{\operatorname{ad} Y}.
\end{aligned}$$

If X, Y, and t are small enough we get $\operatorname{ad} C(t) = \log(e^{t.\operatorname{ad} X}.e^{\operatorname{ad} Y})$, where $\log(z) = \sum_{n\geq 1} \frac{(-1)^{n+1}}{n}(z-1)^n$, thus we have

$$X = g(\operatorname{ad} C(t)).\dot{C}(t) = g(\log(e^{t.\operatorname{ad} X}.e^{\operatorname{ad} Y})).\dot{C}(t).$$

For z near 1 we put $f(z) := \frac{\log(z)}{z-1} = \sum_{n\geq 0} \frac{(-1)^n}{n+1}(z-1)^n$, which satisfies $g(\log(z)).f(z) = 1$. So we have

$$X = g(\log(e^{t.\operatorname{ad} X}.e^{\operatorname{ad} Y})).\dot{C}(t) = f(e^{t.\operatorname{ad} X}.e^{\operatorname{ad} Y})^{-1}.\dot{C}(t),$$

$$\begin{cases} \dot{C}(t) = f(e^{t.\operatorname{ad} X}.e^{\operatorname{ad} Y}).X, \\ C(0) = Y. \end{cases}$$

Passing to the definite integral we get the desired formula

$$\begin{aligned} C(X,Y) &= C(1) = C(0) + \int_0^1 \dot{C}(t)\, dt \\ &= Y + \int_0^1 f(e^{t.\operatorname{ad} X}.e^{\operatorname{ad} Y}).X\, dt \\ &= X + Y + \sum_{n\geq 1} \frac{(-1)^n}{n+1} \int_0^1 \Big(\sum_{\substack{k,\ell\geq 0\\ k+\ell\geq 1}} \frac{t^k}{k!\,\ell!} (\operatorname{ad} X)^k (\operatorname{ad} Y)^\ell \Big)^n X\, dt. \quad \square \end{aligned}$$

Remark. If G is a Lie group of differentiability class C^2, then we may define TG and the Lie bracket of vector fields. The proof above then makes sense and the theorem shows, that in the chart given by $\exp^{-1}$ the multiplication $\mu : G \times G \to G$ is C^ω near e, hence everywhere. So in this case G is a real analytic Lie group. See also remark 5.6 below.

4.30. Convention. We will use the following convention for the rest of the book. If we write a symbol of a classical group from this section without indicating the ground field, then we always mean the field $\mathbb{R}$ — except $Sp(n)$. In particular $GL(n) = GL(n,\mathbb{R})$, and $O(n) = O(n,\mathbb{R})$ from now on.

5. Lie subgroups and homogeneous spaces

5.1. Definition. Let G be a Lie group. A subgroup H of G is called a *Lie subgroup*, if H is itself a Lie group (so it is separable) and the inclusion $i : H \to G$ is smooth.

In this case the inclusion is even an immersion. For that it suffices to check that $T_e i$ is injective: If $X \in \mathfrak{h}$ is in the kernel of $T_e i$, then $i \circ \exp^H(tX) = \exp^G(t.T_e i.X) = e$. Since i is injective, $X = 0$.

From the next result it follows that $H \subset G$ is then an initial submanifold in the sense of 2.14: If H_0 is the connected component of H, then $i(H_0)$ is the Lie subgroup of G generated by $i'(\mathfrak{h}) \subset \mathfrak{g}$, which is an initial submanifold, and this is true for all components of H.

5.2. Theorem. *Let G be a Lie group with Lie algebra $\mathfrak{g}$. If $\mathfrak{h} \subset \mathfrak{g}$ is a Lie subalgebra, then there is a unique connected Lie subgroup H of G with Lie algebra $\mathfrak{h}$. H is an initial submanifold.*

Proof. Put $E_x := \{T_e(\lambda_x).X : X \in \mathfrak{h}\} \subset T_x G$. Then $E := \bigsqcup_{x\in G} E_x$ is a distribution of constant rank on G, in the sense of 3.18. The set $\{L_X : X \in \mathfrak{h}\}$

is an involutive set in the sense of 3.23 which spans E. So by theorem 3.25 the distribution E is integrable and by theorem 3.22 the leaf H through e is an initial submanifold. It is even a subgroup, since for $x \in H$ the initial submanifold $\lambda_x H$ is again a leaf (since E is left invariant) and intersects H (in x), so $\lambda_x(H) = H$. Thus $H.H = H$ and consequently $H^{-1} = H$. The multiplication $\mu : H \times H \to G$ is smooth by restriction, and smooth as a mapping $H \times H \to H$, since H is an initial submanifold, by lemma 2.17. □

5.3. Theorem. *Let $\mathfrak{g}$ be a finite dimensional real Lie algebra. Then there exists a connected Lie group G whose Lie algebra is $\mathfrak{g}$.*

Sketch of Proof. By the theorem of Ado (see [Jacobson, 62] or [Varadarajan, 74, p. 237]) $\mathfrak{g}$ has a faithful (i.e. injective) representation on a finite dimensional vector space V, i.e. $\mathfrak{g}$ can be viewed as a Lie subalgebra of $\mathfrak{gl}(V) = L(V, V)$. By theorem 5.2 above there is a Lie subgroup G of $GL(V)$ with $\mathfrak{g}$ as its Lie algebra. □

This is a rather involved proof, since the theorem of Ado needs the structure theory of Lie algebras for its proof. There are simpler proofs available, starting from a neighborhood of e in G (a neighborhood of 0 in $\mathfrak{g}$ with the Baker-Campbell-Hausdorff formula 4.29 as multiplication) and extending it.

5.4. Theorem. *Let G and H be Lie groups with Lie algebras $\mathfrak{g}$ and $\mathfrak{h}$, respectively. Let $f : \mathfrak{g} \to \mathfrak{h}$ be a homomorphism of Lie algebras. Then there is a Lie group homomorphism φ, locally defined near e, from G to H, such that $\varphi' = T_e\varphi = f$. If G is simply connected, then there is a globally defined homomorphism of Lie groups $\varphi : G \to H$ with this property.*

Proof. Let $\mathfrak{k} := \operatorname{graph}(f) \subset \mathfrak{g} \times \mathfrak{h}$. Then $\mathfrak{k}$ is a Lie subalgebra of $\mathfrak{g} \times \mathfrak{h}$, since f is a homomorphism of Lie algebras. $\mathfrak{g} \times \mathfrak{h}$ is the Lie algebra of $G \times H$, so by theorem 5.2 there is a connected Lie subgroup $K \subset G \times H$ with algebra $\mathfrak{k}$. We consider the homomorphism $g := pr_1 \circ incl : K \to G \times H \to G$, whose tangent mapping satisfies $T_e g(X, f(X)) = T_{(e,e)}pr_1.T_e incl.(X, f(X)) = X$, so is invertible. Thus g is a local diffeomorphism, so $g : K \to G_0$ is a covering of the connected component G_0 of e in G. If G is simply connected, g is an isomorphism. Now we consider the homomorphism $\psi := pr_2 \circ incl : K \to G \times H \to H$, whose tangent mapping satisfies $T_e\psi.(X, f(X)) = f(X)$. We see that $\varphi := \psi \circ (g|U)^{-1} : G \supset U \to H$ solves the problem, where U is an e-neighborhood in K such that $g|U$ is a diffeomorphism. If G is simply connected, $\varphi = \psi \circ g^{-1}$ is the global solution. □

5.5. Theorem. *Let H be a closed subgroup of a Lie group G. Then H is a Lie subgroup and a submanifold of G.*

Proof. Let $\mathfrak{g}$ be the Lie algebra of G. We consider the subset $\mathfrak{h} := \{c'(0) : c \in C^\infty(\mathbb{R}, G), c(\mathbb{R}) \subset H, c(0) = e\}$.

Claim 1. $\mathfrak{h}$ is a linear subspace.

If $c_i'(0) \in \mathfrak{h}$ and $t_i \in \mathbb{R}$, we define $c(t) := c_1(t_1.t).c_2(t_2.t)$. Then $c'(0) = T_{(e,e)}\mu.(t_1.c_1'(0), t_2.c_2'(0)) = t_1.c_1'(0) + t_2.c_2'(0) \in \mathfrak{h}$.

Claim 2. $\mathfrak{h} = \{X \in \mathfrak{g} : \exp(tX) \in H \text{ for all } t \in \mathbb{R}\}$.

Clearly we have '$\supseteq$'. To check the other inclusion, let $X = c'(0) \in \mathfrak{h}$ and consider

$v(t) := (\exp^G)^{-1}c(t)$ for small t. Then we get $X = c'(0) = \frac{d}{dt}|_0 \exp(v(t)) = v'(0) = \lim_{n\to\infty} n.v(\frac{1}{n})$. We put $t_n = \frac{1}{n}$ and $X_n = n.v(\frac{1}{n})$, so that $\exp(t_n.X_n) = \exp(v(\frac{1}{n})) = c(\frac{1}{n}) \in H$. By claim 3 below we then get $\exp(tX) \in H$ for all t.

Claim 3. Let $X_n \to X$ in $\mathfrak{g}$, $0 < t_n \to 0$ in $\mathbb{R}$ with $\exp(t_nX_n) \in H$. Then $\exp(tX) \in H$ for all $t \in R$.

Let $t \in \mathbb{R}$ and take $m_n \in (\frac{t}{t_n} - 1, \frac{t}{t_n}] \cap \mathbb{Z}$. Then $t_n.m_n \to t$ and $m_n.t_n.X_n \to tX$, and since H is closed we may conclude that $\exp(tX) = \lim_n \exp(m_n.t_n.X_n) = \lim_n \exp(t_n.X_n)^{m_n} \in H$.

Claim 4. Let $\mathfrak{k}$ be a complementary linear subspace for $\mathfrak{h}$ in $\mathfrak{g}$. Then there is an open 0-neighborhood W in $\mathfrak{k}$ such that $\exp(W) \cap H = \{e\}$.

If not there are $0 \neq Y_k \in \mathfrak{k}$ with $Y_k \to 0$ such that $\exp(Y_k) \in H$. Choose a norm $|\quad|$ on $\mathfrak{g}$ and let $X_n = Y_n/|Y_n|$. Passing to a subsequence we may assume that $X_n \to X$ in $\mathfrak{k}$, then $|X| = 1$. But $\exp(|Y_n|.X_n) = \exp(Y_n) \in H$ and $0 < |Y_n| \to 0$, so by claim 3 we have $\exp(tX) \in H$ for all $t \in \mathbb{R}$. So by claim 2 $X \in \mathfrak{h}$, a contradiction.

Claim 5. Put $\varphi : \mathfrak{h} \times \mathfrak{k} \to G$, $\varphi(X,Y) = \exp X.\exp Y$. Then there are 0-neighborhoods V in $\mathfrak{h}$, W in $\mathfrak{k}$, and an e-neighborhood U in G such that $\varphi : V \times W \to U$ is a diffeomorphism and $U \cap H = \exp(V)$.

Choose V, W, and U so small that φ becomes a diffeomorphism. By claim 4 W may be chosen so small that $\exp(W) \cap H = \{e\}$. By claim 2 we have $\exp(V) \subseteq H \cap U$. Let $x \in H \cap U$. Since $x \in U$ we have $x = \exp X.\exp Y$ for unique $(X,Y) \in V \times W$. Then x and $\exp X \in H$, so $\exp Y \in H \cap \exp(W)$, thus $Y = 0$. So $x = \exp X \in \exp(V)$.

Claim 6. H is a submanifold and a Lie subgroup.

$(U, (\varphi|V \times W)^{-1} =: u)$ is a submanifold chart for H centered at e by claim 5. For $x \in H$ the pair $(\lambda_x(U), u \circ \lambda_{x^{-1}})$ is a submanifold chart for H centered at x. So H is a closed submanifold of G, and the multiplication is smooth since it is a restriction. □

5.6. Remark. The following stronger results on subgroups and the relation between topological groups and Lie groups in general are available.

Any arc wise connected subgroup of a Lie group is a connected Lie subgroup, [Yamabe, 50].

Let G be a separable locally compact topological group. If it has an e-neighborhood which does not contain a proper subgroup, then G is a Lie group. This is the solution of the 5-th problem of Hilbert, see the book [Montgomery-Zippin, 55, p. 107].

Any subgroup H of a Lie group G has a coarsest Lie group structure, but it might be non separable. To indicate a proof of this statement, consider all continuous curves $c : \mathbb{R} \to G$ with $c(\mathbb{R}) \subset H$, and equip H with the final topology with respect to them. Then the component of the identity satisfies the conditions of the Gleason-Yamabe theorem cited above.

5.7. Let $\mathfrak{g}$ be a Lie algebra. An *ideal* $\mathfrak{k}$ in $\mathfrak{g}$ is a linear subspace $\mathfrak{k}$ such that $[\mathfrak{k}, \mathfrak{g}] \subset \mathfrak{k}$. Then the quotient space $\mathfrak{g}/\mathfrak{k}$ carries a unique Lie algebra structure such that $\mathfrak{g} \to \mathfrak{g}/\mathfrak{k}$ is a Lie algebra homomorphism.

Lemma. *A connected Lie subgroup H of a connected Lie group G is a normal subgroup if and only if its Lie algebra $\mathfrak{h}$ is an ideal in $\mathfrak{g}$.*

Proof. H normal in G means $xHx^{-1} = \operatorname{conj}_x(H) \subset H$ for all $x \in G$. By remark 4.20 this is equivalent to $T_e(\operatorname{conj}_x)(\mathfrak{h}) \subset \mathfrak{h}$, i.e. $\operatorname{Ad}(x)\mathfrak{h} \subset \mathfrak{h}$, for all $x \in G$. But this in turn is equivalent to $\operatorname{ad}(X)\mathfrak{h} \subset \mathfrak{h}$ for all $X \in \mathfrak{g}$, so to the fact that $\mathfrak{h}$ is an ideal in $\mathfrak{g}$. □

5.8. Let G be a connected Lie group. If $A \subset G$ is an arbitrary subset, the *centralizer* of A in G is the closed subgroup $Z_A := \{x \in G : xa = ax \text{ for all } a \in A\}$.

The Lie algebra $\mathfrak{z}_A$ of Z_A then consists of all $X \in \mathfrak{g}$ such that $a.\exp(tX).a^{-1} = \exp(tX)$ for all $a \in A$, i.e. $\mathfrak{z}_A = \{X \in \mathfrak{g} : \operatorname{Ad}(a)X = X \text{ for all } a \in A\}$.

If A is itself a connected Lie subgroup of G, then $\mathfrak{z}_A = \{X \in \mathfrak{g} : \operatorname{ad}(Y)X = 0 \text{ for all } Y \in \mathfrak{a}\}$. This set is also called the *centralizer* of $\mathfrak{a}$ in $\mathfrak{g}$. If $A = G$ then Z_G is called the *center* of G and $\mathfrak{z}_G = \{X \in \mathfrak{g} : [X,Y] = 0 \text{ for all } Y \in \mathfrak{g}\}$ is then the *center* of the Lie algebra $\mathfrak{g}$.

5.9. The *normalizer* of a subset A of a connected Lie group G is the subgroup $N_A = \{x \in G : \lambda_x(A) = \rho_x(A)\} = \{x \in G : \operatorname{conj}_x(A) = A\}$. If A is closed then N_A is also closed.

If A is a connected Lie subgroup of G then $N_A = \{x \in G : \operatorname{Ad}(x)\mathfrak{a} \subset \mathfrak{a}\}$ and its Lie algebra is $\mathfrak{n}_A = \{X \in \mathfrak{g} : \operatorname{ad}(X)\mathfrak{a} \subset \mathfrak{a}\}$ is then the *idealizer* of $\mathfrak{a}$ in $\mathfrak{g}$.

5.10. Group actions. A *left action* of a Lie group G on a manifold M is a smooth mapping $\ell : G \times M \to M$ such that $\ell_x \circ \ell_y = \ell_{xy}$ and $\ell_e = \operatorname{Id}_M$, where $\ell_x(z) = \ell(x,z)$.

A *right action* of a Lie group G on a manifold M is a smooth mapping $r : M \times G \to M$ such that $r^x \circ r^y = r^{yx}$ and $r^e = \operatorname{Id}_M$, where $r^x(z) = r(z,x)$.

A G-space is a manifold M together with a right or left action of G on M.

We will describe the following notions only for a left action of G on M. They make sense also for right actions.

The *orbit* through $z \in M$ is the set $G.z = \ell(G,z) \subset M$. The action is called *transitive*, if M is one orbit, i.e. for all $z, w \in M$ there is some $g \in G$ with $g.z = w$. The action is called *free*, if $g_1.z = g_2.z$ for some $z \in M$ implies already $g_1 = g_2$. The action is called *effective*, if $\ell_x = \ell_y$ implies $x = y$, i.e. if $\ell : G \to \operatorname{Diff}(M)$ is injective, where $\operatorname{Diff}(M)$ denotes the group of all diffeomorphisms of M.

More generally, a continuous transformation group of a topological space M is a pair (G, M) where G is a topological group and where to each element $x \in G$ there is given a homeomorphism ℓ_x of M such that $\ell : G \times M \to M$ is continuous, and $\ell_x \circ \ell_y = \ell_{xy}$. The continuity is an obvious geometrical requirement, but in accordance with the general observation that group properties often force more regularity than explicitly postulated (cf. 5.6), differentiability follows in many situations. So, if G is locally compact, M is a smooth or real analytic manifold, all ℓ_x are smooth or real analytic homeomorphisms and the action is effective, then G is a Lie group and ℓ is smooth or real analytic, respectively, see [Montgomery, Zippin, 55, p. 212]. The latter result is deeply reflected in the theory of bundle functors and will be heavily used in chapter V.

5.11. Homogeneous spaces. Let G be a Lie group and let $H \subset G$ be a closed subgroup. By theorem 5.5 H is a Lie subgroup of G. We denote by G/H the space of all right cosets of G, i.e. $G/H = \{xH : x \in G\}$. Let $p : G \to G/H$ be the projection. We equip G/H with the quotient topology, i.e. $U \subset G/H$ is open if and only if $p^{-1}(U)$ is open in G. Since H is closed, G/H is a Hausdorff space.

G/H is called a *homogeneous space* of G. We have a left action of G on G/H, which is induced by the left translation and is given by $\bar{\lambda}_x(zH) = xzH$.

Theorem. *If H is a closed subgroup of G, then there exists a unique structure of a smooth manifold on G/H such that $p : G \to G/H$ is a submersion. So* $\dim G/H = \dim G - \dim H$.

Proof. Surjective submersions have the universal property 2.4, thus the manifold structure on G/H is unique, if it exists. Let $\mathfrak{h}$ be the Lie algebra of the Lie subgroup H. We choose a complementary linear subspace $\mathfrak{k}$ such that $\mathfrak{g} = \mathfrak{h} \oplus \mathfrak{k}$.

Claim 1. We consider the mapping $f : \mathfrak{k} \times H \to G$, given by $f(X, h) := \exp X.h$. Then there is an open 0-neighborhood W in $\mathfrak{k}$ and an open e-neighborhood U in G such that $f : W \times H \to U$ is a diffeomorphism.

By claim 5 in the proof of theorem 5.5 there are open 0-neighborhoods V in $\mathfrak{h}$, W' in $\mathfrak{k}$, and an open e-neighborhood U' in G such that $\varphi : W' \times V \to U'$ is a diffeomorphism, where $\varphi(X, Y) = \exp X.\exp Y$, and such that $U' \cap H = \exp V$. Now we choose W in $\mathfrak{k}$ so small that $\exp(W)^{-1}.\exp(W) \subset U'$. We will check that this W satisfies claim 1.

Claim 2. $f|W \times H$ is injective.

$f(X_1, h_1) = f(X_2, h_2)$ means $\exp X_1.h_1 = \exp X_2.h_2$, consequently we have $h_2 h_1^{-1} = (\exp X_2)^{-1} \exp X_1 \in \exp(W)^{-1} \exp(W) \cap H \subset U' \cap H = \exp V$. So there is a unique $Y \in V$ with $h_2 h_1^{-1} = \exp Y$. But then $\varphi(X_1, 0) = \exp X_1 = \exp X_2.h_2.h_1^{-1} = \exp X_2.\exp Y = \varphi(X_2, Y)$. Since φ is injective, $X_1 = X_2$ and $Y = 0$, so $h_1 = h_2$.

Claim 3. $f|W \times H$ is a local diffeomorphism.

The diagram

$$\begin{array}{ccc} W \times V & \xrightarrow{\text{Id} \times \exp} & W \times (U' \cap H) \\ {\scriptstyle\varphi}\downarrow & & \downarrow{\scriptstyle f} \\ \varphi(W \times V) & \xrightarrow{incl} & U' \end{array}$$

commutes, and $\text{Id}_W \times \exp$ and φ are diffeomorphisms. So $f|W \times (U' \cap H)$ is a local diffeomorphism. Since $f(X, h) = f(X, e).h$ we conclude that $f|W \times H$ is everywhere a local diffeomorphism. So finally claim 1 follows, where $U = f(W \times H)$.

Now we put $g := p \circ (\exp|W) : \mathfrak{k} \supset W \to G/H$. Then the following diagram commutes:

$$\begin{array}{ccc} W \times H & \xrightarrow{f} & U \\ {\scriptstyle pr_1}\downarrow & & \downarrow{\scriptstyle p} \\ W & \xrightarrow{g} & G/H. \end{array}$$

Claim 4. g is a homeomorphism onto $p(U) =: \bar{U} \subset G/H$.

Clearly g is continuous, and g is open, since p is open. If $g(X_1) = g(X_2)$ then $\exp X_1 = \exp X_2.h$ for some $h \in H$, so $f(X_1, e) = f(X_2, h)$. By claim 1 we get $X_1 = X_2$, so g is injective. Finally $g(W) = \bar{U}$, so claim 4 follows.

For $a \in G$ we consider $\bar{U}_a = \bar{\lambda}_a(\bar{U}) = a.\bar{U}$ and the mapping $u_a := g^{-1} \circ \bar{\lambda}_{a^{-1}} : \bar{U}_a \to W \subset \mathfrak{k}$.

Claim 5. $(\bar{U}_a, u_a = g^{-1} \circ \bar{\lambda}_{a^{-1}} : \bar{U}_a \to W)_{a \in G}$ is a smooth atlas for G/H.

Let $a, b \in G$ such that $\bar{U}_a \cap \bar{U}_b \neq \emptyset$. Then

$$\begin{aligned} u_a \circ u_b^{-1} &= g^{-1} \circ \bar{\lambda}_{a^{-1}} \circ \bar{\lambda}_b \circ g : u_b(\bar{U}_a \cap \bar{U}_b) \to u_a(\bar{U}_a \cap \bar{U}_b) \\ &= g^{-1} \circ \bar{\lambda}_{a^{-1}b} \circ p \circ (\exp|W) \\ &= g^{-1} \circ p \circ \lambda_{a^{-1}b} \circ (\exp|W) \\ &= pr_1 \circ f^{-1} \circ \lambda_{a^{-1}b} \circ (\exp|W) \quad \text{is smooth.} \quad \square \end{aligned}$$

5.12. Let $\ell : G \times M \to M$ be a left action. Then we have partial mappings $\ell_a : M \to M$ and $\ell^x : G \to M$, given by $\ell_a(x) = \ell^x(a) = \ell(a, x) = a.x$.

For any $X \in \mathfrak{g}$ we define the *fundamental vector field* $\zeta_X = \zeta_X^M \in \mathfrak{X}(M)$ by $\zeta_X(x) = T_e(\ell^x).X = T_{(e,x)}\ell.(X, 0_x)$.

Lemma. *In this situation the following assertions hold:*

(1) *$\zeta : \mathfrak{g} \to \mathfrak{X}(M)$ is a linear mapping.*
(2) *$T_x(\ell_a).\zeta_X(x) = \zeta_{\mathrm{Ad}(a)X}(a.x)$.*
(3) *$R_X \times 0_M \in \mathfrak{X}(G \times M)$ is ℓ-related to $\zeta_X \in \mathfrak{X}(M)$.*
(4) *$[\zeta_X, \zeta_Y] = -\zeta_{[X,Y]}$.*

Proof. (1) is clear.

(2) $\ell_a \ell^x(b) = abx = aba^{-1}ax = \ell^{ax}\,\mathrm{conj}_a(b)$, so we get $T_x(\ell_a).\zeta_X(x) = T_x(\ell_a).T_e(\ell^x).X = T_e(\ell_a \circ \ell^x).X = T_e(\ell^{ax}).\mathrm{Ad}(a).X = \zeta_{\mathrm{Ad}(a)X}(ax)$.

(3) $\ell \circ (\mathrm{Id} \times \ell_a) = \ell \circ (\rho_a \times \mathrm{Id}) : G \times M \to M$, so we get $\zeta_X(\ell(a, x)) = T_{(e,ax)}\ell.(X, 0_{ax}) = T\ell.(\mathrm{Id} \times T(\ell_a)).(X, 0_x) = T\ell.(T(\rho_a) \times \mathrm{Id}).(X, 0_x) = T\ell.(R_X \times 0_M)(a, x)$.

(4) $[R_X \times 0_M, R_Y \times 0_M] = [R_X, R_Y] \times 0_M = -R_{[X,Y]} \times 0_M$ is ℓ-related to $[\zeta_X, \zeta_Y]$ by (3) and by 3.10. On the other hand $-R_{[X,Y]} \times 0_M$ is ℓ-related to $-\zeta_{[X,Y]}$ by (3) again. Since ℓ is surjective we get $[\zeta_X, \zeta_Y] = -\zeta_{[X,Y]}$. $\square$

5.13. Let $r : M \times G \to M$ be a right action, so $\check{r} : G \to \mathrm{Diff}(M)$ is a group anti homomorphism. We will use the following notation: $r^a : M \to M$ and $r_x : G \to M$, given by $r_x(a) = r^a(x) = r(x, a) = x.a$.

For any $X \in \mathfrak{g}$ we define the *fundamental vector field* $\zeta_X = \zeta_X^M \in \mathfrak{X}(M)$ by $\zeta_X(x) = T_e(r_x).X = T_{(x,e)}r.(0_x, X)$.

Lemma. *In this situation the following assertions hold:*

(1) $\zeta : \mathfrak{g} \to \mathfrak{X}(M)$ *is a linear mapping.*
(2) $T_x(r^a).\zeta_X(x) = \zeta_{\mathrm{Ad}(a^{-1})X}(x.a)$.
(3) $0_M \times L_X \in \mathfrak{X}(M \times G)$ *is r-related to* $\zeta_X \in \mathfrak{X}(M)$.
(4) $[\zeta_X, \zeta_Y] = \zeta_{[X,Y]}$. □

5.14. Theorem. *Let $\ell : G \times M \to M$ be a smooth left action. For $x \in M$ let $G_x = \{a \in G : ax = x\}$ be the isotropy subgroup of x in G, a closed subgroup of G. Then $\ell^x : G \to M$ factors over $p : G \to G/G_x$ to an injective immersion $i^x : G/G_x \to M$, which is G-equivariant, i.e. $\ell_a \circ i^x = i^x \circ \bar{\lambda}_a$ for all $a \in G$. The image of i^x is the orbit through x.*

The fundamental vector fields span an integrable distribution on M in the sense of 3.20. Its leaves are the connected components of the orbits, and each orbit is an initial submanifold.

Proof. Clearly ℓ^x factors over p to an injective mapping $i^x : G/G_x \to M$; by the universal property of surjective submersions i^x is smooth, and obviously it is equivariant. Thus $T_{p(a)}(i^x).T_{p(e)}(\bar{\lambda}_a) = T_{p(e)}(i^x \circ \bar{\lambda}_a) = T_{p(e)}(\ell_a \circ i^x) = T_x(\ell_a).T_{p(e)}(i^x)$ for all $a \in G$ and it suffices to show that $T_{p(e)}(i^x)$ is injective.

Let $X \in \mathfrak{g}$ and consider its fundamental vector field $\zeta_X \in \mathfrak{X}(M)$. By 3.14 and 5.12.3 we have

$$(1) \qquad \ell(\exp(tX), x) = \ell(\mathrm{Fl}_t^{R_X \times 0_M}(e, x)) = \mathrm{Fl}_t^{\zeta_X}(\ell(e, x)) = \mathrm{Fl}_t^{\zeta_X}(x).$$

So $\exp(tX) \in G_x$, i.e. $X \in \mathfrak{g}_x$, if and only if $\zeta_X(x) = 0_x$. In other words, $0_x = \zeta_X(x) = T_e(\ell^x).X = T_{p(e)}(i^x).T_e p.X$ if and only if $T_e p.X = 0_{p(e)}$. Thus i^x is an immersion.

Since the connected components of the orbits are integral manifolds, the fundamental vector fields span an integrable distribution in the sense of 3.20; but also the condition 3.25.2 is satisfied. So by theorem 3.22 each orbit is an initial submanifold in the sense of 2.14. □

5.15. A mapping $f : M \to \bar{M}$ between two manifolds with left (or right) actions ℓ and $\bar{\ell}$ of a Lie group G is called *G-equivariant* if $f \circ \ell_a = \bar{\ell}_a \circ f$ (or $f \circ r^a = \bar{r}^a \circ f$) for all $a \in G$. Sometimes we say in short that f is a G-mapping. From formula 5.14.(1) we get

Lemma. *If G is connected, then f is G-equivariant if and only if the fundamental field mappings are $f$related, i.e. $Tf \circ \zeta_X = \bar{\zeta}_X \circ f$ for all $X \in \mathfrak{g}$.*

Proof. The image of the exponential mapping generates the connected component of the unit. □

5.16. Semidirect products of Lie groups. Let H and K be two Lie groups and let $\ell : H \times K \to K$ be a left action of H in K such that each $\ell_h : K \to K$ is a group homomorphism. So the associated mapping $\check{\ell} : H \to \mathrm{Aut}(K)$ is a homomorphism into the automorphism group of K. Then we can introduce the following multiplication on $K \times H$

$$(1) \qquad (k, h)(k', h') := (k\ell_h(k'), hh').$$

It is easy to see that this defines a Lie group $G = K \ltimes_\ell H$ called the *semidirect product* of H and K with respect to ℓ. If the action ℓ is clear from the context we write $G = K \ltimes H$ only. The second projection $pr_2 : K \ltimes H \to H$ is a surjective smooth homomorphism with kernel $K \times \{e\}$, and the insertion $\operatorname{ins}_e : H \to K \ltimes H$, $\operatorname{ins}_e(h) = (e, h)$ is a smooth group homomorphism with $pr_2 \circ \operatorname{ins}_e = \operatorname{Id}_H$.

Conversely we consider an exact sequence of Lie groups and homomorphisms

$$\{e\} \to K \xrightarrow{j} G \xrightarrow{p} H \to \{e\}. \tag{2}$$

So j is injective, p is surjective, and the kernel of p equals the image of j. We suppose furthermore that the sequence splits, so that there is a smooth homomorphism $i : H \to G$ with $p \circ i = \operatorname{Id}_H$. Then the rule $\ell_h(k) = i(h)ki(h^{-1})$ (where we suppress j) defines a left action of H on K by automorphisms. It is easily seen that the mapping $K \ltimes_\ell H \to G$ given by $(k, h) \mapsto ki(h)$ is an isomorphism of Lie groups. So we see that semidirect products of Lie groups correspond exactly to splitting short exact sequences.

Semidirect products will appear naturally also in another form, starting from right actions: Let H and K be two Lie groups and let $r : K \times H \to K$ be a right action of H in K such that each $r^h : K \to K$ is a group homomorphism. Then the multiplication on $H \times K$ is given by

$$(h, k)(\bar h, \bar k) := (h\bar h, r^{\bar h}(k)\bar k). \tag{3}$$

This defines a Lie group $G = H \rtimes_r K$, also called the *semidirect product* of H and K with respect to r. If the action r is clear from the context we write $G = H \rtimes K$ only. The first projection $pr_1 : H \rtimes K \to H$ is a surjective smooth homomorphism with kernel $\{e\} \times K$, and the insertion $\operatorname{ins}_e : H \to H \rtimes K$, $\operatorname{ins}_e(h) = (h, e)$ is a smooth group homomorphism with $pr_1 \circ \operatorname{ins}_e = \operatorname{Id}_H$.

Conversely we consider again a splitting exact sequence of Lie groups and homomorphisms

$$\{e\} \to K \xrightarrow{j} G \xrightarrow{p} H \to \{e\}.$$

The splitting is given by a homomorphism $i : H \to G$ with $p \circ i = \operatorname{Id}_H$. Then the rule $r^h(k) = i(h^{-1})ki(h)$ (where we suppress j) defines now a right action of H on K by automorphisms. It is easily seen that the mapping $H \rtimes_r K \to G$ given by $(h, k) \mapsto i(h)k$ is an isomorphism of Lie groups.

Remarks

The material in this chapter is standard. The concept of initial submanifolds in 2.14–2.17 is due to Pradines, the treatment given here follows [Albert, Molino]. The proof of theorem 3.16 is due to [Mauhart, 90]. The main results on distributions of non constant rank (3.18–3.25) are due to [Sussman, 73] and [Stefan, 74], the treatment here follows [Lecomte]. The proof of the Baker-Campbell-Hausdorff formula 4.29 is adapted from [Sattinger, Weaver, 86], see also [Hilgert, Neeb, 91].

CHAPTER II.
DIFFERENTIAL FORMS

This chapter is still devoted to the fundamentals of differential geometry, but here the deviation from the standard presentations is already large. In the section on vector bundles we treat the Lie derivative for natural vector bundles, i.e. functors which associate vector bundles to manifolds and vector bundle homomorphisms to local diffeomorphisms. We give a formula for the Lie derivative of the form of a commutator, but it involves the tangent bundle of the vector bundle in question. So we also give a careful treatment to this situation. The Lie derivative will be discussed in detail in chapter XI; here it is presented in a somewhat special situation as an illustration of the categorical methods we are going to apply later on. It follows a standard presentation of differential forms and a thorough treatment of the Frölicher-Nijenhuis bracket via the study of all graded derivations of the algebra of differential forms. This bracket is a natural extension of the Lie bracket from vector fields to tangent bundle valued differential forms. We believe that this bracket is one of the basic structures of differential geometry (see also section 30), and in chapter III we will base nearly all treatment of curvature and the Bianchi identity on it.

6. Vector bundles

6.1. Vector bundles. Let $p : E \to M$ be a smooth mapping between manifolds. By a *vector bundle chart* on (E, p, M) we mean a pair (U, ψ), where U is an open subset in M and where ψ is a fiber respecting diffeomorphism as in the following diagram:

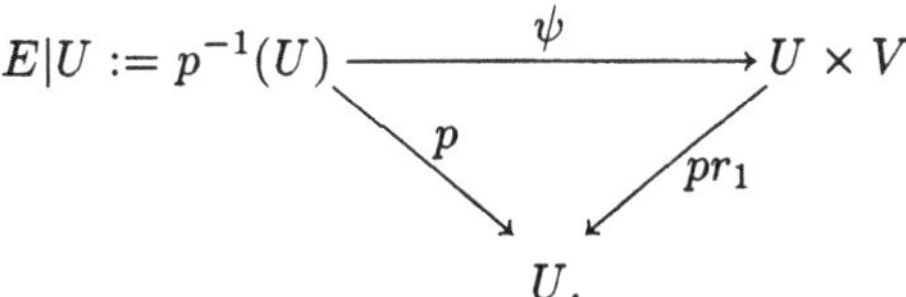

Here V is a fixed finite dimensional vector space, called the *standard fiber* or the *typical fiber*, real as a rule, unless otherwise specified.

Two vector bundle charts (U_1, ψ_1) and (U_2, ψ_2) are called *compatible*, if $\psi_1 \circ \psi_2^{-1}$ is a fiber linear isomorphism, i.e. $(\psi_1 \circ \psi_2^{-1})(x, v) = (x, \psi_{1,2}(x)v)$ for some mapping $\psi_{1,2} : U_{1,2} := U_1 \cap U_2 \to GL(V)$. The mapping $\psi_{1,2}$ is then unique and smooth, and it is called the *transition function* between the two vector bundle charts.

A *vector bundle atlas* $(U_\alpha, \psi_\alpha)_{\alpha\in A}$ for (E,p,M) is a set of pairwise compatible vector bundle charts (U_α, ψ_α) such that $(U_\alpha)_{\alpha\in A}$ is an open cover of M. Two vector bundle atlases are called *equivalent*, if their union is again a vector bundle atlas.

A *vector bundle* (E,p,M) consists of manifolds E (the *total space*), M (the *base*), and a smooth mapping $p: E \to M$ (the *projection*) together with an equivalence class of vector bundle atlases; so we must know at least one vector bundle atlas. The projection p turns out to be a surjective submersion.

The tangent bundle (TM, π_M, M) of a manifold M is the first example of a vector bundle.

6.2. Let us fix a vector bundle (E,p,M) for the moment. On each *fiber* $E_x := p^{-1}(x)$ (for $x \in M$) there is a unique structure of a real vector space, induced from any vector bundle chart (U_α, ψ_α) with $x \in U_\alpha$. So $0_x \in E_x$ is a special element and $0: M \to E$, $0(x) = 0_x$, is a smooth mapping, the *zero section*.

A *section* u of (E,p,M) is a smooth mapping $u: M \to E$ with $p \circ u = \mathrm{Id}_M$. The *support* of the section u is the closure of the set $\{x \in M : u(x) \neq 0_x\}$ in M. The space of all smooth sections of the bundle (E,p,M) will be denoted by either $C^\infty(E) = C^\infty(E,p,M) = C^\infty(E \to M)$. Clearly it is a vector space with fiber wise addition and scalar multiplication.

If $(U_\alpha, \psi_\alpha)_{\alpha\in A}$ is a vector bundle atlas for (E,p,M), then any smooth mapping $f_\alpha : U_\alpha \to V$ (the standard fiber) defines a local section $x \mapsto \psi_\alpha^{-1}(x, f_\alpha(x))$ on U_α. If $(g_\alpha)_{\alpha\in A}$ is a partition of unity subordinated to (U_α), then a global section can be formed by $x \mapsto \sum_\alpha g_\alpha(x) \cdot \psi_\alpha^{-1}(x, f_\alpha(x))$. So a smooth vector bundle has 'many' smooth sections.

6.3. Let (E,p,M) and (F,q,N) be vector bundles. A *vector bundle homomorphism* $\varphi: E \to F$ is a fiber respecting, fiber linear smooth mapping

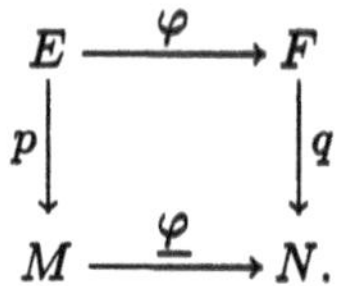

$$\begin{array}{ccc} E & \xrightarrow{\varphi} & F \\ {\scriptstyle p}\downarrow & & \downarrow{\scriptstyle q} \\ M & \xrightarrow{\underline{\varphi}} & N. \end{array}$$

So we require that $\varphi_x : E_x \to F_{\underline{\varphi}(x)}$ is linear. We say that φ covers $\underline{\varphi}$. If φ is invertible, it is called a *vector bundle isomorphism*.

The smooth vector bundles together with their homomorphisms form a category $\mathcal{VB}$.

6.4. We will now give a formal description of the amount of vector bundles with fixed base M and fixed standard fiber V, up to isomorphisms which cover the identity on M.

Let us first fix an open cover $(U_\alpha)_{\alpha\in A}$ of M. If (E,p,M) is a vector bundle which admits a vector bundle atlas (U_α, ψ_α) with the given open cover, then we have $\psi_\alpha \circ \psi_\beta^{-1}(x,v) = (x, \psi_{\alpha\beta}(x)v)$ for transition functions $\psi_{\alpha\beta}: U_{\alpha\beta} = U_\alpha \cap U_\beta \to GL(V)$, which are smooth. This family of transition functions satisfies

$$(1)\qquad \begin{cases} \psi_{\alpha\beta}(x)\cdot\psi_{\beta\gamma}(x)=\psi_{\alpha\gamma}(x) & \text{for each } x\in U_{\alpha\beta\gamma}=U_\alpha\cap U_\beta\cap U_\gamma,\\ \psi_{\alpha\alpha}(x)=e & \text{for all } x\in U_\alpha. \end{cases}$$

Condition (1) is called a *cocycle condition* and thus we call the family $(\psi_{\alpha\beta})$ the *cocycle of transition functions* for the vector bundle atlas (U_α,ψ_α).

Let us suppose now that the same vector bundle (E,p,M) is described by an equivalent vector bundle atlas $(U_\alpha,\varphi_\alpha)$ with the same open cover (U_α). Then the vector bundle charts (U_α,ψ_α) and $(U_\alpha,\varphi_\alpha)$ are compatible for each α, so $\varphi_\alpha\circ\psi_\alpha^{-1}(x,v)=(x,\tau_\alpha(x)v)$ for some $\tau_\alpha:U_\alpha\to GL(V)$. But then we have

$$\begin{aligned}(x,\tau_\alpha(x)\psi_{\alpha\beta}(x)v)&=(\varphi_\alpha\circ\psi_\alpha^{-1})(x,\psi_{\alpha\beta}(x)v)=\\ &=(\varphi_\alpha\circ\psi_\alpha^{-1}\circ\psi_\alpha\circ\psi_\beta^{-1})(x,v)=(\varphi_\alpha\circ\psi_\beta^{-1})(x,v)=\\ &=(\varphi_\alpha\circ\varphi_\beta^{-1}\circ\varphi_\beta\circ\psi_\beta^{-1})(x,v)=(x,\varphi_{\alpha\beta}(x)\tau_\beta(x)v).\end{aligned}$$

So we get

$$(2)\qquad \tau_\alpha(x)\psi_{\alpha\beta}(x)=\varphi_{\alpha\beta}(x)\tau_\beta(x)\quad\text{for all } x\in U_{\alpha\beta}.$$

We say that the two cocycles $(\psi_{\alpha\beta})$ and $(\varphi_{\alpha\beta})$ of transition functions over the cover (U_α) are *cohomologous*. The *cohomology classes* of cocycles $(\psi_{\alpha\beta})$ over the open cover (U_α) (where we identify cohomologous ones) form a set $\check H^1((U_\alpha),\underline{GL}(V))$, the first *Čech cohomology set* of the open cover (U_α) with values in the sheaf $C^\infty(\ \ ,GL(V))=:\underline{GL}(V)$.

Now let $(W_i)_{i\in I}$ be an open cover of M that refines (U_α) with $W_i\subset U_{\varepsilon(i)}$, where $\varepsilon:I\to A$ is some refinement mapping. Then for any cocycle $(\psi_{\alpha\beta})$ over (U_α) we define the cocycle $\varepsilon^*(\psi_{\alpha\beta})=:(\varphi_{ij})$ by the prescription $\varphi_{ij}:=\psi_{\varepsilon(i),\varepsilon(j)}|W_{ij}$. The mapping ε^* respects the cohomology relations and induces therefore a mapping $\varepsilon^\sharp:\check H^1((U_\alpha),\underline{GL}(V))\to\check H^1((W_i),\underline{GL}(V))$. One can show that the mapping ε^* depends on the choice of the refinement mapping ε only up to cohomology (use $\tau_i=\psi_{\varepsilon(i),\eta(i)}|W_i$ if ε and η are two refinement mappings), so we may form the inductive limit $\varinjlim\check H^1(\mathcal U,\underline{GL}(V))=:\check H^1(M,\underline{GL}(V))$ over all open covers of M directed by refinement.

Theorem. *There is a bijective correspondence between $\check H^1(M,\underline{GL}(V))$ and the set of all isomorphism classes of vector bundles over M with typical fiber V.*

Proof. Let $(\psi_{\alpha\beta})$ be a cocycle of transition functions $\psi_{\alpha\beta}:U_{\alpha\beta}\to GL(V)$ over some open cover (U_α) of M. We consider the disjoint union $\bigsqcup_{\alpha\in A}\{\alpha\}\times U_\alpha\times V$ and the following relation on it: $(\alpha,x,v)\sim(\beta,y,w)$ if and only if $x=y$ and $\psi_{\beta\alpha}(x)v=w$.

By the cocycle property (1) of $(\psi_{\alpha\beta})$ this is an equivalence relation. The space of all equivalence classes is denoted by $E=VB(\psi_{\alpha\beta})$ and it is equipped with the quotient topology. We put $p:E\to M$, $p[(\alpha,x,v)]=x$, and we define the vector bundle charts (U_α,ψ_α) by $\psi_\alpha[(\alpha,x,v)]=(x,v)$, $\psi_\alpha:p^{-1}(U_\alpha)=:E|U_\alpha\to$

$U_\alpha \times V$. Then the mapping $\psi_\alpha \circ \psi_\beta^{-1}(x,v) = \psi_\alpha[(\beta,x,v)] = \psi_\alpha[(\alpha,x,\psi_{\alpha\beta}(x)v)] = (x,\psi_{\alpha\beta}(x)v)$ is smooth, so E becomes a smooth manifold. E is Hausdorff: let $u \neq v$ in E; if $p(u) \neq p(v)$ we can separate them in M and take the inverse image under p; if $p(u) = p(v)$, we can separate them in one chart. So (E,p,M) is a vector bundle.

Now suppose that we have two cocycles $(\psi_{\alpha\beta})$ over (U_α), and (φ_{ij}) over (V_i). Then there is a common refinement (W_γ) for the two covers (U_α) and (V_i). The construction described a moment ago gives isomorphic vector bundles if we restrict the cocycle to a finer open cover. So we may assume that $(\psi_{\alpha\beta})$ and $(\varphi_{\alpha\beta})$ are cocycles over the same open cover (U_α). If the two cocycles are cohomologous, so $\tau_\alpha \cdot \psi_{\alpha\beta} = \varphi_{\alpha\beta} \cdot \tau_\beta$ on $U_{\alpha\beta}$, then a fiber linear diffeomorphism $\tau : VB(\psi_{\alpha\beta}) \to VB(\varphi_{\alpha\beta})$ is given by $\varphi_\alpha \tau[(\alpha,x,v)] = (x,\tau_\alpha(x)v)$. By relation (2) this is well defined, so the vector bundles $VB(\psi_{\alpha\beta})$ and $VB(\varphi_{\alpha\beta})$ are isomorphic.

Most of the converse direction was already shown in the discussion before the theorem, and the argument can be easily refined to show also that isomorphic bundles give cohomologous cocycles. □

Remark. If $GL(V)$ is an abelian group, i.e. if V is of real or complex dimension 1, then $\check{H}^1(M, \underline{GL}(V))$ is a usual cohomology group with coefficients in the sheaf $\underline{GL}(V)$ and it can be computed with the methods of algebraic topology. If $GL(V)$ is not abelian, then the situation is rather mysterious: there is no clear definition for $\check{H}^2(M, \underline{GL}(V))$ for example. So $\check{H}^1(M, \underline{GL}(V))$ is more a notation than a mathematical concept.

A coarser relation on vector bundles (stable isomorphism) leads to the concept of topological K-theory, which can be handled much better, but is only a quotient of the whole situation.

6.5. Let (U_α, ψ_α) be a vector bundle atlas on a vector bundle (E,p,M). Let $(e_j)_{j=1}^k$ be a basis of the standard fiber V. We consider the section $s_j(x) := \psi_\alpha^{-1}(x,e_j)$ for $x \in U_\alpha$. Then the $s_j : U_\alpha \to E$ are local sections of E such that $(s_j(x))_{j=1}^k$ is a basis of E_x for each $x \in U_\alpha$: we say that $s = (s_1, \dots, s_k)$ is a local *frame field* for E over U_α.

Now let conversely $U \subset M$ be an open set and let $s_j : U \to E$ be local sections of E such that $s = (s_1, \dots, s_k)$ is a local frame field of E over U. Then s determines a unique vector bundle chart (U,ψ) of E such that $s_j(x) = \psi^{-1}(x,e_j)$, in the following way. We define $f : U \times \mathbb{R}^k \to E|U$ by $f(x,v^1,\dots,v^k) := \sum_{j=1}^k v^j s_j(x)$. Then f is smooth, invertible, and a fiber linear isomorphism, so $(U, \psi = f^{-1})$ is the vector bundle chart promised above.

6.6. A *vector sub bundle* (F,p,M) of a vector bundle (E,p,M) is a vector bundle and a vector bundle homomorphism $\tau : F \to E$, which covers Id_M, such that $\tau_x : E_x \to F_x$ is a linear embedding for each $x \in M$.

Lemma. *Let $\varphi : (E,p,M) \to (E',q,N)$ be a vector bundle homomorphism such that $\operatorname{rank}(\varphi_x : E_x \to E'_{\underline{\varphi}(x)})$ is constant in $x \in M$. Then $\ker\varphi$, given by $(\ker\varphi)_x = \ker(\varphi_x)$, is a vector sub bundle of (E,p,M).*

Proof. This is a local question, so we may assume that both bundles are trivial:

let $E = M \times \mathbb{R}^p$ and let $F = N \times \mathbb{R}^q$, then $\varphi(x, v) = (\underline{\varphi}(x), \overline{\varphi}(x).v)$, where $\overline{\varphi} : M \to L(\mathbb{R}^p, \mathbb{R}^q)$. The matrix $\overline{\varphi}(x)$ has rank k, so by the elimination procedure we can find $p-k$ linearly independent solutions $v_i(x)$ of the equation $\overline{\varphi}(x).v = 0$. The elimination procedure (with the same lines) gives solutions $v_i(y)$ for y near x, so near x we get a local frame field $v = (v_1, \dots, v_{p-k})$ for $\ker \varphi$. By 6.5 $\ker \varphi$ is then a vector sub bundle. □

6.7. Constructions with vector bundles. Let $\mathcal{F}$ be a covariant functor from the category of finite dimensional vector spaces and linear mappings into itself, such that $\mathcal{F} : L(V, W) \to L(\mathcal{F}(V), \mathcal{F}(W))$ is smooth. Then $\mathcal{F}$ will be called a *smooth functor* for shortness sake. Well known examples of smooth functors are $\mathcal{F}(V) = \Lambda^k(V)$ (the k-th exterior power), or $\mathcal{F}(V) = \bigotimes^k V$, and the like.

If (E, p, M) is a vector bundle, described by a vector bundle atlas with cocycle of transition functions $\varphi_{\alpha\beta} : U_{\alpha\beta} \to GL(V)$, where (U_α) is an open cover of M, then we may consider the smooth functions $\mathcal{F}(\varphi_{\alpha\beta}) : x \mapsto \mathcal{F}(\varphi_{\alpha\beta}(x))$, $U_{\alpha\beta} \to GL(\mathcal{F}(V))$. Since $\mathcal{F}$ is a covariant functor, $\mathcal{F}(\varphi_{\alpha\beta})$ satisfies again the cocycle condition 6.4.1, and cohomology of cocycles 6.4.2 is respected, so there exists a unique vector bundle $(\mathcal{F}(E) := VB(\mathcal{F}(\varphi_{\alpha\beta})), p, M)$, the value at the vector bundle (E, p, M) of the canonical extension of the functor $\mathcal{F}$ to the category of vector bundles and their homomorphisms.

If $\mathcal{F}$ is a contravariant smooth functor like duality functor $\mathcal{F}(V) = V^*$, then we have to consider the new cocycle $\mathcal{F}(\varphi_{\alpha\beta}^{-1})$ instead of $\mathcal{F}(\varphi_{\alpha\beta})$.

If $\mathcal{F}$ is a contra-covariant smooth bifunctor like $L(V, W)$, then the rule

$$\mathcal{F}(VB(\psi_{\alpha\beta}), VB(\varphi_{\alpha\beta})) := VB(\mathcal{F}(\psi_{\alpha\beta}^{-1}, \varphi_{\alpha\beta}))$$

describes the induced canonical vector bundle construction, and similarly in other constructions.

So for vector bundles (E, p, M) and (F, q, M) we have the following vector bundles with base M: $\Lambda^k E$, $E \oplus F$, E^*, $\Lambda E = \bigoplus_{k \geq 0} \Lambda^k E$, $E \otimes F$, $L(E, F) \cong E^* \otimes F$, and so on.

6.8. Pullbacks of vector bundles. Let (E, p, M) be a vector bundle and let $f : N \to M$ be smooth. Then the *pullback vector bundle* (f^*E, f^*p, N) with the same typical fiber and a vector bundle homomorphism

$$\begin{array}{ccc} f^*E & \xrightarrow{p^*f} & E \\ {\scriptstyle f^*p}\downarrow & & \downarrow{\scriptstyle p} \\ N & \xrightarrow{f} & M \end{array}$$

are defined as follows. Let E be described by a cocycle $(\psi_{\alpha\beta})$ of transition functions over an open cover (U_α) of M, $E = VB(\psi_{\alpha\beta})$. Then $(\psi_{\alpha\beta} \circ f)$ is a cocycle of transition functions over the open cover $(f^{-1}(U_\alpha))$ of N and the bundle is given by $f^*E := VB(\psi_{\alpha\beta} \circ f)$. As a manifold we have $f^*E = N \underset{(f,M,p)}{\times} E$ in the sense of 2.19.

The vector bundle f^*E has the following universal property: For any vector bundle (F, q, P), vector bundle homomorphism $\varphi : F \to E$ and smooth $g : P \to N$ such that $f \circ g = \underline{\varphi}$, there is a unique vector bundle homomorphism $\psi : F \to f^*E$ with $\underline{\psi} = g$ and $p^*f \circ \psi = \varphi$.

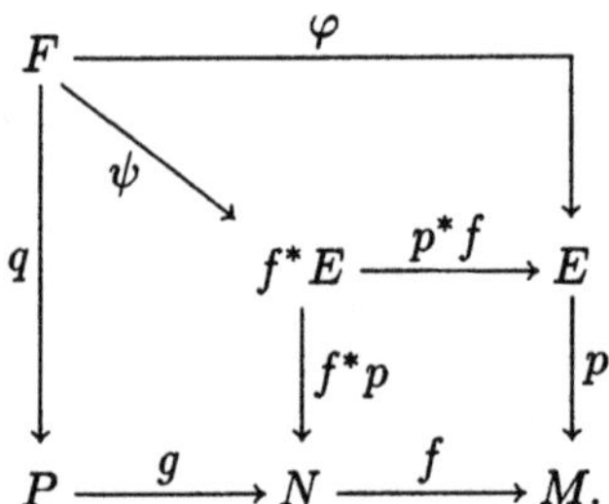

6.9. Theorem. *Any vector bundle admits a finite vector bundle atlas.*

Proof. Let (E, p, M) be the vector bundle in question, let $\dim M = m$. Let $(U_\alpha, \psi_\alpha)_{\alpha \in A}$ be a vector bundle atlas. Since M is separable, by topological dimension theory there is a refinement of the open cover $(U_\alpha)_{\alpha \in A}$ of the form $(V_{ij})_{i=1,\dots,m+1; j \in \mathbb{N}}$, such that $V_{ij} \cap V_{ik} = \emptyset$ for $j \neq k$, see the remarks at the end of 1.1. We define the set $W_i := \bigcup_{j \in \mathbb{N}} V_{ij}$ (a disjoint union) and $\psi_i | V_{ij} = \psi_{\alpha(i,j)}$, where $\alpha : \{1, \dots, m+1\} \times \mathbb{N} \to A$ is a refining map. Then $(W_i, \psi_i)_{i=1,\dots,m+1}$ is a finite vector bundle atlas of E. □

6.10. Theorem. *For any vector bundle (E, p, M) there is a second vector bundle (F, p, M) such that $(E \oplus F, p, M)$ is a trivial vector bundle, i.e. isomorphic to $M \times \mathbb{R}^N$ for some $N \in \mathbb{N}$.*

Proof. Let $(U_i, \psi_i)_{i=1}^n$ be a finite vector bundle atlas for (E, p, M). Let (g_i) be a smooth partition of unity subordinated to the open cover (U_i). Let $\ell_i : \mathbb{R}^k \to (\mathbb{R}^k)^n = \mathbb{R}^k \times \dots \times \mathbb{R}^k$ be the embedding on the i-th factor, where $\mathbb{R}^k$ is the typical fiber of E. Let us define $\psi : E \to M \times \mathbb{R}^{nk}$ by

$$\psi(u) = \left(p(u), \sum_{i=1}^n g_i(p(u)) \, (\ell_i \circ pr_2 \circ \psi_i)(u) \right),$$

then ψ is smooth, fiber linear, and an embedding on each fiber, so E is a vector sub bundle of $M \times \mathbb{R}^{nk}$ via ψ. Now we define $F_x = E_x^\perp$ in $\{x\} \times \mathbb{R}^{nk}$ with respect to the standard inner product on $\mathbb{R}^{nk}$. Then $F \to M$ is a vector bundle and $E \oplus F \cong M \times \mathbb{R}^{nk}$. □

6.11. The tangent bundle of a vector bundle. Let (E, p, M) be a vector bundle with fiber addition $+_E : E \times_M E \to E$ and fiber scalar multiplication $m_t^E : E \to E$. Then (TE, π_E, E), the tangent bundle of the manifold E, is itself a vector bundle, with fiber addition denoted by $+_{TE}$ and scalar multiplication denoted by m_t^{TE}.

If $(U_\alpha, \psi_\alpha : E|U_\alpha \to U_\alpha \times V)_{\alpha\in A}$ is a vector bundle atlas for E, such that (U_α, u_α) is also a manifold atlas for M, then $(E|U_\alpha, \psi'_\alpha)_{\alpha\in A}$ is an atlas for the manifold E, where

$$\psi'_\alpha := (u_\alpha \times \mathrm{Id}_V) \circ \psi_\alpha : E|U_\alpha \to U_\alpha \times V \to u_\alpha(U_\alpha) \times V \subset \mathbb{R}^m \times V.$$

Hence the family $(T(E|U_\alpha), T\psi'_\alpha : T(E|U_\alpha) \to T(u_\alpha(U_\alpha) \times V) = u_\alpha(U_\alpha) \times V \times \mathbb{R}^m \times V)_{\alpha\in A}$ is the atlas describing the canonical vector bundle structure of (TE, π_E, E). The transition functions are in turn:

$$\begin{aligned}
(\psi_\alpha \circ \psi_\beta^{-1})(x, v) &= (x, \psi_{\alpha\beta}(x)v) \quad \text{for } x \in U_{\alpha\beta}\\
(u_\alpha \circ u_\beta^{-1})(y) &= u_{\alpha\beta}(y) \quad \text{for } y \in u_\beta(U_{\alpha\beta})\\
(\psi'_\alpha \circ (\psi'_\beta)^{-1})(y, v) &= (u_{\alpha\beta}(y), \psi_{\alpha\beta}(u_\beta^{-1}(y))v)\\
(T\psi'_\alpha \circ T(\psi'_\beta)^{-1})(y, v; \xi, w) &= \big(u_{\alpha\beta}(y), \psi_{\alpha\beta}(u_\beta^{-1}(y))v; d(u_{\alpha\beta})(y)\xi,\\
&\qquad (d(\psi_{\alpha\beta} \circ u_\beta^{-1})(y))\xi)v + \psi_{\alpha\beta}(u_\beta^{-1}(y))w\big).
\end{aligned}$$

So we see that for fixed (y, v) the transition functions are linear in $(\xi, w) \in \mathbb{R}^m \times V$. This describes the vector bundle structure of the tangent bundle (TE, π_E, E).

For fixed (y, ξ) the transition functions of TE are also linear in $(v, w) \in V \times V$. This gives a vector bundle structure on (TE, Tp, TM). Its fiber addition will be denoted by $T(+_E) : T(E \times_M E) = TE \times_{TM} TE \to TE$, since it is the tangent mapping of $+_E$. Likewise its scalar multiplication will be denoted by $T(m_t^E)$. One may say that the second vector bundle structure on TE, that one over TM, is the derivative of the original one on E.

The space $\{\Xi \in TE : Tp.\Xi = 0 \text{ in } TM\} = (Tp)^{-1}(0)$ is denoted by VE and is called the *vertical bundle* over E. The local form of a vertical vector Ξ is $T\psi'_\alpha.\Xi = (y, v; 0, w)$, so the transition function looks like $(T\psi'_\alpha \circ T(\psi'_\beta)^{-1})(y, v; 0, w) = (u_{\alpha\beta}(y), \psi_{\alpha\beta}(u_\beta^{-1}(y))v; 0, \psi_{\alpha\beta}(u_\beta^{-1}(y))w)$. They are linear in $(v, w) \in V \times V$ for fixed y, so VE is a vector bundle over M. It coincides with $0_M^*(TE, Tp, TM)$, the pullback of the bundle $TE \to TM$ over the zero section. We have a canonical isomorphism $vl_E : E \times_M E \to VE$, called the *vertical lift*, given by $vl_E(u_x, v_x) := \frac{d}{dt}|_0(u_x + tv_x)$, which is fiber linear over M. The local representation of the vertical lift is $(T\psi'_\alpha \circ vl_E \circ (\psi'_\alpha \times \psi'_\alpha)^{-1})((y, u), (y, v)) = (y, u; 0, v)$.

If (and only if) $\varphi : (E, p, M) \to (F, q, N)$ is a vector bundle homomorphism, then we have $vl_F \circ (\varphi \times_M \varphi) = T\varphi \circ vl_E : E \times_M E \to VF \subset TF$. So vl is a natural transformation between certain functors on the category of vector bundles and their homomorphisms.

The mapping $vpr_E := pr_2 \circ vl_E^{-1} : VE \to E$ is called the *vertical projection*. Note also the relation $pr_1 \circ vl_E^{-1} = \pi_E|VE$.

6.12. The second tangent bundle of a manifold. All of 6.11 is valid for the second tangent bundle $T^2M = TTM$ of a manifold, but here we have one more natural structure at our disposal. The *canonical flip* or *involution* $\kappa_M : T^2M \to T^2M$ is defined locally by

$$(T^2u \circ \kappa_M \circ T^2u^{-1})(x, \xi; \eta, \zeta) = (x, \eta; \xi, \zeta),$$

where (U, u) is a chart on M. Clearly this definition is invariant under changes of charts (Tu_α equals ψ'_α from 6.11).

The flip κ_M has the following properties:

(1) $\kappa_N \circ T^2 f = T^2 f \circ \kappa_M$ for each $f \in C^\infty(M, N)$.
(2) $T(\pi_M) \circ \kappa_M = \pi_{TM}$.
(3) $\pi_{TM} \circ \kappa_M = T(\pi_M)$.
(4) $\kappa_M^{-1} = \kappa_M$.
(5) κ_M is a linear isomorphism from the bundle $(TTM, T(\pi_M), TM)$ to (TTM, π_{TM}, TM), so it interchanges the two vector bundle structures on TTM.
(6) It is the unique smooth mapping $TTM \to TTM$ which satisfies $\frac{\partial}{\partial t}\frac{\partial}{\partial s}c(t,s) = \kappa_M \frac{\partial}{\partial s}\frac{\partial}{\partial t}c(t,s)$ for each $c : \mathbb{R}^2 \to M$.

All this follows from the local formula given above. We will come back to the flip later on in chapter VIII from a more advanced point of view.

6.13. Lemma. *For vector fields $X, Y \in \mathfrak{X}(M)$ we have*

$$[X,Y] = vpr_{TM} \circ (TY \circ X - \kappa_M \circ TX \circ Y).$$

We will give global proofs of this result later on: the first one is 6.19. Another one is 37.13.

Proof. We prove this locally, so we assume that M is open in $\mathbb{R}^m$, $X(x) = (x, \bar{X}(x))$, and $Y(x) = (x, \bar{Y}(x))$. By 3.4 we get $[X,Y](x) = (x, d\bar{Y}(x).\bar{X}(x) - d\bar{X}(x).\bar{Y}(x))$, and

$$\begin{aligned}
&vpr_{TM} \circ (TY \circ X - \kappa_M \circ TX \circ Y)(x) = \\
&\quad = vpr_{TM} \circ (TY.(x, \bar{X}(x)) - \kappa_M \circ TX.(x, \bar{Y}(x))) = \\
&\quad = vpr_{TM}\big((x, \bar{Y}(x); \bar{X}(x), d\bar{Y}(x).\bar{X}(x)) - \\
&\qquad\qquad - \kappa_M((x, \bar{X}(x); \bar{Y}(x), d\bar{X}(x).\bar{Y}(x))\big) = \\
&\quad = vpr_{TM}(x, \bar{Y}(x); 0, d\bar{Y}(x).\bar{X}(x) - d\bar{X}(x).\bar{Y}(x)) = \\
&\quad = (x, d\bar{Y}(x).\bar{X}(x) - d\bar{X}(x).\bar{Y}(x)). \quad \square
\end{aligned}$$

6.14. Natural vector bundles. Let $\mathcal{M}f_m$ denote the category of all m-dimensional smooth manifolds and local diffeomorphisms (i.e. immersions) between them. A *vector bundle functor* or *natural vector bundle* is a functor F which associates a vector bundle $(F(M), p_M, M)$ to each m-manifold M and a vector bundle homomorphism

$$\begin{array}{ccc}
F(M) & \xrightarrow{F(f)} & F(N) \\
{\scriptstyle p_M}\downarrow & & \downarrow{\scriptstyle p_N} \\
M & \xrightarrow{f} & N
\end{array}$$

to each $f : M \to N$ in $\mathcal{M}f_m$, which covers f and is fiberwise a linear isomorphism. We also require that for smooth $f : \mathbb{R} \times M \to N$ the mapping $(t,x) \mapsto F(f_t)(x)$ is also smooth $\mathbb{R} \times F(M) \to F(N)$. We will say that F maps smoothly parametrized families to smoothly parametrized families. We shall see later that this last requirement is automatically satisfied. For a characterization of all vector bundle functors see 14.8.

Examples. 1. TM, the tangent bundle. This is even a functor on the category $\mathcal{M}f$.

2. T^*M, the cotangent bundle, where by 6.7 the action on morphisms is given by $(T^*f)_x := ((T_xf)^{-1})^* : T_x^*M \to T_{f(x)}^*N$. This functor is defined on $\mathcal{M}f_m$ only.

3. $\Lambda^k T^*M$, $\Lambda T^*M = \bigoplus_{k\geq 0} \Lambda^k T^*M$.

4. $\bigotimes^k T^*M \otimes \bigotimes^\ell TM = T^*M \otimes \cdots \otimes T^*M \otimes TM \otimes \cdots \otimes TM$, where the action on morphisms involves Tf^{-1} in the T^*M-parts and Tf in the TM-parts.

5. $\mathcal{F}(TM)$, where $\mathcal{F}$ is any smooth functor on the category of finite dimensional vector spaces and linear mappings, as in 6.7.

6.15. Lie derivative. Let F be a vector bundle functor on $\mathcal{M}f_m$ as described in 6.14. Let M be a manifold and let $X \in \mathfrak{X}(M)$ be a vector field on M. Then the flow Fl_t^X, for fixed t, is a diffeomorphism defined on an open subset of M, which we do not specify. The mapping

$$\begin{array}{ccc} F(M) & \xrightarrow{F(\mathrm{Fl}_t^X)} & F(M) \\ {\scriptstyle p_M}\downarrow & & \downarrow{\scriptstyle p_M} \\ M & \xrightarrow{\mathrm{Fl}_t^X} & M \end{array}$$

is then a vector bundle isomorphism, defined over an open subset of M.

We consider a section $s \in C^\infty(F(M))$ of the vector bundle $(F(M), p_M, M)$ and we define for $t \in \mathbb{R}$

$$(\mathrm{Fl}_t^X)^*s := F(\mathrm{Fl}_{-t}^X) \circ s \circ \mathrm{Fl}_t^X .$$

This is a local section of the vector bundle $F(M)$. For each $x \in M$ the value $((\mathrm{Fl}_t^X)^*s)(x) \in F(M)_x$ is defined, if t is small enough. So in the vector space $F(M)_x$ the expression $\frac{d}{dt}|_0((\mathrm{Fl}_t^X)^*s)(x)$ makes sense and therefore the section

$$\mathcal{L}_X s := \tfrac{d}{dt}|_0 (\mathrm{Fl}_t^X)^*s$$

is globally defined and is an element of $C^\infty(F(M))$. It is called the *Lie derivative* of s along X.

Lemma. *In this situation we have*

(1) $(\mathrm{Fl}_t^X)^*(\mathrm{Fl}_r^X)^*s = (\mathrm{Fl}_{t+r}^X)^*s$, *whenever defined.*

(2) $\frac{d}{dt}(\mathrm{Fl}_t^X)^*s = (\mathrm{Fl}_t^X)^*\mathcal{L}_X s = \mathcal{L}_X(\mathrm{Fl}_t^X)^*s$, *so*
$[\mathcal{L}_X, (\mathrm{Fl}_t^X)^*] := \mathcal{L}_X \circ (\mathrm{Fl}_t^X)^* - (\mathrm{Fl}_t^X)^* \circ \mathcal{L}_X = 0$, *whenever defined.*

(3) $(\mathrm{Fl}_t^X)^*s = s$ *for all relevant* t *if and only if* $\mathcal{L}_X s = 0$.

Proof. (1) is clear. (2) is seen by the following computations.

$$\tfrac{d}{dt}(\mathrm{Fl}^X_t)^* s = \tfrac{d}{dr}|_0 (\mathrm{Fl}^X_r)^*(\mathrm{Fl}^X_t)^* s = \mathcal{L}_X (\mathrm{Fl}^X_t)^* s.$$

$$\begin{aligned}\tfrac{d}{dt}((\mathrm{Fl}^X_t)^* s)(x) &= \tfrac{d}{dr}|_0 ((\mathrm{Fl}^X_t)^*(\mathrm{Fl}^X_r)^* s)(x)\\ &= \tfrac{d}{dr}|_0 F(\mathrm{Fl}^X_{-t})(F(\mathrm{Fl}^X_{-r})\circ s\circ \mathrm{Fl}^X_r)(\mathrm{Fl}^X_t(x))\\ &= F(\mathrm{Fl}^X_{-t})\tfrac{d}{dr}|_0 (F(\mathrm{Fl}^X_{-r})\circ s\circ \mathrm{Fl}^X_r)(\mathrm{Fl}^X_t(x))\\ &= ((\mathrm{Fl}^X_t)^*\mathcal{L}_X s)(x),\end{aligned}$$

since $F(\mathrm{Fl}^X_{-t}) : F(M)_{\mathrm{Fl}^X_t(x)} \to F(M)_x$ is linear.

(3) follows from (2). □

6.16. Let F_1, F_2 be two vector bundle functors on $\mathcal{M}f_m$. Then the tensor product $(F_1 \otimes F_2)(M) := F_1(M) \otimes F_2(M)$ is again a vector bundle functor and for $s_i \in C^\infty(F_i(M))$ there is a section $s_1 \otimes s_2 \in C^\infty((F_1 \otimes F_2)(M))$, given by the pointwise tensor product.

Lemma. *In this situation, for $X \in \mathfrak{X}(M)$ we have*

$$\mathcal{L}_X(s_1 \otimes s_2) = \mathcal{L}_X s_1 \otimes s_2 + s_1 \otimes \mathcal{L}_X s_2.$$

In particular, for $f \in C^\infty(M, \mathbb{R})$ we have $\mathcal{L}_X(fs) = df(X)\, s + f\, \mathcal{L}_X s$.

Proof. Using the bilinearity of the tensor product we have

$$\begin{aligned}\mathcal{L}_X(s_1 \otimes s_2) &= \tfrac{d}{dt}|_0 (\mathrm{Fl}^X_t)^*(s_1 \otimes s_2)\\ &= \tfrac{d}{dt}|_0 ((\mathrm{Fl}^X_t)^* s_1 \otimes (\mathrm{Fl}^X_t)^* s_2)\\ &= \tfrac{d}{dt}|_0 (\mathrm{Fl}^X_t)^* s_1 \otimes s_2 + s_1 \otimes \tfrac{d}{dt}|_0 (\mathrm{Fl}^X_t)^* s_2\\ &= \mathcal{L}_X s_1 \otimes s_2 + s_1 \otimes \mathcal{L}_X s_2. \quad \square\end{aligned}$$

6.17. Let $\varphi : F_1 \to F_2$ be a linear *natural transformation* between vector bundle functors on $\mathcal{M}f_m$, i.e. for each $M \in \mathcal{M}f_m$ we have a vector bundle homomorphism $\varphi_M : F_1(M) \to F_2(M)$ covering the identity on M, such that $F_2(f) \circ \varphi_M = \varphi_N \circ F_1(f)$ holds for any $f : M \to N$ in $\mathcal{M}f_m$ (we shall see in 14.11 that for every natural transformation $\varphi : F_1 \to F_2$ in the purely categorical sense each morphism $\varphi_M : F_1(M) \to F_2(M)$ covers Id_M).

Lemma. *In this situation, for $s \in C^\infty(F_1(M))$ and $X \in \mathfrak{X}(M)$, we have $\mathcal{L}_X(\varphi_M\, s) = \varphi_M(\mathcal{L}_X s)$.*

Proof. Since φ_M is fiber linear and natural we can compute as follows.

$$\begin{aligned}\mathcal{L}_X(\varphi_M\, s)(x) &= \tfrac{d}{dt}|_0 ((\mathrm{Fl}^X_t)^*(\varphi_M\, s))(x) = \tfrac{d}{dt}|_0 (F_2(\mathrm{Fl}^X_{-t}) \circ \varphi_M \circ s \circ \mathrm{Fl}^X_t)(x)\\ &= \varphi_M \circ \tfrac{d}{dt}|_0 (F_1(\mathrm{Fl}^X_{-t}) \circ s \circ \mathrm{Fl}^X_t)(x) = (\varphi_M\, \mathcal{L}_X s)(x). \quad \square\end{aligned}$$

6.18. A *tensor field* of type $\binom{p}{q}$ is a smooth section of the natural bundle $\bigotimes^q T^*M \otimes \bigotimes^p TM$. For such tensor fields, by 6.15 the Lie derivative along any vector field is defined, by 6.16 it is a derivation with respect to the tensor product, and by 6.17 it commutes with any kind of contraction or 'permutation of the indices'. For functions and vector fields the Lie derivative was already defined in section 3.

6.19. Let F be a vector bundle functor on $\mathcal{M}f_m$ and let $X \in \mathfrak{X}(M)$ be a vector field. We consider the local vector bundle homomorphism $F(\mathrm{Fl}_t^X)$ on $F(M)$. Since $F(\mathrm{Fl}_t^X) \circ F(\mathrm{Fl}_s^X) = F(\mathrm{Fl}_{t+s}^X)$ and $F(\mathrm{Fl}_0^X) = \mathrm{Id}_{F(M)}$ we have $\frac{d}{dt}F(\mathrm{Fl}_t^X) = \frac{d}{ds}|_0 F(\mathrm{Fl}_s^X) \circ F(\mathrm{Fl}_t^X) = X^F \circ F(\mathrm{Fl}_t^X)$, so we get $F(\mathrm{Fl}_t^X) = \mathrm{Fl}_t^{X^F}$, where $X^F = \frac{d}{ds}|_0 F(\mathrm{Fl}_s^X) \in \mathfrak{X}(F(M))$ is a vector field on $F(M)$, which is called the *flow prolongation* or the *canonical lift* of X to $F(M)$. If it is desirable for technical reasons we shall also write $X^F = \mathcal{F}X$.

Lemma.

(1) $X^T = \kappa_M \circ TX$.
(2) $[X,Y]^F = [X^F, Y^F]$.
(3) $X^F : (F(M), p_M, M) \to (TF(M), T(p_M), TM)$ *is a vector bundle homomorphism for the* $T(+)$*-structure.*
(4) *For* $s \in C^\infty(F(M))$ *and* $X \in \mathfrak{X}(M)$ *we have* $\mathcal{L}_X s = vpr_{F(M)}(Ts \circ X - X^F \circ s)$.
(5) $\mathcal{L}_X s$ *is linear in* X *and* s.

Proof. (1) is an easy computation. $F(\mathrm{Fl}_t^X)$ is fiber linear and this implies (3). (4) is seen as follows:

$$\begin{aligned}(\mathcal{L}_X s)(x) &= \tfrac{d}{dt}|_0 (F(\mathrm{Fl}_{-t}^X) \circ s \circ \mathrm{Fl}_t^X)(x) \quad \text{in } F(M)_x \\ &= vpr_{F(M)}(\tfrac{d}{dt}|_0 (F(\mathrm{Fl}_{-t}^X) \circ s \circ \mathrm{Fl}_t^X)(x) \text{ in } VF(M)) \\ &= vpr_{F(M)}(-X^F \circ s \circ \mathrm{Fl}_0^X(x) + T(F(\mathrm{Fl}_0^X)) \circ Ts \circ X(x)) \\ &= vpr_{F(M)}(Ts \circ X - X^F \circ s)(x).\end{aligned}$$

(5) $\mathcal{L}_X s$ is homogeneous of degree 1 in X by formula (4), and it is smooth as a mapping $\mathfrak{X}(M) \to C^\infty(F(M))$, so it is linear. See [Frölicher, Kriegl, 88] for the convenient calculus in infinite dimensions.

(2) Note first that F induces a smooth mapping between appropriate spaces of local diffeomorphisms which are infinite dimensional manifolds (see [Kriegl, Michor, 91]). By 3.16 we have

$$\begin{aligned}0 &= \tfrac{\partial}{\partial t}\big|_0 (\mathrm{Fl}_{-t}^Y \circ \mathrm{Fl}_{-t}^X \circ \mathrm{Fl}_t^Y \circ \mathrm{Fl}_t^X), \\ [X,Y] &= \tfrac{1}{2}\tfrac{\partial^2}{\partial t^2}|_0 (\mathrm{Fl}_{-t}^Y \circ \mathrm{Fl}_{-t}^X \circ \mathrm{Fl}_t^Y \circ \mathrm{Fl}_t^X) \\ &= \tfrac{\partial}{\partial t}\big|_0 \mathrm{Fl}_t^{[X,Y]}.\end{aligned}$$

Applying F to these curves (of local diffeomorphisms) we get

$$\begin{aligned} 0 &= \tfrac{\partial}{\partial t}\big|_0 (\mathrm{Fl}^{Y^F}_{-t} \circ \mathrm{Fl}^{X^F}_{-t} \circ \mathrm{Fl}^{Y^F}_{t} \circ \mathrm{Fl}^{X^F}_{t}),\\ [X^F, Y^F] &= \tfrac{1}{2}\tfrac{\partial^2}{\partial t^2}|_0 (\mathrm{Fl}^{Y^F}_{-t} \circ \mathrm{Fl}^{X^F}_{-t} \circ \mathrm{Fl}^{Y^F}_{t} \circ \mathrm{Fl}^{X^F}_{t})\\ &= \tfrac{1}{2}\tfrac{\partial^2}{\partial t^2}|_0 F(\mathrm{Fl}^{Y}_{-t} \circ \mathrm{Fl}^{X}_{-t} \circ \mathrm{Fl}^{Y}_{t} \circ \mathrm{Fl}^{X}_{t})\\ &= \tfrac{\partial}{\partial t}\big|_0 F(\mathrm{Fl}^{[X,Y]}_t) = [X,Y]^F. \end{aligned}$$

See also section 50 for a purely finite dimensional proof of a much more general result. □

6.20. Proposition. *For any vector bundle functor F on $\mathcal{M}f_m$ and $X, Y \in \mathfrak{X}(M)$ we have*

$$[\mathcal{L}_X, \mathcal{L}_Y] := \mathcal{L}_X \circ \mathcal{L}_Y - \mathcal{L}_Y \circ \mathcal{L}_X = \mathcal{L}_{[X,Y]} : C^\infty(F(M)) \to C^\infty(F(M)).$$

So $\mathcal{L} : \mathfrak{X}(M) \to \operatorname{End} C^\infty(F(M))$ is a Lie algebra homomorphism.

Proof. See section 50 for a proof of a much more general formula. □

6.21. Theorem. *Let M be a manifold, let $\varphi^i : \mathbb{R} \times M \supset U_{\varphi^i} \to M$ be smooth mappings for $i = 1, \dots, k$ where each U_{φ^i} is an open neighborhood of $\{0\} \times M$ in $\mathbb{R} \times M$, such that each φ^i_t is a diffeomorphism on its domain, $\varphi^i_0 = \mathrm{Id}_M$, and $\frac{\partial}{\partial t}\big|_0 \varphi^i_t = X_i \in \mathfrak{X}(M)$. We put $[\varphi^i, \varphi^j]_t = [\varphi^i_t, \varphi^j_t] := (\varphi^j_t)^{-1} \circ (\varphi^i_t)^{-1} \circ \varphi^j_t \circ \varphi^i_t$. Let F be a vector bundle functor, let $s \in C^\infty(F(M))$ be a section. Then for each formal bracket expression P of lenght k we have*

$$\begin{aligned} 0 &= \tfrac{\partial^\ell}{\partial t^\ell}|_0 P(\varphi^1_t, \dots, \varphi^k_t)^* s \quad \text{for } 1 \le \ell < k,\\ \mathcal{L}_{P(X_1,\dots,X_k)} s &= \tfrac{1}{k!}\tfrac{\partial^k}{\partial t^k}|_0 P(\varphi^1_t, \dots, \varphi^k_t)^* s \in C^\infty(F(M)). \end{aligned}$$

Proof. This can be proved with similar methods as in the proof of 3.16. A concise proof can be found in [Mauhart, Michor, 92] □

6.22. Affine bundles. Given a finite dimensional affine space A modelled on a vector space $V = \vec{A}$, we denote by $+$ the canonical mapping $A \times \vec{A} \to A$, $(p, v) \mapsto p + v$ for $p \in A$ and $v \in \vec{A}$. If A_1 and A_2 are two affine spaces and $f : A_1 \to A_2$ is an affine mapping, then we denote by $\vec{f} : \vec{A_1} \to \vec{A_2}$ the linear mapping given by $f(p + v) = f(p) + \vec{f}(v)$.

Let $p : E \to M$ be a vector bundle and $q : Z \to M$ be a smooth mapping such that each fiber $Z_x = q^{-1}(x)$ is an affine space modelled on the vector space $E_x = p^{-1}(x)$. Let A be an affine space modelled on the standard fiber V of E. We say that Z is an *affine bundle* with standard fiber A modelled on the vector bundle E, if for each vector bundle chart $\psi : E|U = p^{-1}(U) \to U \times V$ on E there exists a fiber respecting diffeomorphism $\varphi : Z|U = q^{-1}(U) \to U \times A$ such that $\varphi_x : Z_x \to A$ is an affine morphism satisfying $\vec{\varphi}_x = \psi_x : E_x \to V$ for each $x \in U$. We also write $E = \vec{Z}$ to have a functorial assignment of the modelling vector bundle.

Let $Z \to M$ and $Y \to N$ be two affine bundles. An affine bundle morphism $f : Z \to Y$ is a fiber respecting mapping such that each $f_x : Z_x \to Y_{\underline{f}(x)}$ is an affine mapping, where $\underline{f} : M \to N$ is the underlying base mapping of f. Clearly the rule $x \mapsto \vec{f}_x : \vec{Z}_x \to \vec{Y}_{\underline{f}(x)}$ induces a vector bundle homomorphism $\vec{f} : \vec{Z} \to \vec{Y}$ over the same base mapping $\underline{f}$.

7. Differential forms

7.1. The *cotangent bundle* of a manifold M is the vector bundle $T^*M := (TM)^*$, the (real) dual of the tangent bundle.

If (U, u) is a chart on M, then $(\frac{\partial}{\partial u^1}, \dots, \frac{\partial}{\partial u^m})$ is the associated frame field over U of TM. Since $\frac{\partial}{\partial u^i}|_x(u^j) = du^j(\frac{\partial}{\partial u^i}|_x) = \delta^j_i$ we see that $(du^1, \dots, du^m)$ is the dual frame field on T^*M over U. It is also called a *holonomous* frame field. A section of T^*M is also called a *1-form.*

7.2. According to 6.18 a *tensor field* of type $\binom{p}{q}$ on a manifold M is a smooth section of the vector bundle

$$\bigotimes^p TM \otimes \bigotimes^q T^*M = TM \overbrace{\otimes \cdots \otimes}^{p \text{ times}} TM \otimes T^*M \overbrace{\otimes \cdots \otimes}^{q \text{ times}} T^*M.$$

The position of p (up) and q (down) can be explained as follows: If (U, u) is a chart on M, we have the holonomous frame field

$$\left(\frac{\partial}{\partial u^{i_1}} \otimes \frac{\partial}{\partial u^{i_2}} \otimes \cdots \otimes \frac{\partial}{\partial u^{i_p}} \otimes du^{j_1} \otimes \cdots \otimes du^{j_q}\right)_{i \in \{1,\dots,m\}^p, j \in \{1,\dots,m\}^q}$$

over U of this tensor bundle, and for any $\binom{p}{q}$-tensor field A we have

$$A \mid U = \sum_{i,j} A^{i_1 \dots i_p}_{j_1 \dots j_q} \frac{\partial}{\partial u^{i_1}} \otimes \cdots \otimes \frac{\partial}{\partial u^{i_p}} \otimes du^{j_1} \otimes \cdots \otimes du^{j_q}.$$

The coefficients have p indices up and q indices down, they are smooth functions on U. From a strictly categorical point of view the position of the indices should be exchanged, but this convention has a long tradition.

7.3 Lemma. *Let $\Phi : \mathfrak{X}(M) \times \cdots \times \mathfrak{X}(M) = \mathfrak{X}(M)^k \to C^\infty(\bigotimes^\ell TM)$ be a mapping which is k-linear over $C^\infty(M, \mathbb{R})$ then Φ is given by a $\binom{\ell}{k}$-tensor field.*

Proof. For simplicity's sake we put $k = 1$, $\ell = 0$, so $\Phi : \mathfrak{X}(M) \to C^\infty(M, \mathbb{R})$ is a $C^\infty(M, \mathbb{R})$-linear mapping: $\Phi(f.X) = f.\Phi(X)$.

CLAIM 1. If $X \mid U = 0$ for some open subset $U \subset M$, then we have $\Phi(X) \mid U = 0$.

Let $x \in U$. We choose $f \in C^\infty(M, \mathbb{R})$ with $f(x) = 0$ and $f \mid M \setminus U = 1$. Then $f.X = X$, so $\Phi(X)(x) = \Phi(f.X)(x) = f(x).\Phi(X)(x) = 0$.

CLAIM 2. If $X(x) = 0$ then also $\Phi(X)(x) = 0$.

Let (U, u) be a chart centered at x, let V be open with $x \in V \subset \bar{V} \subset U$. Then

$X \mid U = \sum X^i \frac{\partial}{\partial u^i}$ and $X^i(x) = 0$. We choose $g \in C^\infty(M, \mathbb{R})$ with $g \mid V \equiv 1$ and $\operatorname{supp} g \subset U$. Then $(g^2.X) \mid V = X \mid V$ and by claim 1 $\Phi(X) \mid V$ depends only on $X \mid V$ and $g^2.X = \sum_i (g.X^i)(g.\frac{\partial}{\partial u^i})$ is a decomposition which is globally defined on M. Therefore we have $\Phi(X)(x) = \Phi(g^2.X)(x) = \Phi\left(\sum_i (g.X^i)(g.\frac{\partial}{\partial u^i})\right)(x) = \sum (g.X^i)(x).\Phi(g.\frac{\partial}{\partial u^i})(x) = 0$.

So we see that for a general vector field X the value $\Phi(X)(x)$ depends only on the value $X(x)$, for each $x \in M$. So there is a linear map $\varphi_x : T_xM \to \mathbb{R}$ for each $x \in M$ with $\Phi(X)(x) = \varphi_x(X(x))$. Then $\varphi : M \to T^*M$ is smooth since $\varphi \mid V = \sum_i \Phi(g.\frac{\partial}{\partial u^i}).du^i$ in the setting of claim 2. □

7.4. Definition. A *differential form* or an *exterior form* of degree k or a *k-form* for short is a section of the vector bundle $\Lambda^k T^*M$. The space of all k-forms will be denoted by $\Omega^k(M)$. It may also be viewed as the space of all skew symmetric $\binom{0}{k}$-tensor fields, i.e. (by 7.3) the space of all mappings

$$\Phi : \mathfrak{X}(M) \times \cdots \times \mathfrak{X}(M) = \mathfrak{X}(M)^k \to C^\infty(M, \mathbb{R}),$$

which are k-linear over $C^\infty(M, \mathbb{R})$ and are skew symmetric:

$$\Phi(X_{\sigma 1}, \ldots, X_{\sigma k}) = \operatorname{sign} \sigma \cdot \Phi(X_1, \ldots, X_k)$$

for each permutation $\sigma \in S_k$.

We put $\Omega^0(M) := C^\infty(M, \mathbb{R})$. Then the space

$$\Omega(M) := \bigoplus_{k=0}^{\dim M} \Omega^k(M)$$

is an algebra with the following product. For $\varphi \in \Omega^k(M)$ and $\psi \in \Omega^\ell(M)$ and for X_i in $\mathfrak{X}(M)$ (or in T_xM) we put

$$(\varphi \wedge \psi)(X_1, \ldots, X_{k+\ell}) = \\ = \frac{1}{k!\,\ell!} \sum_{\sigma \in S_{k+\ell}} \operatorname{sign} \sigma \cdot \varphi(X_{\sigma 1}, \ldots, X_{\sigma k}).\psi(X_{\sigma(k+1)}, \ldots, X_{\sigma(k+\ell)}).$$

This product is defined fiber wise, i.e. $(\varphi \wedge \psi)_x = \varphi_x \wedge \psi_x$ for each $x \in M$. It is also associative, i.e. $(\varphi \wedge \psi) \wedge \tau = \varphi \wedge (\psi \wedge \tau)$, and graded commutative, i.e. $\varphi \wedge \psi = (-1)^{k\ell} \psi \wedge \varphi$. These properties are proved in multilinear algebra.

7.5. If $f : N \to M$ is a smooth mapping and $\varphi \in \Omega^k(M)$, then the pullback $f^*\varphi \in \Omega^k(N)$ is defined for $X_i \in T_xN$ by

$$(1) \qquad (f^*\varphi)_x(X_1, \ldots, X_k) := \varphi_{f(x)}(T_xf.X_1, \ldots, T_xf.X_k).$$

Then we have $f^*(\varphi \wedge \psi) = f^*\varphi \wedge f^*\psi$, so the linear mapping $f^* : \Omega(M) \to \Omega(N)$ is an algebra homomorphism. Moreover we have $(g \circ f)^* = f^* \circ g^* : \Omega(P) \to \Omega(N)$ if $g : M \to P$, and $(\mathrm{Id}_M)^* = \mathrm{Id}_{\Omega(M)}$.

So $M \mapsto \Omega(M) = C^\infty(\Lambda T^*M)$ is a contravariant functor from the category $\mathcal{M}f$ of all manifolds and all smooth mappings into the category of real graded commutative algebras, whereas $M \mapsto \Lambda T^*M$ is a covariant vector bundle functor defined only on $\mathcal{M}f_m$, the category of m-dimensional manifolds and local diffeomorphisms, for each m separately.

7.6. The Lie derivative of differential forms. Since $M \mapsto \Lambda^k T^*M$ is a vector bundle functor on $\mathcal{M}f_m$, by 6.15 for $X \in \mathfrak{X}(M)$ the *Lie derivative* of a k-form φ along X is defined by

$$\mathcal{L}_X\varphi = \tfrac{d}{dt}|_0(\mathrm{Fl}_t^X)^*\varphi.$$

Lemma. *The Lie derivative has the following properties.*

(1) $\mathcal{L}_X(\varphi \wedge \psi) = \mathcal{L}_X\varphi \wedge \psi + \varphi \wedge \mathcal{L}_X\psi$, *so $\mathcal{L}_X$ is a derivation.*

(2) *For $Y_i \in \mathfrak{X}(M)$ we have*

$$(\mathcal{L}_X\varphi)(Y_1,\dots,Y_k) = X(\varphi(Y_1,\dots,Y_k)) - \sum_{i=1}^{k} \varphi(Y_1,\dots,[X,Y_i],\dots,Y_k).$$

(3) $[\mathcal{L}_X,\mathcal{L}_Y]\varphi = \mathcal{L}_{[X,Y]}\varphi$.

Proof. The mapping $Alt : \bigotimes^k T^*M \to \Lambda^k T^*M$, given by

$$(Alt A)(Y_1,\dots,Y_k) := \tfrac{1}{k!}\sum_{\sigma} \mathrm{sign}(\sigma) A(Y_{\sigma 1},\dots,Y_{\sigma k}),$$

is a linear natural transformation in the sense of 6.17 and induces an algebra homomorphism from the tensor algebra $\bigoplus_{k\ge 0} C^\infty(\bigotimes^k T^*M)$ onto $\Omega(M)$. So (1) follows from 6.16.

(2) Again by 6.16 and 6.17 we may compute as follows, where Trace is the full evaluation of the form on all vector fields:

$$\begin{aligned} X(\varphi(Y_1,\dots,Y_k)) &= \mathcal{L}_X \circ \mathrm{Trace}(\varphi\otimes Y_1\otimes\dots\otimes Y_k)\\ &= \mathrm{Trace}\circ\mathcal{L}_X(\varphi\otimes Y_1\otimes\dots\otimes Y_k)\\ &= \mathrm{Trace}\big(\mathcal{L}_X\varphi\otimes(Y_1\otimes\dots\otimes Y_k) + \varphi\otimes(\textstyle\sum_i Y_1\otimes\dots\otimes\mathcal{L}_X Y_i\otimes\dots\otimes Y_k)\big). \end{aligned}$$

Now we use $\mathcal{L}_X Y_i = [X,Y_i]$.

(3) is a special case of 6.20. □

7.7. The insertion operator. For a vector field $X \in \mathfrak{X}(M)$ we define the *insertion operator* $i_X = i(X) : \Omega^k(M) \to \Omega^{k-1}(M)$ by

$$(i_X\varphi)(Y_1,\dots,Y_{k-1}) := \varphi(X,Y_1,\dots,Y_{k-1}).$$

Lemma.

(1) *i_X is a graded derivation of degree -1 of the graded algebra $\Omega(M)$, so we have $i_X(\varphi\wedge\psi) = i_X\varphi\wedge\psi + (-1)^{\deg\varphi}\varphi\wedge i_X\psi$.*

(2) $[\mathcal{L}_X, i_Y] := \mathcal{L}_X\circ i_Y - i_Y\circ\mathcal{L}_X = i_{[X,Y]}$.

Proof. (1) For $\varphi \in \Omega^k(M)$ and $\psi \in \Omega^\ell(M)$ we have

$$(i_{X_1}(\varphi\wedge\psi))(X_2,\dots,X_{k+\ell}) = (\varphi\wedge\psi)(X_1,\dots,X_{k+\ell}) =$$
$$= \tfrac{1}{k!\,\ell!}\sum_\sigma \operatorname{sign}(\sigma)\,\varphi(X_{\sigma 1},\dots,X_{\sigma k})\psi(X_{\sigma(k+1)},\dots,X_{\sigma(k+\ell)}).$$

$$(i_{X_1}\varphi\wedge\psi + (-1)^k\varphi\wedge i_{X_1}\psi)(X_2,\dots,X_{k+\ell}) =$$
$$= \tfrac{1}{(k-1)!\,\ell!}\sum_\sigma \operatorname{sign}(\sigma)\,\varphi(X_1,X_{\sigma 2},\dots,X_{\sigma k})\psi(X_{\sigma(k+1)},\dots,X_{\sigma(k+\ell)})+$$
$$+\frac{(-1)^k}{k!\,(\ell-1)!}\sum_\sigma \operatorname{sign}(\sigma)\,\varphi(X_{\sigma 2},\dots,X_{\sigma(k+1)})\psi(X_1,X_{\sigma(k+2)},\dots).$$

Using the skew symmetry of φ and ψ we may distribute X_1 to each position by adding an appropriate sign. These are $k+\ell$ summands. Since $\frac{1}{(k-1)!\,\ell!}+\frac{1}{k!\,(\ell-1)!} = \frac{k+\ell}{k!\,\ell!}$, and since we can generate each permutation in $\mathcal{S}_{k+\ell}$ in this way, the result follows.

(2) By 6.16 and 6.17 we have:

$$\mathcal{L}_X i_Y\varphi = \mathcal{L}_X \operatorname{Trace}_1(Y\otimes\varphi) = \operatorname{Trace}_1 \mathcal{L}_X(Y\otimes\varphi)$$
$$= \operatorname{Trace}_1(\mathcal{L}_X Y\otimes\varphi + Y\otimes\mathcal{L}_X\varphi) = i_{[X,Y]}\varphi + i_Y\mathcal{L}_X\varphi. \quad \square$$

7.8. The exterior differential. We want to construct a differential operator $\Omega^k(M)\to\Omega^{k+1}(M)$ which is natural. We will show that the simplest choice will work and (later) that it is essentially unique.

So let U be open in $\mathbb{R}^n$, let $\varphi\in\Omega^k(\mathbb{R}^n)$. Then we may view φ as an element of $C^\infty(U, L^k_{alt}(\mathbb{R}^n,\mathbb{R}))$. We consider $D\varphi \in C^\infty(U, L(\mathbb{R}^n, L^k_{alt}(\mathbb{R}^n,\mathbb{R})))$, and we take its canonical image $\operatorname{Alt}(D\varphi)$ in $C^\infty(U, L^{k+1}_{alt}(\mathbb{R}^n,\mathbb{R}))$. Here we write D for the derivative in order to distinguish it from the exterior differential, which we define as $d\varphi := (k+1)\operatorname{Alt}(D\varphi)$, more explicitly as

$$(1)\qquad (d\varphi)_x(X_0,\dots,X_k) = \tfrac{1}{k!}\sum_\sigma \operatorname{sign}(\sigma)\,D\varphi(x)(X_{\sigma 0})(X_{\sigma 1},\dots,X_{\sigma k})$$
$$= \sum_{i=0}^k (-1)^i D\varphi(x)(X_i)(X_0,\dots,\widehat{X_i},\dots,X_k),$$

where the hat over a symbol means that this is to be omitted, and where $X_i\in\mathbb{R}^n$.

Now we pass to an arbitrary manifold M. For a k-form $\varphi\in\Omega^k(M)$ and vector fields $X_i\in\mathfrak{X}(M)$ we try to replace $D\varphi(x)(X_i)(X_0,\dots)$ in formula (1) by Lie derivatives. We differentiate $X_i(\varphi(x)(X_0,\dots)) = D\varphi(x)(X_i)(X_0,\dots)+\sum_{0\le j\le k, j\ne i}\varphi(x)(X_0,\dots,DX_j(x)X_i,\dots)$, so inserting this expression into formula (1) we get (cf. 3.4) our working definition

$$(2)\qquad d\varphi(X_0,\dots,X_k) := \sum_{i=0}^k (-1)^i X_i(\varphi(X_0,\dots,\widehat{X_i},\dots,X_k))$$
$$+\sum_{i<j}(-1)^{i+j}\varphi([X_i,X_j],X_0,\dots,\widehat{X_i},\dots,\widehat{X_j},\dots,X_k).$$

$d\varphi$, given by this formula, is $(k+1)$-linear over $C^\infty(M,\mathbb{R})$, as a short computation involving 3.4 shows. It is obviously skew symmetric, so by 7.3 $d\varphi$ is a $(k+1)$-form, and the operator $d:\Omega^k(M)\to\Omega^{k+1}(M)$ is called the *exterior derivative*.

If (U,u) is a chart on M, then we have

$$\varphi|U=\sum_{i_1<\dots<i_k}\varphi_{i_1,\dots,i_k}du^{i_1}\wedge\dots\wedge du^{i_k},$$

where $\varphi_{i_1,\dots,i_k}=\varphi(\frac{\partial}{\partial u^{i_1}},\dots,\frac{\partial}{\partial u^{i_k}})$. An easy computation shows that (2) leads to

$$(3)\qquad d\varphi|U=\sum_{i_1<\dots<i_k}d\varphi_{i_1,\dots,i_k}\wedge du^{i_1}\wedge\dots\wedge du^{i_k},$$

so that formulas (1) and (2) really define the same operator.

7.9. Theorem. *The exterior derivative $d:\Omega^k(M)\to\Omega^{k+1}(M)$ has the following properties:*

(1) *$d(\varphi\wedge\psi)=d\varphi\wedge\psi+(-1)^{\deg\varphi}\varphi\wedge d\psi$, so d is a graded derivation of degree 1.*
(2) *$\mathcal{L}_X=i_X\circ d+d\circ i_X$ for any vector field X.*
(3) *$d^2=d\circ d=0$.*
(4) *$f^*\circ d=d\circ f^*$ for any smooth $f:N\to M$.*
(5) *$\mathcal{L}_X\circ d=d\circ\mathcal{L}_X$ for any vector field X.*

Remark. In terms of the graded commutator

$$[D_1,D_2]:=D_1\circ D_2-(-1)^{\deg(D_1)\deg(D_2)}D_2\circ D_1$$

for graded homomorphisms and graded derivations (see 8.1) the assertions of this theorem take the following form:

(2) $\mathcal{L}_X=[i_X,d]$.
(3) $\frac{1}{2}[d,d]=0$.
(4) $[f^*,d]=0$.
(5) $[\mathcal{L}_X,d]=0$.

This point of view will be developed in section 8 below.

Proof. (2) For $\varphi\in\Omega^k(M)$ and $X_i\in\mathfrak{X}(M)$ we have

$$(\mathcal{L}_{X_0}\varphi)(X_1,\dots,X_k)=X_0(\varphi(X_1,\dots,X_k))+$$
$$+\sum_{j=1}^k(-1)^{0+j}\varphi([X_0,X_j],X_1,\dots,\widehat{X_j},\dots,X_k)\text{ by 7.6.2,}$$
$$(i_{X_0}d\varphi)(X_1,\dots,X_k)=d\varphi(X_0,\dots,X_k)$$
$$=\sum_{i=0}^k(-1)^iX_i(\varphi(X_0,\dots,\widehat{X_i},\dots,X_k))+$$

$$+\sum_{0\le i<j}(-1)^{i+j}\varphi([X_i,X_j],X_0,\dots,\widehat{X_i},\dots,\widehat{X_j},\dots,X_k).$$

$$(di_{X_0}\varphi)(X_1,\dots,X_k)=\sum_{i=1}^k(-1)^{i-1}X_i((i_{X_0}\varphi)(X_1,\dots,\widehat{X_i},\dots,X_k))+$$
$$+\sum_{1\le i<j}(-1)^{i+j-1}(i_{X_0}\varphi)([X_i,X_j],X_1,\dots,\widehat{X_i},\dots,\widehat{X_j},\dots,X_k)$$
$$=-\sum_{i=1}^k(-1)^iX_i(\varphi(X_0,X_1,\dots,\widehat{X_i},\dots,X_k))-$$
$$-\sum_{1\le i<j}(-1)^{i+j}\varphi([X_i,X_j],X_0,X_1,\dots,\widehat{X_i},\dots,\widehat{X_j},\dots,X_k).$$

By summing up the result follows.

(1) Let $\varphi\in\Omega^p(M)$ and $\psi\in\Omega^q(M)$. We prove the result by induction on $p+q$.
$p+q=0$: $d(f\cdot g)=df\cdot g+f\cdot dg$.
Suppose that (1) is true for $p+q<k$. Then for $X\in\mathfrak{X}(M)$ we have by part (2) and 7.6, 7.7 and by induction

$$\begin{aligned}
i_X\,d(\varphi\wedge\psi)&=\mathcal{L}_X(\varphi\wedge\psi)-d\,i_X(\varphi\wedge\psi)\\
&=\mathcal{L}_X\varphi\wedge\psi+\varphi\wedge\mathcal{L}_X\psi-d(i_X\varphi\wedge\psi+(-1)^p\varphi\wedge i_X\psi)\\
&=i_Xd\varphi\wedge\psi+di_X\varphi\wedge\psi+\varphi\wedge i_Xd\psi+\varphi\wedge di_X\psi-di_X\varphi\wedge\psi\\
&\quad-(-1)^{p-1}i_X\varphi\wedge d\psi-(-1)^pd\varphi\wedge i_X\psi-\varphi\wedge di_X\psi\\
&=i_X(d\varphi\wedge\psi+(-1)^p\varphi\wedge d\psi).
\end{aligned}$$

Since X is arbitrary, (1) follows.

(3) By (1) d is a graded derivation of degree 1, so $d^2=\frac{1}{2}[d,d]$ is a graded derivation of degree 2 (see 8.1), and is obviously local. Since $\Omega(M)$ is locally generated as an algebra by $C^\infty(M,\mathbb{R})$ and $\{df:f\in C^\infty(M,\mathbb{R})\}$, it suffices to show that $d^2f=0$ for each $f\in C^\infty(M,\mathbb{R})$ ($d^3f=0$ is a consequence). But this is easy: $d^2f(X,Y)=Xdf(Y)-Ydf(X)-df([X,Y])=XYf-YXf-[X,Y]f=0$.

(4) $f^*:\Omega(M)\to\Omega(N)$ is an algebra homomorphism by 7.6, so $f^*\circ d$ and $d\circ f^*$ are both graded derivations over f^* of degree 1. By the same argument as in the proof of (3) above it suffices to show that they agree on g and dg for all $g\in C^\infty(M,\mathbb{R})$. We have $(f^*dg)_y(Y)=(dg)_{f(y)}(T_yf.Y)=(T_yf.Y)(g)=Y(g\circ f)(y)=(df^*g)_y(Y)$, thus also $df^*dg=ddf^*g=0$, and $f^*ddg=0$.

(5) $d\mathcal{L}_X=d\,i_X\,d+ddi_X=di_Xd+i_Xdd=\mathcal{L}_Xd$. □

7.10. A differential form $\omega\in\Omega^k(M)$ is called *closed* if $d\omega=0$, and it is called *exact* if $\omega=d\varphi$ for some $\varphi\in\Omega^{k-1}(M)$. Since $d^2=0$, any exact form is closed. The quotient space

$$H^k(M):=\frac{\ker(d:\Omega^k(M)\to\Omega^{k+1}(M))}{\operatorname{im}(d:\Omega^{k-1}(M)\to\Omega^k(M))}$$

is called the k-th De Rham cohomology space of M. We will not treat cohomology in this book, and we finish with the

Lemma of Poicaré. *A closed differential form is locally exact. More precisely: let $\omega \in \Omega^k(M)$ with $d\omega = 0$. Then for any $x \in M$ there is an open neighborhood U of x in M and a $\varphi \in \Omega^{k-1}(U)$ with $d\varphi = \omega|U$.*

Proof. Let (U, u) be chart on M centered at x such that $u(U) = \mathbb{R}^m$. So we may just assume that $M = \mathbb{R}^m$.

We consider $\alpha : \mathbb{R} \times \mathbb{R}^m \to \mathbb{R}^m$, given by $\alpha(t, x) = \alpha_t(x) = tx$. Let $I \in \mathfrak{X}(\mathbb{R}^m)$ be the vector field $I(x) = x$, then $\alpha(e^t, x) = \mathrm{Fl}^I_t(x)$. So for $t > 0$ we have

$$\begin{aligned}\tfrac{d}{dt}\alpha_t^*\omega &= \tfrac{d}{dt}(\mathrm{Fl}^I_{\log t})^*\omega = \tfrac{1}{t}(\mathrm{Fl}^I_{\log t})^*\mathcal{L}_I\omega \\ &= \tfrac{1}{t}\alpha_t^*(i_I d\omega + d i_I\omega) = \tfrac{1}{t} d\alpha_t^* i_I\omega.\end{aligned}$$

Note that $T_x(\alpha_t) = t.\mathrm{Id}$. Therefore

$$\begin{aligned}(\tfrac{1}{t}\alpha_t^* i_I\omega)_x(X_2, \dots, X_k) &= \tfrac{1}{t}(i_I\omega)_{tx}(tX_2, \dots, tX_k) \\ &= \tfrac{1}{t}\omega_{tx}(tx, tX_2, \dots, tX_k) = \omega_{tx}(x, tX_2, \dots, tX_k).\end{aligned}$$

So if $k \geq 1$, the $(k-1)$-form $\frac{1}{t}\alpha_t^* i_I\omega$ is defined and smooth in (t, x) for all $t \in \mathbb{R}$. Clearly $\alpha_1^*\omega = \omega$ and $\alpha_0^*\omega = 0$, thus

$$\begin{aligned}\omega = \alpha_1^*\omega - \alpha_0^*\omega &= \int_0^1 \tfrac{d}{dt}\alpha_t^*\omega dt \\ - \int_0^1 d(\tfrac{1}{t}\alpha_t^* i_I\omega)dt &= d\left(\int_0^1 \tfrac{1}{t}\alpha_t^* i_I\omega dt\right) = d\varphi. \quad \square\end{aligned}$$

7.11. Vector bundle valued differential forms. Let (E, p, M) be a vector bundle. The space of smooth sections of the bundle $\Lambda^k T^*M \otimes E$ will be denoted by $\Omega^k(M; E)$. Its elements will be called *E-valued k-forms*.

If V is a finite dimensional or even a suitable infinite dimensional vector space, $\Omega^k(M; V)$ will denote the space of all V-valued differential forms of degree k. The exterior differential extends to this case, if V is complete in some sense.

8. Derivations on the algebra of differential forms and the Frölicher-Nijenhuis bracket

8.1. In this section let M be a smooth manifold. We consider the graded commutative algebra $\Omega(M) = \bigoplus_{k=0}^{\dim M} \Omega^k(M) = \bigoplus_{k=-\infty}^{\infty} \Omega^k(M)$ of differential forms on M, where we put $\Omega^k(M) = 0$ for $k < 0$ and $k > \dim M$. We denote by $\mathrm{Der}_k\,\Omega(M)$ the space of all *(graded) derivations* of degree k, i.e. all linear mappings $D : \Omega(M) \to \Omega(M)$ with $D(\Omega^\ell(M)) \subset \Omega^{k+\ell}(M)$ and $D(\varphi \wedge \psi) = D(\varphi) \wedge \psi + (-1)^{k\ell}\varphi \wedge D(\psi)$ for $\varphi \in \Omega^\ell(M)$.

Lemma. *Then the space* $\operatorname{Der}\Omega(M) = \bigoplus_k \operatorname{Der}_k \Omega(M)$ *is a graded Lie algebra with the graded commutator* $[D_1, D_2] := D_1 \circ D_2 - (-1)^{k_1 k_2} D_2 \circ D_1$ *as bracket. This means that the bracket is graded anticommutative,* $[D_1, D_2] = -(-1)^{k_1 k_2}[D_2, D_1]$, *and satisfies the graded Jacobi identity* $[D_1, [D_2, D_3]] = [[D_1, D_2], D_3] + (-1)^{k_1 k_2}[D_2, [D_1, D_3]]$ *(so that* $ad(D_1) = [D_1, \quad]$ *is itself a derivation of degree* k_1*).*

Proof. Plug in the definition of the graded commutator and compute. □

In section 7 we have already met some graded derivations: for a vector field X on M the derivation i_X is of degree -1, $\mathcal{L}_X$ is of degree 0, and d is of degree 1. Note also that the important formula $\mathcal{L}_X = d\,i_X + i_X\,d$ translates to $\mathcal{L}_X = [i_X, d]$.

8.2. A derivation $D \in \operatorname{Der}_k \Omega(M)$ is called *algebraic* if $D \mid \Omega^0(M) = 0$. Then $D(f.\omega) = f.D(\omega)$ for $f \in C^\infty(M, \mathbb{R})$, so D is of tensorial character by 7.3. So D induces a derivation $D_x \in \operatorname{Der}_k \Lambda T_x^* M$ for each $x \in M$. It is uniquely determined by its restriction to 1-forms $D_x | T_x^* M : T_x^* M \to \Lambda^{k+1} T^* M$ which we may view as an element $K_x \in \Lambda^{k+1} T_x^* M \otimes T_x M$ depending smoothly on $x \in M$. To express this dependence we write $D = i_K = i(K)$, where $K \in C^\infty(\Lambda^{k+1} T^* M \otimes TM) =: \Omega^{k+1}(M; TM)$. Note the defining equation: $i_K(\omega) = \omega \circ K$ for $\omega \in \Omega^1(M)$. We call $\Omega(M, TM) = \bigoplus_{k=0}^{\dim M} \Omega^k(M, TM)$ the space of all *vector valued differential forms.*

Theorem. (1) *For* $K \in \Omega^{k+1}(M, TM)$ *the formula*

$$(i_K\omega)(X_1, \dots, X_{k+\ell}) = \\ = \frac{1}{(k+1)!\,(\ell-1)!} \sum_{\sigma \in \mathcal{S}_{k+\ell}} \operatorname{sign}\sigma \,.\omega(K(X_{\sigma 1}, \dots, X_{\sigma(k+1)}), X_{\sigma(k+2)}, \dots)$$

for $\omega \in \Omega^\ell(M)$, $X_i \in \mathfrak{X}(M)$ *(or* $T_x M$*) defines an algebraic graded derivation* $i_K \in \operatorname{Der}_k \Omega(M)$ *and any algebraic derivation is of this form.*

(2) *By* $i([K, L]^\wedge) := [i_K, i_L]$ *we get a bracket* $[\quad,\quad]^\wedge$ *on* $\Omega^{*+1}(M, TM)$ *which defines a graded Lie algebra structure with the grading as indicated, and for* $K \in \Omega^{k+1}(M, TM)$, $L \in \Omega^{\ell+1}(M, TM)$ *we have*

$$[K, L]^\wedge = i_K L - (-1)^{k\ell} i_L K,$$

where $i_K(\omega \otimes X) := i_K(\omega) \otimes X$.

$[\quad,\quad]^\wedge$ is called the *algebraic bracket* or the *Nijenhuis-Richardson bracket*, see [Nijenhuis-Richardson, 67].

Proof. Since $\Lambda T_x^* M$ is the free graded commutative algebra generated by the vector space $T_x^* M$ any $K \in \Omega^{k+1}(M, TM)$ extends to a graded derivation. By applying it to an exterior product of 1-forms one can derive the formula in (1). The graded commutator of two algebraic derivations is again algebraic, so the injection $i : \Omega^{*+1}(M, TM) \to \operatorname{Der}_*(\Omega(M))$ induces a graded Lie bracket on $\Omega^{*+1}(M, TM)$ whose form can be seen by applying it to a 1-form. □

8.3. The exterior derivative d is an element of $\operatorname{Der}_1 \Omega(M)$. In view of the formula $\mathcal{L}_X = [i_X, d] = i_X\, d + d\, i_X$ for vector fields X, we define for $K \in \Omega^k(M;TM)$ the *Lie derivation* $\mathcal{L}_K = \mathcal{L}(K) \in \operatorname{Der}_k \Omega(M)$ by $\mathcal{L}_K := [i_K, d]$.

Then the mapping $\mathcal{L} : \Omega(M,TM) \to \operatorname{Der} \Omega(M)$ is injective, since $\mathcal{L}_K f = i_K df = df \circ K$ for $f \in C^\infty(M,\mathbb{R})$.

Theorem. *For any graded derivation $D \in \operatorname{Der}_k \Omega(M)$ there are unique $K \in \Omega^k(M;TM)$ and $L \in \Omega^{k+1}(M;TM)$ such that*

$$D = \mathcal{L}_K + i_L.$$

We have $L = 0$ if and only if $[D,d] = 0$. D is algebraic if and only if $K = 0$.

Proof. Let $X_i \in \mathfrak{X}(M)$ be vector fields. Then $f \mapsto (Df)(X_1,\dots,X_k)$ is a derivation $C^\infty(M,\mathbb{R}) \to C^\infty(M,\mathbb{R})$, so by 3.3 there is a unique vector field $K(X_1,\dots,X_k) \in \mathfrak{X}(M)$ such that

$$(Df)(X_1,\dots,X_k) = K(X_1,\dots,X_k)f = df(K(X_1,\dots,X_k)).$$

Clearly $K(X_1,\dots,X_k)$ is $C^\infty(M,\mathbb{R})$-linear in each X_i and alternating, so K is tensorial by 7.3, $K \in \Omega^k(M;TM)$.

The defining equation for K is $Df = df \circ K = i_K df = \mathcal{L}_K f$ for $f \in C^\infty(M,\mathbb{R})$. Thus $D - \mathcal{L}_K$ is an algebraic derivation, so $D - \mathcal{L}_K = i_L$ by 8.2 for unique $L \in \Omega^{k+1}(M;TM)$.

Since we have $[d,d] = 2d^2 = 0$, by the graded Jacobi identity we obtain $0 = [i_K,[d,d]] = [[i_K,d],d] + (-1)^{k-1}[d,[i_K,d]] = 2[\mathcal{L}_K,d]$. The mapping $K \mapsto [i_K,d] = \mathcal{L}_K$ is injective, so the last assertions follow. $\square$

8.4. Applying $i(\mathrm{Id}_{TM})$ on a k-fold exterior product of 1-forms we see that $i(\mathrm{Id}_{TM})\omega = k\omega$ for $\omega \in \Omega^k(M)$. Thus we have $\mathcal{L}(\mathrm{Id}_{TM})\omega = i(\mathrm{Id}_{TM})d\omega - d\,i(\mathrm{Id}_{TM})\omega = (k+1)d\omega - kd\omega = d\omega$. Thus $\mathcal{L}(\mathrm{Id}_{TM}) = d$.

8.5. Let $K \in \Omega^k(M;TM)$ and $L \in \Omega^\ell(M;TM)$. Then obviously $[[\mathcal{L}_K,\mathcal{L}_L],d] = 0$, so we have

$$[\mathcal{L}(K),\mathcal{L}(L)] = \mathcal{L}([K,L])$$

for a uniquely defined $[K,L] \in \Omega^{k+\ell}(M;TM)$. This vector valued form $[K,L]$ is called the *Frölicher-Nijenhuis bracket* of K and L.

Theorem. *The space $\Omega(M;TM) = \bigoplus_{k=0}^{\dim M} \Omega^k(M;TM)$ with its usual grading is a graded Lie algebra for the Frölicher-Nijenhuis bracket. So we have*

$$[K,L] = -(-1)^{k\ell}[L,K]$$
$$[K_1,[K_2,K_3]] = [[K_1,K_2],K_3] + (-1)^{k_1k_2}[K_2,[K_1,K_3]]$$

$\mathrm{Id}_{TM} \in \Omega^1(M;TM)$ is in the center, i.e. $[K,\mathrm{Id}_{TM}] = 0$ for all K.

$\mathcal{L} : (\Omega(M;TM),[\ ,\ ]) \to \operatorname{Der} \Omega(M)$ is an injective homomorphism of graded Lie algebras. For vector fields the Frölicher-Nijenhuis bracket coincides with the Lie bracket.

Proof. $df \circ [X,Y] = \mathcal{L}([X,Y])f = [\mathcal{L}_X,\mathcal{L}_Y]f$. The rest is clear. $\square$

8.6. Lemma. *For $K \in \Omega^k(M;TM)$ and $L \in \Omega^{\ell+1}(M;TM)$ we have*

$$[\mathcal{L}_K, i_L] = i([K,L]) - (-1)^{k\ell}\mathcal{L}(i_L K), \text{ or}$$
$$[i_L, \mathcal{L}_K] = \mathcal{L}(i_L K) - (-1)^k\, i([L,K]).$$

This generalizes 7.7.2.

Proof. For $f \in C^\infty(M,\mathbb{R})$ we have $[i_L, \mathcal{L}_K]f = i_L\, i_K\, df - 0 = i_L(df \circ K) = df \circ (i_L K) = \mathcal{L}(i_L K)f$. So $[i_L, \mathcal{L}_K] - \mathcal{L}(i_L K)$ is an algebraic derivation.

$$\begin{aligned}[][[i_L, \mathcal{L}_K], d] &= [i_L, [\mathcal{L}_K, d]] - (-1)^{k\ell}[\mathcal{L}_K, [i_L, d]] = \\ &= 0 - (-1)^{k\ell}\mathcal{L}([K,L]) = (-1)^k[i([L,K]), d].\end{aligned}$$

Since $[\quad, d]$ kills the $\mathcal{L}$'s and is injective on the i's, the algebraic part of $[i_L, \mathcal{L}_K]$ is $(-1)^k\, i([L,K])$. □

8.7. The space $\operatorname{Der}\Omega(M)$ is a graded module over the graded algebra $\Omega(M)$ with the action $(\omega \wedge D)\varphi = \omega \wedge D(\varphi)$, because $\Omega(M)$ is graded commutative.

Theorem. *Let the degree of ω be q, of φ be k, and of ψ be ℓ. Let the other degrees be as indicated. Then we have:*

$$[\omega \wedge D_1, D_2] = \omega \wedge [D_1, D_2] - (-1)^{(q+k_1)k_2} D_2(\omega) \wedge D_1. \tag{1}$$
$$i(\omega \wedge L) = \omega \wedge i(L) \tag{2}$$
$$\omega \wedge \mathcal{L}_K = \mathcal{L}(\omega \wedge K) + (-1)^{q+k-1} i(d\omega \wedge K). \tag{3}$$
$$\begin{aligned}[][\omega \wedge L_1, L_2]^\wedge &= \omega \wedge [L_1, L_2]^\wedge - \\ &\quad - (-1)^{(q+\ell_1-1)(\ell_2-1)} i(L_2)\omega \wedge L_1.\end{aligned} \tag{4}$$
$$\begin{aligned}[][\omega \wedge K_1, K_2] &= \omega \wedge [K_1, K_2] - (-1)^{(q+k_1)k_2}\mathcal{L}(K_2)\omega \wedge K_1 \\ &\quad + (-1)^{q+k_1} d\omega \wedge i(K_1)K_2.\end{aligned} \tag{5}$$
$$\begin{aligned}[][\varphi \otimes X, \psi \otimes Y] &= \varphi \wedge \psi \otimes [X,Y] \\ &\quad - \left(i_Y d\varphi \wedge \psi \otimes X - (-1)^{k\ell} i_X d\psi \wedge \varphi \otimes Y\right) \\ &\quad - \left(d(i_Y\varphi \wedge \psi) \otimes X - (-1)^{k\ell} d(i_X\psi \wedge \varphi) \otimes Y\right) \\ &= \varphi \wedge \psi \otimes [X,Y] + \varphi \wedge \mathcal{L}_X\psi \otimes Y - \mathcal{L}_Y\varphi \wedge \psi \otimes X \\ &\quad + (-1)^k \left(d\varphi \wedge i_X\psi \otimes Y + i_Y\varphi \wedge d\psi \otimes X\right).\end{aligned} \tag{6}$$

Proof. For (1), (2), (3) write out the definitions. For (4) compute $i([\omega\wedge L_1, L_2]^\wedge)$. For (5) compute $\mathcal{L}([\omega \wedge K_1, K_2])$. For (6) use (5) . □

8.8. Theorem. *For $K \in \Omega^k(M;TM)$ and $\omega \in \Omega^\ell(M)$ the Lie derivative of ω along K is given by the following formula, where the X_i are vector fields on M.*

$$\begin{aligned}
&(\mathcal{L}_K\omega)(X_1,\dots,X_{k+\ell}) = \\
&= \tfrac{1}{k!\,\ell!}\sum_\sigma \operatorname{sign}\sigma\ \mathcal{L}(K(X_{\sigma 1},\dots,X_{\sigma k}))(\omega(X_{\sigma(k+1)},\dots,X_{\sigma(k+\ell)})) \\
&+ \tfrac{-1}{k!\,(\ell-1)!}\sum_\sigma \operatorname{sign}\sigma\ \omega([K(X_{\sigma 1},\dots,X_{\sigma k}),X_{\sigma(k+1)}],X_{\sigma(k+2)},\dots) \\
&+ \tfrac{(-1)^{k-1}}{(k-1)!\,(\ell-1)!\,2!}\sum_\sigma \operatorname{sign}\sigma\ \omega(K([X_{\sigma 1},X_{\sigma 2}],X_{\sigma 3},\dots),X_{\sigma(k+2)},\dots).
\end{aligned}$$

Proof. It suffices to consider $K = \varphi \otimes X$. Then by 8.7.3 we have $\mathcal{L}(\varphi\otimes X) = \varphi\wedge\mathcal{L}_X - (-1)^{k-1}d\varphi\wedge i_X$. Now use the global formulas of section 7 to expand this. □

8.9. Theorem. *For $K \in \Omega^k(M;TM)$ and $L \in \Omega^\ell(M;TM)$ we have for the Frölicher-Nijenhuis bracket $[K,L]$ the following formula, where the X_i are vector fields on M.*

$$\begin{aligned}
&[K,L](X_1,\dots,X_{k+\ell}) = \\
&= \tfrac{1}{k!\,\ell!}\sum_\sigma \operatorname{sign}\sigma\ [K(X_{\sigma 1},\dots,X_{\sigma k}),L(X_{\sigma(k+1)},\dots,X_{\sigma(k+\ell)})] \\
&+ \tfrac{-1}{k!\,(\ell-1)!}\sum_\sigma \operatorname{sign}\sigma\ L([K(X_{\sigma 1},\dots,X_{\sigma k}),X_{\sigma(k+1)}],X_{\sigma(k+2)},\dots) \\
&+ \tfrac{(-1)^{k\ell}}{(k-1)!\,\ell!}\sum_\sigma \operatorname{sign}\sigma\ K([L(X_{\sigma 1},\dots,X_{\sigma \ell}),X_{\sigma(\ell+1)}],X_{\sigma(\ell+2)},\dots) \\
&+ \tfrac{(-1)^{k-1}}{(k-1)!\,(\ell-1)!\,2!}\sum_\sigma \operatorname{sign}\sigma\ L(K([X_{\sigma 1},X_{\sigma 2}],X_{\sigma 3},\dots),X_{\sigma(k+2)},\dots) \\
&+ \tfrac{(-1)^{(k-1)\ell}}{(k-1)!\,(\ell-1)!\,2!}\sum_\sigma \operatorname{sign}\sigma\ K(L([X_{\sigma 1},X_{\sigma 2}],X_{\sigma 3},\dots),X_{\sigma(\ell+2)},\dots).
\end{aligned}$$

Proof. It suffices to consider $K = \varphi\otimes X$ and $L = \psi\otimes Y$, then for $[\varphi\otimes X,\psi\otimes Y]$ we may use 8.7.6 and evaluate that at $(X_1,\dots,X_{k+\ell})$. After some combinatorial computation we get the right hand side of the above formula for $K = \varphi\otimes X$ and $L = \psi\otimes Y$. □

There are more illuminating ways to prove this formula, see [Michor, 87].

8.10. Local formulas. In a local chart (U,u) on the manifold M we put $K \mid U = \sum K^i_\alpha d^\alpha \otimes \partial_i$, $L \mid U = \sum L^j_\beta d^\beta \otimes \partial_j$, and $\omega \mid U = \sum \omega_\gamma d^\gamma$, where $\alpha = (1 \le \alpha_1 < \alpha_2 < \dots < \alpha_k \le \dim M)$ is a form index, $d^\alpha = du^{\alpha_1}\wedge\dots\wedge du^{\alpha_k}$, $\partial_i = \frac{\partial}{\partial u^i}$ and so on.

Plugging $X_j = \partial_{i_j}$ into the global formulas 8.2, 8.8, and 8.9, we get the following local formulas:

$$i_K\omega \mid U = \sum K^i_{\alpha_1\dots\alpha_k}\omega_{i\alpha_{k+1}\dots\alpha_{k+\ell-1}}\, d^\alpha$$

$$
\begin{aligned}
[K,L]^\wedge \mid U = \sum \Big(& K^i_{\alpha_1\dots\alpha_k} L^j_{i\alpha_{k+1}\dots\alpha_{k+\ell}} \\
& - (-1)^{(k-1)(\ell-1)} L^i_{\alpha_1\dots\alpha_\ell} K^j_{i\alpha_{\ell+1}\dots\alpha_{k+\ell}} \Big) d^\alpha \otimes \partial_j \\
\mathcal{L}_K\omega \mid U = \sum \Big(& K^i_{\alpha_1\dots\alpha_k} \partial_i \omega_{\alpha_{k+1}\dots\alpha_{k+\ell}} \\
& + (-1)^k (\partial_{\alpha_1} K^i_{\alpha_2\dots\alpha_{k+1}}) \omega_{i\alpha_{k+2}\dots\alpha_{k+\ell}} \Big) d^\alpha \\
[K,L] \mid U = \sum \Big(& K^i_{\alpha_1\dots\alpha_k} \partial_i L^j_{\alpha_{k+1}\dots\alpha_{k+\ell}} \\
& - (-1)^{k\ell} L^i_{\alpha_1\dots\alpha_\ell} \partial_i K^j_{\alpha_{\ell+1}\dots\alpha_{k+\ell}} \\
& - k K^j_{\alpha_1\dots\alpha_{k-1} i} \partial_{\alpha_k} L^i_{\alpha_{k+1}\dots\alpha_{k+\ell}} \\
& + (-1)^{k\ell} \ell L^j_{\alpha_1\dots\alpha_{\ell-1} i} \partial_{\alpha_\ell} K^i_{\alpha_{\ell+1}\dots\alpha_{k+\ell}} \Big) d^\alpha \otimes \partial_j
\end{aligned}
$$

8.11. Theorem. *For $K_i \in \Omega^{k_i}(M;TM)$ and $L_i \in \Omega^{k_i+1}(M;TM)$ we have*

$$
\begin{aligned}
(1) \qquad & [\mathcal{L}_{K_1} + i_{L_1}, \mathcal{L}_{K_2} + i_{L_2}] = \\
& = \mathcal{L}\left([K_1,K_2] + i_{L_1}K_2 - (-1)^{k_1k_2} i_{L_2}K_1\right) \\
& \quad + i\left([L_1,L_2]^\wedge + [K_1,L_2] - (-1)^{k_1k_2}[K_2,L_1]\right).
\end{aligned}
$$

Each summand of this formula looks like a semidirect product of graded Lie algebras, but the mappings

$$
\begin{aligned}
i &: \Omega(M;TM) \to \operatorname{End}(\Omega(M;TM), [\ ,\ ]) \\
\operatorname{ad} &: \Omega(M;TM) \to \operatorname{End}(\Omega(M;TM), [\ ,\ ]^\wedge)
\end{aligned}
$$

do not take values in the subspaces of graded derivations. We have instead for $K \in \Omega^k(M;TM)$ and $L \in \Omega^{\ell+1}(M;TM)$ the following relations:

$$
\begin{aligned}
(2) \qquad & i_L[K_1,K_2] = [i_LK_1,K_2] + (-1)^{k_1\ell}[K_1,i_LK_2] \\
& \quad - \left((-1)^{k_1\ell} i([K_1,L])K_2 - (-1)^{(k_1+\ell)k_2} i([K_2,L])K_1\right) \\
(3) \qquad & [K,[L_1,L_2]^\wedge] = [[K,L_1],L_2]^\wedge + (-1)^{kk_1}[L_1,[K,L_2]]^\wedge - \\
& \quad - \left((-1)^{kk_1}[i(L_1)K,L_2] - (-1)^{(k+k_1)k_2}[i(L_2)K,L_1]\right)
\end{aligned}
$$

The algebraic meaning of the relations of this theorem and its consequences in group theory have been investigated in [Michor, 89]. The corresponding product of groups is well known to algebraists under the name 'Zappa-Szep'-product.

Proof. Equation (1) is an immediate consequence of 8.6. Equations (2) and (3) follow from (1) by writing out the graded Jacobi identity, or as follows: Consider $\mathcal{L}(i_L[K_1,K_2])$ and use 8.6 repeatedly to obtain $\mathcal{L}$ of the right hand side of (2). Then consider $i([K,[L_1,L_2]^\wedge])$ and use again 8.6 several times to obtain i of the right hand side of (3). □

8.12. Corollary (of 8.9). *For K, $L \in \Omega^1(M;TM)$ we have*

$$\begin{aligned}[K,L](X,Y) = [KX,LY] - [KY,LX] \\ - L([KX,Y] - [KY,X]) \\ - K([LX,Y] - [LY,X]) \\ + (LK + KL)[X,Y].\end{aligned}$$

8.13. Curvature. Let $P \in \Omega^1(M;TM)$ satisfy $P \circ P = P$, i.e. P is a projection in each fiber of TM. This is the most general case of a (first order) *connection*. We may call $\ker P$ the *horizontal space* and $\operatorname{im} P$ the *vertical space* of the connection. If P is of constant rank, then both are sub vector bundles of TM. If $\operatorname{im} P$ is some primarily fixed sub vector bundle or (tangent bundle of) a foliation, P can be called a connection for it. Special cases of this will be treated extensively later on. The following result is immediate from 8.12.

Lemma. *We have*

$$[P,P] = 2R + 2\bar{R},$$

where R, $\bar{R} \in \Omega^2(M;TM)$ are given by $R(X,Y) = P[(\mathrm{Id} - P)X, (\mathrm{Id} - P)Y]$ and $\bar{R}(X,Y) = (\mathrm{Id} - P)[PX,PY]$.

If P has constant rank, then R is the obstruction against integrability of the horizontal bundle $\ker P$, and $\bar{R}$ is the obstruction against integrability of the vertical bundle $\operatorname{im} P$. Thus we call R the *curvature* and $\bar{R}$ the *cocurvature* of the connection P. We will see later, that for a principal fiber bundle R is just the negative of the usual curvature.

8.14. Lemma (Bianchi identity). *If $P \in \Omega^1(M;TM)$ is a connection (fiber projection) with curvature R and cocurvature $\bar{R}$, then we have*

$$\begin{aligned}[P, R + \bar{R}] &= 0 \\ [R,P] &= i_R\bar{R} + i_{\bar{R}}R.\end{aligned}$$

Proof. We have $[P,P] = 2R + 2\bar{R}$ by 8.13 and $[P,[P,P]] = 0$ by the graded Jacobi identity. So the first formula follows. We have $2R = P \circ [P,P] = i_{[P,P]}P$. By 8.11.2 we get $i_{[P,P]}[P,P] = 2[i_{[P,P]}P,P] - 0 = 4[R,P]$. Therefore $[R,P] = \frac{1}{4}i_{[P,P]}[P,P] = i(R+\bar{R})(R+\bar{R}) = i_R\bar{R} + i_{\bar{R}}R$ since R has vertical values and kills vertical vectors, so $i_RR = 0$; likewise for $\bar{R}$. □

8.15. f-relatedness of the Frölicher-Nijenhuis bracket. Let $f : M \to N$ be a smooth mapping between manifolds. Two vector valued forms $K \in \Omega^k(M;TM)$ and $K' \in \Omega^k(N;TN)$ are called *f-related* or *f-dependent*, if for all $X_i \in T_xM$ we have

$$K'_{f(x)}(T_xf \cdot X_1, \dots, T_xf \cdot X_k) = T_xf \cdot K_x(X_1, \dots, X_k). \tag{1}$$

Theorem.

(2) *If K and K' as above are f-related then $i_K \circ f^* = f^* \circ i_{K'} : \Omega(N) \to \Omega(M)$.*

(3) *If $i_K \circ f^* \mid B^1(N) = f^* \circ i_{K'} \mid B^1(N)$, then K and K' are f-related, where B^1 denotes the space of exact 1-forms.*

(4) *If K_j and K'_j are f-related for $j = 1,2$, then $i_{K_1}K_2$ and $i_{K'_1}K'_2$ are f-related, and also $[K_1, K_2]^\wedge$ and $[K'_1, K'_2]^\wedge$ are f-related.*

(5) *If K and K' are f-related then $\mathcal{L}_K \circ f^* = f^* \circ \mathcal{L}_{K'} : \Omega(N) \to \Omega(M)$.*

(6) *If $\mathcal{L}_K \circ f^* \mid \Omega^0(N) = f^* \circ \mathcal{L}_{K'} \mid \Omega^0(N)$, then K and K' are f-related.*

(7) *If K_j and K'_j are f-related for $j = 1,2$, then their Frölicher-Nijenhuis brackets $[K_1, K_2]$ and $[K'_1, K'_2]$ are also f-related.*

Proof. (2) By 8.2 we have for $\omega \in \Omega^q(N)$ and $X_i \in T_xM$:

$$\begin{aligned}
&(i_K f^*\omega)_x(X_1, \dots, X_{q+k-1}) = \\
&= \tfrac{1}{k!\,(q-1)!} \sum_\sigma \operatorname{sign}\sigma\, (f^*\omega)_x(K_x(X_{\sigma 1}, \dots, X_{\sigma k}), X_{\sigma(k+1)}, \dots) \\
&= \tfrac{1}{k!\,(q-1)!} \sum_\sigma \operatorname{sign}\sigma\, \omega_{f(x)}(T_xf \cdot K_x(X_{\sigma 1}, \dots), T_xf \cdot X_{\sigma(k+1)}, \dots) \\
&= \tfrac{1}{k!\,(q-1)!} \sum_\sigma \operatorname{sign}\sigma\, \omega_{f(x)}(K'_{f(x)}(T_xf \cdot X_{\sigma 1}, \dots), T_xf \cdot X_{\sigma(k+1)}, \dots) \\
&= (f^* i_{K'}\omega)_x(X_1, \dots, X_{q+k-1})
\end{aligned}$$

(3) follows from this computation, since the df, $f \in C^\infty(M, \mathbb{R})$ separate points.

(4) follows from the same computation for K_2 instead of ω, the result for the bracket then follows 8.2.2.

(5) The algebra homomorphism f^* intertwines the operators i_K and $i_{K'}$ by (2), and f^* commutes with the exterior derivative d. Thus f^* intertwines the commutators $[i_K, d] = \mathcal{L}_K$ and $[i_{K'}, d] = \mathcal{L}_{K'}$.

(6) For $g \in \Omega^0(N)$ we have $\mathcal{L}_K\, f^*g = i_K\, d\, f^*g = i_K\, f^*\, dg$ and $f^*\, \mathcal{L}_{K'}\, g = f^*\, i_{K'}\, dg$. By (3) the result follows.

(7) The algebra homomorphism f^* intertwines $\mathcal{L}_{K_j}$ and $\mathcal{L}_{K'_j}$, thus also their graded commutators, which are equal to $\mathcal{L}([K_1, K_2])$ and $\mathcal{L}([K'_1, K'_2])$, respectively. Then use (6). □

8.16. Let $f : M \to N$ be a local diffeomorphism. Then we can consider the pullback operator $f^* : \Omega(N; TN) \to \Omega(M; TM)$, given by

$$(1) \qquad (f^*K)_x(X_1, \dots, X_k) = (T_xf)^{-1} K_{f(x)}(T_xf \cdot X_1, \dots, T_xf \cdot X_k).$$

Note that this is a special case of the pullback operator for sections of natural vector bundles in 6.15. Clearly K and f^*K are then f-related.

Theorem. *In this situation we have:*

(2) $f^*[K, L] = [f^*K, f^*L]$.

(3) $f^* i_K L = i_{f^*K} f^* L$.
(4) $f^* [K, L]^\wedge = [f^*K, f^*L]^\wedge$.
(5) *For a vector field* $X \in \mathfrak{X}(M)$ *and* $K \in \Omega(M; TM)$ *by 6.15 the Lie derivative* $\mathcal{L}_X K = \frac{\partial}{\partial t}\big|_0 (\mathrm{Fl}_t^X)^* K$ *is defined. Then we have* $\mathcal{L}_X K = [X, K]$*, the Frölicher-Nijenhuis-bracket.*

This is sometimes expressed by saying that the Frölicher-Nijenhuis bracket, $[\ ,\ ]^\wedge$, etc. are ***natural bilinear concomitants.***

Proof. (2) – (4) are obvious from 8.15. They also follow directly from the geometrical constructions of the operators in question. (5) Obviously $\mathcal{L}_X$ is $\mathbb{R}$-linear, so it suffices to check this formula for $K = \psi \otimes Y$, $\psi \in \Omega(M)$ and $Y \in \mathfrak{X}(M)$. But then

$$\begin{aligned}
\mathcal{L}_X(\psi \otimes Y) &= \mathcal{L}_X\psi \otimes Y + \psi \otimes \mathcal{L}_X Y \quad \text{by 6.16}\\
&= \mathcal{L}_X\psi \otimes Y + \psi \otimes [X, Y]\\
&= [X, \psi \otimes Y] \quad \text{by 8.7.6.} \quad \square
\end{aligned}$$

8.17. Remark. At last we mention the best known application of the Frölicher-Nijenhuis bracket, which also led to its discovery. A vector valued 1-form $J \in \Omega^1(M; TM)$ with $J \circ J = -\mathrm{Id}$ is called a ***almost complex structure***; if it exists, $\dim M$ is even and J can be viewed as a fiber multiplication with $\sqrt{-1}$ on TM. By 8.12 we have

$$[J, J](X, Y) = 2([JX, JY] - [X, Y] - J[X, JY] - J[JX, Y]).$$

The vector valued form $\frac{1}{2}[J, J]$ is also called the ***Nijenhuis tensor*** of J, because we have the following result:

> A manifold M with an almost complex structure J is a complex manifold (i.e., there exists an atlas for M with holomorphic chart-change mappings) if and only if $[J, J] = 0$. See [Newlander-Nirenberg, 57].

Remarks

The material on the Lie derivative on natural vector bundles 6.14–6.20 appears here for the first time. Most of section 8 is due to [Frölicher-Nijenhuis, 56], the formula in 8.9 was proved by [Mangiarotti-Modugno, 84] and [Michor, 87]. The Bianchi identity 8.14 is from [Michor, 89a].

CHAPTER III.
BUNDLES AND CONNECTIONS

We begin our treatment of connections in the general setting of fiber bundles (without structure group). A connection on a fiber bundle is just a projection onto the vertical bundle. Curvature and the Bianchi identity is expressed with the help of the Frölicher-Nijenhuis bracket. The parallel transport for such a general connection is not defined along the whole of the curve in the base in general - if this is the case for all curves, the connection is called complete. We show that every fiber bundle admits complete connections. For complete connections we treat holonomy groups and the holonomy Lie algebra, a subalgebra of the Lie algebra of all vector fields on the standard fiber.

Then we present principal bundles and associated bundles in detail together with the most important examples. Finally we investigate principal connections by requiring equivariance under the structure group. It is remarkable how fast the usual structure equations can be derived from the basic properties of the Frölicher-Nijenhuis bracket. Induced connections are investigated thoroughly - we describe tools to recognize induced connections among general ones.

If the holonomy Lie algebra of a connection on a fiber bundle is finite dimensional and consists of complete vector fields on the fiber, we are able to show, that in fact the fiber bundle is associated to a principal bundle and the connection is induced from an irreducible principal connection (theorem 9.11). This is a powerful generalization of the theorem of Ambrose and Singer.

Connections will be treated once again from the point of view of jets, when we have them at our disposal in chapter IV.

We think that the treatment of connections presented here offers some didactical advantages besides presenting new results: the geometric content of a connection is treated first, and the additional requirement of equivariance under a structure group is seen to be additional and can be dealt with later - so the reader is not required to grasp all the structures at the same time. Besides that it gives new results and new insights. There are naturally appearing connections in differential geometry which are not principal or induced connections: The universal connection on the bundle J^1P/G of all connections of a principal bundle, and also the Cartan connections.

9. General fiber bundles and connections

9.1. Definition. A *(fiber) bundle* (E, p, M, S) consists of manifolds E, M, S, and a smooth mapping $p : E \to M$; furthermore it is required that each $x \in M$ has an open neighborhood U such that $E \mid U := p^{-1}(U)$ is diffeomorphic to

$U \times S$ via a fiber respecting diffeomorphism:

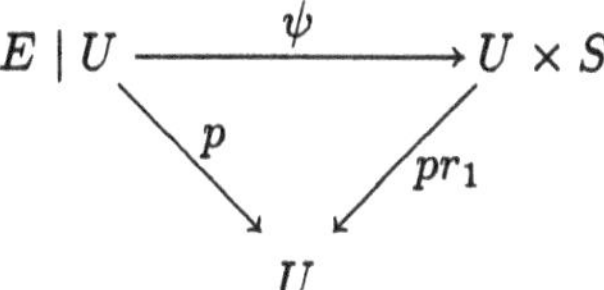

E is called the *total space*, M is called the *base space*, p is a surjective submersion, called the *projection*, and S is called *standard fiber*. (U, ψ) as above is called a *fiber chart* or a *local trivialization* of E.

A collection of fiber charts (U_α, ψ_α), such that (U_α) is an open cover of M, is called a *(fiber) bundle atlas*. If we fix such an atlas, then $(\psi_\alpha \circ \psi_\beta{}^{-1})(x, s) = (x, \psi_{\alpha\beta}(x, s))$, where $\psi_{\alpha\beta} : (U_\alpha \cap U_\beta) \times S \to S$ is smooth and $\psi_{\alpha\beta}(x, \ )$ is a diffeomorphism of S for each $x \in U_{\alpha\beta} := U_\alpha \cap U_\beta$. We may thus consider the mappings $\psi_{\alpha\beta} : U_{\alpha\beta} \to \mathrm{Diff}(S)$ with values in the group $\mathrm{Diff}(S)$ of all diffeomorphisms of S; their differentiability is a subtle question, which will not be discussed in this book, but see [Michor, 88]. In either form these mappings $\psi_{\alpha\beta}$ are called the *transition functions* of the bundle. They satisfy the *cocycle condition*: $\psi_{\alpha\beta}(x) \circ \psi_{\beta\gamma}(x) = \psi_{\alpha\gamma}(x)$ for $x \in U_{\alpha\beta\gamma}$ and $\psi_{\alpha\alpha}(x) = \mathrm{Id}_S$ for $x \in U_\alpha$. Therefore the collection $(\psi_{\alpha\beta})$ is called a *cocycle of transition functions*.

Given an open cover (U_α) of a manifold M and a cocycle of transition functions $(\psi_{\alpha\beta})$ we may construct a fiber bundle (E, p, M, S) similarly as in 6.4.

9.2. Lemma. *Let $p : N \to M$ be a proper surjective submersion (a fibered manifold) which is proper (i.e. compact sets have compact inverse images) and let M be connected. Then (N, p, M) is a fiber bundle.*

Proof. We have to produce a fiber chart at each $x_0 \in M$. So let (U, u) be a chart centered at x_0 on M such that $u(U) \cong \mathbb{R}^m$. For each $x \in U$ let $\xi_x(y) := (T_y u)^{-1}.u(x)$, then $\xi_x \in \mathfrak{X}(U)$, depending smoothly on $x \in U$, such that $u(\mathrm{Fl}^{\xi_x}_t u^{-1}(z)) = z + t.u(x)$, so each ξ_x is a complete vector field on U. Since p is a submersion, with the help of a partition of unity on $p^{-1}(U)$ we may construct vector fields $\eta_x \in \mathfrak{X}(p^{-1}(U))$ which depend smoothly on $x \in U$ and are p-related to ξ_x: $Tp.\eta_x = \xi_x \circ p$. Thus $p \circ \mathrm{Fl}^{\eta_x}_t = \mathrm{Fl}^{\xi_x}_t \circ p$ by 3.14, so $\mathrm{Fl}^{\eta_x}_t$ is fiber respecting, and since p is proper and ξ_x is complete, η_x has a global flow too. Denote $p^{-1}(x_0)$ by S. Then $\varphi : U \times S \to p^{-1}(U)$, defined by $\varphi(x, y) = \mathrm{Fl}^{\eta_x}_1(y)$, is a diffeomorphism and is fiber respecting, so (U, φ^{-1}) is a fiber chart. Since M is connected, the fibers $p^{-1}(x)$ are all diffeomorphic.

9.3. Let (E, p, M, S) be a fiber bundle; we consider the tangent mapping $Tp : TE \to TM$ and its kernel $\ker Tp =: VE$ which is called the *vertical bundle* of E. The following is special case of 8.13.

Definition. A *connection* on the fiber bundle (E, p, M, S) is a vector valued 1-form $\Phi \in \Omega^1(E; VE)$ with values in the vertical bundle VE such that $\Phi \circ \Phi = \Phi$ and $\mathrm{Im}\Phi = VE$; so Φ is just a projection $TE \to VE$.

If we intend to contrast this general concept of connection with some special cases which will be discussed later, we will say that Φ is a *general connection*.

Since $\ker \Phi$ is of constant rank, by 6.6 $\ker \Phi$ is a sub vector bundle of TE, it is called the space of *horizontal vectors* or the *horizontal bundle* and it is denoted by HE. Clearly $TE = HE \oplus VE$ and $T_uE = H_uE \oplus V_uE$ for $u \in E$.

Now we consider the mapping $(Tp, \pi_E) : TE \to TM \times_M E$. We have by definition $(Tp, \pi_E)^{-1}(0_{p(u)}, u) = V_uE$, so $(Tp, \pi_E) \mid HE : HE \to TM \times_M E$ is fiber linear over E and injective, so by reason of dimensions it is a fiber linear isomorphism: Its inverse is denoted by

$$C := ((Tp, \pi_E) \mid HE)^{-1} : TM \times_M E \to HE \hookrightarrow TE.$$

So $C : TM \times_M E \to TE$ is fiber linear over E and is a right inverse for (Tp, π_E). C is called the *horizontal lift* associated to the connection Φ.

Note the formula $\Phi(\xi_u) = \xi_u - C(Tp.\xi_u, u)$ for $\xi_u \in T_uE$. So we can equally well describe a connection Φ by specifying C. Then we call Φ the *vertical projection* (no confusion with 6.11 will arise) and $\chi := \mathrm{id}_{TE} - \Phi = C \circ (Tp, \pi_E)$ will be called the *horizontal projection*.

9.4. Curvature. Suppose that $\Phi : TE \to VE$ is a connection on a fiber bundle (E, p, M, S), then as in 8.13 the curvature R of Φ is given by

$$2R = [\Phi, \Phi] = [\mathrm{Id} - \Phi, \mathrm{Id} - \Phi] = [\chi, \chi] \in \Omega^2(E; VE)$$

(The cocurvature $\bar{R}$ vanishes since the vertical bundle VE is integrable). We have $R(X,Y) = \frac{1}{2}[\Phi, \Phi](X,Y) = \Phi[\chi X, \chi Y]$, so R is an obstruction against integrability of the horizontal subbundle. Note that for vector fields $\xi, \eta \in \mathfrak{X}(M)$ and their horizontal lifts $C\xi, C\eta \in \mathfrak{X}(E)$ we have $R(C\xi, C\eta) = [C\xi, C\eta] - C([\xi, \eta])$.

Since the vertical bundle VE is integrable, by 8.14 we have the *Bianchi identity* $[\Phi, R] = 0$.

9.5. Pullback. Let (E, p, M, S) be a fiber bundle and consider a smooth mapping $f : N \to M$. Since p is a submersion, f and p are transversal in the sense of 2.18 and thus the pullback $N \times_{(f,M,p)} E$ exists. It will be called the *pullback* of the fiber bundle E by f and we will denote it by f^*E. The following diagram sets up some further notation for it:

$$\begin{array}{ccc} f^*E & \xrightarrow{p^*f} & E \\ {\scriptstyle f^*p}\downarrow & & \downarrow{\scriptstyle p} \\ N & \xrightarrow{f} & M. \end{array}$$

Proposition. *In the situation above we have:*

(1) *(f^*E, f^*p, N, S) is again a fiber bundle, and p^*f is a fiber wise diffeomorphism.*

(2) *If $\Phi \in \Omega^1(E; TE)$ is a connection on the bundle E, then the vector valued form $f^*\Phi$, given by $(f^*\Phi)_u(X) := T_u(p^*f)^{-1}.\Phi.T_u(p^*f).X$ for $X \in T_uE$, is a connection on the bundle f^*E. The forms $f^*\Phi$ and Φ are p^*f-related in the sense of 8.15.*

(3) *The curvatures of $f^*\Phi$ and Φ are also p^*f-related.*

Proof. (1) If (U_α, ψ_α) is a fiber bundle atlas of (E, p, M, S) in the sense of 9.1, then $(f^{-1}(U_\alpha), (f^*p, pr_2 \circ \psi_\alpha \circ p^*f))$ is visibly a fiber bundle atlas for (f^*E, f^*p, N, S), by the formal universal properties of a pullback 2.19. (2) is obvious. (3) follows from (2) and 8.15.7. □

9.6. Let us suppose that a connection Φ on the bundle (E, p, M, S) has zero curvature. Then by 9.4 the horizontal bundle is integrable and gives rise to the *horizontal foliation* by 3.25.2. Each point $u \in E$ lies on a unique leaf $L(u)$ such that $T_vL(u) = H_vE$ for each $v \in L(u)$. The restriction $p \mid L(u)$ is locally a diffeomorphism, but in general it is neither surjective nor is it a covering onto its image. This is seen by devising suitable horizontal foliations on the trivial bundle $pr_2 : \mathbb{R} \times S^1 \to S^1$.

9.7. Local description. Let Φ be a connection on (E, p, M, S). Let us fix a fiber bundle atlas (U_α) with transition functions $(\psi_{\alpha\beta})$, and let us consider the connection $((\psi_\alpha)^{-1})^*\Phi \in \Omega^1(U_\alpha \times S; U_\alpha \times TS)$, which may be written in the form

$$((\psi_\alpha)^{-1})^*\Phi)(\xi_x, \eta_y) =: -\Gamma^\alpha(\xi_x, y) + \eta_y \text{ for } \xi_x \in T_xU_\alpha \text{ and } \eta_y \in T_yS,$$

since it reproduces vertical vectors. The Γ^α are given by

$$(0_x, \Gamma^\alpha(\xi_x, y)) := -T(\psi_\alpha).\Phi.T(\psi_\alpha)^{-1}.(\xi_x, 0_y).$$

We consider Γ^α as an element of the space $\Omega^1(U_\alpha; \mathfrak{X}(S))$, a 1-form on U^α with values in the infinite dimensional Lie algebra $\mathfrak{X}(S)$ of all vector fields on the standard fiber. The Γ^α are called the *Christoffel forms* of the connection Φ with respect to the bundle atlas (U_α, ψ_α).

Lemma. *The transformation law for the Christoffel forms is*

$$T_y(\psi_{\alpha\beta}(x, \quad)).\Gamma^\beta(\xi_x, y) = \Gamma^\alpha(\xi_x, \psi_{\alpha\beta}(x, y)) - T_x(\psi_{\alpha\beta}(\quad , y)).\xi_x.$$

The curvature R of Φ satisfies

$$(\psi_\alpha^{-1})^*R = d\Gamma^\alpha + [\Gamma^\alpha, \Gamma^\alpha]_{\mathfrak{X}(S)}.$$

Here $d\Gamma^\alpha$ is the exterior derivative of the 1-form $\Gamma^\alpha \in \Omega^1(U_\alpha; \mathfrak{X}(S))$ with values in the complete locally convex space $\mathfrak{X}(S)$. We will later also use the Lie derivative of it and the usual formulas apply: consult [Frölicher, Kriegl, 88] for calculus in infinite dimensional spaces.

The formula for the curvature is the *Maurer-Cartan* formula which in this general setting appears only in the level of local description.

Proof. From $(\psi_\alpha \circ (\psi_\beta)^{-1})(x, y) = (x, \psi_{\alpha\beta}(x, y))$ we get that $T(\psi_\alpha \circ (\psi_\beta)^{-1}).(\xi_x, \eta_y) = (\xi_x, T_{(x,y)}(\psi_{\alpha\beta}).(\xi_x, \eta_y))$ and thus:

$$T(\psi_\beta^{-1}).(0_x, \Gamma^\beta(\xi_x, y)) = -\Phi(T(\psi_\beta^{-1})(\xi_x, 0_y)) =$$

$$= -\Phi(T(\psi_\alpha^{-1}).T(\psi_\alpha \circ \psi_\beta^{-1}).(\xi_x, 0_y)) =$$
$$= -\Phi(T(\psi_\alpha^{-1})(\xi_x, T_{(x,y)}(\psi_{\alpha\beta})(\xi_x, 0_y))) =$$
$$= -\Phi(T(\psi_\alpha^{-1})(\xi_x, 0_{\psi_{\alpha\beta}(x,y)})) - \Phi(T(\psi_\alpha^{-1})(0_x, T_{(x,y)}\psi_{\alpha\beta}(\xi_x, 0_y))) =$$
$$= T(\psi_\alpha^{-1}).(0_x, \Gamma^\alpha(\xi_x, \psi_{\alpha\beta}(x,y))) - T(\psi_\alpha^{-1})(0_x, T_x(\psi_{\alpha\beta}(\quad, y)).\xi_x).$$

This implies the transformation law.

For the curvature R of Φ we have by 9.4 and 9.5.3

$$(\psi_\alpha^{-1})^* R((\xi^1, \eta^1), (\xi^2, \eta^2)) =$$
$$= (\psi_\alpha^{-1})^*\Phi\,[(\operatorname{Id} - (\psi_\alpha^{-1})^*\Phi)(\xi^1, \eta^1), (\operatorname{Id} - (\psi_\alpha^{-1})^*\Phi)(\xi^2, \eta^2)] =$$
$$= (\psi_\alpha^{-1})^*\Phi[(\xi^1, \Gamma^\alpha(\xi^1)), (\xi^2, \Gamma^\alpha(\xi^2))] =$$
$$= (\psi_\alpha^{-1})^*\Phi\left([\xi^1, \xi^2], \xi^1\Gamma^\alpha(\xi^2) - \xi^2\Gamma^\alpha(\xi^1) + [\Gamma^\alpha(\xi^1), \Gamma^\alpha(\xi^2)]\right) =$$
$$= -\Gamma^\alpha([\xi^1, \xi^2]) + \xi^1\Gamma^\alpha(\xi^2) - \xi^2\Gamma^\alpha(\xi^1) + [\Gamma^\alpha(\xi^1), \Gamma^\alpha(\xi^2)] =$$
$$= d\Gamma^\alpha(\xi^1, \xi^2) + [\Gamma^\alpha(\xi^1), \Gamma^\alpha(\xi^2)]_{\mathfrak{X}(S)}. \quad \square$$

9.8. Theorem (Parallel transport). *Let Φ be a connection on a bundle (E, p, M, S) and let $c : (a, b) \to M$ be a smooth curve with $0 \in (a, b)$, $c(0) = x$.*

Then there is a neighborhood U of $E_x \times \{0\}$ in $E_x \times (a, b)$ and a smooth mapping $\operatorname{Pt}_c : U \to E$ such that:

1. *$p(\operatorname{Pt}(c, u_x, t)) = c(t)$ if defined, and $\operatorname{Pt}(c, u_x, 0) = u_x$.*
2. *$\Phi(\frac{d}{dt}\operatorname{Pt}(c, u_x, t)) = 0$ if defined.*
3. *Reparametrisation invariance: If $f : (a', b') \to (a, b)$ is smooth with $0 \in (a', b')$, then $\operatorname{Pt}(c, u_x, f(t)) = \operatorname{Pt}(c \circ f, \operatorname{Pt}(c, u_x, f(0)), t)$ if defined.*
4. *U is maximal for properties (1) and (2).*
5. *If the curve c depends smoothly on further parameters then $\operatorname{Pt}(c, u_x, t)$ depends also smoothly on those parameters.*

First proof. In local bundle coordinates $\Phi(\frac{d}{dt}\operatorname{Pt}(c, u_x, t)) = 0$ is an ordinary differential equation of first order, nonlinear, with initial condition $\operatorname{Pt}(c, u_x, 0) = u_x$. So there is a maximally defined local solution curve which is unique. All further properties are consequences of uniqueness.

Second proof. Consider the pullback bundle $(c^*E, c^*p, (a, b), S)$ and the pullback connection $c^*\Phi$ on it. It has zero curvature, since the horizontal bundle is 1-dimensional. By 9.6 the horizontal foliation exists and the parallel transport just follows a leaf and we may map it back to E, in detail: $\operatorname{Pt}(c, u_x, t) = p^*c((c^*p \mid L(u_x))^{-1}(t))$.

Third proof. Consider a fiber bundle atlas (U_α, ψ_α) as in 9.7. Then we have $\psi_\alpha(\operatorname{Pt}(c, \psi_\alpha^{-1}(x, y), t)) = (c(t), \gamma(y, t))$, where

$$0 = ((\psi_\alpha^{-1})^*\Phi)\left(\tfrac{d}{dt}c(t), \tfrac{d}{dt}\gamma(y, t)\right) = -\Gamma^\alpha\left(\tfrac{d}{dt}c(t), \gamma(y, t)\right) + \tfrac{d}{dt}\gamma(y, t),$$

so $\gamma(y, t)$ is the integral curve (evolution line) through $y \in S$ of the time dependent vector field $\Gamma^\alpha\left(\frac{d}{dt}c(t)\right)$ on S. This vector field visibly depends smoothly on c. Clearly local solutions exist and all properties follow. For (5) we refer to [Michor, 83]. $\square$

9.9. A connection Φ on (E, p, M, S) is called a *complete connection*, if the parallel transport Pt_c along any smooth curve $c : (a, b) \to M$ is defined on the whole of $E_{c(0)} \times (a, b)$. The third proof of theorem 9.8 shows that on a fiber bundle with compact standard fiber any connection is complete.

The following is a sufficient condition for a connection Φ to be complete:

> There exists a fiber bundle atlas (U_α, ψ_α) and complete Riemannian metrics g_α on the standard fiber S such that each Christoffel form $\Gamma^\alpha \in \Omega^1(U_\alpha, \mathfrak{X}(S))$ takes values in the linear subspace of g_α-bounded vector fields on S.

For in the third proof of theorem 9.8 above the time dependent vector field $\Gamma^\alpha(\frac{d}{dt}c(t))$ on S is g_α-bounded for compact time intervals. So by continuation the solution exists over $c^{-1}(U_\alpha)$, and thus globally.

A complete connection is called an *Ehresmann connection* in [Greub, Halperin, Vanstone I, 72, p. 314], where it is also indicated how to prove the following result.

Theorem. *Each fiber bundle admits complete connections.*

Proof. Let $\dim M = m$. Let (U_α, ψ_α) be a fiber bundle atlas as in 9.1. By topological dimension theory [Nagata, 65] the open cover (U_α) of M admits a refinement such that any $m + 2$ members have empty intersection, see also 1.1. Let (U_α) itself have this property. Choose a smooth partition of unity (f_α) subordinated to (U_α). Then the sets $V_\alpha := \{x : f_\alpha(x) > \frac{1}{m+2}\} \subset U_\alpha$ form still an open cover of M since $\sum f_\alpha(x) = 1$ and at most $m + 1$ of the $f_\alpha(x)$ can be nonzero. By renaming assume that each V_α is connected. Then we choose an open cover (W_α) of M such that $\overline{W_\alpha} \subset V_\alpha$.

Now let g_1 and g_2 be complete Riemannian metrics on M and S, respectively (see [Nomizu - Ozeki, 61] or [Morrow, 70]). For not connected Riemannian manifolds complete means that each connected component is complete. Then $g_1|U_\alpha \times g_2$ is a Riemannian metric on $U_\alpha \times S$ and we consider the metric $g := \sum f_\alpha \psi_\alpha^*(g_1|U_\alpha \times g_2)$ on E. Obviously $p : E \to M$ is a Riemannian submersion for the metrics g and g_1. We choose now the connection $\Phi : TE \to VE$ as the orthonormal projection with respect to the Riemannian metric g.

Claim. Φ is a complete connection on E.

Let $c : [0, 1] \to M$ be a smooth curve. We choose a partition $0 = t_0 < t_1 < \dots < t_k = 1$ such that $c([t_i, t_{i+1}]) \subset V_{\alpha_i}$ for suitable α_i. It suffices to show that $\operatorname{Pt}(c(t_i + \ \ ), u_{c(t_i)}, t)$ exists for all $0 \le t \le t_{i+1} - t_i$ and all $u_{c(t_i)}$, for all i — then we may piece them together. So we may assume that $c : [0, 1] \to V_\alpha$ for some α. Let us now assume that for some $(x, y) \in V_\alpha \times S$ the parallel transport $\operatorname{Pt}(c, \psi_\alpha(x, y), t)$ is defined only for $t \in [0, t')$ for some $0 < t' < 1$. By the third proof of 9.8 we have $\operatorname{Pt}(c, \psi_\alpha(x, y), t) = \psi_\alpha^{-1}(c(t), \gamma(t))$, where $\gamma : [0, t') \to S$ is the maximally defined integral curve through $y \in S$ of the time dependent vector field $\Gamma^\alpha(\frac{d}{dt}c(t), \ \ )$ on S. We put $g_\alpha := (\psi_\alpha^{-1})^* g$, then $(g_\alpha)_{(x,y)} = (g_1)_x \times (\sum_\beta f_\beta(x) \psi_{\beta\alpha}(x, \ \ )^* g_2)_y$. Since $pr_1 : (V_\alpha \times S, g_\alpha) \to (V_\alpha, g_1|V_\alpha)$ is a Riemannian submersion and since the connection $(\psi_\alpha^{-1})^*\Phi$ is also given by orthonormal projection onto the vertical bundle, we get

$$\infty > g_1\text{-length}_0^{t'}(c) = g_\alpha\text{-length}(c,\gamma) = \int_0^{t'} |(c'(t), \tfrac{d}{dt}\gamma(t))|_{g_\alpha}\, dt =$$

$$= \int_0^{t'} \sqrt{|c'(t)|_{g_1}^2 + \textstyle\sum_\beta f_\beta(c(t))(\psi_{\alpha\beta}(c(t),-)^* g_2)(\tfrac{d}{dt}\gamma(t), \tfrac{d}{dt}\gamma(t))}\, dt \geq$$

$$\geq \int_0^{t'} \sqrt{f_\alpha(c(t))}\, |\tfrac{d}{dt}\gamma(t)|_{g_2}\, dt \geq \frac{1}{\sqrt{m+2}} \int_0^{t'} |\tfrac{d}{dt}\gamma(t)|_{g_2} dt.$$

So g_2-lenght(γ) is finite and since the Riemannian metric g_2 on S is complete, $\lim_{t\to t'} \gamma(t) =: \gamma(t')$ exists in S and the integral curve γ can be continued. □

9.10. Holonomy groups and Lie algebras. Let (E, p, M, S) be a fiber bundle with a complete connection Φ, and let us assume that M is connected. We choose a fixed base point $x_0 \in M$ and we identify E_{x_0} with the standard fiber S. For each closed piecewise smooth curve $c : [0,1] \to M$ through x_0 the parallel transport $\operatorname{Pt}(c, \quad, 1) =: \operatorname{Pt}(c, 1)$ (pieced together over the smooth parts of c) is a diffeomorphism of S. All these diffeomorphisms form together the group $\operatorname{Hol}(\Phi, x_0)$, the *holonomy group* of Φ at x_0, a subgroup of the diffeomorphism group $\operatorname{Diff}(S)$. If we consider only those piecewise smooth curves which are homotopic to zero, we get a subgroup $\operatorname{Hol}_0(\Phi, x_0)$, called the *restricted holonomy group* of the connection Φ at x_0.

Now let $C : TM \times_M E \to TE$ be the horizontal lifting as in 9.3, and let R be the curvature (9.4) of the connection Φ. For any $x \in M$ and $X_x \in T_xM$ the horizontal lift $C(X_x) := C(X_x, \quad) : E_x \to TE$ is a vector field along E_x. For X_x and $Y_x \in T_xM$ we consider $R(CX_x, CY_x) \in \mathfrak{X}(E_x)$. Now we choose any piecewise smooth curve c from x_0 to x and consider the diffeomorphism $\operatorname{Pt}(c,t) : S = E_{x_0} \to E_x$ and the pullback $\operatorname{Pt}(c,1)^* R(CX_x, CY_x) \in \mathfrak{X}(S)$. Let us denote by $\operatorname{hol}(\Phi, x_0)$ the closed linear subspace, generated by all these vector fields (for all $x \in M$, X_x, $Y_x \in T_xM$ and curves c from x_0 to x) in $\mathfrak{X}(S)$ with respect to the compact C^∞-topology (see [Hirsch, 76]), and let us call it the *holonomy Lie algebra* of Φ at x_0.

Lemma. $\operatorname{hol}(\Phi, x_0)$ *is a Lie subalgebra of* $\mathfrak{X}(S)$.

Proof. For $X \in \mathfrak{X}(M)$ we consider the local flow Fl_t^{CX} of the horizontal lift of X. It restricts to parallel transport along any of the flow lines of X in M. Then for vector fields X, Y, U, V on M the expression

$$\begin{aligned}
&\tfrac{d}{dt}|_0 (\operatorname{Fl}_s^{CX})^* (\operatorname{Fl}_t^{CY})^* (\operatorname{Fl}_{-s}^{CX})^* (\operatorname{Fl}_z^{CZ})^* R(CU, CV)|E_{x_0} \\
&= (\operatorname{Fl}_s^{CX})^* [CY, (\operatorname{Fl}_{-s}^{CX})^* (\operatorname{Fl}_z^{CZ})^* R(CU, CV)]|E_{x_0} \\
&= [(\operatorname{Fl}_s^{CX})^* CY, (\operatorname{Fl}_z^{CZ})^* R(CU, CV)]|E_{x_0}
\end{aligned}$$

is in $\operatorname{hol}(\Phi, x_0)$, since it is closed in the compact C^∞-topology and the derivative can be written as a limit. Thus

$$[(\operatorname{Fl}_s^{CX})^* [CY_1, CY_2], (\operatorname{Fl}_z^{CZ})^* R(CU, CV)]|E_{x_0} \in \operatorname{hol}(\Phi, x_0)$$

by the Jacobi identity and

$$[(\mathrm{Fl}^{CX}_s)^* C[Y_1, Y_2], (\mathrm{Fl}^{CZ}_z)^* R(CU, CV)]|E_{x_0} \in \mathrm{hol}(\Phi, x_0),$$

so also their difference

$$[(\mathrm{Fl}^{CX}_s)^* R(CY_1, CY_2), (\mathrm{Fl}^{CZ}_z)^* R(CU, CV)]|E_{x_0}$$

is in $\mathrm{hol}(\Phi, x_0)$. □

9.11. The following theorem is a generalization of the theorem of Ambrose and Singer on principal connections. The reader who does not know principal connections is advised to read parts of sections 10 and 11 first. We include this result here in order not to disturb the development in section 11 later.

Theorem. *Let Φ be a complete connection on the fibre bundle (E, p, M, S) and let M be connected. Suppose that for some (hence any) $x_0 \in M$ the holonomy Lie algebra $\mathrm{hol}(\Phi, x_0)$ is finite dimensional and consists of complete vector fields on the fiber E_{x_0}*

Then there is a principal bundle (P, p, M, G) with finite dimensional structure group G, an irreducible connection ω on it and a smooth action of G on S such that the Lie algebra $\mathfrak{g}$ of G equals the holonomy Lie algebra $\mathrm{hol}(\Phi, x_0)$, the fibre bundle E is isomorphic to the associated bundle $P[S]$, and Φ is the connection induced by ω. The structure group G equals the holonomy group $\mathrm{Hol}(\Phi, x_0)$. P and ω are unique up to isomorphism.

By a theorem of [Palais, 57] a finite dimensional Lie subalgebra of $\mathfrak{X}(E_{x_0})$ like $\mathrm{hol}(\Phi, x_0)$ consists of complete vector fields if and only if it is generated by complete vector fields as a Lie algebra.

Proof. Let us again identify E_{x_0} and S. Then $\mathfrak{g} := \mathrm{hol}(\Phi, x_0)$ is a finite dimensional Lie subalgebra of $\mathfrak{X}(S)$, and since each vector field in it is complete, there is a finite dimensional connected Lie group G_0 of diffeomorphisms of S with Lie algebra $\mathfrak{g}$, see [Palais, 57].

Claim 1. G_0 contains $\mathrm{Hol}_0(\Phi, x_0)$, the restricted holonomy group.

Let $f \in \mathrm{Hol}_0(\Phi, x_0)$, then $f = \mathrm{Pt}(c, 1)$ for a piecewise smooth closed curve c through x_0, which is nullhomotopic. Since the parallel transport is essentially invariant under reparametrisation, 9.8, we can replace c by $c \circ g$, where g is smooth and flat at each corner of c. So we may assume that c itself is smooth. Since c is homotopic to zero, by approximation we may assume that there is a smooth homotopy $H : \mathbb{R}^2 \to M$ with $H_1|[0,1] = c$ and $H_0|[0,1] = x_0$. Then $f_t := \mathrm{Pt}(H_t, 1)$ is a curve in $\mathrm{Hol}_0(\Phi, x_0)$ which is smooth as a mapping $\mathbb{R} \times S \to S$. The rest of the proof of claim 1 will follow.

Claim 2. $(\frac{d}{dt} f_t) \circ f_t^{-1} =: Z_t$ is in $\mathfrak{g}$ for all t.

To prove claim 2 we consider the pullback bundle $H^*E \to \mathbb{R}^2$ with the induced connection $H^*\Phi$. It is sufficient to prove claim 2 there. Let $X = \frac{d}{ds}$ and $Y = \frac{d}{dt}$

be the constant vector fields on $\mathbb{R}^2$, so $[X,Y]=0$. Then $\operatorname{Pt}(c,s)=\operatorname{Fl}_s^{CX}|S$ and so on. We put

$$f_{t,s}=\operatorname{Fl}_{-s}^{CX}\circ\operatorname{Fl}_{-t}^{CY}\circ\operatorname{Fl}_s^{CX}\circ\operatorname{Fl}_t^{CY}:S\to S,$$

so $f_{t,1}=f_t$. Then we have in the vector space $\mathfrak{X}(S)$

$$(\tfrac{d}{dt}f_{t,s})\circ f_{t,s}^{-1}=-(\operatorname{Fl}_s^{CX})^*CY+(\operatorname{Fl}_s^{CX})^*(\operatorname{Fl}_t^{CY})^*(\operatorname{Fl}_{-s}^{CX})^*CY,$$

$$(\tfrac{d}{dt}f_{t,1})\circ f_{t,1}^{-1}=\int_0^1\tfrac{d}{ds}\left((\tfrac{d}{dt}f_{t,s})\circ f_{t,s}^{-1}\right)ds$$

$$=\int_0^1\Big(-(\operatorname{Fl}_s^{CX})^*[CX,CY]+(\operatorname{Fl}_s^{CX})^*[CX,(\operatorname{Fl}_t^{CY})^*(\operatorname{Fl}_{-s}^{CX})^*CY]$$
$$-(\operatorname{Fl}_s^{CX})^*(\operatorname{Fl}_t^{CY})^*(\operatorname{Fl}_{-s}^{CX})^*[CX,CY]\Big)\,ds.$$

Since $[X,Y]=0$ we have $[CX,CY]=\Phi[CX,CY]=R(CX,CY)$ and

$$(\operatorname{Fl}_t^{CX})^*CY=C\left((\operatorname{Fl}_t^{X})^*Y\right)+\Phi\left((\operatorname{Fl}_t^{CX})^*CY\right)$$
$$=CY+\int_0^t\tfrac{d}{dt}\Phi(\operatorname{Fl}_t^{CX})^*CY\,dt$$
$$=CY+\int_0^t\Phi(\operatorname{Fl}_t^{CX})^*[CX,CY]\,dt$$
$$=CY+\int_0^t\Phi(\operatorname{Fl}_t^{CX})^*R(CX,CY)\,dt$$
$$=CY+\int_0^t(\operatorname{Fl}_t^{CX})^*R(CX,CY)\,dt.$$

The flows $(\operatorname{Fl}^{C}X_s)^*$ and its derivative at 0 $\mathcal{L}_{CX}=[CX,\ \ ]$ do not lead out of $\mathfrak{g}$, thus all parts of the integrand above are in $\mathfrak{g}$. So $(\frac{d}{dt}f_{t,1})\circ f_{t,1}^{-1}$ is in $\mathfrak{g}$ for all t and claim 2 follows.

Now claim 1 can be shown as follows. There is a unique smooth curve $g(t)$ in G_0 satisfying $T_e(\rho_{g(t)})Z_t=Z_t.g(t)=\frac{d}{dt}g(t)$ and $g(0)=e$; via the action of G_0 on S the curve $g(t)$ is a curve of diffeomorphisms on S, generated by the time dependent vector field Z_t, so $g(t)=f_t$ and $f=f_1$ is in G_0. So we get $\operatorname{Hol}_0(\Phi,x_0)\subseteq G_0$.

Claim 3. $\operatorname{Hol}_0(\Phi,x_0)$ equals G_0.

In the proof of claim 1 we have seen that $\operatorname{Hol}_0(\Phi,x_0)$ is a smoothly arcwise connected subgroup of G_0, so it is a connected Lie subgroup by the results cited in 5.6. It suffices thus to show that the Lie algebra $\mathfrak{g}$ of G_0 is contained in the Lie algebra of $\operatorname{Hol}_0(\Phi,x_0)$, and for that it is enough to show, that for each ξ in a linearly spanning subset of $\mathfrak{g}$ there is a smooth mapping $f:[-1,1]\times S\to S$ such that the associated curve $\check{f}$ lies in $\operatorname{Hol}_0(\Phi,x_0)$ with $\check{f}'(0)=0$ and $\check{f}''(0)=\xi$.

By definition we may assume $\xi=\operatorname{Pt}(c,1)^*R(CX_x,CY_x)$ for X_x, $Y_x\in T_xM$ and a smooth curve c in M from x_0 to x. We extend X_x and Y_x to vector fields

X and $Y \in \mathfrak{X}(M)$ with $[X,Y]=0$ near x. We may also suppose that $Z \in \mathfrak{X}(M)$ is a vector field which extends $c'(t)$ along $c(t)$: if c is simple we approximate it by an embedding and can consequently extend $c'(t)$ to such a vector field. If c is not simple we do this for each simple piece of c and have then several vector fields Z instead of one below. So we have

$$\begin{aligned}\xi &= (\mathrm{Fl}_1^{CZ})^* R(CX,CY) = (\mathrm{Fl}_1^{CZ})^*[CX,CY] \quad \text{since } [X,Y](x)=0\\ &= (\mathrm{Fl}_1^{CZ})^* \tfrac{1}{2}\tfrac{d^2}{dt^2}|_{t=0}(\mathrm{Fl}_{-t}^{CY}\circ \mathrm{Fl}_{-t}^{CX}\circ \mathrm{Fl}_t^{CY}\circ \mathrm{Fl}_t^{CX}) \quad \text{by 3.16}\\ &= \tfrac{1}{2}\tfrac{d^2}{dt^2}|_{t=0}(\mathrm{Fl}_{-1}^{CZ}\circ \mathrm{Fl}_{-t}^{CY}\circ \mathrm{Fl}_{-t}^{CX}\circ \mathrm{Fl}_t^{CY}\circ \mathrm{Fl}_t^{CX}\circ \mathrm{Fl}_1^{CZ}),\end{aligned}$$

where the parallel transport in the last equation first follows c from x_0 to x, then follows a small closed parallelogram near x in M (since $[X,Y]=0$ near x) and then follows c back to x_0. This curve is clearly nullhomotopic.

Step 4. Now we make $\mathrm{Hol}(\Phi,x_0)$ into a Lie group which we call G, by taking $\mathrm{Hol}_0(\Phi,x_0)=G_0$ as its connected component of the identity. Then the quotient group $\mathrm{Hol}(\Phi,x_0)/\mathrm{Hol}_0(\Phi,x_0)$ is countable, since the fundamental group $\pi_1(M)$ is countable (by Morse theory M is homotopy equivalent to a countable CW-complex).

Step 5. Construction of a cocycle of transition functions with values in G. Let $(U_\alpha, u_\alpha : U_\alpha \to \mathbb{R}^m)$ be a locally finite smooth atlas for M such that each $u_\alpha : U_\alpha \to \mathbb{R}^m)$ is surjective. Put $x_\alpha := u_\alpha^{-1}(0)$ and choose smooth curves $c_\alpha : [0,1] \to M$ with $c_\alpha(0)=x_0$ and $c_\alpha(1)=x_\alpha$. For each $x\in U_\alpha$ let $c_\alpha^x : [0,1]\to M$ be the smooth curve $t\mapsto u_\alpha^{-1}(t.u_\alpha(x))$, then c_α^x connects x_α and x and the mapping $(x,t)\mapsto c_\alpha^x(t)$ is smooth $U_\alpha\times[0,1]\to M$. Now we define a fibre bundle atlas $(U_\alpha, \psi_\alpha : E|U_\alpha \to U_\alpha\times S)$ by $\psi_\alpha^{-1}(x,s)=\mathrm{Pt}(c_\alpha^x,1)\,\mathrm{Pt}(c_\alpha,1)\,s$. Then ψ_α is smooth since $\mathrm{Pt}(c_\alpha^x,1)=\mathrm{Fl}_1^{CX_x}$ for a local vector field X_x depending smoothly on x. Let us investigate the transition functions.

$$\begin{aligned}\psi_\alpha\psi_\beta^{-1}(x,s) &= \big(x, \mathrm{Pt}(c_\alpha,1)^{-1}\,\mathrm{Pt}(c_\alpha^x,1)^{-1}\,\mathrm{Pt}(c_\beta^x,1)\,\mathrm{Pt}(c_\beta,1)\,s\big)\\ &= \big(x, \mathrm{Pt}(c_\beta.c_\beta^x.(c_\alpha^x)^{-1}.(c_\alpha)^{-1},4)\,s\big)\\ &=: (x,\psi_{\alpha\beta}(x)\,s), \text{ where } \psi_{\alpha\beta} : U_{\alpha\beta}\to G.\end{aligned}$$

Clearly $\psi_{\beta\alpha} : U_{\beta\alpha}\times S\to S$ is smooth which implies that $\psi_{\beta\alpha} : U_{\beta\alpha}\to G$ is also smooth. $(\psi_{\alpha\beta})$ is a cocycle of transition functions and we use it to glue a principal bundle with structure group G over M which we call (P,p,M,G). From its construction it is clear that the associated bundle $P[S]=P\times_G S$ equals (E,p,M,S).

Step 6. Lifting the connection Φ to P.
For this we have to compute the Christoffel symbols of Φ with respect to the atlas of step 5. To do this directly is quite difficult since we have to differentiate the parallel transport with respect to the curve. Fortunately there is another

way. Let $c : [0,1] \to U_\alpha$ be a smooth curve. Then we have

$$\begin{aligned}
&\psi_\alpha(\operatorname{Pt}(c,t)\psi_\alpha^{-1}(c(0),s)) = \\
&= \Big(c(t), \operatorname{Pt}((c_\alpha)^{-1},1)\operatorname{Pt}((c_\alpha^{c(0)})^{-1},1)\operatorname{Pt}(c,t)\operatorname{Pt}(c_\alpha^{c(0)},1)\operatorname{Pt}(c_\alpha,1)s\Big) \\
&= (c(t), \gamma(t).s),
\end{aligned}$$

where $\gamma(t)$ is a smooth curve in the holonomy group G. Let $\Gamma^\alpha \in \Omega^1(U_\alpha, \mathfrak{X}(S))$ be the Christoffel symbol of the connection Φ with respect to the chart (U_α, ψ_α). From the third proof of theorem 9.8 we have

$$\psi_\alpha(\operatorname{Pt}(c,t)\psi_\alpha^{-1}(c(0),s)) = (c(t), \bar\gamma(t,s)),$$

where $\bar\gamma(t,s)$ is the integral curve through s of the time dependent vector field $\Gamma^\alpha(\frac{d}{dt}c(t))$ on S. But then we get

$$\begin{aligned}
\Gamma^\alpha(\tfrac{d}{dt}c(t))(\bar\gamma(t,s)) = \tfrac{d}{dt}\bar\gamma(t,s) = \tfrac{d}{dt}(\gamma(t).s) = (\tfrac{d}{dt}\gamma(t)).s, \\
\Gamma^\alpha(\tfrac{d}{dt}c(t)) = (\tfrac{d}{dt}\gamma(t)) \circ \gamma(t)^{-1} \in \mathfrak{g}.
\end{aligned}$$

So Γ^α takes values in the Lie sub algebra of fundamental vector fields for the action of G on S. By theorem 11.9 below the connection Φ is thus induced by a principal connection ω on P. Since by 11.8 the principal connection ω has the 'same' holonomy group as Φ and since this is also the structure group of P, the principal connection ω is irreducible, see 11.7. □

10. Principal fiber bundles and G-bundles

10.1. Definition. Let G be a Lie group and let (E,p,M,S) be a fiber bundle as in 9.1. A *G-bundle structure* on the fiber bundle consists of the following data:

(1) A left action $\ell : G \times S \to S$ of the Lie group on the standard fiber.
(2) A fiber bundle atlas (U_α, ψ_α) whose transition functions $(\psi_{\alpha\beta})$ act on S via the G-action: There is a family of smooth mappings $(\varphi_{\alpha\beta} : U_{\alpha\beta} \to G)$ which satisfies the cocycle condition $\varphi_{\alpha\beta}(x)\varphi_{\beta\gamma}(x) = \varphi_{\alpha\gamma}(x)$ for $x \in U_{\alpha\beta\gamma}$ and $\varphi_{\alpha\alpha}(x) = e$, the unit in the group, such that $\psi_{\alpha\beta}(x,s) = \ell(\varphi_{\alpha\beta}(x),s) = \varphi_{\alpha\beta}(x).s$.

A fiber bundle with a G-bundle structure is called a *G-bundle.* A fiber bundle atlas as in (2) is called a *G-atlas* and the family $(\varphi_{\alpha\beta})$ is also called a cocycle of transition functions, but now for the G-bundle. G is called the structure group of the bundle.

To be more precise, two G-atlases are said to be equivalent (to describe the same G-bundle), if their union is also a G-atlas. This translates as follows to the two cocycles of transition functions, where we assume that the two coverings of M are the same (by passing to the common refinement, if necessary): $(\varphi_{\alpha\beta})$

and $(\varphi'_{\alpha\beta})$ are called *cohomologous* if there is a family $(\tau_\alpha : U_\alpha \to G)$ such that $\varphi_{\alpha\beta}(x) = \tau_\alpha(x)^{-1}.\varphi'_{\alpha\beta}(x).\tau_\beta(x)$ holds for all $x \in U_{\alpha\beta}$, compare with 6.4.

In (2) one should specify only an equivalence class of G-bundle structures or only a cohomology class of cocycles of G-valued transition functions. The proof of 6.4 now shows that from any open cover (U_α) of M, some cocycle of transition functions $(\varphi_{\alpha\beta} : U_{\alpha\beta} \to G)$ for it, and a left G-action on a manifold S, we may construct a G-bundle, which depends only on the cohomology class of the cocycle. By some abuse of notation we write (E, p, M, S, G) for a fiber bundle with specified G-bundle structure.

Examples. The tangent bundle of a manifold M is a fiber bundle with structure group $GL(m)$. More general a vector bundle (E, p, M, V) as in 6.1 is a fiber bundle with standard fiber the vector space V and with $GL(V)$-structure.

10.2. Definition. A *principal (fiber) bundle* (P, p, M, G) is a G-bundle with typical fiber a Lie group G, where the left action of G on G is just the left translation.

So by 10.1 we are given a bundle atlas $(U_\alpha, \varphi_\alpha : P|U_\alpha \to U_\alpha \times G)$ such that we have $\varphi_\alpha\varphi_\beta^{-1}(x, a) = (x, \varphi_{\alpha\beta}(x).a)$ for the cocycle of transition functions $(\varphi_{\alpha\beta} : U_{\alpha\beta} \to G)$. This is now called a *principal bundle atlas*. Clearly the principal bundle is uniquely specified by the cohomology class of its cocycle of transition functions.

Each principal bundle admits a unique right action $r : P \times G \to P$, called the *principal right action*, given by $\varphi_\alpha(r(\varphi_\alpha^{-1}(x, a), g)) = (x, ag)$. Since left and right translation on G commute, this is well defined. As in 5.10 we write $r(u, g) = u.g$ when the meaning is clear. The principal right action is visibly free and for any $u_x \in P_x$ the partial mapping $r_{u_x} = r(u_x, \quad) : G \to P_x$ is a diffeomorphism onto the fiber through u_x, whose inverse is denoted by $\tau_{u_x} : P_x \to G$. These inverses together give a smooth mapping $\tau : P \times_M P \to G$, whose local expression is $\tau(\varphi_\alpha^{-1}(x, a), \varphi_\alpha^{-1}(x, b)) = a^{-1}.b$. This mapping is also uniquely determined by the implicit equation $r(u_x, \tau(u_x, v_x)) = v_x$, thus we also have $\tau(u_x.g, u'_x.g') = g^{-1}.\tau(u_x, u'_x).g'$ and $\tau(u_x, u_x) = e$.

When considering principal bundles the reader should think of frame bundles as the foremost examples for this book. They will be treated in 10.11 below.

10.3. Lemma. *Let $p : P \to M$ be a surjective submersion (a fibered manifold), and let G be a Lie group which acts freely on P from the right such that the orbits of the action are exactly the fibers $p^{-1}(x)$ of p. Then (P, p, M, G) is a principal fiber bundle.*

If the action is a left one we may turn it into a right one by using the group inversion if necessary.

Proof. Let $s_\alpha : U_\alpha \to P$ be local sections (right inverses) for $p : P \to M$ such that (U_α) is an open cover of M. Let $\varphi_\alpha^{-1} : U_\alpha \times G \to P|U_\alpha$ be given by $\varphi_\alpha^{-1}(x, a) = s_\alpha(x).a$, which is obviously injective with invertible tangent mapping, so its inverse $\varphi_\alpha : P|U_\alpha \to U_\alpha \times G$ is a fiber respecting diffeomorphism. So $(U_\alpha, \varphi_\alpha)$ is already a fiber bundle atlas. Let $\tau : P \times_M P \to G$ be given by the implicit equation $r(u_x, \tau(u_x, u'_x)) = u'_x$, where r is the right G-action. τ is smooth by the implicit function theorem and clearly we have $\tau(u_x, u'_x.g) = \tau(u_x, u'_x).g$ and $\varphi_\alpha(u_x) = (x, \tau(s_\alpha(x), u_x))$. Thus we have $\varphi_\alpha\varphi_\beta^{-1}(x,g) = \varphi_\alpha(s_\beta(x).g) = (x, \tau(s_\alpha(x), s_\beta(x).g)) = (x, \tau(s_\alpha(x), s_\beta(x)).g)$ and $(U_\alpha, \varphi_\alpha)$ is a principal bundle atlas. □

10.4. Remarks. In the proof of lemma 10.3 we have seen, that a principal bundle atlas of a principal fiber bundle (P, p, M, G) is already determined if we specify a family of smooth sections of P, whose domains of definition cover the base M.

Lemma 10.3 can serve as an equivalent definition for a principal bundle. But this is true only if an implicit function theorem is available, so in topology or in infinite dimensional differential geometry one should stick to our original definition.

From the lemma itself it follows, that the pullback f^*P over a smooth mapping $f : M' \to M$ is again a principal fiber bundle.

10.5. Homogeneous spaces. Let G be a Lie group with Lie algebra $\mathfrak{g}$. Let K be a closed subgroup of G, then by theorem 5.5 K is a closed Lie subgroup whose Lie algebra will be denoted by $\mathfrak{k}$. By theorem 5.11 there is a unique structure of a smooth manifold on the quotient space G/K such that the projection $p : G \to G/K$ is a submersion, so by the implicit function theorem p admits local sections.

Theorem. *$(G, p, G/K, K)$ is a principal fiber bundle.*

Proof. The group multiplication of G restricts to a free right action $\mu : G \times K \to G$, whose orbits are exactly the fibers of p. By lemma 10.3 the result follows. □

For the convenience of the reader we discuss now the best known homogeneous spaces.

The group $SO(n)$ acts transitively on $S^{n-1} \subset \mathbb{R}^n$. The isotropy group of the 'north pole' $(1, 0, \dots, 0)$ is the subgroup

$$\begin{pmatrix} 1 & 0 \\ 0 & SO(n-1) \end{pmatrix}$$

which we identify with $SO(n-1)$. So $S^{n-1} = SO(n)/SO(n-1)$ and we have a principal fiber bundle $(SO(n), p, S^{n-1}, SO(n-1))$. Likewise
$(O(n), p, S^{n-1}, O(n-1))$,
$(SU(n), p, S^{2n-1}, SU(n-1))$,
$(U(n), p, S^{2n-1}, U(n-1))$, and
$(Sp(n), p, S^{4n-1}, Sp(n-1))$ are principal fiber bundles.

The *Grassmann manifold* $G(k, n; \mathbb{R})$ is the space of all k-planes containing 0 in $\mathbb{R}^n$. The group $O(n)$ acts transitively on it and the isotropy group of the k-plane $\mathbb{R}^k \times \{0\}$ is the subgroup

$$\begin{pmatrix} O(k) & 0 \\ 0 & O(n-k) \end{pmatrix},$$

therefore $G(k,n;\mathbb{R}) = O(n)/O(k) \times O(n-k)$ is a compact manifold and we get the principal fiber bundle $(O(n), p, G(k,n;\mathbb{R}), O(k) \times O(n-k))$. Likewise
$(SO(n), p, \tilde{G}(k,n;\mathbb{R}), SO(k) \times SO(n-k))$,
$(U(n), p, G(k,n;\mathbb{C}), U(k) \times U(n-k))$, and
$(Sp(n), p, G(k,n;\mathbb{H}), Sp(k) \times Sp(n-k))$ are principal fiber bundles.

The *Stiefel manifold* $V(k,n;\mathbb{R})$ is the space of all orthonormal k-frames in $\mathbb{R}^n$. Clearly the group $O(n)$ acts transitively on $V(k,n;\mathbb{R})$ and the isotropy subgroup of $(e_1,\dots,e_k)$ is $\mathbb{I}_k \times O(n-k)$, so $V(k,n;\mathbb{R}) = O(n)/O(n-k)$ is a compact manifold and $(O(n), p, V(k,n;\mathbb{R}), O(n-k))$ is a principal fiber bundle. But $O(k)$ also acts from the right on $V(k,n;\mathbb{R})$, its orbits are exactly the fibers of the projection $p : V(k,n;\mathbb{R}) \to G(k,n;\mathbb{R})$. So by lemma 10.3 we get a principal fiber bundle $(V(k,n,\mathbb{R}), p, G(k,n;\mathbb{R}), O(k))$. Indeed we have the following diagram where all arrows are projections of principal fiber bundles, and where the respective structure groups are written on the arrows:

$$\text{(a)}\qquad \begin{array}{ccc} O(n) & \xrightarrow{O(n-k)} & V(k,n;\mathbb{R}) \\ {\scriptstyle O(k)}\big\downarrow & & \big\downarrow{\scriptstyle O(k)} \\ V(n-k,n;\mathbb{R}) & \xrightarrow[O(n-k)]{} & G(k,n;\mathbb{R}) \end{array}$$

It is easy to see that $V(k,n)$ is also diffeomorphic to the space $\{ A \in L(\mathbb{R}^k,\mathbb{R}^n) : A^t.A = \mathbb{I}_k \}$, i.e. the space of all linear isometries $\mathbb{R}^k \to \mathbb{R}^n$. There are furthermore complex and quaternionic versions of the Stiefel manifolds.

Further examples will be given by means of jets in section 12.

10.6. Homomorphisms. Let $\chi : (P,p,M,G) \to (P',p',M',G)$ be a *principal fiber bundle homomorphism*, i.e. a smooth G-equivariant mapping $\chi : P \to P'$. Then obviously the diagram

$$\text{(a)}\qquad \begin{array}{ccc} P & \xrightarrow{\chi} & P' \\ {\scriptstyle p}\big\downarrow & & \big\downarrow{\scriptstyle p'} \\ M & \xrightarrow[\underline{\chi}]{} & M' \end{array}$$

commutes for a uniquely determined smooth mapping $\underline{\chi} : M \to M'$. For each $x \in M$ the mapping $\chi_x := \chi|P_x : P_x \to P'_{\underline{\chi}(x)}$ is G-equivariant and therefore a diffeomorphism, so diagram (a) is a pullback diagram. We denote by $\mathcal{PB}(G)$ the category of principal G-bundles and their homomorphisms.

But the most general notion of a homomorphism of principal bundles is the following. Let $\Phi : G \to G'$ be a homomorphism of Lie groups. $\chi : (P,p,M,G) \to (P',p',M',G')$ is called a *homomorphism over Φ of principal bundles*, if $\chi : P \to P'$ is smooth and $\chi(u.g) = \chi(u).\Phi(g)$ holds for all $u \in P$ and $g \in G$. Then χ is fiber respecting, so diagram (a) makes again sense, but it is no longer a pullback diagram in general. Thus we obtain the category $\mathcal{PB}$ of principal bundles and their homomorphisms.

If χ covers the identity on the base, it is called a *reduction of the structure group* G' *to* G for the principal bundle (P', p', M', G') — the name comes from the case, when Φ is the embedding of a subgroup.

By the universal property of the pullback any general homomorphism χ of principal fiber bundles over a group homomorphism can be written as the composition of a reduction of structure groups and a pullback homomorphism as follows, where we also indicate the structure groups:

$$\text{(b)}\qquad \begin{array}{ccccc} (P,G) & \longrightarrow & (\underline{\chi}^*P', G') & \longrightarrow & (P',G') \\ & \searrow^{p} & \downarrow & & \downarrow p' \\ & & M & \xrightarrow{\underline{\chi}} & M'. \end{array}$$

10.7. Associated bundles. Let (P, p, M, G) be a principal bundle and let $\ell : G \times S \to S$ be a left action of the structure group G on a manifold S. We consider the right action $R : (P \times S) \times G \to P \times S$, given by $R((u, s), g) = (u.g, g^{-1}.s)$.

Theorem. *In this situation we have:*

(1) *The space $P \times_G S$ of orbits of the action R carries a unique smooth manifold structure such that the quotient map $q : P \times S \to P \times_G S$ is a submersion.*

(2) *$(P \times_G S, \bar{p}, M, S, G)$ is a G-bundle in a canonical way, where $\bar{p} : P \times_G S \to M$ is given by*

$$\text{(a)}\qquad \begin{array}{ccc} P \times S & \xrightarrow{q} & P \times_G S \\ \downarrow pr_1 & & \bar{p} \downarrow \\ P & \xrightarrow{p} & M. \end{array}$$

In this diagram $q_u : \{u\} \times S \to (P \times_G S)_{p(u)}$ is a diffeomorphism for each $u \in P$.

(3) *$(P \times S, q, P \times_G S, G)$ is a principal fiber bundle with principal action R.*

(4) *If $(U_\alpha, \varphi_\alpha : P|U_\alpha \to U_\alpha \times G)$ is a principal bundle atlas with cocycle of transition functions $(\varphi_{\alpha\beta} : U_{\alpha\beta} \to G)$, then together with the left action $\ell : G \times S \to S$ this cocycle is also one for the G-bundle $(P \times_G S, \bar{p}, M, S, G)$.*

Notation. $(P \times_G S, \bar{p}, M, S, G)$ is called the *associated bundle* for the action $\ell : G \times S \to S$. We will also denote it by $P[S, \ell]$ or simply $P[S]$ and we will write p for $\bar{p}$ if no confusion is possible. We also define the smooth mapping $\tau^S = \tau : P \times_M P[S, \ell] \to S$ by $\tau(u_x, v_x) := q_{u_x}^{-1}(v_x)$. It satisfies $\tau(u, q(u, s)) = s$, $q(u_x, \tau(u_x, v_x)) = v_x$, and $\tau(u_x.g, v_x) = g^{-1}.\tau(u_x, v_x)$. In the special situation, where $S = G$ and the action is left translation, so that $P[G] = P$, this mapping coincides with $\tau = \tau^G$ considered in 10.2. We denote by $\{u, s\} \in P \times_G S$ the G-orbit through $(u, s) \in P \times S$.

Proof. In the setting of the diagram in (2) the mapping $p \circ pr_1$ is constant on the R-orbits, so $\bar{p}$ exists as a mapping. Let $(U_\alpha, \varphi_\alpha : P|U_\alpha \to U_\alpha \times G)$ be a principal bundle atlas with transition functions $(\varphi_{\alpha\beta} : U_{\alpha\beta} \to G)$. We define $\psi_\alpha^{-1} : U_\alpha \times S \to \bar{p}^{-1}(U_\alpha) \subset P \times_G S$ by $\psi_\alpha^{-1}(x,s) = q(\varphi_\alpha^{-1}(x,e),s)$, which is fiber respecting. For each orbit in $\bar{p}^{-1}(x) \subset P \times_G S$ there is exactly one $s \in S$ such that this orbit passes through $(\varphi_\alpha^{-1}(x,e),s)$, namely $s = \tau^G(u_x, \varphi_\alpha^{-1}(x,e))^{-1}.s'$ if (u_x, s') is the orbit, since the principal right action is free. Thus $\psi_\alpha^{-1}(x, \quad) : S \to \bar{p}^{-1}(x)$ is bijective. Furthermore

$$\begin{aligned}
\psi_\beta^{-1}(x,s) &= q(\varphi_\beta^{-1}(x,e),s) \\
&= q(\varphi_\alpha^{-1}(x,\varphi_{\alpha\beta}(x).e),s) = q(\varphi_\alpha^{-1}(x,e).\varphi_{\alpha\beta}(x),s) \\
&= q(\varphi_\alpha^{-1}(x,e),\varphi_{\alpha\beta}(x).s) = \psi_\alpha^{-1}(x,\varphi_{\alpha\beta}(x).s),
\end{aligned}$$

so $\psi_\alpha \psi_\beta^{-1}(x,s) = (x, \varphi_{\alpha\beta}(x).s)$ So (U_α, ψ_α) is a G-atlas for $P \times_G S$ and makes it into a smooth manifold and a G-bundle. The defining equation for ψ_α shows that q is smooth and a submersion and consequently the smooth structure on $P \times_G S$ is uniquely defined, and $\bar{p}$ is smooth by the universal properties of a submersion.

By the definition of ψ_α the diagram

$$\text{(b)} \qquad \begin{array}{ccc}
p^{-1}(U_\alpha) \times S & \xrightarrow{\varphi_\alpha \times \mathrm{Id}} & U_\alpha \times G \times S \\
\Big\downarrow q & & \Big\downarrow \mathrm{Id} \times \ell \\
\bar{p}^{-1}(U_\alpha) & \xrightarrow{\quad \psi_\alpha \quad} & U_\alpha \times S
\end{array}$$

commutes; since its lines are diffeomorphisms we conclude that $q_u : \{u\} \times S \to \bar{p}^{-1}(p(u))$ is a diffeomorphism. So (1), (2), and (4) are checked.
(3) follows directly from lemma 10.3. □

10.8. Corollary. *Let (E,p,M,S,G) be a G-bundle, specified by a cocycle of transition functions $(\varphi_{\alpha\beta})$ with values in G and a left action ℓ of G on S. Then from the cocycle of transition functions we may glue a unique principal bundle (P,p,M,G) such that $E = P[S,\ell]$.* □

This is the usual way a differential geometer thinks of an associated bundle. He is given a bundle E, a principal bundle P, and the G-bundle structure then is described with the help of the mappings τ and q. We remark that in standard differential geometric situations, the elements of the principal fiber bundle P play the role of certain frames for the individual fibers of each associated fiber bundle $E = P[S,\ell]$. Every frame $u \in P_x$ is interpreted as the above diffeomorphism $q_u : S \to E_x$.

10.9. Equivariant mappings and associated bundles.

1. Let (P,p,M,G) be a principal fiber bundle and consider two left actions of G, $\ell : G \times S \to S$ and $\ell' : G \times S' \to S'$. Let furthermore $f : S \to S'$ be a G-equivariant smooth mapping, so $f(g.s) = g.f(s)$ or $f \circ \ell_g = \ell'_g \circ f$. Then

$\mathrm{Id}_P \times f : P \times S \to P \times S'$ is equivariant for the actions $R : (P \times S) \times G \to P \times S$ and $R' : (P \times S') \times G \to P \times S'$ and is thus a homomorphism of principal bundles, so there is an induced mapping

$$\text{(a)} \qquad \begin{array}{ccc} P \times S & \xrightarrow{\mathrm{Id} \times f} & P \times S' \\ \Big\downarrow q & & \Big\downarrow q' \\ P \times_G S & \xrightarrow{\mathrm{Id} \times_G f} & P \times_G S', \end{array}$$

which is fiber respecting over M, and a homomorphism of G-bundles in the sense of the definition 10.10 below.

2. Let $\chi : (P, p, M, G) \to (P', p', M', G)$ be a homomorphism of principal fiber bundles as in 10.6. Furthermore we consider a smooth left action $\ell : G \times S \to S$. Then $\chi \times \mathrm{Id}_S : P \times S \to P' \times S$ is G-equivariant (a homomorphism of principal fiber bundles) and induces a mapping $\chi \times_G \mathrm{Id}_S : P \times_G S \to P' \times_G S$, which is fiber respecting over M, fiber wise a diffeomorphism, and again a homomorphism of G-bundles in the sense of definition 10.10 below.

3. Now we consider the situation of 1 and 2 at the same time. We have two associated bundles $P[S, \ell]$ and $P'[S', \ell']$. Let $\chi : (P, p, M, G) \to (P', p', M', G)$ be a homomorphism principal fiber bundles and let $f : S \to S'$ be an G-equivariant mapping. Then $\chi \times f : P \times S \to P' \times S'$ is clearly G-equivariant and therefore induces a mapping $\chi \times_G f : P[S, \ell] \to P'[S', \ell']$ which again is a homomorphism of G-bundles.

4. Let S be a point. Then $P[S] = P \times_G S = M$. Furthermore let $y \in S'$ be a fixed point of the action $\ell' : G \times S' \to S'$, then the inclusion $i : \{y\} \hookrightarrow S'$ is G-equivariant, thus $\mathrm{Id}_P \times i$ induces the mapping $\mathrm{Id}_P \times_G i : M = P[\{y\}] \to P[S']$, which is a global section of the associated bundle $P[S']$.

If the action of G on S is trivial, so $g.s = s$ for all $s \in S$, then the associated bundle is trivial: $P[S] = M \times S$. For a trivial principal fiber bundle any associated bundle is trivial.

10.10. Definition. In the situation of 10.9, a smooth fiber respecting mapping $\gamma : P[S, \ell] \to P'[S', \ell']$ covering a smooth mapping $\underline{\gamma} : M \to M'$ of the bases is called a *homomorphism of G-bundles*, if the following conditions are satisfied: P is isomorphic to the pullback $\underline{\gamma}^* P'$, and the local representations of γ in pullback-related fiber bundle atlases belonging to the two G-bundles are fiber wise G-equivariant.

Let us describe this in more detail now. Let $(U'_\alpha, \psi'_\alpha)$ be a G-atlas for $P'[S', \ell']$ with cocycle of transition functions $(\varphi'_{\alpha\beta})$, belonging to the principal fiber bundle atlas $(U'_\alpha, \varphi'_\alpha)$ of (P', p', M', G). Then the pullback-related principal fiber bundle atlas $(U_\alpha = \underline{\gamma}^{-1}(U'_\alpha), \varphi_\alpha)$ for $P = \underline{\gamma}^* P'$ as described in the proof of 9.5 has the cocycle of transition functions $(\varphi_{\alpha\beta} = \varphi'_{\alpha\beta} \circ \underline{\gamma})$; it induces the G-atlas (U_α, ψ_α) for $P[S, \ell]$. Then $(\psi'_\alpha \circ \gamma \circ \psi_\alpha^{-1})(x, s) = (\underline{\gamma}(x), \gamma_\alpha(x, s))$ and $\gamma_\alpha(x, \quad) : S \to S'$ is required to be G-equivariant for all α and all $x \in U_\alpha$.

Lemma. *Let $\gamma : P[S,\ell] \to P'[S',\ell']$ be a homomorphism of G-bundles as defined above. Then there is a homomorphism $\chi : (P,p,M,G) \to (P',p',M',G)$ of principal bundles and a G-equivariant mapping $f : S \to S'$ such that $\gamma = \chi \times_G f : P[S,\ell] \to P'[S',\ell']$.*

Proof. The homomorphism $\chi : (P,p,M,G) \to (P',p',M',G)$ of principal fiber bundles is already determined by the requirement that $P = \underline{\gamma}^* P'$, and we have $\underline{\gamma} = \underline{\chi}$. The G-equivariant mapping $f : S \to S'$ can be read off the following diagram which by the assumptions is seen to be well defined in the right column:

$$\text{(a)} \qquad \begin{array}{ccc} P \times_M P[S] & \xrightarrow{\tau^S} & S \\ {\scriptstyle \chi \times_M \gamma}\downarrow & & \downarrow{\scriptstyle f} \\ P' \times_{M'} P'[S'] & \xrightarrow{\tau^{S'}} & S' \end{array} \qquad \square$$

So a homomorphism of G-bundles is described by the whole triple $(\chi : P \to P', f : S \to S'$ (G-equivariant)$, \gamma : P[S] \to P'[S'])$, such that diagram (a) commutes.

10.11. Associated vector bundles. Let (P,p,M,G) be a principal fiber bundle, and consider a representation $\rho : G \to GL(V)$ of G on a finite dimensional vector space V. Then $P[V,\rho]$ is an associated fiber bundle with structure group G, but also with structure group $GL(V)$, for in the canonically associated fiber bundle atlas the transition functions have also values in $GL(V)$. So by section 6 $P[V,\rho]$ is a vector bundle.

Now let $\mathcal{F}$ be a covariant smooth functor from the category of finite dimensional vector spaces and linear mappings into itself, as considered in section 6.7. Then clearly $\mathcal{F} \circ \rho : G \to GL(V) \to GL(\mathcal{F}(V))$ is another representation of G and the associated bundle $P[\mathcal{F}(V), \mathcal{F} \circ \rho]$ coincides with the vector bundle $\mathcal{F}(P[V,\rho])$ constructed with the method of 6.7, but now it has an extra G-bundle structure. For contravariant functors $\mathcal{F}$ we have to consider the representation $\mathcal{F} \circ \rho \circ \nu$, similarly for bifunctors. In particular the bifunctor $L(V,W)$ may be applied to two different representations of two structure groups of two principal bundles over the same base M to construct a vector bundle $L(P[V,\rho], P'[V',\rho']) = (P \times_M P')[L(V,V'), L \circ ((\rho \circ \nu) \times \rho')]$.

If (E,p,M) is a vector bundle with n-dimensional fibers we may consider the open subset $GL(\mathbb{R}^n, E) \subset L(M \times \mathbb{R}^n, E)$, a fiber bundle over the base M, whose fiber over $x \in M$ is the space $GL(\mathbb{R}^n, E_x)$ of all invertible linear mappings. Composition from the right by elements of $GL(n)$ gives a free right action on $GL(\mathbb{R}^n, E)$ whose orbits are exactly the fibers, so by lemma 10.3 we have a principal fiber bundle $(GL(\mathbb{R}^n, E), p, M, GL(n))$. The associated bundle $GL(\mathbb{R}^n, E)[\mathbb{R}^n]$ for the standard representation of $GL(n)$ on $\mathbb{R}^n$ is isomorphic to the vector bundle (E,p,M) we started with, for the evaluation mapping $ev : GL(\mathbb{R}^n, E) \times \mathbb{R}^n \to E$ is invariant under the right action R of $GL(n)$, and locally in the image there are smooth sections to it, so it factors to a fiber linear diffeomorphism $GL(\mathbb{R}^n, E)[\mathbb{R}^n] = GL(\mathbb{R}^n, E) \times_{GL(n)} \mathbb{R}^n \to E$. The principal bundle $GL(\mathbb{R}^n, E)$ is called the *linear frame bundle of* E. Note that local sections of $GL(\mathbb{R}^n, E)$ are exactly the local frame fields of the vector bundle E as discussed in 6.5.

To illustrate the notion of reduction of structure group, we consider now a vector bundle $(E, p, M, \mathbb{R}^n)$ equipped with a *Riemannian metric* g, that is a section $g \in C^\infty(S^2E^*)$ such that g_x is a positive definite inner product on E_x for each $x \in M$. Any vector bundle admits Riemannian metrics: local existence is clear and we may glue with the help of a partition of unity on M, since the positive definite sections form an open convex subset. Now let $s' = (s'_1, \dots, s'_n) \in C^\infty(GL(\mathbb{R}^n, E)|U)$ be a local frame field of the bundle E over $U \subset M$. Now we may apply the Gram-Schmidt orthonormalization procedure to the basis $(s_1(x), \dots, s_n(x))$ of E_x for each $x \in U$. Since this procedure is smooth (even real analytic), we obtain a frame field $s = (s_1, \dots, s_n)$ of E over U which is orthonormal with respect to g. We call it an *orthonormal frame field.* Now let (U_α) be an open cover of M with orthonormal frame fields $s^\alpha = (s^\alpha_1, \dots, s^\alpha_n)$, where s^α is defined on U_α. We consider the vector bundle charts $(U_\alpha, \psi_\alpha : E|U_\alpha \to U_\alpha \times \mathbb{R}^n)$ given by the orthonormal frame fields: $\psi_\alpha^{-1}(x, v^1, \dots, v^n) = \sum s^\alpha_i(x).v^i =: s^\alpha(x).v$. For $x \in U_{\alpha\beta}$ we have $s^\alpha_i(x) = \sum s^\beta_j(x).g_{\beta\alpha}{}^j_i(x)$ for C^∞-functions $g_{\alpha\beta}{}^j_i : U_{\alpha\beta} \to \mathbb{R}$. Since $s^\alpha(x)$ and $s^\beta(x)$ are both orthonormal bases of E_x, the matrix $g_{\alpha\beta}(x) = (g_{\alpha\beta}{}^j_i(x))$ is an element of $O(n)$. We write $s^\alpha = s^\beta.g_{\beta\alpha}$ for short. Then we have $\psi_\beta^{-1}(x, v) = s^\beta(x).v = s^\alpha(x).g_{\alpha\beta}(x).v = \psi_\alpha^{-1}(x, g_{\alpha\beta}(x).v)$ and consequently $\psi_\alpha\psi_\beta^{-1}(x, v) = (x, g_{\alpha\beta}(x).v)$. So the $(g_{\alpha\beta} : U_{\alpha\beta} \to O(n))$ are the cocycle of transition functions for the vector bundle atlas (U_α, ψ_α). So we have constructed an $O(n)$-structure on E. The corresponding principal fiber bundle will be denoted by $O(\mathbb{R}^n, (E, g))$; it is usually called the *orthonormal frame bundle* of E. It is derived from the linear frame bundle $GL(\mathbb{R}^n, E)$ by reduction of the structure group from $GL(n)$ to $O(n)$. The phenomenon discussed here plays a prominent role in the theory of *classifying spaces.*

10.12. Sections of associated bundles. Let (P, p, M, G) be a principal fiber bundle and $\ell : G \times S \to S$ a left action. Let $C^\infty(P, S)^G$ denote the space of all smooth mappings $f : P \to S$ which are G-equivariant in the sense that $f(u.g) = g^{-1}.f(u)$ holds for $g \in G$ and $u \in P$.

Theorem. *The sections of the associated bundle $P[S, \ell]$ correspond exactly to the G-equivariant mappings $P \to S$; we have a bijection $C^\infty(P, S)^G \cong C^\infty(P[S])$.*

Proof. If $f \in C^\infty(P, S)^G$ we get $s_f \in C^\infty(P[S])$ by the following diagram:

$$\text{(a)} \qquad \begin{array}{ccc} P & \xrightarrow{(\mathrm{Id}, f)} & P \times S \\ {\scriptstyle p}\downarrow & & \downarrow{\scriptstyle q} \\ M & \xrightarrow{s_f} & P[S] \end{array}$$

which exists by 10.9 since $\mathrm{graph}(f) = (\mathrm{Id}, f) : P \to P \times S$ is G-equivariant: $(\mathrm{Id}, f)(u.g) = (u.g, f(u.g)) = (u.g, g^{-1}.f(u)) = ((\mathrm{Id}, f)(u)).g$.

If conversely $s \in C^\infty(P[S])$ we define $f_s \in C^\infty(P,S)^G$ by $f_s := \tau \circ (\mathrm{Id}_P \times_M s) : P = P \times_M M \to P \times_M P[S] \to S$. This is G-equivariant since $f_s(u_x.g) = \tau(u_x.g, s(x)) = g^{-1}.\tau(u_x, s(x)) = g^{-1}.f_s(u_x)$ by 10.7. The two constructions are inverse to each other since we have $f_{s(f)}(u) = \tau(u, s_f(p(u))) = \tau(u, q(u, f(u))) = f(u)$ and $s_{f(s)}(p(u)) = q(u, f_s(u)) = q(u, \tau(u, s(p(u))) = s(p(u))$. □

The G-mapping $f_s : P \to S$ determined by a section s of $P[S]$ will be called the *frame form* of the section s.

10.13. Theorem. *Consider a principal fiber bundle (P, p, M, G) and a closed subgroup K of G. Then the reductions of structure group from G to K correspond bijectively to the global sections of the associated bundle $P[G/K, \bar\lambda]$ in a canonical way, where $\bar\lambda : G \times G/K \to G/K$ is the left action on the homogeneous space from 5.11.*

Proof. By theorem 10.12 the section $s \in C^\infty(P[G/K])$ corresponds to $f_s \in C^\infty(P, G/K)^G$, which is a surjective submersion since the action $\bar\lambda : G \times G/K \to G/K$ is transitive. Thus $P_s := f_s^{-1}(\bar e)$ is a submanifold of P which is stable under the right action of K on P. Furthermore the K-orbits are exactly the fibers of the mapping $p : P_s \to M$, so by lemma 10.3 we get a principal fiber bundle (P_s, p, M, K). The embedding $P_s \hookrightarrow P$ is then a reduction of structure groups as required.

If conversely we have a principal fiber bundle (P', p', M, K) and a reduction of structure groups $\chi : P' \to P$, then χ is an embedding covering the identity of M and is K-equivariant, so we may view P' as a sub fiber bundle of P which is stable under the right action of K. Now we consider the mapping $\tau : P \times_M P \to G$ from 10.2 and restrict it to $P \times_M P'$. Since we have $\tau(u_x, v_x.k) = \tau(u_x, v_x).k$ for $k \in K$ this restriction induces $f : P \to G/K$ by

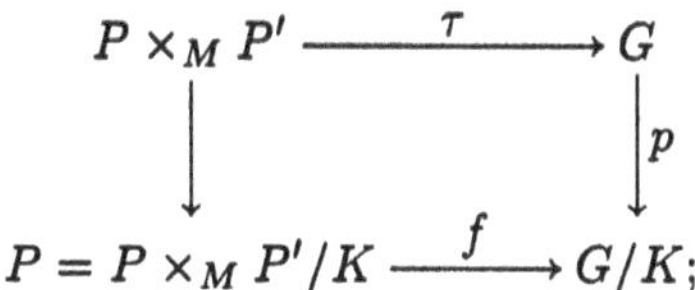

$$\begin{array}{ccc} P \times_M P' & \xrightarrow{\tau} & G \\ \downarrow & & \downarrow p \\ P = P \times_M P'/K & \xrightarrow{f} & G/K; \end{array}$$

and from $\tau(u_x.g, v_x) = g^{-1}.\tau(u_x, v_x)$ it follows that f is G-equivariant as required. Finally $f^{-1}(\bar e) = \{u \in P : \tau(u, P'_{p(u)}) \subseteq K\} = P'$, so the two constructions are inverse to each other. □

10.14. The bundle of gauges. If (P, p, M, G) is a principal fiber bundle we denote by $\mathrm{Aut}(P)$ the group of all G-equivariant diffeomorphisms $\chi : P \to P$. Then $p \circ \chi = \underline{\chi} \circ p$ for a unique diffeomorphism $\underline{\chi}$ of M, so there is a group homomorphism from $\mathrm{Aut}(P)$ into the group $\mathrm{Diff}(M)$ of all diffeomorphisms of M. The kernel of this homomorphism is called $\mathrm{Gau}(P)$, the group of *gauge transformations*. So $\mathrm{Gau}(P)$ is the space of all $\chi : P \to P$ which satisfy $p \circ \chi = p$ and $\chi(u.g) = \chi(u).g$.

Theorem. *The group $\mathrm{Gau}(P)$ of gauge transformations is equal to the space $C^\infty(P, (G, \mathrm{conj}))^G \cong C^\infty(P[G, \mathrm{conj}])$.*

Proof. We use again the mapping $\tau : P \times_M P \to G$ from 10.2. For $\chi \in \operatorname{Gau}(P)$ we define $f_\chi \in C^\infty(P,(G,\operatorname{conj}))^G$ by $f_\chi := \tau \circ (\operatorname{Id},\chi)$. Then $f_\chi(u.g) = \tau(u.g,\chi(u.g)) = g^{-1}.\tau(u,\chi(u)).g = \operatorname{conj}_{g^{-1}} f_\chi(u)$, so f_χ is indeed G-equivariant.

If conversely $f \in C^\infty(P,(G,\operatorname{conj}))^G$ is given, we define $\chi_f : P \to P$ by $\chi_f(u) := u.f(u)$. It is easy to check that χ_f is indeed in $\operatorname{Gau}(P)$ and that the two constructions are inverse to each other. □

10.15. The tangent bundles of homogeneous spaces. Let G be a Lie group and K a closed subgroup, with Lie algebras $\mathfrak{g}$ and $\mathfrak{k}$, respectively. We recall the mapping $\operatorname{Ad}_G : G \to \operatorname{Aut}_{\text{Lie}}(\mathfrak{g})$ from 4.24 and put $\operatorname{Ad}_{G,K} := \operatorname{Ad}_G|K : K \to \operatorname{Aut}_{\text{Lie}}(\mathfrak{g})$. For $X \in \mathfrak{k}$ and $k \in K$ we have $\operatorname{Ad}_{G,K}(k)X = \operatorname{Ad}_G(k)X = \operatorname{Ad}_K(k)X \in \mathfrak{k}$, so $\mathfrak{k}$ is an invariant subspace for the representation $\operatorname{Ad}_{G,K}$ of K in $\mathfrak{g}$, and we have the factor representation $\operatorname{Ad}^\perp : K \to GL(\mathfrak{g}/\mathfrak{k})$. Then

(a) $$0 \to \mathfrak{k} \to \mathfrak{g} \to \mathfrak{g}/\mathfrak{k} \to 0$$

is short exact and K-equivariant.

Now we consider the principal fiber bundle $(G,p,G/K,K)$ and the associated vector bundles $G[\mathfrak{g}/\mathfrak{k},\operatorname{Ad}^\perp]$ and $G[\mathfrak{k},\operatorname{Ad}_K]$.

Theorem. *In these circumstances we have*
$$T(G/K) = G[\mathfrak{g}/\mathfrak{k},\operatorname{Ad}^\perp] = (G \times_K \mathfrak{g}/\mathfrak{k}, p, G/K, \mathfrak{g}/\mathfrak{k}).$$
The left action $g \mapsto T(\bar\lambda_g)$ of G on $T(G/K)$ corresponds to the canonical left action of G on $G \times_K \mathfrak{g}/\mathfrak{k}$. Furthermore $G[\mathfrak{g}/\mathfrak{k},\operatorname{Ad}^\perp] \oplus G[\mathfrak{k},\operatorname{Ad}_K]$ is a trivial vector bundle.

Proof. For $p : G \to G/K$ we consider the tangent mapping $T_e p : \mathfrak{g} \to T_{\bar e}(G/K)$ which is linear and surjective and induces a linear isomorphism $\overline{T_e p} : \mathfrak{g}/\mathfrak{k} \to T_{\bar e}(G/K)$. For $k \in K$ we have $p \circ \operatorname{conj}_k = p \circ \lambda_k \circ \rho_{k^{-1}} = \bar\lambda_k \circ p$ and consequently $T_e p \circ \operatorname{Ad}_{G,K}(k) = T_e p \circ T_e(\operatorname{conj}_k) = T_{\bar e}\bar\lambda_k \circ T_e p$. Thus the isomorphism $\overline{T_e p} : \mathfrak{g}/\mathfrak{k} \to T_{\bar e}(G/K)$ is K-equivariant for the representations $\operatorname{Ad}^\perp$ and $T_{\bar e}\bar\lambda : k \mapsto T_{\bar e}\bar\lambda_k$.

Now we consider the associated vector bundle $G[T_{\bar e}(G/K), T_{\bar e}\bar\lambda] = (G \times_K T_{\bar e}(G/K), p, G/K, T_{\bar e}(G/K))$, which is isomorphic to $G[\mathfrak{g}/\mathfrak{k},\operatorname{Ad}^\perp]$, since the representation spaces are isomorphic. The mapping $T_2\bar\lambda : G \times T_{\bar e}(G/K) \to T(G/K)$ (where T_2 is the second partial tangent functor) is K-invariant and therefore induces a mapping ψ as in the following diagram:

(b)
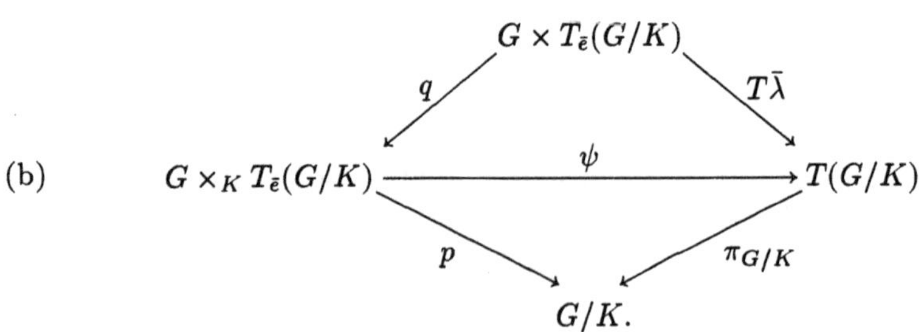

This mapping ψ is an isomorphism of vector bundles.

It remains to show the last assertion. The short exact sequence (a) induces a sequence of vector bundles over G/K:

$$G/K \times 0 \to G[\mathfrak{k}, \mathrm{Ad}_K] \to G[\mathfrak{g}, \mathrm{Ad}_{G,K}] \to G[\mathfrak{g}/\mathfrak{k}, \mathrm{Ad}^{\perp}] \to G/K \times 0$$

This sequence splits fiberwise thus also locally over G/K, so $G[\mathfrak{g}/\mathfrak{k}, \mathrm{Ad}^{\perp}] \oplus G[\mathfrak{k}, \mathrm{Ad}_K] \cong G[\mathfrak{g}, \mathrm{Ad}_{G,K}]$ and it remains to show that $G[\mathfrak{g}, \mathrm{Ad}_{G,K}]$ is a trivial vector bundle. Let $\varphi : G \times \mathfrak{g} \to G \times \mathfrak{g}$ be given by $\varphi(g, X) = (g, \mathrm{Ad}_G(g)X)$. Then for $k \in K$ we have $\varphi((g, X).k) = \varphi(gk, \mathrm{Ad}_{G,K}(k^{-1})X) = (gk, \mathrm{Ad}_G(g.k.k^{-1})X) = (gk, \mathrm{Ad}_G(g)X)$. So φ is K-equivariant from the 'joint' K-action to the 'on the left' K-action and therefore induces a mapping $\bar{\varphi}$ as in the diagram:

(c)

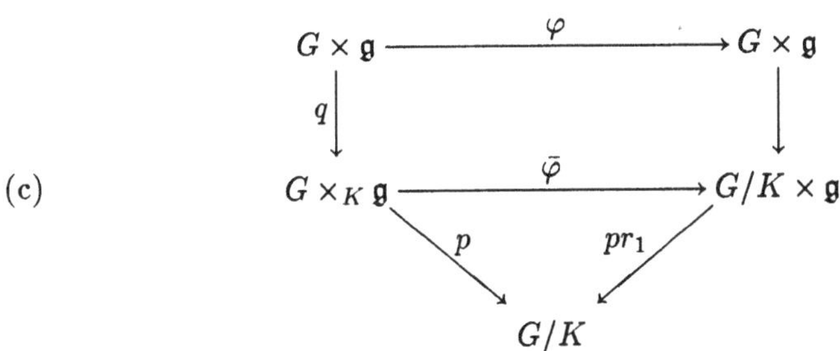

The map $\bar{\varphi}$ is a vector bundle isomorphism. □

10.16. Tangent bundles of Grassmann manifolds. From 10.5 we know that $(V(k,n) = O(n)/O(n-k), p, G(k,n), O(k))$ is a principal fiber bundle. Using the standard representation of $O(k)$ we consider the associated vector bundle $(E_k := V(k,n)[\mathbb{R}^k], p, G(k,n))$. It is called the *universal vector bundle* over $G(k,n)$. Recall from 10.5 the description of $V(k,n)$ as the space of all linear isometries $\mathbb{R}^k \to \mathbb{R}^n$; we get from it the evaluation mapping $ev : V(k,n) \times \mathbb{R}^k \to \mathbb{R}^n$. The mapping (p, ev) in the diagram

(a)

$$\begin{array}{ccc} V(k,n) \times \mathbb{R}^k & & \\ \downarrow q & \searrow^{(p,ev)} & \\ V(k,n) \times_{O(k)} \mathbb{R}^k & \xrightarrow{\psi} & G(k,n) \times \mathbb{R}^n \end{array}$$

is $O(k)$-invariant for the action R and factors therefore to an embedding of vector bundles $\psi : E_k \to G(k,n) \times \mathbb{R}^n$. So the fiber $(E_k)_W$ over the k-plane W in $\mathbb{R}^n$ is just the linear subspace W. Note finally that the fiber wise orthogonal complement ${E_k}^{\perp}$ of E_k in the trivial vector bundle $G(k,n) \times \mathbb{R}^n$ with its standard Riemannian metric is isomorphic to the universal vector bundle E_{n-k} over $G(n-k,n)$, where the isomorphism covers the diffeomorphism $G(k,n) \to G(n-k,n)$ given also by the orthogonal complement mapping.

Corollary. *The tangent bundle of the Grassmann manifold is*

$$TG(k,n) \cong L(E_k, {E_k}^{\perp}).$$

Proof. We have $G(k,n) = O(n)/(O(k)\times O(n-k))$, so by theorem 10.15 we get

$$TG(k,n) = O(n) \underset{O(k)\times O(n-k)}{\times} (\mathfrak{so}(n)/(\mathfrak{so}(k)\times\mathfrak{so}(n-k))).$$

On the other hand we have $V(k,n) = O(n)/O(n-k)$ and the right action of $O(k)$ commutes with the right action of $O(n-k)$ on $O(n)$, therefore

$$V(k,n)[\mathbb{R}^k] = (O(n)/O(n-k)) \underset{O(k)}{\times} \mathbb{R}^k = O(n) \underset{O(k)\times O(n-k)}{\times} \mathbb{R}^k,$$

where $O(n-k)$ acts trivially on $\mathbb{R}^k$. Finally

$$\begin{aligned} L(E_k, E_k{}^\perp) &= L\left(O(n) \underset{O(k)\times O(n-k)}{\times} \mathbb{R}^k, O(n) \underset{O(k)\times O(n-k)}{\times} \mathbb{R}^{n-k}\right) \\ &= O(n) \underset{O(k)\times O(n-k)}{\times} L(\mathbb{R}^k, \mathbb{R}^{n-k}), \end{aligned}$$

where the left action of $O(k)\times O(n-k)$ on $L(\mathbb{R}^k,\mathbb{R}^{n-k})$ is given by $(A,B)(C) = B.C.A^{-1}$. Finally we have an $O(k)\times O(n-k)$ - equivariant linear isomorphism $L(\mathbb{R}^k,\mathbb{R}^{n-k}) \to \mathfrak{so}(n)/(\mathfrak{so}(k)\times\mathfrak{so}(n-k))$, as follows:

$$\mathfrak{so}(n)/(\mathfrak{so}(k)\times\mathfrak{so}(n-k)) =$$

$$\frac{(\text{skew})}{\begin{pmatrix}\text{skew} & 0\\ 0 & \text{skew}\end{pmatrix}} = \left\{\begin{pmatrix} 0 & A \\ -A^t & 0\end{pmatrix} : \quad A\in L(\mathbb{R}^k,\mathbb{R}^{n-k})\right\} \qquad \square$$

10.17. The tangent group of a Lie group. Let G be a Lie group with Lie algebra $\mathfrak{g}$. We will use the notation from 4.1. First note that TG is also a Lie group with multiplication $T\mu$ and inversion $T\nu$, given by (see 4.2) $T_{(a,b)}\mu.(\xi_a,\eta_b) = T_a(\rho_b).\xi_a + T_b(\lambda_a).\eta_b$ and $T_a\nu.\xi_a = -T_e(\lambda_{a^{-1}}).T_a(\rho_{a^{-1}}).\xi_a$.

Lemma. *Via the isomomorphism $T\rho : \mathfrak{g}\times G \to TG$, $T\rho.(X,g) = T_e(\rho_g).X$, the group structure on TG looks as follows: $(X,a).(Y,b) = (X + \operatorname{Ad}(a)Y, a.b)$ and $(X,a)^{-1} = (-\operatorname{Ad}(a^{-1})X, a^{-1})$. So TG is isomorphic to the semidirect product $\mathfrak{g}\ltimes G$, see 5.16.*

Proof. $T_{(a,b)}\mu.(T\rho_a.X, T\rho_b.Y) = T\rho_b.T\rho_a.X + T\lambda_a.T\rho_b.Y =$
$\quad = T\rho_{ab}.X + T\rho_b.T\rho_a.T\rho_{a^{-1}}.T\lambda_a.Y = T\rho_{ab}(X + \operatorname{Ad}(a)Y).$
$T_a\nu.T\rho_a.X = -T\rho_{a^{-1}}.T\lambda_{a^{-1}}.T\rho_a.X = -T\rho_{a^{-1}}.\operatorname{Ad}(a^{-1})X.$ $\square$

Remark. In the left trivialisation $T\lambda : G\times\mathfrak{g}\to TG$, $T\lambda.(g,X) = T_e(\lambda_g).X$, the semidirect product structure is: $(a,X).(b,Y) = (ab, \operatorname{Ad}(b^{-1})X + Y)$ and $(a,X)^{-1} = (a^{-1}, -\operatorname{Ad}(a)X)$.

Lemma 10.17 is a special case of 37.16 and also 38.10 below.

10.18. Tangent bundles and vertical bundles. Let (E,p,M,S) be a fiber bundle. The subbundle $VE = \{\xi\in TE : \quad Tp.\xi = 0\}$ of TE is called the *vertical bundle* and is denoted by (VE, π_E, E).

Theorem. *Let (P, p, M, G) be a principal fiber bundle with principal right action $r : P \times G \to P$. Let $\ell : G \times S \to S$ be a left action. Then the following assertions hold:*

(1) *(TP, Tp, TM, TG) is again a principal fiber bundle with principal right action $Tr : TP \times TG \to TP$.*
(2) *The vertical bundle $(VP, \pi, P, \mathfrak{g})$ of the principal bundle is trivial as a vector bundle over P: $VP \cong P \times \mathfrak{g}$.*
(3) *The vertical bundle of the principal bundle as bundle over M is again a principal bundle: $(VP, p \circ \pi, M, TG)$.*
(4) *The tangent bundle of the associated bundle $P[S, \ell]$ is given by $T(P[S, \ell]) = TP[TS, T\ell]$.*
(5) *The vertical bundle of the associated bundle $P[S, \ell]$ is given by $V(P[S, \ell]) = P[TS, T_2\ell] = P \times_G TS$, where T_2 is the second partial tangent functor.*

Proof. Let $(U_\alpha, \varphi_\alpha : P|U_\alpha \to U_\alpha \times G)$ be a principal fiber bundle atlas with cocycle of transition functions $(\varphi_{\alpha\beta} : U_{\alpha\beta} \to G)$. Since T is a functor which respects products, $(TU_\alpha, T\varphi_\alpha : TP|TU_\alpha \to TU_\alpha \times TG)$ is again a principal fiber bundle atlas with cocycle of transition functions $(T\varphi_{\alpha\beta} : TU_{\alpha\beta} \to TG)$, describing the principal fiber bundle (TP, Tp, TM, TG). The assertion about the principal action is obvious. So (1) follows. For completeness sake we include here the transition formula for this atlas in the right trivialization of TG:

$$T(\varphi_\alpha \circ \varphi_\beta^{-1})(\xi_x, T_e(\rho_g).X) = (\xi_x, T_e(\rho_{\varphi_{\alpha\beta}(x).g}).(\delta\varphi_{\alpha\beta}(\xi_x) + \mathrm{Ad}(\varphi_{\alpha\beta}(x))X)),$$

where $\delta\varphi_{\alpha\beta} \in \Omega^1(U_{\alpha\beta}; \mathfrak{g})$ is the right logarithmic derivative of $\varphi_{\alpha\beta}$, see 4.26.
(2) The mapping $(u, X) \mapsto T_e(r_u).X = T_{(u,e)}r.(0_u, X)$ is a vector bundle isomorphism $P \times \mathfrak{g} \to VP$ over P.
(3) Obviously $Tr : TP \times TG \to TP$ is a free right action which acts transitive on the fibers of $Tp : TP \to TM$. Since $VP = (Tp)^{-1}(0_M)$, the bundle $VP \to M$ is isomorphic to $TP|0_M$ and Tr restricts to a free right action, which is transitive on the fibers, so by lemma 10.3 the result follows.
(4) The transition functions of the fiber bundle $P[S, \ell]$ are given by the expression $\ell \circ (\varphi_{\alpha\beta} \times \mathrm{Id}_S) : U_{\alpha\beta} \times S \to G \times S \to S$. Then the transition functions of $T(P[S, \ell])$ are $T(\ell \circ (\varphi_{\alpha\beta} \times \mathrm{Id}_S)) = T\ell \circ (T\varphi_{\alpha\beta} \times \mathrm{Id}_{TS}) : TU_{\alpha\beta} \times TS \to TG \times TS \to TS$, from which the result follows.
(5) Vertical vectors in $T(P[S, \ell])$ have local representations $(0_x, \eta_s) \in TU_{\alpha\beta} \times TS$. Under the transition functions of $T(P[S, \ell])$ they transform as $T(\ell \circ (\varphi_{\alpha\beta} \times \mathrm{Id}_S)).(0_x, \eta_s) = T\ell.(0_{\varphi_{\alpha\beta}(x)}, \eta_s) = T(\ell_{\varphi_{\alpha\beta}(x)}).\eta_s = T_2\ell.(\varphi_{\alpha\beta}(x), \eta_s)$ and this implies the result. □

11. Principal and induced connections

11.1. Principal connections. Let (P, p, M, G) be a principal fiber bundle. Recall from 9.3 that a (general) connection on P is a fiber projection $\Phi : TP \to$

VP, viewed as a 1-form in $\Omega^1(P;TP)$. Such a connection Φ is called a *principal connection* if it is G-equivariant for the principal right action $r : P \times G \to P$, so that $T(r^g).\Phi = \Phi.T(r^g)$, i.e. Φ is r^g-related to itself, or $(r^g)^*\Phi = \Phi$ in the sense of 8.16, for all $g \in G$. By theorem 8.15.7 the curvature $R = \frac{1}{2}.[\Phi, \Phi]$ is then also r^g-related to itself for all $g \in G$.

Recall from 10.18.2 that the vertical bundle of P is trivialized as a vector bundle over P by the principal action. So we have $\omega(X_u) := T_e(r_u)^{-1}.\Phi(X_u) \in \mathfrak{g}$ and in this way we get a $\mathfrak{g}$-valued 1-form $\omega \in \Omega^1(P;\mathfrak{g})$, which is called the *(Lie algebra valued) connection form* of the connection Φ. Recall from 5.13 the fundamental vector field mapping $\zeta : \mathfrak{g} \to \mathfrak{X}(P)$ for the principal right action. The defining equation for ω can be written also as $\Phi(X_u) = \zeta_{\omega(X_u)}(u)$.

Lemma. *If $\Phi \in \Omega^1(P;VP)$ is a principal connection on the principal fiber bundle (P,p,M,G) then the connection form has the following properties:*

1. *ω reproduces the generators of fundamental vector fields, so we have $\omega(\zeta_X(u)) = X$ for all $X \in \mathfrak{g}$.*
2. *ω is G-equivariant, so $((r^g)^*\omega)(X_u) = \omega(T_u(r^g).X_u) = \mathrm{Ad}(g^{-1}).\omega(X_u)$ for all $g \in G$ and $X_u \in T_uP$.*
3. *For the Lie derivative we have $\mathcal{L}_{\zeta_X}\omega = -\operatorname{ad}(X).\omega$.*

Conversely a 1-form $\omega \in \Omega^1(P,\mathfrak{g})$ satisfying (1) defines a connection Φ on P by $\Phi(X_u) = T_e(r_u).\omega(X_u)$, which is a principal connection if and only if (2) is satisfied.

Proof. (1) $T_e(r_u).\omega(\zeta_X(u)) = \Phi(\zeta_X(u)) = \zeta_X(u) = T_e(r_u).X$. Since $T_e(r_u) : \mathfrak{g} \to V_uP$ is an isomorphism, the result follows.

(2) Both directions follow from

$$\begin{aligned} T_e(r_{ug}).\omega(T_u(r^g).X_u) &= \zeta_{\omega(T_u(r^g).X_u)}(ug) = \Phi(T_u(r^g).X_u) \\ T_e(r_{ug}).\mathrm{Ad}(g^{-1}).\omega(X_u) &= \zeta_{\mathrm{Ad}(g^{-1}).\omega(X_u)}(ug) = \\ &= T_u(r^g).\zeta_{\omega(X_u)}(u) = T_u(r^g).\Phi(X_u). \end{aligned}$$

(3) is a consequence of (2). □

11.2. Curvature. Let Φ be a principal connection on the principal fiber bundle (P,p,M,G) with connection form $\omega \in \Omega^1(P;\mathfrak{g})$. We already noted in 11.1 that the curvature $R = \frac{1}{2}[\Phi, \Phi]$ is then also G-equivariant, $(r^g)^*R = R$ for all $g \in G$. Since R has vertical values we may again define a $\mathfrak{g}$-valued 2-form $\Omega \in \Omega^2(P;\mathfrak{g})$ by $\Omega(X_u, Y_u) := -T_e(r_u)^{-1}.R(X_u, Y_u)$, which is called the *(Lie algebra-valued) curvature form* of the connection. We also have $R(X_u, Y_u) = -\zeta_{\Omega(X_u,Y_u)}(u)$. We take the negative sign here to get the usual curvature form as in [Kobayashi-Nomizu I, 63].

We equip the space $\Omega(P;\mathfrak{g})$ of all $\mathfrak{g}$-valued forms on P in a canonical way with the structure of a graded Lie algebra by

$$\begin{aligned} &[\Psi, \Theta]_\wedge(X_1, \dots, X_{p+q}) = \\ &\qquad = \frac{1}{p!\,q!}\sum_\sigma \operatorname{sign}\sigma\, [\Psi(X_{\sigma 1}, \dots, X_{\sigma p}), \Theta(X_{\sigma(p+1)}, \dots, X_{\sigma(p+q)})]_\mathfrak{g} \end{aligned}$$

or equivalently by $[\psi\otimes X,\theta\otimes Y]_\wedge := \psi\wedge\theta\otimes[X,Y]_{\mathfrak g}$. From the latter description it is clear that $d[\Psi,\Theta]_\wedge = [d\Psi,\Theta]_\wedge + (-1)^{\deg\Psi}[\Psi,d\Theta]_\wedge$. In particular for $\omega\in\Omega^1(P;\mathfrak g)$ we have $[\omega,\omega]_\wedge(X,Y) = 2[\omega(X),\omega(Y)]_{\mathfrak g}$.

Theorem. *The curvature form Ω of a principal connection with connection form ω has the following properties:*

(1) *Ω is horizontal, i.e. it kills vertical vectors.*
(2) *Ω is G-equivariant in the following sense: $(r^g)^*\Omega = \mathrm{Ad}(g^{-1}).\Omega$. Consequently $\mathcal L_{\zeta_X}\Omega = -\,\mathrm{ad}(X).\Omega$.*
(3) *The Maurer-Cartan formula holds: $\Omega = d\omega + \frac12[\omega,\omega]_\wedge$.*

Proof. (1) is true for R by 9.4. For (2) we compute as follows:

$$\begin{aligned}
T_e(r_{ug}).((r^g)^*\Omega)(X_u,Y_u) &= T_e(r_{ug}).\Omega(T_u(r^g).X_u, T_u(r^g).Y_u) = \\
&= -R_{ug}(T_u(r^g).X_u, T_u(r^g).Y_u) = -T_u(r^g).((r^g)^*R)(X_u,Y_u) = \\
&= -T_u(r^g).R(X_u,Y_u) = T_u(r^g).\zeta_{\Omega(X_u,Y_u)}(u) = \\
&= \zeta_{\mathrm{Ad}(g^{-1}).\Omega(X_u,Y_u)}(ug) = \\
&= T_e(r_{ug}).\,\mathrm{Ad}(g^{-1}).\Omega(X_u,Y_u), \quad \text{by 5.13.}
\end{aligned}$$

(3) For $X\in\mathfrak g$ we have $i_{\zeta_X}R = 0$ by (1), and using 11.1.(3) we get

$$\begin{aligned}
i_{\zeta_X}(d\omega + \frac12[\omega,\omega]_\wedge) &= i_{\zeta_X}d\omega + \frac12[i_{\zeta_X}\omega,\omega]_\wedge - \frac12[\omega, i_{\zeta_X}\omega]_\wedge = \\
&= \mathcal L_{\zeta_X}\omega + [X,\omega]_\wedge = -ad(X)\omega + ad(X)\omega = 0.
\end{aligned}$$

So the formula holds for vertical vectors, and for horizontal vector fields $X,Y\in C^\infty(H(P))$ we have

$$\begin{aligned}
&R(X,Y) = \Phi[X-\Phi X, Y-\Phi Y] = \Phi[X,Y] = \zeta_{\omega([X,Y])} \\
&(d\omega + \frac12[\omega,\omega])(X,Y) = X\omega(Y) - Y\omega(X) - \omega([X,Y]) = -\omega([X,Y]). \quad \square
\end{aligned}$$

11.3. Lemma. *Any principal fiber bundle (P,p,M,G) admits principal connections.*

Proof. Let $(U_\alpha,\varphi_\alpha : P|U_\alpha \to U_\alpha\times G)_\alpha$ be a principal fiber bundle atlas. Let us define $\gamma_\alpha(T\varphi_\alpha^{-1}(\xi_x, T_e\lambda_g.X)) := X$ for $\xi_x\in T_xU_\alpha$ and $X\in\mathfrak g$. An easy computation involving lemma 5.13 shows that $\gamma_\alpha\in\Omega^1(P|U_\alpha;\mathfrak g)$ satisfies the requirements of lemma 11.1 and thus is a principal connection on $P|U_\alpha$. Now let (f_α) be a smooth partition of unity on M which is subordinated to the open cover (U_α), and let $\omega := \sum_\alpha (f_\alpha\circ p)\gamma_\alpha$. Since both requirements of lemma 11.1 are invariant under convex linear combinations, ω is a principal connection on P. $\square$

11.4. Local descriptions of principal connections. We consider a principal fiber bundle (P, p, M, G) with some principal fiber bundle atlas $(U_\alpha, \varphi_\alpha : P|U_\alpha \to U_\alpha \times G)$ and corresponding cocycle $(\varphi_{\alpha\beta} : U_{\alpha\beta} \to G)$ of transition functions. We consider the sections $s_\alpha \in C^\infty(P|U_\alpha)$ which are given by $\varphi_\alpha(s_\alpha(x)) = (x, e)$ and satisfy $s_\alpha.\varphi_{\alpha\beta} = s_\beta$.

(1) Let $\Theta \in \Omega^1(G, \mathfrak{g})$ be the left logarithmic derivative of the identity, i.e. $\Theta(\eta_g) := T_g(\lambda_{g^{-1}}).\eta_g$. We will use the forms $\Theta_{\alpha\beta} := {\varphi_{\alpha\beta}}^*\Theta \in \Omega^1(U_{\alpha\beta}; \mathfrak{g})$.

Let $\Phi = \zeta \circ \omega \in \Omega^1(P; VP)$ be a principal connection with connection form $\omega \in \Omega^1(P; \mathfrak{g})$. We may associate the following local data to the connection:

(2) $\omega_\alpha := {s_\alpha}^*\omega \in \Omega^1(U_\alpha; \mathfrak{g})$, the physicists version of the connection.
(3) The Christoffel forms $\Gamma^\alpha \in \Omega^1(U_\alpha; \mathfrak{X}(G))$ from 9.7, which are given by $(0_x, \Gamma^\alpha(\xi_x, g)) = -T(\varphi_\alpha).\Phi.T(\varphi_\alpha)^{-1}(\xi_x, 0_g)$.
(4) $\gamma_\alpha := (\varphi_\alpha^{-1})^*\omega \in \Omega^1(U_\alpha \times G; \mathfrak{g})$, the local expressions of ω.

Lemma. *These local data have the following properties and are related by the following formulas.*

(5) *The forms $\omega_\alpha \in \Omega^1(U_\alpha; \mathfrak{g})$ satisfy the transition formulas*
$$\omega_\alpha = \operatorname{Ad}(\varphi_{\beta\alpha}^{-1})\omega_\beta + \Theta_{\beta\alpha},$$
and any set of forms like that with this transition behavior determines a unique principal connection.
(6) *We have $\gamma_\alpha(\xi_x, T\lambda_g.X) = \gamma_\alpha(\xi_x, 0_g) + X = \operatorname{Ad}(g^{-1})\omega_\alpha(\xi_x) + X$.*
(7) *We have $\Gamma^\alpha(\xi_x, g) = -T_e(\lambda_g).\gamma_\alpha(\xi_x, 0_g) = -T_e(\lambda_g).\operatorname{Ad}(g^{-1})\omega_\alpha(\xi_x) = -T(\rho_g)\omega_\alpha(\xi_x)$, so $\Gamma^\alpha(\xi_x) = R_{\omega_\alpha(\xi_x)}$, a right invariant vector field.*

Proof. From the definition of the Christoffel forms we have
$$\begin{aligned}(0_x, \Gamma^\alpha(\xi_x, g)) &= -T(\varphi_\alpha).\Phi.T(\varphi_\alpha)^{-1}(\xi_x, 0_g)\\ &= -T(\varphi_\alpha).T_e(r_{\varphi_\alpha^{-1}(x,g)})\omega.T(\varphi_\alpha)^{-1}(\xi_x, 0_g)\\ &= -T_e(\varphi_\alpha \circ r_{\varphi_\alpha^{-1}(x,g)})\omega.T(\varphi_\alpha)^{-1}(\xi_x, 0_g)\\ &= -(0_x, T_e(\lambda_g)\omega.T(\varphi_\alpha)^{-1}(\xi_x, 0_g)) = -(0_x, T_e(\lambda_g)\gamma_\alpha(\xi_x, 0_g)).\end{aligned}$$
This is the first part of (7). The second part follows from (6).
$$\begin{aligned}\gamma_\alpha(\xi_x, T\lambda_g.X) &= \gamma_\alpha(\xi_x, 0_g) + \gamma_\alpha(0_x, T\lambda_g.X)\\ &= \gamma_\alpha(\xi_x, 0_g) + \omega(T(\varphi_\alpha)^{-1}(0_x, T\lambda_g.X))\\ &= \gamma_\alpha(\xi_x, 0_g) + \omega(\zeta_X(\varphi_\alpha^{-1}(x, g))) = \gamma_\alpha(\xi_x, 0_g) + X.\end{aligned}$$
So the first part of (6) holds. The second part is seen from
$$\begin{aligned}\gamma_\alpha(\xi_x, 0_g) &= \gamma_\alpha(\xi_x, T_e(\rho_g)0_e) = (\omega \circ T(\varphi_\alpha)^{-1} \circ T(\operatorname{Id}_X \times \rho_g))(\xi_x, 0_e) =\\ &= (\omega \circ T(r^g \circ \varphi_\alpha^{-1}))(\xi_x, 0_e) = \operatorname{Ad}(g^{-1})\omega(T(\varphi_\alpha^{-1})(\xi_x, 0_e))\\ &= \operatorname{Ad}(g^{-1})({s_\alpha}^*\omega)(\xi_x) = \operatorname{Ad}(g^{-1})\omega_\alpha(\xi_x).\end{aligned}$$
Via (7) the transition formulas for the ω_α are easily seen to be equivalent to the transition formulas for the Christoffel forms in lemma 9.7. □

11.5. The covariant derivative. Let (P, p, M, G) be a principal fiber bundle with principal connection $\Phi = \zeta \circ \omega$. We consider the horizontal projection $\chi = \mathrm{Id}_{TP} - \Phi : TP \to HP$, cf. 9.3, which satisfies $\chi \circ \chi = \chi$, $\operatorname{im} \chi = HP$, $\ker \chi = VP$, and $\chi \circ T(r^g) = T(r^g) \circ \chi$ for all $g \in G$.

If W is a finite dimensional vector space, we consider the mapping $\chi^* : \Omega(P; W) \to \Omega(P; W)$ which is given by

$$(\chi^*\varphi)_u(X_1, \dots, X_k) = \varphi_u(\chi(X_1), \dots, \chi(X_k)).$$

The mapping χ^* is a projection onto the subspace of *horizontal differential forms*, i.e. the space $\Omega_{hor}(P; W) := \{\psi \in \Omega(P; W) : i_X\psi = 0 \text{ for } X \in VP\}$. The notion of horizontal form is independent of the choice of a connection.

The projection χ^* has the following properties where in the first assertion one of the two forms has values in $\mathbb{R}$:

$$\begin{aligned}
\chi^*(\varphi \wedge \psi) &= \chi^*\varphi \wedge \chi^*\psi, \\
\chi^* \circ \chi^* &= \chi^*, \\
\chi^* \circ (r^g)^* &= (r^g)^* \circ \chi^* \qquad \text{for all } g \in G, \\
\chi^*\omega &= 0 \\
\chi^* \circ \mathcal{L}(\zeta_X) &= \mathcal{L}(\zeta_X) \circ \chi^*.
\end{aligned}$$

They follow easily from the corresponding properties of χ, the last property uses that $\mathrm{Fl}_t^{\zeta(X)} = r^{\exp tX}$.

Now we define the *covariant exterior derivative* $d_\omega : \Omega^k(P; W) \to \Omega^{k+1}(P; W)$ by the prescription $d_\omega := \chi^* \circ d$.

Theorem. *The covariant exterior derivative d_ω has the following properties.*

(1) $d_\omega(\varphi \wedge \psi) = d_\omega(\varphi) \wedge \chi^*\psi + (-1)^{\deg \varphi}\chi^*\varphi \wedge d_\omega(\psi)$ *if φ or ψ is real valued.*
(2) $\mathcal{L}(\zeta_X) \circ d_\omega = d_\omega \circ \mathcal{L}(\zeta_X)$ *for each $X \in \mathfrak{g}$.*
(3) $(r^g)^* \circ d_\omega = d_\omega \circ (r^g)^*$ *for each $g \in G$.*
(4) $d_\omega \circ p^* = d \circ p^* = p^* \circ d : \Omega(M; W) \to \Omega_{hor}(P; W)$.
(5) $d_\omega\omega = \Omega$, *the curvature form.*
(6) $d_\omega\Omega = 0$, *the Bianchi identity.*
(7) $d_\omega \circ \chi^* - d_\omega = \chi^* \circ i(R)$, *where R is the curvature.*
(8) $d_\omega \circ d_\omega = \chi^* \circ i(R) \circ d$.
(9) *Let $\Omega_{\mathrm{hor}}(P, \mathfrak{g})^G$ be the algebra of all horizontal G-equivariant $\mathfrak{g}$-valued forms, i.e. $(r^g)^*\psi = \mathrm{Ad}(g^{-1})\psi$. Then for any $\psi \in \Omega_{\mathrm{hor}}(P, \mathfrak{g})^G$ we have $d_\omega\psi = d\psi + [\omega, \psi]_\wedge$.*
(10) *The mapping $\psi \mapsto \zeta_\psi$, where $\zeta_\psi(X_1, \dots, X_k)(u) = \zeta_{\psi(X_1, \dots, X_k)(u)}(u)$, is an isomorphism between $\Omega_{\mathrm{hor}}(P, \mathfrak{g})^G$ and the algebra $\Omega_{\mathrm{hor}}(P, VP)^G$ of all horizontal G-equivariant forms with values in the vertical bundle VP. Then we have $\zeta_{d_\omega\psi} = [\Phi, \zeta_\omega]$.*

Proof. (1) through (4) follow from the properties of χ^*.

(5) We have

$$\begin{aligned}(d_\omega\omega)(\xi,\eta) &= (\chi^*d\omega)(\xi,\eta) = d\omega(\chi\xi,\chi\eta)\\ &= (\chi\xi)\omega(\chi\eta) - (\chi\eta)\omega(\chi\xi) - \omega([\chi\xi,\chi\eta])\\ &= -\omega([\chi\xi,\chi\eta]) \text{ and}\\ -\zeta(\Omega(\xi,\eta)) &= R(\xi,\eta) = \Phi[\chi\xi,\chi\eta] = \zeta_{\omega([\chi\xi,\chi\eta])}.\end{aligned}$$

(6) Using 11.2 we have

$$\begin{aligned}d_\omega\Omega &= d_\omega(d\omega + \tfrac{1}{2}[\omega,\omega]_\wedge)\\ &= \chi^*dd\omega + \tfrac{1}{2}\chi^*d[\omega,\omega]_\wedge\\ &= \tfrac{1}{2}\chi^*([d\omega,\omega]_\wedge - [\omega,d\omega]_\wedge) = \chi^*[d\omega,\omega]_\wedge\\ &= [\chi^*d\omega,\chi^*\omega]_\wedge = 0, \text{ since } \chi^*\omega = 0.\end{aligned}$$

(7) For $\varphi \in \Omega(P;W)$ we have

$$\begin{aligned}&(d_\omega\chi^*\varphi)(X_0,\dots,X_k) = (d\chi^*\varphi)(\chi(X_0),\dots,\chi(X_k))\\ &= \sum_{0\le i\le k}(-1)^i\chi(X_i)((\chi^*\varphi)(\chi(X_0),\dots,\widehat{\chi(X_i)},\dots,\chi(X_k)))\\ &\quad + \sum_{i<j}(-1)^{i+j}(\chi^*\varphi)([\chi(X_i),\chi(X_j)],\chi(X_0),\dots\\ &\qquad\qquad \dots,\widehat{\chi(X_i)},\dots,\widehat{\chi(X_j)},\dots)\\ &= \sum_{0\le i\le k}(-1)^i\chi(X_i)(\varphi(\chi(X_0),\dots,\widehat{\chi(X_i)},\dots,\chi(X_k)))\\ &\quad + \sum_{i<j}(-1)^{i+j}\varphi([\chi(X_i),\chi(X_j)] - \Phi[\chi(X_i),\chi(X_j)],\chi(X_0),\dots\\ &\qquad\qquad \dots,\widehat{\chi(X_i)},\dots,\widehat{\chi(X_j)},\dots)\\ &= (d\varphi)(\chi(X_0),\dots,\chi(X_k)) + (i_R\varphi)(\chi(X_0),\dots,\chi(X_k))\\ &= (d_\omega + \chi^*i_R)(\varphi)(X_0,\dots,X_k).\end{aligned}$$

(8) $d_\omega d_\omega = \chi^*d\chi^*d = (\chi^*i_R + \chi^*d)d = \chi^*i_R d$ holds by (7).

(9) If we insert one vertical vector field, say ζ_X for $X \in \mathfrak{g}$, into $d_\omega\psi$, we get 0 by definition. For the right hand side we use $i_{\zeta_X}\psi = 0$ and $\mathcal{L}_{\zeta_X}\psi = \frac{\partial}{\partial t}\big|_0 (\mathrm{Fl}^{\zeta_X}_t)^*\psi = \frac{\partial}{\partial t}\big|_0 (r^{\exp tX})^*\psi = \frac{\partial}{\partial t}\big|_0 \mathrm{Ad}(\exp(-tX))\psi = -\operatorname{ad}(X)\psi$ to get

$$\begin{aligned}i_{\zeta_X}(d\psi + [\omega,\psi]_\wedge) &= i_{\zeta_X}d\psi + di_{\zeta_X}\psi + [i_{\zeta_X}\omega,\psi] - [\omega,i_{\zeta_X}\psi]\\ &= \mathcal{L}_{\zeta_X}\psi + [X,\psi] = -\operatorname{ad}(X)\psi + [X,\psi] = 0.\end{aligned}$$

Let now all vector fields ξ_i be horizontal, then we get

$$\begin{aligned}(d_\omega\psi)(\xi_0,\dots,\xi_k) &= (\chi^*d\psi)(\xi_0,\dots,\xi_k) = d\psi(\xi_0,\dots,\xi_k),\\ (d\psi + [\omega,\psi]_\wedge)(\xi_0,\dots,\xi_k) &= d\psi(\xi_0,\dots,\xi_k).\end{aligned}$$

So the first formula holds.

(10) We proceed in a similar manner. Let Ψ be in the space $\Omega^\ell_{\text{hor}}(P, VP)^G$ of all horizontal G-equivariant forms with vertical values. Then for each $X \in \mathfrak{g}$ we have $i_{\zeta_X}\Psi = 0$; furthermore the G-equivariance $(r^g)^*\Psi = \Psi$ implies that $\mathcal{L}_{\zeta_X}\Psi = [\zeta_X, \Psi] = 0$ by 8.16.(5). Using formula 8.11.(2) we have

$$\begin{aligned} i_{\zeta_X}[\Phi, \Psi] &= [i_{\zeta_X}\Phi, \Psi] - [\Phi, i_{\zeta_X}\Psi] + i([\Phi, \zeta_X])\Psi + i([\Psi, \zeta_X])\Phi \\ &= [\zeta_X, \Psi] - 0 + 0 + 0 = 0. \end{aligned}$$

Let now all vector fields ξ_i again be horizontal, then from the huge formula 8.9 for the Frölicher-Nijenhuis bracket only the following terms in the third and fifth line survive:

$$\begin{aligned} [\Phi, \Psi](\xi_1, \dots, \xi_{\ell+1}) &= \tfrac{(-1)^\ell}{\ell!} \sum_\sigma \operatorname{sign}\sigma \; \Phi([\Psi(\xi_{\sigma 1}, \dots, \xi_{\sigma\ell}), \xi_{\sigma(\ell+1)}]) \\ &\quad + \tfrac{1}{(\ell-1)!\,2!} \sum_\sigma \operatorname{sign}\sigma \; \Phi(\Psi([\xi_{\sigma 1}, \xi_{\sigma 2}], \xi_{\sigma 3}, \dots, \xi_{\sigma(\ell+1)})). \end{aligned}$$

For $f : P \to \mathfrak{g}$ and horizontal ξ we have $\Phi[\xi, \zeta_f] = \zeta_{\xi(f)} = \zeta_{df(\xi)}$ since it is $C^\infty(P, \mathbb{R})$-linear in ξ. So the last expression becomes

$$\zeta(d\psi(\xi_0, \dots, \xi_k)) = \zeta(d_\omega\psi(\xi_0, \dots, \xi_k)) = \zeta((d\psi + [\omega, \psi]_\wedge)(\xi_0, \dots, \xi_k))$$

as required. □

11.6. Theorem. *Let (P, p, M, G) be a principal fiber bundle with principal connection ω. Then the parallel transport for the principal connection is globally defined and G-equivariant.*

In detail: For each smooth curve $c : \mathbb{R} \to M$ there is a smooth mapping $\operatorname{Pt}_c : \mathbb{R} \times P_{c(0)} \to P$ such that the following holds:

(1) $\operatorname{Pt}(c, t, u) \in P_{c(t)}$, $\operatorname{Pt}(c, 0) = \operatorname{Id}_{P_{c(0)}}$, *and* $\omega(\frac{d}{dt}\operatorname{Pt}(c, t, u)) = 0$.
(2) $\operatorname{Pt}(c, t) : P_{c(0)} \to P_{c(t)}$ *is G-equivariant, i.e.* $\operatorname{Pt}(c, t, u.g) = \operatorname{Pt}(c, t, u).g$ *holds for all $g \in G$ and $u \in P$. Moreover we have* $\operatorname{Pt}(c, t)^*(\zeta_X | P_{c(t)}) = \zeta_X | P_{c(0)}$ *for all $X \in \mathfrak{g}$.*
(3) *For any smooth function $f : \mathbb{R} \to \mathbb{R}$ we have* $\operatorname{Pt}(c, f(t), u) = \operatorname{Pt}(c \circ f, t, \operatorname{Pt}(c, f(0), u))$.

Proof. By 11.4 the Christoffel forms $\Gamma^\alpha \in \Omega^1(U_\alpha, \mathfrak{X}(G))$ of the connection ω with respect to a principal fiber bundle atlas $(U_\alpha, \varphi_\alpha)$ are given by $\Gamma^\alpha(\xi_x) = R_{\omega_\alpha(\xi_x)}$, so they take values in the Lie subalgebra $\mathfrak{X}_R(G)$ of all right invariant vector fields on G, which are bounded with respect to any right invariant Riemannian metric on G. Each right invariant metric on a Lie group is complete. So the connection is complete by the remark in 9.9.

Properties (1) and (3) follow from theorem 9.8, and (2) is seen as follows:

$$\omega(\tfrac{d}{dt}\operatorname{Pt}(c, t, u).g) = \operatorname{Ad}(g^{-1})\omega(\tfrac{d}{dt}\operatorname{Pt}(c, t, u)) = 0$$

implies that $\operatorname{Pt}(c,t,u).g = \operatorname{Pt}(c,t,u.g)$. For the second assertion we compute for $u \in P_{c(0)}$:

$$\begin{aligned}
\operatorname{Pt}(c,t)^*(\zeta_X|P_{c(t)})(u) &= T\operatorname{Pt}(c,t)^{-1}\zeta_X(\operatorname{Pt}(c,t,u)) = \\
&= T\operatorname{Pt}(c,t)^{-1}\tfrac{d}{ds}|_0 \operatorname{Pt}(c,t,u).\exp(sX) = \\
&= T\operatorname{Pt}(c,t)^{-1}\tfrac{d}{ds}|_0 \operatorname{Pt}(c,t,u.\exp(sX)) = \\
&= \tfrac{d}{ds}|_0 \operatorname{Pt}(c,t)^{-1}\operatorname{Pt}(c,t,u.\exp(sX)) \\
&= \tfrac{d}{ds}|_0 u.\exp(sX) = \zeta_X(u). \quad \square
\end{aligned}$$

11.7. Holonomy groups. Let (P,p,M,G) be a principal fiber bundle with principal connection $\Phi = \zeta \circ \omega$. We assume that M is connected and we fix $x_0 \in M$.

In 9.10 we defined the holonomy group $\operatorname{Hol}(\Phi,x_0) \subset \operatorname{Diff}(P_{x_0})$ as the group of all $\operatorname{Pt}(c,1) : P_{x_0} \to P_{x_0}$ for c any piecewise smooth closed loop through x_0. (Reparametrizing c by a function which is flat at each corner of c we may assume that any c is smooth.) If we consider only those curves c which are nullhomotopic, we obtain the restricted holonomy group $\operatorname{Hol}_0(\Phi,x_0)$.

Now let us fix $u_0 \in P_{x_0}$. The elements $\tau(u_0,\operatorname{Pt}(c,t,u_0)) \in G$ form a subgroup of the structure group G which is isomorphic to $\operatorname{Hol}(\Phi,x_0)$; we denote it by $\operatorname{Hol}(\omega,u_0)$ and we call it also the *holonomy group* of the connection. Considering only nullhomotopic curves we get the *restricted holonomy group* $\operatorname{Hol}_0(\omega,u_0)$ a normal subgroup of $\operatorname{Hol}(\omega,u_0)$.

Theorem. *The main results for the holonomy are as follows:*

(1) *We have* $\operatorname{Hol}(\omega,u_0.g) = \operatorname{conj}(g^{-1})\operatorname{Hol}(\omega,u_0)$ *and* $\operatorname{Hol}_0(\omega,u_0.g) = \operatorname{conj}(g^{-1})\operatorname{Hol}_0(\omega,u_0)$.

(2) *For each curve c in M with $c(0) = x_0$ we have* $\operatorname{Hol}(\omega,\operatorname{Pt}(c,t,u_0)) = \operatorname{Hol}(\omega,u_0)$ *and* $\operatorname{Hol}_0(\omega,\operatorname{Pt}(c,t,u_0)) = \operatorname{Hol}_0(\omega,u_0)$.

(3) $\operatorname{Hol}_0(\omega,u_0)$ *is a connected Lie subgroup of G and the quotient group* $\operatorname{Hol}(\omega,u_0)/\operatorname{Hol}_0(\omega,u_0)$ *is at most countable, so* $\operatorname{Hol}(\omega,u_0)$ *is also a Lie subgroup of G.*

(4) *The Lie algebra* $\operatorname{hol}(\omega,u_0) \subset \mathfrak{g}$ *of* $\operatorname{Hol}(\omega,u_0)$ *is linearly generated by* $\{\Omega(X_u,Y_u) : X_u,Y_u \in T_uP\}$, *and it is isomorphic to the holonomy Lie algebra* $\operatorname{hol}(\Phi,x_0)$ *we considered in 9.10.*

(5) *For $u_0 \in P_{x_0}$ let $P(\omega,u_0)$ be the set of all* $\operatorname{Pt}(c,t,u_0)$ *for c any (piecewise) smooth curve in M with $c(0) = x_0$ and for $t \in \mathbb{R}$. Then $P(\omega,u_0)$ is a sub fiber bundle of P which is invariant under the right action of* $\operatorname{Hol}(\omega,u_0)$; *so it is itself a principal fiber bundle over M with structure group* $\operatorname{Hol}(\omega,u_0)$ *and we have a reduction of structure group, cf. 10.6 and 10.13. The pullback of ω to $P(\omega,u_0)$ is then again a principal connection form* $i^*\omega \in \Omega^1(P(\omega,u_0);\operatorname{hol}(\omega,u_0))$.

(6) *P is foliated by the leaves $P(\omega,u)$, $u \in P_{x_0}$.*

(7) *If the curvature $\Omega = 0$ then* $\operatorname{Hol}_0(\omega,u_0) = \{e\}$ *and each $P(\omega,u)$ is a covering of M.*

(8) *If one uses piecewise C^k-curves for $1 \le k < \infty$ in the definition, one gets the same holonomy groups.*

In view of assertion 5 a principal connection ω is called *irreducible* if $\operatorname{Hol}(\omega, u_0)$ equals the structure group G for some (equivalently any) $u_0 \in P_{x_0}$.

Proof. 1. This follows from the properties of the mapping τ from 10.2 and from the G-equivariance of the parallel transport.

The rest of this theorem is a compilation of well known results, and we refer to [Kobayashi-Nomizu I, 63, p. 83ff] for proofs. □

11.8. Inducing connections on associated bundles. Let (P, p, M, G) be a principal bundle with principal right action $r : P \times G \to P$ and let $\ell : G \times S \to S$ be a left action of the structure group G on some manifold S. Then we consider the associated bundle $P[S] = P[S, \ell] = P \times_G S$, constructed in 10.7. Recall from 10.18 that its tangent and vertical bundle are given by $T(P[S, \ell]) = TP[TS, T\ell] = TP \times_{TG} TS$ and $V(P[S, \ell]) = P[TS, T_2\ell] = P \times_G TS$.

Let $\Phi = \zeta \circ \omega \in \Omega^1(P; TP)$ be a principal connection on the principal bundle P. We construct the *induced connection* $\bar{\Phi} \in \Omega^1(P[S], T(P[S]))$ by the following diagram:

$$\begin{array}{ccccc}
TP \times TS & \xrightarrow{\Phi \times \mathrm{Id}} & TP \times TS & \xrightarrow{=} & T(P \times S) \\
{\scriptstyle Tq = q'}\downarrow & & {\scriptstyle q'}\downarrow & & {\scriptstyle Tq}\downarrow \\
TP \times_{TG} TS & \xrightarrow{\bar{\Phi}} & TP \times_{TG} TS & \xrightarrow{=} & T(P \times_G S).
\end{array}$$

Let us first check that the top mapping $\Phi \times \mathrm{Id}$ is TG-equivariant. For $g \in G$ and $X \in \mathfrak{g}$ the inverse of $T_e(\lambda_g)X$ in the Lie group TG is denoted by $(T_e(\lambda_g)X)^{-1}$, see lemma 10.17. Furthermore by 5.13 we have

$$\begin{aligned}
Tr(\xi_u, T_e(\lambda_g)X) &= T_u(r^g)\xi_u + Tr((0_P \times L_X)(u, g)) \\
&= T_u(r^g)\xi_u + T_g(r_u)(T_e(\lambda_g)X) \\
&= T_u(r^g)\xi_u + \zeta_X(ug).
\end{aligned}$$

We may compute

$$\begin{aligned}
&(\Phi \times \mathrm{Id})(Tr(\xi_u, T_e(\lambda_g)X), T\ell((T_e(\lambda_g)X)^{-1}, \eta_s)) \\
&= (\Phi(T_u(r^g)\xi_u + \zeta_X(ug)), T\ell((T_e(\lambda_g)X)^{-1}, \eta_s)) \\
&= (\Phi(T_u(r^g)\xi_u) + \Phi(\zeta_X(ug)), T\ell((T_e(\lambda_g)X)^{-1}, \eta_s)) \\
&= ((T_u(r^g)\Phi\xi_u) + \zeta_X(ug), T\ell((T_e(\lambda_g)X)^{-1}, \eta_s)) \\
&= (Tr(\Phi(\xi_u), T_e(\lambda_g)X), T\ell((T_e(\lambda_g)X)^{-1}, \eta_s)).
\end{aligned}$$

So the mapping $\Phi \times \mathrm{Id}$ factors to $\bar{\Phi}$ as indicated in the diagram, and we have $\bar{\Phi} \circ \bar{\Phi} = \bar{\Phi}$ from $(\Phi \times \mathrm{Id}) \circ (\Phi \times \mathrm{Id}) = \Phi \times \mathrm{Id}$. The mapping $\bar{\Phi}$ is fiberwise linear, since $\Phi \times \mathrm{Id}$ and $q' = Tq$ are. The image of $\bar{\Phi}$ is

$$\begin{aligned}
q'(VP \times TS) &= q'(\ker(Tp : TP \times TS \to TM)) \\
&= \ker(Tp : TP \times_{TG} TS \to TM) = V(P[S, \ell]).
\end{aligned}$$

Thus $\bar{\Phi}$ is a connection on the associated bundle $P[S]$. We call it the *induced connection.*

From the diagram it also follows, that the vector valued forms $\Phi \times \mathrm{Id} \in \Omega^1(P \times S; TP \times TS)$ and $\bar{\Phi} \in \Omega^1(P[S]; T(P[S]))$ are $(q : P \times S \to P[S])$-related. So by 8.15 we have for the curvatures

$$\begin{aligned} R_{\Phi\times\mathrm{Id}} &= \tfrac{1}{2}[\Phi \times \mathrm{Id}, \Phi \times \mathrm{Id}] = \tfrac{1}{2}[\Phi, \Phi] \times 0 = R_\Phi \times 0, \\ R_{\bar{\Phi}} &= \tfrac{1}{2}[\bar{\Phi}, \bar{\Phi}], \end{aligned}$$

that they are also q-related, i.e. $Tq \circ (R_\Phi \times 0) = R_{\bar{\Phi}} \circ (Tq \times_M Tq)$.

By uniqueness of the solutions of the defining differential equation we also get that

$$\mathrm{Pt}_{\bar{\Phi}}(c, t, q(u, s)) = q(\mathrm{Pt}_\Phi(c, t, u), s).$$

11.9. Recognizing induced connections. We consider again a principal fiber bundle (P, p, M, G) and a left action $\ell : G \times S \to S$. Suppose that $\Psi \in \Omega^1(P[S]; T(P[S]))$ is a connection on the associated bundle $P[S] = P[S, \ell]$. Then the following question arises: When is the connection Ψ induced from a principal connection on P? If this is the case, we say that Ψ is compatible with the G-bundle structure on $P[S]$. The answer is given in the following

Theorem. *Let Ψ be a (general) connection on the associated bundle $P[S]$. Let us suppose that the action ℓ is infinitesimally effective, i.e. the fundamental vector field mapping $\zeta : \mathfrak{g} \to \mathfrak{X}(S)$ is injective.*

Then the connection Ψ is induced from a principal connection ω on P if and only if the following condition is satisfied:

> *In some (equivalently any) fiber bundle atlas (U_α, ψ_α) of $P[S]$ belonging to the G-bundle structure of the associated bundle the Christoffel forms $\Gamma^\alpha \in \Omega^1(U_\alpha; \mathfrak{X}(S))$ have values in the sub Lie algebra $\mathfrak{X}_{fund}(S)$ of fundamental vector fields for the action ℓ.*

Proof. Let $(U_\alpha, \varphi_\alpha : P|U_\alpha \to U_\alpha \times G)$ be a principal fiber bundle atlas for P. Then by the proof of theorem 10.7 it is seen that the induced fiber bundle atlas $(U_\alpha, \psi_\alpha : P[S]|U_\alpha \to U_\alpha \times S)$ is given by

$$\psi_\alpha^{-1}(x, s) = q(\varphi_\alpha^{-1}(x, e), s), \tag{1}$$

$$(\psi_\alpha \circ q)(\varphi_\alpha^{-1}(x, g), s) = (x, g.s). \tag{2}$$

Let $\Phi = \zeta \circ \omega$ be a principal connection on P and let $\bar{\Phi}$ be the induced connection on the associated bundle $P[S]$. By 9.7 its Christoffel symbols are given by

$$\begin{aligned} (0_x, \Gamma^\alpha_{\bar{\Phi}}(\xi_x, s)) &= -(T(\psi_\alpha) \circ \bar{\Phi} \circ T(\psi_\alpha^{-1}))(\xi_x, 0_s) \\ &= -(T(\psi_\alpha) \circ \bar{\Phi} \circ Tq \circ (T(\varphi_\alpha^{-1}) \times \mathrm{Id}))(\xi_x, 0_e, 0_s) \quad \text{by (1)} \\ &= -(T(\psi_\alpha) \circ Tq \circ (\Phi \times \mathrm{Id}))(T(\varphi_\alpha^{-1})(\xi_x, 0_e), 0_s) \quad \text{by 11.8} \\ &= -(T(\psi_\alpha) \circ Tq)(\Phi(T(\varphi_\alpha^{-1})(\xi_x, 0_e)), 0_s) \end{aligned}$$

$$
\begin{aligned}
&= (T(\psi_\alpha)\circ Tq)(T(\varphi_\alpha^{-1})(0_x,\Gamma^\alpha_\Phi(\xi_x,e)),0_s) && \text{by 11.4.(3)}\\
&= -T(\psi_\alpha\circ q\circ(\varphi_\alpha^{-1}\times \mathrm{Id}))(0_x,\omega_\alpha(\xi_x),0_s) && \text{by 11.4.(7)}\\
&= -T_e(\ell^s)\omega_\alpha(\xi_x) && \text{by (2)}\\
&= -\zeta_{\omega_\alpha(\xi_x)}(s).
\end{aligned}
$$

So the condition is necessary. Now let us conversely suppose that a connection Ψ on $P[S]$ is given such that the Christoffel forms Γ^α_Ψ with respect to a fiber bundle atlas of the G-structure have values in $\mathfrak{X}_{fund}(S)$. Then unique $\mathfrak{g}$-valued forms $\omega_\alpha \in \Omega^1(U_\alpha;\mathfrak{g})$ are given by the equation $\Gamma^\alpha_\Psi(\xi_x) = \zeta(\omega_\alpha(\xi_x))$, since the action is infinitesimally effective. From the transition formulas 9.7 for the Γ^α_Ψ follow the transition formulas 11.4.(5) for the ω^α, so that they give a unique principal connection on P, which by the first part of the proof induces the given connection Ψ on $P[S]$. □

11.10. Inducing connections on associated vector bundles. Let (P,p,M,G) be a principal fiber bundle and let $\rho : G \to GL(W)$ be a representation of the structure group G on a finite dimensional vector space W. We consider the associated vector bundle $(E := P[W,\rho],p,M,W)$, from 10.11. Recall from 6.11 that $T(E) = TP\times_{TG} TW$ has two vector bundle structures with the projections

$$
\begin{aligned}
\pi_E : T(E) = TP\times_{TG} TW \to P\times_G W = E,\\
Tp\circ pr_1 : T(E) = TP\times_{TG} TW \to TM.
\end{aligned}
$$

Now let $\Phi = \zeta\circ\omega \in \Omega^1(P;TP)$ be a principal connection on P. We consider the induced connection $\bar\Phi \in \Omega^1(E;T(E))$ from 11.8. Inserting the projections of both vector bundle structures on $T(E)$ into the diagram in 11.8 we get the following diagram

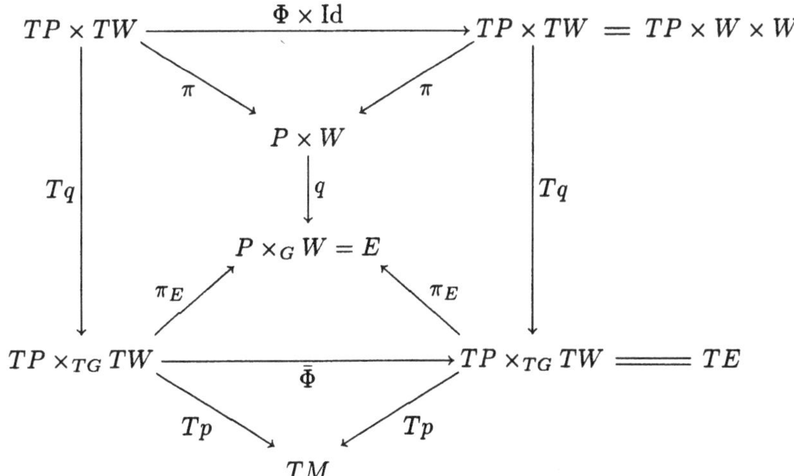

and from it one easily sees that the induced connection is linear in both vector bundle structures. We say that it is a *linear connection* on the associated bundle.

Recall now from 6.11 the vertical lift $vl_E : E \times_M E \to VE$, which is an isomorphism, pr_1–π_E–fiberwise linear and also p–Tp–fiberwise linear.

Now we define the *connector* K of the linear connection $\bar{\Phi}$ by

$$K := pr_2 \circ (vl_E)^{-1} \circ \bar{\Phi} : TE \to VE \to E \times_M E \to E.$$

Lemma. *The connector $K : TE \to E$ is a vector bundle homomorphism for both vector bundle structures on TE and satisfies $K \circ vl_E = pr_2 : E \times_M E \to TE \to E$.*

So K is π_E–p–fiberwise linear and Tp–p–fiberwise linear.

Proof. This follows from the fiberwise linearity of the components of K and from its definition. □

11.11. Linear connections. If (E, p, M) is a vector bundle, a connection $\Psi \in \Omega^1(E; TE)$ such that $\Psi : TE \to VE \to TE$ is also Tp–Tp–fiberwise linear is called a *linear connection.* An easy check with 11.9 or a direct construction shows that Ψ is then induced from a unique principal connection on the linear frame bundle $GL(\mathbb{R}^n, E)$ of E (where n is the fiber dimension of E).

Equivalently a linear connection may be specified by a connector $K : TE \to E$ with the three properties of lemma 11.10. For then $HE := \{\xi_u : K(\xi_u) = 0_{p(u)}\}$ is a complement to VE in TE which is Tp–fiberwise linearly chosen.

11.12. Covariant derivative on vector bundles. Let (E, p, M) be a vector bundle with a linear connection, given by a connector $K : TE \to E$ with the properties in lemma 11.10.

For any manifold N, smooth mapping $s : N \to E$, and vector field $X \in \mathfrak{X}(N)$ we define the *covariant derivative* of s along X by

$$(1) \qquad \nabla_X s := K \circ Ts \circ X : N \to TN \to TE \to E.$$

If $f : N \to M$ is a fixed smooth mapping, let us denote by $C^\infty_f(N, E)$ the vector space of all smooth mappings $s : N \to E$ with $p \circ s = f$ – they are called sections of E along f. From the universal property of the pullback it follows that the vector space $C^\infty_f(N, E)$ is canonically linearly isomorphic to the space $C^\infty(f^*E)$ of sections of the pullback bundle. Then the covariant derivative may be viewed as a bilinear mapping

$$(2) \qquad \nabla : \mathfrak{X}(N) \times C^\infty_f(N, E) \to C^\infty_f(N, E).$$

Lemma. *This covariant derivative has the following properties:*

(3) *$\nabla_X s$ is $C^\infty(N, \mathbb{R})$-linear in $X \in \mathfrak{X}(N)$. So for a tangent vector $X_x \in T_x N$ the mapping $\nabla_{X_x} : C^\infty_f(N, E) \to E_{f(x)}$ makes sense and we have $(\nabla_X s)(x) = \nabla_{X(x)} s$.*

(4) *$\nabla_X s$ is $\mathbb{R}$-linear in $s \in C^\infty_f(N, E)$.*

(5) *$\nabla_X(h.s) = dh(X).s + h.\nabla_X s$ for $h \in C^\infty(N, \mathbb{R})$, the derivation property of ∇_X.*

(6) *For any manifold Q and smooth mapping $g : Q \to N$ and $Y_y \in T_y Q$ we have $\nabla_{Tg.Y_y} s = \nabla_{Y_y}(s \circ g)$. If $Y \in \mathfrak{X}(Q)$ and $X \in \mathfrak{X}(N)$ are g-related, then we have $\nabla_Y(s \circ g) = (\nabla_X s) \circ g$.*

Proof. All these properties follow easily from the definition (1). □

For vector fields X, $Y \in \mathfrak{X}(M)$ and a section $s \in C^\infty(E)$ an easy computation shows that

$$\begin{aligned} R^E(X,Y)s :&= \nabla_X \nabla_Y s - \nabla_Y \nabla_X s - \nabla_{[X,Y]} s \\ &= ([\nabla_X, \nabla_Y] - \nabla_{[X,Y]})s \end{aligned}$$

is $C^\infty(M, \mathbb{R})$-linear in X, Y, and s. By the method of 7.3 it follows that R^E is a 2-form on M with values in the vector bundle $L(E,E)$, i.e. $R^E \in \Omega^2(M; L(E,E))$. It is called the *curvature* of the covariant derivative.

For $f : N \to M$, vector fields X, $Y \in \mathfrak{X}(N)$ and a section $s \in C^\infty_f(N,E)$ along f one may prove that

$$\nabla_X \nabla_Y s - \nabla_Y \nabla_X s - \nabla_{[X,Y]} s = (f^* R^E)(X,Y)s = R^E(Tf.X, Tf.Y)s.$$

We will do this in 37.15.(2).

11.13. Covariant exterior derivative. Let (E,p,M) be a vector bundle with a linear connection, given by a connector $K : TE \to E$.

For a smooth mapping $f : N \to M$ let $\Omega(N; f^*E)$ be the vector space of all forms on N with values in the vector bundle f^*E. We can also view them as forms on N with values along f in E, but we do not introduce an extra notation for this.

The graded space $\Omega(N; f^*E)$ is a graded $\Omega(N)$-module via

$$\begin{aligned} &(\varphi \wedge \Phi)(X_1, \dots, X_{p+q}) = \\ &\qquad = \tfrac{1}{p!\,q!} \sum_\sigma \operatorname{sign}(\sigma)\, \varphi(X_{\sigma 1}, \dots, X_{\sigma p}) \Phi(X_{\sigma(p+1)}, \dots, X_{\sigma(p+q)}). \end{aligned}$$

It can be shown that the graded module homomorphisms $H : \Omega(N; f^*E) \to \Omega(N; f^*E)$ (so that $H(\varphi \wedge \Phi) = (-1)^{\deg H . \deg \varphi} \varphi \wedge H(\Phi))$ are exactly the mappings $\mu(A)$ for $A \in \Omega^p(N; f^* L(E,E))$, which are given by

$$\begin{aligned} &(\mu(A)\Phi)(X_1, \dots, X_{p+q}) = \\ &\qquad = \tfrac{1}{p!\,q!} \sum_\sigma \operatorname{sign}(\sigma)\, A(X_{\sigma 1}, \dots, X_{\sigma p})(\Phi(X_{\sigma(p+1)}, \dots, X_{\sigma(p+q)})). \end{aligned}$$

The *covariant exterior derivative* $d_\nabla : \Omega^p(N; f^*E) \to \Omega^{p+1}(N; f^*E)$ is defined by (where the X_i are vector fields on N)

$$\begin{aligned} (d_\nabla \Phi)(X_0, \dots, X_p) &= \sum_{i=0}^{p} (-1)^i \nabla_{X_i} \Phi(X_0, \dots, \widehat{X_i}, \dots, X_p) \\ &\quad + \sum_{0 \le i < j \le p} (-1)^{i+j} \Phi([X_i, X_j], X_0, \dots, \widehat{X_i}, \dots, \widehat{X_j}, \dots, X_p). \end{aligned}$$

Lemma. *The covariant exterior derivative is well defined and has the following properties.*

(1) *For $s \in C^\infty(f^*E) = \Omega^0(N; f^*E)$ we have $(d_\nabla s)(X) = \nabla_X s$.*
(2) $d_\nabla(\varphi \wedge \Phi) = d\varphi \wedge \Phi + (-1)^{\deg \varphi}\varphi \wedge d_\nabla \Phi$.
(3) *For smooth $g : Q \to N$ and $\Phi \in \Omega(N; f^*E)$ we have $d_\nabla(g^*\Phi) = g^*(d_\nabla \Phi)$.*
(4) $d_\nabla d_\nabla \Phi = \mu(f^*R^E)\Phi$.

Proof. It suffices to investigate decomposable forms $\Phi = \varphi \otimes s$ for $\varphi \in \Omega^p(N)$ and $s \in C^\infty(f^*E)$. Then from the definition we have $d_\nabla(\varphi \otimes s) = d\varphi \otimes s + (-1)^p \varphi \wedge d_\nabla s$. Since by 11.12.(3) $d_\nabla s \in \Omega^1(N; f^*E)$, the mapping d_∇ is well defined. This formula also implies (2) immediately. (3) follows from 11.12.(6). (4) is checked as follows:

$$\begin{aligned} d_\nabla d_\nabla(\varphi \otimes s) &= d_\nabla(d\varphi \otimes s + (-1)^p \varphi \wedge d_\nabla s) \text{ by (2)} \\ &= 0 + (-1)^{2p} \varphi \wedge d_\nabla d_\nabla s \\ &= \varphi \wedge \mu(f^*R^E)s \text{ by the definition of } R^E \\ &= \mu(f^*R^E)(\varphi \otimes s). \quad \square \end{aligned}$$

11.14. Let (P, p, M, G) be a principal fiber bundle and let $\rho : G \to GL(W)$ be a representation of the structure group G on a finite dimensional vector space W.

Theorem. *There is a canonical isomorphism from the space of $P[W,\rho]$-valued differential forms on M onto the space of horizontal G-equivariant W-valued differential forms on P:*

$$\begin{aligned} q^\sharp : \Omega(M; P[W,\rho]) \to \Omega_{hor}(P; W)^G := \{\varphi \in \Omega(P; W) : i_X \varphi = 0 \\ \text{for all } X \in VP, (r^g)^*\varphi = \rho(g^{-1}) \circ \varphi \text{ for all } g \in G\}. \end{aligned}$$

In particular for $W = \mathbb{R}$ with trivial representation we see that

$$p^* : \Omega(M) \to \Omega_{hor}(P)^G = \{\varphi \in \Omega_{hor}(P) : (r^g)^*\varphi = \varphi\}$$

is also an isomorphism. The isomorphism

$$q^\sharp : \Omega^0(M; P[W]) = C^\infty(P[W]) \to \Omega^0_{hor}(P; W)^G = C^\infty(P, W)^G$$

is a special case of the one from 10.12.

Proof. Recall the smooth mapping $\tau^G : P \times_M P \to G$ from 10.2, which satisfies $r(u_x, \tau^G(u_x, v_x)) = v_x$, $\tau^G(u_x.g, u'_x.g') = g^{-1}.\tau^G(u_x, u'_x).g'$, and $\tau^G(u_x, u_x) = e$.

Let $\varphi \in \Omega^k_{hor}(P; W)^G$, $X_1, \dots, X_k \in T_uP$, and $X'_1, \dots, X'_k \in T_{u'}P$ such that $T_up.X_i = T_{u'}p.X'_i$ for each i. Then we have for $g = \tau^G(u, u')$, so that $ug = u'$:

$$\begin{aligned} q(u, \varphi_u(X_1, \dots, X_k)) &= q(ug, \rho(g^{-1})\varphi_u(X_1, \dots, X_k)) \\ &= q(u', ((r^g)^*\varphi)_u(X_1, \dots, X_k)) \\ &= q(u', \varphi_{ug}(T_u(r^g).X_1, \dots, T_u(r^g).X_k)) \\ &= q(u', \varphi_{u'}(X'_1, \dots, X'_k)), \text{ since } T_u(r^g)X_i - X'_i \in V_{u'}P. \end{aligned}$$

By this prescription a vector bundle valued form $\Phi \in \Omega^k(M; P[W])$ is uniquely determined.

For the converse recall the smooth mapping $\tau^W : P \times_M P[W, \rho] \to W$ from 10.7, which satisfies $\tau^W(u, q(u, w)) = w$, $q(u_x, \tau^W(u_x, v_x)) = v_x$, and $\tau^W(u_x g, v_x) = \rho(g^{-1})\tau^W(u_x, v_x)$.

For $\Phi \in \Omega^k(M; P[W])$ we define $q^\sharp\Phi \in \Omega^k(P; W)$ as follows. For $X_i \in T_uP$ we put

$$(q^\sharp\Phi)_u(X_1, \dots, X_k) := \tau^W(u, \Phi_{p(u)}(T_up.X_1, \dots, T_up.X_k)).$$

Then $q^\sharp\Phi$ is smooth and horizontal. For $g \in G$ we have

$$\begin{aligned}((r^g)^*(q^\sharp\Phi))_u(X_1, \dots, X_k) &= (q^\sharp\Phi)_{ug}(T_u(r^g).X_1, \dots, T_u(r^g).X_k)\\ &= \tau^W(ug, \Phi_{p(ug)}(T_{ug}p.T_u(r^g).X_1, \dots, T_{ug}p.T_u(r^g).X_k))\\ &= \rho(g^{-1})\tau^W(u, \Phi_{p(u)}(T_up.X_1, \dots, T_up.X_k))\\ &= \rho(g^{-1})(q^\sharp\Phi)_u(X_1, \dots, X_k).\end{aligned}$$

Clearly the two constructions are inverse to each other. □

11.15. Let (P, p, M, G) be a principal fiber bundle with a principal connection $\Phi = \zeta \circ \omega$, and let $\rho : G \to GL(W)$ be a representation of the structure group G on a finite dimensional vector space W. We consider the associated vector bundle $(E := P[W, \rho], p, M, W)$, the induced connection $\bar{\Phi}$ on it and the corresponding covariant derivative.

Theorem. *The covariant exterior derivative d_ω from 11.5 on P and the covariant exterior derivative for $P[W]$-valued forms on M are connected by the mapping $q^\sharp$ from 11.14, as follows:*

$$q^\sharp \circ d_\nabla = d_\omega \circ q^\sharp : \Omega(M; P[W]) \to \Omega_{hor}(P; W)^G.$$

Proof. Let first $f \in \Omega^0_{hor}(P; W)^G = C^\infty(P, W)^G$, then we have $f = q^\sharp s$ for $s \in C^\infty(P[W])$. Moreover we have $f(u) = \tau^W(u, s(p(u)))$ and $s(p(u)) = q(u, f(u))$ by 11.14 and 10.12. Therefore $Ts.Tp.X_u = Tq(X_u, Tf.X_u)$, where $Tf.X_u = (f(u), df(X_u)) \in TW = W \times W$. If $\chi : TP \to HP$ is the horizontal projection as in 11.5, we have $Ts.Tp.X_u = Ts.Tp.\chi.X_u = Tq(\chi.X_u, Tf.\chi.X_u)$. So we get

$$\begin{aligned}(q^\sharp d_\nabla s)(X_u) &= \tau^W(u, (d_\nabla s)(Tp.X_u))\\ &= \tau^W(u, \nabla_{Tp.X_u} s) && \text{by 11.13.(1)}\\ &= \tau^W(u, K.Ts.Tp.X_u) && \text{by 11.12.(1)}\\ &= \tau^W(u, K.Tq(\chi.X_u, Tf.\chi.X_u)) && \text{from above}\\ &= \tau^W(u, pr_2.vl^{-1}_{P[W]}.\bar{\Phi}.Tq(\chi.X_u, Tf.\chi.X_u)) && \text{by 11.10}\\ &= \tau^W(u, pr_2.vl^{-1}_{P[W]}.Tq.(\Phi \times \mathrm{Id})(\chi.X_u, Tf.\chi.X_u))) && \text{by 11.8}\end{aligned}$$

$$= \tau^W(u, pr_2.vl^{-1}_{P[W]}.Tq(0_u, Tf.\chi.X_u))) \qquad \text{since } \Phi.\chi = 0$$
$$= \tau^W(u, q.pr_2.vl^{-1}_{P\times W}.(0_u, Tf.\chi.X_u))) \qquad \text{since } q \text{ is fiber linear}$$
$$= \tau^W(u, q(u, df.\chi.X_u)) = (\chi^* df)(X_u)$$
$$= (d_\omega q^\sharp s)(X_u).$$

Now we turn to the general case. It suffices to check the formula for a decomposable $P[W]$-valued form $\Psi = \psi \otimes s \in \Omega^k(M, P[W])$, where $\psi \in \Omega^k(M)$ and $s \in C^\infty(P[W])$. Then we have

$$\begin{aligned}
d_\omega q^\sharp(\psi \otimes s) &= d_\omega(p^*\psi \cdot q^\sharp s)\\
&= d_\omega(p^*\psi)\cdot q^\sharp s + (-1)^k \chi^* p^*\psi \wedge d_\omega q^\sharp s \qquad \text{by 11.5.(1)}\\
&= \chi^* p^* d\psi \cdot q^\sharp s + (-1)^k p^*\psi \wedge q^\sharp d_\nabla s \qquad \text{from above and 11.5.(4)}\\
&= p^* d\psi \cdot q^\sharp s + (-1)^k p^*\psi \wedge q^\sharp d_\nabla s\\
&= q^\sharp (d\psi \otimes s + (-1)^k \psi \wedge d_\nabla s)\\
&= q^\sharp d_\nabla(\psi \otimes s). \quad \square
\end{aligned}$$

11.16. Corollary. *In the situation of theorem 11.15 above we have for the curvature form $\Omega \in \Omega^2_{hor}(P;\mathfrak{g})$ and the curvature $R^{P[W]} \in \Omega^2(M; L(P[W], P[W]))$ the relation*

$$q^\sharp_{L(P[W],P[W])} R^{P[W]} = \rho' \circ \Omega,$$

where $\rho' = T_e\rho : \mathfrak{g} \to L(W,W)$ is the derivative of the representation ρ.

Proof. We use the notation of the proof of theorem 11.15. By this theorem we have for $X, Y \in T_uP$

$$\begin{aligned}
(d_\omega d_\omega q^\sharp_{P[W]} s)_u(X,Y) &= (q^\sharp d_\nabla d_\nabla s)_u(X,Y)\\
&= (q^\sharp R^{P[W]} s)_u(X,Y)\\
&= \tau^W(u, R^{P[W]}(T_u p.X, T_u p.Y)s(p(u)))\\
&= (q^\sharp_{L(P[W],P[W])} R^{P[W]})_u(X,Y)(q^\sharp_{P[W]} s)(u).
\end{aligned}$$

On the other hand we have by theorem 11.5.(8)

$$\begin{aligned}
(d_\omega d_\omega q^\sharp s)_u(X,Y) &= (\chi^* i_R d q^\sharp s)_u(X,Y)\\
&= (dq^\sharp s)_u(R(X,Y)) \qquad \text{since } R \text{ is horizontal}\\
&= (dq^\sharp s)(-\zeta_{\Omega(X,Y)}(u)) \qquad \text{by 11.2}\\
&= \tfrac{\partial}{\partial t}\big|_0 (q^\sharp s)(\mathrm{Fl}^{\zeta_{\Omega(X,Y)}}_{-t}(u))\\
&= \tfrac{\partial}{\partial t}\big|_0 \tau^W(u.\exp(-t\Omega(X,Y)), s(p(u.\exp(-t\Omega(X,Y)))))\\
&= \tfrac{\partial}{\partial t}\big|_0 \tau^W(u.\exp(-t\Omega(X,Y)), s(p(u)))\\
&= \tfrac{\partial}{\partial t}\big|_0 \rho(\exp t\Omega(X,Y))\tau^W(u, s(p(u))) \qquad \text{by 10.7}\\
&= \rho'(\Omega(X,Y))(q^\sharp s)(u). \quad \square
\end{aligned}$$

Remarks

The concept of connections on general fiber bundles was formulated at about 1970, see e. g. [Libermann, 73]. The theorem 9.9 that each fiber bundle admits a complete connection is contained in [Wolf, 67], with an incorrect proof. It is an exercise in [Greub-Halperin-Vanstone I, 72, p 314]. The proof given here and the generalization 9.11 of the Ambrose Singer theorem are from [Michor, 88], see also [Michor, 91], which are also the source for 11.8 and 11.9. The results 11.15 and 11.16 appear here for the first time.

CHAPTER IV.
JETS AND NATURAL BUNDLES

In this chapter we start our systematic treatment of geometric objects and operators. It has become commonplace to think of geometric objects on a manifold M as forming fiber bundles over the base M and to work with sections of these bundles. The concrete objects were frequently described in coordinates by their behavior under the coordinate changes. Stressing the change of coordinates, we can say that local diffeomorphisms on the base manifold operate on the bundles of geometric objects. Since a further usual assumption is that the resulting transformations depend only on germs of the underlying morphisms, we actually deal with functors defined on all open submanifolds of M and local diffeomorphisms between them (let us recall that local diffeomorphisms are globally defined locally invertible maps), see the preface. This is the point of view introduced by [Nijenhuis, 72] and worked out later by [Terng, 78], [Palais, Terng, 77], [Epstein, Thurston, 79] and others. These functors are fully determined by their restriction to any open submanifold and therefore they extend to the whole category $\mathcal{M}f_m$ of m-dimensional manifolds and local diffeomorphisms. An important advantage of such a definition of bundles of geometric objects is that we get a precise definition of geometric operators in the concept of natural operators. These are rules transforming sections of one natural bundle into sections of another one and commuting with the induced actions of local diffeomorphisms between the base manifolds.

In the theory of natural bundles and operators, a prominent role is played by jets. Roughly speaking, jets are certain equivalence classes of smooth maps between manifolds, which are represented by Taylor polynomials. We start this chapter with a thorough treatment of jets and jet bundles, and we investigate the so called jet groups. Then we give the definition of natural bundles and deduce that the r-th order natural bundles coincide with the associated fibre bundles to r-th order frame bundles. So they are in bijection with the actions of the r-th order jet group G^r_m on manifolds. Moreover, natural transformations between the r-th order natural bundles bijectively correspond to G^r_m-equivariant maps. Let us note that in chapter V we deduce a rather general theory of functors on categories over manifolds and we prove that both the finiteness of the order and the regularity of natural bundles are consequences of the other axioms, so that actually we describe all natural bundles here. Next we treat the basic properties of natural operators. In particular, we show that k-th order natural operators are described by natural transformations of the k-th order jet prolongations of the bundles in question. This reduces even the problem of finding finite order natural operators to determining G^r_m-equivariant maps, which will be discussed in chapter VI.

Further we present the procedure of principal prolongation of principal fiber bundles based on an idea of [Ehresmann, 55] and we show that the jet prolongations of associated bundles are associated bundles to the principal prolongations of the corresponding principal bundles. This fact is of basic importance for the theory of gauge natural bundles and operators, the foundations of which will be presented in chapter XII. The canonical form on first order principal prolongation of a principal bundle generalizes the well known canonical form on an r-th order frame bundle. These canonical forms are used in a formula for the first jet prolongation of sections of arbitrary associated fiber bundles, which represents a common basis for several algorithms in different branches of differential geometry. At the end of the chapter, we reformulate a part of the theory of connections from the point of view of jets, natural bundles and natural operators. This is necessary for our investigation of natural operations with connections, but we believe that this also demonstrates the power of the jet approach to give a clear picture of geometric concepts.

12. Jets

12.1. Roughly speaking, two maps f, $g: M \to N$ are said to determine the same r-jet at $x \in M$, if they have the r-th order contact at x. To make this idea precise, we first define the r-th order contact of two curves on a manifold. We recall that a smooth function $\mathbb{R} \to \mathbb{R}$ is said to vanish to r-th order at a point, if all its derivatives up to order r vanish at this point.

Definition. Two curves $\gamma, \delta: \mathbb{R} \to M$ have the r-th contact at zero, if for every smooth function φ on M the difference $\varphi \circ \gamma - \varphi \circ \delta$ vanishes to r-th order at $0 \in \mathbb{R}$.

In this case we write $\gamma \sim_r \delta$. Obviously, $\sim_r$ is an equivalence relation. For $r = 0$ this relation means $\gamma(0) = \delta(0)$.

Lemma. *If $\gamma \sim_r \delta$, then $f \circ \gamma \sim_r f \circ \delta$ for every map $f: M \to N$.*

Proof. If φ is a function on N, then $\varphi \circ f$ is a function on M. Hence $\varphi \circ f \circ \gamma - \varphi \circ f \circ \delta$ has r-th order zero at 0. □

12.2. Definition. Two maps f, $g: M \to N$ are said to determine the same r-jet at $x \in M$, if for every curve $\gamma: \mathbb{R} \to M$ with $\gamma(0) = x$ the curves $f \circ \gamma$ and $g \circ \gamma$ have the r-th order contact at zero.

In such a case we write $j^r_x f = j^r_x g$ or $j^r f(x) = j^r g(x)$.

An equivalence class of this relation is called an r-jet of M into N. Obviously, $j^r_x f$ depends on the germ of f at x only. The set of all r-jets of M into N is denoted by $J^r(M,N)$. For $X = j^r_x f \in J^r(M,N)$, the point $x =: \alpha X$ is the *source* of X and the point $f(x) =: \beta X$ is the *target* of X. We denote by π^r_s, $0 \leq s \leq r$, the projection $j^r_x f \mapsto j^s_x f$ of r-jets into s-jets. By $J^r_x(M,N)$ or $J^r(M,N)_y$ we mean the set of all r-jets of M into N with source $x \in M$ or target $y \in N$, respectively, and we write $J^r_x(M,N)_y = J^r_x(M,N) \cap J^r(M,N)_y$. The map $j^r f: M \to J^r(M,N)$ is called the r-th jet prolongation of $f: M \to N$.

12.3. Proposition. *If two pairs of maps f, $\bar{f}: M \to N$ and g, $\bar{g}: N \to Q$ satisfy $j^r_x f = j^r_x \bar{f}$ and $j^r_y g = j^r_y \bar{g}$, $f(x) = y = \bar{f}(x)$, then $j^r_x(g \circ f) = j^r_x(\bar{g} \circ \bar{f})$.*

Proof. Take a curve γ on M with $\gamma(0) = x$. Then $j^r_x f = j^r_x \bar{f}$ implies $f \circ \gamma \sim_r \bar{f} \circ \gamma$, lemma 12.1 gives $\bar{g} \circ f \circ \gamma \sim_r \bar{g} \circ \bar{f} \circ \gamma$ and $j^r_y g = j^r_y \bar{g}$ yields $g \circ f \circ \gamma \sim_r \bar{g} \circ f \circ \gamma$. Hence $g \circ f \circ \gamma \sim_r \bar{g} \circ \bar{f} \circ \gamma$. □

In other words, r-th order contact of maps is preserved under composition. If $X \in J^r_x(M,N)_y$ and $Y \in J^r_y(N,Q)_z$ are of the form $X = j^r_x f$ and $Y = j^r_y g$, we can define the composition $Y \circ X \in J^r_x(M,Q)_z$ by

$$Y \circ X = j^r_x(g \circ f).$$

By the above proposition, $Y \circ X$ does not depend on the choice of f and g. We remark that we find it useful to denote the composition of r-jets by the same symbol as the composition of maps. Since the composition of maps is associative, the same holds for r-jets. Hence all r-jets form a category, the units of which are the r-jets of the identity maps of manifolds. An element $X \in J^r_x(M,N)_y$ is invertible, if there exists $X^{-1} \in J^r_y(N,M)_x$ such that $X^{-1} \circ X = j^r_x(\mathrm{id}_M)$ and $X \circ X^{-1} = j^r_y(\mathrm{id}_N)$. By the implicit function theorem, $X \in J^r(M,N)$ is invertible if and only if the underlying 1-jet $\pi^r_1 X$ is invertible. The existence of such a jet implies $\dim M = \dim N$. We denote by $\mathrm{inv} J^r(M,N)$ the set of all invertible r-jets of M into N.

12.4. Let $f: M \to \bar{M}$ be a local diffeomorphism and $g: N \to \bar{N}$ be a map. Then there is an induced map $J^r(f,g): J^r(M,N) \to J^r(\bar{M},\bar{N})$ defined by

$$J^r(f,g)(X) = (j^r_y g) \circ X \circ (j^r_x f)^{-1}$$

where $x = \alpha X$ and $y = \beta X$ are the source and target of $X \in J^r(M,N)$. Since the jet composition is associative, J^r is a functor defined on the product category $\mathcal{M}f_m \times \mathcal{M}f$. (We shall see in 12.6 that the values of J^r lie in the category $\mathcal{FM}$.)

12.5. We are going to describe the coordinate expression of r-jets. We recall that a *multiindex* of range m is a m-tuple $\alpha = (\alpha_1, \dots, \alpha_m)$ of non-negative integers. We write $|\alpha| = \alpha_1 + \cdots + \alpha_m$, $\alpha! = \alpha_1! \cdots \alpha_m!$ (with $0! = 1$), $x^\alpha = (x^1)^{\alpha_1} \dots (x^m)^{\alpha_m}$ for $x = (x^1, \dots, x^m) \in \mathbb{R}^m$. We denote by

$$D_\alpha f = \frac{\partial^{|\alpha|} f}{(\partial x^1)^{\alpha_1} \dots (\partial x^m)^{\alpha_m}}$$

the partial derivative with respect to the multiindex α of a function $f: U \subset \mathbb{R}^m \to \mathbb{R}$.

Proposition. *Given a local coordinate system x^i on M in a neighborhood of x and a local coordinate system y^p on N in a neighborhood of $f(x)$, two maps f, $g: M \to N$ satisfy $j^r_x f = j^r_x g$ if and only if all the partial derivatives up to order r of the components f^p and g^p of their coordinate expressions coincide at x.*

Proof. We first deduce that two curves $\gamma(t)$, $\delta(t): \mathbb{R} \to N$ satisfy $\gamma \sim_r \delta$ if and only if

$$\frac{d^k(y^p \circ \gamma)(0)}{dt^k} = \frac{d^k(y^p \circ \delta)(0)}{dt^k} \tag{1}$$

$k = 0, 1, \ldots, r$, for all coordinate functions y^p. On one hand, if $\gamma \sim_r \delta$, then $y^p \circ \gamma - y^p \circ \delta$ vanishes to order r, i.e. (1) is true. On the other hand, let (1) hold. Given a function φ on N with coordinate expression $\varphi(y^1, \ldots, y^n)$, we find by the chain rule that all derivatives up to order r of $\varphi \circ \delta$ depend only on the partial derivatives up to order r of φ at $\gamma(0)$ and on (1). Hence $\varphi \circ \gamma - \varphi \circ \delta$ vanishes to order r at 0.

If the partial derivatives up to the order r of f^p and g^p coincide at x, then the chain rule implies $f \circ \gamma \sim_r g \circ \gamma$ by (1). This means $j^r_x f = j^r_x g$. Conversely, assume $j^r_x f = j^r_x g$. Consider the curves $x^i = a^i t$ with arbitrary a^i. Then the coordinate condition for $f \circ \gamma \sim_r g \circ \gamma$ reads

$$\sum_{|\alpha|=k} (D_\alpha f^p(x)) a^\alpha = \sum_{|\alpha|=k} (D_\alpha g^p(x)) a^\alpha \tag{2}$$

$k = 0, 1, \ldots, r$. Since a^i are arbitrary, (2) implies that all partial derivatives up to order r of f^p and g^p coincide at x. $\square$

Now we can easily prove that the auxiliary relation $\gamma \sim_r \delta$ can be expressed in terms of r-jets.

Corollary. *Two curves $\gamma, \delta \colon \mathbb{R} \to M$ satisfy $\gamma \sim_r \delta$ if and only if $j^r_0 \gamma = j^r_0 \delta$.*

Proof. Since $x^i \circ \gamma$ and $x^i \circ \delta$ are the coordinate expressions of γ and δ, (1) is equivalent to $j^r_0 \gamma = j^r_0 \delta$. $\square$

12.6. Write $L^r_{m,n} = J^r_0(\mathbb{R}^m, \mathbb{R}^n)_0$. By proposition 12.5, the elements of $L^r_{m,n}$ can be identified with the r-th order Taylor expansions of the generating maps, i.e. with the n-tuples of polynomials of degree r in m variables without absolute term. Such an expression

$$\sum_{1 \le |\alpha| \le r} a^p_\alpha x^\alpha$$

will be called the polynomial representative of an r-jet. Hence $L^r_{m,n}$ is a numerical space of the variables a^p_α. Standard combinatorics yields $\dim L^r_{m,n} = n\left[\binom{m+r}{m} - 1\right]$. The coordinates on $L^r_{m,n}$ will sometimes be denoted more explicitly by $a^p_i, a^p_{ij}, \ldots, a^p_{i_1 \ldots i_r}$, symmetric in all subscripts. The projection $\pi^r_s \colon L^r_{m,n} \to L^s_{m,n}$ consists in suppressing all terms of degree $> s$.

The jet composition $L^r_{m,n} \times L^r_{n,q} \to L^r_{m,q}$ is evaluated by taking the composition of the polynomial representatives and suppressing all terms of degree higher than r. Some authors call it the truncated polynomial composition. Hence the jet composition $L^r_{m,n} \times L^r_{n,q} \to L^r_{m,q}$ is a polynomial map of the numerical spaces in question. The sets $L^r_{m,n}$ can be viewed as the sets of morphisms of a category L^r over non-negative integers, the composition in which is the jet composition.

The set of all invertible elements of $L^r_{m,m}$ with the jet composition is a Lie group G^r_m called the *r-th differential group* or the *r-th jet group in dimension m*. For $r = 1$ the group G^1_m is identified with $GL(m, \mathbb{R})$. That is why some authors use $GL^r(m, \mathbb{R})$ for G^r_m.

In the case $M = \mathbb{R}^m$, we can identify every $X \in J^r(\mathbb{R}^m, \mathbb{R}^n)$ with a triple $(\alpha X, (j^r_{\beta X} t^{-1}_{\beta X}) \circ X \circ (j^r_0 t_{\alpha X}), \beta X) \in \mathbb{R}^m \times L^r_{m,n} \times \mathbb{R}^n$, where t_x means the translation on $\mathbb{R}^m$ transforming 0 into x. This product decomposition defines the structure of a smooth manifold on $J^r(\mathbb{R}^m,\mathbb{R}^n)$ as well as the structure of a fibered manifold $\pi_0^r\colon J^r(\mathbb{R}^m,\mathbb{R}^n)\to\mathbb{R}^m\times\mathbb{R}^n$. Since the jet composition in L^r is polynomial, the induced map $J^r(f,g)$ of every pair of diffeomorphisms $f\colon\mathbb{R}^m\to\mathbb{R}^m$ and $g\colon\mathbb{R}^n\to\mathbb{R}^n$ is a fibered manifold isomorphism over (f,g). Having two manifolds M and N, every local charts $\varphi\colon U\to\mathbb{R}^m$ and $\psi\colon V\to\mathbb{R}^n$ determine an identification $(\pi_0^r)^{-1}(U\times V)\cong J^r(\mathbb{R}^m,\mathbb{R}^n)$. Since the chart changings are smooth maps, this defines the structure of a smooth fibered manifold on $\pi_0^r\colon J^r(M,N)\to M\times N$. Now we see that J^r is a functor $\mathcal{M}f_m\times\mathcal{M}f\to\mathcal{FM}$. Obviously, all jet projections π_s^r are surjective submersions.

12.7. Remark. In definition 12.2 we underlined the geometrical approach to the concept of r-jets. We remark that there exists a simple algebraic approach as well. Consider the ring $C_x^\infty(M,\mathbb{R})$ of all germs of smooth functions on a manifold M at a point x and its subset $\mathcal{M}(M,x)$ of all germs with zero value at x, which is the unique maximal ideal of $C_x^\infty(M,\mathbb{R})$. Let $\mathcal{M}(M,x)^k$ be the k-th power of the ideal $\mathcal{M}(M,x)$ in the algebraic sense. Using coordinates one verifies easily that two maps f, $g\colon M\to N$, $f(x)=y=g(x)$, determine the same r-jet if and only if $\varphi\circ f-\varphi\circ g\in\mathcal{M}(M,x)^{r+1}$ for every $\varphi\in C_y^\infty(N,\mathbb{R})$.

12.8. Velocities and covelocities. The elements of the manifold $T_k^rM:=J_0^r(\mathbb{R}^k,M)$ are said to be the k-dimensional velocities of order r on M, in short (k,r)-*velocities*. The inclusion $T_k^rM\subset J^r(\mathbb{R}^m,M)$ defines the structure of a smooth fiber bundle on $T_k^rM\to M$. Every smooth map $f\colon M\to N$ is extended into an $\mathcal{FM}$-morphism $T_k^rf\colon T_k^rM\to T_k^rN$ defined by $T_k^rf(j_0^rg)=j_0^r(f\circ g)$. Hence T_k^r is a functor $\mathcal{M}f\to\mathcal{FM}$. Since every map $\mathbb{R}^k\to M_1\times M_2$ coincides with a pair of maps $\mathbb{R}^k\to M_1$ and $\mathbb{R}^k\to M_2$, functor T_k^r preserves products. For $k=r=1$ we obtain another definition of the tangent functor $T=T_1^1$.

We remark that we can now express the contents of definition 12.2 by saying that $j_x^rf=j_x^rg$ holds if and only if the restrictions of both T_1^rf and T_1^rg to $(T_1^rM)_x$ coincide.

The space $T_k^{r*}M=J^r(M,\mathbb{R}^k)_0$ is called the space of all (k,r)-*covelocities* on M. In the most important case $k=1$ we write in short $T_1^{r*}=T^{r*}$. Since $\mathbb{R}^k$ is a vector space, $T_k^{r*}M\to M$ is a vector bundle with $j_x^r\varphi(u)+j_x^r\psi(u)=j_x^r(\varphi(u)+\psi(u))$, $u\in M$, and $kj_x^r\varphi(u)=j_x^rk\varphi(u)$, $k\in\mathbb{R}$. Every local diffeomorphism $f\colon M\to N$ is extended to a vector bundle morphism $T_k^{r*}f\colon T_k^{r*}M\to T_k^{r*}N$, $j_x^r\varphi\mapsto j_{f(x)}^r(\varphi\circ f^{-1})$, where f^{-1} is constructed locally. In this sense T_k^{r*} is a functor on $\mathcal{M}f_m$. For $k=r=1$ we obtain the construction of the cotangent bundles as a functor $T_1^{1*}=T^*$ on $\mathcal{M}f_m$. We remark that the behavior of T_k^{r*} on arbitrary smooth maps will be reflected in the concept of star bundle functors we shall introduce in 41.2.

12.9. Jets as algebra homomorphisms. The multiplication of reals induces a multiplication in every vector space $T_x^{r*}M$ by

$$(j_x^r\varphi(u))(j_x^r\psi(u))=j_x^r(\varphi(u)\psi(u)),$$

which turns $T_x^{r*}M$ into an algebra. Every $j_x^rf\in J_x^r(M,N)_y$ defines an algebra homomorphism $\hom(j_x^rf)\colon T_y^{r*}N\to T_x^{r*}M$ by $j_y^r\varphi\mapsto j_x^r(\varphi\circ f)$. To deduce the converse assertion, consider some local coordinates x^i on M and y^p on N centered at x and y. The algebra $T_y^{r*}N$ is generated by $j_0^r y^p$. If we prescribe quite arbitrarily the images $\Phi(j_0^r y^p)$ in $T_x^{r*}M$, this is extended into a unique algebra homomorphism $\Phi\colon T_y^{r*}N \to T_x^{r*}M$. The n-tuple $\Phi(j_0^r y^p)$ represents the coordinate expression of a jet $X \in J_x^r(M,N)_y$ and one verifies easily $\Phi = \operatorname{hom}(X)$. Thus we have proved

Proposition. *There is a canonical bijection between $J_x^r(M,N)_y$ and the set of all algebra homomorphisms $Hom(T_y^{r*}N, T_x^{r*}M)$.*

For $r=1$ the product of any two elements in T_x^*M is zero. Hence the algebra homomorphisms coincide with the linear maps $T_y^*N \to T_x^*M$. This gives an identification $J^1(M,N) = TN \otimes T^*M$ (which can be deduced by several other ways as well).

12.10. Kernel descriptions. The projection $\pi_{r-1}^r\colon T^{r*}M \to T^{r-1*}M$ is a linear morphism of vector bundles. Its kernel is described by the following exact sequence of vector bundles over M

$$0 \to S^rT^*M \to T^{r*}M \xrightarrow{\pi_{r-1}^r} T^{r-1*}M \to 0 \tag{1}$$

where S^r indicates the r-th symmetric tensor power. To prove it, we first construct a map $p\colon \overset{r}{\times}\, T^*M \to T^{r*}M$. Take r functions $f_1,\dots,f_r$ on M with values zero at x and construct the r-jet at x of their product. One sees directly that $j_x^r(f_1\dots f_r)$ depends on $j_x^1f_1,\dots,j_x^1f_r$ only and lies in $\ker(\pi_{r-1}^r)$. We have $j_x^r(f_1\dots f_r) = j_x^1f_1 \bigcirc \dots \bigcirc j_x^1f_r$, where $\bigcirc$ means the symmetric tensor product, so that p is uniquely extended into a linear isomorphism of S^rT^*M into $\ker(\pi_{r-1}^r)$.

Next we shall use a similar idea for a geometrical construction of an identification, which is usually justified by the coordinate evaluations only. Let $\hat{y}$ denote the constant map of M into $y \in N$.

Proposition. *The subspace $(\pi_{r-1}^r)^{-1}(j_x^{r-1}\hat{y}) \subset J_x^r(M,N)_y$ is canonically identified with $T_yN \otimes S^rT_x^*M$.*

Proof. Let $B \in T_yN$ and $j_x^1f_p \in T_x^*M$, $p=1,\dots,r$. For every $j_y^r\varphi \in T_y^{r*}N$, take the value $B\varphi \in \mathbb{R}$ of the derivative of φ in direction B and construct a function $(B\varphi)f_1(u)\dots f_r(u)$ on M. It is easy to see that $j_y^r\varphi \mapsto j_x^r((B\varphi)f_1\dots f_r)$ is an algebra homomorphism $T_y^{r*}N \to T_x^{r*}M$. This defines a map $p\colon T_yN \times T_x^*M \overset{r\text{-times}}{\times \dots \times} T_x^*M \to J_x^r(M,N)_y$. Using coordinates one verifies that p generates linearly the required identification. □

For $r=1$ we have a distinguished element $j_x^1\hat{y}$ in every fiber of $J^1(M,N) \to M \times N$. This identifies $J^1(M,N)$ with $TN \otimes T^*M$.

In particular, if we apply the above proposition to the projection $\pi_{r-1}^r\colon (T_k^rM)_x \to (T_k^{r-1}M)_x$, $x \in M$, we find

$$(\pi_{r-1}^r)^{-1}(j_0^{r-1}\hat{x}) = T_xM \otimes S^r\mathbb{R}^{k*}. \tag{2}$$

12.11. Proposition. *$\pi^r_{r-1}: J^r(M,N) \to J^{r-1}(M,N)$ is an affine bundle, the modelling vector bundle of which is the pullback of $TN \otimes S^rT^*M$ over $J^{r-1}(M,N)$.*

Proof. Interpret $X \in J^r_x(M,N)_y$ and $A \in T_yN \otimes S^rT^*_xM \subset J^r_x(M,N)_y$ as algebra homomorphisms $T^{r*}_yN \to T^{r*}_xM$. For every $\Phi \in T^{r*}_yN$ we have $\pi^r_{r-1}(A(\Phi)) = 0$ and $\pi^r_0(X(\Phi)) = 0$. This implies $X(\Phi)A(\Psi) = 0$ and $A(\Phi)A(\Psi) = 0$ for any other $\Psi \in T^{r*}_yN$. Hence $X(\Phi\Psi) + A(\Phi\Psi) = X(\Phi)X(\Psi) = (X(\Phi) + A(\Phi))(X(\Psi) + A(\Psi))$, so that $X + A$ is also an algebra homomorphism $T^{r*}_yN \to T^{r*}_xM$. Using coordinates we find easily that the map $(X, A) \mapsto X + A$ gives rise to the required affine bundle structure. □

Since the tangent space to an affine space is the modelling vector space, we obtain immediately the following property of the tangent map $T\pi^r_{r-1}: TJ^r(M,N) \to TJ^{r-1}(M,N)$.

Corollary. *For every $X \in J^r_x(M,N)_y$, the kernel of the restriction of $T\pi^r_{r-1}$ to $T_XJ^r(M,N)$ is $T_yN \otimes S^rT^*_xM$.*

12.12. The frame bundle of order r. The set P^rM of all r-jets with source 0 of the local diffeomorphisms of $\mathbb{R}^m$ into M is called the r-th order frame bundle of M. Obviously, $P^rM = \operatorname{inv}T^r_m(M)$ is an open subset of $T^r_m(M)$, which defines a structure of a smooth fiber bundle on $P^rM \to M$. The group G^r_m acts smoothly on P^rM on the right by the jet composition. Since for every $j^r_0\varphi, j^r_0\psi \in P^r_xM$ there is a unique element $j^r_0(\varphi^{-1} \circ \psi) \in G^r_m$ satisfying $(j^r_0\varphi)\circ(j^r_0(\varphi^{-1}\circ\psi)) = j^r_0\psi$, P^rM is a principal fiber bundle with structure group G^r_m. For $r = 1$, the elements of $\operatorname{inv}J^1_0(\mathbb{R}^m, M)_x$ are identified with the linear isomorphisms $\mathbb{R}^m \to T_xM$ and $G^1_m = GL(m)$, so that P^1M coincides with the bundle of all linear frames in TM, i.e. with the classical frame bundle of M.

Every velocities space T^r_kM is a fiber bundle associated with P^rM with standard fiber $L^r_{k,m}$. The basic idea consists in the fact that for every $j^r_0f \in (T^r_kM)_x$ and $j^r_0\varphi \in P^r_xM$ we have $j^r_0(\varphi^{-1} \circ f) \in L^r_{k,m}$, and conversely, every $j^r_0g \in L^r_{k,m}$ and $j^r_0\varphi \in P^r_xM$ determine $j^r_0(\varphi\circ g) \in (T^r_kM)_x$. Thus, if we formally define a left action $G^r_m \times L^r_{k,m} \to L^r_{k,m}$ by $(j^r_0h, j^r_0g) \mapsto j^r_0(h \circ g)$, then T^r_kM is canonically identified with the associated fiber bundle $P^rM[L^r_{k,m}]$.

Quite similarly, every covelocities space T^{r*}_kM is a fiber bundle associated with P^rM with standard fiber $L^r_{m,k}$ with respect to the left action $G^r_m \times L^r_{m,k} \to L^r_{m,k}$, $(j^r_0h, j^r_0g) \mapsto j^r_0(g \circ h^{-1})$. Furthermore, $P^rM \times P^rN$ is a principal fiber bundle over $M \times N$ with structure group $G^r_m \times G^r_n$. The space $J^r(M,N)$ is a fiber bundle associated with $P^rM \times P^rN$ with standard fiber $L^r_{m,n}$ with respect to the left action $(G^r_m \times G^r_n) \times L^r_{m,n} \to L^r_{m,n}$, $((j^r_0\varphi, j^r_0\psi), j^r_0f) \mapsto j^r_0(\psi \circ f \circ \varphi^{-1})$.

Every local diffeomorphism $f: M \to N$ induces a map $P^rf: P^rM \to P^rN$ by $P^rf(j^r_0\varphi) = j^r_0(f \circ \varphi)$. Since G^r_m acts on the right on both P^rM and P^rN, P^rf is a local principal fiber bundle isomorphism. Hence P^r is a functor from $\mathcal{M}f_m$ into the category $\mathcal{PB}(G^r_m)$.

Given a left action of G^r_m on a manifold S, we have an induced map

$$\{P^rf, \mathrm{id}_S\}: P^rM[S] \to P^rN[S]$$

between the associated fiber bundles with standard fiber S, see 10.9. The rule $M \mapsto P^rM[S]$, $f \mapsto \{P^rf, \mathrm{id}_S\}$ is a bundle functor on $\mathcal{M}f_m$ as defined in 14.1. A very interesting result is that every bundle functor on $\mathcal{M}f_m$ is of this type. This will be proved in section 22, but the proof involves some rather hard analytical results.

12.13. For every Lie group G, T^r_kG is also a Lie group with multiplication $(j^r_0f(u))(j^r_0g(u)) = j^r_0(f(u)g(u))$, $u \in \mathbb{R}^k$, where $f(u)g(u)$ is the product in G. Clearly, if we consider the multiplication map $\mu\colon G \times G \to G$, then the multiplication map of T^r_kG is $T^r_k\mu\colon T^r_kG \times T^r_kG \to T^r_kG$. The jet projections $\pi^r_s\colon T^r_kG \to T^s_kG$ are group homomorphisms. For $s = 0$, there is a splitting $\iota\colon G \to T^r_kG$ of $\pi^r_0 = \beta\colon T^r_kG \to G$ defined by $\iota(g) = j^r_0\hat{g}$, where $\hat{g}$ means the constant map of $\mathbb{R}^k$ into $g \in G$. Hence T^r_kG is a semidirect product of G and of the kernel of $\beta\colon T^r_kG \to G$.

If G acts on the left on a manifold M, then T^r_kG acts on T^r_kM by

$$(j^r_0f(u))(j^r_0g(u)) = j^r_0\big(f(u)(g(u))\big),$$

where $f(u)(g(u))$ means the action of $f(u) \in G$ on $g(u) \in M$. If we consider the action map $\ell\colon G \times M \to M$, then the action map of the induced action is $T^r_k\ell\colon T^r_kG \times T^r_kM \to T^r_kM$. The same is true for right actions.

12.14. ***r*-th order tangent vectors.** In general, consider the dual vector bundle $T^{r\square}_kM = (T^{r*}_kM)^*$ of the (k,r)-covelocities bundle on M. For every map $f\colon M \to N$ the jet composition $A \mapsto A \circ (j^r_xf)$, $x \in M$, $A \in (T^{r*}_kN)_{f(x)}$ defines a linear map $\lambda(j^r_xf)\colon (T^{r*}_kN)_{f(x)} \to (T^{r*}_kM)_x$. The dual map $(\lambda(j^r_xf))^* =: (T^{r\square}_kf)_x\colon (T^{r\square}_kM)_x \to (T^{r\square}_kN)_{f(x)}$ determines a functor $T^{r\square}_k$ on $\mathcal{M}f$ with values in the category of vector bundles. For $r > 1$ these functors do not preserve products by the dimension argument. In the most important case $k = 1$ we shall write $T^{r\square}_1 = T^{(r)}$ (in order to distinguish from the r-th iteration of T). The elements of $T^{(r)}M$ are called r-th order tangent vectors on M. We remark that for $r = 1$ the formula $TM = (T^*M)^*$ can be used for introducing the vector bundle structure on TM.

Dualizing the exact sequence 12.10.(1), we obtain

$$0 \to T^{(r-1)}M \to T^{(r)}M \to S^rTM \to 0. \tag{1}$$

This shows that there is a natural injection of the $(r-1)$-st order tangent vectors into the r-th order ones. Analyzing the proof of 12.10.(1), one finds easily that (1) has functorial character, i.e. for every map $f\colon M \to N$ the following diagram commutes

$$\begin{array}{ccccccccc} 0 & \longrightarrow & T^{(r-1)}M & \longrightarrow & T^{(r)}M & \longrightarrow & S^rTM & \longrightarrow & 0 \\ & & \downarrow{\scriptstyle T^{(r-1)}f} & & \downarrow{\scriptstyle T^{(r)}f} & & \downarrow{\scriptstyle S^rTf} & & \\ 0 & \longrightarrow & T^{(r-1)}N & \longrightarrow & T^{(r)}N & \longrightarrow & S^rTN & \longrightarrow & 0 \end{array} \tag{2}$$

12.15. Contact elements. Let N be an n-dimensional submanifold of a manifold M. For every local chart $\varphi\colon N \to \mathbb{R}^n$, the rule $x \mapsto \varphi^{-1}(x)$ considered as a map $\mathbb{R}^n \to M$ is called a local parametrization of N. The concept of the contact of submanifolds of the same dimension can be reduced to the concept of r-jets.

Definition. Two n-dimensional submanifolds N and $\bar{N}$ of M are said to have r-th order contact at a common point x, if there exist local parametrizations $\psi\colon \mathbb{R}^n \to M$ of N and $\bar{\psi}\colon \mathbb{R}^n \to M$ of $\bar{N}$, $\psi(0) = x = \bar{\psi}(0)$, such that $j_0^r\psi = j_0^r\bar{\psi}$.

An equivalence class of n-dimensional submanifolds of M will be called an n-dimensional contact element of order r on M, in short a *contact* (n,r)-*element* on M. We denote by $K_n^r M$ the set of all contact (n,r)-elements on M. We have a canonical projection 'point of contact' $K_n^r M \to M$.

An (n,r)-velocity $A \in (T_n^r M)_x$ is called regular, if its underlying 1-jet corresponds to a linear map $\mathbb{R}^n \to T_x M$ of rank n. For every local parametrization ψ of an n-dimensional submanifold, $j_0^r\psi$ is a regular (n,r)-velocity. Since in the above definition we can reparametrize ψ and $\bar{\psi}$ in the same way (i.e. we compose them with the same origin preserving diffeomorphism of $\mathbb{R}^m$), every contact (n,r)-element on M can be identified with a class $A \circ G_n^r$, where A is a regular (n,r)-velocity on M. There is a unique structure of a smooth fibered manifold on $K_n^r M \to M$ with the property that the factor projection from the subbundle $\operatorname{reg}T_n^r M \subset T_n^r M$ of all regular (n,r)-velocities into $K_n^r M$ is a surjective submersion. (The simplest way how to check it is to use the identification of an open subset in $K_n^r \mathbb{R}^m$ with the r-th jet prolongation of fibered manifold $\mathbb{R}^n \times \mathbb{R}^{m-n} \to \mathbb{R}^n$, which will be described in the end of 12.16.)

Every local diffeomorphism $f\colon M \to \bar{M}$ preserves the contact of submanifolds. This induces a map $K_n^r f\colon K_n^r M \to K_n^r \bar{M}$, which is a fibered manifold morphism over f. Hence K_n^r is a bundle functor on $\mathcal{M}f_m$. For $r = 1$ each fiber $(K_n^1 M)_x$ coincides with the Grassmann manifold of n-planes in $T_x M$, see 10.5. That is why $K_n^1 M$ is also called the Grassmannian n-bundle of M.

12.16. Jet prolongations of fibered manifolds. Let $p\colon Y \to M$ be a fibered manifold, $\dim M = m$, $\dim Y = m+n$. The set $J^r Y$ (also written as $J^r(Y \to M)$ or $J^r(p\colon Y \to M)$, if we intend to stress the base or the bundle projection) of all r-jets of the local sections of Y will be called the r-th jet prolongation of Y. Using polynomial representatives we find easily that an element $X \in J_x^r(M,Y)$ belongs to $J^r Y$ if and only if $(j_{\beta X}^r p) \circ X = j_x^r(\mathrm{id}_M)$. Hence $J^r Y \subset J^r(M,Y)$ is a closed submanifold. For every section s of $Y \to M$, $j^r s$ is a section of $J^r Y \to M$.

Let x^i or y^p be the canonical coordinates on $\mathbb{R}^m$ or $\mathbb{R}^n$, respectively. Every local fiber chart $\varphi\colon U \to \mathbb{R}^{m+n}$ on Y identifies $(\pi_0^r)^{-1}(U)$ with $J^r(\mathbb{R}^m, \mathbb{R}^n)$. This defines the induced local coordinates y_α^p on $J^r Y$, $1 \le |\alpha| \le r$, where α is any multi index of range m.

Let $q\colon Z \to N$ be another fibered manifold and $f\colon Y \to Z$ be an $\mathcal{FM}$-morphism with the property that the base map $f_0\colon M \to N$ is a local diffeomorphism. Then the map $J^r(f, f_0)\colon J^r(M,Y) \to J^r(N,Z)$ constructed in 12.4 transforms $J^r Y$ into $J^r Z$. Indeed, $X \in J^r Y$, $\beta X = y$ is characterized by $(j_y^r p) \circ X = j_x^r \mathrm{id}_M$, $x = p(y)$, and $q \circ f = f_0 \circ p$ implies $(j_{f(y)}^r q) \circ ((j_y^r f) \circ X \circ (j_{f_0(x)}^r f_0^{-1})) = (j_x^r f_0) \circ (j_y^r p) \circ X \circ j_{f_0(x)}^r f_0^{-1} = j_{f_0(x)}^r \mathrm{id}_N$. The restricted map will be denoted by $J^r f\colon J^r Y \to J^r Z$ and called the r-th jet prolongation of f. Let $\mathcal{FM}_m$ denote the category of fibered manifolds with m-dimensional bases and their morphisms with the additional property that the base maps are local diffeomorphisms. Then the construction of the r-th jet prolongations can be interpreted as a functor $J^r\colon \mathcal{FM}_m \to \mathcal{FM}$. (If there will be a danger of confusion with the bifunctor J^r of spaces of r-jets between pairs of manifolds, we shall write J^r_{fib} for the fibered manifolds case.)

By proposition 12.11, $\pi^r_{r-1}\colon J^r(M,Y) \to J^{r-1}(M,Y)$ is an affine bundle, the associated vector bundle of which is the pullback of $TY \otimes S^r T^*M$ over $J^{r-1}(M,Y)$. Taking into account the local trivializations of Y, we find that $\pi^r_{r-1}\colon J^r Y \to J^{r-1}Y$ is an affine subbundle of $J^r(M,Y)$ and its modelling vector bundle is the pullback of $VY \otimes S^r T^*M$ over $J^{r-1}Y$, where VY denotes the vertical tangent bundle of Y. For $r = 1$ it is useful to give a direct description of the affine bundle structure on $J^1 Y \to Y$ because of its great importance in the theory of connections. The space $J^1(M,Y)$ coincides with the vector bundle $TY \otimes T^*M = L(TM, TY)$. A 1-jet $X\colon T_x M \to T_y Y$, $x = p(y)$, belongs to $J^1 Y$ if and only if $Tp \circ X = \mathrm{id}_{T_x M}$. The kernel of such a projection induced by Tp is $V_y Y \otimes T^*_x M$, so that the pre-image of $\mathrm{id}_{T_x M}$ in $T_y Y \otimes T^*_x M$ is an affine subspace with modelling vector space $V_y Y \otimes T^*_x M$.

If we specialize corollary 12.11 to the case of a fibered manifold Y, we deduce that for every $X \in J^r Y$ the kernel of the restriction of $T\pi^r_{r-1}\colon TJ^r Y \to TJ^{r-1}Y$ to $T_X J^r Y$ is $V_{\beta X} Y \otimes S^r T^*_{\alpha X} M$.

In conclusion we describe the relation between the contact (n,r)-elements on a manifold M and the elements of the r-th jet prolongation of a suitable local fibration on M. In a sufficiently small neighborhood U of an arbitrary $x \in M$ there exists a fibration $p\colon U \to N$ over an n-dimensional manifold N. By the definition of contact elements, every $X \in K^r_n M$ transversal to p (i.e. the underlying contact 1-element of X is transversal to p) is identified with an element of $J^r(U \to N)$ and vice versa. In particular, if we take $U \cong \mathbb{R}^n \times \mathbb{R}^{m-n}$, then the latter identification induces some simple local coordinates on $K^r_n M$.

12.17. If $E \to M$ is a vector bundle, then $J^r E \to M$ is also a vector bundle, provided we define $j^r_x s_1(u) + j^r_x s_2(u) = j^r_x(s_1(u) + s_2(u))$, where u belongs to a neighborhood of $x \in M$, and $k j^r_x s(u) = j^r_x ks(u)$, $k \in \mathbb{R}$.

Let $Z \to M$ be an affine bundle with the modelling vector bundle $E \to M$. Then $J^r Z \to M$ is an affine bundle with the modelling vector bundle $J^r E \to M$. Given $j^r_x s \in J^r Z$ and $j^r_x \sigma \in J^r E$, we set $j^r_x s(u) + j^r_x \sigma(u) = j^r_x(s(u) + \sigma(u))$, where the sum $s(u) + \sigma(u)$ is defined by the canonical map $Z \times_M E \to Z$.

12.18. Infinite jets. Consider an infinite sequence

$$A_1, A_2, \dots, A_r, \dots \tag{1}$$

of jets $A_i \in J^i(M,N)$ satisfying $A_i = \pi^{i+1}_i(A_{i+1})$ for all $i = 1, \dots$. Such a sequence is called a *jet of order* ∞ or an *infinite jet* of M into N. Hence the set $J^\infty(M,N)$ of all infinite jets of M into N is the projective limit of the sequence

$$J^1(M,N) \xleftarrow{\pi^2_1} J^2(M,N) \xleftarrow{\pi^3_2} \dots \xleftarrow{\pi^r_{r-1}} J^r(M,N) \xleftarrow{\pi^{r+1}_r} \dots$$

We denote by $\pi_r^\infty\colon J^\infty(M,N) \to J^r(M,N)$ the projection transforming the sequence (1) into its r-th term. In this book we usually treat $J^\infty(M,N)$ as a set only, i.e. we consider no topological or smooth structure on $J^\infty(M,N)$. (For the latter subject the reader can consult e.g. [Michor, 80].)

Given a smooth map $f\colon M \to N$, the sequence

$$j_x^1 f \leftarrow j_x^2 f \leftarrow \cdots \leftarrow j_x^r f \leftarrow \dots$$

$x \in M$, which is denoted by $j_x^\infty f$ or $j^\infty f(x)$, is called the *infinite jet of f at x*. The classical Borel theorem, see 19.4, implies directly that every element of $J^\infty(M,N)$ is the infinite jet of a smooth map of M into N, see also 19.4.

The spaces $T_k^\infty M$ of all k-dimensional velocities of infinite order and the infinite differential group G_m^∞ in dimension m are defined in the same way. Having a fibered manifold $Y \to M$, the infinite jets of its sections form the *infinite jet prolongation $J^\infty Y$ of Y*.

12.19. Jets of fibered manifold morphisms. If we consider the jets of morphisms of fibered manifolds, we can formulate additional conditions concerning the restrictions to the fibers or the induced base maps. In the first place, if we have two maps f, g of a fibered manifold Y into another manifold, we say they determine the same (r,s)-jet at $y \in Y$, $s \geq r$, if

$$j_y^r f = j_y^r g \text{ and } j_y^s(f|Y_x) = j_y^s(g|Y_x), \tag{1}$$

where Y_x is the fiber passing through y. The corresponding equivalence class will be denoted by $j_y^{r,s} f$. Clearly (r,s)-jets of $\mathcal{FM}$-morphisms form a category, and the bundle projection determines a functor from this category into the category of r-jets. We denote by $J^{r,s}(Y,\bar{Y})$ the space of all (r,s)-jets of the fibered manifold morphisms of Y into another fibered manifold $\bar{Y}$.

Moreover, let $q \geq r$ be another integer. We say that two $\mathcal{FM}$-morphisms $f, g\colon Y \to \bar{Y}$ determine the same (r,s,q)-jet at y, if it holds (1) and

$$j_x^q Bf = j_x^q Bg, \tag{2}$$

where Bf and Bg are the induced base maps and x is the projection of y to the base BY of Y. We denote by $j_y^{r,s,q} f$ such an equivalence class and by $J^{r,s,q}(Y,\bar{Y})$ the space of all (r,s,q)-jets of the fibered manifold morphisms between Y and $\bar{Y}$. The bundle projection determines a functor from the category of (r,s,q)-jets of $\mathcal{FM}$-morphisms into the category of q-jets. Obviously, it holds

$$J^{r,s,q}(Y,\bar{Y}) = J^{r,s}(Y,\bar{Y}) \times_{J^r(BY,B\bar{Y})} J^q(BY,B\bar{Y}) \tag{3}$$

where we consider the above mentioned projection $J^{r,s}(Y,\bar{Y}) \to J^r(BY,B\bar{Y})$ and the jet projection $\pi_r^q\colon J^q(BY,B\bar{Y}) \to J^r(BY,B\bar{Y})$.

12.20. An abstract characterization of the jet spaces. We remark that [Kolář, to appear c] has recently deduced that the r-th order jets can be characterized as homomorphic images of germs of smooth maps in the following way. According to 12.3, the rule j^r defined by

$$j^r(\text{germ}_x f) = j^r_x f$$

transforms germs of smooth maps into r-jets and preserves the compositions. By 12.6, $J^r(M,N)$ is a fibered manifold over $M \times N$ for every pair of manifolds M, N. So if we denote by $G(M,N)$ the set of all germs of smooth maps of M into N, j^r can be interpreted as a map

$$j^r = j^r_{M,N}\colon G(M,N) \to J^r(M,N).$$

More generally, consider a rule F transforming every pair M, N of manifolds into a fibered manifold $F(M,N)$ over $M \times N$ and a system φ of maps $\varphi_{M,N}\colon G(M,N) \to F(M,N)$ commuting with the projections $G(M,N) \to M \times N$ and $F(M,N) \to M \times N$ for all M, N. Let us formulate the following requirements I–IV.

I. *Every $\varphi_{M,N}\colon G(M,N) \to F(M,N)$ is surjective.*

II. *For every pairs of composable germs B_1, B_2 and $\bar{B}_1$, $\bar{B}_2$, $\varphi(B_1) = \varphi(\bar{B}_1)$ and $\varphi(B_2) = \varphi(\bar{B}_2)$ imply $\varphi(B_2 \circ B_1) = \varphi(\bar{B}_2 \circ \bar{B}_1)$.*

By I and II we have a well defined composition (denoted by the same symbol as the composition of germs and maps)

$$X_2 \circ X_1 = \varphi(B_2 \circ B_1)$$

for every $X_1 = \varphi(B_1) \in F_x(M,N)_y$ and $X_2 = \varphi(B_2) \in F_y(N,P)_z$. Every local diffeomorphism $f\colon M \to \bar{M}$ and every smooth map $g\colon N \to \bar{N}$ induces a map $F(f,g)\colon F(M,N) \to F(\bar{M},\bar{N})$ defined by

$$F(f,g)(X) = \varphi(\text{germ}_y g) \circ X \circ \varphi((\text{germ}_x f)^{-1}), \qquad X \in F_x(M,N)_y.$$

III. *Each map $F(f,g)$ is smooth.*

Consider the product $N_1 \xleftarrow{p_1} N_1 \times N_2 \xrightarrow{p_2} N_2$ of two manifolds. Then we have the induced maps $F(\text{id}_M, p_1) : F(M, N_1 \times N_2) \to F(M,N_1)$ and $F(\text{id}_M, p_2)\colon F(M, N_1 \times N_2) \to F(M,N_2)$. Both $F(M,N_1)$ and $F(M,N_2)$ are fibered manifolds over M.

IV. *$F(M, N_1 \times N_2)$ coincides with the fibered product $F(M,N_1) \times_M F(M,N_2)$ and $F(id_M, p_1)$, $F(id_M, p_2)$ are the induced projections.*

Then it holds: *For every pair (F,φ) satisfying* I–IV *there exists an integer $r \geq 0$ such that $(F,\varphi) = (J^r, j^r)$.* (The proof is heavily based on the theory of Weil functors presented in chapter VIII below.)

13. Jet groups

In spite of the fact that the jet groups lie at the core of considerations concerning geometric objects and operations, they have not been studied very extensively. The paper [Terng, 78] is one of the exceptions and many results presented in this section appeared there for the first time.

13.1. Let us recall the jet groups $G^k_m = \operatorname{inv} J^k_0(\mathbb{R}^m, \mathbb{R}^m)_0$ with the multiplication defined by the composition of jets, cf. 12.6. The jet projections π^{l+1}_l define the sequence

$$G^k_m \to G^{k-1}_m \to \cdots \to G^1_m \to 1 \tag{1}$$

and the normal subgroups $B_l = \ker \pi^k_l$ (or B^k_l if more suitable) form the filtration

$$G^k_m = B_0 \supset B_1 \supset \cdots \supset B_{k-1} \supset B_k = 1. \tag{2}$$

Since we identify $J^k_0(\mathbb{R}^m, \mathbb{R}^m)$ with the space of polynomial maps $\mathbb{R}^m \to \mathbb{R}^m$ of degree less then or equal to k, we can write $G^k_m = \{f = f_1 + f_2 + \cdots + f_k\,;\ f_i \in L^i_{\text{sym}}(\mathbb{R}^m, \mathbb{R}^m),\ 1 \le i \le k, \text{ and } f_1 \in GL(m) = G^1_m\}$, where $L^i_{\text{sym}}(\mathbb{R}^m, \mathbb{R}^n)$ is the space of all homogeneous polynomial maps $\mathbb{R}^m \to \mathbb{R}^n$ of degree i. Hence G^k_m is identified with an open subset of an Euclidean space consisting of two connected components. The connected component of the unit, i.e. the space of all invertible jets of orientation preserving diffeomorphisms, will be denoted by $G^k_m{}^+$. It follows that the Lie algebra $\mathfrak{g}^k_m$ is identified with the whole space $J^k_0(\mathbb{R}^m, \mathbb{R}^m)_0$, or equivalently with the space of k-jets of vector fields on $\mathbb{R}^m$ at the origin that vanish at the origin. Since each $j^k_0 X$, $X \in \mathfrak{X}(\mathbb{R}^m)$, has a canonical polynomial representative, the elements of $\mathfrak{g}^k_m$ can also be viewed as polynomial vector fields $X = \sum a^i_\mu x^\mu \frac{\partial}{\partial x_i}$. Here the sum goes over i and all multi indices μ with $1 \le |\mu| \le k$.

For technical reasons, we shall not use any summation convention in the rest of this section and we shall use only subscripts for the indices of the space variables $x \in \mathbb{R}^n$, i.e. if $(x_1, \dots, x_n) \in \mathbb{R}^n$, then x_1^2 always means $x_1.x_1$, etc.

13.2. The tangent maps to the jet projections turn out to be jet projections as well. Hence the sequence 13.1.(1) gives rise to the sequence of Lie algebra homomorphisms

$$\mathfrak{g}^k_m \xrightarrow{\pi^k_{k-1}} \mathfrak{g}^{k-1}_m \xrightarrow{\pi^{k-1}_{k-2}} \cdots \xrightarrow{\pi^2_1} \mathfrak{g}^1_m \to 0$$

and we get the filtration by ideals $\mathfrak{b}_l = \ker \pi^k_l$ (or $\mathfrak{b}^k_l$ if more suitable)

$$\mathfrak{g}^k_m = \mathfrak{b}_0 \supset \mathfrak{b}_1 \supset \cdots \supset \mathfrak{b}_{k-1} \supset \mathfrak{b}_k = 0.$$

Let us define $\mathfrak{g}_p \subset \mathfrak{g}^k_m$, $0 \le p \le k-1$, as the space of all homogeneous polynomial vector fields of degree $p+1$, i.e. $\mathfrak{g}_p = L^{p+1}_{\text{sym}}(\mathbb{R}^m, \mathbb{R}^m)$. By definition, $\mathfrak{g}_p$ is identified with the quotient $\mathfrak{b}_p/\mathfrak{b}_{p+1}$ and at the level of vector spaces we have

$$\mathfrak{g}^k_m = \mathfrak{g}_0 \oplus \mathfrak{g}_1 \oplus \cdots \oplus \mathfrak{g}_{k-1}. \tag{1}$$

For any two subsets L_1, L_2 in a Lie algebra $\mathfrak{g}$ we write $[L_1, L_2]$ for the linear subspace generated by the brackets $[l_1, l_2]$ of elements $l_1 \in L_1$, $l_2 \in L_2$. A decomposition $\mathfrak{g} = \mathfrak{g}_0 \oplus \mathfrak{g}_1 \oplus \dots$ of a Lie algebra is called a *grading* if $[\mathfrak{g}_i, \mathfrak{g}_j] \subset \mathfrak{g}_{i+j}$ for all $0 \le i, j < \infty$. In our decomposition of $\mathfrak{g}^k_m$ we take $\mathfrak{g}_i = 0$ for all $i \ge k$.

Proposition. *The Lie algebra $\mathfrak{g}^k_m$ of the Lie group G^k_m is the vector space $\{j^k_0 X\ ;\ X \in \mathfrak{X}(\mathbb{R}^m),\ X(0) = 0\}$ with the bracket*

$$[j^k_0 X, j^k_0 Y] = -j^k_0 [X, Y] \tag{2}$$

and with the exponential mapping

$$\exp(j^k_0 X) = j^k_0 \operatorname{Fl}^X_1, \qquad j^k_0 X \in \mathfrak{g}^k_m. \tag{3}$$

The decomposition (1) is a grading and for all indices $0 \le i, j < k$ we have

$$[\mathfrak{g}_i, \mathfrak{g}_j] = \mathfrak{g}_{i+j} \qquad \text{if } m > 1, \text{ or if } m = 1 \text{ and } i \neq j. \tag{4}$$

Proof. For every vector field $X \in \mathfrak{X}(\mathbb{R}^m)$, the map $t \mapsto j^k_0 \operatorname{Fl}^X_t$ is a one-parameter subgroup in G^k_m and the corresponding element in $\mathfrak{g}^k_m$ is

$$\tfrac{\partial}{\partial t}\big|_0 j^k_0 \operatorname{Fl}^X_t = j^k_0 \left(\tfrac{\partial}{\partial t}\big|_0 \operatorname{Fl}^X_t\right) = j^k_0 X.$$

Hence $\exp(t.j^k_0 X) = j^k_0 \operatorname{Fl}^X_t$, see 4.18. Now, let us consider vector fields X, Y on $\mathbb{R}^m$ vanishing at the origin and let us write briefly $a := j^k_0 X$, $b := j^k_0 Y$. According to 3.16 and 4.18.(3) we have

$$\begin{aligned}
-2j^k_0[X,Y] = 2j^k_0[Y,X] &= j^k_0 \tfrac{\partial^2}{\partial t^2}\Big|_0 \left(\operatorname{Fl}^X_{-t} \circ \operatorname{Fl}^Y_{-t} \circ \operatorname{Fl}^X_t \circ \operatorname{Fl}^Y_t\right) \\
&= \tfrac{\partial^2}{\partial t^2}\Big|_0 \left(j^k_0 \operatorname{Fl}^X_{-t} \circ j^k_0 \operatorname{Fl}^Y_{-t} \circ j^k_0 \operatorname{Fl}^X_t \circ j^k_0 \operatorname{Fl}^Y_t\right) \\
&= \tfrac{\partial^2}{\partial t^2}\Big|_0 \left(\exp(-ta) \circ \exp(-tb) \circ \exp(ta) \circ \exp(tb)\right) \\
&= \tfrac{\partial^2}{\partial t^2}\Big|_0 \left(\operatorname{Fl}^{L_b}_t \circ \operatorname{Fl}^{L_a}_t \circ \operatorname{Fl}^{L_b}_{-t} \circ \operatorname{Fl}^{L_a}_{-t}\right)(e) = 2[j^k_0 X, j^k_0 Y].
\end{aligned}$$

So we have proved formulas (2) and (3). For all polynomial vector fields $a = \sum a^i_\lambda x^\lambda \frac{\partial}{\partial x_i}$, $b = \sum b^i_\mu x^\mu \frac{\partial}{\partial x_i} \in \mathfrak{g}^k_m$ the coordinate formula for the Lie bracket of vector fields, see 3.4, and formula (2) imply

$$\begin{aligned}
[a, b] &= \sum_{i,\gamma} c^i_\gamma x^\gamma \frac{\partial}{\partial x_i} \qquad \text{where} \\
c^i_\gamma &= \sum_{\substack{1 \le j \le m \\ \mu+\lambda-1_j=\gamma}} (\lambda_j b^j_\mu a^i_\lambda - \mu_j a^j_\lambda b^i_\mu).
\end{aligned} \tag{5}$$

Here 1_j means the multi index α with $\alpha_i = \delta^i_j$ and there is no implicit summation in the brackets. This formula shows that (1) is a grading. Let us evaluate

$$\left[x^\alpha \frac{\partial}{\partial x_i}, x^\beta \frac{\partial}{\partial x_i}\right] = (\alpha_i - \beta_i) x^{\alpha+\beta-1_i} \frac{\partial}{\partial x_i}$$

and consider two degrees p, q, $0 \le p+q \le k-1$. If $p \ne q$ then for every γ with $|\gamma| = p+q+1$ and for every index $1 \le i \le m$, we are able to find some α and β with $|\alpha| = p+1$, $|\beta| = q+1$ and $\alpha + \beta = \gamma + 1_i$, $\beta_i \ne \alpha_i$. Since the vector fields $x^\gamma \frac{\partial}{\partial x_i}$, $1 \le i \le m$, $|\gamma| = p+q+1$, form a linear base of the homogeneous component $\mathfrak{g}_{p+q}$, we get equality (4). If $p = q$, then the above consideration fails only in the case $\gamma_i = |\gamma|$. But if $m > 1$, then we can take the bracket

$$[x_j x_i^p \tfrac{\partial}{\partial x_i}, x_i^{q+1} \tfrac{\partial}{\partial x_j}] = x_i^{p+q+1} \tfrac{\partial}{\partial x_i} - (q+1) x_i^{p+q} x_j \tfrac{\partial}{\partial x_j} \qquad j \ne i.$$

Since the second summand belongs to $[\mathfrak{g}_p, \mathfrak{g}_q]$ this completes the proof. □

13.3. Let us recall some general concepts. The commutator of elements a_1, a_2 of a Lie group G is the element $a_1 a_2 a_1^{-1} a_2^{-1} \in G$. The closed subgroup $K(S_1, S_2)$ generated by all commutators of elements $s_1 \in S_1 \subset G$, $s_2 \in S_2 \subset G$ is called the commutator of the subsets S_1 and S_2. In particular, $G' := K(G, G)$ is called the *derived group* of the Lie group G. We get two sequences of closed subgroups

$$\begin{aligned} G^{(0)} &= G = G_{(0)} \\ G^{(n)} &= (G^{(n-1)})' \qquad n \in \mathbb{N} \\ G_{(n)} &= K(G, G_{(n-1)}) \qquad n \in \mathbb{N}. \end{aligned}$$

A Lie group G is called *solvable* if $G^{(n)} = \{e\}$ and *nilpotent* if $G_{(n)} = \{e\}$ for some $n \in \mathbb{N}$. Since always $G_{(n)} \supset G^{(n)}$, every nilpotent Lie group is solvable.

The Lie bracket determines in each Lie algebra $\mathfrak{g}$ the following two sequences of Lie subalgebras

$$\begin{aligned} \mathfrak{g} &= \mathfrak{g}^{(0)} = \mathfrak{g}_{(0)} \\ \mathfrak{g}^{(n)} &= [\mathfrak{g}^{(n-1)}, \mathfrak{g}^{(n-1)}] \qquad n \in \mathbb{N} \\ \mathfrak{g}_{(n)} &= [\mathfrak{g}, \mathfrak{g}_{(n-1)}] \qquad n \in \mathbb{N}. \end{aligned}$$

The sequence $\mathfrak{g}_{(n)}$ is called the *descending central sequence* of $\mathfrak{g}$. A Lie algebra $\mathfrak{g}$ is called *solvable* if $\mathfrak{g}^{(n)} = 0$ and *nilpotent* if $\mathfrak{g}_{(n)} = 0$ for some $n \in \mathbb{N}$, respectively. Every nilpotent Lie algebra is solvable. If $\mathfrak{b}$ is an ideal in $\mathfrak{g}^{(n)}$ such that the factor $\mathfrak{g}^{(n)}/\mathfrak{b}$ is commutative, then $\mathfrak{b} \supset \mathfrak{g}^{(n+1)}$. Consequently Lie algebra $\mathfrak{g}$ is solvable if and only if there is a sequence of subalgebras $\mathfrak{g} = \mathfrak{b}_0 \supset \mathfrak{b}_1 \supset \cdots \supset \mathfrak{b}_l = 0$ where $\mathfrak{b}_{k+1} \subset \mathfrak{b}_k$ is an ideal, $0 \le k < l$, and all factors $\mathfrak{b}_k/\mathfrak{b}_{k+1}$ are commutative.

Proposition. [Naymark, 76, p. 516] *A connected Lie group is solvable, or nilpotent if and only if its Lie algebra is solvable, or nilpotent, respectively.*

13.4. Let $i\colon GL(m) \to G_m^k$ be the map transforming every matrix $A \in GL(m)$ into the r-jet at zero of the linear isomorphism $x \mapsto A(x)$, $x \in \mathbb{R}^m$. This is a splitting of the short exact sequence of Lie groups

$$e \longrightarrow B_1 \longrightarrow G_m^k \underset{i}{\overset{\pi_1^k}{\rightleftarrows}} G_m^1 \longrightarrow e \tag{1}$$

so that we have the situation of 5.16.

Proposition. *The Lie group G^k_m is the semidirect product $GL(m) \rtimes B_1$ with the action of $GL(m)$ on B_1 given by (1). The normal subgroup B_1 is connected, simply connected and nilpotent. The exponential map $\exp: \mathfrak{b}_1 \to B_1$ is a global diffeomorphism.*

Proof. Since the normal subgroup B_1 is diffeomorphic to a Euclidean space, see 13.1, it is connected and simply connected. Hence B_1 is also nilpotent, for its Lie algebra $\mathfrak{b}_1$ is nilpotent, see 13.2.(4) and 13.3. By a general theorem, see [Naymark, 76, p. 516], the exponential map of a connected and simply connected solvable Lie group is a global diffeomorphism. Since our group is even nilpotent this also follows from the Baker-Campbell-Hausdorff formula, see 4.29. □

13.5. We shall need some very basic concepts from representation theory. A *representation* π of a Lie group G on a finite dimensional vector space V is a Lie group homomorphism $\pi: G \to GL(V)$. Analogously, a representation of a Lie algebra $\mathfrak{g}$ on V is a Lie algebra homomorphism $\mathfrak{g} \to \mathfrak{gl}(V)$. For every representation $\pi: G \to GL(V)$ of a Lie group, the tangent map at the identity $T\pi: \mathfrak{g} \to \mathfrak{gl}(V)$ is a representation of its Lie algebra, cf. 4.24.

Given two representations π_1 on V_1 and π_2 on V_2 of a Lie group G, or a Lie algebra $\mathfrak{g}$, a linear map $f: V_1 \to V_2$ is called a *G-module* or *$\mathfrak{g}$-module homomorphism*, if $f(\pi_1(a)(x)) = \pi_2(a)(f(x))$ for all $a \in G$ or $a \in \mathfrak{g}$ and all $x \in V$, respectively. We say that the representations π_1 and π_2 are *equivalent*, if there is a G-module isomorphism or $\mathfrak{g}$-module isomorphism $f: V_1 \to V_2$, respectively.

A linear subspace $W \subset V$ in the representation space V is called *invariant* if $\pi(a)(W) \subset W$ for all $a \in G$ (or $a \in \mathfrak{g}$) and π is called *irreducible* if there is no proper invariant subspace $W \subset V$. A representation π is said to be *completely reducible* if V decomposes into a direct sum of irreducible invariant subspaces. A decomposition of a completely reducible representation is unique up to the ordering and equivalences. A classical result reads that the standard action of $GL(V)$ on every invariant linear subspace of $\otimes^p V \otimes \otimes^q V^*$ is completely reducible for each p and q, see e.g. [Boerner, 67].

A representation π of a connected Lie group G is irreducible, or completely reducible if and only if the induced representation $T\pi$ of its Lie algebra $\mathfrak{g}$ is irreducible, or completely reducible, respectively, see [Naymark, 76, p. 346].

A representation $\pi: GL(m) \to GL(V)$ is said to have *homogeneous degree* r if $\pi(t.\mathrm{id}_{\mathbb{R}^m}) = t^r \mathrm{id}_V$ for all $t \in \mathbb{R} \setminus \{0\}$. Obviously, two irreducible representations with different homogeneous degrees cannot be equivalent.

13.6. The $GL(m)$-module structure on $\mathfrak{b}_1 \subset \mathfrak{g}^k_m$. Since $B_1 \subset G^k_m$ is a normal subgroup, the corresponding subalgebra $\mathfrak{b}_1 = \mathfrak{g}_1 \oplus \cdots \oplus \mathfrak{g}_{k-1}$ is an ideal. The (lower case) adjoint action ad of $\mathfrak{g}_0 = \mathfrak{gl}(m)$ on $\mathfrak{b}_1$ and the adjoint action Ad of $GL(m) = G^1_m$ on $\mathfrak{b}_1$ determine structures of a $\mathfrak{g}_0$-module and a $GL(m)$-module on $\mathfrak{b}_1$. As we proved in 13.2, all homogeneous components $\mathfrak{g}_r \subset \mathfrak{b}_1$ are $\mathfrak{g}_0$-submodules.

Let us consider the canonical volume form $\omega = dx_1 \wedge \cdots \wedge dx_m$ on $\mathbb{R}^m$ and recall that for every vector field X on $\mathbb{R}^m$ its *divergence* is a function $\mathrm{div}X$ on $\mathbb{R}^m$ defined by $\mathcal{L}_X \omega = (\mathrm{div}X)\omega$.

In coordinates we have $\operatorname{div}(\sum \xi^i \partial/\partial x_i) = \sum \partial\xi^i/\partial x_i$ and so every k-jet $j_0^k X \in \mathfrak{g}_m^k$ determines the $(k-1)$-jet $j_0^{k-1}(\operatorname{div} X)$. Hence we can define $\operatorname{div}(j_0^k X) = j_0^{k-1}(\operatorname{div} X)$ for all $j_0^k X \in \mathfrak{g}_m^k$. If X is the canonical polynomial representative of $j_0^k X$ of degree k, then $\operatorname{div} X$ is a polynomial of degree $k-1$. Let $C_1^r \subset \mathfrak{g}_r$ be the subspace of all elements $j_0^k X \in \mathfrak{g}_r$ with divergence zero. By definition,

$$\begin{aligned} \operatorname{div}[X,Y]\omega &= \mathcal{L}_{[X,Y]}\omega = \mathcal{L}_X\mathcal{L}_Y\omega - \mathcal{L}_Y\mathcal{L}_X\omega \\ &= (X(\operatorname{div}Y) - Y(\operatorname{div}X))\omega. \end{aligned} \tag{1}$$

Since every linear vector field $X \in \mathfrak{g}_0$ has constant divergence, $C_1^r \subset \mathfrak{g}_r$ is a $\mathfrak{gl}(m)$-submodule. In coordinates,

$$\sum a_\lambda^i x^\lambda \frac{\partial}{\partial x_i} \in C_1^r \text{ if and only if } \sum_{i,\lambda} \lambda_i a_\lambda^i x^{\lambda - 1_i} = 0,$$

i.e. $\sum_i (\mu_i + 1) a_{\mu+1_i}^i = 0$ for each μ with $|\mu| = r$.

Further, let us notice that the Lie bracket of the field $Y_0 = \sum_j x_j \frac{\partial}{\partial x_j}$ with any linear field $X \in \mathfrak{g}_0$ is zero. Hence, also the subspace C_2^r of all vector fields $Y \in \mathfrak{g}_r$ of the form $Y = fY_0$ with an arbitrary polynomial $f = \sum f_\alpha x^\alpha$ of degree r is $\mathfrak{g}_0$-invariant. Indeed, it holds $[X, fY_0] = -(Xf)Y_0$.

Since $\operatorname{div}(fY_0) = \sum_j (\alpha_j + 1) f_\alpha x^\alpha$, we see that $\mathfrak{g}_r = C_1^r \oplus C_2^r$. In coordinates, we have linear generators of C_2^r

$$X_\alpha = x^\alpha \Big(\sum_k x_k \tfrac{\partial}{\partial x_k}\Big), \qquad |\alpha| = r, \tag{2}$$

and if $m > 1$ then there are linear generators of C_1^r

$$\begin{aligned} X_{\alpha,k} &= x^\alpha \left((\alpha_k + 1) x_1 \tfrac{\partial}{\partial x_1} - (\alpha_1 + 1) x_k \tfrac{\partial}{\partial x_k} \right), && |\alpha| = r,\ k = 2, \dots, m \\ Y_{\mu,k} &= x^\mu \tfrac{\partial}{\partial x_k}, && k = 1, \dots, m,\ |\mu| = r+1,\ \mu_k = 0. \end{aligned} \tag{3}$$

We shall write $C_1 = C_1^1 \oplus C_1^2 \oplus \dots \oplus C_1^{k-1}$ and $C_2 = C_2^1 \oplus C_2^2 \oplus \dots \oplus C_2^{k-1}$. According to (1), $C_1 \subset \mathfrak{b}_1$ is a Lie subalgebra. Since for smooth functions f, g on $\mathbb{R}^m$ we have $[fX, gX] = (g(Xf) + f(Xg))X$, $C_2 \subset \mathfrak{b}_1$ is a Lie subalgebra as well. So we have got a decomposition $\mathfrak{b}_1 = C_1 \oplus C_2$. According to the general theory this is also a decomposition into $G_m^1{}^+$-submodules, but as all the spaces C_j^r are invariant with respect to the adjoint action of any exchange of two coordinates, the latter spaces are even $GL(m)$-submodules.

Proposition. *If $m > 1$, then the $GL(m)$-submodules C_1^r, C_2^r in $\mathfrak{g}_r$, $1 \le r \le k-1$, are irreducible and inequivalent. For $m = 1$, $C_1^r = 0$, $1 \le r \le k-1$, and all C_2^r are irreducible inequivalent $GL(1)$-modules.*

Proof. Assume first $m > 1$. A reader familiar with linear representation theory could verify that the modules C_2^r are equivalent to the irreducible modules

$\det^{-r} C^m_{(r,r,\dots,r,0)}$, where the symbol $C^m_{(r,\dots,r,0)}$ corresponds to the Young's diagram $(r,\dots,r,0)$, while C^r_1 are equivalent to $\det^{-(r+1)} C^m_{(r+2,r+1,\dots,r+1,0)}$, see e.g. [Dieudonné, Carrell, 71]. We shall present an elementary proof of the proposition.

Let us first discuss the modules C^r_2. Consider one of the linear generators X_α defined in (2) and a linear vector field $x_i\frac{\partial}{\partial x_j} \in \mathfrak{gl}(m)$. We have

$$(4) \qquad [-x_i\tfrac{\partial}{\partial x_j}, x^\alpha(\sum_k x_k\tfrac{\partial}{\partial x_k})] = \alpha_j x_i x^{\alpha-1_j}\sum_k(x_k\tfrac{\partial}{\partial x_k}).$$

If $j=i$, we get a scalar multiplication, but in all other cases the index α_j decreases while α_i increases by one and if $\alpha_j=0$, then the bracket is zero. Hence an iterated action of suitable linear vector fields on an arbitrary linear combination of the base elements X_α yields one of the base elements. Further, formula (4) implies that the submodule generated by any X_α is the whole C^r_2. This proves the irreducibility of the $GL(m)$-modules C^r_2.

In a similar way we shall prove the irreducibility of C^r_1. Let us evaluate the action of $Z_{i,j} = x_i\frac{\partial}{\partial x_j}$ on the linear generators $X_{\alpha,k}$, $Y_{\mu,k}$.

$$\begin{aligned}
[-Z_{i,j}, X_{\alpha,k}] &= (\alpha_k+1)(\alpha_j+\delta^j_1)x^{\alpha+1_1+1_i-1_j}\tfrac{\partial}{\partial x_1} - \\
&\quad - (\alpha_1+1)(\alpha_j+\delta^j_k)x^{\alpha+1_k+1_i-1_j}\tfrac{\partial}{\partial x_k} - \\
&\quad - \delta^i_1(\alpha_k+1)x^{\alpha+1_1}\tfrac{\partial}{\partial x_j} + \delta^i_k(\alpha_1+1)x^{\alpha+1_k}\tfrac{\partial}{\partial x_j} \\
[-Z_{i,j}, Y_{\mu,k}] &= \mu_j x^{\mu-1_j+1_i}\tfrac{\partial}{\partial x_k} - \delta^i_k x^\mu\tfrac{\partial}{\partial x_j}.
\end{aligned}$$

In particular, we get

$$[-Z_{i,1}, Y_{\mu,1}] = 0$$

$$[-Z_{i,1}, X_{\alpha,k}] = \begin{cases} (\alpha_1+1)X_{\alpha+1_i-1_1,k} & \text{if } \alpha_1\neq 0,\ i\neq 1 \\ (\alpha_k+1+\delta^i_k)Y_{\alpha+1_i,1} & \text{if } \alpha_1 = 0,\ i\neq 1 \end{cases}$$

$$[-Z_{i,j}, Y_{\mu,k}] = \begin{cases} \mu_j Y_{\mu-1_j+1_i,k} & \text{if } i\neq k \\ X_{\mu-1_j,j} & \text{if } i=k,\ \mu_j\neq 0 \\ -Y_{\mu,j} & \text{if } i=k,\ \mu_j=0. \end{cases}$$

Hence starting with an arbitrary linear combination of the base elements, an iterated action of suitable vector fields leads to one of the base elements $Y_{\mu,k}$. Then any other base element can be reached by further actions. Therefore also the modules C^r_2 are irreducible.

If $m=1$, then all $C^r_1=0$ by the definition and for all $0\le r\le k-1$ we have $C^r_2 = \mathfrak{g}_r = \mathbb{R}$ with the action of $\mathfrak{g}_0$ given by $[ax\frac{\partial}{\partial x}, bx^{r+1}\frac{\partial}{\partial x}] = -rabx^{r+1}\frac{\partial}{\partial x}$.

The submodules C^r_1 and C^r_2 cannot be equivalent for dimension reasons. The adjoint action Ad of $GL(m)$ on $\mathfrak{g}^k_m$ is given by $Ad(a)(j^k_0X) = j^k_0(a\circ X\circ a^{-1})$. So each irreducible component of $\mathfrak{g}_r$ has homogeneous degree $-r$. Therefore the modules C^r_i with different r are inequivalent. □

13.7. Corollary. *The normal subgroup $B_1 \subset G^k_m$ is generated by two closed Lie subgroups D_1, D_2 invariant under the canonical action of G^1_m. The group D_1 is formed by the jets of volume preserving diffeomorphisms and D_2 consists of the jets of diffeomorphisms keeping all the one-dimensional linear subspaces in $\mathbb{R}^m$. The corresponding Lie subalgebras are the subalgebras with grading $C_1 = C^1_1 \oplus \cdots \oplus C^{k-1}_1$ and $C_2 = C^1_2 \oplus \cdots \oplus C^{k-1}_2$ where all the homogeneous components are irreducible $GL(m)$-modules with respect to the adjoint action and $\mathfrak{b}_1 = C_1 \oplus C_2$.*

Let us point out that an element $j^k_0 f \in G^k_m$ belongs to D_1 or D_2 if and only if its polynomial representative is of the form $f = id_{\mathbb{R}_m} + f_2 + \cdots + f_k$ with $f_i \in C_1 \cap L^i_{\text{sym}}(\mathbb{R}^m, \mathbb{R}^m) = C^{i-1}_1$ or $f_i \in C_2 \cap L^i_{\text{sym}}(\mathbb{R}^m, \mathbb{R}^m) = C^{i-1}_2$, respectively.

13.8. Proposition. *If $m \geq 2$ and $l > 1$, or $m = 1$ and $l > 2$, then there is no splitting in the exact sequence $e \to B_l \to G^k_m \to G^l_m \to e$. In dimension $m = 1$, there is the exceptional projective splitting $G^2_1 \to G^k_1$ defined by*

$$ax + bx^2 \to a\left(x + \frac{b}{a}x^2 + \cdots + \frac{b^{k-1}}{a^{k-1}}x^k\right). \tag{1}$$

Proof. Let us assume there is a splitting j in the exact sequence of Lie algebra homomorphisms $0 \to \mathfrak{b}_l \to \mathfrak{g}^k_m \to \mathfrak{g}^l_m \to 0$, $l > 1$. So $j\colon \mathfrak{g}_0 \oplus \cdots \oplus \mathfrak{g}_{l-1} \to \mathfrak{g}_0 \oplus \cdots \oplus \mathfrak{g}_{k-1}$ and the restrictions $j^p_{t,q}$ of the components $j_q\colon \mathfrak{g}^l_m \to \mathfrak{g}_q$ to the $\mathfrak{g}_0$-submodules C^p_t in the homogeneous component $\mathfrak{g}_p$ are morphisms of $\mathfrak{g}_0$-modules. Hence $j^p_{t,q} = 0$ whenever $p \neq q$. Since j is a splitting the maps $j^p_{t,p}$ are the identities.

Assume now $m > 1$. Since $[\mathfrak{g}_{l-1}, \mathfrak{g}_1]$ equals $\mathfrak{g}_l$ in $\mathfrak{g}^k_m$ but at the same time this bracket equals zero in $\mathfrak{g}^l_m$, we have got a contradiction.

If $m = 1$ and $l > 2$ the same argument applies, but the inclusion $j\colon \mathfrak{g}_0 \oplus \mathfrak{g}_1 \to \mathfrak{g}_0 \oplus \mathfrak{g}_1 \oplus \cdots \oplus \mathfrak{g}_{k-1}$ is a Lie algebra homomorphism, for in $\mathfrak{g}^k_1$ the bracket $[\mathfrak{g}_1, \mathfrak{g}_1]$ equals zero. Let us find the splitting on the Lie group level. The germs of transformations $f_{\alpha,\beta}(x) = \frac{x}{\alpha x + \beta}$, $\beta \neq 0$, are determined by their second jets, so we can view them as elements in G^2_1. Since the composition of two such transformations is a transformation of the same type, they give rise to Lie group homomorphisms $G^2_1 \to G^r_1$ for all $r \in \mathbb{N}$. One computes easily the derivatives $f^{(n)}_{\alpha,\beta}(0) = (-1)^{n-1} n! \alpha^{n-1} \beta^{-n}$. Hence the 2-jet $ax + bx^2$ corresponds to $f_{\alpha,\beta}$ with $\alpha = -ba^{-2}$, $\beta = a^{-1}$. Consequently, the homomorphism $G^2_1 \to G^r_1$ has the form (1) and its tangent at the unit is the inclusion j. □

We remark that a geometric definition of the exceptional splitting (1) is based on the fact that the construction of the second order jets determines a bijection between G^2_1 and the germs at zero of the origine preserving projective transformations of $\mathbb{R}$.

13.9. Proposition. *The Lie group G^k_1 is solvable. Its Lie algebra $\mathfrak{g}^k_1$ can be characterized as a Lie algebra generated by three elements*

$$X_0 = x\tfrac{d}{dx} \in \mathfrak{g}_0, \quad X_1 = x^2\tfrac{d}{dx} \in \mathfrak{g}_1, \quad X_2 = x^3\tfrac{d}{dx} \in \mathfrak{g}_2$$

with relations

$$(1)\qquad [X_0, X_1] = -X_1$$

$$(2)\qquad [X_0, X_2] = -2X_2$$

$$(3)\qquad (ad(X_1))^i X_2 = 0 \qquad \text{for } i \geq k-2.$$

Proof. The filtration $\mathfrak{g}_1^k = \mathfrak{b}_0 \supset \cdots \supset \mathfrak{b}_{k-1} \supset 0$ from 13.2 is a descending chain of ideals with $\dim(\mathfrak{b}_i/\mathfrak{b}_{i+1}) = 1$. Hence $\mathfrak{g}_1^k$ is solvable.

Let us write $X_i = x^{i+1}\frac{d}{dx} \in \mathfrak{g}_i$. Since $[X_1, X_i] = (1-i)X_{i+1}$, we have

$$(4)\qquad X_i = \frac{(-1)^{i-2}}{(i-2)!}(\mathrm{ad}(X_1))^{i-2}X_2 \qquad \text{for } k-1 \geq i \geq 3$$

$$(5)\qquad [X_i, X_j] = (i-j)X_{i+j}\,.$$

Now, let $\mathfrak{g}$ be a Lie algebra generated by $\bar{X}_0$, $\bar{X}_1$, $\bar{X}_2$ which satisfy relations (1)–(3) and let us define $\bar{X}_i$, $i > 2$ by (4). Consider the linear map $\alpha\colon \mathfrak{g}_1^k \to \mathfrak{g}$, $X_i \to \bar{X}_i$, $0 \leq i \leq k-1$. Then $[\bar{X}_1, \bar{X}_i] = (1-i)\bar{X}_{i+1}$ and using Jacobi identity, the induction on i yields $[\bar{X}_0, \bar{X}_i] = -i\bar{X}_i$. A further application of Jacobi identity and induction on i lead to $[\bar{X}_i, \bar{X}_j] = (i-j)\bar{X}_{i+j}$. Hence the map α is an isomorphism. □

13.10. The group G_m^k with $m \geq 2$ has a more complicated structure. In particular G_m^k cannot be solvable, for $[\mathfrak{g}_m^k, \mathfrak{g}_m^k]$ contains the whole homogeneous component $\mathfrak{g}_0$, so that this cannot be nilpotent. But we have

Proposition. *The Lie algebra $\mathfrak{g}_m^k$, $m \geq 2$, $k \geq 2$, is generated by $\mathfrak{g}_0$ and any element $a \in \mathfrak{g}_1$ with $a \notin C_1^1 \cup C_2^1$. In particular, we can take $a = x_1^2\frac{\partial}{\partial x_1}$.*

Proof. Let $\mathfrak{g}$ be the Lie algebra generated by $\mathfrak{g}_0$ and a. Since $\mathfrak{g}_1 = C_1^1 \oplus C_2^1$ is a decomposition into irreducible $\mathfrak{g}_0$-modules, $\mathfrak{g}_1 \subset \mathfrak{g}$. But then 13.2.(4) implies $\mathfrak{g} = \mathfrak{g}_m^k$. □

13.11. Normal subgroup structure. Let us first describe several normal subgroups of G_m^k. For every $r \in \mathbb{N}$, $1 \leq r \leq k-1$, we define $B_{r,1} \subset B_r$, $B_{r,1} = \{j_0^r f;\ f = \mathrm{id}_{\mathbb{R}^m} + f_{r+1} + \cdots + f_k,\ f_{r+1} \in C_1^r,\ f_i \in L^i_{\mathrm{sym}}(\mathbb{R}^m, \mathbb{R}^m)\}$. The corresponding Lie subalgebra in $\mathfrak{g}_m^k$ is the ideal $C_1^r \oplus \mathfrak{g}_{r+1} \oplus \cdots \oplus \mathfrak{g}_{k-1}$ so that $B_{r,1}$ is a normal subgroup. Analogously, we set $B_{r,2} = \{j_0^r f;\ f = \mathrm{id}_{\mathbb{R}^m} + f_{r+1} + \cdots + f_k,\ f_{r+1} \in C_2^r,\ f_i \in L^i_{\mathrm{sym}}(\mathbb{R}^m, \mathbb{R}^m)\}$ with the corresponding Lie subalgebra $C_2^r \oplus \mathfrak{g}_{r+1} \oplus \cdots \oplus \mathfrak{g}_{k-1}$. We can characterize the normal subgroups $B_{r,j}$ as the subgroups in B_r with the projections $\pi_{r+1}^k(B_{r,j})$ belonging to the subgroups $D_j \subset G_m^{r+1}$, $j = 1, 2$, cf. 13.7.

Proposition. *Every connected normal subgroup H of G_m^k, $m \geq 2$, is one of the following:*

(1) $\{e\}$, the identity subgroup,

(2) B_r, $1 \leq r < k$, the kernel of the projection $\pi_r^k\colon G_m^k \to G_m^r$,

(3) $B_{r,1}$, $1 \le r < k$, the subgroup in B_r of jets of diffeomorphisms keeping the standard volume form up to the order $r+1$ at the origin,

(4) $B_{r,2}$, $1 \le r < k$, the subgroup in B_r of jets of diffeomorphisms keeping the linear one-dimensional subspaces in $\mathbb{R}^m$ up to the order $r+1$ at the origin,

(5) $N \rtimes B_1$, where N is a normal subgroup of $GL(m) = G^1_m$.

Proof. Since we deal with connected subgroups $H \subset G^k_m$, we can prove the proposition on the Lie algebra level.

Let us first assume that $H \subset B_1$. Then it suffices to prove that the ideal in $\mathfrak{g}^k_m$ generated by C^r_j, $j = 1, 2$, is the whole $C^r_j \oplus \mathfrak{b}_{r+1}$. But the whole algebra $\mathfrak{g}^k_m$ is generated by $\mathfrak{g}_0$ and $X_1 = x_1^2 \frac{\partial}{\partial x_1}$, and $[\mathfrak{g}_1, \mathfrak{g}_i] = \mathfrak{g}_{i+1}$ for all $2 \le i < k$. That is why we have only to prove that $\mathfrak{g}_{r+1}$ is contained in the subalgebra generated by $\mathfrak{g}_0$, X_1 and C^r_j for both $j = 1$ and $j = 2$. Since C^{r+1}_j are irreducible $\mathfrak{g}_0$-submodules, it suffices to find an element $Y \in C^r_j$ such that $[X_1, Y] \notin C^{r+1}_1$ and at the same time $[X_1, Y] \notin C^{r+1}_2$.

Let us take first $j = 2$, i.e. $Y = fY_0$ for certain polynomial f. Since $[fY_0, X_1] = (X_1 f)Y_0 + f[Y_0, X_1] = (X_1 f)Y_0 - fX_1$, the choice $f(x) = -x_2^r$ gives $[Y, X_1] = x_2^r x_1^2 \frac{\partial}{\partial x_1}$ which does not belong to $C^{r+1}_1 \cup C^{r+1}_2$, for its divergence equals to $2x_1 x_2^r \neq 0$, cf. 13.5.

Further, consider $Y = x_2^{r+1} \frac{\partial}{\partial x_1} \in C^r_1$ and let us evaluate $[x_2^{r+1} \frac{\partial}{\partial x_1}, x_1^2 \frac{\partial}{\partial x_1}] = -2x_1 x_2^{r+1} \frac{\partial}{\partial x_1}$. Since the divergence of the latter field does not vanish, $[Y, X_2] \notin C^{r+1}_1 \cup C^{r+1}_2$ as required. Hence we have proved that all connected normal subgroups $H \subset G^k_m$ contained in B_1 are of the form (1)–(4).

Consider now an arbitrary ideal $\mathfrak{h}$ in $\mathfrak{g}^k_m$ and let us denote $\mathfrak{n} = \mathfrak{h} \cap \mathfrak{g}_0 \subset \mathfrak{g}_0$. By virtue of 13.2.(4), if $\mathfrak{h}$ contains a vector which generates $\mathfrak{g}_1$ as a $\mathfrak{g}_0$-module, then $\mathfrak{b}_1 \subset \mathfrak{h}$. We shall prove that for every $X \in \mathfrak{g}_0$ any of the equalities $[X, C^1_1] = 0$ and $[X, C^1_2] = 0$ implies $X = 0$. Therefore either $\mathfrak{h} \supset \mathfrak{b}_1$ or $\mathfrak{n} = 0$ which concludes the proof of the proposition.

Let $X = \sum_{i,j} b_{ij} x_j \frac{\partial}{\partial x_i} \in \mathfrak{g}_0$ and $Y = x_k \sum_j x_j \frac{\partial}{\partial x_j} \in C^1_2$. Then $[X, Y] = -(\sum_j b_{kj} x_j) Y_0$. Hence $[X, C^1_2] = 0$ implies $X = 0$. Similarly, for $Y = x_l^2 \frac{\partial}{\partial x_k} \in C^1_1$ and $X \in \mathfrak{g}_0$, the equalities $[X, Y] = 0$ for all $k \neq l$ yield $X = 0$. The simple computation is left to the reader. □

13.12. G^k_m-modules. In the next sections we shall see that the actions of the jet groups on manifolds correspond to bundles of geometric objects. In particular, the vector bundle functors on m-dimensional manifolds correspond to linear representations of G^k_m, i.e. to G^k_m-modules. Since there is a well known representation theory of $GL(m)$ which is a subgroup in G^k_m, we should try to describe possible extensions of a given representation of $GL(m)$ on a vector space V to a representation of G^k_m. A step towards such description was done in [Terng, 78], we shall present only an observation showing that the study of geometric operations on irreducible vector bundles restricts in fact to the case of irreducible $GL(m)$-modules (with trivial action of the normal subgroup B_1). According to 5.4, there is a bijective correspondence between Lie group homomorphisms from B_1 to $GL(V)$ and Lie algebra homomorphisms from $\mathfrak{b}_1$ to $\mathfrak{gl}(V)$, for B_1 is connected and simply connected. Further, there is the semidirect product structure $\mathfrak{g}_m^k = \mathfrak{gl}(m) \rtimes \mathfrak{b}_1$ with the adjoint action of $\mathfrak{gl}(m)$ on $\mathfrak{b}_1$ which is tangent to the adjoint action of $GL(m)$ and every representation of $GL(m)$ on V induces a $GL(m)$-module structure on $\mathfrak{gl}(V)$ via the adjoint action of $GL(V)$ on $\mathfrak{gl}(V)$. This implies immediately

Proposition. *For every representation $\rho\colon GL(m) \to GL(V)$ there is a bijection between the representations $\bar{\rho}\colon G_m^k \to GL(V)$ with $\bar{\rho}|GL(m) = \rho$ and the set of mappings $T\colon \mathfrak{b}_1 \to \mathfrak{gl}(V)$ which are both Lie algebra homomorphisms and homomorphisms of $GL(m)$-modules.*

13.13. A G-module is called primary if it is equivalent to a direct sum of copies of a single irreducible G-module.

Proposition. *If V is a G_m^k-module such that the induced $GL(m)$-module is primary, then the action of the normal subgroup $B_1 \subset G_m^k$ is trivial.*

Proof. Assume that the $GL(m)$-module V equals sW, where W is an irreducible $GL(m)$-module. Then each irreducible component of the $GL(m)$-module $\mathfrak{gl}(V) = V \otimes V^*$ has homogeneous degree zero. But all the irreducible components of $\mathfrak{b}_1$ have negative homogeneous degrees. So there are no non-zero homomorphisms between the $GL(m)$-modules $\mathfrak{b}_1$ and $\mathfrak{gl}(V)$ and 13.12 implies the proposition. □

13.14. Proposition. *Let $\rho\colon G_m^k \to GL(V)$ be a linear representation such that the corresponding $GL(m)$-module is completely reducible and let $V = \sum_{i=1}^r n_i V_i$, where V_i are inequivalent irreducible $GL(m)$-modules ordered by their homogeneous degrees, i.e. the homogeneous degree of V_i is less than or equal to the homogeneous degree of V_j whenever $i \le j$. Then $W = (\sum_{i=1}^{l-1} n_i V_i) \oplus n V_l$ is a G_m^k-submodule of V for all $1 \le l \le r$ and $n \le n_l$.*

Proof. By definition, $(\sum_{i=1}^{l-1} n_i V_i) \oplus n V_l$ is a $GL(m)$-submodule. Since every irreducible component of the $GL(m)$-module $\mathfrak{b}_1$ has negative homogeneous degree and for all $1 \le i \le l$ the homogeneous degree of $L(V_i, V_l)$ is non-negative, we get

$$T_e\rho(X)\Big(\big(\sum_{i=1}^{l-1} n_i V_i\big) \oplus n V_l\Big) \subset \sum_{i=1}^{l-1} n_i V_i$$

for all $n \le n_l$ and for every $X \in \mathfrak{b}_1$. Now the proposition follows from 13.12 and 13.5. □

13.15. Corollary. *Every irreducible G_m^k-module which is completely reducible as a $GL(m)$-module is an irreducible $GL(m)$-module with a trivial action of the normal nilpotent subgroup $B_1 \subset G_m^k$.*

Proof. Let V be an irreducible G_m^k-module. Then V is irreducible when viewed as a $GL(m)$-module, cf. proposition 13.14. But then B_1 acts trivially on V by virtue of proposition 13.13. □

13.16. Remark. In the sequel we shall often work with various subgroups in the group of all diffeomorphisms $\mathbb{R}^m \to \mathbb{R}^m$ which determine Lie subgroups in the jet groups G^k_m. Proposition 13.2 describes the bracket and the exponential map in the corresponding Lie algebras and also their gradings $\mathfrak{g} = \mathfrak{g}_0 \oplus \cdots \oplus \mathfrak{g}_{k-1}$. Let us mention at least volume preserving diffeomorphisms, symplectic diffeomorphisms, isometries and fibered isomorphisms on the fibrations $\mathbb{R}^{m+n} \to \mathbb{R}^m$. We shall essentially need the latter case in the next chapter, see 18.8. The r-th jet group of the category $\mathcal{FM}_{m,n}$ is $G^r_{m,n} \subset G^r_{m+n}$ and the corresponding Lie subalgebra $\mathfrak{g}^k_{m,n} \subset \mathfrak{g}^k_{m+n}$ consists of all polynomial vector fields $\sum_{i,\mu} a^i_\mu x^\mu \frac{\partial}{\partial x_i}$ with $a^i_\mu = 0$ whenever $i \leq m$ and $\mu_j \neq 0$ for some $j > m$. The arguments from the end of the proof of proposition 13.2 imply that even 13.2.(4) remains valid in the following formulation.

The decomposition $\mathfrak{g}^k_{m,n} = \mathfrak{g}_0 \oplus \cdots \oplus \mathfrak{g}_{k-1}$ *is a grading and for every indices* $0 \leq i, j < k$ *it holds*

$$(1) \qquad [\mathfrak{g}_i, \mathfrak{g}_j] = \mathfrak{g}_{i+j} \qquad \textit{if } m > 1,\ n > 1, \textit{ or if } i \neq j.$$

14. Natural bundles and operators

In the preface and in the introduction to this chapter, we mentioned that geometric objects are in fact functors defined on a category of manifolds with values in category $\mathcal{FM}$ of fibered manifolds. Therefore we shall use the name bundle functors, in general. But the best known among them are defined on category $\mathcal{M}f_m$ of m-dimensional manifolds and local diffeomorphisms and in this case many authors keep the traditional name natural bundles. Throughout this section, we shall use the original definition of natural bundles including the regularity assumption, see [Nijenhuis, 72], [Terng, 78], [Palais, Terng, 77], but we shall prove in chapter V that every bundle functor on $\mathcal{M}f_m$ is of finite order and that the regularity condition 14.1.(iii) follows from the other axioms. Since the presentation of these results needs rather long and technical analytical considerations, we prefer to derive first geometric properties of bundle functors in the best known situations under stronger assumptions. In fact the material of this section presents a model for the more general situation treated in the next chapter.

14.1. Definition. A *bundle functor on* $\mathcal{M}f_m$ or a *natural bundle* over m-manifolds, is a covariant functor $F\colon \mathcal{M}f_m \to \mathcal{FM}$ satisfying the following conditions

(i) (Prolongation) $B \circ F = \mathrm{Id}_{\mathcal{M}f_m}$, where $B\colon \mathcal{FM} \to \mathcal{M}f$ is the base functor. Hence the induced projections form a natural transformation $p\colon F \to \mathrm{Id}_{\mathcal{M}f_m}$.

(ii) (Locality) If $i\colon U \to M$ is an inclusion of an open submanifold, then $FU = p_M^{-1}(U)$ and Fi is the inclusion of $p_M^{-1}(U)$ into FM.

(iii) (Regularity) If $f\colon P \times M \to N$ is a smooth map such that for all $p \in P$ the maps $f_p = f(p,\ )\colon M \to N$ are local diffeomorphisms, then $\tilde{F}f\colon P \times FM \to FN$, defined by $\tilde{F}f(p, \ ) = Ff_p, p \in P$, is smooth, i.e. smoothly parameterized systems of local diffeomorphisms are transformed into smoothly parameterized systems of fibered local isomorphisms.

In sections 10 and 12 we met several bundle functors on $\mathcal{M}f_m$.

14.2. Now let F be a natural bundle. We shall denote by $t_x\colon \mathbb{R}^m \to \mathbb{R}^m$ the translation $y \mapsto y + x$ and for any manifold M and point $x \in M$ we shall write F_xM for the pre image $p_M^{-1}(x)$. In particular, $F_0\mathbb{R}^m$ will be called the *standard fiber of the bundle functor* F. Every bundle functor $F\colon \mathcal{M}f_m \to \mathcal{FM}$ determines an action τ of the abelian group $\mathbb{R}^m$ on $F\mathbb{R}^m$ via $\tau_x = Ft_x$.

Proposition. *Let $F\colon \mathcal{M}f_m \to \mathcal{FM}$ be a bundle functor on $\mathcal{M}f_m$ and let $S := F_0\mathbb{R}^m$ be the standard fiber of F. Then there is a canonical isomorphism $\mathbb{R}^m \times S \cong F\mathbb{R}^m$, $(x, z) \mapsto Ft_x(z)$, and for every m-dimensional manifold M the value FM is a locally trivial fiber bundle with standard fiber S.*

Proof. The map $\psi\colon F\mathbb{R}^m \to \mathbb{R}^m \times S$ defined by $z \mapsto (x, Ft_{-x}(z))$, $x = p(z)$, is the inverse to the map defined in the proposition and both maps are smooth according to the regularity condition 14.1.(iii). The rest of the proposition follows from the locality condition 14.1.(ii). Indeed, a fibered atlas of FM is formed by the values of F on the charts of any atlas of M. □

14.3. Definition. A natural bundle $F\colon \mathcal{M}f_m \to \mathcal{FM}$ is said to be of *finite order* r, $0 \le r < \infty$, if for all local diffeomorphisms f, $g\colon M \to N$ and every point $x \in M$, the equality $j_x^r f = j_x^r g$ implies $Ff|F_xM = Fg|F_xM$.

14.4. Associated maps. Let us consider a natural bundle $F\colon \mathcal{M}f_m \to \mathcal{FM}$ of order r. For all m-dimensional manifolds M, N we define the mapping $F_{M,N}\colon \operatorname{inv}J^r(M, N) \times_M FM \to FN$, $(j_x^r f, y) \mapsto Ff(y)$. The mappings $F_{M,N}$ are called the *associated maps of the bundle functor* F.

Proposition. *The associated maps are smooth.*

Proof. For $m = 0$ the assertion is trivial. Let us assume $m > 0$. Since smoothness is a local property, we may restrict ourselves to $M = N = \mathbb{R}^m$. Indeed, chosen local charts on M and N we get local trivializations on FM and FN and the induced local chart on $\operatorname{inv}J^r(M, N)$. Hence we have

$$\operatorname{inv}J^r(\mathbb{R}^m, \mathbb{R}^m) \times_{\mathbb{R}^m} F\mathbb{R}^m \xrightarrow{\cong} \operatorname{inv}J^r(U, V) \times_U FU \xrightarrow{F_{U,V}} FV \xrightarrow{\cong} F\mathbb{R}^m$$

and we can apply the locality condition.

Now, let us recall that every jet in $J^r(\mathbb{R}^m, \mathbb{R}^n)$ has a canonical polynomial representative and that this space coincides with the cartesian product of $\mathbb{R}^m$ and the Euclidean space of coefficients of these polynomials, as a smooth manifold. If we consider the map $\operatorname{ev}\colon \operatorname{inv}J^r(\mathbb{R}^m, \mathbb{R}^m) \times \mathbb{R}^m \to \mathbb{R}^m$, $\operatorname{ev}_x(j_0^r f) = f(x)$, then the associated map $F_{\mathbb{R}^m, \mathbb{R}^m}$ coincides with the map $\tilde{F}(\operatorname{ev})$ appearing in the regularity condition. □

14.5. Induced action. According to proposition 14.4 the restriction $\ell = F_{\mathbb{R}^m,\mathbb{R}^m}|G^r_m \times S$ is a smooth left action of the jet group G^r_m on the standard fiber S.

Let us define $q_M = F_{\mathbb{R}^m,M}|\operatorname{inv}J^r_0(\mathbb{R}^m,M) \times S\colon P^rM \times S \to FM$. For every $u = j^r_0 g \in \operatorname{inv}J^r_0(\mathbb{R}^m,M)$, $s \in S$ and $j^r_0 f \in G^r_m$ we have

$$(1) \qquad q_M(j^r_0 g \circ j^r_0 f, \ell(j^r_0 f^{-1}, s)) = q_M(j^r_0 g, s)$$

and the restriction $(q_M)_u := q_M(j^r_0 g, \ )$ is a diffeomorphism. Hence q determines the structure of the associated fiber bundle $P^rM[S;\ell]$ on FM, cf. 10.7.

Proposition. *For every bundle functor $F\colon \mathcal{M}f_m \to \mathcal{FM}$ of order r and every m-dimensional manifold M there is a canonical structure of an associated bundle $P^rM[S;\ell]$ on FM given by the map q_M and the values of the functor F lie in the category of bundles with structure group G^r_m and standard fiber S.*

Proof. The first part was already proved. Consider a local diffeomorphism $f\colon M \to N$. For every $j^r_0 g \in P^rM$, $s \in S$ we have

$$Ff \circ q_M(j^r_0 g, s) = Ff \circ Fg(s) = q_N(j^r_0(f \circ g), s).$$

So we identify Ff with $\{P^rf, \mathrm{id}_S\}\colon P^rM \times_{G^r_m} S \to P^rN \times_{G^r_m} S$. □

14.6. Description of r-th order natural bundles. Every smooth left action ℓ of G^r_m on a manifold S determines a covariant functor $L\colon \mathcal{PB}(G^r_m) \to \mathcal{FM}_m$, $LP = P[S;\ell]$, $Lf = \{f, \mathrm{id}_S\}$. An r-th order bundle functor F with standard fiber S induces an action ℓ of G^r_m on S and we can construct a natural bundle $G = L \circ P^r\colon \mathcal{M}f_m \to \mathcal{FM}$.

We claim that F is naturally equivalent to G. For every $u = j^r_0 g \in P^r_x M$ there is the diffeomorphism $(q_M)_u\colon S \to F_xM$ which we shall denote Fu. Hence we can define maps $\chi_M\colon GM \to FM$ by

$$\chi_M(\{u,s\}) = Fu(s) = q_M(j^r_0 g, s) = Fg(s).$$

According to 14.5.(1), this is a correct definition, and by the construction, the maps χ_M are fibered isomorphisms. Since $Gf = \{P^rf, \mathrm{id}_S\}$ for every local diffeomorphism $f\colon M \to N$, we have $Ff \circ \chi_M(\{j^r_0 g, s\}) = F(f \circ g)(s) = \chi_N \circ Gf(\{j^r_0 g, s\})$.

From the geometrical point of view, naturally equivalent functors can be identified. Hence we have proved

Theorem. *There is a bijective correspondence between the set of all r-th order natural bundles on m-dimensional manifolds and the set of smooth left actions of the jet group G^r_m on smooth manifolds.*

In the next examples, we demonstrate on well known natural bundles, that the identification in the theorem is exactly what the geometers usually do.

14.7. Examples.

1. The reader should reconsider that in the case of frame bundles P^r the identification used in 14.6, i.e. the relation of the functor P^r to the functor G constructed from the induced action, is exactly the usual identification of principal fiber bundles (P, p, M, G) with their associated bundles $P[G, \lambda]$, where λ is the left action of G on itself.

2. For the tangent bundle T, the map $(q_M)_u$ with $u = j^1_0 g \in P^1_x M$ is just the linear map $T_0 g \colon T_0\mathbb{R}^m \to T_x M$ determined by $j^1_0 g$, i.e. the linear coordinates on $T_x M$ induced by local chart g. Hence the tangent bundle corresponds to the canonical action of $G^1_m = GL(m, \mathbb{R})$ on $\mathbb{R}^m$.

3. Further well known natural bundles are the functors T^r_k of r-th order k-velocities. More precisely, we consider the restrictions of the functors defined in 12.8 to the category $\mathcal{M}f_m$. Let us recall that $T^r_k M = J^r_0(\mathbb{R}^k, M)$ and the action on morphisms is given by the composition of jets. Hence, in this case, for every $u = j^r_0 g \in P^r_x M$ the map $(q_M)_u$ transforms the classes of r-equivalent maps $(\mathbb{R}^k, 0) \to (M, x)$ into their induced coordinate expressions in the local chart g, i.e. $(q_M)^{-1}_u(j^r_0 f) = j^r_0(g^{-1} \circ f)$.

14.8. Vector bundle functors. In accordance with 6.14, a bundle functor $F \colon \mathcal{M}f_m \to \mathcal{FM}$ is called a *vector bundle functor on* $\mathcal{M}f_m$, or *natural vector bundle*, if there is a canonical vector bundle structure on each value FM and the values Ff on morphisms are morphisms of vector bundles. Let F be an r-th order natural vector bundle with standard fiber V and with induced action $\ell : G^r_m \times V \to V$. Then ℓ is a group homomorphism $G^r_m \to GL(V)$ and so V carries a structure of G^r_m-module. On the other hand, every G^r_m module V gives rise to a natural bundle F, see the construction in 14.6, and an application of F to charts of any atlas on a manifold M yields a vector bundle atlas on the value $FM \to M$. Therefore proposition 14.6 implies

Proposition. *There is a bijective correspondence between r-th order vector bundle functors on $\mathcal{M}f_m$ and G^r_m-modules.*

14.9. Examples.

1. In our setting, the p-covariant and q-contravariant tensor fields on a manifold M are just the smooth global sections of $FM \to M$, where F is the vector bundle functor corresponding to the $GL(m)$-module $\otimes^p\mathbb{R}^{m*} \otimes \otimes^q\mathbb{R}^m$, cf. 7.2.

2. In 6.7 we discussed constructions with vector bundles corresponding to a smooth covariant functor $\mathcal{F}$ on the category of finite dimensional vector spaces and these constructions can be applied to the values of any natural vector bundle to get new natural vector bundles, cf. 6.14. There we applied $\mathcal{F}$ to the cocycle of transition functions. Let us look what happens on the level of the corresponding G^r_m-modules. If we apply $\mathcal{F}$ to a G^r_m-module V with action $\ell \colon G^r_m \to GL(V)$, we get a vector space $\mathcal{F}V$ with action $\tilde{\ell} \colon G^r_m \to GL(\mathcal{F}V)$, $\tilde{\ell}(g) = \mathcal{F}(\ell(g))$, i.e. a new G^r_m-module $\mathcal{F}V$. Let us assume that G and $\mathcal{F}G$ are the natural vector bundles corresponding to V and $\mathcal{F}V$. The canonical vector bundle structure on $(\mathcal{F}G)M = P^r M \times_{G^r_m} \mathcal{F}V$ coincides with that on $\mathcal{F}(GM)$ by 10.7.(4). Similarly, we can handle contravariant functors and bifunctors on the category of vector spaces, cf. 6.7. In particular, the values of natural vector bundles corresponding to direct sums of the modules are just fibered products over the base manifolds of the individual bundles. Let us also note that $C^\infty(\oplus^i F_i M) = \oplus^i (C^\infty(F_i M))$.

3. There are also well known examples of higher order natural vector bundles. First of all, we recall the functor of r-th order k-dimensional covelocities $J^r(\ ,\mathbb{R}^k)_0 = T_k^{r*}$ introduced in 12.8. If $r, k = 1$, we get the dual bundles to the tangent bundles $J_0^1(\mathbb{R}, M) = TM$. So the vector bundle structure on the cotangent bundle is natural and the tangent spaces are the duals, from our point of view. But we can apply the construction of a dual module to any G_m^r-module and this leads to *dual natural vector bundles* according to 14.6. In this way we get the r-th order tangent bundles $T^{(r)} := (T^{r*})^*$ or, more general the bundle functors $T_k^{r\Box} = (T_k^{r*})^*$, see 12.14.

14.10. Affine bundle functors. A bundle functor $F\colon \mathcal{M}f_m \to \mathcal{FM}$ is called an *affine bundle functor on* $\mathcal{M}f_m$, or *natural affine bundle*, if each value $FM \to M$ is an affine bundle and the values on morphisms are affine maps. Hence the standard fiber V of an r-th order natural affine bundle is an affine space and the induced action ℓ is a representation of G_m^r in the group of affine transformations of V. So for each $g \in G_m^r$ there is a unique linear map $\vec{\ell}(g)\colon \vec{V} \to \vec{V}$ satisfying $\ell(g)(y) = \ell(g)(x) + \vec{\ell}(g)(y - x)$ for all $x, y \in V$. It follows that $\vec{\ell}$ is a linear representation of G_m^r on the vector space $\vec{V}$ and there is the corresponding natural vector bundle $\vec{F}$. By the construction, for every m-dimensional manifold M the value $\vec{F}M$ is just the modelling vector bundle to FM and for every morphism $f\colon M \to N$, $\vec{F}f$ is the modelling linear map to Ff. Hence two arbitrary sections of FM 'differ' by a section of $\vec{F}M$. The best known example of a second order natural affine bundle is the bundle of elements of linear connections QP^1 which we shall study in section 17. The modelling natural vector bundle $\overrightarrow{QP^1}$ is the tensor bundle $T \otimes T^* \otimes T^*$ corresponding to $GL(m)$-module $\mathbb{R}^m \otimes \mathbb{R}^{m*} \otimes \mathbb{R}^{m*}$.

Next we shall describe all natural transformations between natural bundles in the terms of G_m^r-equivariant maps.

14.11. Lemma. *For every natural transformation $\chi\colon F \to G$ between two natural bundles on $\mathcal{M}f_m$ all mappings $\chi_M\colon FM \to GM$ cover the identities* id_M.

Proof. Let $\chi\colon F \to G$ be a natural transformation and let us write $p\colon FM \to M$ and $q\colon GM \to M$ for the canonical projections onto an m-dimensional manifold M. If $y \in FM$ is a point with $z := q(\chi_M(y)) \neq p(y)$, then there is a local diffeomorphism $f\colon M \to M$ such that $\mathrm{germ}_{p(y)} f = \mathrm{germ}_{p(y)} \mathrm{id}_M$ and $f(z) = \bar{z}$, $\bar{z} \neq z$. But now the localization condition implies $q \circ \chi_M \circ Ff(y) \neq q \circ Gf \circ \chi_M(y)$, for $q \circ Gf = f \circ q$. This is a contradiction. □

14.12. Theorem. *There is a bijective correspondence between the set of all natural transformations between two r-th order natural bundles on $\mathcal{M}f_m$ and the set of smooth G_m^r-equivariant maps between their standard fibers.*

Proof. Let F and G be natural bundles with standard fibers S and Q and let $\chi\colon F \to G$ be a natural transformation. According to 14.11, we have the restriction $\chi_{\mathbb{R}^m}|S\colon S \to Q$ and we claim that this is G^r_m-equivariant with respect to the induced actions. Indeed, for any $j^r_0 f \in G^r_m$ we get $(\chi_{\mathbb{R}^m}|S) \circ Ff = Gf \circ (\chi_{\mathbb{R}^m}|S)$, but $Ff\colon S \to S$ and $Gf\colon Q \to Q$ are just the induced actions of $j^r_0 f$ on S and Q. Now we have to show that the whole transformation χ is determined by the map $\chi_{\mathbb{R}^m}|S$. First, using translations $t_x\colon \mathbb{R}^m \to \mathbb{R}^m$ we see this for the map $\chi_{\mathbb{R}^m}$. Then, if we choose any atlas (U_α, u_α) on a manifold M, the maps Fu_α form a fiber bundle atlas on FM and we know $\chi_M \circ Fu_\alpha = Gu_\alpha \circ \chi_{\mathbb{R}^m}$. Hence the locality of bundle functors implies $\chi_M|(p^F_M)^{-1}(U_\alpha) = Gu_\alpha \circ \chi_{\mathbb{R}^m} \circ (Fu_\alpha)^{-1}$.

On the other hand, let $\chi_0\colon S \to Q$ be an arbitrary G^r_m-equivariant smooth map. According to 14.6, the functors F or G are canonically naturally equivalent to the functors $L \circ P^r$ or $K \circ P^r$, where L or K are the functors corresponding to the induced G^r_m-actions ℓ or k on the standard fibers S or Q, respectively. So it suffices to define a natural transformation $\chi\colon L \circ P^r \to K \circ P^r$. We set $\chi_M = \{\mathrm{id}_{P^r M}, \chi_0\}$. It is an easy exercise to verify that χ is a natural transformation. Moreover, we have $\chi_{\mathbb{R}^m}|S = \chi_0$. □

In general, an operator is a rule transforming sections of a fibered manifold $Y \to M$ into sections of another fibered manifold $\bar{Y} \to \bar{M}$. We shall deal with the case $M = \bar{M}$ in this section. Let us recall that $C^\infty Y$ means the set of all smooth sections of a fibered manifold $Y \to M$.

14.13. Definition. Let $Y \xrightarrow{p} M$, $\bar{Y} \xrightarrow{\bar{p}} M$ be fibered manifolds. A *local operator* $A\colon C^\infty Y \to C^\infty \bar{Y}$ is a map such that for every section $s\colon M \to Y$ and every point $x \in M$ the value $As(x)$ depends on the germ of s at x only. If, moreover, for certain $k \in \mathbb{N}$ or $k = \infty$ the condition $j^k_x s = j^k_x q$ implies $As(x) = Aq(x)$, then A is said to be *of order* k. An operator $A\colon C^\infty Y \to C^\infty \bar{Y}$ is called a *regular operator* if every smoothly parameterized family of sections of Y is transformed into a smoothly parameterized family of sections of $\bar{Y}$.

14.14. Associated maps to an k-th order operator. Consider an operator $A\colon C^\infty Y \to C^\infty \bar{Y}$ of order k. We define a map $\mathcal{A}\colon J^k Y \to \bar{Y}$ by $\mathcal{A}(j^k_x s) = As(x)$ which is called the *associated map to the k-th order operator A*.

Proposition. *The associated map to any finite order operator A is smooth if and only if A is regular.*

Proof. Let $A\colon C^\infty Y \to C^\infty \bar{Y}$ be an operator of order k. If we choose local fibered coordinates on Y, we also get the induced fibered coordinates on $J^k Y$. But in these local coordinates, the jets of sections are identified with (polynomial) sections. Thus, a chart on $J^k Y$ can be viewed as a smoothly parameterized family of sections in $C^\infty Y$ and so the smoothness of $\mathcal{A}$ follows from the regularity. The converse implication is obvious. □

14.15. Natural operators. A *natural operator* $A\colon F \rightsquigarrow G$ between two natural bundles F and G is a system of regular operators $A_M\colon C^\infty(FM) \to C^\infty(GM)$, $M \in \mathrm{Ob}\mathcal{M}f_m$, satisfying

(i) for every section $s \in C^\infty(FM \to M)$ and every diffeomorphism $f\colon M \to N$ it holds

$$A_N(Ff \circ s \circ f^{-1}) = Gf \circ A_M s \circ f^{-1}$$

(ii) $A_U(s|U) = (A_M s)|U$ for every $s \in C^\infty(FM)$ and every open submanifold $U \subset M$.

In particular, condition (ii) implies that natural operators are formed by local operators.

A natural operator $A\colon F \rightsquigarrow G$ is said to be of order k, $0 \le k \le \infty$, if all operators A_M are of order k. The system of associated maps $\mathcal{A}_M\colon J^kFM \to GM$ to the k-th order operators A_M is called the system of associated maps to the natural operator A. The associated maps to finite order natural operators are smooth.

We can look at condition (i) even from the viewpoint of the local coordinates on a manifold M. Given a local chart $u\colon U \subset M \to V \subset \mathbb{R}^m$, the diffeomorphisms $f\colon V \to W \subset \mathbb{R}^m$ correspond to the changes of coordinates on U. Combining this observation with localization property (ii), we conclude that the natural operators coincide, in fact, with those operators, the local descriptions of which do not depend on the changes of coordinates.

14.16. Proposition. *For every r-th order bundle functor F on $\mathcal{M}f_m$ its composition with the functor of k-th jet prolongations of fibered manifolds $J^k\colon \mathcal{FM} \to \mathcal{FM}$ is a natural bundle of order $r+k$.*

Proof. Let $f\colon M \to N$ be a local diffeomorphism. Then, by definition of the associated maps $F_{M,N}$, we have

$$Ff = F_{M,N} \circ \big((j^r f \circ p_M) \times \mathrm{id}_{FM}\big)\colon FM \to FN.$$

Hence $J^k(Ff)$ depends on $(k+r)$-jets of f in the underlying points in M only. It is an easy exercise to verify the axioms of natural bundles. □

14.17. Proposition. *There is a bijective correspondence between the set of k-th order natural operators $A\colon F \rightsquigarrow G$ between two natural bundles on $\mathcal{M}f_m$ and the set of all natural transformations $\alpha\colon J^k \circ F \to G$.*

Proof. Let $\mathcal{A}_M$ be the associated maps of an k-th order natural operator $A\colon F \rightsquigarrow G$. We claim that these maps form a natural transformation $\alpha\colon J^kF \to G$. They are smooth by virtue of 14.14 and we have to verify $Gf \circ \mathcal{A}_M = \mathcal{A}_N \circ J^kFf$ for an arbitrary local diffeomorphism $f\colon M \to N$. We have

$$\begin{aligned}\mathcal{A}_N((J^kFf)(j^k_x s)) &= \mathcal{A}_N(j^k(Ff \circ s \circ f^{-1})(f(x))) \\ &= A_N(Ff \circ s \circ f^{-1})(f(x)) = Gf \circ A_M s(x) \\ &= Gf \circ \mathcal{A}_M(j^k_x s).\end{aligned}$$

On the other hand, consider a natural transformation $\alpha\colon J^kF \to G$. We define operators $A_M\colon FM \rightsquigarrow GM$ by $A_M s(x) = \alpha_M(j^k_x s)$ for all sections $s \in C^\infty(FM)$. Since the maps α_M are smooth fibered morphisms and according to lemma 14.11 they all cover the identities id_M, the maps $A_M s$ are smooth sections of GM. The straightforward verification of the axioms of natural operators is left to the reader. □

14.18. Let $F\colon \mathcal{M}f_m \to \mathcal{FM}$ be an r-th order natural bundle with standard fiber S and let $\ell\colon G^r_m \times S \to S$ be the induced action. The identification $\mathbb{R}^m \times S \cong F\mathbb{R}^m$, $(x,s) \mapsto F(t_x)(s)$, induces the identification $C^\infty(\mathbb{R}^m, S) \cong C^\infty(F\mathbb{R}^m)$, $(\tilde{s}\colon \mathbb{R}^m \to S) \mapsto (s(x) = Ft_x(\tilde{s}(x))) \in C^\infty(F\mathbb{R}^m)$. Hence the standard fiber of the natural bundle J^kF equals $T^k_m S$. Under these identifications, the action of F on an arbitrary local diffeomorphism is of the form

$$Fg(x,s) = (g(x), F(t_{-g(x)} \circ g \circ t_x)(s))$$

and the induced action $\ell^k\colon G^{r+k}_m \times T^k_m S \to T^k_m S$ determined by the functor J^kF is expressed by the following formula

$$\begin{aligned} (1)\qquad \ell^k(j^{r+k}_0 g, j^k_0 \tilde{s}) &= \ell^k(j^{r+k}_0 g, j^k_0(Ft_x \circ \tilde{s}(x))) \\ &= j^k_0(Fg \circ Ft_{g^{-1}(x)} \circ \tilde{s}(g^{-1}(x))) \qquad \in J^k_0 F\mathbb{R}^m \\ &= j^k_0(Ft_{-x} \circ Fg \circ Ft_{g^{-1}(x)} \circ \tilde{s}(g^{-1}(x))) \qquad \in T^k_m S \\ &= j^k_0\big(\ell(j^r_0(t_{-x} \circ g \circ t_{g^{-1}(x)}), \tilde{s}(g^{-1}(x)))\big). \end{aligned}$$

In particular, if $a = j^{r+k}_0 g \in G^1_m \subset G^{r+k}_m$, i.e. g is linear, then

$$(2)\qquad \ell^k(a, j^k_0 \tilde{s}) = j^k_0(\ell(j^r_0 g, \tilde{s} \circ g^{-1}(x))) = j^k_0(\ell_a \circ \tilde{s} \circ g^{-1}).$$

As a consequence of the last two propositions we get the basic result for finding natural operators of prescribed types. Consider natural bundles F or F' on $\mathcal{M}f_m$ of finite orders r or r', with standard fibers S or S' and induced actions ℓ or ℓ' of G^r_m or $G^{r'}_m$, respectively. If $q = \max\{r+k, r'\}$ with some fixed $k \in \mathbb{N}$ then the actions ℓ^k and ℓ' trivially extend to actions of G^q_m on both $T^k_m S$ and S' and we have

Theorem. *There is a canonical bijective correspondence between the set of all k-th order natural operators $A\colon F \rightsquigarrow F'$ and the set of all smooth G^q_m-equivariant maps between the left G^q_m-spaces $T^k_m S$ and S'.*

14.19. Examples.

1. By the construction in 3.4, the Lie bracket of vector fields is a bilinear natural operator $[\ ,\ ]\colon T \oplus T \rightsquigarrow T$ of order one, see also corollary 3.11. The corresponding bilinear G^2_m-equivariant map is

$$\begin{aligned} b = (b^1, \dots, b^m)\colon T^1_m\mathbb{R}^m \times T^1_m\mathbb{R}^m \to \mathbb{R}^m \\ b^j(X^i, X^k_\ell; Y^m, Y^n_p) = X^i Y^j_i - Y^i X^j_i. \end{aligned}$$

Later on we shall be able to prove that every bilinear equivariant map $b'\colon T^r_m\mathbb{R}^m \times T^r_m\mathbb{R}^m \to \mathbb{R}^m$ is a constant multiple of b composed with the jet projections and, moreover, every natural bilinear operator is of a finite order, so that all bilinear natural operators on vector fields are the constant multiples of the Lie bracket. On the other hand, if we drop the bilinearity, then we can iterate the Lie bracket to get operators of higher orders. But nevertheless, one can prove that there are no other G^2_m-equivariant maps $b'\colon T^1_m\mathbb{R}^m \times T^1_m\mathbb{R}^m \to \mathbb{R}^m$ beside the constant multiples of b and the projections $T^1_m\mathbb{R}^m \times \mathbb{R}^m \to \mathbb{R}^m$. This implies, that the constant multiples of the Lie bracket are essentially the only natural operators $T \oplus T \rightsquigarrow T$ of order 1.

2. The exterior derivative introduced in 7.8 is a first order natural operator $d\colon \Lambda^k T^* \rightsquigarrow \Lambda^{k+1}T^*$. Formula 7.8.(1) expresses the corresponding G^2_m-equivariant map

$$T^1_m(\Lambda^k\mathbb{R}^{m*}) \to \Lambda^{k+1}\mathbb{R}^{m*}$$

$$(\varphi_{i_1\dots i_k}, \varphi_{i_1\dots i_k,i_{k+1}}) \mapsto \sum_j (-1)^{j+1}\varphi_{i_1\dots\hat{i_j}\dots i_{k+1},i_j}$$

where the hat denotes that the index is omitted. We shall derive in 25.4 that for $k > 0$ this is the only G^2_m-equivariant map up to constant multiples. Consequently, the constant multiples of the exterior derivative are the only natural operators of the type in question.

14.20. In concrete problems we often meet a situation where the representations of G^r_m are linear, or at least their restrictions to $G^1_m \subset G^r_m$ turn the standard fibers into $GL(m)$-modules. Then the linear equivariant maps between the standard fibers are $GL(m)$-module homomorphisms and so the structure of the modules in question is often a very useful information for finding all equivariant maps. Given a G^1_m-module V and linear coordinates y^p on V, there are the induced coordinates $y^p_\alpha = \frac{\partial^{|\alpha|}y^p}{\partial x^\alpha}$ on $T^k_m V$, where x^i are the canonical coordinates on $\mathbb{R}^m$ and $0 \le |\alpha| \le k$. Then the linear subspace in $T^k_m V$ defined by $y^p_\alpha = 0$, $|\alpha| \ne i$, coincides with $V \otimes S^i\mathbb{R}^{m*}$. Clearly, these identifications do not depend on our choice of the linear coordinates y^p. Formula 14.18.(2) shows that $T^k_m V = V \oplus \dots \oplus V \otimes S^k\mathbb{R}^{m*}$ is a decomposition of $T^k_m V$ into G^1_m-submodules and the same formula implies the following result.

Proposition. *Let V be a G^1_m-invariant subspace in $\otimes^p\mathbb{R}^m \otimes \otimes^q\mathbb{R}^{m*}$ and let us consider a representation $\ell\colon G^r_m \to \mathrm{Diff}(V)$ such that its restriction to $G^1_m \subset G^r_m$ is the canonical tensorial action. Then the restriction of the induced action ℓ^k of G^{r+k}_m on $T^k_m V = V \oplus \dots \oplus V \otimes S^k\mathbb{R}^{m*}$ to $G^1_m \subset G^{r+k}_m$ is also the canonical tensorial action.*

14.21. Some geometric constructions are performed on the whole category $\mathcal{M}f$ of smooth manifolds and smooth maps. Similarly to natural bundles, the bundle functors on the category $\mathcal{M}f$ present a special case of the more general concept of bundle functors.

Definition. A *bundle functor on the category* $\mathcal{M}f$ is a covariant functor $F\colon \mathcal{M}f \to \mathcal{FM}$ satisfying the following conditions

(i) $B \circ F = \mathrm{Id}_{\mathcal{M}f}$, so that the fiber projections form a natural transformation $p\colon F \to \mathrm{Id}_{\mathcal{M}f}$.

(ii) If $i\colon U \to M$ is an inclusion of an open submanifold, then $FU = p_M^{-1}(U)$ and Fi is the inclusion of $p_M^{-1}(U)$ into FM.

(iii) If $f\colon P \times M \to N$ is a smooth map, then $\tilde{F}f\colon P \times FM \to FN$, defined by $\tilde{F}f(p, \;) = Ff_p$, $p \in P$, is smooth.

For every non-negative integer m the restriction F_m of a bundle functor F on $\mathcal{M}f$ to the subcategory $\mathcal{M}f_m \subset \mathcal{M}f$ is a natural bundle. Let us call the sequence $\mathcal{S} = \{S_0, S_1, \dots, S_m, \dots\}$ of the standard fibers of the natural bundles F_m the *system of standard fibers of the bundle functor* F. Proposition 14.2 implies that for every m there is the canonical isomorphism $\mathbb{R}^m \times S_m \cong F\mathbb{R}^m$, $(x, s) \mapsto Ft_x(s)$, and given an m-dimensional manifold M, $p_M\colon FM \to M$ is a locally trivial bundle with standard fiber S_m.

Analogously to 14.3 and 14.4, a bundle functor F on $\mathcal{M}f$ is said to be of order r if for every smooth map $f\colon M \to N$ and point $x \in M$ the restriction $Ff|F_xM$ depends only on j^r_xf. Then the maps $F_{M,N}\colon J^r(M,N) \times_M FM \to FN$, $F_{M,N}(j^r_xf, y) = Ff(y)$ are called the *associated maps to the r-th order functor* F. Since in the proof of proposition 14.3 we never used the invertibility of the jets in question, the same proof applies to the present situation and so the associated maps to any finite order bundle functor on $\mathcal{M}f$ are smooth. For every m-dimensional manifold M, there is the canonical structure of the associated bundle $FM \cong P^rM[S_m]$, cf. 14.5.

Let $\mathcal{S} = \{S_0, S_1, \dots\}$ be the system of standard fibers of an r-th order bundle functor F on $\mathcal{M}f$. The restrictions $\ell_{m,n}$ of the associated maps $F_{\mathbb{R}^m,\mathbb{R}^n}$ to $J^r_0(\mathbb{R}^m, \mathbb{R}^n)_0 \times S_m$ have the following property. For every $A \in J^r_0(\mathbb{R}^m, \mathbb{R}^n)_0$, $B \in J^r_0(\mathbb{R}^n, \mathbb{R}^p)_0$ and $s \in S_m$

$$\ell_{m,p}(B \circ A, s) = \ell_{n,p}(B, \ell_{m,n}(A, s)). \tag{1}$$

Hence instead of the action of one group G^r_m on the standard fiber in the case of bundle functors on $\mathcal{M}f_m$, we get an action of the category L^r on $\mathcal{S}$, see below and 12.6 for the definitions. We recall that the objects of L^r are the non-negative integers and the set of morphisms between m and n is the set $L^r_{m,n} = J^r_0(\mathbb{R}^m, \mathbb{R}^n)_0$.

Let $\mathcal{S} = \{S_0, S_1, \dots\}$ be a system of manifolds. A left *action ℓ of the category* L^r on $\mathcal{S}$ is defined as a system of maps $\ell_{m,n}\colon L^r_{m,n} \times S_m \to S_n$ satisfying (1). The action is called smooth if all maps $\ell_{m,n}$ are smooth. The canonical action of L^r on the system of standard fibers of a bundle functor F is called the *induced action*. Every induced action of a finite order bundle functor is smooth.

14.22. Consider a system of smooth manifolds $\mathcal{S} = \{S_0, S_1, \dots\}$ and a smooth action ℓ of the category L^r on $\mathcal{S}$. We shall construct a bundle functor L determined by this action. The restrictions ℓ_m of the maps $\ell_{m,m}$ to invertible jets form smooth left actions of the jet groups G^r_m on manifolds S_m. Hence for every m-dimensional manifold M we can define $LM = P^rM[S_m; \ell_m]$. Let us recall the notation $\{u, s\}$ for the elements in $P^rM \times_{G^r_m} S_m$, i.e. $\{u, s\} = \{u \circ A, \ell_m(A^{-1}, s)\}$ for all $u \in P^rM$, $A \in G^r_m$, $s \in S_m$. For every smooth map $f\colon M \to N$ we define $Lf\colon FM \to FN$ by

$$Lf(\{u, s\}) = \{v, \ell_{m,n}(v^{-1} \circ A \circ u, s)\}$$

where $m = \dim M$, $n = \dim N$, $u \in P^r_x M$, $A = j^r_x f$, and $v \in P^r_{f(x)} N$ is an arbitrary element. We claim that this is a correct definition. Indeed, chosen another representative for $\{u, s\}$ and another frame $v' \in P^r_{f(x)}$, say $\{u \circ B, \ell_m(B^{-1}, s)\}$, and $v' = v \circ C$, formula 14.21.(1) implies

$$\begin{aligned} Lf(\{u \circ B, \ell_m(B^{-1}, s)\} = \\ &= \{v \circ C, \ell_{m,n}(C^{-1} \circ v^{-1} \circ A \circ u \circ B, \ell_m(B^{-1}, s))\} = \\ &= \{v \circ C, \ell_n(C^{-1}, \ell_{m,n}(v^{-1} \circ A \circ u, s))\} = \\ &= \{v, \ell_{m,n}(v^{-1} \circ A \circ u, s)\}. \end{aligned}$$

One verifies easily all the axioms of bundle functors, this is left to the reader.

On the other hand, consider an r-th order bundle functor F on $\mathcal{M}f$ and its induced action ℓ. Let L be the corresponding bundle functor, we have just constructed. Analogously to 14.6, there is a canonical natural equivalence $\chi \colon L \to F$. In fact, we have the restrictions of χ to manifolds of any fixed dimension which consists of the maps q_M determining the canonical structures of associated bundles on the values FM, see 14.6. It remains only to show that $Ff \circ \chi_M = \chi_N \circ Ff$ for all smooth maps $f \colon M \to N$. But given $j^r_0 g \in P^r_x M$, $j^r_0 h \in P^r_{f(x)} N$ and $s \in S_m$, we have

$$\begin{aligned} Ff \circ \chi_M(\{j^r_0 g, s\}) &= Ff \circ Fg(s) = Fh \circ F(h^{-1} \circ f \circ g)(s) \\ &= \chi_N(j^r_0 h, \ell_{m,n}(j^r_0(h^{-1} \circ f \circ g), s)) = \chi_N \circ Lf(\{j^r_0 g, s\}). \end{aligned}$$

Since in geometry we usually identify naturally equivalent functors, we have proved

Theorem. *There is a bijective correspondence between the set of r-th order bundle functors on $\mathcal{M}f$ and the set of smooth left actions of the category L^r on systems $\mathcal{S} = \{S_0, S_1, \dots\}$ of smooth manifolds.*

14.23. Natural transformations. Consider a smooth action ℓ or ℓ' of the category L^r on a system $\mathcal{S} = \{S_0, S_1, \dots\}$ or $\mathcal{S}' = \{S'_0, S'_1, \dots\}$ of smooth manifolds, respectively. A sequence φ of smooth maps $\varphi_i \colon S_i \to S'_i$ is called a smooth L^r-equivariant map between ℓ and ℓ' if for every $s \in S_m$, $A \in L^r_{m,n}$ it holds

$$\varphi_n(\ell_{m,n}(A, s)) = \ell'_{m,n}(A, \varphi_m(s)).$$

Theorem. *There is a bijective correspondence between the set of natural transformations of two r-th order bundle functors on $\mathcal{M}f$ and the set of smooth L^r-equivariant maps between the induced actions of L^r on the systems of standard fibers.*

Proof. Let $\chi \colon F \to G$ be a natural transformation, ℓ or k be the induced action on the system of standard fibers $\mathcal{S} = \{S_0, S_1, \dots\}$ or $\mathcal{Q} = \{Q_0, Q_1, \dots\}$, respectively. As we proved in 14.11, all maps $\chi_M \colon FM \to GM$ are over identities. Let us define $\varphi_n : S_n \to Q_n$ as the restriction of $\chi_{\mathbb{R}^n}$ to S_n. If $j_0^r f \in L^r_{m,n}$, $s \in S_m$, then

$$\varphi_n(\ell_{m,n}(j_0^r f, s)) = \chi_{\mathbb{R}^n} \circ Ff(s) = Gf \circ \chi_{\mathbb{R}^m}(s) = k_{m,n}(j_0^r f, \varphi_m(s)),$$

so that the maps φ_m form a smooth L^r-equivariant map between ℓ and k. Moreover, the arguments used in 14.11 imply that χ is completely determined by the maps φ_m.

Conversely, by virtue of 14.22, we may assume that the functors F and G coincide with the functors L and K constructed from the induced actions. Consider a smooth L^r-equivariant map φ between ℓ and k. Then we can define for all m-dimensional manifolds M maps $\chi_M : FM \to GM$ by

$$\chi_M := \{\mathrm{id}_{P^r M}, \varphi_m\}.$$

The reader should verify easily that the maps χ_M form a natural transformation. □

14.24. Remark. Let F be an r-th order bundle functor on $\mathcal{M}f$. Its induced action can be interpreted as a smooth functor $F_{\mathrm{Inf}} : L^r \to \mathcal{M}f$, where the smoothness means that all the maps $L^r_{m,n} \times F_{\mathrm{Inf}}(m) \to F_{\mathrm{Inf}}(n)$ defined by $(A, x) \mapsto F_{\mathrm{Inf}}A(x)$ are smooth. Then the concept of smooth L^r-equivariant maps between the actions coincides with that of a natural transformation. Hence we can reformulate theorems 14.22 and 14.23 as follows. The full subcategory of r-th order bundle functors on $\mathcal{M}f$ in the category of functors and natural transformations is naturally equivalent to the full subcategory of smooth functors $L^r \to \mathcal{M}f$. Let us also remark, that the L^r-objects can be viewed as numerical spaces $\mathbb{R}^m$, $0 \le m < \infty$, with distinguished origins. Then every $\mathcal{M}f$-object is locally isomorphic to exactly one L^r-object and, up to local diffeomorphisms, L^r contains all r-jets of smooth maps. Therefore, we can call L^r the r-th order skeleton of $\mathcal{M}f$. We shall work out this point of view in our treatment of general bundle functors in the next chapter. Let us mention that the bundle functors on $\mathcal{M}f_m$ also admit such a description. Indeed, the r-th order skeleton then consists of the group G^r_m only.

15. Prolongations of principal fiber bundles

15.1. In the present section, we shall mostly deal with the category $\mathcal{PB}_m(G)$ consisting of principal fiber bundles with m-dimensional bases and a fixed structure group G, with $\mathcal{PB}(G)$-morphisms which cover local diffeomorphisms between the base manifolds. So a $\mathcal{PB}_m(G)$-morphism $\varphi : (P, p, M) \to (P', p', M')$ is a smooth fibered map over a local diffeomorphism $\varphi_0 : M \to M'$ satisfying $\varphi \circ \rho_g = \rho'_g \circ \varphi$ for all $g \in G$, where ρ and ρ' are the principal actions on P and P'. In particular, every automorphism $\varphi : \mathbb{R}^m \times G \to \mathbb{R}^m \times G$ is fully determined by its restriction $\bar\varphi : \mathbb{R}^m \to G$, $\bar\varphi(x) = \mathrm{pr}_2 \circ \varphi(x, e)$, where $e \in G$ is the unit, and by the underlying map $\varphi_0\colon \mathbb{R}^m \to \mathbb{R}^m$. We shall identify the morphism φ with the couple $(\varphi_0, \bar\varphi)$, i.e. we have

$$(1) \qquad \varphi(x,a) = (\varphi_0(x), \bar\varphi(x).a).$$

Analogously, every morphism $\psi\colon \mathbb{R}^m \times G \to P$, i.e. every local trivialization of P, is determined by ψ_0 and $\tilde\psi := \psi|(\mathbb{R}^m \times \{e\})\colon \mathbb{R}^m \to P$ covering ψ_0. Further we define $\psi_1 = \tilde\psi \circ \psi_0^{-1}$, so that ψ_1 is a local section of the principal bundle P, and we identify the morphism ψ with the couple (ψ_0, ψ_1). We have

$$(2) \qquad \psi(x,a) = (\psi_1 \circ \psi_0(x)).a\ .$$

Of course, for an automorphism φ on $\mathbb{R}^m \times G$ we have $\bar\varphi = \mathrm{pr}_2 \circ \tilde\varphi$.

15.2. Principal prolongations of Lie groups. We shall apply the construction of r-jets to such a situation. Since all $\mathcal{PB}_m(G)$-objects are locally isomorphic to the trivial principal bundle $\mathbb{R}^m \times G$ and all $\mathcal{PB}_m(G)$-morphisms are local isomorphisms, we first have to consider the group $W^r_m G$ of r-jets at $(0,e)$ of all automorphisms $\varphi\colon \mathbb{R}^m \times G \to \mathbb{R}^m \times G$ with $\varphi_0(0) = 0$, where the multiplication μ is defined by the composition of jets,

$$\mu(j^r\varphi(0,e), j^r\psi(0,e)) = j^r(\psi \circ \varphi)(0,e).$$

This is a correct definition according to 15.1.(1) and the inverse elements are the jets of inverse maps (which always exist locally). The identification 15.1 of automorphisms on $\mathbb{R}^m \times G$ with couples $(\varphi_0, \bar\varphi)$ determines the identification

$$(1) \qquad W^r_m G \cong G^r_m \times T^r_m G, \qquad j^r\varphi(0,e) \mapsto (j^r_0\varphi_0, j^r_0\bar\varphi).$$

Let us describe the multiplication μ in this identification. For every φ, $\psi \in \mathcal{PB}_m(G)(\mathbb{R}^m \times G, \mathbb{R}^m \times G)$ we have

$$\psi \circ \varphi(x,a) = \psi(\varphi_0(x), \bar\varphi(x).a) = (\psi_0 \circ \varphi_0(x), \bar\psi(\varphi_0(x)).\bar\varphi(x).a)$$

so that given any (A,B), $(A',B') \in G^r_m \times T^r_m G$ we get

$$(2) \qquad \mu((A,B),(A',B')) = (A \circ A', (B \circ A').B').$$

Here the dot means the multiplication in the Lie group $T^r_m G$, cf. 12.13. Hence there is the structure of a semi direct product of Lie groups on $W^r_m G$. The Lie group $W^r_m G = G^r_m \rtimes T^r_m G$ is called the *(m,r)-principal prolongation of Lie group G*.

15.3. Principal prolongations of principal bundles. For every principal fiber bundle $(P,p,M,G) \in \mathrm{Ob}\mathcal{PB}_m(G)$ we define

$$W^r P := \{j^r\psi(0,e);\ \psi \in \mathcal{PB}_m(G)(\mathbb{R}^m \times G, P)\}.$$

In particular, $W^r(\mathbb{R}^m \times G)$ is identified with $\mathbb{R}^m \times W^r_m G$ by the rule

$$\mathbb{R}^m \times W^r_m G \ni (x, j^r\varphi(0,e)) \mapsto j^r(\tau_x \circ \varphi)(0,e) \in W^r(\mathbb{R}^m \times G)$$

where $\tau_x = t_x \times \mathrm{id}_G$, and so there is a well defined structure of a smooth manifold on $W^r(\mathbb{R}^m \times G)$. Furthermore, if we define the action of W^r on $\mathcal{PB}_m(G)$-morphisms by the composition of jets, i.e.

$$W^r\chi(j^r\psi(0,e)) := j^r(\chi \circ \psi)(0,e),$$

W^r becomes a functor. Now, taking any principle atlas on a principal bundle P, the application of the functor W^r to the local trivializations yields a fibered atlas on W^r. Finally, there is the right action of $W^r_m G$ on $W^r P$ defined for every $j^r\varphi(0,e) \in W^r_m G$ and $j^r\psi(0,e) \in W^r P$ by $(j^r\psi(0,e))(j^r\varphi(0,e)) = j^r(\psi \circ \varphi)(0,e)$. Since all the jets in question are invertible, this action is free and transitive on the individual fibers and therefore we have got principal bundle $(W^r P, p \circ \beta, M, W^r_m G)$ called the r-th *principal prolongation* of the principal bundle (P, p, M, G). By the definition, for a morphism φ the mapping $W^r\varphi$ always commutes with the right principal action of $W^r_m G$ and we have defined the functor $W^r \colon \mathcal{PB}_m(G) \to \mathcal{PB}_m(W^r_m G)$ of r-th principal prolongation of principal bundles.

15.4. Every $\mathcal{PB}_m(G)$-morphism $\psi \colon \mathbb{R}^m \times G \to P$ is identified with a couple (ψ_0, ψ_1), see 15.1.(2). This yields the identification

$$W^r P = P^r M \times_M J^r P \tag{1}$$

and also the smooth structures on both sides coincide. Let us express the corresponding action of $G^r_m \rtimes T^r_m G$ on $P^r M \times_M J^r P$. If $(u,v) = (j^r_0\psi_0, j^r\psi_1(\psi_0(0))) \in P^r M \times_M J^r P$ and $(A,B) = (j^r_0\varphi_0, j^r_0\bar\varphi) \in G^r_m \rtimes T^r_m G$, then 15.2.(2) implies

$$\begin{aligned}\psi \circ \varphi(x,a) &= \psi(\varphi_0(x), \bar\varphi(x).a) = \psi_1(\psi_0 \circ \varphi_0(x)).\bar\varphi(x).a \\ &= (\rho \circ (\psi_1, \bar\varphi \circ \varphi_0^{-1} \circ \psi_0^{-1}) \circ (\psi_0 \circ \varphi_0)(x)).a\end{aligned}$$

where ρ is the principal right action on P. Hence we have

$$(u,v)(A,B) = (u \circ A, v.(B \circ A^{-1} \circ u^{-1})) \tag{2}$$

where '.' is the multiplication

$$m \colon J^r P \times_M J^r(M,G) \to J^r P, \quad (j^r_x\sigma, j^r_x s) \mapsto j^r_x(\rho \circ (\sigma, s)).$$

The decomposition (1) is natural in the following sense. For every $\mathcal{PB}_m(G)$-morphism $\psi \colon (P,p,M,G) \to (P',p',M,'G)$, the $\mathcal{PB}_m(W^r_m G)$-morphism $W^r\psi$ has the form $(P^r\psi_0, J^r\psi)$. Indeed, given $\varphi \colon \mathbb{R}^m \times G \to P$, we have $(\psi \circ \varphi)_0 = \psi_0\varphi_0$, $(\psi \circ \varphi)_1 = \psi \circ \tilde\varphi \circ (\psi_0 \circ \varphi_0)^{-1} = \psi \circ \varphi_1 \circ \psi_0^{-1}$. Therefore, in the category of functors and natural transformations, the following diagram is a pullback

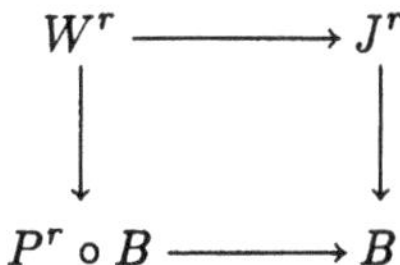

Here $B \colon \mathcal{PB}_m(G) \to \mathcal{M}f_m$ is the base functor, the upper and left-hand natural transformations are given by the above decomposition and the right-hand and bottom arrows are the usual projections.

15.5. For every associated bundle $E = P[S;\ell]$ to a principal bundle (P,p,M,G) there is a canonical left action $\ell^r\colon W^r_mG \times T^r_mS \to T^r_mS$ of $W^r_mG = G^r_m \rtimes T^r_mG$ on T^r_mS. We simply compose the prolonged action $T^r_m\ell$ of T^r_mG on T^r_mS, see 12.13, with the canonical left action of G^r_m on both T^r_mG and T^r_mS, i.e. we set

$$\ell^r(j^r\varphi(0,e), j^r_0 s) = j^r_0(\ell \circ (\bar\varphi \circ \varphi_0^{-1}, s \circ \varphi_0^{-1})) \tag{1}$$

for every $j^r\varphi(0,e) = (j^r_0\varphi_0, j^r_0\bar\varphi) \in G^r_m \rtimes T^r_mG$.

Proposition. *For every associated bundle $E = P[S;\ell]$, there is a canonical structure of the associated bundle $W^rP[T^r_mS;\ell^r]$ on the r-th jet prolongation J^rE.*

Proof. Similarly to 14.6, every action $\ell\colon G \times S \to S$ determines the functor L on $\mathcal{PB}_m(G)$, $P \mapsto P[S,\ell]$ and $\varphi \mapsto \{\varphi, \mathrm{id}_S\}$, with values in the category of the associated bundles with standard fiber S and structure group G. We shall essentially use the identification

$$\begin{aligned} T^r_mS &\cong J^r_0(\mathbb{R}^m \times S) \cong J^r_0((\mathbb{R}^m \times G)[S;\ell]) \\ j^r_0 s &\mapsto j^r_0(\mathrm{id}_{\mathbb{R}^m}, s) \mapsto j^r_0\{\hat e, s\} \end{aligned} \tag{2}$$

where $\hat e\colon \mathbb{R}^m \to \mathbb{R}^m \times G$, $\hat e(x) = (x,e)$. Then the action ℓ^r becomes the form

$$\begin{aligned} \ell^r(j^r\varphi(0,e), j^r_0\{\hat e, s\}) &= j^r_0\{\hat e, \ell \circ (\bar\varphi \circ \varphi_0^{-1}, s \circ \varphi_0^{-1})\} \\ &= J^r(L\varphi)(j^r_0\{\hat e, s\}). \end{aligned} \tag{3}$$

Now we can define a map $q\colon W^rP \times T^r_mS \to J^rE$ determining the required structure on J^rE. Given $u = j^r\psi(0,e) \in W^rP$ and $B = j^r_0 s \in T^r_mS$, we set

$$q(u,B) = J^r(L\psi)(j^r_0\{\hat e, s\}).$$

Since the map ψ is a local trivialization of the principal bundle P, the restriction $q_u = q(u,\ )\colon T^r_mS \to J^r_{\psi_0(0)}E$ is a diffeomorphism. Moreover, for every $A = j^r\varphi(0,e) \in W^r_mG$, formula (3) implies

$$q(u.A, \ell^r(A^{-1}, B)) = J^r(L(\psi \circ \varphi))(J^r(L\varphi^{-1})(j^r_0\{\hat e, s\})) = q(u,B)$$

and the proposition is proved. □

For later purposes, let us express the corresponding map $\tau\colon W^rP \times_M J^rE \to T^r_mS$. It holds

$$\tau(u, j^r_x s) = j^r_0(\tau_E \circ (\psi \circ \hat e, s \circ \psi_0))$$

where $\tau_E\colon P \times_M E \to S$ is the canonical map of E and $u = j^r\psi(0,e) \in W^r_xP$.

15.6. First order principal prolongation. We shall point out some special properties of the groups $W^1_m G$ and the bundles $W^1 P$. Let us start with the group $T^r_m G$. Every map $s\colon \mathbb{R}^m \to G$ can be identified with the couple $(s(0), \lambda_{s(0)^{-1}} \circ s)$, and for a second map $s'\colon \mathbb{R}^m \to G$ we have (we recall that λ_a and ρ^a are the left and right translations by a in G, μ is the multiplication on G)

$$\begin{aligned}(1)\qquad \mu \circ (s', s)(x) &= s'(0)s'(0)^{-1}s'(x)s(0)s(0)^{-1}s(x)\\ &= \big(s'(0)s(0)\big)\big(\mathrm{conj}_{s(0)^{-1}}(s'(0)^{-1}s'(x))\big)\big(s(0)^{-1}s(x)\big).\end{aligned}$$

It follows that $T^r_m G$ is the semi direct product $G \rtimes J^r_0(\mathbb{R}^m, G)_e$. This can be described easily in more details in the case $r = 1$. Namely, the first order jets are identified with linear maps between the tangent spaces, so that (1) implies $T^1_m G = G \rtimes (\mathfrak{g} \otimes \mathbb{R}^{m*})$ with the multiplication

$$(2)\qquad (a', Z').(a, Z) = (a'a, \mathrm{Ad}(a^{-1})(Z') + Z),$$

where a, $a' \in G$, Z, $Z' \in \mathrm{Hom}(\mathbb{R}^m, \mathfrak{g})$. Taking into account the decomposition 15.2.(1) and formula 15.2.(2), we get

$$W^1_m G = (GL(m) \times G) \rtimes (\mathfrak{g} \otimes \mathbb{R}^{m*})$$

with multiplication

$$(3)\qquad (A', a', Z').(A, a, Z) = (A' \circ A, a'a, \mathrm{Ad}(a^{-1})(Z') \circ A + Z).$$

Now, let us view fibers $P^1_x M$ as subsets in $\mathrm{Hom}(\mathbb{R}^m, T_x M)$ and elements in $J^1_x P$ as homomorphisms in $\mathrm{Hom}(T_x M, T_y P)$, $y \in P_x$. Given any $(u, v) \in P^1 M \times_M J^1 P = W^1 P$ and $(A, a, Z) \in (G^1_m \times G) \rtimes (\mathfrak{g} \otimes \mathbb{R}^{m*})$, 15.4.(2) implies

$$(4)\qquad (u, v)(A, a, Z) = (u \circ A, T\rho(v, T\lambda_a \circ Z \circ A^{-1} \circ u^{-1}))$$

where ρ is the principal right action on P.

15.7. Principal prolongations of frame bundles. Consider the r-th principal prolongation $W^r(P^s M)$ of the s-th order frame bundle $P^s M$ of a manifold M. Every local diffeomorphism $\varphi\colon \mathbb{R}^m \to M$ induces a principal fiber bundle morphism $P^s\varphi\colon P^s\mathbb{R}^m \to P^s M$ and we can construct $j^r_{(0,e_s)}(P^s\varphi) \in W^r(P^s M)$, where e_s denotes the unit of G^s_m. One sees directly that this element depends on the $(r+s)$-jet $j^{r+s}_0\varphi$ only. Hence the map $j^{r+s}_0\varphi \mapsto j^r_{(0,e_s)}(P^s\varphi)$ defines an injection $i_M\colon P^{r+s}M \to W^r(P^s M)$. Since the group multiplication in both G^{r+s}_m and $W^r_m G^s_m$ is defined by the composition of jets, the restriction $i_0\colon G^{r+s}_m \to W^r_m G^s_m$ of $i_{\mathbb{R}^m}$ to the fibers over $0 \in \mathbb{R}^m$ is a group homomorphism. Thus, the *$(r+s)$-order frames on a manifold M form a natural reduction $i_M\colon P^{r+s}M \to W^r(P^s M)$ of the r-th principal prolongation of the s-th order frame bundle of M to the subgroup $i_0(G^{r+s}_m) \subset W^r_m G^s_m$.*

15.8. Coordinate expression of $i_0\colon G^{r+s}_m \to W^r_m G^s_m$. The canonical coordinates x^i on $\mathbb{R}^m$ induce coordinates a^i_α, $0 < |\alpha| \le r+s$, on G^{r+s}_m, $a^i_\alpha(j^{r+s}_0 f) = \frac{1}{\alpha!}\frac{\partial^{|\alpha|} f^i}{\partial x^\alpha}(0)$, and the following coordinates on $W^r_m G^s_m$: Any element $j^r\varphi(0,e) \in W^r_m G^s_m$ is given by $j^r_0\varphi_0 \in G^r_m$ and $j^r_0\bar\varphi \in T^r_m G^s_m$, see 15.2. Let us denote the coordinate expression of $\bar\varphi$ by $b^i_\gamma(x)$, $0 < |\gamma| \le s$, so that we have the coordinates $b^i_{\gamma,\delta}$, $0 < |\gamma| \le s$, $0 \le |\delta| \le r$ on $T^r_m G^s_m$, $b^i_{\gamma,\delta}(j^r_0\bar\varphi) = \frac{1}{\delta!}\frac{\partial^{|\delta|} b^i_\gamma}{\partial x^\delta}(0)$, and the coordinates $(a^i_\beta; b^i_{\gamma,\delta})$, $0 < |\beta| \le r$, $0 < |\gamma| \le s$, $0 \le |\delta| \le r$, on $W^r_m G^s_m$. By definition, we have

$$i_0(a^i_\alpha) = (a^i_\beta; a^i_{\gamma+\delta}). \tag{1}$$

In the first order case, i.e. for $r = 1$, we have to take into account a further structure, namely $T^1_m G^s_m = G^s_m \rtimes (\mathfrak{g}^s_m \otimes \mathbb{R}^{m*})$, cf. 15.6. So given $i_0(j^{s+1}_0 f) = (j^1_0 f, j^1_0 q)$, where $q\colon \mathbb{R}^m \to G^s_m$, we are looking for $b = q(0) \in G^s_m$ and $Z = T\lambda_{b^{-1}} \circ T_0 q \in \mathfrak{g}^s_m \otimes \mathbb{R}^{m*}$. Let us perform this explicitly for $s = 2$.

In G^2_m we have $(a^i_j, a^i_{jk})^{-1} = (\tilde a^i_j, \tilde a^i_{jk})$ with $a^i_j \tilde a^j_k = \delta^i_k$ and $\tilde a^i_{jk} = -\tilde a^i_l a^l_{ps} \tilde a^s_k \tilde a^p_j$. Let $X = (a^i_k, a^i_{jk}, A^i_j, A^i_{jk}) \in TG^2_m$ and $b = (b^i_k, b^i_{jk}) \in G^2_m$. It is easy to compute

$$T\lambda_b(X) = (b^i_k a^k_j, b^i_l a^l_{jk} + b^i_{ps} a^p_j a^s_k, b^i_p A^p_j, b^i_p A^p_{jk} + b^i_{ps} A^p_j a^s_k + b^i_{ps} a^p_j A^s_k).$$

Taking into account all our identifications we get a formula for $i_0\colon G^3_m \to W^1_m G^2_m$

$$i_0(a^i_j, a^i_{jk}, a^i_{jkl}) = (a^i_j; a^i_j, a^i_{jk}; \tilde a^i_p a^p_{jl}, \tilde a^i_p a^p_{jkl} + \tilde a^i_{ps} a^p_{jl} a^s_k + \tilde a^i_{ps} a^p_j a^s_{kl}).$$

If we perform the above consideration up to the first order terms only, we get $i_0\colon G^2_m \to W^1_m G^1_m$, $i_0(a^i_j, a^i_{jk}) = (a^i_j; a^i_j; \tilde a^i_p a^p_{jl})$.

16. Canonical differential forms

16.1. Consider a vector bundle $E = P[V, \ell]$ associated to a principal bundle (P, p, M, G) and the space of all E-valued differential forms $\Omega(M; E)$. By theorem 11.14, there is the canonical isomorphism $q^\sharp$ between $\Omega(M; E)$ and the space of horizontal G-equivariant V-valued differential forms on P. According to 10.12, the image $\Phi = q^\sharp(\varphi) \in \Omega^k_{hor}(P; V)^G$ is called the *frame form of* $\varphi \in \Omega^k(M; E)$. We have

$$\Phi(X_1, \dots, X_k) = \tau(u, \ ) \circ \varphi(TpX_1, \dots, TpX_k) \tag{1}$$

where $X_i \in T_u P$ and $\tau\colon P \times_M E \to V$ is the canonical map. Conversely, for every $X_1, \dots, X_k \in T_x M$, we can choose arbitrary vectors $\bar X_1, \dots, \bar X_k \in T_u P$ with $u \in P_x$ and $Tp\bar X_i = X_i$ to get

$$\varphi(X_1, \dots, X_k) = q(u, \ ) \circ \Phi(\bar X_1, \dots, \bar X_k) \tag{2}$$

where $q\colon P \times V \to E$ is the other canonical map. The elements $\Phi \in \Omega_{hor}(P; V)^G$ are sometimes called the *tensorial forms of type* ℓ, while the differential forms in $\Omega(P; V)^G$ are called *pseudo tensorial forms of type* ℓ.

16.2. The canonical form on P^1M. We define an $\mathbb{R}^m$-valued one-form $\theta = \theta_M$ on P^1M for every m-dimensional manifold M as follows. Given $u = j_0^1 g \in P^1M$ and $X = j_0^1 c \in T_u P^1 M$ we set

$$\theta_M(X) = u^{-1} \circ Tp(X) = j_0^1(g^{-1} \circ p \circ c) \in T_0\mathbb{R}^m = \mathbb{R}^m.$$

In words, the choice of $u \in P^1M$ determines a local chart at $x = p(u)$ up to the first order and the form θ_M transforms $X \in T_u P^1 M$ into the induced coordinates of TpX. If we insert $\varphi = \mathrm{id}_{TM}$ into 16.1.(1) we get immediately

Proposition. *The canonical form $\theta_M \in \Omega^1(P^1M; \mathbb{R}^m)$ is a tensorial form which is the frame form of the 1-form $\mathrm{id}_{TM} \in \Omega^1(M; TM)$.*

Consider further a principal connection Γ on P^1M. Then the covariant exterior differential $d_\Gamma \theta_M$ is called the *torsion form* of Γ. By 11.15, $d_\Gamma\theta_M$ is identified with a section of $TM \otimes \Lambda^2 T^*M$, which is called the *torsion tensor* of Γ. If $d_\Gamma\theta_M = 0$, connection Γ is said to be *torsion-free*.

16.3. The canonical form on W^1P. For every principal bundle (P, p, M, G) we can generalize the above construction to an $(\mathbb{R}^m \oplus \mathfrak{g})$-valued one-form on W^1P. Consider the target projection $\beta\colon W^1P \to P$, an element $u = j^1\psi(0,e) \in W^1P$ and a tangent vector $X = j_0^1 c \in T_u(W^1P)$. We define the form $\theta = \theta_P$ by

$$\theta(X) = u^{-1} \circ T\beta(X) = j_0^1(\psi^{-1} \circ \beta \circ c) \in T_{(0,e)}(\mathbb{R}^m \times G) = \mathbb{R}^m \oplus \mathfrak{g}.$$

Let us notice that if $G = \{e\}$ is the trivial structure group, then we get $P = M$, $W^1P = P^1M$ and $\theta_P = \theta_M$.

The principal action ρ on P induces an action of G on the tangent space TP. We claim that the space of orbits TP/G is the associated vector bundle $E = W^1P[\mathbb{R}^m \oplus \mathfrak{g}; \ell]$ with the left action ℓ of $W_m^1 G$ on $T_{(0,e)}(\mathbb{R}^m \times G) = \mathbb{R}^m \oplus \mathfrak{g}$,

$$\ell(j^1\varphi(0,e), j_0^1 c) = j_0^1(\rho^{\bar\varphi(0)^{-1}} \circ \varphi \circ c).$$

Indeed, every $\mathcal{PB}_m(G)$-morphism commutes with the principal actions, so that ℓ is a left action which is obviously linear and the map $q\colon W^1P \times T_{(0,e)}(\mathbb{R}^m \times G) \to E$ transforming every couple $j^1\psi(0,e) \in W^1P$ and $j_0^1 c \in T_{(0,e)}(\mathbb{R}^m \times G)$ into the orbit in TP/G determined by $j_0^1(\psi \circ c)$ describes the associated bundle structure on E.

Proposition. *The canonical form θ_P on W^1P is a pseudo tensorial one-form of type ℓ.*

Proof. We have to prove $\theta_P \in \Omega^1(W^1P; \mathbb{R}^m \oplus \mathfrak{g})^{W_m^1 G}$. Let ρ and $\bar\rho$ be the principal actions on P and W^1P, $X = j_0^1 c \in T_u W^1 P$, $u = j^1\psi(0,e)$, $A = j^1\varphi(0,e) \in W_m^1 G$, $a = \mathrm{pr}_2 \circ \beta(A)$. We have

$$\beta \circ \bar\rho^A = \rho^a \circ \beta$$

$$(\bar\rho^A)_* X = j_0^1(\bar\rho^A \circ c) \in T_{uA} W^1 P$$

$$\theta_P \circ (\bar\rho^A)_* X = j_0^1(\varphi^{-1} \circ \psi^{-1} \circ \beta \circ \bar\rho^A \circ c) = j_0^1(\rho^a \circ \varphi^{-1} \circ \psi^{-1} \circ \beta \circ c).$$

Hence

$$\ell_{A^{-1}} \circ \theta_P(X) = \ell_{A^{-1}}(j_0^1(\psi^{-1} \circ \beta \circ c) = \theta_P \circ (\bar{\rho}^A)_*(X). \quad \square$$

Unfortunately, θ_P is not horizontal since the principal bundle projection on W^1P is $p \circ \beta$.

16.4. Lemma. *Let (P, p, M, G) be a principal bundle and $q\colon W^1P = P^1M \times_M J^1P \to P^1M$ be the projection onto the first factor. Then the following diagram commutes*

$$\begin{array}{ccc} \mathbb{R}^m \oplus \mathfrak{g} & \xleftarrow{\theta_P} & TW^1P \\ \Big\downarrow{\scriptstyle pr_1} & & \Big\downarrow{\scriptstyle Tq} \\ \mathbb{R}^m & \xleftarrow{\theta_M} & TP^1M \end{array}$$

Proof. Consider $X = j_0^1 c \in T_u W^1P$, $u = j^1\psi(0, e)$. Then $Tq(X) = j^1(q \circ c)$ and $q(u) = j_0^1 \psi_0$. It holds

$$\begin{aligned} \mathrm{pr}_1 \circ \theta_P(X) &= \mathrm{pr}_1(j_0^1(\psi^{-1} \circ \beta \circ c)) = j_0^1(\psi_0^{-1} \circ p \circ \beta \circ c) \\ &= j_0^1(\psi_0^{-1} \circ \bar{p} \circ q \circ c) = \theta_M \circ Tq(X) \end{aligned}$$

where $\bar{p}\colon P^1M \to M$ is the canonical projection. $\square$

16.5. Canonical forms on frame bundles. Let us consider a frame bundle P^rM and the first order principal prolongation $W^1(P^{r-1}M)$. We know the canonical form $\theta \in \Omega^1(W^1(P^{r-1}M); \mathbb{R}^m \oplus \mathfrak{g}_m^{r-1})^{W_m^1 G_m^{r-1}}$ and the reduction $i_M\colon P^rM \to W^1(P^{r-1}M)$ to the structure group G_m^r, see 15.7. So we can define the canonical form θ^r on P^rM to be the pullback $i_M^*\theta \in \Omega^1(P^rM, \mathbb{R}^m \oplus \mathfrak{g}_m^{r-1})$. By virtue of 16.3 there is the linear action $\bar{\ell} = \ell \circ \kappa$ where κ is the group homomorphism corresponding to i_M, see 15.7, and θ^r is a pseudo tensorial form of type $\bar{\ell}$. The form θ^r can also be described directly. Given $X \in T_uP^rM$, we set $\bar{u} = \pi_{r-1}^r u$, $\bar{X} = T\pi_{r-1}^r(X) \in T_{\bar{u}}P^{r-1}M$. Since every $u = j_0^r f \in P^rM$ determines a linear map $\tilde{u} = T_{(0,e)}P^{r-1}f\colon \mathbb{R}^m \oplus \mathfrak{g}_m^{r-1} \to T_{j_0^{r-1}f}P^{r-1}M$ we get $\theta^r(X) = \tilde{u}^{-1}(\bar{X})$.

16.6. Coordinate functions of sections of associated bundles. Let us fix an associated bundle $E = P[S; \ell]$ to a principal bundle (P, p, M, G). The canonical map $\tau_E\colon P \times_M E \to S$ determines the so called *frame form* $\sigma\colon P \to S$ of a section $s\colon M \to E$, $\sigma(u) = \tau_E(u, s(p(u)))$. As we proved in 15.5, $J^rE = W^rP[T_m^rS; \ell^r]$, $m = \dim M$, and so for every fixed section $s\colon M \to E$ the frame form σ^r of its r-th prolongation j^rs is a map $\sigma^r\colon W^rP \to T_m^rS$. If we choose some local coordinates (U, φ), $\varphi = (y^p)$, on S, then there are the induced local coordinates y_α^p on $(\pi_0^r)^{-1}(U) \subset T_m^rS$, $0 \le |\alpha| \le r$, and for every section $s\colon M \to E$ the compositions $y_\alpha^p \circ \sigma^r$ define (on the corresponding preimages) the *coordinate functions* a_α^p of j^rs induced by the local chart (U, φ). We deduced in 15.5 that for every $u = j^r\psi(0, e) = (j_0^r\psi_0, j^r\psi_1(\psi_0(0))) \in W^rP$

$$\sigma^r(u) = j_0^r \tau_E(\psi_1 \circ \psi_0, s \circ \psi_0).$$

In particular, for the first order case we get

$$a^p(u) = y^p \circ \sigma(u)$$
$$a_i^p(u) = dy^p(j_0^1 \tau_E(\psi_1 \circ \psi_0, s \circ \psi_0) \circ c)$$

where $c\colon \mathbb{R} \to \mathbb{R}^m$ is the curve $t \mapsto tx^i$.

We shall describe the first order prolongation in more details. Let us denote e_i, $i = 1, \dots, m$, the canonical basis in $\mathbb{R}^m$ and let e_α, $\alpha = m+1, \dots, m+\dim G$, be a linear basis of the Lie algebra $\mathfrak{g}$. So the canonical form θ on W^1P decomposes into $\theta = \theta^i e_i + \theta^\alpha e_\alpha$. Let us further write Y_α for the fundamental vector fields on S determined by e_α and let ω^α be the dual basis to that induced from e_α on VP. Hence if the coordinate formulas for Y_α are $Y_\alpha = \eta_\alpha^p(y)\frac{\partial}{\partial y^p}$, then for $z \in E_x$, $u \in P_x$, $X \in V_uP$, $y = \tau_E(u, z)$ we get

$$\tau_E(\ ,z)_*X = -Y_\alpha(y)\omega^\alpha(X) = -\eta_\alpha^p(y)\omega^\alpha(X)\tfrac{\partial}{\partial y^p}.$$

The next proposition describes the coordinate functions of j^1s on W^1P by means of the canonical form θ and the coordinate functions a^p of s on P.

Proposition. *Let $\bar{a}^p$ be the coordinate functions of a geometric object field $s\colon M \to E$ and let a_i^p, a^p be the coordinate functions of j^1s. Then $a^p = \bar{a}^p \circ \beta$, where $\beta\colon W^1P \to P$ is the target projection, and*

$$da^p + \eta_\alpha^p(a^q)\theta^\alpha = a_i^p\theta^i.$$

Proof. The equality $a^p = \bar{a}^p \circ \beta$ follows directly from the definition. We shall evaluate $da^p(X)$ with arbitrary $X \in T_uW^1P$, where $u \in W^1P$, $u = j^1\psi(0,e) = (j_0^1\psi_0, j^1\psi_1(\psi_0(0)))$. The frame u determines the linear isomorphism

$$\tilde{u} = T_{(0,e)}\psi\colon \mathbb{R}^m \oplus \mathfrak{g} \to T_{\bar{u}}P,$$

$\bar{u} = \beta(u)$. We shall denote $\theta^i(X) = \xi^i$, $\theta^\alpha(X) = \xi^\alpha$, so that $\theta(X) = \tilde{u}^{-1}(\beta_*X) = \xi^i e_i + \xi^\alpha e_\alpha$. Let us write $\bar{X} = \beta_*X = \bar{X}_1 + \bar{X}_2$ with $\bar{X}_1 = \tilde{u}(\xi^i e_i)$, $\bar{X}_2 = \tilde{u}(\xi^\alpha e_\alpha)$ and let c be the curve $t \mapsto t\xi^i e_i$ on $\mathbb{R}^m$. We have

$$\begin{aligned} d\bar{a}^p(\bar{X}_1) &= dy^p(j_0^1(\sigma \circ \psi_1 \circ \psi_0 \circ c)) \\ &= dy^p(j_0^1(\tau_E(\psi_1 \circ \psi_0, s \circ \psi_0) \circ c)) = a_i^p(u)\xi^i \\ d\bar{a}^p(\bar{X}_2) &= dy^p(\tau_E(\ , s(p(\bar{u})))_*\bar{X}_2) = -\eta_\alpha^p(a^q(\bar{u}))\xi^\alpha\tfrac{\partial}{\partial y^p}. \end{aligned}$$

Hence

$$da^p(X) = d\bar{a}^p(\beta_*X) = a_i^p(u)\theta^i(X) - \eta_\alpha^p(a^q(u))\theta^\alpha(X). \quad \square$$

17. Connections and the absolute differentiation

17.1. Jet approach to general connections. The (general) connections on any fiber bundle (Y, p, M, S) were introduced in 9.3 as the vector valued 1-forms $\Phi \in \Omega^1(Y; VY)$ with $\Phi \circ \Phi = \Phi$ and $\mathrm{Im}\Phi = VY$. Equivalently, any connection is determined by the horizontal projection $\chi = \mathrm{id}_{TY} - \Phi$, or by the horizontal subspaces $\chi(T_yY) \subset T_yY$ in the individual tangent spaces, i.e. by the horizontal distribution. But every horizontal subspace $\chi(T_yY)$ is complementary to the vertical subspace V_yY and therefore it is canonically identified with a unique element $j^1_y s \in J^1_y Y$. On the other hand, each $j^1_y s \in J^1_y Y$ determines a subspace in T_yY complementary to V_yY. This leads us to the following equivalent definition.

Definition. A (general) connection Γ on a fiber bundle (Y, p, M) is a section $\Gamma: Y \to J^1Y$ of the first jet prolongation $\beta: J^1Y \to Y$.

Now, the horizontal lifting $\gamma: TM \times_M Y \to TY$ corresponding to a connection Γ is given by the composition of jets, i.e. for every $\xi_x = j^1_0 c \in T_xM$ and $y \in Y$, $p(y) = x$, we have $\gamma(\xi_x, y) = \Gamma(y) \circ \xi_x$. Given a vector field ξ, we get the Γ-lift $\Gamma\xi \in \mathfrak{X}(Y)$, $\Gamma\xi(y) = \Gamma(y) \circ \xi(p(y))$ which is a projectable vector field on $Y \to M$. Note that for every connection Γ on $p: Y \to M$ and $\xi \in T_yY$ it holds $\chi(\xi) = \gamma(Tp(\xi), y)$ and $\Phi = \mathrm{id}_{TY} - \chi$.

Since the first jet prolongations carry a natural affine structure, we can consider J^1 as an affine bundle functor on the category $\mathcal{FM}_{m,n}$ of fibered manifolds with m-dimensional bases and n-dimensional fibers and their local fibered manifold isomorphisms. The corresponding vector bundle functor is $V \otimes T^*B$, where $B: \mathcal{FM}_{m,n} \to \mathcal{M}f_m$ is the base functor, see 12.11. The choice of a (general) connection Γ on $p: Y \to M$ yields an identification of $J^1Y \to Y$ with $VY \otimes T^*M$. Chosen any fibered atlas $\varphi_\alpha: (\mathbb{R}^{m+n} \to \mathbb{R}^m) \to (Y \to M)$ with $\varphi_\alpha(R^{m+n}) = U_\alpha$, we can use the canonical flat connection on $\mathbb{R}^{m+n}$ to get such identifications on J^1U_α. In this way we obtain the local sections $\gamma_\alpha: U_\alpha \to (V \otimes T^*B)(U_\alpha)$ which correspond to the Christoffel forms introduced in 9.7. More explicitly, if we pull back the sections γ_α to $\mathbb{R}^{m+n} \to \mathbb{R}^m$ and use the product structure, then we obtain exactly the Christoffel forms.

In 9.4 we defined the curvature R of a (general) connection Γ by means of the Frölicher-Nijenhuis bracket, $2R = [\Phi, \Phi]$. It holds $R[X_1, X_2] = \Phi([\chi X_1, \chi X_2])$ for all vector fields X_1, X_2 on Y. In other words, given two vectors A_1, $A_2 \in T_yY$, we extend them to arbitrary vector fields X_1 and X_2 on Y and we have $R(A_1, A_2) = \Phi([\chi X_1, \chi X_2](y))$. Clearly, we can take for X_1 and X_2 projectable vector fields over some vector fields ξ_1, ξ_2 on M. Then $\chi X_i = \gamma\xi_i$, $i = 1, 2$. This implies that R can be interpreted as a map $R(y, \xi_1, \xi_2) = \Phi([\gamma\xi_1, \gamma\xi_2](y))$. Such a map is identified with a section $Y \to VY \otimes \Lambda^2T^*M$. Obviously, the latter formula can be rewritten as

$$R(y, \xi_1, \xi_2) = [\gamma\xi_1, \gamma\xi_2](y) - \gamma([\xi_1, \xi_2])(y).$$

This relation is usually expressed by saying that the curvature is the obstruction against lifting the bracket of vector fields.

17.2. Principal connections. Consider a principal fiber bundle (P, p, M, G) with the principal action $r: P \times G \to P$. We shall also denote by r the canonical right action $r: J^1P \times G \to J^1P$ given by $r^g(j_x^1 s) = j_x^1(r^g \circ s)$ for all $g \in G$ and $j_x^1 s \in J^1P$. In accordance with 11.1 we define a *principal connection* Γ on a principal fiber bundle P with a principal action r as an r-equivariant section $\Gamma: P \to J^1P$ of the first jet prolongation $J^1P \to P$.

Let us recall that for every principal bundle, there are the canonical right actions of the structure group on its tangent bundle and vertical tangent bundle. By definition, for every vector field $\xi \in \mathfrak{X}(M)$ and principal connection Γ the Γ-lift $\Gamma\xi$ is a right invariant projectable vector field on P. Furthermore, a principal connection induces an identification $J^1P \cong VP \otimes T^*M$ which maps principal connections into right invariant sections.

17.3. Induced connections on associated fiber bundles. Let us consider an associated fiber bundle $E = P[S; \ell]$. Every local section σ of P determines a local trivialization of E. Hence the idea of the definition of induced connections used in 11.8 gets the following simple form. For any principal connection Γ on P we define the section $\Gamma_E: E \to J^1E$ by $\Gamma_E\{u, s\} = j_x^1\{\sigma, \hat{s}\}$, where $u \in P_x$ and $s \in S$ are arbitrary, $\Gamma(u) = j_x^1\sigma$ and $\hat{s}$ means the constant map $M \to S$ with value s. It follows immediately that the parallel transport $\mathrm{Pt}_E(c, \{u, s\})$ of an element $\{u, s\} \in E$ along a curve $c: \mathbb{R} \to M$ is the curve $t \mapsto \{\mathrm{Pt}(c, u, t), s\}$ where Pt is the G-equivariant parallel transport with respect to the principal connection on P.

We recall the canonical principal bundle structure (TP, Tp, TM, TG) on TP and $TE = TP[TS, T\ell]$, see 10.18. The horizontal lifting determined by the induced connection Γ_E is given for every $\xi \in \mathfrak{X}(M)$ by

$$\Gamma_E\xi(\{u, s\}) = \{\Gamma\xi(u), 0_s\} \in (TE)_{\xi(p(u))}, \tag{1}$$

where $0_s \in T_sS$ is the zero tangent vector. Let us now consider an arbitrary general connection Γ_E on E. Chosen an auxiliary principal connection Γ_P on P, we can express the horizontal lifting γ_E in the form $\Gamma_E\xi(\{u, s\}) = \{\Gamma_P\xi(u), \bar{\gamma}(\xi(p(u)), s)\}$. The map $\bar{\gamma}$ is uniquely determined if the action ℓ is infinitesimally effective, i.e. the fundamental field mapping $\mathfrak{g} \to \mathfrak{X}(S)$ is injective. Then it is not difficult to check that the horizontal lifting γ_E can be expressed in the form (1) with certain principal connection Γ on P if and only if the map $\bar{\gamma}$ takes values in the fundamental fields on S. This is equivalent to 11.9.

17.4. The bundle of (principal) connections. We intend to treat principal connections as sections of an appropriate bundle. We have defined them as right invariant sections of the first jet prolongation of principal bundles, so that given a principal connection Γ on (P, p, M, G) and a point $x \in M$, its value on the whole fiber P_x is determined by the value in any point from P_x. We define QP to be the set of orbits J^1P/G. Since the source projection $\alpha: J^1P \to M$ is G-invariant, we have the projection $QP \to M$, also denoted by α. Furthermore, for every morphism of principal fiber bundles $(\varphi, \varphi_1): (P, p, M, G) \to (\bar{P}, \bar{p}, \bar{M}, \bar{G})$ over $\varphi_1: G \to \bar{G}$ it holds

$$J^1\varphi(j_x^1(r^a \circ s)) = j_{\varphi_0(x)}^1(r^{\varphi_1(a)} \circ \varphi \circ s \circ \varphi_0^{-1})$$

for all $j^1_x s \in J^1P$, $a \in G$. Hence the map $J^1\varphi\colon J^1P \to J^1\bar{P}$ factors to a map $Q\varphi\colon QP \to Q\bar{P}$ and Q becomes a functor with values in fibered sets. More explicitly, for every $j^1_x s$ in an orbit $A \in QP$ the value $Q\varphi(A)$ is the orbit in $J^1\bar{P}$ going through $J^1\varphi(j^1_x s)$. By the construction, we have a bijective correspondence between the sections of the fibered set $QP \to M$ and the G-equivariant sections of $J^1P \to P$ which are smooth along the individual fibers of P. It remains to define a suitable smooth structure on QP.

Let us first assume $P = \mathbb{R}^m \times G$. Then there is a canonical representative in each orbit $J^1(\mathbb{R}^m \times G)/G$, namely $j^1_x s$ with $s(x) = (x,e)$, $e \in G$ being the unit. Moreover, $J^1(\mathbb{R}^m \times G)$ is identified with $\mathbb{R}^m \times J^1_0(\mathbb{R}^m, G)$, $j^1_x s \mapsto (x, j^1_0(\mathrm{pr}_2 \circ s \circ t_x))$. Hence there is the induced smooth structure $Q(\mathbb{R}^m \times G) \cong \mathbb{R}^m \times J^1_0(\mathbb{R}^m, G)_e$ and the canonical projection $J^1(\mathbb{R}^m \times G) \to Q(\mathbb{R}^m \times G)$ becomes a surjective submersion. Let $\mathcal{PB}_m$ be the category of principal fiber bundles over m-manifolds and their morphisms covering local diffeomorphisms on the base manifolds. For every $\mathcal{PB}_m$-morphism $\varphi\colon \mathbb{R}^m \times G \to \mathbb{R}^m \times \bar{G}$ and element $j^1_x s \in A \in Q(\mathbb{R}^m \times G)$ with $s(x) = (x,e)$, the orbit $Q\varphi(A)$ is determined by $J^1\varphi(j^1_x s)$. This means that

$$Q\varphi(j^1_x s) = j^1_{\varphi_0(x)}(r^{a^{-1}} \circ \varphi \circ s \circ \varphi_0^{-1})$$

where $a = \mathrm{pr}_2 \circ \varphi(x,e)$ and consequently $Q\varphi$ is smooth.

Now for every principal fiber bundle atlas $(U_\alpha, \varphi_\alpha)$ on a principal fiber bundle P the maps $Q\varphi_\alpha$ form a fiber bundle atlas $(U_\alpha, Q\varphi_\alpha)$ on $QP \to M$. Let us summarize.

Proposition. *The functor $Q\colon \mathcal{PB}_m \to \mathcal{FM}_m$ associates with each principal fiber bundle (P,p,M,G) the fiber bundle QP over the base M with standard fiber $J^1_0(\mathbb{R}^m, G)_e$. The smooth sections of QP are in bijection with the principle connections on P.*

The functor Q is a typical example of the so called gauge natural bundles which will be studied in detail in chapter XII. On replacing the first jets by k-jets in the above construction, we get the functor $Q^k\colon \mathcal{PB}_m \to \mathcal{FM}_m$ of k-th order (principal) connections.

17.5. The structure of an associated bundle on QP. Let us consider a principal fiber bundle (P,p,M,G) and a local trivialization $\psi\colon \mathbb{R}^m \times G \to P$. By the definition, the restriction of $Q\psi$ to the fiber $S := (Q(\mathbb{R}^m \times G))_0$ is a diffeomorphism onto the fiber $QP_{\psi_0(0)}$. Since the functor Q is of order one, this diffeomorphism is determined by $j^1\psi(0,e) \in W^1P$, cf. 15.3. For the same reason, every element $j^1\varphi(0,e) \in W^1_m G$ determines a diffeomorphism $Q\varphi|S\colon S \to S$. By the definition of the Lie group structure on $W^1_m G$, this defines a left action ℓ of $W^1_m G$ on S. We define a mapping $q\colon W^1P \times S \to QP$ by

$$q(j^1\psi(0,e), A) = Q\psi(A).$$

Since $q(j^1(\psi\circ\varphi)(0,e), Q\varphi^{-1}(A)) = Q\psi \circ Q\varphi \circ Q\varphi^{-1}(A)$, the map q identifies QP with $W^1P[S;\ell]$. We shall see in chapter XII that the map q is an analogy to our identifications of the values of bundle functors on $\mathcal{M}f_m$ with associated bundles to frame bundles and that this construction goes through for every gauge natural bundle.

We are going to describe the action ℓ in more details. We know that

$$S = J_0^1(R^m \times G)/G \cong (\mathbb{R}^m \times T_m^1 G)_0/G \cong J_0^1(\mathbb{R}^m, G)_e \cong \mathfrak{g} \otimes \mathbb{R}^{m*},$$

see 17.4, and $W_m^1 G = G_m^1 \rtimes T_m^1 G$. Moreover, we have introduced the identification $T_m^1 G = G \rtimes (\mathfrak{g} \otimes \mathbb{R}^{m*})$ with the multiplication $(a, Z)(\bar{a}, \bar{Z}) = (a\bar{a}, \mathrm{Ad}(\bar{a}^{-1})Z + \bar{Z})$, see 15.6. Let us now express the action ℓ of $W_m^1 G = (G_m^1 \times G) \rtimes (\mathfrak{g} \otimes \mathbb{R}^{m*})$ on $S = (\mathfrak{g} \otimes \mathbb{R}^{m*})$. Given $(A, a, Z) \cong j^1\varphi(0, e) \in W_m^1 G$, and $Y \cong j_0^1 s \in J_0^1(\mathbb{R}^m \times G)$, $s(0) = (0, e)$, we have $A = j_0^1\varphi_0$, $a = \mathrm{pr}_2 \circ \varphi(0, e)$, $Z = T\lambda_{a^{-1}} \circ T_0\bar{\varphi}$ and $Y = T_0\tilde{s}$, where $\tilde{s} = \mathrm{pr}_2 \circ s$. By definition, $Q\varphi(j_0^1 s) = j_0^1 q$ and if we require $\tilde{q}(0) := \mathrm{pr}_2 \circ q(0) = e$ we have $q = \rho^{a^{-1}} \circ \varphi \circ s \circ \varphi_0^{-1}$, where ρ denotes the principal right action of G. Then we evaluate

$$\tilde{q} = \rho^{a^{-1}} \circ \mu \circ (\bar{\varphi}, \tilde{s}) \circ \varphi_0^{-1} = \mathrm{conj}(a) \circ \mu \circ (\lambda_{a^{-1}} \circ \bar{\varphi}, \tilde{s}) \circ \varphi_0^{-1}.$$

Hence by applying the tangent functor we get the action ℓ in form

$$(A, a, Z)(Y) = \mathrm{Ad}(a)(Y + Z) \circ A^{-1}. \tag{1}$$

Proposition. *For every principal bundle (P, p, M, G) the bundle of principal connections QP is the associated fiber bundle $W^1P[\mathfrak{g} \otimes \mathbb{R}^{m*}, \ell]$ with the action ℓ given by (1).*

Since the standard fiber of QP is a Euclidean space, there are always global sections of QP and so we have reproved in this way that every principal fiber bundle admits principal connections.

17.6. The affine structure on QP. In 17.2 and 17.3 we deduced that every principal connection on P determines a bijection between principal connections on P and the right invariant sections in $C^\infty(VP \otimes T^*M \to P)$. For every principal fiber bundle (P, p, M, G), let us denote by LP the associated vector bundle $P[\mathfrak{g}, \mathrm{Ad}]$. Since the fundamental field mapping $(u, A) \mapsto \zeta_A(u) \in V_uP$ identifies VP with $P \times \mathfrak{g}$ and $(ua, \mathrm{Ad}(a^{-1})(A)) \mapsto TR^a \circ \zeta_A(u)$, there is the induced identification $P[\mathfrak{g}, \mathrm{Ad}] \cong VP/G$. Hence every element in LP can be viewed as a right invariant vertical vector field on a fiber of P. Let us now consider $\mathfrak{g} \otimes \mathbb{R}^{m*}$ as a standard fiber of the vector bundle $LP \otimes T^*M$ with the left action of the product of Lie groups $G \times G_m^1$ given by

$$(a, A)(Y) = \mathrm{Ad}(a)(Y) \circ A^{-1}. \tag{1}$$

At the same time, we can view $\mathfrak{g} \otimes \mathbb{R}^{m*}$ as the standard fiber of QP with the action ℓ of $W_m^1 G$ given in 17.5.(1). Using the canonical affine structure on the vector space $\mathfrak{g} \otimes \mathbb{R}^{m*}$, we get for every two elements $Y_1, Y_2 \in \mathfrak{g} \otimes \mathbb{R}^{m*}$

$$\ell((A, a, Z), Y_1) - \ell((A, a, Z), Y_2) = \mathrm{Ad}(a)(Y_1 - Y_2) \circ A^{-1},$$

cf. 15.6.(3). Hence QP is an associated affine bundle to W^1P with the modelling vector bundle $LP \otimes T^*M = W^1P[\mathfrak{g} \otimes \mathbb{R}^{m*}]$ corresponding to the action (1) of the Lie subgroup $G^1_m \times G \subset W^1_mG$ via the canonical homomorphism $W^1_mG \to G^1_m \times G$. Since the curvature R of a principal connection is a right invariant section in $C^\infty(VP \otimes \Lambda^2T^*M \to P)$, we can view the curvature as an operator $R\colon C^\infty(QP \to M) \to C^\infty(LP \otimes \Lambda^2T^*M \to M)$. By the definition, R commutes with the action of the $\mathcal{PB}_m(G)$-morphisms, so that this is a typical example of the so called gauge natural operators which will be treated in chapter XII.

17.7. Principal connections on higher order frame bundles. Let us consider a frame bundle P^rM and the bundle of principal connections QP^rM. The composition $Q \circ P^r$ is a bundle functor on $\mathcal{M}f_m$ of order $r+1$, so that there is the canonical structure $QP^rM \cong P^{r+1}M[\mathfrak{g}^r_m \otimes \mathbb{R}^{m*}]$, but there also is the identification $QP^rM \cong W^1P^r[\mathfrak{g}^r_m \otimes \mathbb{R}^{m*}; \ell]$ described in 17.6. It is an easy exercise to verify that the former structure of an associated bundle is obtained from the latter one by the natural reduction $i_M\colon P^{r+1}M \to W^1P^rM$, see proposition 15.7.

The most important case is $r = 1$, since the functor QP^1 associates to each manifold M the bundle of linear connections on M. Let us deduce the coordinate expressions of the actions of $W^1_mG^1_m$ and G^2_m on $(\mathfrak{g}^1_m \otimes \mathbb{R}^{m*}) = \operatorname{Hom}(\mathbb{R}^m, \mathfrak{gl}(m))$. Given $(A, B, Z) \in W^1_mG^1_m$, $A = (a^i_j) \in G^1_m$, $B = (b^i_j) \in G^1_m$, $Z = (z^i_{jk}) \in (\mathfrak{g}^1_m \otimes \mathbb{R}^{m*})$, $\Gamma = (\Gamma^i_{jk}) \in (\mathfrak{g}^1_m \otimes \mathbb{R}^{m*})$, we have $\operatorname{Ad}(B)(Z) = (b^i_m z^m_{nj} \tilde b^n_k)$, so that 17.5.(1) implies

$$(A, B, Z)(\Gamma^i_{jk}) = (b^i_m(\Gamma^m_{nl} + z^m_{nl})\tilde a^l_k \tilde b^n_j).$$

The coordinate expression of the homomorphism $i_0\colon G^2_m \to W^1_mG^1_m$ deduced in 15.8 yields the formula

$$(a^i_j, a^i_{jk})(\Gamma^i_{jk}) = (a^i_m \Gamma^m_{nl} \tilde a^l_k \tilde a^n_j + a^i_{nl} \tilde a^l_k \tilde a^n_j).$$

We remark that the Γ^i_{jk} introduced in this way differ from the classical Christoffel symbols, [Kobayashi, Nomizu, 69], by sign and by the order of subscripts, see 17.15.

Let us mention briefly the second order case. We have to deal with $(A, B, Z) \in W^1_mG^2_m$, $A = (a^i_j) \in G^1_m$, $B = (b^i_j, b^i_{jk}) \in G^2_m$, $Z = (z^i_{jk}, z^i_{jkl}) \in (\mathfrak{g}^2_m \otimes \mathbb{R}^{m*})$. We compute

$$\begin{aligned}\operatorname{Ad}(B)(Z) \circ A^{-1} = (&b^i_p z^p_{sm} \tilde a^m_k \tilde b^s_j, b^i_p z^p_{sm} \tilde a^m_l \tilde b^s_{jk} \\ &+ b^i_p z^p_{mnq} \tilde a^q_l \tilde b^n_j \tilde b^m_k + b^i_{ps} z^p_{mn} \tilde a^n_l \tilde b^m_j \tilde b^s_k + b^i_{ps} z^s_{mn} \tilde a^n_l \tilde b^p_j \tilde b^m_k)\end{aligned}$$

and we have to compose this action with the homomorphism $i_0\colon G^3_m \to W^1_mG^2_m$. For every $a = (a^i_j, a^i_{jk}, a^i_{jkl}) \in G^3_m$, the formula derived in 15.8 implies

$$\begin{aligned}a.(\Gamma^i_{jk}, \Gamma^i_{jkl}) = \big(&a^i_m \Gamma^m_{nl} \tilde a^l_k \tilde a^n_j + a^i_{nl} \tilde a^l_k \tilde a^n_j, \\ &a^i_p \Gamma^p_{mnq} \tilde a^q_l \tilde a^n_k \tilde a^m_j + a^i_p \Gamma^p_{sm} \tilde a^m_l \tilde a^s_{jk} + a^i_{ps} \Gamma^p_{mn} \tilde a^n_l \tilde a^m_j \tilde a^s_k \\ &+ a^i_{ps} \Gamma^s_{nm} \tilde a^m_l \tilde a^p_j \tilde a^n_k + a^i_{mnq} \tilde a^q_l \tilde a^m_k \tilde a^n_j + a^i_{sm} \tilde a^s_{kj} \tilde a^m_l\big).\end{aligned}$$

17.8. The absolute differential. Let us consider a fixed principal connection $\Gamma\colon P \to J^1P$ on a principal fiber bundle (P, p, M, G) and an associated fiber bundle $E = P[S; \ell]$. We recall the maps $q\colon P \times S \to E$ and $\tau\colon P \times_M E \to S$, see 10.7, and we denote $\tilde{u}\colon = q(u, \ )\colon S \to E_{p(u)}$. Hence given local sections $\sigma\colon M \to P$ and $s\colon M \to E$ with a common domain $U \subset M$ and a point $x \in U$, there is the map $\varphi_{\sigma,s}\colon U \ni y \mapsto \widetilde{\sigma(x)} \circ \widetilde{\sigma(y)}^{-1} \circ s(y) \in E_x$, i.e. $\varphi_{\sigma,s} = q(\sigma(x), \ ) \circ \tau \circ (\sigma, s)$. In fact we use the local trivialization of E induced by σ to describe the local behavior of s in a single fiber. If P and (consequently) also E are trivial bundles and $\sigma(x) = (0, e)$, then we get just the projection onto the standard fiber. Since the principal connection Γ associates to every $u \in P_x$ a 1-jet $\Gamma(u) = j^1_x\sigma$ of a section σ, for every local section $s\colon M \to E$ and point x in its domain the one jet of $\varphi_{\sigma,s}$ at x describes the local behavior of s at x up to the first order. Our construction does not depend on the choice of $u \in P_x$, for Γ is right invariant. So we define the *absolute* (or *covariant*) *differential* $\nabla s(x)$ of s at x with respect to the principal connection Γ by

$$\nabla s(x) = j^1_x\varphi_{\sigma,s} \in J^1_x(M, E_x)_{s(x)} \cong \mathrm{Hom}(T_xM, V_{s(x)}E).$$

If E is an associated vector bundle, then there is the canonical identification $V_{s(x)}E = E_x$. Then we have $\nabla s(x) \in \mathrm{Hom}(T_xM, E_x)$ and we shall see that this coincides with the values of the covariant derivative ∇ as defined in section 11.

We can define a structure of an associated bundle on the union of the manifolds $J^1_x(M, E_x)$, $x \in M$, where the mappings ∇s take their values. Let us consider the principal fiber bundle $P^1M \times_M P$ with the principal action $r^{(a_1,a_2)}(u_1, u_2) = (u_1.a_1, u_2.a_2)$ of the Lie group $G^1_m \times G$ (here the dots mean the obvious principal actions). We define

$$\tau\colon (P^1M \times_M P) \times_M (\cup_{x\in M} J^1_x(M, E_x)) \to T^1_mS$$
$$\tau((j^1_0f, u), j^1_x\varphi) = j^1_0(\tilde{u}^{-1} \circ \varphi \circ f).$$

Let us further define a left action $\bar{\ell}$ of $G^1_m \times G$ on T^1_mS by (remember $E = P[S; \ell]$)

$$\bar{\ell}((j^1_0h, a_2), j^1_0q) = j^1_0(\ell_{a_2} \circ q) \circ j^1_0h^{-1}.$$

One verifies easily that τ determines the structure of the associated bundle $E_1 = (P^1M \times_M P)[T^1_mS; \bar{\ell}]$ and that for every section $s\colon M \to E$ its absolute differential ∇s with respect to a fixed principal connection Γ on P is a smooth section of E_1. Hence ∇ can be viewed as an operator

$$\nabla\colon C^\infty(E) \to C^\infty((P^1M \times_M P)[T^1_mS; \bar{\ell}]).$$

17.9. Absolute differentiation along vector fields. Let E, P, Γ be as in 17.8. Given a tangent vector $X_x \in T_xM$, we define the absolute differentiation in the direction X_x of a section $s\colon M \to E$ to be the value $\nabla s(x)(X_x) \in V_{s(x)}E$.

Applying this procedure to a vector field $X \in \mathfrak{X}(M)$ we get a map $\nabla_X s\colon M \to VE$ with the following properties

(1) $$\pi_E \circ \nabla_X s = s$$

(2) $$\nabla_{fX+gY} s = f\nabla_X s + g\nabla_Y s$$

for all vector fields X, $Y \in \mathfrak{X}(M)$ and smooth functions f, g on M, $\pi_E\colon VE \subset TE \to E$ being the canonical projection.

So every $X \in \mathfrak{X}(M)$ determines an operator $\nabla_X\colon C^\infty(E) \to C^\infty(VE)$ and the whole procedure of the absolute differentiation can be viewed as an operator $\nabla\colon C^\infty(TM \times_M E) \to C^\infty(VE)$.

By the definition of the connection form Φ_E of the induced connection Γ_E, it holds

(3) $$\nabla_X s = \Phi_E \circ Ts \circ X$$

(4) $$\nabla_X s = Ts \circ X - (\Gamma_E X) \circ s.$$

17.10. The frame forms. For every vector field $X \in \mathfrak{X}(M)$ and every map $\bar{s}\colon P \to S$ we define

$$\nabla_X \bar{s}\colon P \to TS, \quad \nabla_X \bar{s} = T\bar{s} \circ \Gamma_E X$$
$$\nabla \bar{s}\colon P^1 M \times_M P \to T^1_m S, \quad \nabla \bar{s}(v,u) = T\bar{s} \circ T\sigma \circ v,$$

where $\Gamma(u) = j^1_x \sigma$, $x = p(u)$. We call $\nabla \bar{s}$ the *absolute differential* of $\bar{s}$ while $\nabla_X \bar{s}$ is called the *absolute differential along* X.

Proposition. *Let $\bar{s}\colon P \to S$ be the frame form of a section $s\colon M \to E$. Then $\nabla \bar{s}$ is the frame form of ∇s and for every $X \in \mathfrak{X}(M)$, $\nabla_X \bar{s}$ is the frame form of $\nabla_X s$.*

Proof. The map $\nabla_X s$ is a section of $VE = P[TS]$ and $\bar{s}(u) = \tau_E(u, s \circ p(u))$, $u \in P$. Further, for every $u \in P_x$ with $\Gamma(u) = j^1_x \sigma$, we have $\nabla s(x) = j^1_x(\tilde{u} \circ \bar{s} \circ \sigma) \in \operatorname{Hom}(T_x M, V_{s(x)} E)$. Hence for every $X \in \mathfrak{X}(M)$ we get $\nabla_X s = T\tilde{u} \circ T(\bar{s} \circ \sigma) \circ X$ and since the diffeomorphism $TS \to (VE)_x$ determined by $u \in P$ is just $T\tilde{u}$, the frame form of $\nabla_X s$ is $\nabla_X \bar{s}$.

In order to prove the other equality, let us evaluate

$$\nabla s(x) = \{(v,u), (j^1_x(\tilde{u}^{-1} \circ \varphi)) \circ v\}.$$

Since $\varphi = \tilde{u} \circ \bar{s} \circ \sigma$, where $\Gamma(u) = j^1_x \sigma$, the frame form of ∇s is $\nabla \bar{s}$. □

17.11. If $E = P[S; \ell]$ is an associated vector bundle, then we can use the canonical identification $S \cong T_y S$ for each point $y \in S$. Consider a section $s\colon M \to E$ and its frame form $\bar{s}\colon P \to S$. Then $\nabla s(x) \in J^1_x(M, E_x)$ can be viewed as a value of a form $Ds \in \Omega^1(M; E)$. The corresponding S-valued tensorial 1-form $D\bar{s}\colon TP \to S$ is defined by $D\bar{s} = d\bar{s} \circ \chi = (\chi^* d)(\bar{s})$, where χ is the horizontal projection of Γ_E. Of course, this formula defines the absolute differentiation $D\colon \Omega^k(P; S) \to \Omega^{k+1}(P; S)$ for all $k \geq 0$, cf. section 11. The absolute differentials of higher order can also be defined in the nonlinear case. However, this requires an inductive procedure and we refer the reader to [Kolář, 73 b].

17.12. We are going to deduce a general coordinate formula for the absolute differentiation of sections of an arbitrary associated fiber bundle. We shall do it in a geometric way, which reduces the problem to the proposition 16.6. For every principal connection $\Gamma\colon P \to J^1P$ the image of the map Γ defines a reduction

$$R(\Gamma)\colon P^1M \times_M P \xrightarrow{\Gamma} P^1M \times_M \Gamma(P) \hookrightarrow P^1M \times_M J^1P = W^1P$$

of the principal bundle W^1P to the structure group

$$G^1_m \times G \hookrightarrow G^1_m \rtimes T^1_mG = G^1_m \rtimes (G \rtimes (\mathfrak{g} \otimes \mathbb{R}^{m*})).$$

Let us write $\tilde\theta$ for the restriction of the canonical form θ on W^1P to $P^1M \times_M \Gamma(P)$, let ω be the connection form of Γ and θ_M will denote the canonical form $\theta_M \in \Omega^1(P^1M; \mathbb{R}^m)$.

Lemma. *The following diagram is commutative*

$$\begin{array}{ccccc}
TP & \xleftarrow{\beta_*} & T(P^1M \times_M \Gamma(P)) & \xrightarrow{pr_1} & TP^1M \\
\downarrow{\scriptstyle\omega} & & \downarrow{\scriptstyle\tilde\theta} & & \downarrow{\scriptstyle\theta_M} \\
\mathfrak{g} & \xleftarrow{pr_2} & \mathbb{R}^m \oplus \mathfrak{g} & \xrightarrow{pr_1} & \mathbb{R}^m
\end{array}$$

Proof. For every $u \in W^1P$, $u = j^1\psi(0,e)$, $\beta(u) = \bar u$, we have the isomorphism $\tilde u\colon \mathbb{R}^m \oplus \mathfrak{g} \to T_{\bar u}P$ and for every $X \in T_uW^1P$, $\theta(X) = \tilde u^{-1}(\beta_*X)$. If $X \in T(P^1M \times_M \Gamma(P))$, we denote $\theta(X) = Y_1 + Y_2 \in \mathbb{R}^m \oplus \mathfrak{g}$. Then $\tilde u(Y_1) = T(\psi_1 \circ \psi_0)Y_1 = \chi(\beta_*X)$ and $\tilde u(Y_2) = \beta_*X - \tilde u(Y_1) = \Phi(\beta_*X)$, where Φ and χ are the vertical and horizontal projections determined by Γ. Since the restriction of $\tilde u$ to the second factor in $\mathbb{R}^m \oplus \mathfrak{g}$ coincides with the fundamental vector field mapping, the commutativity of the left-hand square follows.

The commutativity of the right-hand one was proved in 16.4. □

17.13. Lemma. *Let $s\colon M \to E$ be a section, $\bar s\colon P \to S$ its frame form and let $\bar s^1\colon W^1P \to T^1_mS$ be the frame form of j^1s. Then for all $u \in P^1M \times_M P \cong P^1M \times_M \Gamma(P) \subset W^1P$ it holds $\bar s^1(u) = \nabla\bar s(u)$.*

Proof. If $u = j^1\psi(0,e)$, $\bar u = \beta(u)$, then $\Gamma(\bar u) = j^1\psi_1(\psi_0(0))$. Since we know $\bar s^1(u) = j^1_0(\tau_E(\psi_1 \circ \psi_0, s \circ \psi_0))$, we get $\nabla\bar s(u) = j^1_0(\bar s \circ \psi_1 \circ \psi_0) = \bar s^1(u)$. □

17.14. Proposition. *Let E, S, P, Γ, ω be as before and consider a local chart (U,φ), $\varphi = (y^p)$, on S. Let e_i, $i = 1,\dots,m$ be the canonical basis in $\mathbb{R}^m$ and e_α, $\alpha = m+1,\dots,m+\dim G$ be a base of Lie algebra $\mathfrak{g}$. Let us denote $\theta_M = \theta^i_M e_i$ the canonical form on P^1M, $\omega = \omega^\alpha e_\alpha$, j_1 and j_2 be the canonical projections on $P^1M \times_M P$. Further, let us write $\bar\omega^\alpha = j_2^*\omega^\alpha$, $\bar\theta^i_M = j_1^*\theta^i_M$ and let $\eta^p_\alpha(y)\frac{\partial}{\partial y^p}$ be the fundamental vector fields corresponding to e_α. For a section $s\colon M \to E$ let a^p, a^p_i be the coordinate functions of ∇s on $P^1M \times_M P$ while $\bar a^p$ be those of s. Then it holds*

$$da^p + \eta^p_\alpha(a^q)\bar\omega^\alpha = a^p_i\bar\theta^i_M.$$

Proof. In 16.6 we described the coordinate functions b^p, b^p_i of j^1s defined on W^1P, $b^p = \beta^*\bar{a}^p$, $db^p + \eta^p_\alpha(b^q)\theta^\alpha = b^p_i\theta^i$. According to 17.13, the functions a^p, a^p_i are restrictions of b^p, b^p_i to $P^1M \times_M P$. But then the proposition follows from lemma 17.12. □

17.15. Example. We find it instructive to apply this general formula to the simplest case of the absolute differential of a vector field ξ on a manifold M with respect to a classical linear connection Γ on M. Since we consider the standard action $\bar{y}^i = a^i_j y^j$ of $GL(m)$ on $\mathbb{R}^m$, the fundamental vector fields η^i_j on $\mathbb{R}^m$ corresponding to the canonical basis of the Lie algebra of $GL(m)$ are of the form $\delta^k_i y^j \frac{\partial}{\partial y^k}$. Every local coordinates (x^i) on an open subset $U \subset M$ define a section $\rho\colon U \to P^1M$ formed by the coordinate frames $(\frac{\partial}{\partial x^1}, \dots, \frac{\partial}{\partial x^m})$ and it holds $\rho^*\theta^i_M = dx^i$. On the other hand, from the explicit equation 25.2.(2) of Γ we deduce easily that the restriction of the connection form $\omega = (\omega^i_j)$ of Γ to ρ is $(-\Gamma^i_{jk}(x)dx^k)$. Thus, if we consider the coordinate expression $\xi^i(x)\frac{\partial}{\partial x^i}$ of ξ in our coordinate system and we write $\nabla_j\xi^i$ for the additional coordinates of $\nabla\xi$, we obtain from 17.14

$$\nabla_j\xi^i = \frac{\partial\xi^i}{\partial x^j} - \Gamma^i_{kj}\xi^k.$$

Comparing with the classical formula in [Kobayashi, Nomizu, 63, p. 144], we conclude that our quantities Γ^i_{jk} differ from the classical Christoffel symbols by sign and by the order of subscripts.

Remarks

The development of the theory of natural bundles and operators is described in the preface and in the introduction to this chapter. But let us come back to the jet groups. As mentioned in [Reinhart, 83], it is remarkable how very little of existing Lie group theory applies to them. The results deduced in our exposition are mainly due to [Terng, 78] where the reader can find some more information on the classification of G^r_m-modules. For the first order jet groups, it is very useful to study in detail the properties of irreducible representations, cf. section 34. But in view of 13.15 it is not interesting to extend this approach to the higher orders. The bundle functors on the whole category $\mathcal{M}f$ were first studied by [Janyška, 83]. We shall continue the study of such functors in chapter IX.

The basic ideas from section 15 were introduced in a slightly modified situation by [Ehresmann, 55]. Every principal fiber bundle $p\colon P \to M$ with structure group G determines the associated groupoid PP^{-1} which can be defined as the factor space $P \times P/\sim$ with respect to the equivalence relation $(u,v) \sim (ug, vg)$, u, $v \in P$, $g \in G$. Writing uv^{-1} for such an equivalence class, we have two projections a, $b\colon PP^{-1} \to M$, $a(uv^{-1}) = p(v)$, $b(uv^{-1}) = p(u)$. If E is a fiber bundle associated with P with standard fiber S, then every $\theta = uv^{-1} \in PP^{-1}$ determines a diffeomorphism $q_u \circ (q_v)^{-1}\colon E_{a\theta} \to E_{b\theta}$, where $q_v\colon S \to E_{a\theta}$ and $q_u\colon S \to E_{b\theta}$ are the 'frame maps' introduced in 10.7. This defines an action of groupoid PP^{-1} on fiber bundle E. The space PP^{-1} is a prototype of a smooth groupoid over M. In [Ehresmann, 55] the r-th prolongation Φ^r of an arbitrary smooth groupoid Φ over M is defined and every action of Φ on a fiber bundle $E \to M$ is extended into an action of Φ^r on the r-th jet prolongation J^rE of $E \to M$. This construction was modified to the principal fiber bundles by [Libermann, 71], [Virsik, 69] and [Kolář, 71b].

The canonical $\mathbb{R}^m$-valued form on the first order frame bundle P^1M is one of the basic concepts of modern differential geometry. Its generalization to r-th order frame bundles was introduced by [Kobayashi, 61]. The canonical form on W^1P (as well as on W^rP) was defined in [Kolář, 71b] in connection with some local considerations by [Laptev, 69] and [Gheorghiev, 68]. Those canonical forms play an important role in a generalization of the Cartan method of moving frames, see [Kolář, 71c, 73a, 73b, 77].

CHAPTER V. FINITE ORDER THEOREMS

The purpose of this chapter is to develop a general framework for the theory of geometric objects and operators and to reduce local geometric considerations to finite order problems. In general, the latter is a hard analytical problem and its solution essentially depends on the category in question. Roughly speaking, our methods are efficient when we deal with a sufficiently large class of smooth maps, but they fail e.g. for analytic maps.

We first extend the concepts and results from section 14 to a wider class of categories. Then we present our important analytical tool, a nonlinear generalization of well known Peetre theorem. In section 20 we prove the regularity of bundle functors for a class of categories which includes $\mathcal{M}f$, $\mathcal{M}f_m$, $\mathcal{FM}$, $\mathcal{FM}_m$, $\mathcal{FM}_{m,n}$, and we get near to the finiteness of the order of bundle functors. It remains to deduce estimates on the possible orders of jet groups acting on manifolds. We derive such estimates for the actions of jet groups in the category $\mathcal{FM}_{m,n}$ so that we describe all bundle functors on $\mathcal{FM}_{m,n}$. For $n = 0$ this reproves in a different way the classical results due to [Palais, Terng, 77] and [Epstein, Thurston, 79] on the regularity and the finiteness of the order of natural bundles.

The end of the chapter is devoted to a discussion on the order of natural operators. Also here we essentially profit from the nonlinear Peetre theorem. First of all, its trivial consequence is that every (even not natural) local operator depends on infinite jets only. So instead of natural transformations between the infinite dimensional spaces of sections of the bundles in question, we have to deal with natural transformations between the (infinite) jet prolongations. The full version of Peetre theorem implies that in fact the order is finite on large subsets of the infinite jet spaces and, by naturality, the order is invariant under the action of local isomorphisms on the infinite jets. In many concrete situations the whole infinite jet prolongation happens to be the orbit of such a subset. Then all natural operators from the bundle in question are of finite order and the problem of finding a full list of them can be attacked by the methods developed in the next chapter.

18. Bundle functors and natural operators

Roughly speaking, the objects of a differential geometric category should be manifolds with an additional structure and the morphisms should be smooth maps. The following approach is somewhat abstract, but this is a direct modification of the contemporary point of view to the concept of a concrete category, which is defined as a category over the category of sets.

18.1. Definition. A *category over manifolds* is a category $\mathcal{C}$ endowed with a faithful functor $m\colon \mathcal{C} \to \mathcal{M}f$. The manifold mA is called the *underlying manifold* of $\mathcal{C}$-object A and A is said to be a $\mathcal{C}$-object *over* mA.

The assumption that the functor m is faithful means that every induced map $m_{A,B}\colon \mathcal{C}(A,B) \to C^\infty(mA,mB)$, A, $B \in \mathrm{Ob}\mathcal{C}$, is injective. Taking into account this inclusion $\mathcal{C}(A,B) \subset C^\infty(mA,mB)$, we shall use the standard abuse of language identifying every smooth map $f\colon mA \to mB$ in $m_{A,B}(\mathcal{C}(A,B))$ with a $\mathcal{C}$-morphism $f\colon A \to B$.

The best known examples of categories over manifolds are the categories $\mathcal{M}f_m$ or $\mathcal{M}f$, the categories $\mathcal{FM}$, $\mathcal{FM}_m$, $\mathcal{FM}_{m,n}$ of fibered manifolds, oriented manifolds, symplectic manifolds, manifolds with fixed volume forms, Riemannian manifolds, etc., with appropriate morphisms.

For a category over manifolds $m\colon \mathcal{C} \to \mathcal{M}f$, we can define a bundle functor on $\mathcal{C}$ as a functor $F\colon \mathcal{C} \to \mathcal{FM}$ satisfying $B \circ F = m$ where $B\colon \mathcal{FM} \to \mathcal{M}f$ is the base functor. However, we have seen that the localization property of a natural bundle over m-dimensional manifolds plays an important role. To incorporate it into our theory, we adapt the general concept of a local category by [Eilenberg, 57] and [Ehresmann, 57] to the case of a category over manifolds.

18.2. Definition. A category over manifolds $m\colon \mathcal{C} \to \mathcal{M}f$ is said to be *local*, if every $A \in \mathrm{Ob}\mathcal{C}$ and every open subset $U \subset mA$ determine a $\mathcal{C}$-subobject $L(A,U)$ of A over U, called the *localization of* A *over* U, such that

(a) $L(A,mA) = A$, $L(L(A,U),V) = L(A,V)$ for every $A \in \mathrm{Ob}\mathcal{C}$ and every open subsets $V \subset U \subset mA$,

(b) (aggregation of morphisms) if (U_α), $\alpha \in I$, is an open cover of mA and $f \in C^\infty(mA,mB)$ has the property that every $f \circ i_{U_\alpha}$ is a $\mathcal{C}$-morphism $L(A,U_\alpha) \to B$, then f is a $\mathcal{C}$-morphism $A \to B$,

(c) (aggregation of objects) if (U_α), $\alpha \in I$, is an open cover of a manifold M and (A_α), $\alpha \in I$, is a system of $\mathcal{C}$-objects such that $mA_\alpha = U_\alpha$ and $L(A_\alpha, U_\alpha \cap U_\beta) = L(A_\beta, U_\alpha \cap U_\beta)$ for all α, $\beta \in I$, then there exists a unique $\mathcal{C}$-object A over M such that $A_\alpha = L(A,U_\alpha)$.

We recall that the requirement $L(A,U)$ is a $\mathcal{C}$-subobject of A means

(i) the inclusion $i_U\colon U \to mA$ is a $\mathcal{C}$-morphism $L(A,U) \to A$,

(ii) if for a smooth map $f\colon mB \to U$ the composition $i_U \circ f$ is a $\mathcal{C}$-morphism $B \to A$, then f is a $\mathcal{C}$-morphism $B \to L(A,U)$.

There are categories like the category $\mathcal{VB}$ of vector bundles with no localization of the above type, i.e. we cannot localize to an arbitrary open subset of the total space. From our point of view it is more appropriate to consider $\mathcal{VB}$ (and other similar categories) as a category over fibered manifolds, see 51.4.

18.3. Definition. Given a local category $\mathcal{C}$ over manifolds, a *bundle functor on* $\mathcal{C}$ is a functor $F\colon \mathcal{C} \to \mathcal{FM}$ satisfying $B \circ F = m$ and the localization condition:

(i) for every inclusion of an open subset $i_U\colon U \hookrightarrow mA$, $F(L(A,U))$ is the restriction $p_A^{-1}(U)$ of the value $p_A\colon FA \to mA$ over U and Fi_U is the inclusion $p_A^{-1}(U) \hookrightarrow FA$.

In particular, the projections p_A, $A \in \mathrm{Ob}\mathcal{C}$, form a natural transformation $p\colon F \to m$. We shall see later on that for a large class of categories one can equivalently define bundle functors as functors $F\colon \mathcal{C} \to \mathcal{M}f$ endowed with such a natural transformation and satisfying the above localization condition.

18.4. Definition. A *locally defined $\mathcal{C}$-morphism* of A into B is a $\mathcal{C}$-morphism $f\colon L(A,U) \to L(B,V)$ for some open subsets $U \subset mA$, $V \subset mB$. A $\mathcal{C}$-object A is said to be *locally homogeneous*, if for every x, $y \in mA$ there exists a locally defined $\mathcal{C}$-isomorphism f of A into A such that $f(x) = y$. The category $\mathcal{C}$ is called *locally homogeneous*, if each $\mathcal{C}$-object is locally homogeneous. A *local skeleton* of a locally homogeneous category $\mathcal{C}$ is a system (C_α), $\alpha \in I$, of $\mathcal{C}$-objects such that locally every $\mathcal{C}$-object A is isomorphic to a unique C_α. In such a case we say that A is an *object of type α*. The set I is called the *type set* of $\mathcal{C}$. A *pointed local skeleton* of a locally homogeneous category $\mathcal{C}$ is a local skeleton (C_α), $\alpha \in I$, with a distinguished point $0_\alpha \in mC_\alpha$ for each $\alpha \in I$.

A $\mathcal{C}$-morphism $f\colon A \to B$ is said to be a *local isomorphism*, if for every $x \in mA$ there are neighborhoods U of x and V of $f(x)$ such that the restricted map $U \to V$ is a $\mathcal{C}$-isomorphism $L(A,U) \to L(B,V)$. We underline that a local isomorphism is a globally defined map, which should be carefully distinguished from a locally defined isomorphism.

18.5. Examples. All the categories $\mathcal{M}f_m$, $\mathcal{M}f$, $\mathcal{FM}_{m,n}$, $\mathcal{FM}_m$, $\mathcal{FM}$ are locally homogeneous. A pointed local skeleton of the category $\mathcal{M}f$ is the sequence $(\mathbb{R}^m, 0)$, $m = 0,1,2,\dots$, while a pointed local skeleton of the category $\mathcal{FM}$ is the double sequence $(\mathbb{R}^{m+n} \to \mathbb{R}^m, 0)$, m, $n = 0,1,2\dots$.

18.6. Definition. The space $J^r(A,B)$ of all r-jets of a $\mathcal{C}$-object A into a $\mathcal{C}$-object B is the subset of the space $J^r(mA,mB)$ of all r-jets of mA into mB generated by the locally defined $\mathcal{C}$-morphisms of A into B. If it is useful to underline the category $\mathcal{C}$, we write $\mathcal{C}J^r(A,B)$ for $J^r(A,B)$.

18.7. Definition. A locally homogeneous category $\mathcal{C}$ is called *infinitesimally admissible*, if we have

(a) $J^r(A,B)$ is a submanifold of $J^r(mA,mB)$,

(b) the jet projections $\pi^r_k\colon J^r(A,B) \to J^k(A,B)$, $0 \le k < r$, are surjective submersions,

(c) if $X \in J^r(A,B)$ is an invertible r-jet of mA into mB, then X is generated by a locally defined $\mathcal{C}$-isomorphism.

Taking into account (c), we write

$$\mathrm{inv}J^r(A,B) = J^r(A,B) \cap \mathrm{inv}J^r(mA,mB).$$

18.8. Assume $\mathcal{C}$ is infinitesimally admissible and fix a pointed local skeleton $(C_\alpha, 0_\alpha)$, $\alpha \in I$. Let us write $\mathcal{C}^r(\alpha,\beta) = J^r_{0_\alpha}(C_\alpha, C_\beta)_{0_\beta}$ for the set of all r-jets of C_α into C_β with source 0_α and target 0_β. Definition 18.7 implies that every $\mathcal{C}^r(\alpha,\beta)$ is a smooth manifold, so that the restrictions of the jet composition $\mathcal{C}^r(\alpha,\beta) \times \mathcal{C}^r(\beta,\gamma) \to \mathcal{C}^r(\alpha,\gamma)$ are smooth maps. Thus we obtain a category $\mathcal{C}^r$ over I called the *r-th order skeleton of* $\mathcal{C}$.

By definition 18.7, $G^r_\alpha := \operatorname{inv} J^r_{0_\alpha}(C_\alpha, C_\alpha)_{0_\alpha}$ is a Lie group with respect to the jet composition, which is called the *r-th jet group* (or the r-th differential group) *of type* α. Moreover, if A is a $\mathcal{C}$-object of type α, then $P^r A := \operatorname{inv} J^r_{0_\alpha}(C_\alpha, A)$ is a principal fiber bundle over mA with structure group G^r_α, which is called the *r-th order frame bundle of* A. Let us remark that every jet group G^r_α is a Lie subgroup in the usual jet group G^r_m, $m = \dim C_\alpha$.

For example, all objects of the category $\mathcal{FM}_{m,n}$ are of the same type, so that $\mathcal{FM}_{m,n}$ determines a unique r-th jet group $G^r_{m,n} \subset G^r_{m+n}$ in every order r. In other words, $G^r_{m,n}$ is the group of all r-jets at $0 \in \mathbb{R}^{m+n}$ of fibered manifold isomorphisms $f\colon (\mathbb{R}^{m+n} \to \mathbb{R}^m) \to (\mathbb{R}^{m+n} \to \mathbb{R}^m)$ satisfying $f(0) = 0$.

18.9. The following assumption, which deals with the local skeleton of $\mathcal{C}$ only, has purely technical character.

A category $\mathcal{C}$ is said to have the *smooth splitting property*, if for every smooth curve $\gamma\colon \mathbb{R} \to J^r(C_\alpha, C_\beta)$, α, $\beta \in I$, there exists a smooth map $\Gamma\colon \mathbb{R} \times mC_\alpha \to mC_\beta$ such that $\gamma(t) = j^r_{c(t)}\Gamma(t,\ )$, where $c(t)$ is the source of r-jet $\gamma(t)$.

Since $\gamma(t)$ is a curve on $J^r(C_\alpha, C_\beta)$, we know that $\gamma(t)$ is generated by a system of locally defined $\mathcal{C}$-morphisms. So we require that on the local skeleton this can be done globally and in a smooth way. In all our concrete examples the underlying manifolds of the objects of the canonical skeleton are numerical spaces and each polynomial map determined by a jet of $J^r(C_\alpha, C_\beta)$ belongs to $\mathcal{C}$. This implies immediately that $\mathcal{C}$ has the smooth splitting property.

Definition. An infinitesimally admissible category $\mathcal{C}$ with the smooth splitting property is called *admissible*.

18.10. Regularity. From now on we assume that $\mathcal{C}$ is an admissible category. A family of $\mathcal{C}$-morphisms $f\colon M \to \mathcal{C}(A,B)$ parameterized by a manifold M is said to be *smoothly parameterized*, if the map $M \times mA \to mB$, $(u,x) \mapsto f(u)(x)$, is smooth.

Definition. A bundle functor $F\colon \mathcal{C} \to \mathcal{FM}$ is called *regular*, if F transforms every smoothly parameterized family of $\mathcal{C}$-morphisms into a smoothly parameterized family of $\mathcal{FM}$-morphisms.

18.11. Definition. A bundle functor $F\colon \mathcal{C} \to \mathcal{FM}$ is said to be of order r, $r \in \mathbb{N}$, if for any two locally defined $\mathcal{C}$-morphisms f and g of A into B, the equality $j^r_x f = j^r_x g$ implies that the restrictions of Ff and Fg to the fiber $F_x A$ of FA over $x \in mA$ coincide.

18.12. Associated maps. An r-th order bundle functor F defines the so-called *associated maps*

$$F_{A,B}\colon J^r(A,B) \times_{mA} FA \to FB, \quad (j^r_x f, y) \mapsto Ff(y)$$

where the fibered product is constructed with respect to the source projection $J^r(A,B) \to mA$.

Proposition. *The associated maps of an r-th order bundle functor F on an admissible category $\mathcal{C}$ are smooth if and only if F is regular.*

Proof. By locality, it suffices to discuss

$$F_{C_\alpha,C_\beta}\colon J^r(C_\alpha,C_\beta)\times_{mC_\alpha} FC_\alpha \to FC_\beta.$$

Consider a smooth curve $(\gamma(t),\delta(t))$ on $J^r(C_\alpha,C_\beta)\times_{mC_\alpha} FC_\alpha$, so that $p_{C_\alpha}\delta(t) = c(t)$, where $c(t)$ is the source of r-jet $\gamma(t)$. Since $\mathcal{C}$ has the smooth splitting property, there exists a smooth map $\Gamma\colon \mathbb{R}\times mC_\alpha \to mC_\beta$ such that $\gamma(t) = j^r_{c(t)}\Gamma(t,\ )$. The regularity of F implies $\varepsilon(t) := F(\Gamma(t,\ ))(\delta(t))$ is a smooth curve on FC_β. By the definition of the associated map, it holds $F_{C_\alpha,C_\beta}(\gamma(t),\delta(t)) = \varepsilon(t)$. Hence F_{C_α,C_β} transforms smooth curves into smooth curves. Now, we can use the following theorem due to [Boman, 67]

A mapping $f\colon \mathbb{R}^m \to \mathbb{R}^n$ is smooth if and only if for every smooth curve $c\colon \mathbb{R}\to\mathbb{R}^m$ the composition $f\circ c$ is smooth.

Then we conclude F_{C_α,C_β} is a smooth map. The other implication is obvious. □

18.13. The induced action. Consider an r-th order regular bundle functor F on an admissible category $\mathcal{C}$. The fibers $S_\alpha = F_{0_\alpha}C_\alpha$, $\alpha\in I$, will be called the *standard fibers* of F. Write $F_{\alpha\beta}$ for the restriction of F_{C_α,C_β} to $C^r(\alpha,\beta)\times S_\alpha \to S_\beta$. In the following definition we consider an arbitrary system (S_α), $\alpha\in I$, of manifolds with indices from the type set of $\mathcal{C}$.

Definition. A smooth action of C^r on a system (S_α), $\alpha\in I$, of manifolds is a system $\varphi_{\alpha\beta}\colon C^r(\alpha,\beta)\times S_\alpha \to S_\beta$ of smooth maps satisfying

$$\varphi_{\beta\gamma}(b,\varphi_{\alpha\beta}(a,s)) = \varphi_{\alpha\gamma}(b\circ a, s)$$

for all α, β, $\gamma\in I$, $a\in C^r(\alpha,\beta)$, $b\in C^r(\beta,\gamma)$, $s\in S_\alpha$.

By proposition 18.12, $F_{\alpha\beta}$ are smooth maps so that they form a smooth action of C^r on the system of standard fibers.

18.14. Theorem. *There is a canonical bijection between the regular r-th order bundle functors on $\mathcal{C}$ and the smooth actions of the r-th order skeleton of $\mathcal{C}$.*

Proof. For every regular r-th order bundle functor F on $\mathcal{C}$, $F_{\alpha\beta}$ is a smooth action of C^r on $(F_{0_\alpha}C_\alpha)$, $\alpha\in I$. Conversely, let $(\varphi_{\alpha\beta})$ be a smooth action of C^r on a system of manifolds (S_α), $\alpha\in I$. The inclusion $G^r_\alpha \hookrightarrow C^r(\alpha,\alpha)$ gives a smooth left action of G^r_α on S_α. For a $\mathcal{C}$-object A of type α we define GA to be the fiber bundle associated to P^rA with standard fiber S_α. For a $\mathcal{C}$-morphism $f\colon A\to B$ we define $Gf\colon GA\to GB$ by

$$Gf(\{u,s\}) = \{v,\varphi_{\alpha\beta}(v^{-1}\circ j^r_x f\circ u, s)\}$$

$x\in mA$, $u\in P^r_xA$, $v\in P^r_{f(x)}B$, $s\in S_\alpha$. One verifies easily that G is a well-defined regular r-th order bundle functor on $\mathcal{C}$, cf. 14.22. Clearly, if we apply the latter construction to the action $F_{\alpha\beta}$, we get a bundle functor naturally equivalent to the original functor F. □

18.15. Natural transformations. Given two bundle functors F, $G: \mathcal{C} \to \mathcal{FM}$, by a natural transformation $T: F \to G$ we shall mean a system of *base-preserving* morphisms $T_A: FA \to GA$, $A \in \mathrm{Ob}\mathcal{C}$, satisfying $Gf \circ T_A = T_B \circ Ff$ for every $\mathcal{C}$-morphism $f: A \to B$. (We remark that for a large class of admissible categories every natural transformation between any two bundle functors is formed by base-preserving morphisms, see 14.11.)

Given two smooth actions $(\varphi_{\alpha\beta}, S_\alpha)$ and $(\psi_{\alpha\beta}, Z_\alpha)$, a $\mathcal{C}^r$*-map*

$$\tau: (\varphi_{\alpha\beta}, S_\alpha) \to (\psi_{\alpha\beta}, Z_\alpha)$$

is a system of smooth maps $\tau_\alpha: S_\alpha \to Z_\alpha$, $\alpha \in I$, satisfying

$$\tau_\beta(\varphi_{\alpha\beta}(a, s)) = \psi_{\alpha\beta}(a, \tau_\alpha(s))$$

for all $s \in S_\alpha$, $a \in \mathcal{C}^r(\alpha, \beta)$.

Theorem. *Natural transformations $F \to G$ between two r-th order regular bundle functors on $\mathcal{C}$ are in a canonical bijection with the $\mathcal{C}^r$-maps between the corresponding actions of $\mathcal{C}^r$.*

Proof. Given $T: F \to G$, we define $\tau_\alpha: F_{0_\alpha} C_\alpha \to G_{0_\alpha} C_\alpha$ by $\tau_\alpha(s) = T_{C_\alpha}(s)$. One verifies directly that (τ_α) is a $\mathcal{C}^r$-map $(F_{\alpha\beta}, F_{0_\alpha} C_\alpha) \to (G_{\alpha\beta}, G_{0_\alpha} C_\alpha)$. Conversely, let $(\tau_\alpha): (\varphi_{\alpha\beta}, S_\alpha) \to (\psi_{\alpha\beta}, Z_\alpha)$ be a $\mathcal{C}^r$-map between two smooth actions of $\mathcal{C}^r$. Then the induced bundle functors transform $A \in \mathrm{Ob}\mathcal{C}$ of type α into the fiber bundle associated with $P^r A$ with standard fibers S_α and Z_α and we define $T_A = (\mathrm{id}_{P^r A}, \tau_\alpha)$. One verifies easily that T is a natural transformation between the induced bundle functors. □

18.16. Morphism operators. We are going to generalize the concept of natural operator from 14.15 in the following three directions: 1. We replace the category $\mathcal{M}f_m$ by an admissible category $\mathcal{C}$ over manifolds. 2. We consider the operators defined on morphisms of fibered manifolds. 3. We study an operator defined on some morphisms only, not on all of them. We start with the general concept of a morphism operator.

If $Y_1 \to M$ and $Y_2 \to M$ are two fibered manifolds, we denote by $C^\infty_M(Y_1, Y_2)$ the space of all base-preserving morphisms $Y_1 \to Y_2$. Given another pair $Z_1 \to M$ and $Z_2 \to M$ of fibered manifolds, a *morphism operator* D is a map $D: E \subset C^\infty_M(Y_1, Y_2) \to C^\infty_M(Z_1, Z_2)$. In the case Z_1 is a fibered manifold over Y_1, i.e. we have a surjective submersion $q: Z_1 \to Y_1$, we also say that D is a *base extending* operator.

In general, if we have four manifolds N_1, N_2, N_3, N_4, a map $\pi: N_3 \to N_1$ and a subset $E \subset C^\infty(N_1, N_2)$, an operator $A: E \to C^\infty(N_3, N_4)$ is called π-local, if the value $As(x)$ depends only on the germ of s at $\pi(x)$ for all $s \in E$, $x \in N_3$. Such an operator is said to be of *order* k, $0 \le k \le \infty$, if $j^k_{\pi(x)} s_1 = j^k_{\pi(x)} s_2$ implies $As_1(x) = As_2(x)$ for all s_1, $s_2 \in E$, $x \in N_3$. We call A regular if smoothly parameterized families in E are transformed into smoothly parameterized families in $C^\infty(N_3, N_4)$.

Assume we have a surjective submersion $q\colon Z_1 \to Y_1$. Then we have defined both local and k-th order operators $C^\infty_M(Y_1, Y_2) \to C^\infty_M(Z_1, Z_2)$ with respect to q. Such a k-th order operator D determines the *associated map*

$$\text{(1)} \qquad \mathcal{D}\colon J^k_M(Y_1, Y_2) \times_{Y_1} Z_1 \to Z_2, \quad (j^k_y s, z) \mapsto Ds(z), \quad y = q(z),$$

where $J^k_M(Y_1, Y_2)$ means the space of all k-jets of the maps of $C^\infty_M(Y_1, Y_2)$. If D is regular, then $\mathcal{D}$ is smooth. Conversely, every smooth map (1) defines a regular operator $C^\infty_M(Y_1, Y_2) \to C^\infty_M(Z_1, Z_2)$, $s \mapsto \mathcal{D}((j^k s) \circ q, \ )\colon Z_1 \to Z_2$, $s \in C^\infty_M(Y_1, Y_2)$.

18.17. Natural morphism operators. Let F_1, F_2, G_1, G_2 be bundle functors on an admissible category $\mathcal{C}$. A *natural operator* $D\colon (F_1, F_2) \rightsquigarrow (G_1, G_2)$ is a system of regular operators $D_A\colon C^\infty_{mA}(F_1A, F_2A) \to C^\infty_{mA}(G_1A, G_2A)$, $A \in \mathrm{Ob}\mathcal{C}$, such that for all $s_1 \in C^\infty_{mA}(F_1A, F_2A)$, $s_2 \in C^\infty_{mB}(F_1B, F_2B)$ and $f \in \mathcal{C}(A, B)$ the right-hand diagram commutes whenever the left-hand one does.

$$\begin{array}{ccc} F_2A & \xleftarrow{s_1} & F_1A \\ \Big\downarrow F_2f & & \Big\downarrow F_1f \\ F_2B & \xleftarrow{s_2} & F_1B \end{array} \qquad\qquad \begin{array}{ccc} G_1A & \xrightarrow{D_As_1} & G_2A \\ \Big\downarrow G_1f & & \Big\downarrow G_2f \\ G_1B & \xrightarrow{D_Bs_2} & G_2B \end{array}$$

This implies the localization property

$$D_{L(A,U)}(s|(p^{F_1})^{-1}(U)) = (D_As)|(p^{G_1})^{-1}(U)$$

for every $A \in \mathrm{Ob}\mathcal{C}$ and every open subset $U \subset mA$. If $q\colon G_1 \to F_1$ is a natural transformation formed by surjective submersions q_A and if all operators D_A are q_A-local, then we say that D is *q-local.*

In the special case $F_1 = m$ we have $C^\infty_{mA}(mA, F_2A) = C^\infty(F_2A)$, so that D_A transforms sections of F_2A into base-preserving morphisms $G_1A \to G_2A$; in this case we write $D\colon F_2 \rightsquigarrow (G_1, G_2)$. Then D is always p^{G_1}-local by definition. If we have a natural surjective submersion $q_M\colon G_2M \to G_1M$ and we require the values of operator D to be sections of q, we write $D\colon (F_1, F_2) \rightsquigarrow (G_2 \to G_1)$ and $D\colon F_2 \rightsquigarrow (G_2 \to G_1)$ in the special case $F_1 = m$. In particular, if G_2 is of the form $G_2 = H \circ G_1$, where H is a bundle functor on a suitable category, and $q = p^H$ is the bundle projection of H, we write $D\colon (F_1, F_2) \rightsquigarrow HG_1$ and $D\colon F_2 \rightsquigarrow HG_1$ for $F_1 = m$. In the case $F_1 = m = G_1$, we have an operator $D\colon F_2 \rightsquigarrow G_2$ transforming sections of F_2A into sections of G_2A for all $A \in \mathrm{Ob}\mathcal{C}$. The classical natural operators from 14.15 correspond to the case $\mathcal{C} = \mathcal{M}f_m$.

Example 1. The tangent functor T is defined on the whole category $\mathcal{M}f$. The Lie bracket of vector fields is a natural operator $[\ ,\ ]\colon T \oplus T \rightsquigarrow T$, see 3.10 for the verification. Let us remark that the naturality of the bracket with respect to local diffeomorphisms follows directly from the fact that its definition does not depend on any coordinate construction.

Example 2. Let F be a natural bundle over m-manifolds and X be a vector field on an m-manifold M. If we apply F to the flow of X, we obtain the flow of a vector field $\mathcal{F}_MX$ on FM. This defines a natural operator $\mathcal{F}\colon T \rightsquigarrow TF$.

18.18. Natural domains. This concept reflects the situation when the operators are defined on some morphisms only.

Definition. A system of subsets $E_A \subset C^\infty_{mA}(F_1A, F_2A)$, $A \in \mathrm{Ob}\mathcal{C}$, is called a natural domain, if

(i) the restriction of every $s \in E_A$ to $L(A,U)$ belongs to $E_{L(A,U)}$ for every open subset $U \subset mA$,

(ii) for every $\mathcal{C}$-isomorphism $f: A \to B$ it holds $f_*(E_A) = E_B$, where $f_*(s) = F_2f \circ s \circ (F_1f)^{-1}$, $s \in E_A$.

If we replace $C^\infty_{mA}(F_1A, F_2A)$ by a natural domain E_A in 18.17, we obtain the definition of a natural operator $E \rightsquigarrow (G_1, G_2)$.

Example 1. For every admissible category $m: \mathcal{C} \to \mathcal{M}f$ we define the *$\mathcal{C}$-fields* on the $\mathcal{C}$-objects as those vector fields on the underlying manifolds, the flows of which are formed by local $\mathcal{C}$-morphisms. For every regular bundle functor on $\mathcal{C}$ there is the flow operator $\mathcal{F}: T \rightsquigarrow TF$ defined on all $\mathcal{C}$-fields. Indeed, if we apply F to the flow of a $\mathcal{C}$-field $X \in \mathfrak{X}(mA)$, we get a flow of a vector field $\mathcal{F}X$ on FA. The naturality of $\mathcal{F}$ follows from 3.14. In particular, if $\mathcal{C}$ is the category of symplectic $2m$-dimensional manifolds, then the $\mathcal{C}$-fields are the locally Hamiltonian vector fields. For the category $\mathcal{C}$ of Riemannian manifolds and isometries, the $\mathcal{C}$-fields are the Killing vector fields. If $\mathcal{C} = \mathcal{FM}$, we obtain the projectable vector fields.

Example 2. The Frölicher-Nijenhuis bracket is a natural operator $[\ ,\ ]: T \otimes \Lambda^kT^* \oplus T \otimes \Lambda^lT^* \to T \otimes \Lambda^{k+l}T^*$ with respect to local diffeomorphisms by the definition. The functors in question do not act on the whole category $\mathcal{M}f$. However, we have proved more than this naturality in section 8. Let us consider $E^k_M = \Omega^k(M;TM) \subset C^\infty_M(\oplus^kTM, TM)$. Then we can view the bracket as an operator $[\ ,\ ]: (\oplus^kT \oplus \oplus^lT, T \oplus T) \rightsquigarrow (\oplus^{k+l}T, T)$ with the natural domain $(E_M = E^k_M \times E^l_M)_{M \in \mathrm{Ob}\mathcal{M}f}$ and its naturality follows from 8.15. We remark that even the Schouten-Nijenhuis bracket satisfies such a kind of naturality, [Michor, 87b].

18.19. To deduce a result analogous to 14.17 for natural morphism operators, we shall assume that all $\mathcal{C}$-objects are of the same type and all $\mathcal{C}$-morphisms are local isomorphisms. Hence the r-th order skeleton of $\mathcal{C}$ is one Lie group $G^r \subset G^r_m$, where m is the dimension of the only object C of a local skeleton of $\mathcal{C}$.

Consider four bundle functors F_1, F_2, G_1, G_2 on $\mathcal{C}$ and a q-local natural operator $D: (F_1, F_2) \rightsquigarrow (G_1, G_2)$. Then the rule

$$A \mapsto J^k_{mA}(F_1A, F_2A) \times_{F_1A} G_1A =: HA$$

with its canonical extension to the $\mathcal{C}$-morphisms defines a bundle functor H on $\mathcal{C}$. Using 18.16.(1), we deduce quite similarly to 14.15 the following assertion

Proposition. *k-th order natural operators $D: (F_1, F_2) \rightsquigarrow (G_1, G_2)$ are in bijection with the natural transformations $H \to G_2$.* □

By 18.15, these natural transformations are in bijective correspondence with the G^s-equivariant maps $H_0 \to (G_2)_0$ between the standard fibers, where s is the maximum of the orders of G_2 and H.

If we pose some additional natural conditions on such an operator D, they are reflected directly in our model. For example, in the case $F_1 = m$ assume we have a natural surjective submersion $p\colon G_2 \to G_1$ and require every $D_A s$ to be a section of p_A. Then the k-order operators of this type are in bijection with the G^s-maps $f\colon (J^k F_2)_0 \times (G_1)_0 \to (G_2)_0$ satisfying $p_0 \circ f = \mathrm{pr}_2$, where $p_0\colon (G_2)_0 \to (G_1)_0$ is the map induced by p.

18.20. We are going to extend 18.19 to the case of a natural domain $E \subset (F_1, F_2)$. Such a domain will be called *k-admissible*, if

(i) the space $E^k_A \subset J^k_{mA}(F_1 A, F_2 A)$ of all k-jets of the maps from E_A is a fibered submanifold of $J^k_{mA}(F_1 A, F_2 A) \to F_1 A$,

(ii) for every smooth curve $\gamma(t)\colon \mathbb{R} \to E^k_C$ there is a smoothly parametrized family $s_t \in E_C$ such that $\gamma(t) = j^k_{c(t)} s_t$, where $c(t)$ is the source of $\gamma(t)$.

The second condition has a similar technical character as the smooth splitting property in 18.9.

Then the rule

$$A \mapsto E^k_A \times_{F_1 A} G_1 A =: HA$$

with its canonical extension to the $\mathcal{C}$-morphisms defines a bundle functor H on $\mathcal{C}$. Analogously to 18.19 we deduce

Proposition. *If E is a k-admissible natural domain, then k-th order natural operators $E \rightsquigarrow (G_1, G_2)$ are in bijection with the natural transformations $H \to G_2$.*

19. Peetre-like theorems

We first present the well known Peetre theorem on the finiteness of the order of linear support non-increasing operators. After sketching a non-traditional proof of this theorem, we discuss the way to its generalization and the most of this section is occupied by the proof and corollaries of a nonlinear version of the Peetre theorem formulated in 19.7.

19.1. Let us recall that the *support* $\operatorname{supp} s$ of a section $s\colon M \to L$ of a vector bundle L over M is the closure of the set $\{x \in M; s(x) \neq 0\}$ and for every operator $D\colon C^\infty(L_1) \to C^\infty(L_2)$ support non-increasing means $\operatorname{supp} Ds \subset \operatorname{supp} s$ for all sections $s \in C^\infty(L_1)$.

Theorem, [Peetre, 60]. *Consider vector bundles $L_1 \to M$ and $L_2 \to M$ over the same base M and a linear support non-increasing operator $D\colon C^\infty(L_1) \to C^\infty(L_2)$. Then for every compact set $K \subset M$ there is a natural number r such that for all sections s_1, $s_2 \in C^\infty(L_1)$ and every point $x \in K$ the condition $j^r s_1(x) = j^r s_2(x)$ implies $Ds_1(x) = Ds_2(x)$.*

Briefly, for any compact set $K \subset M$, D is a differential operator of some finite order r on K.

We shall see later that the theorem follows easily from more general results. However the following direct (but rather sketched) proof based on lemma 19.2.

contains the basic ideas of the forthcoming generalization. By the standard compactness argument, we may restrict ourselves to $M = \mathbb{R}^m$, $L_1 = \mathbb{R}^m \times \mathbb{R}^n$, $L_2 = \mathbb{R}^m \times \mathbb{R}^p$ and to view D as a linear map $D\colon C^\infty(\mathbb{R}^m, \mathbb{R}^n) \to C^\infty(\mathbb{R}^m, \mathbb{R}^p)$.

19.2. Lemma. *Let $D\colon C^\infty(\mathbb{R}^m, \mathbb{R}^n) \to C^\infty(\mathbb{R}^m, \mathbb{R}^p)$ be a support non increasing linear operator. Then for every point $x \in \mathbb{R}^m$ and every real constant $C > 0$, there is a neighborhood V of x and an order $r \in \mathbb{N}$, such that for all $y \in V \setminus \{x\}$, $s \in C^\infty(\mathbb{R}^m, \mathbb{R}^n)$ the condition $j^r s(y) = 0$ implies $|Ds(y)| \le C$.*

Proof. Let us assume the lemma is not true for some x and C. Then we can construct sequences $s_k \in C^\infty(\mathbb{R}^m, \mathbb{R}^n)$ and $x_k \to x$, $x_k \ne x$ with $j^k s_k(x_k) = 0$ and $|Ds_k(x_k)| > C$ and we can even require $|x_k - x_j| \ge 4|x_k - x|$ for all $k > j$. Further, let us choose maps $q_k \in C^\infty(\mathbb{R}^m, \mathbb{R}^n)$ in such a way that $q_k(y) = 0$ for $|y - x_k| > \frac{1}{2}|x_k - x|$, $\operatorname{germ} s_k(x_k) = \operatorname{germ} q_k(x_k)$, and $\max_{y \in \mathbb{R}^m} |\partial^\alpha q_k(y)| \le 2^{-k}$, $0 \le |\alpha| \le k$. This is possible since $j^k s_k(x_k) = 0$ for all $k \in \mathbb{N}$ and we shall not verify this in detail. Now one can show that the map

$$q(y) := \textstyle\sum_{k=0}^\infty q_{2k}(y), \qquad y \in \mathbb{R}^m,$$

is well defined and smooth (note that the supports of the maps q_k are disjoint). It holds $\operatorname{germ} q(x_{2k}) = \operatorname{germ} s_{2k}(x_{2k})$ and $\operatorname{germ} q(x_{2k+1}) = 0$. Since the operator D is support non-increasing and linear, its values depend on germs only. Therefore

$$|Dq(x_{2k+1})| = 0 \text{ and } |Dq(x_{2k})| = |Ds_{2k}(x_{2k})| > C > 0$$

which is a contradiction with $x_k \to x$ and $Dq \in C^\infty(\mathbb{R}^m, \mathbb{R}^p)$. □

Proof of theorem 19.1. Given a compact subset K we choose $C = 1$ and apply lemma 19.2. We get an open cover of K by neighborhoods V_x, $x \in K$, so we can choose a finite cover $V_{x_1}, \dots, V_{x_k}$. Let r be the maximum of the corresponding orders. Then the condition $j^r s(x) = 0$ implies $|Ds(x)| \le 1$ for all $x \in K$, $s \in C^\infty(\mathbb{R}^m, \mathbb{R}^n)$, with a possible exception of points $x_1, \dots, x_k \in K$. But if $|Ds(x)| = \varepsilon > 0$, then $|D(\frac{2}{\varepsilon}s)(x)| = 2$. Hence for all $x \in K \setminus \{x_1, \dots, x_k\}$, $Ds(x) = 0$ whenever $j^r s(x) = 0$. The linearity expressed in local coordinates implies, that this is true for the points $x_1, \dots, x_k$ as well. □

If we look carefully at the proof of lemma 19.2, we see that the result does not essentially depend on the linearity of the operator. Dealing with a nonlinear operator, the assertion can be formulated as follows. For all sections s, q, each point x and real constant $\varepsilon > 0$, there is a neighborhood V of the point x and an order $r \in \mathbb{N}$ such that the values $Dq(y)$ and $Ds(y)$ do not differ more then by ε for all $y \in V \setminus \{x\}$ with $j^r q(y) = j^r s(y)$. At the same time, there are two essential assumptions in the proof only. First, the operator D depends on germs, and second, the domain of D is the whole $C^\infty(\mathbb{R}^m, \mathbb{R}^n)$. Moreover, let us note that we have used only the continuity of the values in the proof of 19.2. But the next example shows, that having no additional assumptions on the values of the operators, there is no reason for any finiteness of the order.

19.3. Example. We define an operator $D\colon C^\infty(\mathbb{R},\mathbb{R}) \to C^0(\mathbb{R},\mathbb{R})$. For all $f \in C^\infty(\mathbb{R},\mathbb{R})$ we put

$$Df(x) = \sum_{k=0}^{\infty} 2^{-k}\left(\operatorname{arctg}\circ \frac{d^k f}{dx^k}(x)\right), \qquad x \in \mathbb{R}.$$

The value $Df(x)$ depends essentially on $j^\infty f(x)$.

That is why in the rest of this section we shall deal with operators with smooth values, only. The technique used in 19.2 can be applied to more general types of operators. We will study the π-local operators $D\colon E \subset C^\infty(X,Y) \to C^\infty(Z,W)$ with a continuous map $\pi\colon Z \to X$, see 18.19 for the definition.

In the nonlinear case we need a general tool for extending a sequence of germs of sections to one globally defined section. In our considerations, this role will be played by the Whitney extension theorem:

19.4. Theorem. *Let $K \subset \mathbb{R}^m$ be a compact set and let f_α be continuous functions defined on K for all multi-indices α, $0 \le |\alpha| < \infty$. There exists a function $f \in C^\infty(\mathbb{R}^m)$ satisfying $\partial^\alpha f|K = f_\alpha$ for all α if and only if for every natural number m*

$$f_\alpha(b) = \textstyle\sum_{|\beta|\le m} \frac{1}{\beta!} f_{\alpha+\beta}(a)(b-a)^\beta + o(|b-a|^m) \tag{1}$$

holds uniformly for $|b-a| \to 0$, b, $a \in K$.

Let us recall that $f(x) = o(|x|^m)$ means $\lim_{x\to 0} f(x)x^{-m} = 0$.

The proof is rather complicated and technical and can be found in [Whitney, 34], [Malgrange, 66] or [Tougeron, 72]. If K is a one-point set, we obtain the classical Borel theorem. We shall work with a special case of this theorem where the compact set K consists of a convergent sequence of points in $\mathbb{R}^m$. Therefore we shall use the following assumptions on the domains of the operators.

19.5. Definition. A subset $E \subset C^\infty(X,Y)$ is said to be *Whitney-extendible*, or briefly *W-extendible*, if for every map $f \in C^\infty(X,Y)$, every convergent sequence $x_k \to x$ in X and each sequence $f_k \in E$ and $f_0 \in E$, satisfying $\operatorname{germ} f(x_k) = \operatorname{germ} f_k(x_k)$, $k \in \mathbb{N}$, $j^\infty f_0(x) = j^\infty f(x)$, there exists a map $g \in E$ and a natural number k_0 satisfying $\operatorname{germ} g(x_k) = \operatorname{germ} f_k(x_k)$ for all $k \ge k_0$.

19.6. Examples.

1. By definition $E = C^\infty(X,Y)$ is Whitney-extendible.

2. Let $E \subset C^\infty(\mathbb{R}^m,\mathbb{R}^m)$ be the subset of all local diffeomorphisms. Then E is W-extendible. Indeed, we need to join given germs on some neighborhood of x only, but the original map f itself has to be a local diffeomorphism around x, for $j^\infty f(x) = j^\infty f_0(x)$ and every germ of a locally defined diffeomorphism on $\mathbb{R}^m$ is a germ of a globally defined local diffeomorphism. So every bundle functor F on $\mathcal{M}f_m$ defines a map $F\colon E \to C^\infty(F\mathbb{R}^m, F\mathbb{R}^m)$ which is a $p_{\mathbb{R}^m}$-local operator with W-extendible domain.

3. Consider a fibered manifold $p\colon Y \to M$. The set of all sections $E = C^\infty(Y)$ is W-extendible. Indeed, since we require the extension of given germs on an

arbitrary neighborhood of the limit point x only, we may restrict ourselves to a local chart $\mathbb{R}^m \times \mathbb{R}^n \hookrightarrow Y$. Now, we can work with the coordinate expressions of the given germs of sections, i.e. with germs of functions. The existence of the 'extension' f of given germs implies that the germs of coordinate functions satisfy condition 19.4.(1), and so there are functions joining these germs. But these functions represent a coordinate expression of the required section. Therefore the operators dealt with in 19.1 are id_M-local linear operators with W-extendible domains.

19.7. Nonlinear Peetre theorem. Now we can formulate the main result of this section. The last technical assumption is that for our π-local operators, the map π should be *locally non-constant*, i.e. there are at least two different points in the image $\pi(U)$ of any open set U.

Theorem. *Let $\pi\colon Z \to \mathbb{R}^m$ be a locally non-constant continuous map and let $D\colon E \subset C^\infty(\mathbb{R}^m, \mathbb{R}^n) \to C^\infty(Z, W)$ be a π-local operator with a Whitney-extendible domain. Then for every fixed map $f \in E$ and for every compact subset $K \subset Z$ there exist a natural number r and a smooth function $\varepsilon\colon \pi(K) \to \mathbb{R}$ which is strictly positive, with a possible exception of a finite set of points in $\pi(K)$, such that the following statement holds.*

For every point $z \in K$ and all maps $g_1, g_2 \in E$ satisfying $|\partial^\alpha(g_i - f)(\pi(z))| \le \varepsilon(\pi(z))$, $i = 1, 2$, $0 \le |\alpha| \le r$, the condition

$$j^r g_1(\pi(z)) = j^r g_2(\pi(z))$$

implies

$$Dg_1(z) = Dg_2(z).$$

Before going into details of the proof, we present some remarks and corollaries.

19.8. Corollary. *Let X, Y, Z, W be manifolds, $\pi\colon Z \to X$ a locally non-constant continuous map and let $D\colon E \subset C^\infty(X, Y) \to C^\infty(Z, W)$ be a π-local operator with Whitney-extendible domain. Then for every fixed map $f \in E$ and for every compact set $K \subset Z$, there exists $r \in \mathbb{N}$ such that for every $x \in \pi(K)$, $g \in E$ the condition $j^r f(x) = j^r g(x)$ implies*

$$Df|(\pi^{-1}(x) \cap K) = Dg|(\pi^{-1}(x) \cap K).$$

19.9. Multilinear version of Peetre theorem. Let us note that the classical Peetre theorem 19.1 follows easily from 19.8. Indeed, id_M-locality is equivalent to the condition on supports in 19.1, the sections of a fibration form a W-extendible domain (see 19.6), so we can apply 19.8 to the zero section of the vector bundle $L_1 \to M$. Hence for every compact set $K \subset M$ there is an order $r \in \mathbb{N}$ such that $Ds(x) = 0$ whenever $j^r s(x) = 0$, $x \in K$, $s \in C^\infty(L)$, and the classical Peetre theorem follows.

But applying the full formulation of theorem 19.7, we can prove in a similar way a 'multilinear base-extending' Peetre theorem.

Theorem. *Let $L_1, \dots, L_k$ be vector bundles over the same base M, $L \to N$ be another vector bundle and let $\pi\colon N \to M$ be continuous and locally non-constant. If $D\colon C^\infty(L_1) \times \dots \times C^\infty(L_k) \to C^\infty(L)$ is a k-linear π-local operator, then for every compact set $K \subset N$ there is a natural number r such that for every $x \in \pi(K)$ and all sections s, $q \in C^\infty(L_1 \oplus \dots \oplus L_k)$ the condition $j^r s(x) = j^r q(x)$ implies*

$$Ds|(\pi^{-1}(x) \cap K) = Dq|(\pi^{-1}(x) \cap K).$$

Proof. We may assume $L_i = \mathbb{R}^m \times \mathbb{R}^{n_i}$, $i = 1, \dots, k$. Then all assumptions of 19.7 are satisfied and so, chosen a compact set $K \subset N$ and the zero section of $L_1 \oplus \dots \oplus L_k$, we get some order r and a function $\varepsilon\colon \pi(K) \to \mathbb{R}$. Consider arbitrary sections q, $s \in C^\infty(L_1 \oplus \dots \oplus L_k)$ and a point $x \in \pi(K)$, $\varepsilon(x) > 0$. Using multiplication of sections by positive real constants, we can arrange that all their derivatives up to order r at the point x are less then $\varepsilon(x)$. Hence if $j^r q(x) = j^r s(x)$, then for a suitable $c > 0$, $c \in \mathbb{R}$, it holds

$$c^k \cdot Ds(z) = D(c \cdot s)(z) = D(c \cdot q)(z) = c^k \cdot Dq(z)$$

for all $z \in K \cap \pi^{-1}(x)$. According to 19.7, the function ε can be chosen in such a way that the set $\{x \in \pi(K); \varepsilon(x) = 0\}$ is discrete. So the theorem follows from the multilinearity of the operator and the continuity of its values, what is easily checked looking at the coordinate description of the multilinear operators. □

19.10. One could certainly replace the Whitney extendibility by some other property, but this cannot be completely omitted. To see this, consider the operator constructed in 19.3 and let us restrict its domain to the subset $E \subset C^\infty(\mathbb{R}, \mathbb{R})$ of all polynomials. We get an operator $D\colon E \to C^\infty(\mathbb{R}, \mathbb{R})$ essentially depending on infinite jets. Also the requirement on π is essential because dropping it, any action of the group of germs of maps $f\colon (\mathbb{R}^m, 0) \to (\mathbb{R}^m, 0)$ on a manifold should factorize to an action of some jet group G^r_m.

Let us notice that the assertion of our theorem is near to local finiteness of the order with respect to the topology on Z and to the compact open C^∞-topology on $C^\infty(\mathbb{R}^m, \mathbb{R}^n)$, see e.g. [Hirsch, 76] for definition. It would be sufficient if we might always choose a strictly positive function $\varepsilon\colon \pi(K) \to \mathbb{R}$ in the conclusion of the theorem. However, example 19.15 shows that this need not be possible in general. On the other hand, if we add a suitable regularity condition, then the mentioned local finiteness can be proved. Regularity will mean that smoothly parameterized families of maps in the domain are transformed into smoothly parameterized families. The idea of the proof is to define a new operator $\tilde{D}$ with domain $\tilde{E}$ formed by all one-parameter families of maps, then to perform a similar construction as in the proof of 19.7 and to apply theorem 19.7 to $\tilde{D}$ to get a contradiction, see [Slovák, 88]. Therefore, beside the regularity, we need that $\tilde{E}$ is also W-extendible. This is not obvious in general, but it is evident if E consists of all sections of a fibration. Since we shall mostly deal with regular operators defined on all sections of a fibration, we present the full formulation.

Theorem. *Let Z, W be manifolds, $Y \to X$ a fibration, $\pi\colon Z \to X$ a locally non-constant map and let $D\colon E = C^\infty(Y \to X) \to C^\infty(Z,W)$ be a regular π-local operator. Then for every fixed map $f \in E$ and for every compact set $K \subset Z$, there exist an order $r \in \mathbb{N}$ and a neighborhood V of f in the compact open C^∞-topology such that for every $x \in \pi(K)$ and all g_1, $g_2 \in V \cap E$ the condition*

$$j^r g_1(x) = j^r g_2(x)$$

implies

$$Dg_1|(\pi^{-1}(x) \cap K) = Dg_2|(\pi^{-1}(x) \cap K).$$

Similar, but essentially weaker, results can also be deduced dealing with operators with continuous values, see [Chrastina, 87], [Slovák, 87 b].

Let us pass to the proof of 19.7. In the sequel, we fix manifolds Z, W, a locally non-constant continuous map $\pi\colon Z \to \mathbb{R}^m$, a Whitney-extendible subset $E \subset C^\infty(\mathbb{R}^m, \mathbb{R}^n)$ and a π-local operator $D\colon E \to C^\infty(Z,W)$. The proof is based on two lemmas.

19.11. Lemma. *Let $z_0 \in Z$ be a point, $x_0 := \pi(z_0)$, $f \in E$, and let us define a function $\varepsilon\colon \mathbb{R}^m \to \mathbb{R}$ by $\varepsilon(x) = \exp(-|x - x_0|^{-1})$ if $x \neq x_0$ and $\varepsilon(x_0) = 0$. Then there is a neighborhood V of the point $z_0 \in Z$ and a natural number r such that for every $z \in V - \pi^{-1}(x_0)$ and all maps g_1, $g_2 \in E$ satisfying $|\partial^\alpha(g_i - f)(\pi(z))| \leq \varepsilon(\pi(z))$, $i = 1,2$, $0 \leq |\alpha| \leq r$, the condition $j^r g_1(\pi(z)) = j^r g_2(\pi(z))$ implies $Dg_1(z) = Dg_2(z)$.*

Proof. We assume the lemma does not hold and we shall find a contradiction. If the assertion is not true, then we can construct sequences $z_k \to z_0$ in Z, $x_k := \pi(z_k) \to x_0$ and maps f_k, $g_k \in E$ satisfying for all $k \in \mathbb{N}$

$$|\partial^\alpha(f_k - f)(x_k)| \leq \varepsilon(x_k) \qquad \text{for all } 0 \leq |\alpha| \leq k \tag{1}$$

$$j^k f_k(x_k) = j^k g_k(x_k) \tag{2}$$

$$Df_k(z_k) \neq Dg_k(z_k). \tag{3}$$

Since all x_k are different from x_0, by passing to subsequences we can assume

$$|x_{k+1} - x_0| < \frac{1}{4}|x_k - x_0|. \tag{4}$$

Let us fix Riemannian metrics ρ_Z or ρ_W on Z or W, respectively, and choose further points $\bar{z}_k \in Z$, $\bar{z}_k \to z_0$, $\bar{x}_k := \pi(\bar{z}_k)$ and neighborhoods U_k or V_k of x_k or $\bar{x}_k$, respectively, in such a way that for all $k \in \mathbb{N}$ the following six conditions hold

$$|x_k - x_0| \leq 2|a - b| \quad \text{for all } a \in U_k \cup V_k,\ b \in U_j \cup V_j,\ k \neq j \tag{5}$$

$$|\partial^\alpha(f_k - f)(a)| \leq 2\varepsilon(x_k) \quad \text{for all } a \in U_k \cup V_k,\ 0 \leq |\alpha| \leq k \tag{6}$$

$$|\partial^\alpha(g_k - f)(a)| \leq 2\varepsilon(x_k) \quad \text{for all } a \in U_k \cup V_k,\ 0 \leq |\alpha| \leq k \tag{7}$$

$$\rho_W(Dg_k(z_k), Df_k(\bar{z}_k)) \geq k\rho_Z(z_k, \bar{z}_k) \tag{8}$$

and for all m, $k \in \mathbb{N}$, and multi-indices α with $|\alpha| + 2m \leq k$, $a \in U_k$, and $b \in V_k$ we require

$$\frac{1}{|b-a|^m}\left|\Big(\sum_{|\beta|\leq m}\frac{1}{\beta!}\partial^{\alpha+\beta}f_k(b)(a-b)^\beta\Big) - \partial^\alpha g_k(a)\right| \leq \frac{1}{k} \tag{9}$$

$$\frac{1}{|b-a|^m}\left|\Big(\sum_{|\beta|\leq m}\frac{1}{\beta!}\partial^{\alpha+\beta}g_k(a)(b-a)^\beta\Big) - \partial^\alpha f_k(b)\right| \leq \frac{1}{k}. \tag{10}$$

All these requirements can be satisfied. Indeed, the equalities (5), (6), (7) are valid for all points a, b from some suitable neighborhoods W_k of the points x_k. By the Taylor formula, for any fixed k and $|\alpha| + m \leq k$, (2) implies $\partial^\alpha g_k(a) = \partial^\alpha f_k(a) + o(|a - x_k|^m)$. Therefore, if we consider only points a, $b \in W_k$ such that

$$|b - x_k| \leq 2|b-a|, \qquad |a - x_k| \leq 2|b-a|, \tag{11}$$

then under the condition $|\alpha| + 2m \leq k$ we get (note that $o(|a - x_k|^m)$ or $o(|b - x_k|^m)$ now implies $o(|a-b|^m)$)

$$\begin{aligned}\sum_{|\beta|\leq m}\frac{1}{\beta!}\partial^{\alpha+\beta}f_k(b)(a-b)^\beta &= \sum_{|\beta|\leq m}\frac{1}{\beta!}\partial^{\alpha+\beta}g_k(b)(a-b)^\beta + o(|a-b|^m)\\ &= \partial^\alpha g_k(a) + o(|b-a|^m)\\ \sum_{|\beta|\leq m}\frac{1}{\beta!}\partial^{\alpha+\beta}g_k(a)(b-a)^\beta &= \sum_{|\beta|\leq m}\frac{1}{\beta!}\partial^{\alpha+\beta}f_k(a)(b-a)^\beta + o(|a-b|^m)\\ &= \partial^\alpha f_k(b) + o(|b-a|^m).\end{aligned}$$

Hence also conditions (9), (10) are realizable if we take U_k, V_k in sufficiently small neighborhoods W_k of x_k in such a way that (11) holds for all $a \in U_k$, $b \in V_k$. By virtue of (3), there are also neighborhoods of the points z_k in Z ensuring (8). Finally, we are able to choose appropriate points $\bar{z}_k$ and neighborhoods U_k, V_k using the fact that π is continuous and locally non-constant.

The aim of conditions (1), (4)–(7), (9), (10) is to guarantee the existence of a map $h \in C^\infty(\mathbb{R}^m, \mathbb{R}^n)$ satisfying

$$\operatorname{germ} h(x_k) = \operatorname{germ} g_k(x_k) \quad \text{and} \quad \operatorname{germ} h(\bar{x}_k) = \operatorname{germ} f_k(\bar{x}_k). \tag{12}$$

Then, by virtue of our requirements on E, we may assume $h \in E$, provided we use (12) for large indices k, only. But applying D to h, the π-locality and (8) imply

$$\rho_W(Dh(z_k), Dh(\bar{z}_k)) \geq k\rho_Z(z_k, \bar{z}_k)$$

for large k's, and this is a contradiction with $Dh \in C^\infty(Z, W)$ and $(z_k, \bar{z}_k) \to (z_0, z_0)$.

So it remains to verify condition 19.4.(1) in the Whitney extension theorem with $K = \bigcup_k(\bar{U}_k \cup \bar{V}_k) \cup \{x_0\}$ and $f_\alpha(x) = \partial^\alpha g_k(x)$ if $x \in \bar{U}_k$, $f_\alpha(x) = \partial^\alpha f_k(x)$ if $x \in \bar{V}_k$ and $f_\alpha(x_0) = \partial^\alpha f(x_0)$. This follows by our construction for all couples $(a,b) \in \bigcup_k(\bar{U}_k \times \bar{V}_k)$, see (9), (10). In all other cases and for all $m \in \mathbb{N}$ we have to use (6) and (7), (5), the Taylor formula, (6) and (7), and (5) to get

$$\begin{aligned}\sum_{|\beta|\le m} \tfrac{1}{\beta!} f_{\alpha+\beta}(a)(b-a)^\beta &= \sum_{|\beta|\le m} \tfrac{1}{\beta!}\left(\partial^{\alpha+\beta} f(a) + o(|x_{k(a)} - x_0|^m)\right)(b-a)^\beta \\ &= \sum_{|\beta|\le m} \tfrac{1}{\beta!}\partial^{\alpha+\beta} f(a)(b-a)^\beta + o(|b-a|^m) \\ &= \partial^\alpha f(b) + o(|b-a|^m) \\ &= f_\alpha(b) + o(|x_{k(b)} - x_0|^m) + o(|b-a|^m) \\ &= f_\alpha(b) + o(|b-a|^m). \quad \square\end{aligned}$$

19.12. Lemma. *Let $z_0 \in Z$ be a point, $x = \pi(z_0)$ and $f \in E$. Then there is a neighborhood V of z_0 in $\pi^{-1}(x)$ and a natural number r such that for every $z \in V$ and all maps $g \in E$ the condition $j^r g(x) = j^r f(x)$ implies $Dg(z) = Df(z)$.*

Proof. The proof is quite similar to that of 19.11, but we first have to prove the dependence on infinite jets. Consider $g_1, g_2 \in E$ with $j^\infty g_1(x) = j^\infty g_2(x)$ and a point $y \in \pi^{-1}(x)$. Let us choose a sequence $y_k \to y$ in Z, $\pi(y_k) =: x_k \neq x$ and neighborhoods U_k of x_k satisfying $|a-x| \ge 2|a-b|$ for all $a \in U_k$, $b \in U_j$, $k \neq j$. Using the Whitney extension theorem 19.4, the Taylor formula, and our assumptions on E we find a map $h \in E$ satisfying for all large k's

$$\operatorname{germ} h(x_{2k}) = \operatorname{germ} g_1(x_{2k}) \text{ and } \operatorname{germ} h(x_{2k+1}) = \operatorname{germ} g_2(x_{2k+1}).$$

This implies $Dh(y_{2k}) = Dg_1(y_{2k})$, $Dh(y_{2k+1}) = Dg_2(y_{2k+1})$ and consequently $Dg_1(y) = Dg_2(y)$.

Now, we assume the assertion of the lemma is not true. So we can construct a sequence $z_k \to z_0$, $\pi(z_k) = x$ and maps $g_k \in E$ satisfying for all $k \in \mathbb{N}$

$$j^k f(x) = j^k g_k(x) \tag{1}$$

$$Dg_k(z_k) \neq Df(z_k). \tag{2}$$

We choose further points $\bar{z}_k \to z_0$ in Z, $\bar{x}_k := \pi(\bar{z}_k)$, $\bar{x}_k \neq x$, and neighborhoods V_k of $\bar{x}_k$ in such a way that

$$\rho_W(Dg_k(\bar{z}_k), Df(z_k)) \ge k\rho_Z(\bar{z}_k, z_k) \qquad \text{for all } k \in \mathbb{N} \tag{3}$$

$$|a-x| \ge 2|a-b| \qquad \text{for all } a \in V_k,\ b \in V_j,\ k \neq j \tag{4}$$

$$\frac{|\partial^\alpha(g_k - f)(a)|}{|a-x|^m} \le \frac{1}{k} \qquad \text{for all } a \in V_k,\ |\alpha| + m \le k. \tag{5}$$

This is possible by virtue of (1), (2) and the Taylor formula analogously to 19.11. Finally, using (4), (5), the Whitney extension theorem and our assumptions, we get a map $h \in E$ satisfying

$$\operatorname{germ} h(\bar{x}_k) = \operatorname{germ} g_k(\bar{x}_k) \text{ and } j^\infty h(x) = j^\infty f(x)$$

for large k's. Hence (3) and the first part of this proof imply

$$\rho_W(Dh(\bar{z}_k), Dh(z_k)) = \rho_W(Dg_k(\bar{z}_k), Df(z_k)) \geq k\rho_Z(\bar{z}_k, z_k)$$

which is a contradiction with $Dh \in C^\infty(Z, W)$. □

Proof of theorem 19.7. According to lemmas 19.11 and 19.12, for every point $z \in K$ we find a neighborhood V_z of z, an order r_z and a smooth function $\varepsilon_z \colon \pi(V_z) \to \mathbb{R}$ which is strictly positive with a possible exception of the point $\pi(z)$, such that the conclusion of 19.7 is true for these data. The proof is then completed by the standard compactness argument. □

19.13. Let us note that our definition of Whitney-extendibility was not fully exploited in the proof of lemma 19.12. Namely, we dealt with 'fast converging' sequences only. However, we might be unable to verify the W-extendibility for certain domains $E \subset C^\infty(X, Y)$ while the proof of lemma 19.12 might still go through. So we find it profitable to present explicit formulations. For technical reasons, we consider the case $X = \mathbb{R}^m$.

Definition. A subset $E \subset C^\infty(\mathbb{R}^m, Y)$ is said to be *almost Whitney-extendible* if for every map $f \in C^\infty(\mathbb{R}^m, Y)$, sequence $f_k \in E$, $f_0 \in E$ and every convergent sequence $x_k \to x$ satisfying for all $k \in \mathbb{N}$, $|x_k - x| \geq 2|x_{k+1} - x|$, $\operatorname{germ} f(x_k) = \operatorname{germ} f_k(x_k)$, $j^\infty f(x) = j^\infty f_0(x)$, there is a map $g \in E$ and a natural number k_0 satisfying $\operatorname{germ} g(x_k) = \operatorname{germ} f_k(x_k)$ for all $k \geq k_0$.

19.14. Proposition. *Let $\pi \colon Z \to \mathbb{R}^m$ be a locally non-constant continuous map, $E \subset C^\infty(\mathbb{R}^m, Y)$ be an almost Whitney-extendible subset and let $D \colon E \to C^\infty(Z, W)$ be a π-local operator. Then for every fixed map $f \in E$, point $x \in \mathbb{R}^m$, and for every compact subset $K \subset \pi^{-1}(x)$, there exists a natural number r such that for all maps $g \in E$ the condition $j^r g(x) = j^r f(x)$ implies $Dg|K = Df|K$.*

Proof. The proposition is implied by lemma 19.12 and by the standard compactness argument. □

At the end of this section, we present an example showing that the results in 19.7 are the best possible ones in our general setting.

19.15. Example. We shall construct a simple $\mathrm{id}_\mathbb{R}$-local operator

$$D \colon C^\infty(\mathbb{R}, \mathbb{R}) \to C^\infty(\mathbb{R}, \mathbb{R})$$

such that if we take $f = \mathrm{id}_\mathbb{R}$, then for any order r and any compact neighborhood K of $0 \in \mathbb{R}$, every function $\varepsilon \colon \mathbb{R} \to \mathbb{R}$ from 19.7 satisfies $\varepsilon(0) = 0$.

Let $g \colon \mathbb{R}^2 \to \mathbb{R}$ be a function with the following three properties

(1) g is smooth in all points $x \in \mathbb{R}^2 \setminus \{(0,1)\}$
(2) $\limsup_{x \to 1} g(0, x) = \infty$
(3) g is identically zero on the closed unit discs centered in $(-1, 1)$ and $(1, 1)$.

Further, let $a \colon \mathbb{R}^2 \to \mathbb{R}$ be a smooth function satisfying $a(t, x) \neq 0$ if and only if $|x| > t > 0$.

Given $f \in C^\infty(\mathbb{R},\mathbb{R})$, $x \in \mathbb{R}$, we define

$$Df(x) = \sum_{k=0}^{\infty} \left(\left((a(k,-) \circ g \circ (f \times \frac{df}{dx})(x) \right) \frac{d^k f}{dx^k}(x) \right).$$

The sum is locally finite if $g \circ (f \times \frac{df}{dx})$ is locally bounded. Hence Df is well defined and smooth if $g \circ (f \times \frac{df}{dx})$ is smooth. The only difficulty may happen if we deal with some $f \in C^\infty(\mathbb{R},\mathbb{R})$ and $x \in \mathbb{R}$ with $f(x) = 0$, $\frac{df}{dx}(x) = 1$. However, in this case it holds

$$\lim_{y \to x} \frac{\frac{df}{dx}(y) - 1}{f(y)} = \frac{d^2 f}{dx^2}(x)$$

and the property (3) of g implies $g \circ (f \times \frac{df}{dx}) = 0$ on some neighborhood of x.

On the other hand, for $f = \mathrm{id}_{\mathbb{R}}$, arbitrary $\varepsilon > 0$ and order $r \in \mathbb{N}$, there are functions $h_1, h_2 \in C^\infty(\mathbb{R},\mathbb{R})$ such that $j^r h_1(0) = j^r h_2(0)$, $|\frac{d^k}{dx^k}(h_1 - \mathrm{id}_{\mathbb{R}})(0)| < \varepsilon$ for all $0 \le k \le r$, and $Dh_1(0) \ne Dh_2(0)$. This is caused by property (2) of g.

20. The regularity of bundle functors

20.1. Definition. A category $\mathcal{C}$ over manifolds is called *locally flat* if $\mathcal{C}$ admits a local pointed skeleton $(C_\alpha, 0_\alpha)$ where each $\mathcal{C}$-object C_α is over some $\mathbb{R}^{m(\alpha)}$ and if all translations t_x on $\mathbb{R}^{m(\alpha)}$ are $\mathcal{C}$-morphisms.

Each local pointed skeleton of a locally flat category will be assumed to have this property.

Every bundle functor $F : \mathcal{C} \to \mathcal{M}f$ on a locally flat category $\mathcal{C}$ determines the induced action τ of the abelian subgroup $\mathbb{R}^{m(\alpha)} \subset \mathcal{C}(C_\alpha, C_\alpha)$ on the manifold FC_α, $\tau_x = F(t_x)$. In section 14 we used this action and the regularity of the natural bundles to find canonical diffeomorphisms $F\mathbb{R}^m \cong \mathbb{R}^m \times p_{\mathbb{R}^m}^{-1}(0)$. The same consideration applies also in our general case, but we have first to prove the smoothness of τ. The most difficult and rather technical job is to prove that τ is continuous. Therefore we first formulate this result, then we deduce some of its consequences including the regularity of bundle functors and only at the very end of this section we present the proof consisting of several analytical lemmas.

20.2. Proposition. *Let $\mathcal{C}$ be an admissible locally flat category over manifolds with almost Whitney-extendible sets of morphisms and with the faithful functor $m\colon \mathcal{C} \to \mathcal{M}f$. Let $(C_\alpha, 0_\alpha)$ be its local pointed skeleton. Let $F\colon \mathcal{C} \to \mathcal{M}f$ be a functor endowed with a natural transformation $p\colon F \to m$ such that the locality condition 18.3.(i) holds. Then the induced actions of the abelian groups $\mathbb{R}^{m(\alpha)}$ on FC_α are continuous.*

The proof will be given in 20.9–20.12.

20.3. Theorem. *Let $\mathcal{C}$ be an admissible locally flat category over manifolds with almost Whitney-extendible sets of morphisms, $(C_\alpha, 0_\alpha)$ its local pointed* *skeleton, and $m: \mathcal{C} \to \mathcal{M}f$ the faithful functor. Let $F: \mathcal{C} \to \mathcal{M}f$ be a functor endowed with a natural transformation $p: F \to m$ such that the locality condition 18.3.(i) holds. Then there are canonical diffeomorphisms*

$$mC_\alpha \times p_{C_\alpha}^{-1}(0_\alpha) \cong FC_\alpha, \quad (x,z) \mapsto Ft_x(z) \tag{1}$$

and for every $A \in \text{Ob}\mathcal{C}$ of type α the map $p_A: FA \to A$ is a locally trivial fiber bundle with standard fiber $p_{C_\alpha}^{-1}(0_\alpha)$. In particular F is a bundle functor on $\mathcal{C}$.

Proof. Let us fix a type α and write $\mathbb{R}^m$ for mC_α. By proposition 20.2, the action $\tau: \mathbb{R}^m \times FC_\alpha \to FC_\alpha$ is a continuous action and each map $\tau_x: FC_\alpha \to FC_\alpha$ is a diffeomorphism. But then a general theorem, see 5.10, implies that this action is smooth. It follows that for every $z \in p_{C_\alpha}^{-1}(0_\alpha)$ the map $s: \mathbb{R}^m \to FC_\alpha$, $s(x) = \tau_x(z)$ is smooth and $p_{C_\alpha} \circ s = \text{id}_{\mathbb{R}^m}$. Therefore p_{C_α} is a submersion and $p_{C_\alpha}^{-1}(0_\alpha)$ is a manifold. Since both the maps $(x,z) \mapsto \tau(x,z)$ and $y \mapsto \tau(-p_{C_\alpha}(y), y)$ are smooth, (1) is a diffeomorphism. The rest of the theorem follows now from the locality of functor F. □

20.4. Consider a bundle functor F on an admissible category $\mathcal{C}$. Since for every $\mathcal{C}$-object A the action of $\mathcal{C}(A,A)$ on FA determined by F can be viewed as a p_A-local operator, a simple application of our results from section 19 will enable us to get near to the finiteness of the order of bundle functors.

Consider a point $x \in A$ and a compact set $K \subset p_A^{-1}(x) \subset FA$. We define

$$Q_K := \cup_{f \in \text{inv}\mathcal{C}(A,A)} Ff(K).$$

Lemma. *If $\mathcal{C}(A,A) \subset C^\infty(mA, mA)$ is almost Whitney-extendible, then for every compact K as above there is an order $r \in \mathbb{N}$ such that for all invertible $\mathcal{C}$-morphisms f, g and for every point $y \in A$ the equality $j_y^r f = j_y^r g$ implies*

$$Ff|(Q_K \cap p_A^{-1}(y)) = Fg|(Q_K \cap p_A^{-1}(y)).$$

Proof. Let us fix the map $\text{id}_A \in \mathcal{C}(A,A)$ and let us apply proposition 19.14 to $F: \mathcal{C}(A,A) \to C^\infty(FA, FA)$, $\pi = p_A$ and K. We denote by r the resulting order. For every $z \in Q_K$ there are $y \in K$ and $g \in \text{inv}\mathcal{C}(A,A)$ with $Fg(y) = z$. Consider $f_1, f_2 \in \text{inv}\mathcal{C}(A,A)$ such that $j^r f_1(\pi(z)) = j^r f_2(\pi(z))$. Then $j^r(f_1 \circ g)(\pi(y)) = j^r(f_2 \circ g)(\pi(y))$ and therefore $j^r(g^{-1} \circ f_1^{-1} \circ f_2 \circ g)(\pi(y)) = j^r \text{id}_A(\pi(y))$. Hence $Ff_1(z) = Ff_1 \circ Fg(y) = Ff_2 \circ Fg(y) = Ff_2(z)$. □

20.5. Theorem. *Let $\mathcal{C}$ be an admissible locally flat category over manifolds with almost Whitney-extendible sets of morphisms. If all $\mathcal{C}$-morphisms are locally invertible, then every bundle functor F on $\mathcal{C}$ is regular.*

Proof. Since all morphisms are locally invertible and the functors are local, we may restrict ourselves to objects of one fixed type, say α. We shall write $(C, 0)$ for $(C_\alpha, 0_\alpha)$, $mC = \mathbb{R}^m$, $p = p_C$. Let us consider a smoothly parameterized family $g_s \in \mathcal{C}(C,C)$ with parameters in a manifold P. For any $z \in FC$, $x = p(z)$, $f \in \mathcal{C}(C,C)$ we have

$$Ff(z) = \tau_{f(x)} \circ F(t_{-f(x)} \circ f \circ t_x) \circ \tau_{-x}(z) \tag{1}$$

and the mapping in the brackets transforms 0 into 0. Since τ is a smooth action by theorem 20.3, the regularity will follow from (1) if we show that for families with $g_s(0) = 0$ the restrictions of Fg_s to the standard fiber $S = p^{-1}(0)$ are smoothly parameterized. Since the case $m = 0$ is trivial, we may assume $m > 0$. By lemma 20.4 F is of order ∞. We first show that the induced action of the group of infinite jets $G_\alpha^\infty = \text{inv}J_0^\infty(C,C)_0$ on S is continuous with respect to the inverse limit topology.

Consider converging sequences $z_n \to z$ in S and $j_0^\infty f_n \to j_0^\infty f_0$ in G_α^∞. We shall show that any subsequence of $Ff_n(z_n)$ contains a further subsequence converging to the point $Ff_0(z)$. On replacing f_n by $f_n \circ f_0^{-1}$, we may assume $f_0 = \text{id}_C$. By passing to subsequences, we may assume that all absolute values of the derivatives of $(f_n - \text{id}_C)$ at 0 up to order $2n$ are less then e^{-n}. Let us choose positive reals $\varepsilon_n < e^{-n}$ in such a way that on the open balls $B(0,\varepsilon_n)$ centered at 0 with diameters ε_n all the derivatives in question vary at most by e^{-n}. Let $x_n := (2^{-n}, 0, \dots, 0) \in \mathbb{R}^m$. By the Whitney extension theorem there is a local diffeomorphism $f : \mathbb{R}^m \to \mathbb{R}^m$ such that

$$f|B(x_{2n+1}, \varepsilon_{2n+1}) = \text{id}_C \quad \text{and} \quad f|B(x_{2n}, \varepsilon_{2n}) = t_{x_{2n}} \circ f_{2n} \circ t_{-x_{2n}}$$

for large n's. Since the sets of $\mathcal{C}$-morphisms are almost Whitney extendible, there is a $\mathcal{C}$-morphism h satisfying the same equalities for large n's. Now

$$\begin{aligned} &\tau_{-x_n} \circ Fh \circ \tau_{x_n}(z_n) = Ff_n(z_n) && \text{if } n \text{ is even} \\ &\tau_{-x_n} \circ Fh \circ \tau_{x_n}(z_n) = z_n && \text{if } n \text{ is odd.} \end{aligned}$$

Hence, by virtue of proposition 20.2, $Ff_{2n}(z_{2n})$ converges to z and we have proved the continuity of the action of G_α^∞ on S as required.

Now, let us choose a relatively compact open neighborhood V of z and define $Q_V := (\bigcup_{f \in \text{inv}\mathcal{C}(C,C)} Ff(V)) \cap S$. This is an open submanifold in S and the functor F defines an action of the group G_α^∞ on Q_V. According to lemma 20.4 this action factorizes to an action of a jet group G_α^r on Q_V which is continuous by the above part of the proof. Hence this action has to be smooth for the reason discussed in the proof of theorem 20.3 and since smoothness is a local property and all $\mathcal{C}$-morphisms are locally invertible this concludes the proof. □

20.6. Corollary. *Every bundle functor on $\mathcal{FM}_{m,n}$ is regular.*

We can also deduce the regularity for bundle functors on $\mathcal{FM}_m$ using theorems 20.3 and 20.5.

20.7. Corollary. *Every bundle functor on $\mathcal{FM}_m$ is regular.*

Proof. The system $(\mathbb{R}^{m+n} \to \mathbb{R}^m, 0)$, $n \in \mathbb{N}_0$, is a local pointed skeleton of $\mathcal{FM}_m$. Every morphism $f \colon \mathbb{R}^{m+n} \to \mathbb{R}^{m+k}$ is locally of the form $f = h \circ g$ where $g = g_0 \times \text{id}_{\mathbb{R}^n} : \mathbb{R}^{m+n} \to \mathbb{R}^{m+n}$ and h is a morphism over identity on $\mathbb{R}^m$ ($g_0 = f_0$, $h_1(x,y) = f_1(f_0^{-1}(x), y)$). So we can deal separately with this two special types of morphisms.

The restriction F_n of functor F to subcategory $\mathcal{FM}_{m,n}$ is a regular bundle functor according to 20.6 and the morphisms of the type $g_0 \times \mathrm{id}_{\mathbb{R}^n}$ are $\mathcal{FM}_{m,n}$-morphisms.

Hence it remains to discuss the latter type of morphisms. We may restrict ourselves to families $h_p\colon \mathbb{R}^{m+n} \to \mathbb{R}^{m+k}$ parameterized by $p \in \mathbb{R}^q$, for some $q \in \mathbb{N}$. Let us consider $i\colon \mathbb{R}^{m+n} \to \mathbb{R}^{m+n} \times \mathbb{R}^q$, $(x, y) \mapsto (x, y, 0)$, $h\colon \mathbb{R}^{m+n+q} \to \mathbb{R}^{m+k}$, $h(-,-,p) = h_p$. Since all the maps h_p are over the identity, h is a fibered morphism. We have $h_p = h \circ t_{(0,0,p)} \circ i$, so that $Fh_p = Fh \circ Ft_{(0,0,p)} \circ Fi$. According to theorem 20.3 Fh_p is smoothly parameterized. □

20.8. Remarks. Since every bundle functor is completely determined by its restriction to a local pointed skeleton, there must be a bijective correspondence between bundle functors on categories with a common local pointed skeleton. Hence, although the category $\mathcal{FM}_{m,0}$ does not coincide with $\mathcal{M}f_m$ (in the former category, there are coverings of m-dimensional manifolds), the bundle functors on $\mathcal{M}f_m$ and $\mathcal{FM}_{m,0}$ are in fact the same ones. Analogously, the usual local skeleton of $\mathcal{FM}_0$ coincides with that of $\mathcal{M}f$. So corollary 20.6 reproves the classical result on natural bundles due to [Epstein, Thurston, 79] while 20.7 implies that every bundle functor defined on the whole category of manifolds is regular. For the same reason our results also apply to the category of $(m+n)$-dimensional manifolds with a foliation of codimension m and morphisms transforming leafs into leafs.

The rest of this section is devoted to the proof of proposition 20.2. Let us fix a bundle functor F on an admissible locally flat category $\mathcal{C}$ over manifolds with almost W-extendible sets of morphisms and an object $(\mathbb{R}^m, 0)$ in a local pointed skeleton. We shall briefly write p instead of $p_{\mathbb{R}^m}$, τ for the action of $\mathbb{R}^m$ on $F\mathbb{R}^m$ and we denote by $B(x, \varepsilon)$ the open ball $\{y \in \mathbb{R}^m; |y - x| < \varepsilon\} \subset \mathbb{R}^m$.

First the technique used in section 19 will help us to get a lemma that seems to be near to the continuity of τ claimed in proposition 20.2. However, the complete proof of 20.2 will require a lot of other analytical considerations.

20.9. Lemma. *Let $z_i \in F\mathbb{R}^m$, $i = 1, 2, \dots$, be a sequence of points converging to $z \in F\mathbb{R}^m$ such that $p(z_i) \neq p(z)$. Then there is a sequence of real constants $\varepsilon_i > 0$ such that for any point $a \in \mathbb{R}^m$ and any neighborhood W of $\tau_a(z)$ the inclusion $\tau(B(a, \varepsilon_i) \times \{z_i\}) \subset W$ holds for all large i's.*

Proof. Let us assume that the lemma is not true for some sequence $z_i \to z$. Then for any sequence ε_i of positive real numbers there are a point $a \in \mathbb{R}^m$, a neighborhood W of $\tau_a(z)$ and a sequence $a_i \in B(a, \varepsilon_i)$ such that $\tau(a_i, z_i) \notin W$ for an infinite set of indices $i \in I_0 \subset \mathbb{N}$. Let us denote $x_i := p(z_i)$, $x := p(z)$. Passing to a further subset of indices we can arrange that $2|x_i - x_j| > |x_i - x|$ for all i, $j \in I_0$, $i \neq j$. If we construct a smooth map $f : \mathbb{R}^m \to \mathbb{R}^m$ such that

$$\text{germ}\, f(x_i) = \text{germ}\, t_{a_i}(x_i) \tag{1}$$

for an infinite subset of indices $i \in I \subset I_0$ and

$$\text{germ}\, f(x_j) = \text{germ}\, t_a(x_j) \tag{2}$$

for an infinite subset of indices $j \in J \subset I_0$, then using the almost W-extendibility of $\mathcal{C}$-morphisms we find some $g \in \mathcal{C}(\mathbb{R}^m, \mathbb{R}^m)$ satisfying $Fg(z_i) = Ft_{a_i}(z_i) = \tau(a_i, z_i)$ for large $i \in I$ and $Fg(z_j) = \tau(a, z_j)$ for large $j \in J$. Hence $F(t_{-a}) \circ Fg(z_i) = \tau(a_i - a, z_i)$ for large $i \in I$ while $F(t_{-a}) \circ Fg(z_j) = z_j$ for large $j \in J$ and this implies $Fg(z) = \tau_a(z)$ which is in contradiction with $Fg(z_i) = \tau(a_i, z_i) \notin W$ for large $i \in I$.

The existence of a smooth map $f : \mathbb{R}^m \to \mathbb{R}^m$ satisfying (1), (2) is ensured by the Whitney extension theorem (see 19.4) if we choose the numbers ε_i small enough. To see this, let us view (1) and (2) as a prescription of all derivatives of f on some small neighborhoods of the points x_i, $i \in I_0$. Then the condition 19.4.(1) reads

$$\lim_{\substack{j,i\to\infty \\ j,i\in I}} \frac{|a_i - a_j|}{|x_i - x_j|^k} \to 0, \quad \lim_{\substack{i\to\infty \\ i\in I}} \frac{|a - a_i|}{|x_i - x|^k} \to 0 \quad \lim_{\substack{j,i\to\infty \\ i\in I, j\in J}} \frac{|a - a_i|}{|x_i - x_j|^k} \to 0$$

for all $k \in \mathbb{N}$.

Let us choose $0 < \varepsilon_i < e^{-1/(|x_i - x|)}$. Now, if $i < j$ then $|a_i - a_j| < 2\varepsilon_i$ and $|x_i - x_j| > \frac{1}{2}|x_i - x|$ and the first estimate follows. Analogously we get the remaining ones. □

The next lemma is necessary to overcome difficulties with constant sequences in $F\mathbb{R}^m$.

20.10. Lemma. *Let $z_j \in F\mathbb{R}^m$, $j = 1, 2, \dots$, be a sequence of points converging to $z \in F\mathbb{R}^m$. Then there is a sequence of points $a_i \in \mathbb{R}^m$, $a_i \neq 0$, $i = 1, 2, \dots$, converging to $0 \in \mathbb{R}^m$ and a subsequence z_{j_i} such that $Ft_{a_i}(z_{j_i}) \to z$ if $i \to \infty$.*

Proof. Let us recall that $F\mathbb{R}^m$ has a countable basis of open sets and let U_j, $j \in \mathbb{N}$, form a basis of open neighborhoods of the point z satisfying $U_{j+1} \subset U_j$. For each number $j \in \mathbb{N}$, there is a sequence of points $a(j,k) \in \mathbb{R}^m$, $k \in \mathbb{N}$, such that

$$\bigcup_{a\in\mathbb{R}^m} F(t_a)(U_j) = \bigcup_{k\in\mathbb{N}} F(t_{a(j,k)})(U_j).$$

Let $b_j \in \mathbb{R}^m$ be such a sequence that for all $k \in \mathbb{N}$, $b_j \neq a(j,k)$. Passing to subsequences, we may assume $z_j \in U_j$ for all j and consequently we get $F(t_{b_j})(z_j) \in \bigcup_{k\in\mathbb{N}} F(t_{a(j,k)})(U_j)$ for all $j \in \mathbb{N}$. Let us choose a sequence $k_j \in \mathbb{N}$, such that

$$F(t_{b_j})(z_j) \in F(t_{a(j,k_j)})(U_j)$$

for all $j \in \mathbb{N}$, and denote $a_j := b_j - a(j,k_j)$. Then $a_j \neq 0$ and $F(t_{a_j})(z_j) \in U_j$ for all $j \in \mathbb{N}$. Therefore $F(t_{a_j})(z_j) \to z$ and since $a_j = p(F(t_{a_j})(z_j)) - p(z_j)$, we also have $a_j \to 0$ □

A further step we need is to exclude the dependence of the balls $B(a, \varepsilon_i)$ on the indices i in the formulation of 20.9.

20.11. Lemma. *Let $z_i \to z$ be a convergent sequence in $F\mathbb{R}^m$, $p(z_i) \neq p(z)$, and let W be an open neighborhood of z. Then there exist $b \in \mathbb{R}^m$ and $\varepsilon > 0$ such that*

$$\tau\big(B(b,\varepsilon) \times \{z\}\big) \subset W, \quad \tau\big(B(b,\varepsilon) \times \{z_i\}\big) \subset W$$

for large i's.

Proof. We first deduce that there is some open ball $B(y,\eta) \subset \mathbb{R}^m$ satisfying

$$B(y,\eta) \subset \{a \in \mathbb{R}^m; F(t_a)(z) \in W\}. \tag{1}$$

Let us apply lemma 20.10 to a constant sequence $y_j := z$. So there is a sequence $a_i \in \mathbb{R}^m$, $a_i \neq 0$, $a_i \to 0$ such that $\tau_{a_i}(z) \to z$. Now we apply lemma 20.9 to the sequence $w_i := \tau_{a_i}(z)$. Since for $a = 0$ we have $\tau_a(z) \in W$, there is a sequence of positive constants η_i such that $\tau\big(B(0,\eta_i) \times \{w_i\}\big) \subset W$ for large i's. Let us choose one of these indices, say i_0, and put $y := a_{i_0}$, $\eta := \eta_{i_0}$. Now for any $b \in B(y,\eta)$ we have $\tau_b(z) = \tau_{b-y} \circ \tau_y(z) = \tau_{b-y}(w_i) \subset W$, so that (1) holds.

Further, let us apply lemma 20.9 to the sequence $z_i \to z$ and let us fix a neighborhood W of z. Then the conclusion of 20.9 reads as follows. There is a sequence of positive real constants ε_i such that for any $a \in \mathbb{R}^m$ the condition $\tau_a(z) \in W$ implies $\tau\big(B(a,\varepsilon_i) \times \{z_i\}\big) \subset W$ for all large i's. Therefore

$$B(y,\eta) \subset \bigcup_{k\in\mathbb{N}} \bigcap_{i\geq k} \{a \in \mathbb{R}^m; \tau\big(B(a,\varepsilon_i) \times \{z_i\}\big) \subset W\}.$$

For any natural number k we define

$$B_k := \bigcap_{i\geq k} \{a \in B(y,\eta); \tau\big(B(a,\varepsilon_i) \times \{z_i\}\big) \subset W\}.$$

Since $\bigcup_{k\in\mathbb{N}} B_k = B(y,\eta)$, the Baire category theorem implies that there is a natural number k_0 such that $\operatorname{int}(\bar{B}_{k_0}) \cap B(y,\eta) \neq \emptyset$.

Now, let us choose $b \in \mathbb{R}^m$ and $\varepsilon > 0$ such that $B(b,\varepsilon) \subset \operatorname{int}(\bar{B}_{k_0}) \cap B(y,\eta)$. If $x \in B(b,\varepsilon)$ and $i \geq k_0$, then there is $\bar{x} \in B_{k_0}$ with $x \in B(\bar{x},\varepsilon_i)$ so that we have $\tau_x(z_i) \in W$ and (1) implies $\tau_x(z) \in W$. □

20.12. Proof of proposition 20.2. Let $z_i \to z$ be a convergent sequence in $F\mathbb{R}^m$, $x_i \to x$ a convergent sequence in $\mathbb{R}^m$. We have to show

$$\tau_{x_i}(z_i) = F(t_{x_i})(z_i) \to F(t_x)(z) = \tau_x(z). \tag{1}$$

Since we can apply the isomorphism $F(t_{-x})$, we may assume $x = 0$. Moreover, it is sufficient to show that any subsequence of (x_i, z_i) contains a further subsequence satisfying (1). That is why we may assume either $p(z_i) \neq p(z)$ or $p(z_i) = p(z)$ for all $i \in \mathbb{N}$.

Let us first deal with the latter case. According to lemma 20.10 there is a sequence $y_i \in \mathbb{R}^m$ and subsequence z_{i_j} such that $\tau_{y_j}(z_{i_j}) \to z$ and $y_j \to 0$,

$y_j \neq 0$. But $\tau_{x_{i_j}}(z_{i_j}) \to z$ if and only if $\tau_{x_{i_j}-y_j} \circ \tau_{y_j}(z_{i_j}) \to z$, so that if we consider $\bar{z}_j := \tau_{y_j}(z_{i_j})$, $\bar{z} := z$ and $\bar{x}_j := x_{i_j} - y_j$, we transform the problem to the former case.

So we assume $p(z_i) \neq p(z)$ for all $i \in \mathbb{N}$ and $x_i \to 0$. Let us moreover assume that $\tau_{x_i}(z_i)$ does not converge to z. Then, for each $x \in \mathbb{R}^m$, $\tau_{x+x_i}(z_i)$ does not converge to $\tau_x(z)$ as well. Therefore, if we set

$$A := \{x \in \mathbb{R}^m; \tau_{x+x_i}(z_i) \text{ does not converge to } \tau_x(z)\}$$

we find $A = \mathbb{R}^m$. Now we use the separability of $F\mathbb{R}^m$. Let V_s, $s \in \mathbb{N}$, be a basis of open sets in $F\mathbb{R}^m$ and let

$$L_s := \{x \in \mathbb{R}^m; \tau_{x+x_i}(z_i) \in V_s \text{ for large } i\text{'s}\}$$

$$Q_s := \{x \in \mathbb{R}^m; \tau_x(z) \in V_s \text{ and } x \notin L_s\}.$$

We know $A \subset \cup_{s\in\mathbb{N}} Q_s$ and consequently $\cup_{s\in\mathbb{N}} Q_s = \mathbb{R}^m$. By virtue of the Baire category theorem there is a natural number k such that $\operatorname{int}(\overline{Q_k}) \neq \emptyset$.

Let us choose a point $a \in Q_k \cap \operatorname{int}(\overline{Q_k})$. Then $z \in \tau_{-a}(V_k)$ and so

$$W := p^{-1}\left(t_{p(z)-a}(\operatorname{int}(\overline{Q_k}))\right) \bigcap \tau_{-a}(V_k)$$

is an open neighborhood of z. According to lemma 20.11 there is an open ball $B(b,\varepsilon) \subset \mathbb{R}^m$ such that

$$\tau\big(B(b,\varepsilon) \times \{z\}\big) \subset p^{-1}\left(t_{p(z)-a}(\operatorname{int}(\overline{Q_k}))\right) \tag{2}$$

$$\tau\big(B(b,\varepsilon) \times \{z_i\}\big) \subset \tau_{-a}(V_k) \tag{3}$$

for all large i's. Inclusion (2) implies $p(z) + B(b,\varepsilon) \subset p(z) - a + \operatorname{int}(\overline{Q_k})$ or, equivalently,

$$B(b+a,\varepsilon) \subset \operatorname{int}(\overline{Q_k}). \tag{4}$$

Formula (3) is equivalent to

$$\tau\big(B(b+a,\varepsilon) \times \{z_i\}\big) \subset V_k$$

for large i's. Since $x_i \to 0$, we know that for any $x \in B(b+a,\varepsilon)$ also $(x + x_i) \in B(b+a,\varepsilon)$ for large i's and we get the inclusion $B(b+a,\varepsilon) \subset L_k$. Finally, (4) implies

$$B(b+a,\varepsilon) \subset L_k \cap \operatorname{int}(\overline{Q_k}) \subset (\mathbb{R}^m \setminus Q_k) \cap \operatorname{int}(\overline{Q_k}).$$

This is a contradiction. □

21. Actions of jet groups

Let us recall the jet group $G^r_{m,n}$ of the only type in the category $\mathcal{FM}_{m,n}$ which we mentioned in 18.8. In this section, we derive estimates on the possible order of this jet group acting on a manifold S depending only on $\dim S$. In view of lemma 20.4, these estimates will imply the finiteness of the order of bundle functors on $\mathcal{FM}_{m,n}$.

21.1. The whole procedure leading to our estimates is rather technical but the main idea is very simple and can be applied to other categories as well. Consider a jet group G^r_α of an admissible category $\mathcal{C}$ over manifolds acting on a manifold S and write B^r_k for the kernel of the jet projection $\pi^r_k\colon G^r_\alpha \to G^k_\alpha$. For every point $y \in S$, let H_y be the isotropy subgroup at the point y. The action factorizes to an action of a group G^k_α on S if and only if $B^r_k \subset H_y$ for all points $y \in S$. So if we assume that the order r is essential, i.e. the action does not factorize to G^{r-1}_α, then there is a point $y \in S$ such that H_y does not contain B^r_{k-1}. If the action is continuous, then H_y is closed and the homogeneous space G^r_α/H_y is mapped injectively and continuously into S. Hence we have

$$\dim S \geq \dim(G^r_\alpha/H_y) \tag{1}$$

and we see that $\dim S$ is bounded from below by the smallest possible codimension of Lie subgroups in G^r_α which do not contain B^r_k.

A proof of such a bound in the special case $\mathcal{C} = \mathcal{FM}_{m,n}$ will occupy the rest of this section.

21.2. Theorem. *Let a jet group $G^r_{m,n}$, $m \geq 1$, $n \geq 0$, act continuously on a manifold S, $\dim S = s$, $s \geq 0$, and assume that r is essential, i.e. the action does not factorize to an action of $G^k_{m,n}$, $k < r$. Then*

$$r \leq 2s + 1.$$

Moreover, if m, $n > 1$, then

$$r \leq \max\{\frac{s}{m-1},\ \frac{s}{m}+1,\ \frac{s}{n-1},\ \frac{s}{n}+1\}$$

and if $m > 1$, $n = 0$, then

$$r \leq \max\{\frac{s}{m-1},\ \frac{s}{m}+1\}.$$

All these estimates are sharp for all $m \geq 1$, $n \geq 0$, $s \geq 0$.

21.3. Proof of the estimate $r \leq 2s + 1$. Let us first assume $s > 0$. By the general arguments discussed in 21.1, there is a point $y \in S$ such that its isotropy group H_y does not contain the normal closed subgroup B^r_{r-1}. We shall denote $\mathfrak{g}^r_{m,n}$, $\mathfrak{b}^r_{r-1}$ and $\mathfrak{h}$ the Lie algebras of $G^r_{m,n}$, B^r_{r-1} and H_y, respectively. Since B^r_{r-1} is a connected and simply connected nilpotent Lie group, its exponential map is a global diffeomorphism of $\mathfrak{b}^r_{r-1}$ onto B^r_{r-1}, cf. 13.16 and 13.4. Therefore $\mathfrak{h}$ does not contain $\mathfrak{b}^r_{r-1}$. In this way, our problem reduces to the determination of a lower bound of the codimensions of subalgebras of $\mathfrak{g}^r_{m,n}$ that do not contain the whole $\mathfrak{b}^r_{r-1}$.

Since $\mathfrak{g}^r_{m,n}$ is a Lie subalgebra in $\mathfrak{g}^r_{m+n}$, there is the induced grading

$$\mathfrak{g}^r_{m,n} = \mathfrak{g}_0 \oplus \cdots \oplus \mathfrak{g}_{r-1}$$

where homogeneous components $\mathfrak{g}_p$ are formed by jets of homogeneous projectable vector fields of degrees $p+1$, cf. 13.16.

If we consider the intersections of $\mathfrak{h}$ with the filtration defining the grading $\mathfrak{g}^r_{m,n} = \oplus^p \mathfrak{g}_p$, then we get the filtration

$$\mathfrak{h} = \mathfrak{h}^0 \supset \mathfrak{h}^1 \supset \ldots \supset \mathfrak{h}^{r-1} \supset 0$$

and the quotient spaces $\mathfrak{h}_p = \mathfrak{h}^p/\mathfrak{h}^{p+1}$ are subalgebras in $\mathfrak{g}_p$. Therefore we can construct a new algebra $\tilde{\mathfrak{h}} = \mathfrak{h}_0 \oplus \cdots \oplus \mathfrak{h}_{r-1}$ with grading and since

$$\dim \mathfrak{h} = \dim \mathfrak{h}/\mathfrak{h}^1 + \dim \mathfrak{h}^1/\mathfrak{h}^2 + \cdots + \dim \mathfrak{h}^{r-1} = \dim \tilde{\mathfrak{h}},$$

both the algebras $\mathfrak{h}$ and $\tilde{\mathfrak{h}}$ have the same codimension. By the construction, $\mathfrak{b}^r_{r-1} \not\subset \mathfrak{h}$ if and only if $\mathfrak{h}_{r-1} \neq \mathfrak{g}_{r-1}$, so that $\tilde{\mathfrak{h}}$ does not contain $\mathfrak{b}^r_{r-1}$ as well. That is why in the proof of theorem 21.2 we may restrict ourselves to Lie subalgebras $\mathfrak{h} \subset \mathfrak{g}^r_{m,n}$ with grading $\mathfrak{h} = \mathfrak{h}_0 \oplus \cdots \oplus \mathfrak{h}_{r-1}$ satisfying $\mathfrak{h}_i \subset \mathfrak{g}_i$ for all $0 \leq i \leq r-1$, and $\mathfrak{h}_{r-1} \neq \mathfrak{g}_{r-1}$.

Now the proof of the estimate $r \leq 2s+1$ becomes rather easy. To see this, let us fix two degrees $p \neq q$ with $p+q = r-1$ and recall $[\mathfrak{g}_p, \mathfrak{g}_q] = \mathfrak{g}_{r-1}$, see 13.16. Hence there is either $a \in \mathfrak{g}_p$ or $a \in \mathfrak{g}_q$ with $a \notin \mathfrak{h}$, for if not then $[\mathfrak{g}_p, \mathfrak{g}_q] = \mathfrak{g}_{r-1} \subset \mathfrak{h}_{r-1}$. It follows

$$\operatorname{codim} \mathfrak{h} \geq \frac{1}{2}(r-1).$$

According to 21.1.(1) we get $s \geq \frac{1}{2}(r-1)$ and consequently $r \leq 2s+1$.

The remaining case $s = 0$ follows immediately from the fact that given an action $\rho\colon G^r_{m,n} \to \operatorname{Diff}(S)$ on a zero-dimensional manifold S, then its kernel $\ker \rho$ contains the whole connected component of the unit. Since $G^r_{m,n}$ has two components and these can be distinguished by the first order jet projection, we see that the order can be at most one. □

Let us notice, that the only special property of $\mathfrak{g}^r_{m,n}$ among the general jet groups which we used in 21.3 was the equality $[\mathfrak{g}_p, \mathfrak{g}_q] = \mathfrak{g}_{p+q}$. Hence the first estimate from theorem 21.2 can be easily generalized to some other categories.

The proof of the better estimates for higher dimensions is based on the same ideas but supported by some considerations from linear algebra. We choose some non-zero linear form C on $\mathfrak{g}_{r-1}$ with $\ker C \supset \mathfrak{h}_{r-1}$. Then given p, q, $p+q = r-1$, we define a bilinear form $f : \mathfrak{g}_p \times \mathfrak{g}_q \to \mathbb{R}$ by $f(a,b) = C([a,b])$ and we study the dimensions of the annihilators.

21.4. Lemma. *Let V, W, be finite dimensional real vector spaces and let $f: V \times W \to \mathbb{R}$ be a bilinear form. Denote by V^0 or W^0 the annihilators of V or W related to f, respectively. Let $M \subset V$, $N \subset W$ be subspaces satisfying $f|(M \times N) = 0$. Then*

$$\operatorname{codim} M + \operatorname{codim} N \geq \operatorname{codim} V^0.$$

Proof. Consider the associated form $f^* : V/W^0 \times W/V^0 \to \mathbb{R}$ and let $[M]$, $[N]$ be the images of M, N in the projections onto quotient spaces. Since f^* is not degenerated, we have

$$\dim[M] + \dim[M]^0 = \operatorname{codim} W^0. \tag{1}$$

Note that $\operatorname{codim} V^0 = \operatorname{codim} W^0$. We know $\dim[M] = \dim(M/M \cap W^0) = \dim(M + W^0) - \dim W^0$ and similarly for N. Therefore

$$\begin{aligned}\dim[M] + \dim[N] = \\ &= \dim(M + W^0) - \dim W^0 + \dim(N + V^0) - \dim V^0 \\ &= \operatorname{codim} W^0 - \operatorname{codim}(M + W^0) + \operatorname{codim} V^0 - \operatorname{codim}(N + V^0) \\ &\geq \operatorname{codim} W^0 + (\operatorname{codim} V^0 - \operatorname{codim} M - \operatorname{codim} N).\end{aligned} \tag{2}$$

According to our assumptions $N \subset M^0$, so that $\dim[N] \leq \dim[M]^0$. But then (1) implies

$$\dim[M] + \dim[N] \leq \operatorname{codim} W^0$$

and therefore the term in the last bracket in (2) must be less then zero. □

If we fix a basis of the vector space $\mathbb{R}^m$ then there is the induced basis on the vector space $\mathfrak{g}_{r-1}$ and the induced coordinate expressions of linear forms C on $\mathfrak{g}_{r-1}$. By naturality of the Lie bracket, using arbitrary coordinates on $\mathbb{R}^m$ the coordinate formula for the Lie bracket does not change. Since fiber respecting linear transformations of $\mathbb{R}^{m+n} \to \mathbb{R}^m$ preserve the projectability of vector fields, we can use arbitrary affine coordinates on the fibration $\mathbb{R}^{m+n} \to \mathbb{R}^m$ in our discussion on possible codimensions of the subalgebras, which is based on formula 13.2.(5).

The coordinate expression of C will be written like $C = (C_i^\alpha)$, $i = 1, \dots, m+n$, $|\alpha| = r$. This means $C(X) = \sum_{\alpha,i} C_i^\alpha a_\alpha^i$, if $X = \sum_{\alpha,i} a_\alpha^i x^\alpha \frac{\partial}{\partial x^i} \in \mathfrak{g}_{r-1}$, where we sum also over repeated indices. For technical reasons we set $C_i^\alpha = 0$ whenever $i \leq m$ and $\alpha_j > 0$ for some $j > m$.

If suitable, we also write $\alpha = (\alpha_1, \dots \alpha_{m+n})$ in the form $\alpha = i_1 \cdots i_r$, where $r = |\alpha|$, $1 \leq i_j \leq m+n$, so that α_j is the number of indices i_k that equal j. Further we shall use the symbol (j) for a multiindex α with $\alpha_i = 0$ for all $i \neq j$, and its length will be clear from the context. As before, the symbol 1_j denotes a multiindex α with $\alpha_i = \delta_j^i$.

21.5. Lemma. *Let C be a non-zero form on $\mathfrak{g}_{r-1}$, $m \geq 1$, $n \geq 0$. Then in suitable affine coordinates on the fibration $\mathbb{R}^{m+n} \to \mathbb{R}^m$, the induced coordinate expression of C satisfies one of the following conditions:*

(i) $C_m^{(1)} \neq 0$ and $m > 1$.

(ii) $C_1^{(1)} \neq 0$; $C_j^\alpha = 0$ *whenever* $\alpha_j = 0$ *and* $1 \leq j \leq m$; $C_1^{\alpha+1_1} = C_j^{\alpha+1_j}$ *(no summation) for all* $|\alpha| = r - 1$, $1 \leq j \leq m$.

(iii) $C_{m+n}^{(m+1)} \neq 0$, $n > 1$, *and* $C_j^\alpha = 0$ *whenever* $j \leq m$.

(iv) $C_{m+1}^{(m+1)} \neq 0$; $C_j^\alpha = 0$ *if* $j \leq m$ or $\alpha_j = 0$; *and* $C_{m+1}^{\alpha+1_{m+1}} = C_j^{\alpha+1_j}$ *(no summation) for all* $|\alpha| = r - 1$, $j \geq m + 1$.

Proof. Let C be a non-zero form on $\mathfrak{g}_{r-1}$ with coordinates C_j^α in the canonical basis of $\mathbb{R}^{m+n} \to \mathbb{R}^m$. Let us consider a matrix $A \in GL(m+n)$ whose first row consists of arbitrary real parameters $a_1^1 = t_1 \neq 0$, $a_2^1 = t_2, \dots, a_m^1 = t_m$, $a_j^1 = 0$ for $j > m$, and let all the other elements be like in the unit matrix. Let $\tilde{a}_j^i$ be the elements of the inverse matrix A^{-1}. If we perform this linear transformation, we get a new coordinate expression of C, in particular

$$\bar{C}_j^{(1)} = a_{i_1}^1 \cdots a_{i_r}^1 C_s^{i_1 \dots i_r} \tilde{a}_j^s \,. \tag{1}$$

Hence we get

$$\bar{C}_1^{(1)} = t_{i_1} \cdots t_{i_r} C_1^{i_1 \dots i_r} \frac{1}{t_1} \tag{2}$$

$$\bar{C}_j^{(1)} = t_{i_1} \cdots t_{i_r} \left(C_j^{i_1 \dots i_r} - \frac{t_j}{t_1} C_1^{i_1 \dots i_r} \right) \qquad \text{for } 1 < j \leq m. \tag{3}$$

Formula (2) implies that either we can obtain $\bar{C}_1^{(1)} \neq 0$ or $C_1^\alpha = 0$ for all multi indices α, $|\alpha| = r$. Let us assume $m > 1$ and try to get condition (i). According to (3), if (i) does not hold after performing any of our transformations, then the expression on the right hand side of (3) has to be identically zero for all values of the parameters and this implies $C_j^\alpha = 0$ whenever $\alpha_j = 0$, $|\alpha| = r$, and $C_1^{\alpha+1_1} = C_j^{\alpha+1_j}$ for all $|\alpha| = r - 1$, $1 \leq j \leq m$. Hence we can summarize: either (i) can be obtained, or (ii) holds, or $\bar{C}_j^\alpha = 0$ for all $1 \leq j \leq m$, $|\alpha| = r$, in suitable affine coordinates.

Analogously, let us take a matrix $A \in GL(m+n)$ whose $(m+1)$-st row consists of real parameters $t_1, \dots, t_{m+n}$, $t_{m+1} \neq 0$ and let the other elements be like in the unit matrix. The new coordinates of C are obtained as above

$$\bar{C}_{m+1}^{(m+1)} = t_{i_1} \cdots t_{i_r} C_{m+1}^{i_1 \dots i_r} \frac{1}{t_{m+1}} \tag{4}$$

$$\bar{C}_j^{(m+1)} = t_{i_1} \cdots t_{i_r} \left(C_j^{i_1 \dots i_r} - \frac{t_j}{t_{m+1}} C_{m+1}^{i_1 \dots i_r} \right) . \tag{5}$$

Now we may assume $C_j^\alpha = 0$ whenever $1 \leq j \leq m$, for if not then (i) or (ii) could be obtained. As before, either there is a basis relative to which $C_{m+1}^{(m+1)} \neq 0$ or

$C^\alpha_{m+1} = 0$ for all $|\alpha| = r$. Further, according to (5) either we can get (iii) or $C^\alpha_j = 0$ whenever $\alpha_j = 0$, and $C^{\alpha+1_{m+1}}_{m+1} = C^{\alpha+1_j}_j$, for all $|\alpha| = r-1$, $j \geq m+1$. Therefore if both (iii) and (iv) do not hold after arbitrary transformations, then all C^α_j have to be zero, but this is contradictory to the fact that C is non-zero. □

21.6. Lemma. *Let p, q be two degrees with $p+q = r-1 > 0$ and $p \geq q \geq 0$. Let $m > 1$ and $n > 1$ or $n = 0$, and let $\mathfrak{h}_p$, $\mathfrak{h}_q$ be subspaces of $\mathfrak{g}_p$, $\mathfrak{g}_q$. Let C be a non-zero linear form on $\mathfrak{g}_{r-1}$ and suppose $[\mathfrak{h}_p, \mathfrak{h}_q] \subset \ker C$. If C^α_i, $1 \leq i \leq m+n$, $|\alpha| = r$, is a coordinate expression of C satisfying one of the conditions 21.5.(i)-(iv), then*

$$\operatorname{codim} h_p + \operatorname{codim} \mathfrak{h}_q \geq \begin{cases} 2m-2, & \text{if 21.5.(i) holds} \\ 2m, & \text{if 21.5.(ii) holds and } q > 0 \\ m, & \text{if 21.5.(ii) holds and } q = 0 \\ 2n-2, & \text{if 21.5.(iii) holds} \\ 2n, & \text{if 21.5.(iv) holds and } q > 0 \\ n, & \text{if 21.5.(iv) holds and } q = 0. \end{cases}$$

Proof. Define a bilinear form

$$f : \mathfrak{g}_p \times \mathfrak{g}_q \to \mathbb{R} \qquad f(a,b) = C([a,b]) .$$

By our assumptions $f(\mathfrak{h}_p, \mathfrak{h}_q) = \{0\}$. Hence by lemma 21.4 it suffices to prove that the codimension of the f-annihilator of $\mathfrak{g}_q$ in $\mathfrak{g}_p$ has the above lower bounds. Let $\mathfrak{h}_0$ be this annihilator and consider elements $a \in \mathfrak{h}_0$, $b \in \mathfrak{g}_q$. We get

$$C([a,b]) = \sum_{\substack{1\leq i\leq m+n \\ |\alpha|=r}} C^\alpha_i([a,b])^i_\alpha = 0.$$

Using formula for the bracket 13.2.(5) we obtain

$$\begin{aligned} 0 &= \sum_{\substack{1\leq i,j\leq m+n \\ |\mu|=q+1 \\ |\lambda|=p+1}} C^{\mu+\lambda-1_j}_i \left(\lambda_j b^j_\mu a^i_\lambda - \mu_j a^j_\lambda b^i_\mu\right) \\ &= \sum_{\substack{1\leq i,j\leq m+n \\ |\mu|=q+1 \\ |\lambda|=p+1}} \left(C^{\mu+\lambda-1_j}_i \lambda_j - C^{\mu+\lambda-1_i}_j \mu_i\right) a^i_\lambda b^j_\mu . \end{aligned}$$

Since $b \in \mathfrak{g}_q$ is arbitrary, we have got a system of linear equations for the annihilator $\mathfrak{h}_0$ containing one equation for each couple (j,μ), where $1 \leq j \leq m+n$, $|\mu| = q+1$ and $\mu_i = 0$ whenever $i > m$ and $j \leq m$. The (j,μ)-equation reads

$$\sum_{\substack{1\leq i\leq m+n \\ |\lambda|=p+1}} \left(C^{\mu+\lambda-1_j}_i \lambda_j - C^{\mu+\lambda-1_i}_j \mu_i\right) a^i_\lambda = 0. \tag{1}$$

A lower bound of the codimension of $\mathfrak{h}_0$ is given by any number of linearly independent (j,μ)-equations and we have to discuss this separately for the cases 21.5.(i)–(iv).

Let us first assume that 21.5.(i) holds, i.e. $C_m^{(1)} \neq 0$, $m > 1$. We denote by E_s the $(s,(1))$-equation, $1 \leq s < m$ and by F_k the $(m,(1)+1_k)$-equation, $1 \leq k < m$ (note that if $q = 0$ then $(1)+1_k = 1_k$). We claim that this subsystem is of full rank. In order to verify this, consider a linear combination

$$\sum_{s=1}^{m-1} a^s E_s + \sum_{k=1}^{m-1} b^k F_k = 0 \qquad a^s, b^k \in \mathbb{R}.$$

From (1) we get

$$(2) \quad \sum_{\substack{1\leq i\leq m+n \\ |\lambda|=p+1}} \Bigg(\sum_{s=1}^{m-1} \big(C_i^{(1)+\lambda-1_s}\lambda_s - C_s^{(1)+\lambda-1_i}\delta_i^1(q+1)\big)a^s + \\ + \sum_{k=1}^{m-1}\big(C_i^{(1)+1_k+\lambda-1_m}\lambda_m - C_m^{(1)+1_k+\lambda-1_i}(\delta_i^1 q + \delta_i^k)\big)b^k \Bigg) a_\lambda^i = 0.$$

Hence all the coefficients at the variables a_λ^i with $1 \leq i \leq m+n$, $\lambda = p+1$, and $\lambda_j = 0$ whenever $j > m$ and $i \leq m$, have to vanish. Therefore, we get equations on reals a^s, b^k, whenever we choose i and λ. We have to show that all these reals are zero.

First, let us substitute $\lambda = (1)$ and $i = m$. Then (2) implies $C_m^{(1)}(p+1)a^1 = 0$ and consequently $a^1 = 0$. Now we choose $\lambda = (1)+1_v$, $i = m$, with $1 < v < m$, and we get $C_m^{(1)}a^v = 0$ so that $a^s = 0$ for $1 \leq s \leq m-1$. Further, take $\lambda = (1)$ and $1 < i < m$ to obtain $-C_m^{(1)}b^i = 0$. Finally, the choice $i = 1$ and $\lambda = (1)$ leads to $-C_m^{(1)}(q+1)b^1 = 0$. In this way, we have proved that the chosen $2m-2$ equations E_s and F_k are independent and this implies the first lower bound in 21.6.

Now suppose 21.5.(ii) takes place and let us denote E_s the $(s,(1))$-equation, $1 \leq s \leq m$, and if $q > 0$, then F_k will be the $(m,(1)+1_m+1_k)$-equation, $1 \leq k \leq m$. As before, we assume $\sum_{s=1}^m a^s E_s + \sum_{k=1}^m b^k F_k = 0$ for some reals a^s and b^k and we compare the coefficients at a_λ^i to show that all these reals are zero. But before doing this, we can simplify all (j,μ)-equations with $1 \leq j \leq m$ using the relations from 21.5.(ii). Indeed, (1) reduces to

$$\sum_{\substack{1\leq i\leq m \\ |\lambda|=p+1}} C_j^{\mu+\lambda-1_i}(\lambda_j - \mu_i)a_\lambda^i + R = 0.$$

where R involves all terms with indices $i > m$. Consequently E_s and F_k have the forms

$$\sum_{\substack{1\le i\le m\\ |\lambda|=p+1}} C_s^{(1)+\lambda-1_i}(\lambda_s-\delta_i^1(q+1))a_\lambda^i+R=0$$

$$\sum_{\substack{1\le i\le m\\ |\lambda|=p+1}} C_m^{(1)+1_k+1_m+\lambda-1_i}(\lambda_m-\delta_i^1(q-1)-\delta_i^k-\delta_i^m)a_\lambda^i+R=0.$$

Assume first $q>0$. If we choose $1<i\le m$, $\lambda=(1)$, then the variables a_λ^i do not appear in the equations E_s at all. Hence the choice $i=m$, $\lambda=(1)$ gives (see 21.5.(ii)) $0=-2C_m^{(1)+1_m}b^m=-2C_1^{(1)}b^m$; and $1<i<m$, $\lambda=(1)$ now yields $-C_m^{(1)+1_m}b^i=0$. Hence $b^i=0$ for all $1<i\le m$. Further, we take $i=m$, $\lambda=(1)+1_v+1_m$, $v\ne m$ (note $p\ge q>0$), so that all the coefficients in F_1 are zero. In particular, $v=1$ implies $C_1^{(1)}a^1=0$ so that $a^1=0$. Now, if $1<v<m$, then $C_v^{(1)+1_v}a^v=C_1^{(1)}a^v=0$ and what remains are a^m and b^1, only. Taken $\lambda=(1)$, $i=1$, we see $0=-(q+1)C_m^{(1)}a^m-qC_m^{(1)+1_m}b^1=C_1^{(1)}b^1$ and, finally, the choice $i=m$ and $\lambda=(1)+1_m+1_m$ gives $C_m^{(1)+1_m}2a^m=0$. This completes the proof of the second lower bound in 21.6.

But if $q=0$ and 21.5.(ii) holds, we can perform the above procedure after forgetting all the equations F_k which are not defined. We have only to notice $p+q=r-1>0$, so that $|\lambda|=p+1=r\ge 2$.

If $n>1$, then the remaining three parts of the proof are complete recapitulations of the above ones. This becomes clear if we notice, that we have used neither any information on C_j^α, $j>m$, nor the fact that $C_j^\alpha=0$ if $j\le m$ and $\alpha_i\ne 0$ for some $i>m$. That is why we can go step by step through the above proof on replacing 1 or m by $m+1$ or $m+n$, respectively.

If $n=0$, then neither 21.5.(iii) nor 21.5.(iv) can hold. □

21.7. Proposition. *Let $\mathfrak{h}$ be a subalgebra of $\mathfrak{g}^r_{m,n}$, $m\ge 1$, $n\ge 0$, $r\ge 2$, which does not contain $\mathfrak{b}^r_{r-1}$. Then*

$$\operatorname{codim}\mathfrak{h}\ge \frac{1}{2}(r-1). \tag{1}$$

Moreover, if $m>1$, $n>1$, then

$$\operatorname{codim}\mathfrak{h}\ge \min\{r(m-1),\ (r-1)m,\ r(n-1),\ (r-1)n\} \tag{2}$$

and if $m>1$, $n=0$, then

$$\operatorname{codim} h\ge \min\{r(m-1),\ (r-1)m\}. \tag{3}$$

Proof. In 21.3 we deduced that we may suppose $\mathfrak{h}$ is a subalgebra with grading $\mathfrak{h}=\mathfrak{h}_0\oplus\cdots\oplus\mathfrak{h}_{r-1}$, $\mathfrak{h}_i\subset\mathfrak{g}_i$, $\mathfrak{h}_{r-1}\ne\mathfrak{g}_{r-1}$, and we proved the lower bound (1). Let us assume $m>1$, $n=0$ and choose a non-zero form C on $\mathfrak{g}_{r-1}$ with

$\ker C \supset \mathfrak{h}_{r-1}$. Then we know $[\mathfrak{h}_j, \mathfrak{h}_{r-j-1}] \subset \mathfrak{h}_{r-1} \subset \ker C$ and by lemma 21.5 there is a suitable coordinate expression of C satisfying one of the conditions 21.5.(i), 21.5.(ii). Therefore we can apply lemma 21.6.

Assume first C_i^α satisfies 21.5.(i). Then for all j

$$\operatorname{codim} \mathfrak{h}_j + \operatorname{codim} \mathfrak{h}_{r-j-1} \geq 2m - 2$$

and consequently

$$\operatorname{codim} \mathfrak{h} = \sum_{j=0}^{r-1} \operatorname{codim} \mathfrak{h}_j \geq r(m-1).$$

If 21.5.(ii) holds, then

$$\operatorname{codim} \mathfrak{h}_0 + \operatorname{codim} \mathfrak{h}_{r-1} \geq m$$
$$\operatorname{codim} \mathfrak{h}_j + \operatorname{codim} \mathfrak{h}_{r-j-1} \geq 2m$$

for $1 \leq j \leq r-2$, so that $\operatorname{codim} \mathfrak{h} \geq m + (r-2)m = (r-1)m$. This completes the proof of (3) and analogous considerations lead to the estimate (2) if $n > 1$ and the coordinate expression of C satisfies 21.5.(iii) or 21.5.(iv). □

21.8. Examples.

1. Let $\mathfrak{h}_1 \subset \mathfrak{g}_m^r$, $m > 1$, be defined by

$$\mathfrak{h}_1 = \{a_\lambda^i x^\lambda \frac{\partial}{\partial x_i}; a_{(1)}^j = 0 \text{ for } j = 2, \dots, m, \ 1 \leq |(1)| \leq r\}.$$

One sees immediately that the linear subspace $\mathfrak{h}_1$ consists just of polynomial vector fields of degree r tangent to the line $x_2 = x_3 = \dots = x_m = 0$, so that $\mathfrak{h}_1$ clearly is a Lie subalgebra in $\mathfrak{g}_m^r$ of codimension $r(m-1)$. Consider now the subalgebra $\mathfrak{h} \subset \mathfrak{g}_{m,n}^r$ consisting of projectable polynomial vector fields of degree r over polynomial vector fields from $\mathfrak{h}_1$. This is a subalgebra of codimension $r(m-1)$ in $\mathfrak{g}_{m,n}^r$.

2. Consider the algebra $\mathfrak{h}_2 \subset \mathfrak{g}_{m+n}^r$, $n > 1$, defined analogously to the subalgebra $\mathfrak{h}_1$

$$\mathfrak{h}_2 = \{a_\lambda^i x^\lambda \frac{\partial}{\partial x_i}; a_{(m+1)}^j = 0 \text{ for } 1 \leq j \leq m+n, \ j \neq m+1, \ 1 \leq |(m+1)| \leq r\}$$

and define $\mathfrak{h} = \mathfrak{h}_2 \cap \mathfrak{g}_{m,n}^r$. Since every polynomial vector field in $\mathfrak{g}_{m,n}^r$ is tangent to the fiber over zero, this clearly is a Lie subalgebra with coordinate description

$$\mathfrak{h} = \{a_\lambda^i x^\lambda \frac{\partial}{\partial x_i}; a_{(m+1)}^j = 0 \text{ for } m+1 < j \leq m+n, \ 1 \leq |(m+1)| \leq r\}$$

and codimension $r(n-1)$.

3. Let us recall that the divergence $\operatorname{div} X$ of a polynomial vector field $X \in \mathfrak{g}^r_m$ can be viewed as the jet $j^{r-1}_0(\operatorname{div} X)$, see 13.6. So for an element $a = a^i_\lambda x^\lambda \frac{\partial}{\partial x_i}$ we have

$$\operatorname{div} a = \sum_{\substack{1 \le i \le m \\ 1 \le |\lambda| \le r}} \lambda_i a^i_\lambda x^{\lambda - 1_i}.$$

Let M be the line in $\mathbb{R}^m$, $m > 1$, defined by $x_2 = x_3 = \cdots = x_m = 0$ and denote by $\mathfrak{h}_3$ the linear subspace in $\mathfrak{g}_1 \oplus \cdots \oplus \mathfrak{g}_{r-1}$ (note $\mathfrak{g}_0$ is missing!)

$$\mathfrak{h}_3 = \{a^i_\lambda x^\lambda \frac{\partial}{\partial x_i} \in \mathfrak{g}_1 \oplus \cdots \oplus \mathfrak{g}_{r-1}; \left(\operatorname{div}(a^i_\lambda x^\lambda \frac{\partial}{\partial x_i})\right)\big|_M = 0\}.$$

Of course, $\mathfrak{h}_3$ is not a Lie subalgebra in $\mathfrak{g}^r_m$. Let us further consider the Lie subalgebra $\mathfrak{h}_4 \subset \mathfrak{g}^{r-1}_m$ consisting of all polynomial vector fields without absolute terms and tangent to M, cf. example 1. Let $\mathfrak{h}_5 = \left(\pi^r_{r-1}\right)^{-1} \mathfrak{h}_4 \subset \mathfrak{g}^r_m$ and let us define a linear subspace

$$\mathfrak{h}_6 = (\mathfrak{h}_5 \cap \mathfrak{g}_0) \oplus (\mathfrak{h}_3 \cap \mathfrak{h}_5).$$

First we claim that $\mathfrak{h}_3 \cap \mathfrak{h}_5$ is a subalgebra. Indeed, if $X, Y \in \mathfrak{h}_3 \cap \mathfrak{h}_5$, then either the degree of both of them is less then r or their bracket is zero. But in the first case, X and Y are tangent to M and their divergences are zero on M, so that 13.6.(1) implies $\operatorname{div}([X,Y])|_M = 0$.

Now, consider a polynomial vector field X from the subalgebra $\mathfrak{h}_5 \cap \mathfrak{g}_0$ and a field $Y \in \mathfrak{h}_3 \cap \mathfrak{h}_5$. Since every field from $\mathfrak{g}_0$ has constant divergence everywhere and X is tangent to M, 13.6.(1) implies $\operatorname{div}([X,Y])|_M = 0$. So we have proved that $\mathfrak{h}_6$ is a subalgebra. In coordinates, we have

$$\mathfrak{h}_6 = \{a^i_\lambda x^\lambda \frac{\partial}{\partial x_i} \in \mathfrak{g}^r_m;\ a^j_{(1)} = 0,\ \sum_{i=1}^m a^i_{(1)+1_i}(1 + \delta^1_i |(1)|) = 0$$
$$\text{for } j = 2, \dots, m,\ 1 \le |(1)| \le r-1\}.$$

Now, we take the subalgebra $\mathfrak{h}$ in $\mathfrak{g}^r_{m,n}$ consisting of polynomial vector fields over the fields from $\mathfrak{h}_6$. The codimension of $\mathfrak{h}$ is $(r-1)m$.

4. Analogously to example 2, let us consider the subalgebra $\mathfrak{h}_7$ in $\mathfrak{g}^r_{m+n}$, $n > 1$,

$$\mathfrak{h}_7 = \{a^i_\lambda x^\lambda \frac{\partial}{\partial x_i} \in \mathfrak{g}^r_{m+n};\ a^j_{(m+1)} = 0,\ \sum_{i=1}^{m+n} a^i_{(m+1)+1_i}(1 + \delta^{m+1}_i |(m+1)|) = 0$$
$$\text{for } j = 1, \dots, m+n,\ j \ne m+1,\ 1 \le |(m+1)| \le r-1\}$$

and let us define $\mathfrak{h} = \mathfrak{h}_7 \cap \mathfrak{g}^r_{m,n}$. Then

$$\mathfrak{h} = \{a^i_\lambda x^\lambda \frac{\partial}{\partial x_i} \in \mathfrak{g}^r_{m,n};\ a^j_{(m+1)} = 0,\ \sum_{i=m+1}^{m+n} a^i_{(m+1)+1_i}(1 + \delta^{m+1}_i |(m+1)|) = 0$$
$$\text{for } j = m+2, \dots, m+n,\ 1 \le |(m+1)| \le r-1\}$$

and we have found a Lie subalgebra in $\mathfrak{g}^r_{m,n}$ of codimension $(r-1)n$.

Let us look at the subgroups corresponding to the above subalgebras. In the first and the second examples, the groups consist of polynomial fibered isomorphisms keeping invariant the given lines. These are closed subgroups. In the remaining two examples, we have to consider analogous subgroups in $G^{r-1}_{m,n}$, then to take their preimages in the group homomorphism π^r_{r-1}. Further we consider the subgroups of polynomial local isomorphisms at the origin identical in linear terms and without the absolute ones. Their subsets consisting of maps keeping the volume form along the given lines are subgroups. Finally, we take the intersections of the above constructed subgroups. All these subgroups are closed.

21.9. Proof of 21.2. The idea of the proof was explained in 21.1 and 21.3. In particular, we deduced that the dimension of every manifold with an action of $G^r_{m,n}$, $r \geq 2$, which does not factorize to an action of $G^{r-1}_{m,n}$, is bounded from below by the smallest possible codimension of Lie subalgebras $\mathfrak{h} = \mathfrak{h}_0 \oplus \cdots \oplus \mathfrak{h}_{r-1}$, $\mathfrak{h}_i \subset \mathfrak{g}_i$, $\mathfrak{h}_{r-1} \neq \mathfrak{g}_{r-1}$, with grading. We also got the lower bound $\frac{1}{2}(r-1)$ for the codimensions and this implied the estimate $r \leq 2 \dim S + 1$. But now, we can use proposition 21.7 to get a better lower bound for every $m > 1$ and $n > 1$. Indeed,

$$s = \dim S \geq \min\{r(m-1), (r-1)m, r(n-1), (r-1)n\}$$

and consequently

$$r \leq \max\{\frac{s}{m-1}, \frac{s}{m} + 1, \frac{s}{n-1}, \frac{s}{n} + 1\}.$$

If $n = 0$ we get

$$s \geq \min\{r(m-1), (r-1)m\},$$

so that

$$r \leq \max\{\frac{s}{m-1}, \frac{s}{m} + 1\}.$$

Since all the groups determined by the subalgebras we have constructed in 21.8 are closed, the corresponding homogeneous spaces are examples of manifolds with actions of $G^r_{m,n}$ with the extreme values of r.

If $m = 1$, let us consider $\mathfrak{h} = \mathfrak{g}_0 \oplus \mathfrak{g}_s \oplus \mathfrak{g}_{s+1} \oplus \cdots \oplus \mathfrak{g}_{2s-1} \subset \mathfrak{g}^{2s+1}_1$. Since $[\mathfrak{g}_s, \mathfrak{g}_s] = 0$ in dimension one, this is a Lie subalgebra and one can see that the corresponding subgroup H in G^{2s+1}_1 is closed (in general, every connected Lie subgroup in a simply connected Lie group is closed, see e.g. [Hochschild, 68, p. 137]). The homogeneous space G^{2s+1}_1/H has dimension s and G^{2s+1}_1 acts non trivially. Since there are group homomorphisms $G^r_{m,n} \to G^r_m$ and $G^r_{m,n} \to G^r_n$ (the latter one is the restriction of the polynomial maps to the fiber over zero), we have found the two remaining examples. □

22. The order of bundle functors

Now we will collect the results from the previous sections to get a description of bundle functors on fibered manifolds. Let us remark that the bundle functors on categories with the same local skeletons in fact coincide. So we also describe bundle functors on $\mathcal{M}f_m$ and $\mathcal{M}f$ in this way, cf. remark 20.8. In view of the general description of finite order regular bundle functors on admissible categories and natural transformations between them deduced in theorems 18.14 and 18.15, the next theorem presents a rather detailed information. As usual $m\colon \mathcal{FM}_{m,n} \to \mathcal{M}f$ is the faithful functor forgetting the fibrations.

22.1. Theorem. *Let $F : \mathcal{FM}_{m,n} \to \mathcal{M}f$, $m \geq 1$, $n \geq 0$, be a functor endowed with a natural transformation $p\colon F \to m$ and satisfying the localization property 18.3.(i). Then $S := p^{-1}_{\mathbb{R}^{m+n}}(0)$ is a manifold of dimension $s \geq 0$ and for every $(Y \to M)$ in $\mathrm{Ob}\mathcal{FM}_{m,n}$ the mapping $p_Y : FY \to Y$ is a locally trivial fiber bundle with standard fiber S, i.e. $F\colon \mathcal{FM}_{m,n} \to \mathcal{FM}$. The functor F is a regular bundle functor of a finite order $r \leq 2s + 1$. If moreover $m > 1$, $n = 0$, then*

$$r \leq \max\{\frac{s}{m-1}, \frac{s}{m} + 1\},$$

and if $m > 1$, $n > 1$, then

$$r \leq \max\{\frac{s}{m-1}, \frac{s}{m} + 1, \frac{s}{n-1}, \frac{s}{n} + 1\}.$$

All these estimates are sharp.

Proof. Since $\mathcal{FM}_{m,n}$ is a locally flat category with Whitney-extendible sets of morphisms, we have only to prove the assertion concerning the order. The rest of the theorem follows from theorems 20.3 and 20.5. By definition of bundle functors, it suffices to prove that the action of the group G of germs of fibered morphisms $f\colon \mathbb{R}^{m+n} \to \mathbb{R}^{m+n}$ with $f(0) = 0$ on the standard fiber S factorizes to an action of $G^r_{m,n}$ with the above bounds of r depending on s, m, n.

As in the proof of theorem 20.5, let $V \subset S$ be a relatively compact open set and $Q_V \subset S$ be the open submanifold invariant with respect to the action of G, as defined in 20.4. By virtue of lemma 20.4 the action of G on Q_V factorizes to an action of $G^k_{m,n}$ for some $k \in \mathbb{N}$. But then theorem 21.2 yields the necessary estimates. Moreover, if we consider the $G^r_{m,n}$-spaces with the extreme orders from theorem 21.2, then the general construction of a bundle functor from an action of the r-th skeleton yields bundle functors with the extreme orders, cf. 18.14. □

22.2. Example. All objects in the category $\mathcal{FM}_{m,n}$ are of the same type. Now we will show that the order of bundle functors may vary on objects of different types. We shall construct a bundle functor on $\mathcal{M}f$ of infinite order.

Consider the sequence of the r-th order tangent functors $T^{(r)}$ from 12.14. These are bundle functors of orders $r \in \mathbb{N}$ with values in the category $\mathcal{VB}$ of vector bundles. Let us denote d_k the dimension of the standard fiber of $T^{(k)}\mathbb{R}^k$ and define a functor $F\colon \mathcal{M}f \to \mathcal{FM}$ as follows.

Consider the functors Λ^k operating on category $\mathcal{VB}$ of vector bundles. For every manifold M the value FM is defined as the Whitney sum over M

$$FM = \bigoplus_{1\le k\le\infty} \Lambda^{d_k} T^{(k)} M$$

and for every smooth map $f: M \to N$ we set

$$Ff = \bigoplus_{1\le k\le\infty} \Lambda^{d_k} T^{(k)} f: FM \to FN.$$

Since $\Lambda^{d_k} T^{(k)} M = M \times \{0\}$ whenever $k > \dim M$, the value FM is a well defined finite dimensional smooth manifold and Ff is a smooth map. The fiber projections on $T^{(k)}M$ yield a fibration of FM and all the axioms of bundle functors are easily verified. Since the order of $\Lambda^{d_k} T^{(k)}$ is at least k the functor F is of infinite order.

22.3. The order of bundle functors on $\mathcal{FM}_m$. Consider a bundle functor $F: \mathcal{FM}_m \to \mathcal{FM}$ and let F_n be its restriction to the subcategory $\mathcal{FM}_{m,n} \subset \mathcal{FM}_m$. Write S_n for the standard fibers of functors F_n and $s_n := \dim S_n$. We have proved that functors F_n have finite orders r_n bounded by the estimates given in theorem 22.1.

Theorem. *Let $F: \mathcal{FM}_m \to \mathcal{FM}$ be a bundle functor. Then for all fibered manifolds Y with n-dimensional fibers and for all fibered maps f, $g: Y \to \bar{Y}$, the condition $j_x^{r_n+1} f = j_x^{r_n+1} g$ implies $Ff|F_xY = Fg|F_xY$. If $\dim\bar{Y} \le \dim Y$, then even the equality of r_n-jets implies that the values on the corresponding fibers coincide.*

Proof. We may restrict ourselves to the case f, $g: \mathbb{R}^{m+n} \to \mathbb{R}^{m+k}$, $f(0) = g(0) = 0 \in \mathbb{R}^{m+k}$.

(a) First we discuss the case $n = k$. Let us assume $j_0^r f = j_0^r g$, $r = r_n$ and consider families $f_t = f + t\mathrm{id}_{\mathbb{R}^{m+n}}$, $g_t = g + t\mathrm{id}_{\mathbb{R}^{m+n}}$, $t \in \mathbb{R}$. The Jacobians at zero are certain polynomials in t, so that the maps f_t and g_t are local diffeomorphisms at zero except a finite number of values of t. Since $j_0^r f_t = j_0^r g_t$ for all t, we have $Ff_t|S_n = Fg_t|S_n$ except a finite number of values of t. Hence the regularity of F implies $Ff|S_n = Fg|S_n$.

Every fibered map $f \in \mathcal{FM}_m(\mathbb{R}^{m+n}, \mathbb{R}^{m+k})$ over $f_0: \mathbb{R}^m \to \mathbb{R}^m$ locally decomposes as $f = h \circ g$ where $g = f_0 \times \mathrm{id}_{\mathbb{R}^n}: \mathbb{R}^{m+n} \to \mathbb{R}^{m+n}$ and $h = f \circ g^{-1}$ is over the identity on $\mathbb{R}^m$. Hence in the rest of the proof we will restrict ourselves to morphisms over the identity.

(b) Next we assume $n = k + q$, $q > 0$, f, $g: \mathbb{R}^{m+k+q} \to \mathbb{R}^{m+k}$, and let $j_0^r f = j_0^r g$ with $r = r_n$. Consider $\bar{f} = (f, \mathrm{pr}_2)$, $\bar{g} = (g, \mathrm{pr}_2): \mathbb{R}^{m+n} \to \mathbb{R}^{m+n}$, where $\mathrm{pr}_2: \mathbb{R}^{m+k+q} \to \mathbb{R}^q$ is the projection onto the last factor. Since $j_0^r \bar{f} = j_0^r \bar{g}$, $f = \mathrm{pr}_1 \circ \bar{f}$ and $g = \mathrm{pr}_1 \circ \bar{g}$, the functoriality and (a) imply $Ff|S_n = Fg|S_n$.

(c) If $k = n + 1$ and if $j_0^r f = j_0^r g$ with $r = r_{n+1}$, then we consider $\bar{f}$, $\bar{g}: \mathbb{R}^{m+n+1} \to \mathbb{R}^{m+n+1}$ defined by $\bar{f} = f \circ \mathrm{pr}_1$, $\bar{g} = g \circ \mathrm{pr}_1$. Let us write

$i\colon \mathbb{R}^{m+n} \to \mathbb{R}^{m+n+1}$ for the inclusion $x \mapsto (x,0)$. For every $y \in \mathbb{R}^{m+n+1}$ with $\mathrm{pr}_1(y) = 0$ we have $j^r_y\bar{f} = j^r_y\bar{g}$ and since $f = \bar{f}\circ i$, $g = \bar{g}\circ i$, we get $Ff|S_n = Fg|S_n$.

(d) Let $k = n+q$, $q > 0$, and $i\colon \mathbb{R}^{m+n} \to \mathbb{R}^{m+n+q}$, $x \mapsto (x,0)$. Analogously to (a) we may assume that f and g have maximal rank at 0. Hence according to the canonical local form of maps of maximal rank we may assume $g = i$.

(e) Let us write $f = (\mathrm{id}_{\mathbb{R}^m}, f^1, \dots, f^k)\colon \mathbb{R}^{m+n} \to \mathbb{R}^{m+k}$, $k > n$, and assume $j^r_0 f = j^r_0 i$ with $r = r_{n+1}$. We define $h\colon \mathbb{R}^{m+n+1} \to \mathbb{R}^{m+k}$

$$h(x,y) = (\mathrm{id}_{\mathbb{R}^m}, f^1(x), \dots, f^n(x), y, f^{n+2}(x), \dots, f^k(x)).$$

Then we have

$$h \circ (\mathrm{id}_{\mathbb{R}^m}, \mathrm{id}_{\mathbb{R}^n}, f^{n+1}) = f$$
$$h \circ i = (\mathrm{id}_{\mathbb{R}^m}, f^1, \dots, f^n, 0, f^{n+2}, \dots, f^k).$$

Since $j^r_0(\mathrm{id}_{\mathbb{R}^m}, \mathrm{id}_{\mathbb{R}^n}, f^{n+1}) = j^r_0 i$, part (c) of this proof implies

$$F(\mathrm{id}_{\mathbb{R}^{m+n}}, f^{n+1})|S_n = Fi|S_n$$

and we get for every $z \in S_n$

$$Ff(z) = Fh \circ Fi(z) = F(\mathrm{id}_{\mathbb{R}^m}, f^1, \dots, f^n, 0, f^{n+2}, \dots, f^k)(z).$$

Now, we shall proceed by induction. Let us assume

$$Ff(z) = F(\mathrm{id}_{\mathbb{R}^m}, f^1, \dots, f^n, 0, \dots, 0, f^{n+s}, \dots, f^k)(z), \qquad s > 1,$$

for every $z \in S_n$ and $j_0^{r_{n+1}} f = j_0^{r_{n+1}} i$. Let $\sigma\colon \mathbb{R}^{m+n+k} \to \mathbb{R}^{m+n+k}$ be the map which exchanges the coordinates x^{n+1} and x^{n+s}, i.e.

$$\sigma(x, x^1, \dots, x^n, x^{n+1}, \dots, x^{n+s}, \dots, x^k) = \\ = (x, x^1, \dots, x^n, x^{n+s}, \dots, x^{n+1}, \dots, x^k).$$

We get

$$\begin{aligned} F(\mathrm{id}_{\mathbb{R}^m}, &f^1, \dots, f^n, 0, \dots, 0, f^{n+s}, \dots, f^k)(z) = \\ &= F(\sigma \circ (\mathrm{id}_{\mathbb{R}^m}, f^1, \dots, f^n, f^{n+s}, 0, \dots, 0, f^{n+s+1}, \dots, f^k))(z) \\ &= F\sigma \circ F(\mathrm{id}_{\mathbb{R}^m}, f^1, \dots, f^n, 0, \dots, 0, f^{n+s+1}, \dots, f^k)(z) \\ &= F(\mathrm{id}_{\mathbb{R}^m}, f^1, \dots, f^n, 0, \dots, 0, f^{n+s+1}, \dots, f^k)(z). \end{aligned}$$

So the induction yields $Ff(z) = F(\mathrm{id}_{\mathbb{R}^m}, f^1, \dots, f^n, 0, \dots, 0)$. Since we always have $r_{n+1} \ge r_n$, (a) implies

$$F(\mathrm{id}_{\mathbb{R}^m}, f^1, \dots, f^n)|S_n = F\,\mathrm{id}_{\mathbb{R}^{m+n}}|S_n.$$

Finally, we get

$$\begin{aligned} Ff|S_n &= F(\mathrm{id}_{\mathbb{R}^m}, f^1, \dots, f^n, 0, \dots, 0)|S_n \\ &= F(i \circ (\mathrm{id}_{\mathbb{R}^m}, f^1, \dots, f^n))|S_n = Fi|S_n. \qquad \square \end{aligned}$$

Theorem 22.3 reads that every bundle functor on $\mathcal{FM}_m$ is of locally finite order and we also have estimates on these 'local orders'. But there still remains an open question. Namely, all values on morphisms with an m-dimensional source manifold depend on r_{m+1}-jets. It is not clear whether one could get a better estimate.

23. The order of natural operators

In this section, we shall continue the general discussion on natural operators started in 18.16–18.20. Let us fix an admissible category $\mathcal{C}$ over manifolds, its local pointed skeleton $(C_\alpha, 0_\alpha)$, $\alpha \in I$, and consider bundle functors on $\mathcal{C}$.

23.1. The local order. We call a natural domain E of a natural operator $D\colon E \rightsquigarrow (G_1, G_2)$ *W-extendible* (or Whitney-extendible) if all the domains $E_A \subset C^\infty_{mA}(F_1A, F_2A)$, $A \in \mathrm{Ob}\mathcal{C}$, are W-extendible. We recall that the set of all sections of any fibration is W-extendible, so that the classical natural operators between natural bundles always have W-extendible domains.

Let us recall that we can apply corollary 19.8 to each q-local operator $D\colon E \subset C^\infty(Y_1, Y_2) \to C^\infty(Z_1, Z_2)$, where Y_1, Y_2, Z_1, Z_2 are smooth manifolds, $q\colon Z_1 \to Y_1$ is a surjective submersion and E is Whitney-extendible. In particular, D is of some order k, $0 \le k \le \infty$. Let us consider a mapping $s \in E$, $z \in Z_1$ and the compact set $K = \{z\} \subset Z_1$. According to 19.8 applied to K and s, there is the smallest possible order $r =: \chi(j^\infty s(q(z)), z) \in \mathbb{N}$ such that for all $\bar{s} \in E$ the condition $j^r s(q(z)) = j^r \bar{s}(q(z))$ implies $Ds(z) = D\bar{s}(z)$. Let us write $E^k \subset J^k(Y_1, Y_2)$ for the set of all k-jets of mappings from the domain E. The just defined mapping $\chi\colon E^\infty \times_{Y_1} Z_1 \to \mathbb{N}$ is called the *local order* of D.

For every π-local natural operator $D\colon E \rightsquigarrow (G_1, G_2)$ with a natural W-extendible domain E, the operators

$$D_A\colon E_A \subset C^\infty_{mA}(F_1A, F_2A) \to C^\infty_{mA}(G_1A, G_2A)$$

are π_A-local. The system of local orders $(\chi_A)_{A\in\mathrm{Ob}\mathcal{C}}$ is called the *local order* of the natural π-local operator D.

Every locally invertible $\mathcal{C}$-morphism $f\colon A \to B$ acts on $E^\infty_A \times_{F_1A} G_1A$ by

$$f^*(j^\infty_x s, z) = \big(j^\infty(F_2f \circ s \circ F_1f^{-1})(F_1f(x)), G_1f(z)\big).$$

Lemma. *Let $D\colon E \rightsquigarrow (G_1, G_2)$ be a natural operator with a natural Whitney-extendible domain E. For every locally invertible $\mathcal{C}$-morphism $f\colon A \to B$ and every $(j^\infty_x s, z) \in E^\infty_A \times_{F_1A} G_1A$ we have*

$$\chi_B\big(f^*(j^\infty_x s, z)\big) = \chi_A(j^\infty_x s, z).$$

Proof. Since $\mathcal{C}$ is admissible and the domain E is natural, we may restrict ourselves to $A = B = C_\alpha$, for some $\alpha \in I$. Assume $\chi_A(j^\infty_x s, z) = r$ and $j^r q(F_1f(x)) = j^r(F_2f \circ s \circ F_1f^{-1})(F_1f(x))$ for some $x \in F_1C_\alpha$ and $s, q \in E_{C_\alpha}$. Then $j^r((F_2f)^{-1} \circ q \circ F_1f)(x) = j^r s(x)$ and therefore $D_{C_\alpha}((F_2f)^{-1} \circ q \circ F_1f)(z) = D_{C_\alpha}s(z)$. We have locally for each $s \in E_{C_\alpha}$

$$s \circ F_1f^{-1} = F_2f^{-1} \circ (F_2f \circ s \circ F_1f^{-1})$$
$$D_{C_\alpha}s = G_2f^{-1} \circ D_{C_\alpha}(F_2f \circ s \circ F_1f^{-1}) \circ G_1f.$$

Hence $D_{C_\alpha}q(G_1f(z)) = G_2f \circ D_{C_\alpha}s(z)$ and we have proved $\chi_B \circ f^* \le \chi_A$. Applying the action of the inverse f^{-1} we get the converse inequality. □

23.2. Consider the associated maps

$$\mathcal{D}_{C_\alpha}: E^\infty_{C_\alpha} \times_{F_1 C_\alpha} G_1 C_\alpha \to G_2 C_\alpha$$

determined by a natural π-local operator D with a natural W-extendible domain E. We shall write briefly $E^r_\alpha \subset E^r_{C_\alpha}$ for the subset of jets with sources in the fiber S_α over 0_α in $F_1 C_\alpha$, and $Z_\alpha := \pi^{-1}_{C_\alpha}(S_\alpha) \subset G_1 C_\alpha$. By naturality, the whole operator D is determined by the restrictions

$$\mathcal{D}_\alpha: E^\infty_\alpha \times_{S_\alpha} Z_\alpha \to G_2 C_\alpha$$

of the maps $\mathcal{D}_{C_\alpha}$. Let us write $\chi_\alpha: E^\infty_\alpha \times_{S_\alpha} Z_\alpha \to \mathbb{N}$ for the restrictions of χ_{C_α}.

Lemma. *The maps χ_α are G^∞_α-invariant and if $\chi_\alpha \leq r$, then the operator D is of order r on all objects of type α.*

Proof. The lemma follows immediately from the definition of naturality, the homogeneity of category $\mathcal{C}$ and lemma 23.1. □

23.3. The above lemma suggests how to prove finiteness of the order in concrete situations. Namely, theorem 19.7 implies that 'locally' χ_α is bounded and so it must be bounded on each orbit under the action of G^∞_α. Assume now $F_1 = \mathrm{Id}_\mathcal{C}$, i.e. we deal with a natural p^{G_1}-local operator $D: E \rightsquigarrow (G_1, G_2)$ with a natural W-extendible domain ($E_A \subset C^\infty(FA)$). Then $E^\infty_\alpha \subset T^\infty_n Q_\alpha$, where $Q_\alpha = F_0 C_\alpha$ is the standard fiber and $n = \dim(mC_\alpha)$. Further assume that the category $\mathcal{C}$ is locally flat and that the bundle functors F and G_1 have the properties asserted in theorem 20.3 (so this always holds if $\mathcal{C}$ has almost W-extendible sets of morphisms). Consider a section $s \in E_{C_\alpha} \subset C^\infty(FC_\alpha)$ invariant with respect to all translations, i.e. $F(t_x) \circ s \circ t_{-x}(y) = s(y)$ for all $x \in mC_\alpha = \mathbb{R}^n$, $y \in C_\alpha$ and denote Z the standard fiber $(G_1)_0 C_\alpha$.

Lemma. *For every compact set $K \subset Z$ there is an order $r \in \mathbb{N}$ and a neighborhood $V \subset E^\infty_\alpha \subset T^\infty_n Q_\alpha$ of $j^\infty_{0_\alpha} s$ in the C^r-topology such that $\chi_\alpha \leq r$ on $V \times K$.*

Proof. Let us apply theorem 19.7 to the translation invariant section s and a compact set $K' \times K \subset C_\alpha \times Z = G_1 C_\alpha$, where K' is a compact neighborhood of $0_\alpha \in \mathbb{R}^n$. We get an order r and a smooth function $\varepsilon > 0$ except for finitely many points $y \in K'$ where $\varepsilon(y) = 0$. Let us fix x in the interior of K' with $\varepsilon(x) > 0$. Hence there is a neighborhood V of s in the C^r-topology on E_{C_α} and a neighborhood $U \subset C_\alpha$ of x in K' such that $\chi_U(j^\infty_y q, (y, z)) \leq r$ whenever $(y, z) \in U \times K$ and $q \in V$. Now, let W be a neighborhood of $j^\infty_0 s$ in C^r-topology on E^∞_α such that $t_x{}^* W$ is contained in the set of all infinite jets of sections from V. Since we might assume that t_x acts on $G_1 C_\alpha = \mathbb{R}^n \times Z$ by $G_1 t_x = t_x \times \mathrm{id}_Z$ and we have assumed $t_x{}^* s = s$, the lemma follows from lemma 23.2. □

Under the assumptions of 23.3 we get

23.4. Corollary. *Let $s \in E_{C_\alpha}$ be a translation invariant section and $K \subset Z$ a compact set. Assume that for every order $r \in \mathbb{N}$, every neighborhood V of $j^r_{0_\alpha} s$ in $E^r_\alpha \subset T^r_n Q_\alpha$, every relatively compact neighborhood K' of K and every couple $(j^r_{0_\alpha} q, z) \in E^r_\alpha \times Z_\alpha$ there is an element $g \in G^\infty_\alpha$ such that $g^*(j^r_{0_\alpha} q, z) \in V \times K'$. Then every natural operator in question has finite order on all objects of type α.*

Proof. For every relatively compact neighborhood K' of K, there is an $r \in \mathbb{N}$ and a neighborhood of $j^\infty_{0_\alpha} s$ in the C^r-topology such that $\chi_\alpha \leq r$ on $V \times K'$. But the assumptions of the corollary ensure that the orbit of $V \times K'$ coincides with the whole space $E^\infty_\alpha \times Z_\alpha$. □

Next we deduce several simple applications of this procedure.

23.5. Proposition. *Let $F\colon \mathcal{M}f_m \to \mathcal{FM}$ be a bundle functor of order r such that its standard fiber Q together with the induced action of $G^1_m \subset G^r_m$ can be identified with a linear subspace in a finite direct sum*

$$\bigoplus^{i}\Big(\bigotimes^{a_i}\mathbb{R}^m \otimes \bigotimes^{b_i}\mathbb{R}^{m*}\Big)$$

and $b_i > a_i$ for all i. Let $G_1 : \mathcal{M}f_m \to \mathcal{FM}$ be a bundle functor such that either its standard fiber Z together with the induced G^1_m-action can be identified with a linear subspace in a finite direct sum

$$\bigoplus^{j}\Big(\bigotimes^{a'_j}\mathbb{R}^m \otimes \bigotimes^{b'_j}\mathbb{R}^{m*}\Big)$$

and $b'_j > a'_j$ for all j, or Z is compact.

Then every natural operator $D\colon F \rightsquigarrow (G_1, G_2)$ defined on all sections of the bundles FM has finite order.

Proof. Write $\varphi_t\colon \mathbb{R}^m \to \mathbb{R}^m$, $t \in \mathbb{R}$, for the homotheties $x \mapsto tx$. Let us consider the canonical identification $F\mathbb{R}^m = \mathbb{R}^m \times Q$ and the zero section $s = (\mathrm{id}_{\mathbb{R}^m}, 0)$ in $C^\infty(F\mathbb{R}^m)$. Further, consider an arbitrary section $q\colon \mathbb{R}^m \to F\mathbb{R}^m$ and let us denote $q_t = F\varphi_t \circ q \circ \varphi_t^{-1}$ and $q_t(x) = (x, q^i_t(x))$. Under our identification, s is translation invariant and we can use formula 14.18.(2) to study the derivatives of the maps q^i_t at the origin. For all partial derivatives $\partial^\alpha q^i_t$ we get

$$\partial^\alpha q^i_t(0) = t^{a_i - b_i - |\alpha|}\partial^\alpha q^i(0). \tag{1}$$

If the standard fiber Z is compact, then we can use lemma 23.4 with $K = Z$ and the zero section s. Indeed, if we choose an order r and a neighborhood V of $j^r_0 s$ in $T^r_m Q$, then taking t large enough we obtain $j^r_0 q_t \in V$, so that the bound r is valid everywhere. But if Z is not compact, then an analogous equality to (1) holds for the sections of $G_1\mathbb{R}^m$ with $a_i - b_i$ replaced by $a'_j - b'_j$ and these are also negative. Hence we can apply the same procedure taking $K = \{0\}$, where 0 is the zero element in the tensor space Z. □

23.6. Examples. The assumptions of the proposition are satisfied by all tensor bundles with more covariant then contravariant components. But clearly, these are also satisfied for all affine natural bundles with associated natural vector bundles formed by the above tensor bundles. So in particular, F can equal to $QP^1: \mathcal{M}f_m \to \mathcal{FM}$, the bundle functor of elements of linear connections, cf. 17.7, or to the bundle functors of elements of exterior forms. If $G_1 = \mathrm{Id}_C$ then Z is a one-point-manifold, i.e. a compact. Hence we have proved that all natural operators on connections or on exterior forms that do not extend the bases have finite order.

23.7. Let us apply corollary 23.4 to the natural operators on the bundle functor $J^1: \mathcal{FM}_{m,n} \to \mathcal{FM}$, i.e. we want to derive the finiteness of the order for geometric operations with general connections. For this purpose, consider the maps $\varphi_{a,b}: \mathbb{R}^{m+n} \to \mathbb{R}^{m+n}$, $\varphi(x,y) = (ax, by)$. In words, we will use the inclusion $G^r_m \times G^r_n \hookrightarrow G^r_{m,n}$ and the jets of homotheties in the jet groups G^r_m and G^r_n. In canonical coordinates (x^i, y^p, y^p_i) on $J^1(\mathbb{R}^{m+n} \to \mathbb{R}^m)$, $i = 1, \dots, m$, $j = 1, \dots, n$, we get for every section $s = y^p_i(x^i, y^p)$ and every local fibered isomorphism $\varphi = (\varphi^i, \varphi^p)$

$$J^1\varphi \circ s \circ \varphi^{-1} = \left(\frac{\partial \varphi^p}{\partial x^j} \circ \varphi^{-1}\right) \frac{\partial \varphi_0^{-1}}{\partial x^i} + \left(\frac{\partial \varphi^p}{\partial y^q} \circ \varphi^{-1}\right) (y^q_j \circ \varphi^{-1}) \frac{\partial \varphi_0^{-1}}{\partial x^i}.$$

In particular, for $\varphi = \varphi_{a,b}$ we obtain

$$\varphi_{a,b}{}^* s(x^i, y^p) = ba^{-1} y^p_i \circ \varphi^{-1}_{a,b}.$$

Hence for every multi index $\alpha = \alpha_1 + \alpha_2$, where α_1 includes all the derivatives with respect to the indices i while α_2 those with respect to p's, it holds

$$\partial^{\alpha_1+\alpha_2}(\varphi_{a,b}{}^* s)(0) = a^{-1-|\alpha_1|} b^{1-|\alpha_2|} \partial^{\alpha_1+\alpha_2} s(0). \tag{1}$$

Proposition. *Let $H: \mathcal{FM}_{m,n} \to \mathcal{FM}$ be an arbitrary bundle functor while $G: \mathcal{FM}_{m,n} \to \mathcal{FM}$ is either the identity functor or the functor J^1 or the vertical tangent bundle V. Then every natural operator $D: J^1 \rightsquigarrow (G, H)$ defined on all sections of the first jet prolongations has finite order.*

Proof. If $G = \mathrm{Id}_{\mathcal{FM}_{m,n}}$, then we can take $b = 1$, $a > 0$ and corollary 23.4 together with (1) imply the assertion. The same choice of a and b leads also to the case $G = J^1$, for $J^1\varphi_{a,b}(y^p_i) = a^{-1} y^p_i$ on the standard fiber over $0 \in \mathbb{R}^{m+n}$.

In the third case we have to be more careful. On the standard fiber $\mathbb{R}^n$ of $V\mathbb{R}^{m+n}$ we have $V\varphi_{a,b}(\xi^p) = (b\xi^p)$. Let us fix some $r \in \mathbb{N}$ and choose $a = b^{-r}$, $0 < b < 1$ arbitrary. Then

$$|\partial^{\alpha_1+\alpha_2}(\varphi_{a,b}{}^* s)(0)| = b^{r(1+|\alpha_1|)+1-|\alpha_2|} |\partial^{\alpha_1+\alpha_2} s(0)|$$

and so for all $|\alpha| \le r$ we get

$$|\partial^{\alpha}(\varphi_{a,b}{}^* s)(0)| \le b |\partial^{\alpha} s(0)|.$$

Hence also in this case corollary 23.4 implies our assertion. □

At the end of this section, we illustrate on two examples how bad things may be. First we construct a natural operator which essentially depends on infinite jets and the next example presents a non-regular natural operator. This contrasts the results on bundle functors where the regularity follows from the other axioms.

23.8. Example. Consider the bundle functor $F = T \oplus T^*: \mathcal{M}f_m \to \mathcal{M}f$ and let G be the bundle functor defined by $GM = M \times \mathbb{R}$, $Gf = f \times \mathrm{id}_{\mathbb{R}}$, for all m-dimensional manifolds M and local diffeomorphisms f, i.e. 'the bundle of real functions'. The contraction defines a natural function, i.e. a natural operator $F \rightsquigarrow G$, of order zero. The composition with any fixed real function $\mathbb{R} \to \mathbb{R}$ is a natural transformation $G \to G$ and also the addition $G \oplus G \xrightarrow{+} G$ and multiplication $G \oplus G \xrightarrow{\cdot} G$ are natural transformations. Moreover, there is the exterior differential $d: G \rightsquigarrow T^*$, a natural operator of order 1.

By induction, let us define operators $D_k: T \oplus T^* \to G$. We set

$$(D_0)_M(X, \omega) = i_X\omega \quad \text{and} \quad (D_{k+1})_M(X, \omega) = i_X\big(d((D_k)_M(X, \omega))\big)$$

for $k = 0, 1, \dots$. Further, consider a smooth function $a: \mathbb{R}^2 \to \mathbb{R}$ satisfying $a(t, x) \neq 0$ if and only if $|x| > t > 0$. We define

$$D_M(X, \omega) = \sum_{k=0}^{\infty} \big(a(k, -) \circ (i_X\omega)\big).\big((D_k)_M(X, \omega)\big).$$

Since the sum is locally finite for every $(X, \omega) \in C^\infty(FM)$, this is a natural operator of infinite order.

23.9. Example. Consider once more the bundle functors F, G and operators D_k from example 23.8. Let a and $g: \mathbb{R}^2 \to \mathbb{R}$ be the functions used in 19.15. We shall modify operator D from example 19.15 to get a non-regular natural operator. Let us define operators $D_M: C^\infty(FM) \rightsquigarrow C^\infty(GM)$ by

$$D_M(X, \omega) = \sum_{k=0}^{\infty} \big(a(k, -) \circ g \circ ((i_X\omega) \times (i_X d(i_X\omega)))\big).\big((D_k)_M(X, \omega)\big)$$

for all $(X, \omega) \in C^\infty(T \oplus T^*M)$. We have used only natural operators in our construction, but, unfortunately, the values $D_M(X, \omega)$ need not be smooth (or even defined) if dimension m is greater then one. This is caused by the infinite value of $\limsup_{x \to (0,1)} g(x)$. But if $m = 1$, then all values are smooth and the system D_M satisfies all axioms of natural operators except the regularity. Indeed, it suffices to verify the smoothness of the values of $D_{\mathbb{R}}$. But if $(i_X\omega)(t_0) = 0$ and $(i_X d(i_X\omega))(t_0) = 1$, i.e. $X(t_0)\frac{d}{dx}(X\omega)(t_0) = 1$, then $\frac{d}{dx}(X\omega)(t_0) \neq 0$ and therefore the curve

$$t \mapsto \Big((X\omega)(t), X\frac{d}{dx}(X\omega)(t)\Big)$$

lies on a neighborhood of t_0 inside the unit circles centered in $(-1, 1)$ and $(1, 1)$. Hence $D_{\mathbb{R}}(X, \omega) = 0$ on some neighborhood of t_0.

Let us note that our operator D is not only non-regular, but also of infinite order and it shows that the assertion of lemma 23.3 does not hold for all maps $s \in \mathcal{D}_{C_\alpha}$, in general. A non-regular natural operator of order 4 on Riemannian metrics for dimension $m = 2$ can be found in [Epstein, 75].

23.10. If we consider natural operators $D\colon F \rightsquigarrow (G_1, G_2)$ with domains formed by all sections of the bundles $FM \to M$, then we can use the regularity of D and apply the stronger version of nonlinear Peetre theorem 19.10 instead of 19.7 in the proof of 23.3. Hence we do not need the invariance of the section s. Consequently, the assertion of lemma 23.3 holds for all sections $s \in E_{C_\alpha}$. That is why, under the assumptions of corollary 23.4 we can strengthen its assertion.

Corollary. *Let $s \in C^\infty(FC_\alpha)$ be a section and $K \subset Z$ be a compact set. Assume that for every order $r \in \mathbb{N}$, every neighborhood V of $j^r_{0_\alpha} s$ in $E^r_\alpha \subset T^r_n Q_\alpha$, every relatively compact neighborhood K' of K and every couple $(j^r_{0_\alpha} q, z) \in E^r_\alpha \times Z_\alpha$ there is an element $g \in G^\infty_\alpha$ such that $g^*(j^r_{0_\alpha} q, z) \in V \times K'$. Then every natural operator $D\colon F \rightsquigarrow (G_1, G_2)$ has a finite order on all objects of type α.*

Remarks

The general setting for bundle functors and natural operators extends the original categorical approach to geometric objects and operators due to [Nijenhuis, 72] and we follow mainly [Kolář, 90] and partially [Slovák, 91].

The multilinear version of Peetre theorem, proved in [Cahen, De Wilde, Gutt, 80], seems to be the first non-linear generalization of the famous Peetre theorem, [Peetre, 60]. The study of general nonlinear operators started in [Chrastina, 87] and [Slovák, 87b]. The original aim of the nonlinear version 19.7, first proved in [Slovák, 87b], was the reduction of the problem of finding natural operators to a finite order. The pure analytical results were further generalized and completed in a setting of Hölder-continuous maps and metric spaces in [Slovák, 88] and it became clear that they should help to unify the approach to the finiteness of the orders of both natural operators and bundle functors and to avoid the original manipulation with infinite dimensional Lie algebras, see [Palais, Terng, 77]. Let us remark that nearly all categories over manifolds used in differential geometry are admissible and locally flat, however the verification of the Whitney extendibility might present a serious analytical problem in concrete examples. In the most technical part of the description of bundle functors, i.e. in the proof of the regularity, we mainly follow [Mikulski, 85] which generalizes the original proof due to [Epstein, Thurston, 79] to natural bundles with infinite dimensional values. Let us point out that our proof also applies to continuous regularity of bundle functors on the categories in question with values in infinite dimensional manifolds.

Our sharp estimate on the orders of jet groups acting on manifolds is a generalization of [Zajtz, 87], where similar results are obtained for the full group G^r_m.

The results on the order of bundle functors on $\mathcal{FM}_m$ follow some ideas from [Kolář, Slovák, 89] and [Mikulski, 89 a, b]. The methods used in our discussion on the order of natural operators never exploit the regularity of the natural operators which we have incorporated into our definition. So the results of section 23 can be applied to non-regular natural operators which can also be classified in some concrete situations.

CHAPTER VI.
METHODS FOR FINDING NATURAL OPERATORS

We present certain general procedures useful for finding some equivariant maps and we clarify their application by solving concrete geometric problems. The equivariance with respect to the homotheties in $GL(m)$ gives frequently a homogeneity condition. The homogeneous function theorem reads that under certain assumptions a globally defined smooth homogeneous function must be polynomial. In such a case the use of the invariant tensor theorem and the polarization technique can specify the form of the polynomial equivariant map up to such an extend, that all equivariant maps can then be determined by direct evaluation of the equivariance condition with respect to the kernel of the jet projection $G^r_m \to G^1_m$. We first deduce in such a way that all natural operators transforming linear connections into linear connections form a simple 3-parameter family. Then we strengthen a classical result by Palais, who deduced that all linear natural operators $\Lambda^p T^* \to \Lambda^{p+1}T^*$ are the constant multiples of the exterior derivative. We prove that for $p > 0$ even linearity follows from naturality. We underline, as a typical feature of our procedures, that in both cases we first have guaranteed by the results from chapter V that the natural operators in question have finite order. Then the homogeneous function theorem implies that the natural operators have zero order in the first case and first order in the second case. In section 26 we develop the smooth version of the tensor evaluation theorem. As the first application we determine all natural transformations $TT^* \to T^*T$. The result implies that, unlike to the case of cotangent bundle, there is no natural symplectic structure on the tangent bundle.

As an example of a natural operator related with fibered manifolds we discuss the curvature of a general connection. An important tool here is the generalized invariant tensor theorem, which describes all $GL(m) \times GL(n)$-invariant tensors. We deduce that all natural operators of the curvature type are the constant multiples of the curvature and that all such operators on a pair of connections are linear combinations of the curvatures of the individual connections and of the so-called mixed curvature of both connections. The next section is devoted to the orbit reduction. We develop a complete version of the classical reduction theorem for linear symmetric connections and Riemannian metrics, in which the factorization procedure is described in terms of the curvature spaces and the Ricci spaces. The so-called method of differential equations is based on the simple fact that on the Lie algebra level the equivariance condition represents a system of partial differential equations. As an example we deduce that the only first order natural operator transforming Riemannian metrics into linear connections is the Levi-Cività operator. But we apply the method of differential equations only in the first part of the proof, while in the final step a direct geometric consideration is used.

24. Polynomial $GL(V)$-equivariant maps

24.1. We first deduce a result on the globally defined smooth homogeneous functions, which is useful in the theory of natural operators.

Consider a product $V_1 \times \ldots \times V_n$ of finite dimensional vector spaces. Write $x_i \in V_i$, $i = 1, \ldots, n$.

Homogeneous function theorem. *Let $f(x_1, \ldots, x_n)$ be a smooth function defined on $V_1 \times \ldots \times V_n$ and let $a_i > 0$, b be real numbers such that*

$$k^b f(x_1, \ldots, x_n) = f(k^{a_1} x_1, \ldots, k^{a_n} x_n) \tag{1}$$

holds for every real number $k > 0$. Then f is a sum of the polynomials of degree d_i in x_i satisfying the relation

$$a_1 d_1 + \cdots + a_n d_n = b. \tag{2}$$

If there are no non-negative integers $d_1, \ldots, d_n$ with the property (2), then f is the zero function.

Proof. First we remark that if f satisfies (1) with $b < 0$, then f is the zero function. Indeed, if there were $f(x_1, \ldots, x_n) \neq 0$, then the limit of the right-hand side of (1) for $k \to 0_+$ would be $f(0, \ldots, 0)$, while the limit of the left-hand side would be improper.

In the case $b \geq 0$ we write $a = \min(a_1, \ldots, a_n)$ and $r = \left[\frac{b}{a}\right]$(=the integer part of the ratio $\frac{b}{a}$). Consider some linear coordinates x^{j_i} on each V_i. We claim that all partial derivatives of the order $r+1$ of every function f satisfying (1) vanish identically. Differentiating (1) with respect to x^{j_i}, we obtain

$$k^b \frac{\partial f(x_1, \ldots, x_n)}{\partial x^{j_i}} = k^{a_i} \frac{\partial f(k^{a_1} x_1, \ldots, k^{a_n} x_n)}{\partial x^{j_i}}.$$

Hence for $\frac{\partial f}{\partial x^{j_i}}$ we have (1) with b replaced by $b - a_i$. This implies that every partial derivative of the order $r+1$ of f satisfies (1) with a negative exponent on the left-hand side, so that it is the zero function by the above remark.

Since all the partial derivatives of f of order $r+1$ vanish identically, the remainder in the r-th order Taylor expansion of f at the origin vanishes identically as well, so that f is a polynomial of order at most r. For every monomial $x_1^{\alpha_1} \ldots x_n^{\alpha_n}$ of degree $|\alpha_i|$ in x_i, we have

$$(k^{a_1} x_1)^{\alpha_1} \ldots (k^{a_n} x_n)^{\alpha_n} = k^{a_1|\alpha_1| + \cdots + a_n|\alpha_n|} x_1^{\alpha_1} \ldots x_n^{\alpha_n}.$$

Since k is an arbitrary positive real number, a non-zero polynomial satisfies (1) if and only if (2) holds. □

24.2. Remark. The assumption $a_i > 0$, $i = 1, \dots, n$ in the homogeneous function theorem is essential. We shall see in section 26 that e.g. all smooth functions $f(x,y)$ of two independent variables satisfying $f(kx, k^{-1}y) = f(x,y)$ for all $k \neq 0$ are of the form $\varphi(xy)$, where $\varphi(t)$ is any smooth function of one variable. In this case we have $a_1 = 1$, $a_2 = -1$, $b = 0$.

24.3. Invariant tensors. Consider a finite dimensional vector space V with a linear action of a group G. The induced action of G on the dual space V^* is given by

$$\langle av^*, v\rangle = \langle v^*, a^{-1}v\rangle$$

for all $v \in V$, $v^* \in V^*$, $a \in G$. In any linear coordinates, if $av = (a^i_j v^j)$, then $av^* = (\tilde{a}^j_i v^*_j)$, where $\tilde{a}^i_j$ denotes the inverse matrix to a^i_j. Moreover, if we have some linear actions of G on vector spaces $V_1, \dots, V_n$, then there is a unique linear action of G on the tensor product $V_1 \otimes \dots \otimes V_n$ satisfying $g(v_1 \otimes \dots \otimes v_n) = (gv_1) \otimes \dots \otimes (gv_n)$ for all $v_1 \in V_1, \dots, v_n \in V_n$, $g \in G$. The latter action is called the *tensor product* of the original actions.

In particular, every tensor product $\otimes^r V \otimes \otimes^q V^*$ is considered as a $GL(V)$-space with respect to the tensor product of the canonical action of $GL(V)$ on V and the induced action of $GL(V)$ on V^*.

Definition. A tensor $B \in \otimes^r V \otimes \otimes^q V^*$ is said to be invariant, if $aB = B$ for all $a \in GL(V)$.

The invariance of B with respect to the homotheties in $GL(V)$ yields $k^{r-q}B = B$ for all $k \in \mathbb{R} \setminus \{0\}$. This implies that for $r \neq q$ the only invariant tensor is the zero tensor. An invariant tensor from $\otimes^r V \otimes \otimes^r V^*$ will be called an *invariant tensor of degree r*. For every s from the group S_r of all permutations of r letters we define $I^s \in \otimes^r V \otimes \otimes^r V^*$ to be the result of the permutation s of the superscripts of

$$I^{\mathrm{id}} = \mathrm{id}_V \underbrace{\otimes \cdots \otimes}_{r\text{-times}} \mathrm{id}_V. \tag{1}$$

In coordinates, $I^s = (\delta^{i_{s(1)}}_{j_1} \dots \delta^{i_{s(r)}}_{j_r})$. The tensors I^s, which are clearly invariant, are called the *elementary invariant tensors of degree r*. Obviously, if we replace the permutation of superscripts in (1) by the permutation of subscripts, we obtain the same collection of the elementary invariant tensors of degree r.

24.4. Invariant tensor theorem. *Every invariant tensor B of degree r is a linear combination of the elementary invariant tensors of degree r.*

Proof. The condition for $B = (b^{i_1\dots i_r}_{j_1\dots j_r}) \in \otimes^r \mathbb{R}^m \otimes \otimes^r \mathbb{R}^{m*}$ to be invariant reads

$$a^{i_1}_{k_1} \dots a^{i_r}_{k_r} b^{k_1\dots k_r}_{l_1\dots l_r} = b^{i_1\dots i_r}_{j_1\dots j_r} a^{j_1}_{l_1} \dots a^{j_r}_{l_r} \tag{1}$$

for all $a^i_j \in GL(m)$. To delete the a's, we rewrite (1) as

$$a^{j_1}_{k_1} \dots a^{j_r}_{k_r} \delta^{i_1}_{j_1} \dots \delta^{i_r}_{j_r} b^{k_1\dots k_r}_{l_1\dots l_r} = b^{i_1\dots i_r}_{j_1\dots j_r} \delta^{k_1}_{l_1} \dots \delta^{k_r}_{l_r} a^{j_1}_{k_1} \dots a^{j_r}_{k_r}.$$

Comparing the coefficients by the individual monomials in a^i_j, we obtain the following equivalent form of (1)

$$\sum_{s\in S_r} \delta^{i_1}_{j_{s(1)}} \dots \delta^{i_r}_{j_{s(r)}} b^{k_{s(1)}\dots k_{s(r)}}_{l_1\dots l_r} = \sum_{s\in S_r} \delta^{k_{s(1)}}_{l_1} \dots \delta^{k_{s(r)}}_{l_r} b^{i_1\dots i_r}_{j_{s(1)}\dots j_{s(r)}}. \tag{2}$$

The case $r \le m$ is very simple. Set $c_s = b^{1\dots r}_{s(1)\dots s(r)}$. If we put $i_1 = 1,\dots, i_r = r$, $j_1 = 1,\dots, j_r = r$ in (2), then the only non-zero term on the left-hand side corresponds to $s = \mathrm{id}$. This yields

$$b^{k_1\dots k_r}_{l_1\dots l_r} = \sum_{s\in S_r} c_s \delta^{k_{s(1)}}_{l_1} \dots \delta^{k_{s(r)}}_{l_r} \tag{3}$$

which is the coordinate form of our theorem.

For $r > m$ we have to use a more complicated procedure (due to [Gurevich, 48]). In this case, the coefficients c_s in (3) are not uniquely determined. This follows from the fact that for $r > m$ the system of m^{2r} equations in $r!$ variables z_s

$$\sum_{s\in S_r} \delta^{i_1}_{j_{s(1)}} \dots \delta^{i_r}_{j_{s(r)}} z_s = 0 \tag{4}$$

has non-zero solutions. Indeed, in this case e.g. every tensor

$$c\delta^{i_1}_{[j_1} \dots \delta^{i_{m+1}}_{j_{m+1}]} \delta^{i_{m+2}}_{j_{m+2}} \dots \delta^{i_r}_{j_r} \tag{5}$$

(where the square bracket denotes alternation) is the zero tensor, since among every $j_1, \dots, j_{m+1}$ at least two indices coincide. Hence (5) expresses the zero tensor as a non-trivial linear combination of the elementary invariant tensors.

Let z^α_s, $\alpha = 1,\dots,q$ be a basis of the solutions of (4). Consider the linear equations

$$\sum_{s\in S_r} z^\alpha_s z_s = 0 \qquad \alpha = 1,\dots,q. \tag{6}$$

To deduce that the rank of the system (4) and (6) is $r!$, it suffices to prove that this system has the zero solution only. Let z^0_s be a solution of (4) and (6). Since z^0_s satisfy (4), there are $k_\alpha \in \mathbb{R}$ such that

$$z^0_s = \sum_{\alpha=1}^{q} k_\alpha z^\alpha_s. \tag{7}$$

Since z^0_s satisfy (6) as well, they annihilate the linear combination

$$\sum_{\alpha=1}^{q} k_\alpha \Big(\sum_{s\in S_r} z^\alpha_s z^0_s\Big) = 0.$$

By (7) the latter relation means $\sum_{s\in S_r}(z_s^0)^2 = 0$, so that all z_s^0 vanish.

In this situation, we can formulate a lemma:

Let $r!$ tensors $X_s \in \otimes^r\mathbb{R}^m \otimes \otimes^r\mathbb{R}^{m*}$, $s \in S_r$, satisfy the equations

$$\sum_{s\in S_r} \delta^{i_1}_{j_{s(1)}} \dots \delta^{i_r}_{j_{s(r)}} X_s = \sum_{s\in S_r} c^{i_1\dots i_r}_{j_{s(1)}\dots j_{s(r)}} I^s \tag{8}$$

with some real coefficients $c^{i_1\dots i_r}_{j_{s(1)}\dots j_{s(r)}}$ and

$$\sum_{s\in S_r} z_s^\alpha X_s = 0 \qquad \alpha = 1,\dots,q \tag{9}$$

Then every X_s is a linear combination of the elementary invariant tensors.

Indeed, since the system (4) and (6) has rank $r!$ and the equations (6) are linearly independent, there is a subsystem (4') in (4) such that the system (4') and (6) has non-zero determinant. Let (8') be the subsystem in (8) corresponding to (4'). Then we can apply the Cramer rule for modules to the system (8') and (9). This yields that every X_s is a linear combination of the right-hand sides, which are linearly generated by the elementary invariant tensors.

Now we can complete the proof of our theorem. Let B be an invariant tensor and B^s be the result of permutation s on its superscripts. Then (2) can be rewritten as

$$\sum_{s\in S_r} \delta^{i_1}_{j_{s(1)}} \dots \delta^{i_r}_{j_{s(r)}} B^s = \sum_{s\in S_r} b^{i_1\dots i_r}_{j_{s(1)}\dots j_{s(r)}} I^s. \tag{10}$$

Contract the zero tensor $\sum_{s\in S_r} \delta^{i_1}_{j_{s(1)}} \dots \delta^{i_r}_{j_{s(r)}} z_s^\alpha$, $\alpha = 1,\dots,q$, with undetermined $x_{j_1\dots j_r}$. This yields the algebraic relations

$$\sum_{s\in S_r} z_s^\alpha x_{i_{s(1)}\dots i_{s(r)}} = 0. \tag{11}$$

In particular, for $x_{i_1\dots i_r} = b^{i_1\dots i_r}_{j_1\dots j_r}$ with parameters $j_1,\dots,j_r$ we obtain

$$\sum_{s\in S_r} z_s^\alpha B^s = 0 \qquad \alpha = 1,\dots,q. \tag{12}$$

Applying the above lemma to (10) and (12) we deduce that B is a linear combination of the elementary invariant tensors. □

24.5. Remark. The invariant tensor theorem follows directly from the classification of all relative invariants of $GL(m,\Omega)$ with p vectors in Ω^m and q covectors in Ω^{m*} given in section 2.7 of [Dieudonné, Carrell, 71], p. 29. But Ω is assumed to be an algebraically closed field there and the complexification procedure is rather technical in this case. That is why we decided to present a more elementary proof, which fits better to the main line of our book.

24.6. Having two vector spaces V and W, there is a canonical bijection between the linear maps $f\colon V \to W$ and the elements $f^{\otimes} \in W \otimes V^*$ given by $f(v) = \langle f^{\otimes}, v\rangle$ for all $v \in V$. The following assertion is a direct consequence of the definition.

Proposition. *A linear map $f\colon \otimes^p V \otimes \otimes^q V^* \to \otimes^r V \otimes \otimes^t V^*$ is $GL(V)$-equivariant if and only if $f^{\otimes} \in \otimes^{r+q} V \otimes \otimes^{p+t} V^*$ is an invariant tensor.*

24.7. In several cases we can combine the use of the homogeneous function theorem and the invariant tensor theorem to deduce all smooth $GL(V)$-equivariant maps of certain types. As an example we determine all smooth $GL(V)$-equivariant maps of $\otimes^r V$ into itself. Having such a map $f\colon \otimes^r V \to \otimes^r V$, the equivariance with respect to the homotheties in $GL(V)$ gives $k^r f(x) = f(k^r x)$. Since the only solution of $rd = r$ is $d = 1$, the homogeneous function theorem implies f is linear. Then the invariant tensor theorem and 24.6 yield that *all smooth $GL(V)$-equivariant maps $\otimes^r V \to \otimes^r V$ are the linear combinations of the permutations of indices.*

24.8. If we study the symmetric and antisymmetric tensor powers, we can apply the invariant tensor theorem when taking into account that the tensor symmetrization $\mathrm{Sym}\colon \otimes^r V \to S^r V$ and alternation $\mathrm{Alt}\colon \otimes^r V \to \Lambda^r V$ as well as the inclusions $S^r V \hookrightarrow \otimes^r V$ and $\Lambda^r V \hookrightarrow \otimes^r V$ are equivariant maps. We determine in such a way all smooth $GL(V)$-equivariant maps $S^r V \to S^r V$. Consider the diagram

$$\begin{array}{ccc} S^r V & \xrightarrow{\ f\ } & S^r V \\ {\scriptstyle \mathrm{Sym}}\big\uparrow\big\downarrow{\scriptstyle i} & & {\scriptstyle i}\big\downarrow\big\uparrow{\scriptstyle \mathrm{Sym}} \\ \otimes^r V & \xrightarrow{\ \varphi\ } & \otimes^r V \end{array}$$

Then $\varphi = i \circ f \circ \mathrm{Sym}\colon \otimes^r V \to \otimes^r V$ is an equivariant map and it holds $f = \mathrm{Sym} \circ \varphi \circ i$. Using 24.7, we deduce

(1) all smooth $GL(V)$-maps $S^r V \to S^r V$ are the constant multiples of the identity.

Quite similarly one obtains the following simple assertions.

All smooth $GL(V)$-maps

(2) $\Lambda^r V \to \Lambda^r V$ are the constant multiples of the identity,

(3) $\otimes^r V \to S^r V$ are the constant multiples of the symmetrization,

(4) $\otimes^r V \to \Lambda^r V$ are the constant multiples of the alternation,

(5) $S^r V \to \otimes^r V$ and $\Lambda^r V \to \otimes^r V$ are the constant multiples of the inclusion.

24.9. In the next section we shall need all smooth $GL(m)$-equivariant maps of $\mathbb{R}^m \otimes \mathbb{R}^{m*} \otimes \mathbb{R}^{m*}$ into itself. Let $f^i_{jk}(x^l_{mn})$ be the components of such a map f. Consider first the homotheties $\frac{1}{k}\delta^i_j$ in $GL(m)$. The equivariance of f with respect to these homotheties yields $kf(x) = f(kx)$. By the homogeneous function theorem, f is a linear map. The corresponding tensor $f^{\otimes}$ is invariant in $\otimes^3 \mathbb{R}^m \otimes \otimes^3 \mathbb{R}^{m*}$. Hence $f^{\otimes}$ is a linear combination of all six permutations of the tensor products of the identity maps, i.e.

$$f^i_{jk} = (a_1\delta^i_j\delta^m_k\delta^n_l + a_2\delta^i_j\delta^m_l\delta^n_k + a_3\delta^i_k\delta^m_j\delta^n_l + a_4\delta^i_k\delta^m_l\delta^n_j + a_5\delta^i_l\delta^m_j\delta^n_k + a_6\delta^i_l\delta^m_k\delta^n_j)x^l_{mn}$$

$a_1, \dots, a_6 \in \mathbb{R}$. Thus, all smooth $GL(m)$-maps of $\mathbb{R}^m \otimes \mathbb{R}^{m*} \otimes \mathbb{R}^{m*}$ into itself form the following 6-parameter family

$$f^i_{jk} = a_1\delta^i_j x^l_{kl} + a_2\delta^i_j x^l_{lk} + a_3\delta^i_k x^l_{jl} + a_4\delta^i_k x^l_{lj} + a_5 x^i_{jk} + a_6 x^i_{kj}.$$

24.10. The invariant tensor theorem can be used for finding the polynomial equivariant maps, if we add the standard polarization technique. We present the basic general facts according to [Dieudonné, Carrell, 71].

Let V and W be two finite dimensional vector spaces. A map $f\colon V \to W$ is called polynomial, if in its coordinate expression

$$f(x^i v_i) = f^p(x^i)w_p$$

in a basis (v_i) of V and a basis (w_p) of W the functions $f^p(x^i)$ are polynomial. One sees directly that such a definition does not depend on the choice of both bases.

We recall that for a multi index $\alpha = (\alpha_1, \dots, \alpha_m)$ of range $m = \dim V$ we write

$$x^\alpha = (x^1)^{\alpha_1} \dots (x^m)^{\alpha_m}.$$

The degree of monomial x^α is $|\alpha|$. A linear combination of the monomials of the same degree r is called a homogeneous polynomial of degree r. Every polynomial map $f\colon V \to W$ is uniquely decomposed into the homogeneous components

$$f = f_0 + f_1 + \dots + f_r.$$

Consider a group G acting linearly on both V and W.

Proposition. *Each homogeneous component of an equivariant polynomial map $f\colon V \to W$ is also equivariant.*

Proof. This follows directly from the fact that the actions of G on both V and W are linear. □

24.11. In the same way one introduces the notion of a polynomial map

$$f\colon V_1 \times \dots \times V_n \to W$$

of a finite product of finite dimensional vector spaces into W. Let $x_i \in V_i$ and α_i be a multi index of range $m_i = \dim V_i$, $i = 1, \dots, n$. A monomial

$$x_1^{\alpha_1} \dots x_n^{\alpha_n}$$

is said to be of degree $(|\alpha_1|, \dots, |\alpha_n|)$. The multihomogeneous component $f_{(r_1,\dots,r_n)}$ of degree $(r_1, \dots, r_n)$ of a polynomial map $f\colon V_1 \times \dots \times V_n \to W$ consists of all monomials of this degree in f.

Having a group G acting linearly on all $V_1, \dots, V_n$ and W, one deduces quite similarly to 24.10

Proposition. *Each multihomogeneous component of an equivariant polynomial map $f: V_1 \times \ldots \times V_n \to W$ is also equivariant.*

24.12. Let $f: V \to \mathbb{R}$ be a homogeneous polynomial of degree r. Its first polarization $P_1 f: V \times V \to \mathbb{R}$ is defined as the coefficient by t in Taylor's formula

$$f(x+ty) = f(x) + t\, P_1 f(x,y) + \cdots \tag{1}$$

The coordinate expression of $P_1 f(x,y)$ is $\frac{\partial f}{\partial x^i} y^i$. Since f is homogeneous of degree r, Euler's theorem implies

$$P_1 f(x,x) = r f(x).$$

The second polarization $P_2 f(x, y_1, y_2): V \times V \times V \to \mathbb{R}$ is defined as the first polarization of $P_1 f(x, y_1)$ with fixed values of y_1. By induction, the i-th polarization $P_i f(x, y_1, \ldots, y_i)$ of f is the first polarization of $P_{i-1} f(x, y_1, \ldots, y_{i-1})$ with fixed values of $y_1, \ldots, y_{i-1}$. Obviously, the r-th polarization $P_r f$ is independent on x and is linear and symmetric in $y_1, \ldots, y_r$. The induced linear map $Pf: S^r V \to \mathbb{R}$ is called the *total polarization of f*. An iterated application of the Euler formula gives

$$r!\, f(x) = Pf(\underbrace{x \otimes \cdots \otimes x}_{r\text{-times}}).$$

The concept of polarization is extended to a homogeneous polynomial map $f: V \to W$ of degree r by applying this procedure to each component of f with respect to a basis of W. Thus, the i-th polarization of f is a map $P_i f: \overset{i+1}{\times} V \to W$ and the total polarization of f is a linear map $Pf: S^r V \to W$. Let a group G act linearly on both V and W.

Proposition. *If $f: V \to W$ is an equivariant homogeneous polynomial map of degree r, then every polarization $P_i f: \overset{i+1}{\times} V \to W$ as well as the total polarization Pf are also equivariant.*

Proof. The first polarization is given by formula 24.12.(1). Since f is equivariant, we have $f(gx + tgy) = gf(x+ty)$ for all $g \in G$. Then 24.12.(1) implies $g\, P_1 f(x,y) = P_1 f(gx, gy)$. By iteration we deduce the same result for the i-th polarization. The equivariance of the r-th polarization implies the equivariance of the total polarization. □

24.13. The same construction can be applied to a multihomogeneous polynomial map $f: V_1 \times \ldots \times V_n \to W$ of degree $(r_1, \ldots, r_n)$. For any $(i_1, \ldots, i_n)$, $i_1 \leq r_1, \ldots, i_n \leq r_n$, we define the multipolarization $P_{(i_1, \ldots, i_n)} f$ of type $(i_1, \ldots, i_n)$ by constructing the corresponding polarization of f in each component separately. Hence

$$P_{(i_1,\ldots,i_n)} f: \overset{i_1+1}{\times} V_1 \times \ldots \times \overset{i_n+1}{\times} V_n \to W.$$

The multipolarization $P_{(r_1,\ldots,r_n)} f$ induces a linear map

$$Pf: S^{r_1} V_1 \otimes \cdots \otimes S^{r_n} V_n \to W$$

called the total polarization of f.

Given a linear action of a group on $V_1, \ldots, V_n$, W, the following assertion is a direct analogy of proposition 24.12.

Proposition. *If $f\colon V_1 \times \ldots \times V_n \to W$ is an equivariant multihomogeneous polynomial map, then all its multipolarizations $P_{(i_1,\ldots,i_n)}f$ and its total polarization Pf are also equivariant.*

24.14. Example. The simplest example for the polarization technique is the problem of finding all smooth $GL(V)$-equivariant maps $f\colon V \to \otimes^r V$. Using the homotheties in $GL(V)$, we obtain $k^r f(x) = f(kx)$. By the homogeneous function theorem, f is a homogeneous polynomial map of degree r. Its total polarization is an equivariant map $Pf\colon S^r V \to \otimes^r V$. By 24.8.(5), Pf is a constant multiple of the inclusion $S^r V \hookrightarrow \otimes^r V$. Hence all smooth $GL(V)$-equivariant maps $V \to \otimes^r V$ are of the form $x \mapsto k(x \otimes \cdots \otimes x)$, $k \in \mathbb{R}$.

25. Natural operators on linear connections, the exterior differential

25.1. Our first geometrical application of the general methods deals with the natural operators transforming the linear connections on an m-dimensional manifold M into themselves. In 17.7 we denoted by QP^1M the connection bundle of the first order frame bundle P^1M of M. This is an affine bundle modelled on vector bundle $TM \otimes T^*M \otimes T^*M$. The linear connections on M coincide with the sections of QP^1M. Obviously, QP^1 is a second order bundle functor on the category $\mathcal{M}f_m$ of all m-dimensional manifolds and their local diffeomorphisms.

25.2. We determine all natural operators $QP^1 \rightsquigarrow QP^1$. Let S be the torsion tensor of a linear connection $\Gamma \in C^\infty(QP^1M)$, see 16.2, let $\hat{S}$ be the contracted torsion tensor and let I be the identity tensor of $TM \otimes T^*M$. Then S, $I \otimes \hat{S}$ and $\hat{S} \otimes I$ are three sections of $TM \otimes T^*M \otimes T^*M$.

Proposition. *All natural operators $QP^1 \rightsquigarrow QP^1$ form the following 3-parameter family*

$$\Gamma + k_1 S + k_2 I \otimes \hat{S} + k_3 \hat{S} \otimes I, \qquad k_1, k_2,\ k_3 \in \mathbb{R}. \tag{1}$$

Proof. In the canonical coordinates x^i, x^i_j on $P^1\mathbb{R}^m$, the equations of a principal connection Γ are

$$dx^i_j = \Gamma^i_{lk}(x) x^l_j dx^k \tag{2}$$

where Γ^i_{jk} are any smooth functions on $\mathbb{R}^m$. From (2) we obtain the action of G^2_m on the standard fiber $F_0 = (QP^1\mathbb{R}^m)_0$

$$\bar{\Gamma}^i_{jk} = a^i_l \Gamma^l_{mn} \tilde{a}^m_j \tilde{a}^n_k + a^i_{lm} \tilde{a}^l_j \tilde{a}^m_k \tag{3}$$

see 17.7. The proof will be performed in 3 steps, which are typical for a wider class of naturality problems.

Step I. The zero order operators correspond to the G^2_m-equivariant maps $f\colon F_0 \to F_0$. The group G^2_m is a semidirect product of the kernel K of the jet projection $G^2_m \to G^1_m$, the elements of which satisfy $a^i_j = \delta^i_j$, and of the subgroup $i(G^1_m)$, the elements of which are characterized by $a^i_{jk} = 0$. By (3), F_0 with the action of $i(G^1_m)$ coincides with $\mathbb{R}^m \otimes \mathbb{R}^{m*} \otimes \mathbb{R}^{m*}$ with the canonical action of $GL(m)$. We have deduced in 24.9 that all $GL(m)$-equivariant maps of $\mathbb{R}^m \otimes \mathbb{R}^{m*} \otimes \mathbb{R}^{m*}$ into itself form the 6-parameter family

$$f^i_{jk} = a_1\delta^i_j x^l_{kl} + a_2\delta^i_j x^l_{lk} + a_3\delta^i_k x^l_{jl} + a_4\delta^i_k x^l_{lj} + a_5 x^i_{jk} + a_6 x^i_{kj}. \tag{4}$$

The equivariance of (4) with respect to K then yields

$$a^i_{jk} = (a_1 + a_2)\delta^i_j a^l_{lk} + (a_3 + a_4)\delta^i_k a^l_{lj} + (a_5 + a_6)a^i_{jk}. \tag{5}$$

This is a polynomial identity in a^i_{jk}. For $m \geq 2$, (5) is equivalent to $a_1 + a_2 = 0$, $a_3 + a_4 = 0$, $a_5 + a_6 = 1$. From 16.2 we find easily $S = (\Gamma^i_{jk} - \Gamma^i_{kj}) =: (S^i_{jk})$, so that $I \otimes \hat{S} = (\delta^i_j S^l_{lk})$ and $\hat{S} \otimes I = (\delta^i_k S^l_{lj})$. Hence (5) implies (1). For $m = 1$, we have only one quantity a^1_{11}, so that (5) gives $1 = a_1 + a_2 + a_3 + a_4 + a_5 + a_6$. But it is easy to check this leads to the same geometrical result (1).

Step II. The r-th order natural operators $QP^1 \rightsquigarrow QP^1$ correspond to the G^{r+2}_m-equivariant maps from $(J^r QP^1\mathbb{R}^m)_0$ into F_0. Denote by Γ_s the collection of all s-th order partial derivatives $\Gamma^i_{jk,l_1,\dots,l_s}$, $s = 1,\dots,r$. According to 14.20, the action of $i(G^1_m) \subset G^{r+2}_m$ on every Γ_s is tensorial. Using the equivariance with respect to the homotheties in G^1_m, we obtain a homogeneity condition

$$k\, f(\Gamma, \Gamma_1, \dots, \Gamma_r) = f(k\Gamma, k^2\Gamma_1, \dots, k^{r+1}\Gamma_r).$$

By the homogeneous function theorem, f is a polynomial of degree d_0 in Γ and d_s in Γ_s such that

$$1 = d_0 + 2d_1 + \dots + (r+1)d_r.$$

Obviously, the only possibility is $d_0 = 1$, $d_1 = \dots = d_r = 0$. This implies that f is independent of $\Gamma_1, \dots, \Gamma_r$, so that we get the case I.

Step III. In example 23.6 we deduced that every natural operator $QP^1 \rightsquigarrow QP^1$ has finite order. This completes the proof. □

25.3. Rigidity of the torsion-free connections. Let $Q_\tau P^1 M \to M$ be the bundle of all torsion-free (in other words: symmetric) linear connections on M. The symmetrization $\Gamma \mapsto \Gamma - \frac{1}{2}S$ of linear connections is a natural transformation $\sigma\colon QP^1 \to Q_\tau P^1$ satisfying $\sigma \circ i = \mathrm{id}_{Q_\tau P^1}$, where $i\colon Q_\tau P^1 \to QP^1$ is the inclusion. Hence for every natural operator $A\colon Q_\tau P^1 \rightsquigarrow Q_\tau P^1$, $B = i \circ A \circ \sigma$ is a natural operator $QP^1 \rightsquigarrow QP^1$, i.e. one of the list 25.2.(1). By this list, $B(\Gamma) = \Gamma$ for every symmetric connection. This implies that *the only natural operator $Q_\tau P^1 \rightsquigarrow Q_\tau P^1$ is the identity.*

25.4. The exterior differential of p-forms is a natural operator $d\colon \Lambda^pT^* \rightsquigarrow \Lambda^{p+1}T^*$. The oldest result on natural operators is a theorem by Palais, who deduced that all linear natural operators $\Lambda^pT^* \rightsquigarrow \Lambda^{p+1}T^*$ are the constant multiples of the exterior differential only, [Palais, 59]. Using a similar procedure as in the proof of proposition 25.2, we deduce that for $p > 0$ even linearity follows from naturality.

Proposition. *For $p > 0$, all natural operators $\Lambda^pT^* \rightsquigarrow \Lambda^{p+1}T^*$ are the constant multiples kd of the exterior differential d, $k \in \mathbb{R}$.*

Proof. The canonical coordinates on $\Lambda^p\mathbb{R}^{m*}$ are $b_{i_1\dots i_p} =: b$ antisymmetric in all subscripts and the action of $GL(m)$ is

$$\bar{b}_{i_1\dots i_p} = b_{j_1\dots j_p}\tilde{a}^{j_1}_{i_1}\dots\tilde{a}^{j_p}_{i_p}. \tag{1}$$

The induced coordinates on $F_1 = J^1_0\Lambda^pT^*\mathbb{R}^m$ are $b_{i_1\dots i_p,i_{p+1}} =: b_1$. One evaluates easily that the action of G^2_m on F_1 is given by (1) and

$$\begin{aligned}\bar{b}_{i_1\dots i_p,i} = b_{j_1\dots j_p,j}\tilde{a}^{j_1}_{i_1}\dots\tilde{a}^{j_p}_{i_p}\tilde{a}^{j}_{i} + b_{j_1\dots j_p}\tilde{a}^{j_1}_{i_1 i}\dots\tilde{a}^{j_p}_{i_p} + \\ \dots + b_{j_1\dots j_p}\tilde{a}^{j_1}_{i_1}\dots\tilde{a}^{j_p}_{i_p i}.\end{aligned} \tag{2}$$

The action of $GL(m)$ on $\Lambda^{p+1}\mathbb{R}^{m*}$ is

$$\bar{c}_{i_1\dots i_{p+1}} = c_{j_1\dots j_{p+1}}\tilde{a}^{j_1}_{i_1}\dots\tilde{a}^{j_{p+1}}_{i_{p+1}}. \tag{3}$$

Step I. The first order natural operators are in bijection with G^2_m-maps $f\colon F_1 \to \Lambda^{p+1}\mathbb{R}^{m*}$. Consider first the equivariance of f with respect to the homotheties in $i(G^1_m)$. This gives a homogeneity condition

$$k^{p+1}f(b,b_1) = f(k^pb, k^{p+1}b_1). \tag{4}$$

For $p > 0$, f must be a polynomial of degrees d_0 in b and d_1 in b_1 such that $p+1 = pd_0 + (p+1)d_1$. For $p > 1$ the only possibility is $d_0 = 0$, $d_1 = 1$, i.e. f is linear in b_1. By 24.8.(4), the equivariance of f with respect to the whole group $i(G^1_m)$ implies

$$\bar{c}_{i_1\dots i_{p+1}} = k\, b_{[i_1\dots i_p,i_{p+1}]} \qquad k \in \mathbb{R}. \tag{5}$$

For $p = 1$, there is another possibility $d_0 = 2$, $d_1 = 0$. But 24.8 and the polarization technique yield that the only smooth $GL(m)$-map of $S^2\mathbb{R}^{m*}$ into $\Lambda^2\mathbb{R}^{m*}$ is the zero map. Thus all first order natural operators are of the form (5), which is the coordinate expression of kd.

Step II. Every r-th order natural operator is determined by a G^{r+1}_m-map $f\colon F_r := J^r_0\Lambda^pT^*\mathbb{R}^m \to \Lambda^{p+1}\mathbb{R}^{m*}$. Denote by b_s the collection of all s-th order coordinates $b_{i_1\dots i_p,j_1\dots j_s}$ induced on F_r, $s = 1,\dots,r$. According to 14.20 the action of $i(G^1_m) \subset G^{r+1}_m$ on every b_s is tensorial. Using the equivariance with respect to the homotheties in G^1_m, we obtain

$$k^{p+1}f(b,b_1,\dots,b_r) = f(k^pb, k^{p+1}b_1,\dots,k^{p+r}b_r).$$

This implies that f is independent of $b_2,\dots,b_r$. Hence the r-th order natural operators are reduced to the case I for every $r > 1$.

Step III. In example 23.6 we deduced that every natural operator $\Lambda^pT^* \rightsquigarrow \Lambda^{p+1}T^*$ has finite order. □

25.5. Remark. For $p = 0$ the homogeneity condition 25.4.(4) yields $f = \varphi(b)b_1$, b, $b_1 \in \mathbb{R}$, where φ is any smooth function of one variable. Hence *all natural operators $\Lambda^0 T^* \rightsquigarrow \Lambda^1 T^*$ are of the form $g \mapsto \varphi(g)dg$ with an arbitrary smooth function $\varphi\colon \mathbb{R} \to \mathbb{R}$.*

26. The tensor evaluation theorem

26.1. We first formulate an important special case. Consider the product

$$V_{k,l} := V \overbrace{\times \ldots \times}^{k\text{-times}} V \times V^* \overbrace{\times \ldots \times}^{l\text{-times}} V^*$$

of k copies of a vector space V and of l copies of its dual V^*. Let $\langle\ ,\ \rangle\colon V \times V^* \to \mathbb{R}$ be the evaluation map $\langle x, y\rangle = y(x)$. The following assertion gives a very useful description of all smooth $GL(V)$-invariant functions

$$f(x_\alpha, y_\lambda)\colon V_{k,l} \to \mathbb{R}, \qquad \alpha = 1, \ldots, k,\ \lambda = 1, \ldots, l.$$

Proposition. *For every smooth $GL(V)$-invariant function $f\colon V_{k,l} \to \mathbb{R}$ there exists a smooth function $g(z_{\alpha\lambda})\colon \mathbb{R}^{kl} \to \mathbb{R}$ such that*

$$f(x_\alpha, y_\lambda) = g(\langle x_\alpha, y_\lambda\rangle). \tag{1}$$

We remark that this result can easily be proved in the case $k \le m = \dim V$ (or $l \le m$ by duality). Consider first the case $k = m$. Let $e_1, \ldots, e_m$ be a basis of V and $e^1, \ldots, e^m$ be the dual basis of V^*. Write $Z_\lambda = z_{1\lambda}e^1 + \cdots + z_{k\lambda}e^k \in V^*$ and define

$$g(z_{11}, \ldots, z_{kl}) = f(e_1, \ldots, e_k, Z_1, \ldots, Z_l).$$

Assume $x_1, \ldots, x_m$ are linearly independent vectors. Hence there is a linear isomorphism transforming $e_1, \ldots, e_k$ into $x_1, \ldots, x_k$. Since we have

$$y_\lambda = \langle e_1, y_\lambda\rangle e^1 + \cdots + \langle e_m, y_\lambda\rangle e^m,$$

$f(x_i, y_\lambda) = g(\langle x_i, y_\lambda\rangle)$ follows from the invariance of f. But the subset with linearly independent $x_1, \ldots, x_m$ is dense in $V_{m,l}$ and f and g are smooth functions, so that the latter relation holds everywhere. In the case $k < m$, $f\colon V_{k,l} \to \mathbb{R}$ can be interpreted as a function $V_{m,l} \to \mathbb{R}$ independent of $(k+1)$-st up to m-th vector components. This function is also $GL(V)$-invariant. Hence there is a smooth function $G(z_{i\lambda})\colon \mathbb{R}^{ml} \to \mathbb{R}$ satisfying $f(x_i, y_\lambda) = G(\langle x_i, y_\lambda\rangle)$. Put $g(z_{i\lambda}) = G(z_{i\lambda}, 0)$. Since f is independent of $x_{k+1}, \ldots, x_m$, we can set $x_{k+1} = 0, \ldots, x_m = 0$. This implies (1).

However, in the case $m < \min(k, l)$, the function g need not to be uniquely determined. For example, in the extreme case $m = 1$ our proposition asserts that for every smooth function $f(x_1, \ldots, x_k, y_1, \ldots, y_l)$ of $k + l$ scalar variables satisfying

$$f(x_1, \ldots, x_k, y_1, \ldots, y_l) = f(cx_1, \ldots, cx_k, \tfrac{1}{c}y_1, \ldots, \tfrac{1}{c}y_l)$$

for all $0 \ne c \in \mathbb{R}$, there exists a smooth function $g\colon \mathbb{R}^{kl} \to \mathbb{R}$ such that $f(x_1, \ldots, x_k, y_1, \ldots, y_l) = g(x_1y_1, \ldots, x_ky_l)$. Even this is a non-trivial analytical problem.

26.2. In general, consider k copies of V and a finite number of tensor products $\otimes^p V^*, \dots, \otimes^q V^*$ of V^*. (Proposition 26.1 corresponds to the case $p = 1, \dots, q = 1$.) Write x_i for the elements of the i-th copy of V and $a \in \otimes^p V^*, \dots, b \in \otimes^q V^*$. Denote by $a(x_{i_1}, \dots, x_{i_p})$ or ... or $b(x_{j_1}, \dots, x_{j_q})$ the full contraction of a with $x_{i_1}, \dots, x_{i_p}$ or ... or of b with $x_{j_1}, \dots, x_{j_q}$, respectively. Let $y_{i_1 \dots i_p} \in \mathbb{R}^{k^p}, \dots, z_{j_1 \dots j_q} \in \mathbb{R}^{k^q}$ be the canonical coordinates.

Tensor evaluation theorem. *For every smooth $GL(V)$-invariant function $f\colon \otimes^p V^* \times \dots \times \otimes^q V^* \times \times^k V \to \mathbb{R}$ there exists a smooth function*

$$g(y_{i_1 \dots i_p}, \dots, z_{j_1 \dots j_q})\colon \mathbb{R}^{k^p} \times \dots \times \mathbb{R}^{k^q} \to \mathbb{R}$$

such that

$$f(a, \dots, b, x_1, \dots, x_k) = g(a(x_{i_1}, \dots, x_{i_p}), \dots, b(x_{j_1}, \dots, x_{j_q})). \tag{1}$$

To prove this, we shall use a general result by D. Luna.

26.3. Luna's theorem. Consider a completely reducible action of a group G on $\mathbb{R}^n$, see 13.5. Let $P(\mathbb{R}^n)$ be the ring of all polynomials on $\mathbb{R}^n$ and $P(\mathbb{R}^n)^G$ be the subring of all G-invariant polynomials. By the classical Hilbert theorem, $P(\mathbb{R}^n)^G$ is finitely generated. Consider a system $p_1, \dots, p_s$ of its generators (called the Hilbert generators) and denote by $p\colon \mathbb{R}^n \to \mathbb{R}^s$ the mapping with components $p_1, \dots, p_s$. Luna deduced the following theorem, [Luna, 76], which we present without proof.

Theorem. *For every smooth function $f\colon \mathbb{R}^n \to \mathbb{R}$ which is constant on the fibers of p there exists a smooth function $g\colon \mathbb{R}^s \to \mathbb{R}$ satisfying $f = g \circ p$.*

We remark that in the category of sets it is trivial that constant values of f on the pre-images of p form a necessary and sufficient condition for the existence of a map g such that $f = g \circ p$. If some pre-images are empty, then g is not uniquely determined. The proper meaning of the above result by Luna is that smoothness of f implies the existence of a smooth g.

26.4. Remark. In the real analytic case [Luna, 76] deduced an essentially stronger result: *If f is a real analytic G-invariant function on $\mathbb{R}^n$, then there exists a real analytic function g defined on a neighborhood of $p(\mathbb{R}^n) \subset \mathbb{R}^s$ such that $f = g \circ p$.* But the following example shows that the smooth case is really different from the analytic one.

Example. The connected component of unity in $GL(1)$ coincides with the multiplicative group $\mathbb{R}^+$ of all positive real numbers. The formula $(cx, \frac{1}{c}y)$, $c \in \mathbb{R}^+$, $(x, y) \in \mathbb{R}^2$ defines a linear action of $\mathbb{R}^+$ on $\mathbb{R}^2$. The rule $(x, y) \mapsto \operatorname{sgn} x$ is a non-smooth $\mathbb{R}^+$-invariant function on $\mathbb{R}^2$. Take a smooth function $\varphi(t)$ of one variable with infinite order zero at $t = 0$. Then $(\operatorname{sgn} x)\varphi(xy)$ is a smooth $\mathbb{R}^+$-invariant function on $\mathbb{R}^2$. Using homogeneity one finds directly that the ring of $\mathbb{R}^+$-invariant polynomials on $\mathbb{R}^2$ is generated by xy. But $(\operatorname{sgn} x)\varphi(xy)$ cannot be expressed as a function of xy, since it changes sign when replacing (x, y) by $(-x, -y)$.

26.5. Theorem 26.2 can easily be proved in the case $k \leq m$. Assume first $k = m$. Let $a_{i_1 \dots i_p}, \dots, b_{j_1 \dots j_q}$ be the coordinates of $a, \dots, b$. Hence $f = f(a_{i_1 \dots i_p}, \dots, b_{j_1 \dots j_q}, x_1^i, \dots, x_k^j)$ and we define

$$g(y_{i_1 \dots i_p}, \dots, z_{j_1 \dots j_q}) = f(y_{i_1 \dots i_p}, \dots, z_{j_1 \dots j_q}, e_1, \dots, e_k).$$

Obviously, g is a smooth function. Then 26.2.(1) holds on the set of all linearly independent vector k-tuples of V by invariance of f. But the latter set is dense, so that 26.2.(1) holds everywhere by the continuity. In the case $k < m$ we interpret f as a function $\otimes^p V^* \times \dots \times \otimes^q V^* \times \times^m V \to \mathbb{R}$ independent of the $(k+1)$-st up to m-th vector component and we proceed in the same way as in 26.1.

26.6. In the case $m < k$ we have to apply Luna's theorem. First we claim that the set of all contractions $a(x_{i_1}, \dots, x_{i_p}), \dots, b(x_{j_1}, \dots, x_{j_q})$ form the Hilbert generators on $\otimes^p V^* \times \dots \times \otimes^q V^* \times \times^k V$. Indeed, let h be a $GL(V)$-invariant polynomial and $H^{i_1 \dots i_s}_{A \dots B}$ be its component linearly generated by all monomials of degree A in the components of $a, \dots$, of degree B in the components of b and with simple entries of the components of $x_{i_1}, \dots, x_{i_s}$ (repeated indices being allowed). Since h is $GL(V)$-invariant, the total polarization of each $H^{i_1 \dots i_s}_{A \dots B}$ corresponds to an invariant tensor. By the invariant tensor theorem, the latter tensor is a linear combination of the elementary invariant tensors in the case $Ap + \dots + Bq = s$ and vanishes otherwise. But the elementary invariant tensors induce just the contractions we mentioned in our claim.

Then we have to prove that

$$\bar{a}(\bar{x}_{i_1}, \dots, \bar{x}_{i_p}) = a(x_{i_1}, \dots, x_{i_p}), \dots, \bar{b}(\bar{x}_{j_1}, \dots, \bar{x}_{j_q}) = b(x_{j_1}, \dots, x_{j_q}) \tag{1}$$

implies

$$f(\bar{a}, \dots, \bar{b}, \bar{x}_1, \dots, \bar{x}_k) = f(a, \dots, b, x_1, \dots, x_k). \tag{2}$$

Consider first the case that both m-tuples $x_1, \dots, x_m$ and $\bar{x}_1, \dots, \bar{x}_m$ are linearly independent. Hence $x_\lambda = c^i_\lambda x_i$, $\bar{x}_\lambda = \bar{c}^i_\lambda \bar{x}_i$, $i = 1, \dots, m$, $\lambda = m+1, \dots, k$. Then the first collection from (1) yields, for each $\lambda = m+1, \dots, k$,

$$\begin{aligned} &\sum_{i=1}^m (c^i_\lambda - \bar{c}^i_\lambda) a(x_i, x_1, \dots, x_1) = 0 \\ &\qquad\qquad\vdots \\ &\sum_{i=1}^m (c^i_\lambda - \bar{c}^i_\lambda) a(x_1, \dots, x_1, x_i) = 0. \end{aligned} \tag{3}$$

We restrict ourselves to the subset, on which the determinant of linear system (3) does not vanish. (This determinant does not vanish identically, as for $x_i = e_i$ it is a polynomial in the components of the tensor a, whose coefficient by $(a_{1\dots 1})^m$ is 1.) Then (3) yields $c^i_\lambda = \bar{c}^i_\lambda$. Consider now the functions

$$\tilde{f}(a,\dots,b,x_1,\dots,x_m) = f(a,\dots,b,x_1,\dots,x_m,c^i_\lambda x_i). \tag{4}$$

By the first part of the proof, $\tilde{f}$ can be expressed in the form 26.2.(1). This implies (2).

Thus, we have deduced that a dense subset of the solutions of (1) is formed by the solutions of (2). Since both solution sets are closed, this completes the proof of the tensor evaluation theorem.

26.7. Remark. We remark that there are some obstructions to obtain a general result of such a type if we replace the product $\times^k V$ by a product of some tensorial powers of V. Consider the simpliest case of the smooth $GL(1)$-invariant functions on $\otimes^2\mathbb{R} \times \otimes^2\mathbb{R}^*$. Let x or y be the canonical coordinate on $\otimes^2\mathbb{R}$ or $\otimes^2\mathbb{R}^*$, respectively. The action of $GL(1)$ is $(x,y) \mapsto (k^2x, \frac{1}{k^2}y)$, $0 \neq k \in \mathbb{R}$. But this is the situation of example 26.4, so that e.g. $(\operatorname{sgn} x)\varphi(xy)$, where $\varphi(t)$ is a smooth function on $\mathbb{R}$ with infinite zero at $t = 0$, is a smooth $GL(1)$-invariant function on $\otimes^2\mathbb{R} \times \otimes^2\mathbb{R}^*$. Here the smooth case is essentially different from the analytic one.

26.8. Tensor evaluation theorem with parameters. Analyzing the proof of theorem 26.2, one can see that the result depends smoothly on 'constant' parameters in the following sense. Let W be another vector space endowed with the identity action of $GL(V)$.

Theorem. *For every smooth $GL(V)$-invariant function $f\colon \otimes^p V^* \times \dots \times \otimes^q V^* \times \times^k V \times W \to \mathbb{R}$ there exists a smooth function $g(y_{i_1\dots i_p},\dots,z_{j_1\dots j_q},t)\colon \mathbb{R}^{k^p} \times \dots \times \mathbb{R}^{k^q} \times W \to \mathbb{R}$ such that*

$$f(a,\dots,b,x_1,\dots,x_k,t) = g(a(x_{i_1},\dots,x_{i_p}),\dots,b(x_{j_1},\dots,x_{j_q}),t), \quad t \in W.$$

The proof is left to the reader.

26.9. Smooth $GL(V)$-equivariant maps $V_{k,l} \to V$. As the first application of the tensor evaluation theorem we determine all smooth $GL(V)$-equivariant maps $f\colon V_{k,l} \to V$. Let us construct a function $F\colon V_{k,l} \times V^* \to \mathbb{R}$ by

$$F(x_\alpha, y_\lambda, w) = \langle f(x_\alpha, y_\lambda), w\rangle, \qquad w \in V^*.$$

This is a $GL(V)$-invariant function, so that there is a smooth function

$$g(z_{\alpha\lambda}, z_\alpha)\colon \mathbb{R}^{k(l+1)} \to \mathbb{R}$$

such that

$$F(x_\alpha, y_\lambda, w) = g(\langle x_\alpha, y_\lambda\rangle, \langle x_\alpha, w\rangle).$$

Taking the partial differential with respect to w and setting $w = 0$, we obtain

$$f(x_\alpha, y_\lambda) = \sum_\beta \frac{\partial g(\langle x_\alpha, y_\lambda\rangle, 0)}{\partial z_\beta} x_\beta, \qquad \beta = 1,\dots,k.$$

This proves

Proposition. *All $GL(V)$-equivariant maps $V_{k,l} \to V$ are of the form*

$$\sum_{\beta=1}^{k} g_\beta(\langle x_\alpha, y_\lambda \rangle) x_\beta$$

with arbitrary smooth functions $g_\beta \colon \mathbb{R}^{kl} \to \mathbb{R}$.

If we replace vectors and covectors, we obtain

26.10. Proposition. *All $GL(V)$-equivariant maps $V_{k,l} \to V^*$ are of the form*

$$\sum_{\mu=1}^{l} g_\mu(\langle x_\alpha, y_\lambda \rangle) y_\mu$$

with arbitrary smooth functions $g_\mu \colon \mathbb{R}^{kl} \to \mathbb{R}$.

Next we present a simple application of this result in the theory of natural operations.

26.11. Natural transformations $TT^* \to T^*T$. Starting from some problems in analytical mechanics, Modugno and Stefani introduced a geometrical isomorphism between the bundles $TT^*M = T(T^*M)$ and $T^*TM = T^*(TM)$ for every manifold M, [Modugno, Stefani, 78]. From the categorical point of view this is a natural equivalence between bundle functors TT^* and T^*T defined on the category $\mathcal{M}f_m$. Our aim is to determine all natural transformations $TT^* \to T^*T$.

We first give a simple construction of the isomorphism $s_M \colon TT^*M \to T^*TM$ by Modugno and Stefani. Let $q \colon T^*M \to M$ be the bundle projection and $\kappa \colon TTM \to TTM$ be the canonical involution. Every $A \in TT^*M$ is a vector tangent to a curve $\gamma(t) \colon \mathbb{R} \to T^*M$ at $t = 0$. If B is any vector of $T_{Tq(A)}TM$, then κB is tangent to the curve $\delta(t) \colon \mathbb{R} \to TM$ over the curve $q(\gamma(t))$ on M. Hence we can evaluate $\langle \gamma(t), \delta(t) \rangle$ for every t and the derivative $\frac{\partial}{\partial t}\big|_0 \langle \gamma(t), \delta(t) \rangle =: \sigma(A, B)$ depends on A and B only. This determines a linear map $T_{Tq(A)}TM \to \mathbb{R}$, $B \mapsto \sigma(A, B)$, i.e. an element $s_M(A) \in T^*TM$.

In general, for every vector bundle $p \colon E \to M$, the tangent map $Tp \colon TE \to TM$ defines another vector bundle structure on TE. Even on the cotangent bundle $T^*E \to E$ there is another vector bundle structure $\rho \colon T^*E \to E^*$ defined by the restriction of a linear map $T_yE \to \mathbb{R}$ to the vertical tangent space, which is identified with $E_{p(y)}$. This enables us to introduce a sum $Y \dotplus Z$ for every $Y \in T^*_yTM$ and $Z \in T^*_{\pi(y)}M$ as follows. We have $(\rho(Y), Z) \in T^*M \times_M T^*M = VT^*M \hookrightarrow TT^*M$ and we can apply $s_M \colon TT^*M \to T^*TM$. Then $Y \dotplus Z$ is defined as the sum $Y + s_M(\rho(Y), Z)$ with respect to the vector bundle structure ρ.

26.12. For every $X \in TT^*M$ we write $p \in T^*M$ for its point of contact and $\xi = Tq(X) \in TM$. Taking into account both vector bundle structures on T^*TM, we denote by $Y \mapsto (k)_1 Y$ or $Y \mapsto (k)_2 Y$, $k \in \mathbb{R}$, the scalar multiplication with respect to the first or second one, respectively.

Proposition. *All natural transformations $TT^* \to T^*T$ are of the form*

$$(F(\langle p,\xi\rangle))_1 (G(\langle p,\xi\rangle))_2 s_M(X) \dotplus H(\langle p,\xi\rangle)p \tag{1}$$

where $F(t)$, $G(t)$, $H(t)$ are three arbitrary smooth functions of one variable.

Proof. Since TT^* and T^*T are second order bundle functors on $\mathcal{M}f_m$, we have to determine all G^2_m-equivariant maps of $S := TT^*_0\mathbb{R}^m$ into $Z := T^*T_0\mathbb{R}^m$. The canonical coordinates x^i on $\mathbb{R}^m$ induce the additional coordinates p_i on $T^*\mathbb{R}^m$ and $\xi^i = dx^i$, $\pi_i = dp_i$ on $TT^*\mathbb{R}^m$. If we evaluate the effect of a diffeomorphism on $\mathbb{R}^m$ and pass to 2-jets, we find easily that the equations of the action of G^2_m on S are

$$\bar{p}_i = \tilde{a}^j_i p_j, \quad \bar{\xi}^i = a^i_j \xi^j, \quad \bar{\pi}_i = \tilde{a}^j_i \pi_j - a^l_{jk}\tilde{a}^m_l \tilde{a}^j_i p_m \xi^k. \tag{2}$$

Further, if η^i are the induced coordinates on $T\mathbb{R}^m$, then the expression $\rho_i dx^i + \sigma_i d\eta^i$ determines the additional coordinates ρ_i, σ_i on $T^*T\mathbb{R}^m$. Similarly to (2) we obtain the following action of G^2_m on Z

$$\bar{\eta}^i = a^i_j \eta^j, \quad \bar{\sigma}_i = \tilde{a}^j_i \sigma_j, \quad \bar{\rho}_i = \tilde{a}^j_i \rho_j - a^l_{jk}\tilde{a}^m_l \tilde{a}^j_i \sigma_m \eta^k. \tag{3}$$

Any map $\varphi\colon S \to Z$ has the form

$$\eta^i = f^i(p,\xi,\pi), \quad \sigma_i = g_i(p,\xi,\pi), \quad \rho_i = h_i(p,\xi,\pi).$$

The equivariance of f^i is expressed by

$$a^i_j f^j(p,\xi,\pi) = f^i(\tilde{a}^j_i p_j, a^i_j\xi^j, \tilde{a}^j_i\pi_j - a^l_{jk}\tilde{a}^m_l\tilde{a}^j_i p_m \xi^k). \tag{4}$$

Setting $a^i_j = \delta^i_j$, we obtain $f^i(p,\xi,\pi) = f^i(p,\xi,\pi_j - a^l_{jk}p_l\xi^k)$. This implies that the f^i are independent of π_j. Then (4) shows that $f^i(p,\xi)$ is a $GL(m)$-equivariant map $\mathbb{R}^m \times \mathbb{R}^{m*} \to \mathbb{R}^m$. By proposition 26.9,

$$f^i = F(\langle p,\xi\rangle)\xi^i \tag{5}$$

where F is an arbitrary smooth function of one variable. Using the same procedure we obtain that the g_i are independent of π_j. Then proposition 26.10 yields

$$g_i = G(\langle p,\xi\rangle)p_i \tag{6}$$

where G is another smooth function of one variable.

Consider further the difference $k_i = h_i - F(\langle p,\xi\rangle)G(\langle p,\xi\rangle)\pi_i$. Using the fact that $\langle p,\xi\rangle$ is invariant, we express the equivariance of k_i in the form

$$\tilde{a}^j_i k_j(p,\xi,\pi) = k_i(\tilde{a}^j_i p_j, a^i_j\xi^j, \tilde{a}^j_i\pi_j - a^l_{jk}\tilde{a}^m_l\tilde{a}^j_i p_m\xi^k).$$

Quite similarly to (4) and (6) we then deduce $k_i = H(\langle p,\xi\rangle)p_i$, i.e.

$$h_i = F(\langle p,\xi\rangle)G(\langle p,\xi\rangle)\pi_i + H(\langle p,\xi\rangle)p_i. \tag{7}$$

One verifies easily that (5), (6) and (7) is the coordinate form of (1). □

26.13. To interpret all natural transformations of proposition 26.12 geometrically, we first show that for any constant values $F = f$, $G = g$, $H = h$, 26.12.(1) can be determined by a simple modification of the above mentioned construction of s (s corresponds to the case $f = 1$, $g = 1$, $h = 0$). If $A \in TT^*M$ is tangent to a curve $\gamma(t)$, then fA is tangent to $\gamma(ft)$. For every vector $B \in T_{fTq(A)}TM$, κB is tangent to a curve $\delta(t)\colon \mathbb{R} \to TM$ over the curve $q(\gamma(ft))$ on M. Then we define an element $s_{(f,g,h)}A \in T^*TM$ by

$$(1) \qquad \langle s_{(f,g,h)}A, B\rangle = \tfrac{\partial}{\partial t}\big|_0 \langle \gamma(ft), g\delta(t)\rangle + h\langle \gamma(0), \delta(0)\rangle.$$

The coordinate expression of (1) is $(fg\pi_i + hp_i)dx^i + gp_i d\eta^i$ and our construction implies $\eta^i = f\xi^i$. This gives 26.12.(1) with constant coefficients. Moreover, the general case can also be interpreted in such a way. Let $\pi\colon TT^*M \to T^*M$ be the bundle projection. Every $A \in TT^*M$ determines $Tq(A) \in TM$ and $\pi(A) \in T^*M$ over the same base point in M. Then we take the values of F, G and H at $\langle \pi(A), Tq(A)\rangle$ and apply the latter construction.

We remark that the natural transformation s by Modugno and Stefani can be distinguished among all natural transformations $TT^* \to T^*T$ by an interesting geometric construction explained in [Kolář, Radziszewski, 88].

26.14. The functor T^*T^*. The iterated cotangent functor T^*T^* is also a second order bundle functor on $\mathcal{M}f_m$. The problem of finding of all natural transformations between any two of the functors TT^*, T^*T and T^*T^* can be reduced to proposition 26.12, if we take into account a classical geometrical construction of a natural equivalence between TT^* and T^*T^*. Consider the Liouville 1-form $\omega\colon TT^*M \to \mathbb{R}$ defined by $\omega(A) = \langle \pi(A), Tq(A)\rangle$. The exterior differential $d\omega = \Omega$ endows T^*M with a natural symplectic structure. This defines a bijection between the tangent and cotangent bundles of T^*M transforming $X \in TT^*M$ into its inner product with Ω. Hence *the natural transformations between any two of the functors TT^*, T^*T and T^*T^* depend on three arbitrary smooth functions of one variable.* Their coordinate expressions can be found in [Kolář, Radziszewski, 88].

26.15. Non-existence of natural symplectic structure on the tangent bundles. We shall see in 37.4 that the natural transformations of the iterated tangent functor into itself depend on four real parameters. This is related with the fact that TT is defined on the whole category $\mathcal{M}f$ and is product preserving. Since the natural transformations of TT into itself are essentially different from the natural transformations of T^*T into itself, there is no natural equivalence between TT and T^*T. This implies that *there is no natural symplectic structure on the tangent bundles.*

26.16. Remark. Taking into account the natural isomorphism $s\colon TT^* \to T^*T$ and the canonical symplectic structure on the cotangent bundles, one sees easily that any two of the third order functors TTT^*, TT^*T, TT^*T^*, T^*TT, T^*TT^*, T^*T^*T and $T^*T^*T^*$ are naturally equivalent, but TTT is naturally equivalent to none of them. All natural transformations $TTT^* \to TT^*T$ for manifolds of dimension at least two are determined in [Doupovec, to appear].

27. Generalized invariant tensors

To study the natural operators on $\mathcal{FM}_{m,n}$, we need a modification of the Invariant tensor theorem.

27.1. Consider two vector spaces V and W. The tensor product of the standard actions of $GL(V)$ on $\otimes^p V \otimes \otimes^q V^*$ and of $GL(W)$ on $\otimes^r W \otimes \otimes^s W^*$ defines the standard action of $GL(V) \times GL(W)$ on $\otimes^p V \otimes \otimes^q V^* \otimes \otimes^r W \otimes \otimes^s W^*$. A tensor B of the latter space is said to be a *generalized invariant tensor*, if $aB = B$ for all $a \in GL(V) \times GL(W)$. The invariance of B with respect to the homotheties in $GL(V)$ or $GL(W)$ gives $k^{p-q}B = B$ or $k^{r-s}B = B$, respectively. This implies that for $p \neq q$ or $r \neq s$ the only generalized invariant tensor is the zero tensor.

Generalized invariant tensor theorem. *Every generalized invariant tensor* $B \in \otimes^q V \otimes \otimes^q V^* \otimes \otimes^r W \otimes \otimes^r W^*$ *is a linear combination of the tensor products* $I \otimes J$, *where* I *is an elementary* $GL(V)$*-invariant tensor of degree* q *and* J *is an elementary* $GL(W)$*-invariant tensor of degree* r.

Proof. Contracting B with q vectors of V and q covectors of V^*, we obtain a $GL(W)$-invariant tensor. By the invariant tensor theorem 24.4 and by multilinearity, B is of the form

$$B = \sum_{s \in S_r} B_s \otimes J^s \qquad \text{with } B_s \in \otimes^q V \otimes \otimes^q V^*, \tag{1}$$

where J^s are the elementary $GL(W)$-invariant tensors of degree r. If we construct the total contraction of (1) with one tensor J^σ, $\sigma \in S_r$, we obtain $B_{\sigma^{-1}}$. Hence every B_s is a $GL(V)$-invariant tensor. Using theorem 24.4 once again, we prove our assertion. □

27.2. Example. We determine all smooth equivariant maps $W \otimes V^* \times W \otimes W^* \otimes V^* \to W \otimes V^* \otimes V^*$. Let $f^p_{ij}(x^q_k, y^r_{sl})$ be the coordinate expression of such a map. The equivariance of f with respect to the homotheties $\frac{1}{k}\delta^i_j$ in $GL(V)$ gives

$$k^2 f^p_{ij}(x^q_k, y^r_{sl}) = f^p_{ij}(kx^q_k, ky^r_{sl}).$$

By the homogeneous function theorem, we have to discuss the condition $2 = d_1 + d_2$. There are three possibilities: a) $d_1 = 2$, $d_2 = 0$, b) $d_1 = 1$, $d_2 = 1$, c) $d_1 = 0$, $d_2 = 2$. In each case f is a polynomial map. The homotheties $k\delta^p_q$ in $GL(W)$ yield

$$kf^p_{ij}(x^q_k, y^r_{sl}) = f^p_{ij}(kx^q_k, y^r_{sl}).$$

This condition is compatible with the case b) only, so that f is bilinear in x^q_k and y^r_{sl}. Its total polarization corresponds to a generalized invariant tensor in $\otimes^2 V \otimes \otimes^2 V^* \otimes \otimes^2 W^* \otimes \otimes^2 W^*$. By theorem 27.1, the coordinate form of f is

$$f^p_{ij} = (a\delta^p_q\delta^r_s\delta^k_i\delta^l_j + b\delta^p_s\delta^r_q\delta^k_i\delta^l_j + c\delta^p_q\delta^r_s\delta^k_j\delta^l_i + d\delta^p_s\delta^r_q\delta^k_j\delta^l_i)x^q_k y^s_{rl},$$

a, b, c, $d \in \mathbb{R}$. Hence all smooth equivariant maps $W \otimes V^* \times W \otimes W^* \otimes V^* \to W \otimes V^* \otimes V^*$ form the following 4-parameter family

$$ax^p_i y^q_{qj} + bx^q_i y^p_{qj} + cx^p_j y^q_{qi} + dx^q_j y^p_{qi}.$$

27.3. Curvature like operators. Consider a general connection $\Gamma\colon Y \to J^1Y$ on an arbitrary fibered manifold $Y \to BY$, where $B\colon \mathcal{FM} \to \mathcal{M}f$ denotes the base functor. In 17.1 we have deduced that the curvature of Γ is a map $C_Y\Gamma\colon Y \to VY \otimes \Lambda^2T^*BY$. The geometrical definition of curvature implies that C is a natural operator between two bundle functors J^1 and $V \otimes \Lambda^2T^*B$ defined on the category $\mathcal{FM}_{m,n}$. In the following assertion we may replace the second exterior power by the second tensor power (so that the antisymmetry of the curvature operator is a consequence of its naturality).

Proposition. *All natural operators $J^1 \rightsquigarrow V \otimes \otimes^2T^*B$ are the constant multiples kC of the curvature operator, $k \in \mathbb{R}$.*

Proof. We shall proceed in three steps as in the proof of proposition 25.2.

Step I. We first determine the first order operators. The canonical coordinates on the standard fiber $S_1 = J^1_0(J^1(\mathbb{R}^{n+m} \to \mathbb{R}^m) \to \mathbb{R}^{n+m})$ of J^1J^1 are y^p_i, $y^p_{ij} = \partial y^p_i/\partial x^j$, $y^p_{iq} = \partial y^p_i/\partial y^q$. Evaluating the effect of the isomorphisms in $\mathcal{FM}_{m,n}$ and passing to 2-jets, we obtain the following action of $G^2_{m,n}$ on S_1

$$\bar{y}^p_i = a^p_q y^q_j \tilde{a}^j_i + a^p_j \tilde{a}^j_i \tag{1}$$

$$\bar{y}^p_{iq} = a^p_r y^r_{js} \tilde{a}^s_q \tilde{a}^j_i + a^p_{rs} y^r_j \tilde{a}^s_q \tilde{a}^j_i + a^p_{rj} \tilde{a}^r_q \tilde{a}^j_i \tag{2}$$

$$\begin{aligned}\bar{y}^p_{ij} = {} & a^p_q y^q_{kl} \tilde{a}^k_i \tilde{a}^l_j + a^p_q y^q_{kr} \tilde{a}^r_j \tilde{a}^k_i + a^p_{ql} y^q_k \tilde{a}^k_i \tilde{a}^l_j + a^p_{qr} y^q_k \tilde{a}^r_j \tilde{a}^k_i \\ & + a^p_q y^q_k \tilde{a}^k_{ij} + a^p_{kl} \tilde{a}^k_i \tilde{a}^l_j + a^p_{kq} \tilde{a}^q_j \tilde{a}^k_i + a^p_k \tilde{a}^k_{ij}\end{aligned} \tag{3}$$

On the other hand, the standard fiber of $V \otimes \otimes^2T^*B$ is $\mathbb{R}^n \otimes \otimes^2\mathbb{R}^{m*}$ with canonical coordinates z^p_{ij} and the following action

$$\bar{z}^p_{ij} = a^p_q z^q_{kl} \tilde{a}^k_i \tilde{a}^l_j$$

We have to determine all $G^2_{m,n}$- equivariant maps $S_1 \to \mathbb{R}^n \otimes \otimes^2\mathbb{R}^{m*}$. Let $z^p_{ij} = f^p_{ij}(y^q_k, y^r_{\ell s}, y^t_{mn})$ be the coordinate expression of such a map. Consider the canonical injection of $GL(m) \times GL(n)$ into $G^2_{m,n}$ defined by 2-jets of the products of linear transformations of $\mathbb{R}^m$ and $\mathbb{R}^n$. The equivariance with respect to the homotheties in $GL(m)$ gives a homogeneity condition

$$k^2 f^p_{ij}(y^q_k, y^r_{\ell s}, y^t_{mn}) = f^p_{ij}(ky^q_k, ky^r_{\ell s}, k^2 y^t_{mn}).$$

When applying the homogeneous function theorem, we have to discuss the equation $2 = d_1 + d_2 + 2d_3$. Hence f^p_{ij} is a sum $g^p_{ij} + h^p_{ij}$ where g^p_{ij} is a linear map of $\mathbb{R}^n \otimes \mathbb{R}^{m*} \otimes \mathbb{R}^{m*}$ into itself and h^p_{ij} is a polynomial map $\mathbb{R}^n \otimes \mathbb{R}^{m*} \times \mathbb{R}^n \otimes \mathbb{R}^{n*} \otimes \mathbb{R}^{m*} \to \mathbb{R}^n \otimes \mathbb{R}^{m*} \otimes \mathbb{R}^{m*}$. Then we see directly that both g^p_{ij} and h^p_{ij} are $GL(m) \times GL(n)$-equivariant. For h^p_{ij} we have deduced in example 27.2

$$h^p_{ij} = ay^p_i y^q_{jq} + by^q_i y^p_{jq} + cy^p_j y^q_{iq} + dy^q_j y^p_{iq}$$

while for g^p_{ij} a direct use of theorem 27.1 yields

$$g^p_{ij} = ey^p_{ij} + fy^p_{ji}.$$

Moreover, the equivariance with respect to the subgroup $K \subset G^2_{m,n}$ characterized by $a^i_j = \delta^i_j$, $a^p_q = \delta^p_q$ leads to the relations $a = 0 = c$, $e = -f = -b = d$. Hence $f^p_{ij} = e(y^p_{ij} - y^p_{ji} - y^q_i y^p_{qj} + y^q_j y^p_{qi})$, which is the coordinate expression of eC, $e \in \mathbb{R}$.

Step II. Assume we have an r-th order natural operator $A\colon J^1 \rightsquigarrow V \otimes \otimes^2 T^* B$. It corresponds to a $G^{r+1}_{m,n}$-equivariant map from the standard fiber S_r of $J^r J^1$ into $\mathbb{R}^n \otimes \mathbb{R}^{m*} \otimes \mathbb{R}^{m*}$. Denote by $y^p_{i\alpha\beta}$ the partial derivative of y^p_i with respect to a multi index α in x^i and β in y^p. Any map $f\colon S_r \to \mathbb{R}^n \otimes \mathbb{R}^{m*} \otimes \mathbb{R}^{m*}$ is of the form $f(y^p_{i\alpha\beta})$, $\alpha + \beta \le r$. Similarly to the first part of the proof, $GL(m) \times GL(n)$ can be considered as a subgroup of $G^{r+1}_{m,n}$. One verifies easily that the transformation law of $y^p_{i\alpha\beta}$ with respect to $GL(m) \times GL(n)$ is tensorial. Using the homotheties in $GL(m)$, we obtain a homogeneity condition $k^2 f(y^p_{i\alpha\beta}) = f(k^{|\alpha|+1} y^p_{i\alpha\beta})$. This implies that f is a polynomial linear in the coordinates with $|\alpha| = 1$ and bilinear in the coordinates with $|\alpha| = 0$. Using the homotheties in $GL(n)$, we find $kf(y^p_{i\alpha\beta}) = f(k^{1-|\beta|} y^p_{i\alpha\beta})$. This yields that f is independent of all coordinates with $|\alpha| + |\beta| > 1$. Hence A is a first order operator.

Step III. Using 23.7 we conclude that every natural operator $J^1 \rightsquigarrow V \otimes \otimes^2 T^* B$ has finite order. This completes the proof. □

27.4. Curvature-like operators on pairs of connections. The Frölicher-Nijenhuis bracket $[\Gamma, \Delta] =: \kappa(\Gamma, \Delta)$ of two general connections Γ and Δ on Y is a section $Y \to VY \otimes \Lambda^2 T^* BY$, which may be called the *mixed curvature* of Γ and Δ. Since the pair Γ, Δ can be interpreted as a section $Y \to J^1 Y \times_Y J^1 Y$, κ is a natural operator $\kappa\colon J^1 \oplus J^1 \rightsquigarrow V \otimes \Lambda^2 T^* B$ between two bundle functors on $\mathcal{FM}_{m,n}$. Let $C_1\colon J^1 \oplus J^1 \rightsquigarrow V \otimes \Lambda^2 T^* B$ or $C_2\colon J^1 \oplus J^1 \rightsquigarrow V \otimes \Lambda^2 T^* B$ denote the curvature operator of the first or the second connection, respectively. The following assertion can be deduced in the same way as proposition 27.3, see [Kolář, 87a].

Proposition. *All natural operators $J^1 \oplus J^1 \rightsquigarrow V \otimes \otimes^2 T^* B$ form the following 3-parameter family*

$$k_1 C_1 + k_2 C_2 + k_3 \kappa, \qquad k_1,\ k_2,\ k_3 \in \mathbb{R}.$$

From a general point of view, this result enlightens us on the fact that the mixed curvature of two general connections can be defined in an 'essentially unique' way, i.e. the possibility of defining the mixed curvature is limited by the above 3-parameter family with trivial terms C_1 and C_2.

27.5. Remark. [Kurek, 91] deduced that the only natural operator $J^1 \rightsquigarrow V \otimes \Lambda^3 T^* B$ is the zero operator. This result presents an interesting point of view to the Bianchi identity for general connections.

28. The orbit reduction

We are going to explain another general procedure used in the theory of natural operators. From the computational point of view, the orbit reduction is an almost self-evident assertion about independence of the maps in question on some variables. This was already used e.g. for the simplification of (4) in 26.12. But the explicit formulation of such a procedure presented below is useful in several problems. First we discuss a concrete example, in which we obtain a Utiyama-like theorem for general connections. Then we present a complete treatment of the 'classical' reduction theorems from the theory of linear connections and from Riemannian geometry.

28.1. Let $p\colon G \to H$ be a Lie group homomorphism with kernel K, M be a G-space, Q be an H-space and $\pi\colon M \to Q$ be a p-equivariant surjective submersion, i.e. $\pi(gx) = p(g)\pi(x)$ for all $x \in M$, $g \in G$. Having p, we can consider every H-space N as a G-space by $gy = p(g)y$, $g \in G$, $y \in N$.

Proposition. *If each $\pi^{-1}(q)$, $q \in Q$ is a K-orbit in M, then there is a bijection between the G-maps $f\colon M \to N$ and the H-maps $\varphi\colon Q \to N$ given by $f = \varphi \circ \pi$.*

Proof. Clearly, $\varphi \circ \pi$ is a G-map $M \to N$ for every H-map $\varphi\colon Q \to N$. Conversely, let $f\colon M \to N$ be a G-map. Then we define $\varphi\colon Q \to N$ by $\varphi(\pi(x)) = f(x)$. This is a correct definition, since $\pi(\bar{x}) = \pi(x)$ implies $\bar{x} = kx$ with $k \in K$ by the orbit condition, so that $\varphi(\pi(\bar{x})) = f(kx) = p(k)f(x) = ef(x)$. We have $f = \varphi \circ \pi$ by definition and φ is smooth, since π is a surjective submersion. □

28.2. Example. We continue in our study of the standard fiber

$$S_1 = J_0^1(J^1(\mathbb{R}^{m+n} \to \mathbb{R}^m) \to \mathbb{R}^{m+n})$$

corresponding to the first order operators on general connections from 27.3. If we replace the coordinates y_{ij}^p by

$$Y_{ij}^p = y_{ij}^p + y_{iq}^p y_j^q, \tag{1}$$

we find easily that the action of $G_{m,n}^2$ on S_1 is given by 27.3.(1), 27.3.(2) and

$$\begin{aligned}\bar{Y}_{ij}^p = {} & a_q^p Y_{kl}^q \tilde{a}_i^k \tilde{a}_j^l + a_{rs}^p y_k^r y_l^s \tilde{a}_i^k \tilde{a}_j^l + a_{ql}^p y_k^q \tilde{a}_i^l \tilde{a}_j^k + a_{ql}^p y_k^q \tilde{a}_i^k \tilde{a}_j^l \\ & + a_q^p y_k^q \tilde{a}_{ij}^k + a_{kl}^p \tilde{a}_i^k \tilde{a}_j^l + a_k^p \tilde{a}_{ij}^k.\end{aligned} \tag{2}$$

Define further

$$S_{ij}^p = \frac{1}{2}(Y_{ij}^p + Y_{ji}^p), \qquad R_{ij}^p = \frac{1}{2}(Y_{ij}^p - Y_{ji}^p). \tag{3}$$

Since the right-hand side of (2) except the first term is symmetric in i and j, we obtain the action formula for $\bar{S}_{ij}^p$ by replacing Y_{kl}^q by S_{kl}^q on the right-hand side of (2). On the other hand,

$$\bar{R}_{ij}^p = a_q^p R_{kl}^q \tilde{a}_i^k \tilde{a}_j^l.$$

The map $\gamma\colon S_1 \to \mathbb{R}^n \otimes \Lambda^2\mathbb{R}^{m*}$, $\gamma(y^q_k, y^r_{\ell s}, y^t_{mn}) = R^p_{ij}$ will be called the *formal curvature map.*

Let Z be any $(G^2_m \times G^2_n)$-space. The canonical projection $G^2_{m,n} \to G^2_m$ and the group homomorphism $G^2_{m,n} \to G^2_n$ determined by the restriction of local isomorphisms of $\mathbb{R}^{m+n} \to \mathbb{R}^m$ to $\{0\} \times \mathbb{R}^n \subset \mathbb{R}^{m+n}$ define a map $p\colon G^2_{m,n} \to G^2_m \times G^2_n$. The kernel K of p is characterized by $a^i_j = \delta^i_j$, $a^i_{jk} = 0$, $a^p_q = \delta^p_q$, $a^p_{qr} = 0$. The group $G^2_{m,n}$ acts on $\mathbb{R}^n \otimes \Lambda^2\mathbb{R}^{m*}$ by means of the jet homomorphism π^2_1 into $G^1_m \times G^1_n$. One sees directly, that the curvature map γ satisfies the orbit condition with respect to K. Indeed, on K we have

$$(5) \qquad \bar y^p_i = y^p_i + a^p_i, \quad \bar y^p_{iq} = y^p_{iq} + a^p_{iq}, \quad \bar S^p_{ij} = S^p_{ij} + a^p_{qi} y^q_j + a^p_{qj} y^q_i + a^p_{ij}.$$

Using a^p_i, a^q_{jr}, $a^s_{k\ell}$, we can transform every $(y^p_i, y^q_{jr}, S^s_{k\ell})$ into $(0,0,0)$. In this situation, proposition 28.1 yields directly the following assertion.

Proposition. *Every $G^2_{m,n}$-map $S_1 \to Z$ factorizes through the formal curvature map $\gamma\colon S_1 \to \mathbb{R}^n \otimes \Lambda^2\mathbb{R}^{m*}$.*

28.3. The Utiyama theorem and general connections. In general, an r-th order Lagrangian on a fibered manifold $Y \to M$ is defined as a base-preserving morphism $J^rY \to \Lambda^m T^*M$, $m = \dim M$. Roughly speaking, the Utiyama theorem reads that every invariant first order Lagrangian on the connection bundle $QP \to M$ of an arbitrary principal fiber bundle $P \to M$ factorizes through the curvature map. This assertion will be formulated in a precise way in the framework of the theory of gauge natural operators in chapter XII. At this moment we shall apply proposition 28.2 to deduce similar results for the general connections on an arbitrary fibered manifold $Y \to M$.

Since the action 28.2.(5) is simply transitive, proposition 28.2 reflects exactly the possibilities for formulating Utiyama-like theorems for general connections. But the general interpretation of proposition 28.2 in terms of natural operators is beyond the scope of this example and we restrict ourselves to one special case only.

If we let the group $G^2_m \times G^2_n$ act on a manifold S by means of the first product projection, we obtain a G^2_m-space, which corresponds to a second order bundle functor F on $\mathcal{M}f_m$. (In the classical Utiyama theorem we have the first order bundle functor $\Lambda^m T^*$, which is allowed to be viewed as a second order functor as well.) Obviously, F can be interpreted as a bundle functor on $\mathcal{FM}_{m,n}$, if we compose it with the base functor $B\colon \mathcal{FM} \to \mathcal{M}f$ and apply the pullback construction. If we interpret proposition 28.2 in terms of natural operators between bundle functors on $\mathcal{FM}_{m,n}$, we obtain immediately

Proposition. *There is a bijection between the first order natural operators $A\colon J^1 \rightsquigarrow F$ and the zero order natural operators $A_0\colon V \otimes \Lambda^2 T^*B \rightsquigarrow F$ given by $A = A_0 \circ C$, where $C\colon J^1 \rightsquigarrow V \otimes \Lambda^2 T^*B$ is the curvature operator.*

28.4. The general Ricci identity. Before treating the classical tensor fields on manifolds, we deduce a general result for arbitrary vector bundles. Consider a linear connection Γ on a vector bundle $E \to M$ and a classical linear connection

Λ on M, i.e. a linear connection on $TM \to M$. The absolute differential ∇s of a section $s\colon M \to E$ is a section $M \to E \otimes T^*M$. Hence we can use the tensor product $\Gamma \otimes \Lambda^*$ of connection Γ and the dual connection Λ^* of Λ, see 47.14, to construct the absolute differential of ∇s. This is a section $\nabla^2_\Lambda s\colon E \otimes T^*M \otimes T^*M$ called the second absolute differential of s with respect to Γ and Λ. We describe the alternation $\mathrm{Alt}(\nabla^2_\Lambda s)\colon M \to E \otimes \Lambda^2 T^*M$. Let $R\colon M \to E \otimes E^* \otimes \Lambda^2 T^*M$ be the curvature of Γ and $S\colon M \to TM \otimes \Lambda^2 T^*M$ be the torsion of Λ. Then the contractions $\langle R, s\rangle$ and $\langle S, \nabla s\rangle$ are sections of $E \otimes \Lambda^2 T^*M$.

Proposition. *It holds*

$$\mathrm{Alt}(\nabla^2_\Lambda s) = -\langle R, s\rangle + \langle S, \nabla s\rangle. \tag{1}$$

Proof. This follows directly from the coordinate formula for $\nabla^2_\Lambda s$

$$\frac{\partial}{\partial x^j}\Big(\frac{\partial s^p}{\partial x^i} - \Gamma^p_{qi}s^q\Big) - \Gamma^p_{rj}\Big(\frac{\partial s^r}{\partial x^i} - \Gamma^r_{qi}s^q\Big) + \Lambda^k_{ij}\nabla_k s^p. \quad \square$$

The coordinate form of (1) will be called the *general Ricci identity* of E. If E is a vector bundle associated to P^1M and Γ is induced from a principal connection on P^1M, we take for Λ the connection induced from the same principal connection. In this case we write $\nabla^2 s$ only. For the classical tensor fields on M our proposition gives the classical Ricci identity, see e.g. [Lichnerowicz, 76, p. 69].

28.5. Curvature subspaces. We are going to describe some properties of the absolute derivatives of curvature tensors of linear symmetric connections on m-manifolds. Let $Q = (Q_r P^1\mathbb{R}^m)_0$ denote the standard fiber of the connection bundle in question, see 25.3, let $W = \mathbb{R}^m \otimes \mathbb{R}^{m*} \otimes \Lambda^2\mathbb{R}^{m*}$, $W_r = W \otimes \otimes^r \mathbb{R}^{m*}$, $W^r = W \times W_1 \times \ldots \times W_r$. The formal curvature is a map $C\colon T^1_m Q \to W$, its formal r-th absolute differential is $C_r = \nabla^r C\colon T^{r+1}_m Q \to W_r$. We write $C^r = (C, C_1, \ldots, C_r)\colon T^{r+1}_m Q \to W^r$, where the jet projections $T^{r+1}_m Q \to T^s_m Q$, $s < r+1$, are not indicated explicitly. (Such a slight simplification of notation will be used even later in this section.)

We define the *r-th order curvature equations* E_r on W^r as follows.

i) E_0 are the first Bianchi identity

$$W^i_{jkl} + W^i_{klj} + W^i_{ljk} = 0 \tag{1}$$

ii) E_1 are the absolute derivatives of (1)

$$W^i_{jklm} + W^i_{kljm} + W^i_{ljkm} = 0 \tag{2}$$

and the second Bianchi identity

$$W^i_{jklm} + W^i_{jlmk} + W^i_{jmkl} = 0 \tag{3}$$

iii) E_s, $s > 1$, are the absolute derivatives of E_{s-1} and the formal Ricci identity of the product vector bundle $W_{s-2} \times \mathbb{R}^m$. By 28.4, the latter equations are of the form

$$W^i_{jklm_1\cdots[m_{s-1}m_s]} = \mathrm{bilin}(W, W_{s-2}) \tag{4}$$

where the right-hand sides are some bilinear functions on $W \times W_{s-2}$.

Definition. The *r-th order curvature subspace* $K^r \subset W^r$ is defined by

$$E_0 = 0, \dots, E_r = 0.$$

We write $K = K^0 \subset W$. For $r = 1$ we denote by $K_1 \subset W_1$ the subspace defined by $E_1 = 0$. Hence $K^1 = K \times K_1$.

Lemma. *K^r is a submanifold of W^r, it holds $K^r = C^r(T_m^{r+1}Q)$ and the restricted map $C^r\colon T_m^{r+1}Q \to K^r$ is a submersion.*

Proof. To prove K^r is a submanifold we proceed by induction. For $r = 0$ we have a linear subspace. Assume $K^{r-1} \subset W^{r-1}$ is a submanifold. Consider the product bundle $K^{r-1} \times W_r$. Equations E_r consist of the following 3 systems

$$W^i_{\{jkl\}m_1\dots m_r} = 0 \tag{5}$$

$$W^i_{j\{klm_1\}m_2\dots m_r} = 0 \tag{6}$$

$$W^i_{jklm_1\cdots[m_{s-1}m_s]\cdots m_r} + \mathrm{polyn}(W^{r-2}) = 0 \tag{7}$$

where $\{\dots\}$ denotes the cyclic permutation and $\mathrm{polyn}(W^{r-2})$ are some polynomials on W^{r-2}. The map defined by the left-hand sides of (5)–(7) represents an affine bundle morphism $K^{r-1} \times W_r \to K^{r-1} \times \mathbb{R}^N$ of constant rank, $N =$ the number of equations (5)–(7). Analogously to 6.6 we find that its kernel K^r is a subbundle of $K^{r-1} \times W_r$.

To prove $K^r = C^r(T_m^{r+1}Q)$ we also proceed by induction.

Sublemma. *It holds $K = C(T_m^1 Q)$ and $K_1 = C_1(T_m^2 Q)$.*

Proof. The coordinate form of C is

$$W^i_{jkl} = \Gamma^i_{jk,l} - \Gamma^i_{jl,k} + \Gamma^i_{ml}\Gamma^m_{jk} - \Gamma^i_{mk}\Gamma^m_{jl}. \tag{8}$$

This is an affine bundle morphism of affine bundle $T_m^1 Q \to Q$ into W of constant rank. We know that the values of C lie in K, so that it suffices to prove that the image is the whole K at one point $0 \in Q$. The restricted map $\bar{C}\colon \mathbb{R}^m \otimes S^2\mathbb{R}^{m*} \otimes \mathbb{R}^{m*} \to W$ is of the form

$$W^i_{jkl} = \Gamma^i_{jk,l} - \Gamma^i_{jl,k}. \tag{9}$$

Denote by $\dim E_0$ the number of independent equations in E_0, so that $\dim K = \dim W - \dim E_0$. From linear algebra we know that K is the image of $\bar{C}$ if

$$\dim W - \dim E_0 = \dim \mathbb{R}^m \otimes S^2\mathbb{R}^{m*} \otimes \mathbb{R}^{m*} - \dim \operatorname{Ker}\bar{C}. \tag{10}$$

Clearly, $\dim W = m^3(m-1)/2$ and $\dim \mathbb{R}^m \otimes S^2\mathbb{R}^{m*} \otimes \mathbb{R}^{m*} = m^3(m+1)/2$. By (9) we have $\operatorname{Ker}\bar{C} = \mathbb{R}^m \otimes S^3\mathbb{R}^{m*}$, so that $\dim \operatorname{Ker}\bar{C} = m^2(m+1)(m+2)/6$. One finds easily that (1) represents one equation on W for any i and mutually different j, k, l, while (1) holds identically if at least two subscripts coincide. Hence $\dim E_0 = m^2(m-1)(m-2)/6$. Now (10) is verified by simple evaluation.

The absolute differentiation of (8) yields that C_1 is an affine morphism of affine bundle $T^2_m Q \to T^1_m Q$ into W_1 of constant rank. We know that the values of C_1 lie in K_1 so that it suffices to prove that the image is the whole K_1 at one point $0 \in T^1_m Q$. The restricted map $\bar{C}_1 : \mathbb{R}^m \otimes S^2\mathbb{R}^{m*} \otimes S^2\mathbb{R}^{m*} \to W_1$ is

$$W^i_{jklm} = \Gamma^i_{jk,lm} - \Gamma^i_{jl,km}. \tag{11}$$

Analogously to (10) we shall verify the dimension condition

$$\dim W_1 - \dim E_1 = \dim \mathbb{R}^m \otimes S^2\mathbb{R}^{m*} \otimes S^2\mathbb{R}^{m*} - \dim \operatorname{Ker}\bar{C}_1. \tag{12}$$

Clearly, $\dim W_1 = m^4(m-1)/2$, $\dim \mathbb{R}^m \otimes \otimes^2 S^2\mathbb{R}^{m*} = m^3(m+1)^2/4$. We have $\operatorname{Ker}\bar{C}_1 = \mathbb{R}^m \otimes S^4\mathbb{R}^{m*}$, so that $\dim \operatorname{Ker}\bar{C}_1 = m^2(m+1)(m+2)(m+3)/24$. For any i and mutually different j, k, l, m, (2) and (3) represent 8 equations, but one finds easily that only 7 of them are linearly independent. This yields $7m^2(m-1)(m-2)(m-3)/24$ independent equations. If exactly two subscripts coincide, (2) and (3) represent 2 independent equations. This yields another $m^2(m-1)(m-2)$ equations. In the remaining cases (2) and (3) hold identically. Now a direct evaluation proves our sublemma. □

Assume by induction $C^{r-1}: T^r_m Q \to K^{r-1}$ is a surjective submersion. The iterated absolute differentiation of (8) yields the following coordinate form of C_r

$$W^i_{jklm_1 \dots m_r} = \Gamma^i_{j[k,l]m_1 \dots m_r} + \operatorname{polyn}(T^r_m Q) \tag{13}$$

where $\operatorname{polyn}(T^r_m Q)$ are some polynomials on $T^r_m Q$. This implies C^r is an affine bundle morphism

$$\begin{array}{ccc} T^{r+1}_m Q & \xrightarrow{C^r} & K^r \\ \downarrow & & \downarrow \\ T^r_m Q & \xrightarrow{C^{r-1}} & K^{r-1} \end{array}$$

of constant rank. Hence it suffices to prove at one point $0 \in T^r_m Q$ that the image is the whole fiber of $K^r \to K^{r-1}$. The restricted map $\bar{C}_r : \mathbb{R}^m \otimes S^2\mathbb{R}^{m*} \otimes S^{r+1}\mathbb{R}^{m*} \to W_r$ is of the form

$$W^i_{jklm_1 \dots m_r} = \Gamma^i_{jk,lm_1 \dots m_r} - \Gamma^i_{jl,km_1 \dots m_r}. \tag{14}$$

By (7) the values of $\bar{C}_r$ lie in $W \otimes S^r\mathbb{R}^{m*}$. Then (5) and (6) characterize $(K \otimes S^r\mathbb{R}^{m*}) \cap (K_1 \otimes S^{r-1}\mathbb{R}^{m*})$. Consider an element $X = (X^i_{jklm_1 \dots m_r})$ of the latter space. Since $\bar{C}_1(\mathbb{R}^m \otimes S^2\mathbb{R}^{m*} \otimes S^2\mathbb{R}^{m*}) = K_1$ by the sublemma, the tensor product $\bar{C}_1 \otimes \mathrm{id}_{S^{r-1}\mathbb{R}^{m*}} : \mathbb{R}^m \otimes S^2\mathbb{R}^{m*} \otimes S^2\mathbb{R}^{m*} \otimes S^{r-1}\mathbb{R}^{m*} \to K_1 \otimes S^{r-1}\mathbb{R}^{m*}$ is a surjective map. Hence there is a $Y \in \mathbb{R}^m \otimes S^2\mathbb{R}^{m*} \otimes S^2\mathbb{R}^{m*} \otimes S^{r-1}\mathbb{R}^{m*}$ such that

$$X^i_{jklm_1 \dots m_r} = Y^i_{jklm_1 \dots m_r} - Y^i_{jlkm_1 \dots m_r}. \tag{15}$$

Consider the symmetrization $\bar{Y} = (Y^i_{jkl(m_1 m_2) \cdots m_r}) \in \mathbb{R}^m \otimes S^2\mathbb{R}^{m*} \otimes S^{r+1}\mathbb{R}^{m*}$. The second condition $X \in K \otimes S^r\mathbb{R}^{m*}$ implies X is symmetric in m_1 and m_2, so that $\bar{C}_r(\bar{Y}) = X$.

Finally, since $C^{r-1}: T^r_m Q \to K^{r-1}$ is a submersion and $C^r: T^{r+1}_m Q \to K^r$ is an affine bundle morphism surjective on each fiber, C^r is also a submersion. □

28.6. Linear symmetric connections. A fundamental result on the r-th order natural operators on linear symmetric connections with values in a first order natural bundle is that they factorize through the curvature operator and its absolute derivatives up to order $r-1$. We present a formal version of this result, which involves a precise description of the factorization.

Let F be a G^1_m-space, which is considered as a G^{r+2}_m-space by means of the jet homomorphism $G^{r+2}_m \to G^1_m$.

Theorem. *For every G^{r+2}_m-map $f\colon T^r_m Q \to F$ there exists a unique G^1_m-map $g\colon K^{r-1} \to F$ satisfying $f = g \circ C^{r-1}$.*

Proof. We use a recurrence procedure, in the first step of which we apply the orbit reduction with respect to the kernel B^{r+2}_{r+1} of the jet projection $G^{r+2}_m \to G^{r+1}_m$. Let $S_r\colon T^r_m Q \to \mathbb{R}^m \otimes S^{r+2}\mathbb{R}^{m*} =: S^1_{r+2}$ be the symmetrization

$$S^i_{j_1\dots j_{r+2}} = \Gamma^i_{(j_1 j_2, j_3 \dots j_{r+2})} \tag{1}$$

and $\pi^r_{r-1}\colon T^r_m Q \to T^{r-1}_m Q$ be the jet projection. Define

$$\varphi_r = (S_r, \pi^r_{r-1}, C_{r-1})\colon T^r_m Q \to S^1_{r+2} \times T^{r-1}_m Q \times W_{r-1}.$$

The map C_{r-1} is of the form

$$W^i_{jkl_1\dots l_r} = \Gamma^i_{jk,l_1\dots l_r} - \Gamma^i_{jl_1,kl_2\dots l_r} + \operatorname{polyn}(T^{r-1}_m Q). \tag{2}$$

One sees easily that in the formula

$$\Gamma^i_{jk,l_1\dots l_r} = S^i_{jkl_1\dots l_r} + (\Gamma^i_{jk,l_1\dots l_r} - \Gamma^i_{(jk,l_1\dots l_r)}) \tag{3}$$

the expression in brackets can be rewritten as a linear combination of terms of the form $\Gamma^i_{mn,p_1\dots p_r} - \Gamma^i_{mp_1,np_2\dots p_r}$. If we replace each of them by $W^i_{mnp_1\dots p_r} - \operatorname{polyn}(T^{r-1}_m Q)$ according to (2), we obtain a map (not uniquely determined) $\psi_r\colon S^1_{r+2} \times T^{r-1}_m Q \times W_{r-1} \to T^r_m Q$ over $\mathrm{id}_{T^{r-1}_m Q}$ satisfying

$$\psi_r \circ \varphi_r = \mathrm{id}_{T^r_m Q}. \tag{4}$$

Consider the canonical action of Abelian group $B^{r+2}_{r+1} = S^1_{r+2}$ on itself, which is simply transitive. From the transformation laws of Γ^i_{jk} it follows that ψ_r is a B^{r+2}_{r+1}-map. Thus the composed map $f \circ \psi_r\colon S^1_{r+2} \times T^{r-1}_m Q \times W_{r-1} \to F$ satisfies the orbit condition for B^{r+2}_{r+1} with respect to the product projection $p_r\colon S^1_{r+2} \times T^{r-1}_m Q \times W_{r-1} \to T^{r-1}_m Q \times W_{r-1}$. By 28.1 there is a G^{r+1}_m-map $g_r\colon T^{r-1}_m Q \times W_{r-1} \to F$ satisfying $f \circ \psi_r = g_r \circ p_r$. Composing both sides with φ_r, we obtain $f = g_r \circ (\pi^r_{r-1}, C_{r-1})$.

In the second step we define analogously

$$\varphi_{r-1} = (S_{r-1}, \pi^{r-1}_{r-2}, C_{r-2})\colon T^{r-1}_m Q \to S^1_{r+1} \times T^{r-2}_m Q \times W_{r-2}$$

and construct $\psi_{r-1}\colon S^1_{r+1}\times T^{r-2}_m Q\times W_{r-2}\to T^{r-1}_m Q$ satisfying $\psi_{r-1}\circ\varphi_{r-1}=\mathrm{id}_{T^{r-1}_m Q}$. The composed map $g_r\circ(\psi_{r-1}\times\mathrm{id}_{W_{r-1}})\colon S^1_{r+1}\times T^{r-2}_m Q\times W_{r-2}\times W_{r-1}\to F$ is equivariant with respect to the kernel B^{r+1}_r of the jet projection $G^{r+1}_m\to G^r_m$. The product projection of $S^1_{r+1}\times T^{r-2}_m Q\times W_{r-2}\times W_{r-1}$ omitting the first factor satisfies the orbit condition for B^{r+1}_r. This yields a G^r_m-map $g_{r-1}\colon T^{r-2}_m Q\times W_{r-2}\times W_{r-1}\to F$ such that $g_r=g_{r-1}\circ((\pi^{r-1}_{r-2},C_{r-2})\times\mathrm{id}_{W_{r-1}})$, i.e. $f=g_{r-1}\circ(\pi^r_{r-2},C_{r-2},C_{r-1})$.

In the last but one step we construct a G^2_m-map $g_1\colon Q\times W\times\ldots\times W_{r-1}\to F$ such that $f=g_1\circ(\pi^r_0,C,\ldots,C_{r-1})$. The product projection p_1 of $Q\times W\times\ldots\times W_{r-1}$ omitting the first factor satisfies the orbit condition for the kernel B^2_1 of the jet projection $G^2_m\to G^1_m$. By 28.1 there is a G^1_m-map $g_0\colon W\times\ldots\times W_{r-1}\to F$ satisfying $g_1=g_0\circ p_1$. Hence $f=g_0\circ C^{r-1}$. Since $K^{r-1}=C^{r-1}(T^r_m Q)$, the restriction $g=g_0|K^{r-1}$ is uniquely determined.

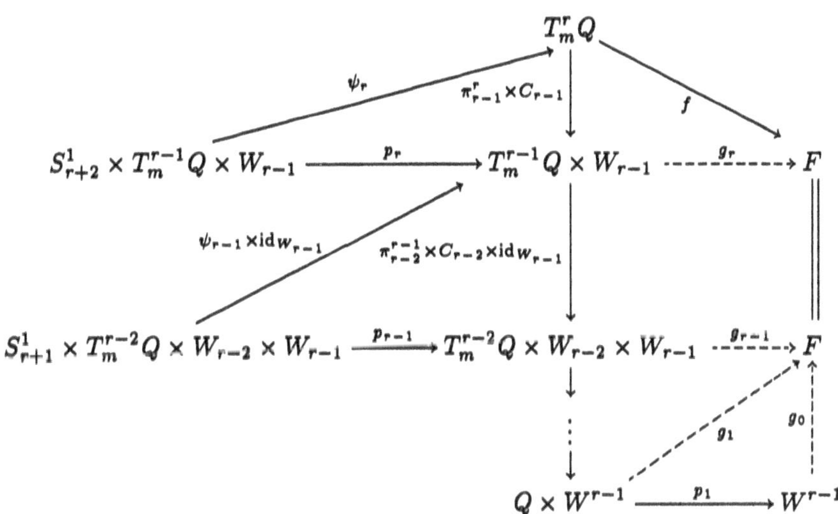

□

28.7. Example. We determine all natural operators $Q_\tau P^1 \rightsquigarrow T^*\otimes T^*$. By 23.5, every such operator has a finite order r. Let

$$u=f(\Gamma_0,\Gamma_1,\ldots,\Gamma_r)$$

$\Gamma_s\in\mathbb{R}^m\otimes S^2\mathbb{R}^{m*}\otimes S^s\mathbb{R}^{m*}$, be its associated map. The equivariance of f with respect to the homotheties in $G^1_m\subset G^{r+2}_m$ yields

$$k^2 f(\Gamma_0,\Gamma_1,\ldots,\Gamma_r)=f(k\Gamma_0,k^2\Gamma_1,\ldots,k^{r+1}\Gamma_r).$$

By the homogeneous function theorem, f is a first order operator. According to 28.6, the first order operators are in bijection with G^1_m-maps $K\to\mathbb{R}^{m*}\otimes\mathbb{R}^{m*}$. Let $u=g(W)$ be such a map. The equivariance with respect to the homotheties yields $k^2 g(W)=g(k^2 W)$, so that g is linear. Consider the injection $i\colon K\to$

$\mathbb{R}^m \otimes \otimes^3\mathbb{R}^{m*}$. Since $\mathbb{R}^m \otimes \otimes^3\mathbb{R}^{m*}$ is a completely reducible $GL(m)$-module, there is an equivariant projection $p\colon \mathbb{R}^m \otimes \otimes^3\mathbb{R}^{m*} \to K$ satisfying $p \circ i = \mathrm{id}_K$. Hence we can proceed analogously to 24.8. By the invariant tensor theorem, all linear G^1_m-maps $\mathbb{R}^m \otimes \otimes^3\mathbb{R}^{m*} \to \mathbb{R}^{m*} \otimes \mathbb{R}^{m*}$ form a 6-parameter family. Its restriction to K gives the following 2-parameter family

$$k_1 W^k_{kij} + k_2 W^k_{ikj}.$$

Let R_1 and R_2 be the corresponding contractions of the curvature tensor. By theorem 28.6, *all natural operators $Q_rP^1 \rightsquigarrow T^*\otimes T^*$ form a two parameter family linearly generated by two contractions R_1 and R_2 of the curvature tensor.*

28.8. Ricci subspaces. Let $V = \mathbb{R}^n$ be a $GL(m)$-module and $\tilde{V}$ denote the corresponding first order natural vector bundle over m-manifolds. Write $V_r = V \otimes \otimes^r\mathbb{R}^{m*}$, $V^r = V \times V_1 \times \ldots \times V_r$. The formal r-th order absolute differentiation defines a map $D^V_r = \nabla^r\colon T^{r-1}_m Q \times T^r_m V \to V_r$, $D^V_0 = \mathrm{id}_V$. If v^p, $v^p_i, \ldots, v^p_{i_1\ldots i_r}$ are the jet coordinates on $T^r_m V$ (symmetric in all subscripts) and $V^p_{i_1\ldots i_r}$ are the canonical coordinates on V_r, then D^V_r is of the form

$$V^p_{i_1\ldots i_r} = v^p_{i_1\ldots i_r} + \mathrm{polyn}(T^{r-1}_m Q \times T^{r-1}_m V). \tag{1}$$

Set $D^r_V = (D^V_0, D^V_1, \ldots, D^V_r)\colon T^{r-1}_m Q \times T^r_m V \to V^r$.

We define the *r-th order Ricci equations* E^V_r, $r \geq 2$, as follows. For $r = 2$, E^V_2 are the formal Ricci identities of $\tilde{V}(\mathbb{R}^m)$. By 28.4, they are of the form

$$V^p_{[ij]} - \mathrm{bilin}(W, V) = 0. \tag{2}$$

For $r > 2$, E^V_r are the absolute derivatives of E^V_{r-1} and the formal Ricci identities of $\tilde{V}(\mathbb{R}^m) \otimes \otimes^{r-2}T^*\mathbb{R}^m$. These equations are of the form

$$V^p_{i_1\cdots[i_{s-1}i_s]\cdots i_r} - \mathrm{bilin}(W^{r-2}, V^{r-2}) = 0. \tag{3}$$

Definition. The *r-th order Ricci subspace* $Z^r_V \subset K^{r-2} \times V^r$ is defined by $E^V_2 = 0, \ldots, E^V_r = 0$, $r \geq 2$. For $r = 0, 1$ we set $Z^0_V = V$ and $Z^1_V = V^1$.

Lemma. *Z^r_V is a submanifold of $K^{r-2} \times V^r$, it holds $Z^r_V = (C^{r-2}, D^r_V)(T^{r-1}_m Q \times T^r_m V)$ and the restricted map $(C^{r-2}, D^r_V)\colon T^{r-1}_m Q \times T^r_m V \to Z^r_V$ is a submersion.*

Proof. For $r = 0$ we have $Z^0_V = V$ and $D^0_V = \mathrm{id}_V$. For $r = 1$, $D^1_V\colon Q \times T^1_m V \to V^1 = Z^1_V$ is of the form

$$V^p = v^p, \qquad V^p_i = v^p_i + \mathrm{bilin}(Q, V)$$

so that our claim is trivial. Assume by induction Z^{r-1}_V is a submanifold and the restriction of the first product projection of $K^{r-3} \times V^{r-1}$ to Z^{r-1}_V is a surjective submersion. Consider the fiber product $K^{r-2} \times_{K^{r-3}} Z^{r-1}_V$ and the product vector bundle $(K^{r-2} \times_{K^{r-3}} Z^{r-1}_V) \times V_r$. By (3) Z^r_V is characterized by affine equations of constant rank. This proves Z^r_V is a subbundle and $Z^r_V \to K^{r-2}$ is a surjective submersion.

Assume by induction $(C^{r-3}, D^{r-1}_V)\colon T^{r-2}_m Q \times T^{r-1}_m V \to Z^{r-1}_V$ is a surjective submersion. We have $T^r_m V = T^{r-1}_m V \times V \otimes S^r\mathbb{R}^{m*}$. By (1) and (3), $(C^{r-2}, D^r_V)\colon (T^{r-1}_m Q \times T^{r-1}_m V) \times V \otimes S^r\mathbb{R}^{m*} \to (Z^r_V \to K^{r-2} \times_{K^{r-3}} Z^{r-1}_V)$ is bijective on each fiber. This proves our lemma. □

28.9. The following result is of technical character, but it covers the core of several applications. Let F be a G^1_m-space.

Proposition. *For every G^{r+1}_m-map $f\colon T^{r-1}_m Q \times T^r_m V \to F$ there exists a unique G^1_m-map $g\colon Z^r_V \to F$ satisfying $f = g \circ (C^{r-2}, D^r_V)$.*

Proof. First we deduce a lemma.

Lemma. *If $y, \bar{y} \in T^{r-1}_m Q$ satisfy $C^{r-2}(y) = C^{r-2}(\bar{y})$, then there is an element $h \in B^{r+1}_1$ of the kernel B^{r+1}_1 of the jet projection $G^{r+1}_m \to G^1_m$ such that $h(\bar{y}) = y$.*

Indeed, consider the orbit set $T^{r-1}_m Q/B^{r+1}_1$. (We shall not need a manifold structure on it, as one checks easily that 28.1 and 28.6 work at the set-theoretical level as well.) This is a G^1_m-set under the action $a(B^{r+1}_1 y) = aB^{r+1}_1(y)$, $y \in T^{r-1}_m Q$, $a \in G^1_m \subset G^{r+1}_m$. Clearly, the factor projection

$$p\colon T^{r-1}_m Q \to T^{r-1}_m Q/B^{r+1}_1$$

is a G^{r+1}_m-map. By 28.6 there is a map $g\colon K^{r-2} \to T^{r-1}_m Q/B^{r+1}_1$ satisfying $p = g \circ C^{r-2}$. If $C^{r-2}(y) = C^{r-2}(\bar{y}) = x$, then $p(y) = p(\bar{y}) = g(x)$. This proves our lemma.

Consider the map $(\mathrm{id}_{T^{r-1}_m Q}, D^r_V)\colon T^{r-1}_m Q \times T^r_m V \to T^{r-1}_m Q \times V^r$ and denote by $\tilde{V}^r \subset T^{r-1}_m Q \times V^r$ its image. By 28.8.(1), the restricted map $\mathcal{D}^r_V\colon T^{r-1}_m Q \times T^r_m V \to \tilde{V}^r$ is bijective for every $y \in T^{r-1}_m Q$, so that $\mathcal{D}^r_V$ is an equivariant diffeomorphism. Define $\tilde{C}^{r-2}\colon \tilde{V}^r \to Z^r_V$, $\tilde{C}^{r-2}(y,z) = (C^{r-2}(y), z)$, $y \in T^{r-1}_m Q$, $z \in V^r$. By lemma 28.5, $\tilde{C}^{r-2}$ is a surjective submersion. By definition, $\tilde{C}^{r-2}(y,z) = \tilde{C}^{r-2}(\bar{y},\bar{z})$ means $C^{r-2}(y) = C^{r-2}(\bar{y})$ and $z = \bar{z}$. Thus, the above lemma implies $\tilde{C}^{r-2}$ satisfies the orbit condition for B^{r+1}_1. By 28.1 there is a G^1_m-map $g\colon Z^r_V \to F$ satisfying $f \circ (\mathcal{D}^r_V)^{-1} = g \circ \tilde{C}^{r-2}$. Composing both sides with $\mathcal{D}^r_V$, we find $f = g \circ (C^{r-2}, D^r_V)$. □

28.10. Remark. The idea of the proof of proposition 28.9 can be applied to suitable invariant subspaces of V as well. We shall need the case $P = \mathrm{Reg}S^2\mathbb{R}^{m*} \subset S^2\mathbb{R}^{m*}$ of the standard fiber of the bundle of pseudoriemannian metrics over m-manifolds. In this case we only have to modify the definition of P_r to $P_r = S^2\mathbb{R}^{m*} \otimes \otimes^r\mathbb{R}^{m*}$, but the rest of 28.8 and 28.9 remains to be unchanged. Thus, *for every G^{r+1}_m-map $f\colon T^{r-1}_m Q \times T^r_m P \to F$ there exists a unique G^1_m-map $g\colon Z^r_P \to F$ satisfying $f = g \circ (C^{r-2}, D^r_P)$.*

28.11. Linear symmetric connection and a general vector field. Let $\tilde{F}$ denote the first order natural bundle over m-manifolds determined by G^1_m-space F. Consider an r-th order natural operator $Q_r P^1 \oplus \tilde{V} \rightsquigarrow \tilde{F}$ with associated G^{r+2}_m-map $f\colon T^r_m Q \times T^r_m V \to F$. Let $\bar{Z}^r_V \subset K^{r-1} \times V^r$ be the pre-image of $Z^r_V \subset K^{r-2} \times V^r$ with respect to the canonical projection $K^{r-1} \to K^{r-2}$.

Take the map $\psi_r\colon S^1_{r+2} \times T^{r-1}_m Q \times W_{r-1} \to T^r_m Q$ from 28.6 and construct $\psi_r \times \mathrm{id}_{T^r_m V}\colon S^1_{r+2} \times T^{r-1}_m Q \times W_{r-1} \times T^r_m V \to T^r_m Q \times T^r_m V$. If we apply the orbit reduction to $f \circ (\psi_r \times \mathrm{id}_{T^r_m V})$ in the previous way, we obtain a G^{r+1}_m-map $h\colon T^{r-1}_m Q \times W_{r-1} \times T^r_m V \to F$ such that $f = h \circ ((\pi^r_{r-1}, C_{r-1}) \times \mathrm{id}_{T^r_m V})$. Applying proposition 28.9 (with 'parameters' from W_{r-1}) to h, we obtain

Proposition. *For every G_m^{r+2}-map $f\colon T_m^r Q \times T_m^r V \to F$ there exists a unique G_m^1-map $g\colon \bar{Z}_V^r \to F$ satisfying $f = g \circ (C^{r-1}, D_V^r)$.*

Roughly speaking, every r-th order natural operator $Q_\tau P^1 \oplus \tilde{V} \rightsquigarrow \tilde{F}$ factorizes through the curvature operator and its absolute derivatives up to order $r-1$ and through the absolute derivatives on vector bundle $\tilde{V}$ up to order r.

28.12. Linear non-symmetric connections. An arbitrary linear connection on TM can be uniquely decomposed into its symmetrization and its torsion tensor. In other words, $QP^1M = Q_\tau P^1 M \oplus TM \otimes \Lambda^2 T^*M$. Hence we have the situation of 28.11, in which the role of standard fiber V is played by $\mathbb{R}^m \otimes \Lambda^2 \mathbb{R}^{m*} =: H$. This proves

Corollary. *For every G_m^{r+2}-map $f\colon J_0^r(QP^1\mathbb{R}^m) \to F$ there exists a unique G_m^1-map $g\colon \bar{Z}_H^r \to F$ satisfying $f = g \circ (C^{r-1}, D_H^r)$.*

28.13. Example. We determine all natural operators $QP^1 \rightsquigarrow T^* \otimes T^*$. In the same way as in 28.7 we deduce that such operators are of the first order. By 28.12 we have to find all G_m^1-maps $f\colon K \times H \times H_1 \to \mathbb{R}^{m*} \otimes \mathbb{R}^{m*}$. The equivariance with respect to the homotheties yields the homogeneity condition

$$k^2 f(W, H, H_1) = f(k^2 W, kH, k^2 H_1).$$

Hence f is linear in W and H_1 and quadratic in H. The term linear in W was determined in 28.7. By the invariant tensor theorem, the term quadratic in H is generated by the permutations of m, n, p, q in

$$\delta_i^m \delta_j^n \delta_k^p \delta_l^q H^k_{mn} H^l_{pq}.$$

This yields the 3 different double contractions $S^k_{ik} S^l_{jl}$, $S^k_{ij} S^l_{kl}$, $S^k_{il} S^l_{jk}$ of the tensor product $S \otimes S$ of the torsion tensor with itself. Finally, the term linear in H_1 corresponds to the permutations of l, m, n in

$$\delta_i^l \delta_j^m \delta_k^n H^k_{lmn}.$$

This gives 3 generators

$$H^k_{ijk}, \quad H^k_{ikj}, \quad H^k_{jki}. \tag{1}$$

Thus, *all natural operators $QP^1 \to T^* \otimes T^*$ form an 8-parameter family linearly generated by 2 different contractions of the curvature tensor of the symmetrized connection, by 3 different double contractions of $S \otimes S$ and by 3 operators constructed from the covariant derivatives of the torsion tensor with respect to the symmetrized connection according to (1).*

We remark that the first author determined all natural operators $QP^1 \to T^* \otimes T^*$ by direct evaluation in [Kolář, 87b]. Some of his generators are geometrically different of our present result, but both 8-parameter families are, of course, linearly equivalent.

28.14. Pseudoriemannian metrics. Using the notation of 28.10, we deduce a reduction theorem for natural operators on pseudoriemannian metrics. Let $\bar{P}^r = Z^r_P \cap (K^{r-2} \times P \times \{0\} \times \ldots \times \{0\})$ be the subspace determined by $0 \in P_1, \ldots, 0 \in P_r$.

Lemma. *$\bar{P}^r$ is a submanifold of Z^r_P.*

Proof. By [Lichnerowicz, 76, p. 69], the Ricci identity in the case of the bundle of pseudoriemannian metrics has the form

$$P_{ij[kl]} + W^m_{ikl} P_{mj} + W^m_{jkl} P_{im} = 0. \tag{1}$$

Thus, for $r = 2$, $\bar{P}^2 \subset W \times P^2$ is characterized by the curvature equations E_2, by $P_{ijk} = 0$, $P_{ijkl} = 0$ and by

$$W^m_{ikl} P_{mj} + W^m_{jkl} P_{im} = 0. \tag{2}$$

Equations (2) are G^1_m-equivariant. We know that P is divided into $m+1$ components P_σ according to the signature σ of the metric in question. Every element in each component can be transformed by a linear isomorphism into a canonical form $\pm\delta_{ij}$. This implies that $\bar{P}^2$ is characterized by linear equations of constant rank over each component P_σ. Assume by induction $\bar{P}^{r-1} \subset W^{r-3} \times P^{r-1}$ is a submanifold. Then $\bar{P}^r \subset \bar{P}^{r-1} \times \{0\} \times W_{r-2}$ is characterized by the curvature equations E_r and by

$$W^n_{iklm_1\ldots m_{r-2}} P_{nj} + W^n_{jklm_1\ldots m_{r-2}} P_{in} = 0. \tag{3}$$

By the above argument we deduce that this is a system of affine equations of constant rank over each P_σ. $\square$

Consider a G^{r+1}_m-map $f\colon T^r_m P \to F$. Applying 28.10 to $f \circ p_2 = T^{r-1}_m Q \times T^r_m P \to F$, where p_2 is the second product projection, we obtain a G^1_m-map $h\colon Z^r_P \to F$ satisfying

$$f \circ p_2 = h \circ (C^{r-2}, D^r_P). \tag{4}$$

Let $\lambda_r\colon T^r_m P \to T^{r-1}_m Q$ be the map determined by constructing the r-jets of the Levi-Cività connection. Composing (4) with $(\lambda_r, \mathrm{id})\colon T^r_m P \to T^{r-1}_m Q \times T^r_m P$, we find

$$f = h \circ (C^{r-2}, D^r_P) \circ (\lambda_r, \mathrm{id}). \tag{5}$$

Let g be the restriction of h to $\bar{P}^r$. Since the Levi-Cività connection is characterized by the fact that the absolute differential of the metric tensor vanishes, the values of $(C^{r-2}, D^r_P) \circ (\lambda_r, \mathrm{id})$ lie in $\bar{P}^r$. Write $L^{r-2} = (C^{r-2}, D^r_P) \circ (\lambda_r, \mathrm{id})\colon T^r_m P \to \bar{P}^r$. Then we can summarize by

Proposition. *For every G_m^{r+1}-map $f\colon T_m^r P \to F$ there exists a G_m^1-map $g\colon \bar P^r \to F$ satisfying $f = g \circ L^{r-2}$.* □

This is the classical assertion that every r-th order natural operator on pseudoriemannian metrics with values in an arbitrary first order natural bundle factorizes through the metric itself and through the absolute derivatives of the curvature tensor of the Levi-Cività connection up to order $r-2$.

We remark that each component P_σ of P can be treated separately in course of the proof of the above proposition. Hence the result holds for any kind of pseudoriemannian metrics (in particular for the proper Riemannian metrics).

28.15. Pseudoriemannian metric and a general vector field. A simple modification of 28.11 and 28.14 leads to a reduction theorem for the r-th order natural operators transforming a pseudoriemannian metric and a general vector field into a section of a first order natural bundle. In the notation from 28.11 and 28.14, let $f\colon T_m^r P \times T_m^r V \to F$ be a G_m^{r+1}-map. Consider the product projection $p\colon T_m^{r-1} Q \times T_m^r P \times T_m^r V \to T_m^r P \times T_m^r V$. Then we can apply 28.9 and 28.10 to the product $P \times V$. Hence there exists a G_m^1-map $h\colon Z_{P\times V}^r \to F$ satisfying

(1) $$f \circ p = h \circ (C^{r-2}, D_{P\times V}^r).$$

Denote by $\bar P_V^r \subset Z_P^r P \times V \subset K^{r-2} \times P^r \times V^r$ the subspace determined by $0 \in P_1, \dots, 0 \in P_r$. Analogously to 28.14 we deduce that $\bar P_V^r$ is a submanifold. Write $L_V^{r-2} = (\lambda_r, \mathrm{id}_{T_m^r P}) \times \mathrm{id}_{T_m^r V}\colon T_m^r P \times T_m^r V \to \bar P_V^r$, i.e. $L_V^{r-2}(u,v) = (C^{r-2}(\lambda_r(u)), u_0, 0, \dots, 0, D_V^r(\lambda_r(u), v))$, $u \in T_m^r P$, $v \in T_m^r V$, $u_0 = \pi_0^r(u)$. Then (1) implies $f = h \circ L_V^{r-2}$. If we denote by g the restriction of h to $\bar P_V^r$, we obtain the following assertion.

Proposition. *For every G_m^{r+1}-map $f\colon T_m^r P \times T_m^r V \to F$ there exists a G_m^1-map $g\colon \bar P_V^r \to F$ satisfying $f = g \circ L_V^{r-2}$.*

Hence every r-th order natural operator transforming a pseudoriemannian metric and a general vector field into a section of a first order natural bundle factorizes through the metric itself, through the absolute derivatives of the curvature tensor of the Levi-Cività connection up to the order $r-2$ and through the absolute derivatives with respect to the Levi-Cività connection of the general vector field up to the order r.

28.16. Remark. Since $Q_r P^1 M \to M$ is an affine bundle, the standard fiber $T_m^r Q$ of its r-th jet prolongation is an affine space by 12.17. In other words, G_m^{r+2} acts on $T_m^r Q$ by affine isomorphisms. Consider an affine action of G_m^1 of F (with the linear action as a special case). Then we can introduce the concept of a polynomial map $T_m^r Q \to F$ analogously to 24.10. Analyzing the proof of theorem 28.6, we observe that all the maps ψ_r and φ_r are polynomial. This implies that *for every polynomial G_m^{r+2} equivariant map $f\colon T_m^r Q \to F$, the unique G_m^1-equivariant map $g\colon K^{r-1} \to F$ from the theorem 28.6 is the restriction of a polynomial map $\bar g\colon W^{r-1} \to F$.*

Consider further a G_m^1-module V as in 28.8 or an invariant open subset of such a module as in 28.10. Then we also have defined the concept of a polynomial map of $T^{r-1}_m Q \times T^r_m V$ into an affine G^1_m-space F. Quite similarly to the first part of this remark we deduce that *for every polynomial G^{r+1}_m-equivariant map $f: T^{r-1}_m Q \times T^r_m V \to F$ the unique G^1_m-equivariant map $g: Z^r_V \to F$ from the proposition 28.9 is the restriction of a polynomial map $\bar{g}: W^{r-2} \times V^r \to F$.*

29. The method of differential equations

29.1. In chapter IV we have clarified that the finite order natural operators between any two bundle functors are in a canonical bijection with the equivariant maps between certain G-spaces. We recall that in 5.15 we deduced the following infinitesimal characterization of G-equivariance. Given a connected Lie group G and two G-spaces S and Z we construct the induced fundamental vector field ζ^S_A and ζ^Z_A on S and Z for every element $A \in \mathfrak{g}$ of the Lie algebra of G. Then $f: S \to Z$ is a G-equivariant map if and only if vector fields ζ^S_A and ζ^Q_A are f-related for every $A \in \mathfrak{g}$, i.e.

$$(1) \qquad Tf \circ \zeta^S_A = \zeta^Z_A \circ f \qquad \text{for all } A \in \mathfrak{g}.$$

The coordinate expression of (1) is a system of partial differential equations for the coordinate components of f. If we can find the general solution of this system, we obtain all G-equivariant maps. This procedure is sometimes called the method of differential equations.

29.2. Remark. If G is not connected and G^+ denotes its connected component of unity, then the solutions of 29.1.(1) determine all G^+-equivariant maps $S \to Z$. Obviously, there is an algebraic procedure how to decide which of these maps are G-equivariant. We select one element g_a in each connected component of G and we check which solutions of 29.1.(1) are invariant with respect to all g_a. However, one usually interprets the solutions of 29.1.(1) geometrically. In practice, if we succeed in finding the geometrical constructions of all solutions of 29.1.(1), it is clear that all of them determine the G-equivariant maps and we are not obliged to discuss the individual connected components of G.

29.3. From 5.12 we have that for each left G-space S the map of the fundamental vector fields $A \mapsto \zeta^S_A$, $A \in \mathfrak{g}$, is a Lie algebra antihomomorphism, i.e. $\zeta^S_{[A,B]} = -[\zeta^S_A, \zeta^S_B]$ for all A, $B \in \mathfrak{g}$, where on the left-hand side is the Lie bracket in $\mathfrak{g}$ and on the right-hand side we have the bracket of vector fields. Hence if some vectors A_α, $\alpha = 1, \dots, q \leq \dim G$ generate $\mathfrak{g}$ as a Lie algebra, i.e. A_α with all their iterated brackets generate $\mathfrak{g}$ as a vector space, then the equations

$$Tf \circ \zeta^S_{A_\alpha} = \zeta^Z_{A_\alpha} \circ f \qquad \alpha = 1, \dots, q$$

imply $Tf \circ \zeta^S_A = \zeta^Z_A \circ f$ for all $A \in \mathfrak{g}$. In particular, for the group G^r_m the generators of its Lie algebra are described in 13.9 and 13.10.

29.4. The Levi-Cività connection. We are going to determine all first order natural operators transforming pseudoriemannian metrics into linear connections. We denote by $\mathrm{Reg}S^2T^*M$ the bundle of all pseudoriemannian metrics over an m-manifold M, so that the standard fiber of the corresponding natural bundle over m-manifolds is the subset $\mathrm{Reg}S^2\mathbb{R}^{m*} \subset S^2\mathbb{R}^{m*}$ of all elements g_{ij} satisfying $\det(g_{ij}) \neq 0$. Since the zero of $S^2\mathbb{R}^{m*}$ does not lie in $\mathrm{Reg}S^2\mathbb{R}^{m*}$, the homogeneous function theorem is of no use for our problem. (Of course, this analytical fact is deeply reflected in the geometry of pseudoriemannian manifolds.) Hence we shall try to apply the method of differential equations. In the canonical coordinates $g_{ij} = g_{ji}$, $g_{ij,k}$ on the standard fiber $S = J_0^1\mathrm{Reg}S^2T^*\mathbb{R}^m$, the action of G_m^2 has the following form

$$\bar{g}_{ij} = g_{kl}\tilde{a}_i^k\tilde{a}_j^l \tag{1}$$

$$\bar{g}_{ij,k} = g_{lm,n}\tilde{a}_i^l\tilde{a}_j^m\tilde{a}_k^n + g_{lm}(\tilde{a}_{ik}^l\tilde{a}_j^m + \tilde{a}_i^l\tilde{a}_{jk}^m). \tag{2}$$

Since we deal with a classical problem, we shall use the classical Christoffel's on the standard fiber $Z = (QP^1\mathbb{R}^m)_0$. In this case we have the following action of G_m^2

$$\bar{\Gamma}^i_{jk} = a_l^i\Gamma^l_{mn}\tilde{a}_j^m\tilde{a}_k^n + a_l^i\tilde{a}_{jk}^l \tag{3}$$

see 17.15.

We shall not need all differential equations of our problem, since we shall proceed in another way in the final step. It is sufficient to deduce the fundamental vector fields $S^{\ jk}_i$ on S and $Z^{\ jk}_i$ on Z corresponding to the one-parameter subgroups $a_j^i = \delta_j^i$, $\tilde{a}^i_{jk} = t$ for $j \neq k$ and $a_j^i = \delta_j^i$, $a^i_{jj} = 2t$. From (1)–(3) we deduce easily

$$S^{\ jk}_i = 2g_{il}\left(\frac{\partial}{\partial g_{lj,k}} + \frac{\partial}{\partial g_{lk,j}}\right) \tag{4}$$

and

$$Z^{\ jk}_i = \frac{\partial}{\partial \Gamma^i_{jk}} + \frac{\partial}{\partial \Gamma^i_{kj}} \tag{5}$$

Hence the corresponding part of the differential equations for a G_m^2-equivariant map $\Gamma\colon S \to Z$ with components $\Gamma^i_{jk}(g_{lm}, g_{lm,n})$ is

$$2g_{lp}\left(\frac{\partial\Gamma^i_{jk}}{\partial g_{pm,n}} + \frac{\partial\Gamma^i_{jk}}{\partial g_{pn,m}}\right) = \delta_l^i\left(\delta_j^m\delta_k^n + \delta_j^n\delta_k^m\right). \tag{6}$$

Multiplying by g^{lq} and replacing q by l, we find

$$\frac{\partial\Gamma^i_{jk}}{\partial g_{lm,n}} + \frac{\partial\Gamma^i_{jk}}{\partial g_{ln,m}} = \frac{1}{2}g^{il}\left(\delta_j^m\delta_k^n + \delta_k^m\delta_j^n\right). \tag{7}$$

Let (7') or (7") be the equations derived from (7) by the permutation $(l,m,n) \mapsto (m,n,l)$ or $(l,m,n) \mapsto (n,l,m)$, respectively. Then the sum (7)+(7')−(7") yields

$$2\frac{\partial \Gamma^i_{jk}}{\partial g_{lm,n}} = \frac{1}{2}\Big(g^{il}(\delta^m_j\delta^n_k + \delta^m_k\delta^n_j) + g^{im}(\delta^n_j\delta^l_k + \delta^n_k\delta^l_j) - g^{in}(\delta^l_j\delta^m_k + \delta^l_k\delta^m_j)\Big). \tag{8}$$

The right-hand sides are independent on $g_{ij,k}$. Since we meet such a situation frequently, it is useful to formulate a simple lemma of general character.

29.5. Lemma. *Let U be an open subset in $\mathbb{R}^a$ with coordinates z^α and let $f(z^\alpha, w^\lambda)$ be a smooth function on $U\times\mathbb{R}^b$, $(w^\lambda)\in\mathbb{R}^b$, satisfying $\frac{\partial f(z,w)}{\partial w^\lambda} = g_\lambda(z)$. Then*

$$f(z,w) = \sum_{\lambda=1}^{b} g_\lambda(z)w^\lambda + h(z) \tag{1}$$

where $h(z)$ is a smooth function on U.

Proof. Notice that the difference $F(z,w) = f(z,w) - \sum_{\lambda=1}^b g_\lambda(z)w^\lambda$ satisfies $\frac{\partial F}{\partial w^\lambda} = 0$. □

Applying lemma 29.5 to 29.4.(8), we find

$$\Gamma^i_{jk} = \frac{1}{2}g^{il}(g_{lj,k} + g_{lk,j} - g_{jk,l}) + \gamma^i_{jk}(g_{lm}).$$

For $\gamma^i_{jk} = 0$ we obtain the coordinate expression of the Levi-Cività connection Λ, which is natural by its standard geometric interpretation. Hence the difference $\Gamma - \Lambda$ is a $GL(m)$-equivariant map $\mathrm{Reg}S^2\mathbb{R}^{m*} \to \mathbb{R}^m \otimes \mathbb{R}^{m*} \otimes \mathbb{R}^{m*}$.

29.6. Lemma. *The only $GL(m)$-equivariant map $f\colon \mathrm{Reg}S^2\mathbb{R}^{m*} \to \mathbb{R}^m\otimes\mathbb{R}^{m*}\otimes\mathbb{R}^{m*}$ is the zero map.*

Proof. Let I_s be the matrix $g_{ii} = 1$ for $i \le s$, $g_{jj} = -1$ for $j > s$ and $g_{ij} = 0$ for $i \neq j$. Since every $g \in \mathrm{Reg}S^2\mathbb{R}^{m*}$ can be transformed into some I_s, it suffices to deduce $f^i_{jk}(I_s) = 0$ for all i, j, k. If $j \neq i \neq k$ or $j = i = k$, the equivariance with respect to the change of orientation on the i-th axis gives $f^i_{jk}(I_s) = -f^i_{jk}(I_s)$. If $j = i \neq k$, we obtain the same result by changing the orientation on both the i-th and k-th axes. □

Lemma 29.6 implies $\Gamma - \Lambda = 0$. This proves

29.7. Proposition. *The only first order natural operator transforming pseudoriemannian metrics into linear connections is the Levi-Cività operator.*

Remarks

The first version of our systematical approach to the problem of finding natural operators was published in [Kolář, 87b]. In the same paper both geometric results from section 25 are deduced. The smooth version of the tensor evaluation theorem is first presented in this book. Proposition 26.12 was proved by [Kolář, Radziszewski, 88]. The generalized invariant tensor theorem was first used in [Kolář, 87b].

The reduction theorems for symmetric linear connections and pseudoriemannian metrics are classical, see e.g. [Schouten, 54]. Some extensions or reformulations of them are presented in [Lubczonok, 72] and [Krupka, 82]. The method of differential equations is used systematically e.g. in the book [Krupka, Janyška, 90]. The complete version of proposition 29.7 was deduced in [Slovák, 89].

CHAPTER VII.
FURTHER APPLICATIONS

In this chapter we discuss some further geometric problems about different types of natural operators. First we deduce that all natural bilinear operators transforming a vector field and a differential k-form into a differential k-form form a 2-parameter family. This further clarifies the well known relation between Lie derivatives and exterior derivatives of k-forms. From the technical point of view this problem can be considered as a preparatory exercise to the problem of finding all bilinear natural operators of the type of the Frölicher-Nijenhuis bracket. We deduce that in general case all such operators form a 10-parameter family. Then we prove that there is exactly one natural operator transforming general connections on a fibered manifold $Y \to M$ into general connections on its vertical tangent bundle $VY \to M$. Furthermore, starting from some geometric problems in analytical mechanics, we deduce that all first-order natural operators transforming second-order differential equations on a manifold M into general connections on its tangent bundle $TM \to M$ form a one parameter family. Further we study the natural transformations of the jet functors. The construction of the bundle of all r-jets between any two manifolds can be interpreted as a functor J^r on the product category $\mathcal{M}f_m \times \mathcal{M}f$. We deduce that for $r \geq 2$ the only natural transformations of J^r into itself are the identity and the contraction, while for $r = 1$ we have a one-parameter family of homotheties. This implies easily that the only natural transformation of the functor of the r-th jet prolongation of fibered manifolds into itself is the identity. For the second iterated jet prolongation $J^1(J^1Y)$ of a fibered manifold Y we look for an analogy of the canonical involution on the second iterated tangent bundle TTM. We prove that such an exchange map depends on a linear connection on the base manifold and we give a simple list of all natural transformations of this type.

The next section is devoted to some problems from Riemannian geometry. Here we complete our study of natural connections on Riemannian manifolds, we prove the Gilkey theorem on natural differential forms and we find all natural lifts of Riemannian metrics to the tangent bundles. We also deduce that all natural operators transforming linear symmetric connections into exterior forms are generated by the Chern forms. Since there are no natural forms of odd degree, all of them are closed.

In the last section, we present a survey of some results concerning the multilinear natural operators which are based heavily on the (linear) representation theory of Lie algebras. First we treat the naturality over the whole category $\mathcal{M}f_m$, where the main tools come from the representation theory of infinite dimensional algebras of vector fields. At the very end we comment briefly on the category of conformal (Riemannian) manifolds, which leads to finite dimensional representation theory of some parabolic subalgebras of the Lie algebras of the pseudo orthogonal groups.

30. The Frölicher-Nijenhuis bracket

The main goal of this section is to determine all bilinear natural operators of the type of the Frölicher-Nijenhuis bracket. But we find it useful to start with a technically simpler problem, which can serve as an introduction.

30.1. Bilinear natural operators $T \oplus \Lambda^p T^* \rightsquigarrow \Lambda^p T^*$. We are going to study the natural operators transforming a vector field and an exterior p-form into an exterior p-form. In order to get results of geometric interest, it is reasonable to restrict ourselves to the bilinear operators. The two simplest examples of such operators are $(X,\omega) \mapsto di_X\omega$ and $(X,\omega) \mapsto i_X d\omega$.

Proposition. *All bilinear natural operators $T \oplus \Lambda^p T^* \rightsquigarrow \Lambda^p T^*$ form the 2-parameter family*

$$k_1 d i_X \omega + k_2 i_X d\omega, \qquad k_1, k_2 \in \mathbb{R}. \tag{1}$$

Proof. First of all, every such operator has finite order r by the bilinear Peetre theorem. The canonical coordinates on the standard fiber $S = J_0^r T\mathbb{R}^m \times J_0^r \Lambda^p T^* \mathbb{R}^m$ are X^i_α, $b_{i_1\dots i_p,\beta}$, $|\alpha| \le r$, $|\beta| \le r$, while the canonical coordinates on the standard fiber $Z = \Lambda^p \mathbb{R}^{m*}$ are $c_{i_1\dots i_p}$. Since we consider the bilinear operators, even the associated maps $f\colon S \to Z$ are bilinear in X^i_α and $b_{i_1\dots i_p,\beta}$. Using the homotheties in $GL(m) \subset G^{r+1}_m$, we obtain

$$k^p f(X^i_\alpha, b_{i_1\dots i_p,\beta}) = f(k^{|\alpha|-1} X^i_\alpha, k^{p+|\beta|} b_{i_1\dots i_p,\beta}). \tag{2}$$

This implies that only the products $X^i b_{i_1\dots i_p,j}$ and $X^i_j b_{i_1\dots i_p}$ can appear in f. (In particular, every natural bilinear operator $T \oplus \Lambda^p T^* \rightsquigarrow \Lambda^p T^*$ is a first order operator.) Denote by $f = f_1 + f_2$ the corresponding decomposition of f.

The transformation laws of $b_{i_1\dots i_p}$, $b_{i_1\dots i_p,j}$ can be found in 25.4 and one deduces easily

$$\bar X^i = a^i_j X^j, \quad \bar X^i_j = a^i_{kl} \tilde a^k_j X^l + a^i_k X^k_l \tilde a^l_j. \tag{3}$$

In particular, the transformation laws with respect to the subgroup $GL(m) \subset G^2_m$ are tensorial in all cases. Hence we first have to determine the $GL(m)$-equivariant bilinear maps $\mathbb{R}^m \times \Lambda^p \mathbb{R}^{m*} \otimes \mathbb{R}^{m*} \to \Lambda^p \mathbb{R}^{m*}$. Consider the following diagram

$$\begin{array}{ccc} \mathbb{R}^m \times \Lambda^p \mathbb{R}^{m*} \otimes \mathbb{R}^{m*} & \xrightarrow{\ f_1\ } & \Lambda^p \mathbb{R}^{m*} \\ {\scriptstyle \mathrm{id} \times \mathrm{Alt}_p \otimes \mathrm{id}} \uparrow\downarrow & & \downarrow\uparrow {\scriptstyle \mathrm{Alt}_p} \\ \mathbb{R}^m \times \otimes^{p+1} \mathbb{R}^{m*} & \longrightarrow & \otimes^p \mathbb{R}^{m*} \end{array} \tag{4}$$

where Alt denotes the alternator of the indicated degree. The vertical maps are also $GL(m)$-equivariant and the $GL(m)$-equivariant map in the bottom row can be determined by the invariant tensor theorem. This implies that f_1 is a linear combination of the contraction of X^i with the derivation entry in $b_{i_1\dots i_p,j}$ and of the contraction of X^i with a non-derivation entry in $b_{i_1\dots i_p,j}$ followed by the alternation. To specify f_2, consider the diagram

$$\begin{array}{ccc} \mathbb{R}^m\otimes\mathbb{R}^{m*}\otimes\Lambda^p\mathbb{R}^{m*} & \xrightarrow{\tilde f_2} & \Lambda^p\mathbb{R}^{m*} \\ {\scriptstyle \mathrm{id}\otimes\mathrm{id}\otimes\mathrm{Alt}_p}\uparrow\downarrow & & \uparrow\downarrow{\scriptstyle \mathrm{Alt}_p} \\ \mathbb{R}^m\times\otimes^{p+1}\mathbb{R}^{m*} & \longrightarrow & \otimes^p\mathbb{R}^{m*} \end{array} \tag{5}$$

where $\tilde f_2$ is the linearization of f_2. Taking into account the maps in the bottom row determined by the invariant tensor theorem, we conclude similarly as above that f_2 is a linear combination of the inner contraction X^j_j multiplied by $b_{i_1\dots i_p}$ and of the contraction $X^j_{i_1}b_{i_2\dots i_p j}$ followed by the alternation. Thus, the equivariance of f with respect to $GL(m)$ leads to the following 4-parameter family

$$f_{i_1\dots i_p} = aX^j b_{i_1\dots i_p,j} + bX^j b_{j[i_2\dots i_p,i_1]} + cX^j_j b_{i_1\dots i_p} + eX^j_{[i_1} b_{i_2\dots i_p]j} \tag{6}$$

$a,\ b,\ c,\ e\in\mathbb{R}$.

The equivariance of f on the kernel $a^i_j=\delta^i_j$ is expressed by the relation

$$\begin{aligned} 0 = & -aX^j(b_{ki_2\dots i_p}a^k_{i_1j}+\dots+b_{i_1\dots i_{p-1}k}a^k_{i_pj})+ \\ & bX^j b_{k[i_2\dots i_p}a^k_{i_1]j} + ca^k_{kj}X^j b_{i_1\dots i_p} + eX^j a^k_{j[i_1}b_{i_2\dots i_p]k}. \end{aligned} \tag{7}$$

This implies

$$c=0 \quad\text{and}\quad a=b+e \tag{8}$$

which gives the coordinate form of (1). □

30.2. The Lie derivative. Proposition 30.1 gives a new look at the well known formula expressing the Lie derivative $\mathcal{L}_X\omega$ of a p-form as the sum of $di_X\omega$ and $i_Xd\omega$. Clearly, the Lie derivative operator on p-forms $(X,\omega)\mapsto\mathcal{L}_X\omega$ is a bilinear natural operator $T\oplus\Lambda^pT^*\rightsquigarrow\Lambda^pT^*$. By proposition 30.1, there exist certain real numbers a_1 and a_2 such that

$$\mathcal{L}_X\omega = a_1 di_X\omega + a_2 i_X d\omega$$

for every vector field X and every p-form ω on m-manifolds. If we evaluate $a_1=1=a_2$ in two suitable special cases, we obtain an interesting proof of the classical formula.

30.3. Bilinear natural operators $T \oplus \Lambda^p T^* \rightsquigarrow \Lambda^q T^*$. These operators can be determined in the same way as in 30.1, see [Kolář, 90b]. That is why we restrict ourselves to the result. *The only natural bilinear operators $T \oplus \Lambda^p T^* \rightsquigarrow \Lambda^{p-1}T^*$ or $\Lambda^{p+1}T^*$ are the constant multiples of $i_X\omega$ or $d(i_X d\omega)$, respectively. In the case $q \neq p-1$, p, $p+1$, we have the zero operator only.*

30.4. Bilinear natural operators of the Frölicher-Nijenhuis type. The wedge product of a differential q-form and a vector valued p-form is a bilinear map $\Omega^q(M) \times \Omega^p(M,TM) \to \Omega^{p+q}(M,TM)$ characterized by $\omega \wedge (\varphi \otimes X) = (\omega \wedge \varphi) \otimes X$ for all $\omega \in \Omega^q(M)$, $\varphi \in \Omega^p(M)$, $X \in \mathfrak{X}(M)$. Further let $C\colon \Omega^p(M,TM) \to \Omega^{p-1}(M)$ be the contraction operator defined by $C(\omega \otimes X) = i(X)\omega$ for all $\omega \in \Omega^p(M)$, $X \in \mathfrak{X}(M)$. In particular, for $P \in \Omega^0(M,TM)$ we have $C(P) = 0$. Clearly $C(i(P)Q)$ is a linear combination of $C(i(Q)P)$, $i(P)(C(Q))$, $i(Q)(C(P))$, $P \in \Omega^p(M,TM)$, $Q \in \Omega^q(M,TM)$. By I we denote Id_{TM}, viewed as an element of $\Omega^1(M,TM)$.

Theorem. *For $\dim M \geq p+q$, all bilinear natural operators $A\colon \Omega^p(M,TM) \times \Omega^q(M,TM) \to \Omega^{p+q}(M,TM)$ form a vector space linearly generated by the following 10 operators*

$$\begin{aligned}
&[P,Q], \quad dC(P)\wedge Q, \quad dC(Q)\wedge P, \quad dC(P)\wedge C(Q)\wedge I,\\
&dC(Q)\wedge C(P)\wedge I, \quad dC(i(P)Q)\wedge I, \quad i(P)dC(Q)\wedge I,\\
&i(Q)dC(P)\wedge I, \quad d(i(P)C(Q))\wedge I, \quad d(i(Q)C(P))\wedge I.
\end{aligned} \tag{1}$$

These operators form a basis if p, $q \geq 2$ and $m \geq p+q+1$.

30.5. Remark. If p or q is ≤ 1, then all bilinear natural operators in question are generated by those terms from 30.4.(1) that make sense. For example, in the extreme case $p = q = 0$ our result reads that the only bilinear natural operators $\mathfrak{X}(M) \times \mathfrak{X}(M) \to \mathfrak{X}(M)$ are the constant multiples of the Lie bracket. This was proved by [van Strien, 80], [Krupka, Mikolášová, 84], and in an 'infinitesimal' sense by [de Wilde, Lecomte, 82]. For a detailed discussion of all special cases we refer the reader to [Cap, 90]. Clearly, for $m < p+q$ we have the zero operator only.

30.6. To prove theorem 30.4, we start with the fact that the bilinear Peetre theorem implies that every A has finite order r. Denote by $P^i_{j_1\dots j_p}$ or $Q^i_{j_1\dots j_q}$ the canonical coordinates on $\mathbb{R}^m \otimes \Lambda^p\mathbb{R}^{m*}$ or $\mathbb{R}^m \otimes \Lambda^q\mathbb{R}^{m*}$, respectively. The associated map A_0 of A is bilinear in P's and Q's and their partial derivatives up to order r. Using equivariance with respect to homotheties in $GL(m)$, we find that A_0 contains only the products $P^i_{j_1\dots j_p,k}Q^m_{n_1\dots n_q}$ and $Q^i_{j_1\dots j_q,k}P^m_{n_1\dots n_p}$, where the first term in both expressions means the partial derivative with respect to x^k. In other words, A is a first order operator and A_0 is a sum $A_1 + A_2$ where

$$A_1\colon \mathbb{R}^m \otimes \Lambda^p\mathbb{R}^{m*} \otimes \mathbb{R}^{m*} \times \mathbb{R}^m \otimes \Lambda^q\mathbb{R}^{m*} \to \mathbb{R}^m \otimes \Lambda^{p+q}\mathbb{R}^{m*}$$
$$A_2\colon \mathbb{R}^m \otimes \Lambda^p\mathbb{R}^{m*} \times \mathbb{R}^m \otimes \Lambda^q\mathbb{R}^{m*} \otimes \mathbb{R}^{m*} \to \mathbb{R}^m \otimes \Lambda^{p+q}\mathbb{R}^{m*}$$

are bilinear maps. One finds easily that the transformation law of $P^i_{j_1\dots j_p,k}$ is

$$\begin{aligned}\bar P^i_{j_1\dots j_p,k} &= P^l_{m_1\dots m_p,n} a^i_l \tilde a^{m_1}_{j_1}\dots \tilde a^{m_p}_{j_p}\tilde a^n_k\\ &+ P^l_{m_1\dots m_p}(a^i_{ln}\tilde a^{m_1}_{j_1}\dots\tilde a^{m_p}_{j_p}\tilde a^n_k + a^i_l\tilde a^{m_1}_{j_1k}\tilde a^{m_2}_{j_2}\dots\tilde a^{m_p}_{j_p}+\dots\\ &+ a^i_l\tilde a^{m_1}_{j_1}\dots\tilde a^{m_p}_{j_pk}).\end{aligned}\tag{1}$$

30.7. Taking into account the canonical inclusion $GL(m)\subset G^2_m$, we see that the linear maps associated with the bilinear maps A_1 and A_2, which will be denoted by the same symbol, are $GL(m)$-equivariant. Consider first the following diagram

$$\begin{array}{ccc}\mathbb{R}^m\otimes\Lambda^p\mathbb{R}^{m*}\otimes\mathbb{R}^{m*}\otimes\mathbb{R}^m\otimes\Lambda^q\mathbb{R}^{m*} & \xrightarrow{A_1} & \mathbb{R}^m\otimes\Lambda^{p+q}\mathbb{R}^{m*}\\ \mathrm{id}\otimes\mathrm{Alt}_p\otimes\mathrm{id}\otimes\mathrm{Alt}_q\ \big\uparrow\big\downarrow & & \big\downarrow\big\uparrow\ \mathrm{id}\otimes\mathrm{Alt}_{p+q}\\ \mathbb{R}^m\otimes\otimes^p\mathbb{R}^{m*}\otimes\mathbb{R}^{m*}\otimes\mathbb{R}^m\otimes\otimes^q\mathbb{R}^{m*} & \longrightarrow & \mathbb{R}^m\otimes\otimes^{p+q}\mathbb{R}^{m*}\end{array}\tag{1}$$

where Alt denotes the alternator of the indicated degree. It suffices to determine all equivariant maps in the bottom row, to restrict them and to take the alternator of the result. By the invariant tensor theorem, all $GL(m)$-equivariant maps $\otimes^2\mathbb{R}^m\otimes\otimes^{p+q+1}\mathbb{R}^{m*}\to\mathbb{R}^m\otimes\otimes^{p+q}\mathbb{R}^{m*}$ are given by all kinds of permutations of the indices, all contractions and tensorizing with the identity. Since we apply this to alternating forms and use the alternator on the result, permutations do not play a role.

In what follows we discuss the case $p\ge 2$, $q\ge 2$ only and we leave the other cases to the reader. (A direct discussion shows that in the remaining cases the list (2) below should be reduced by those terms that do not make sense, but the next procedure leads to theorem 30.4 as well.) Constructing A_1, we may contract the vector field part of P into a non-derivation entry of P or into the derivation entry of P or into Q, and we may contract the vector field part of Q into Q or into a non-derivation entry of P or into the derivation entry of P, and then tensorize with the identity of $\mathbb{R}^m$. This gives 8 possibilities. If we perform only one contraction, we get 6 further possibilities, so that we have a 14-parameter family denoted by the lower case letters in the list (2) below. Constructing A_2, we obtain analogously another 14-parameter family denoted by upper case letters in the list (2) below. Hence $GL(m)$-equivariance yields the following expression for A_0 (we do not indicate alternation in the subscripts and we write α, β for any kind of free form-index on the right hand side)

$$\begin{aligned}&aP^m_{m\alpha,k}Q^n_{n\beta}\delta^i_l + bP^m_{\alpha,m}Q^n_{n\beta}\delta^i_l + cP^m_{\alpha,k}Q^n_{nm\beta}\delta^i_l + dP^m_{mn\alpha,k}Q^n_\beta\delta^i_l+\\ &eP^m_{n\alpha,m}Q^n_\beta\delta^i_l + fP^m_{n\alpha,k}Q^n_{m\beta}\delta^i_l + gP^m_{m\alpha,n}Q^n_\beta\delta^i_l + hP^m_{\alpha,n}Q^n_{m\beta}\delta^i_l+\\ &iP^m_{m\alpha,k}Q^i_\beta + jP^m_{\alpha,m}Q^i_\beta + kP^m_{\alpha,k}Q^i_{m\beta} + lP^i_{\alpha,k}Q^n_{n\beta} + mP^i_{n\alpha,k}Q^n_\beta+\\ &nP^i_{\alpha,n}Q^n_\beta + AP^m_{m\alpha}Q^n_{n\beta,k}\delta^i_l + BP^m_{m\alpha}Q^n_{\beta,n}\delta^i_l + CP^m_{mn\alpha}Q^n_{\beta,k}\delta^i_l+\\ &DP^m_\alpha Q^n_{nm\beta,k}\delta^i_l + EP^m_\alpha Q^n_{m\beta,n}\delta^i_l + FP^m_{n\alpha}Q^n_{m\beta,k}\delta^i_l + GP^m_\alpha Q^n_{n\beta,m}\delta^i_l+\\ &HP^m_{n\alpha}Q^n_{\beta,m}\delta^i_l + IP^i_\alpha Q^n_{n\beta,k} + JP^i_\alpha Q^n_{\beta,n} + KP^i_{n\alpha}Q^n_{\beta,k}+\\ &LP^m_{m\alpha}Q^i_{\beta,k} + MP^m_\alpha Q^i_{m\beta,k} + NP^m_\alpha Q^i_{\beta,m}.\end{aligned}\tag{2}$$

30.8. Then we consider the kernel K of the jet projection $G^2_m \to G^1_m$. Using 30.5.(1) with $a^i_j = \delta^i_j$, we evaluate that A_0 is K-equivariant if and only if the following coordinate expression

(1)
$$\begin{aligned}
&\Big(BP^m_{m\alpha}Q^t_\beta a^n_{tn} + (-1)^q qBP^m_{m\alpha}Q^n_{t\beta}a^t_{nk} + bP^t_\alpha Q^n_{n\beta}a^m_{tm} - \\
&(-1)^{p+q}pbP^m_{t\alpha}Q^n_{n\beta}a^t_{mk} + ((-1)^q c - D - (-1)^q(q-1)G)P^m_\alpha Q^n_{nt\beta}a^t_{mk} + \\
&(C - (-1)^q d - (-1)^{p+q}(p-1)g)P^m_{mt\alpha}Q^n_\beta a^t_{nk} + eP^m_{n\alpha}Q^n_\beta a^t_{mt} + \\
&(H - e)P^m_{t\alpha}Q^n_\beta a^t_{mn} + EP^m_\alpha Q^t_{m\beta}a^n_{tn} + (h - E)P^m_\alpha Q^n_{t\beta}a^t_{mn} + \\
&((-1)^q qH - (-1)^q f - F)P^m_{n\alpha}Q^n_{t\beta}a^t_{mk} + \\
&(F + (-1)^q f - (-1)^{p+q}ph)P^m_{n\alpha}Q^t_{m\beta}a^n_{tk} - (-1)^{p+q}(p-1)eP^m_{nt\alpha}Q^n_\beta a^t_{mk} - \\
&(-1)^q(q-1)EP^m_\alpha Q^n_{mt\beta}a^t_{nk}\Big)\delta^i_l + jP^t_\alpha Q^i_\beta a^m_{tm} + (-1)^{p+q}pjP^m_{t\alpha}Q^i_\beta a^t_{mk} + \\
&JP^i_\alpha Q^t_\beta a^m_{tm} + (-1)^q qJP^i_\alpha Q^m_{t\beta}a^t_{mk} - ((-1)^q k - (-1)^q qN + M)P^m_\alpha Q^i_{t\beta}a^t_{mk} + \\
&(K + (-1)^{p+q}pn - (-1)^q m)P^i_{m\alpha}Q^t_\beta a^m_{tk} - l(-1)^q P^t_\alpha Q^m_{m\beta}a^i_{tk} + \\
&LP^m_{m\alpha}Q^t_\beta a^i_{tk} - (-1)^q mP^t_{m\alpha}Q^m_\beta a^i_{tk} + MP^m_\alpha Q^t_{m\beta}a^i_{tk} + (n + N)P^m_\alpha Q^t_\beta a^i_{mt}
\end{aligned}$$

represents the zero map $\mathbb{R}^m \otimes \Lambda^p\mathbb{R}^{m*} \times \mathbb{R}^m \otimes \Lambda^q\mathbb{R}^{m*} \times \mathbb{R}^m \otimes S^2\mathbb{R}^{m*} \to \mathbb{R}^m \otimes \Lambda^{p+q}\mathbb{R}^{m*}$.

For $\dim M \geq p+q+1$, the individual terms in (1) are linearly independent. Hence (1) is the zero map if and only if all the coefficients vanish. This leads to the following equations

$$\begin{gathered}
b = B = e = E = h = H = j = J = l = L = m = M = 0 \\
c = (q-1)G + (-1)^q D, \qquad C = (-1)^q d + (-1)^{p+q}(p-1)g \\
F = (-1)^{q-1}f, \qquad k = -qn, \qquad K = (-1)^{p+q-1}pn, \qquad N = -n
\end{gathered} \tag{2}$$

while a, A, d, D, f, g, G, n, i, I are independent parameters. This yields the coordinate form of 30.4.(1).

In the case $m = p+q$, p, $q \geq 2$, there are certain linear relations between the individual terms of (1). They are described explicitly in [Cap, 90]. But even in this case we obtain the final result in the form indicated in theorem 30.4. □

30.9. Linear and bilinear natural operators on vector valued forms. Roughly speaking, we can characterize theorem 30.4 by saying that the Frölicher-Nijenhuis bracket is the only non-trivial operator in the list 30.4.(1), since the remaining terms can easily be constructed by means of tensor algebra and exterior differentiation. We remark that the natural operators on vector valued forms were systematically studied by A. Cap. He deduced the complete list of all linear natural operators $\Omega^p(M,TM) \to \Omega^q(M,TM)$ and all bilinear natural operators $\Omega^p(M,TM) \times \Omega^q(M,TM) \to \Omega^r(M,TM)$, which can be found in [Cap, 90]. From a general point of view, the situation is analogous to 30.4: except the Frölicher-Nijenhuis bracket, all other operators in question can easily be constructed by means of tensor algebra and exterior differentiation.

30.10. Remark on the Schouten-Nijenhuis bracket. This is a bilinear operator $C^\infty\Lambda^pTM \times C^\infty\Lambda^qTM \to C^\infty\Lambda^{p+q-1}TM$ introduced geometrically by [Schouten, 40] and further studied by [Nijenhuis, 55]. In [Michor, 87b] the natural operators of this type are studied. The problem is technically much simpler than in the Frölicher-Nijenhuis case and the same holds for the result: *The only natural bilinear operators $\Lambda^pT \oplus \Lambda^qT \rightsquigarrow \Lambda^{p+q-1}T$ are the constant multiples of the Schouten-Nijenhuis bracket.*

31. Two problems on general connections

31.1. Vertical prolongation of connections. Consider a connection $\Gamma\colon Y \to J^1Y$ on a fibered manifold $Y \to M$. If we apply the vertical tangent functor V, we obtain a map $V\Gamma\colon VY \to VJ^1Y$. Let $i_Y\colon VJ^1Y \to J^1VY$ be the canonical involution, see 39.8. Then the composition

$$\mathcal{V}_Y\Gamma := i_Y \circ V\Gamma\colon VY \to J^1VY \tag{1}$$

is a connection on $VY \to M$, which will be called the *vertical prolongation* of Γ. Since this construction has geometrical character, $\mathcal{V}$ is an operator $J^1 \rightsquigarrow J^1V$ natural on the category $\mathcal{FM}_{m,n}$.

Proposition. *The vertical prolongation $\mathcal{V}$ is the only natural operator $J^1 \rightsquigarrow J^1V$.*

We start the proof with finding the equations of $\mathcal{V}\Gamma$. If

$$dy^p = F_i^p(x,y)dx^i \tag{2}$$

is the coordinate form of Γ and $Y^p = dy^p$ are the additional coordinates on VY, then (1) implies that the equations of $\mathcal{V}\Gamma$ are (2) and

$$dY^p = \frac{\partial F_i^p}{\partial y^q}Y^q dx^i. \tag{3}$$

31.2. The standard fiber of V on the category $\mathcal{FM}_{m,n}$ is $\mathbb{R}^n$. Let $S_1 = J_0^1(J^1(\mathbb{R}^m \times \mathbb{R}^n \to \mathbb{R}^m) \to \mathbb{R}^m \times \mathbb{R}^n)$ and $Z = J_0^1(V(\mathbb{R}^m \times \mathbb{R}^n) \to \mathbb{R}^m)$, $0 \in \mathbb{R}^m \times \mathbb{R}^n$. By 18.19, the first order natural operators are in bijection with $G^2_{m,n}$-maps $S_1 \times \mathbb{R}^n \to Z$ over the identity of $\mathbb{R}^n$. The canonical coordinates on S_1 are y_i^p, y_{iq}^p, y_{ij}^p and the action of $G^2_{m,n}$ can be found in 27.3. The action of $G^2_{m,n}$ on $\mathbb{R}^n$ is

$$\bar{Y}^p = a_q^pY^q. \tag{1}$$

The coordinates on Z are Y^p, $z_i^p = \partial y^p/\partial x^i$, $Y_i^p = \partial Y^p/\partial x^i$. By standard evaluation we find the following action of $G^2_{m,n}$

$$\begin{aligned} &\bar{Y}^p = a_q^pY^q, \qquad \bar{z}_i^p = a_q^pz_j^q\tilde{a}_i^j + a_j^p\tilde{a}_i^j \\ &\bar{Y}_i^p = a_q^pY_j^q\tilde{a}_i^j + a_{qr}^pz_j^rY^q\tilde{a}_i^j + a_{qj}^pY^q\tilde{a}_i^j. \end{aligned} \tag{2}$$

The coordinate form of any map $S_1 \times \mathbb{R}^n \to Z$ over the identity of $\mathbb{R}^n$ is $Y^p = Y^p$ and

$$z_i^p = f_i^p(Y^q, y_j^r, y_{kt}^s, y_{lm}^u)$$
$$Y_i^p = g_i^p(Y^q, y_j^r, y_{kt}^s, y_{lm}^u).$$

First we discuss f_i^p. The equivariance with respect to base homotheties yields

$$kf_i^p = f_i^p(Y^q, ky_j^r, ky_{kt}^s, k^2 y_{lm}^u). \tag{3}$$

By the homogeneous function theorem, if we fix Y^q, then f_i^p is linear in y_j^r, y_{kt}^s and independent of y_{lm}^u. The fiber homotheties then give

$$kf_i^p = f_i^p(kY^q, ky_j^r, y_{kt}^s). \tag{4}$$

By (3) and (4), f_i^p is a sum of an expression linear in y_i^p and bilinear in Y^p and y_{iq}^p. Since f_i^p is $GL(m) \times GL(n)$-equivariant, the generalized invariant tensor theorem implies it has the following form

$$ay_i^p + bY^p y_{qi}^q + cY^q y_{qi}^p. \tag{5}$$

The equivariance on the kernel K of the projection $G_{m,n}^2 \to G_m^1 \times G_n^1$ yields

$$a_i^p = aa_i^p + bY^p(a_{qi}^q + a_{qr}^q y_i^r) + cY^q(a_{qi}^p + a_{qr}^p y_i^r). \tag{6}$$

This implies $a = 1$, $b = c = 0$, which corresponds to 31.1.(2).

For g_i^p, the above procedure leads to the same form (5). Then the equivariance with respect to K yields $a = b = 0$, $c = 1$. This corresponds to 31.1.(3). Thus we have proved that $\mathcal{V}$ is the only first order natural operator $J^1 \rightsquigarrow J^1 V$.

31.3. By 23.7, every natural operator $A: J^1 \rightsquigarrow J^1 V$ has a finite order r. Let $f = (f_i^p, g_i^p): S_r \times \mathbb{R}^n \to Z$ be the associated map of A, where $S^r = J_0^r(J^1(\mathbb{R}^{m+n} \to \mathbb{R}^m) \to \mathbb{R}^{m+n})$. Consider first $f_i^p(Y^q, y_{j\alpha\beta}^r)$ with the same notation as in the second step of the proof of proposition 27.3. The base homotheties yield

$$kf_i^p = f_i^p(Y^q, k^{|\alpha|+1} y_{j\alpha\beta}^r). \tag{1}$$

By the homogeneous function theorem, if we fix Y^p, then f_i^p are independent of $y_{i\alpha\beta}^p$ with $|\alpha| \geq 1$ and linear in $y_{i\beta}^p$. Hence the only s-th order term is $\varphi_s = \varphi_{iq}^{pj\beta}(Y^r) y_{j\beta}^q$, $|\beta| = s$, $s \geq 2$. Using fiber homotheties we find that φ_s is of degree s in Y^p. Then the generalized invariant tensor theorem implies that φ_s is of the form

$$a_s y_{iq_1\dots q_s}^p Y^{q_1} \dots Y^{q_s} + b_s Y^p y_{iq_1 q_2 \dots q_s}^{q_1} Y^{q_2} \dots Y^{q_s}.$$

Consider the equivariance with respect to the kernel of the jet projection $G_{m,n}^{r+1} \to G_{m,n}^r$. Using induction we deduce the transformation law

$$\bar{y}_{iq_1\dots q_r}^p = y_{iq_1\dots q_r}^p + a_{tq_1\dots q_r}^p y_i^t + a_{iq_1\dots q_r}^p,$$

while the lower order terms remain unchanged. By direct evaluation we find $a_r = b_r = 0$. The same procedure takes place for g_i^p. Hence A is an operator of order $r - 1$. By recurrence we conclude A is a first order operator. This completes the proof of proposition 31.1.

31.4. Remark. In [Kolář, 81a] it was clarified geometrically that the vertical prolongation $\mathcal{V}\Gamma$ plays an important role in the theory of the original connection Γ. The uniqueness of $\mathcal{V}\Gamma$ proved in proposition 31.1 gives a theoretical justification of this phenomenon.

It is remarkable that there is another construction of $\mathcal{V}\Gamma$ using flow prolongations of vector fields, see 45.4. The equivalence of both definitions is an interesting consequence of the uniqueness of operator $\mathcal{V}$.

31.5. Natural operators transforming second order differential equations into general connections. We recall that a *second order differential equation* on a manifold M is usually defined as a vector field $\xi: TM \to TTM$ on TM satisfying $Tp_M \circ \xi = \mathrm{id}_{TM}$, where $p_M: TM \to M$ is the bundle projection. Let L_M be the Liouville vector field on TM, i.e. the vector field generated by the homotheties. If $[\xi, L_M] = \xi$, then ξ is said to be a *spray*. There is a classical bijection between sprays and linear symmetric connections, which is used in several branches of differential geometry. (We shall obtain it as a special case of a more general construction.)

A. Dekrét, [Dekrét, 88], studied the problem whether an arbitrary second order differential equation on M determines a general connection on TM by means of the naturality approach. He deduced rather quickly a simple analytical expression for all first order natural operators. Only then he looked for the geometrical interpretation. Keeping the style of this book, we first present the geometrical construction and then we discuss the naturality problem.

According to 9.3, the horizontal projection of a connection Γ on an arbitrary fibered manifold Y is a vector valued 1-form on Y, which will be called the horizontal form of Γ.

On the tangent bundle TM, we have the following natural tensor field V_M of type $\binom{1}{1}$. Since TM is a vector bundle, its vertical tangent bundle is identified with $TM \oplus TM$. For every $B \in TTM$ we define

$$V_M(B) = (p_{TM}B, Tp_M B). \tag{1}$$

(A general approach to natural tensor fields of type $\binom{1}{1}$ on an arbitrary Weil bundle is explained in [Kolář, Modugno, 92].)

Given a second order differential equation ξ on M, the Lie derivative $\mathcal{L}_\xi V_M$ is a vector valued 1-form on TM. Let 1_{TTM} be the identity on TTM. The following result gives a construction of a general connection on TM determined by ξ.

Lemma. *For every second order differential equation ξ on M, $\frac{1}{2}(1_{TTM} - \mathcal{L}_\xi V_M)$ is the horizontal form of a connection on TM.*

Proof. Let x^i be local coordinates on M and $y^i = dx^i$ be the induced coordinates on TM. The coordinate expression of the horizontal form of a connection on TM is

$$dx^i \otimes \frac{\partial}{\partial x^i} + F_i^j(x,y)dx^i \otimes \frac{\partial}{\partial y^j}. \tag{2}$$

By (1), the coordinate expression of V_M is

$$(3) \qquad dx^i \otimes \frac{\partial}{\partial y^i}.$$

Having a second order differential equation ξ of the form

$$(4) \qquad y^i \frac{\partial}{\partial x^i} + \xi^i(x,y)\frac{\partial}{\partial y^i}$$

we evaluate directly for $\mathcal{L}_\xi V_M$

$$(5) \qquad -dx^i \otimes \frac{\partial}{\partial x^i} - \frac{\partial \xi^i}{\partial y^j} dx^j \otimes \frac{\partial}{\partial y^i} + dy^i \otimes \frac{\partial}{\partial y^i}.$$

Hence $\frac{1}{2}(1_{TM} - \mathcal{L}_\xi V_M)$ has the required form

$$(6) \qquad dx^i \otimes \frac{\partial}{\partial x^i} + \frac{1}{2}\frac{\partial \xi^i}{\partial y^j} dx^j \otimes \frac{\partial}{\partial y^i}. \quad \square$$

31.6. Denote by A the operator from lemma 31.5. By the general theory, the difference of two general connections on $TM \to M$ is a section $TM \to VTM \otimes T^*M = (TM \oplus TM) \otimes T^*M$. The identity tensor of $TM \otimes T^*M$ determines a natural section $I_M : TM \to VTM \otimes T^*M$. Hence $A + kI$ is a natural operator for every $k \in \mathbb{R}$.

Proposition. *All first order natural operators transforming second order differential equations on a manifold into connections on the tangent bundle form the one-parameter family*

$$A + kI, \qquad k \in \mathbb{R}.$$

Proof. We have the case of a morphism operator from 18.17 with $\mathcal{C} = \mathcal{M}f_m$, $F_1 = G_1 = T$, $q = \mathrm{id}$, $F_2 = T_1^2$, $G_2 = J^1T$ and the additional conditions that we consider the sections of $T_1^2 \to T$ and $J^1T \to T$. Let S be the fiber of $J^1(T_1^2\mathbb{R}^m \to T\mathbb{R}^m)$ over $0 \in \mathbb{R}^m$ and Z be the fiber of $J^1T\mathbb{R}^m$ over $0 \in \mathbb{R}^m$. By 18.19 we have to determine all G_m^3-equivariant maps $f : S \to Z$ over the identity of $T_0\mathbb{R}^m$.

Denote by $X^i = \frac{dx^i}{dt}$, $Y^i = \frac{d^2x^i}{dt^2}$ the induced coordinates on $T_1^2\mathbb{R}^m$ and by $X^i_j = \partial Y^i/\partial x^j$, $Y^i_j = \partial Y^i/\partial X^j$ the induced coordinates on S. By direct evaluation, one finds the following action of G_m^3

$$(1) \qquad \bar{X}^i = a^i_j X^j, \qquad \bar{Y}^i = a^i_{jk} X^j X^k + a^i_j Y^j$$

$$(2) \qquad \bar{X}^i_j = a^i_{klm} \tilde{a}^m_j X^k X^l + a^i_{kl} \tilde{a}^l_j Y^k + 2a^i_{kl} \tilde{a}^k_{mj} a^m_n X^l X^n + a^i_k X^k_l \tilde{a}^l_j + a^i_k \tilde{a}^l_{mj} a^m_n X^n Y^k_l$$

$$(3) \qquad \bar{Y}^i_j = 2a^i_{kl} \tilde{a}^l_j X^k + a^i_k Y^k_l \tilde{a}^l_j$$

The standard coordinates Z^i, Z^i_j on Z have the transformation law

$$\bar{Z}^i = a^i_j Z^j, \qquad \bar{Z}^i_j = a^i_{kl}\tilde{a}^k_j Z^l + a^i_k Z^k_l \tilde{a}^l_j. \tag{4}$$

Let $Z^i = X^i$ and $Z^i_j = f^i_j(X^p, Y^q, X^m_n, Y^k_l)$ be the coordinate expression of f. The equivariance with respect to the homotheties in $GL(m) \subset G^3_m$ yields

$$f^i_j(X^p, Y^q, X^m_n, Y^k_l) = f^i_j(kX^p, kY^q, X^m_n, Y^k_l). \tag{5}$$

Hence f^i_j do not depend on X^p and Y^q. Let $a^i_j = \delta^i_j$, $a^i_{jk} = 0$. Then the equivariance condition reads

$$f^i_j(X^m_n, Y^k_l) = f^i_j(X^m_n + a^m_{npq}X^pX^q, Y^k_l). \tag{6}$$

This implies f^i_j are independent of X^m_n. Putting $a^i_j = \delta^i_j$, we obtain

$$f^i_j(Y^k_l) + a^i_{mj}X^m = f^i_j(Y^k_l + 2a^k_{lm}X^m) \tag{7}$$

with arbitrary a^i_{jk}. Differentiating with respect to Y^k_l, we find $\partial f^i_j/\partial Y^k_l = \text{const}$. Hence f^i_j are affine functions. By the Invariant tensor theorem, we deduce

$$f^i_j = k_1 Y^i_j + k_2\delta^i_j Y^k_k + k_3\delta^i_j. \tag{8}$$

Using (7) once again, we obtain $k_1 = \frac{1}{2}$, $k_2 = 0$. This gives the coordinate form of our assertion. □

31.7. Remark. If X is a spray, then the operator A from lemma 31.5 determines the classical linear symmetric connection induced by X. Indeed, 31.5.(4) satisfies the spray condition if and only if

$$\frac{\partial \xi^i}{\partial y^j} y^j = 2\xi^i.$$

This kind of homogeneity implies $\xi^i = b^i_{jk}(x)y^jy^k$. Then the coordinate form of $A(X)$ is

$$dy^i = b^i_{jk}(x)y^j dx^k.$$

32. Jet functors

32.1. By 12.4, the construction of r-jets of smooth maps can be viewed as a bundle functor J^r on the product category $\mathcal{M}f_m \times \mathcal{M}f$. We are going to determine all natural transformations of J^r into itself. Denote by $\hat{y}\colon M \to N$ the constant map of M into $y \in N$. Obviously, the assignment $X \mapsto j^r_{\alpha X}\widehat{\beta X}$ is a natural transformation of J^r into itself called the *contraction*. For $r = 1$, $J^1(M, N)$ coincides with $\text{Hom}(TM, TN)$, which is a vector bundle over $M \times N$.

Proposition. *For $r \geq 2$ the only natural transformations $J^r \to J^r$ are the identity and the contraction. For $r = 1$, all natural transformations $J^1 \to J^1$ form the one-parametric family of homotheties $X \mapsto cX$, $c \in \mathbb{R}$.*

Proof. Consider first the subcategory $\mathcal{M}f_m \times \mathcal{M}f_n \subset \mathcal{M}f_m \times \mathcal{M}f$. The standard fiber $S = J_0^r(\mathbb{R}^m, \mathbb{R}^n)_0$ is a $G_m^r \times G_n^r$-space and the action of $(A, B) \in G_m^r \times G_n^r$ on $X \in S$ is given by the jet composition

$$\bar{X} = B \circ X \circ A^{-1}. \tag{1}$$

According to the general theory, the natural transformations $J^r \to J^r$ are in bijection with the $G_m^r \times G_n^r$-equivariant maps $f\colon S \to S$.

Write $A^{-1} = (\tilde{a}_j^i, \dots, \tilde{a}_{j_1 \dots j_r}^i)$, $B = (b_q^p, \dots, b_{q_1 \dots q_r}^p)$, $X = (X_i^p, \dots, X_{i_1 \dots i_r}^p) = (X_1, \dots, X_r)$. Consider the equivariance of $f = (f_1, \dots, f_r)$ with respect to the homotheties in $GL(m) \subset G_m^r$. This gives the homogeneity conditions

$$\begin{aligned} k f_1(X_1, \dots, X_s, \dots, X_r) &= f_1(kX_1, \dots, k^s X_s, \dots, k^r X_r) \\ &\vdots \\ k^s f_s(X_1, \dots, X_s, \dots, X_r) &= f_s(kX_1, \dots, k^s X_s, \dots, k^r X_r) \\ &\vdots \\ k^r f_r(X_1, \dots, X_s, \dots, X_r) &= f_r(kX_1, \dots, k^s X_s, \dots, k^r X_r). \end{aligned} \tag{2}$$

Taking into account the homotheties in $GL(n)$, we further find

$$\begin{aligned} k f_1(X_1, \dots, X_r) &= f_1(kX_1, \dots, kX_r) \\ &\vdots \\ k f_r(X_1, \dots, X_r) &= f_r(kX_1, \dots, kX_r). \end{aligned} \tag{3}$$

Applying the homogeneous function theorem to both (2) and (3), we deduce that f_s is linear in X_s and independent of the remaining coordinates, $s = 1, \dots, r$. Consider furthemore the equivariance with respect to the subgroup $GL(m) \times GL(n)$. This yields that f_s corresponds to an equivariant map of $\mathbb{R}^n \otimes S^s\mathbb{R}^{m*}$ into itself. By the generalized invariant tensor theorem, it holds $f_s = c_s X_s$ with any $c_s \in \mathbb{R}$.

For $r = 1$ we have deduced $f_i^p = c_1 X_i^p$. For $r = 2$ consider the equivariance with respect to the kernel of the jet projection $G_m^2 \times G_n^2 \to G_m^1 \times G_n^1$. Taking into account the coordinate form of the jet composition, we find that the action of an element $((\delta_j^i, \tilde{a}_{jk}^i), (\delta_q^p, b_{qr}^p))$ on (X_i^p, X_{ij}^p) is $\bar{X}_i^p = X_i^p$ and

$$\bar{X}_{ij}^p = X_{ij}^p + b_{qr}^p X_i^q X_j^r + X_k^p \tilde{a}_{ij}^k. \tag{4}$$

Then the equivariance condition for f_{ij}^p reads

$$c_2 X_{ij}^p + (c_1)^2 b_{qr}^p X_i^q X_j^r + c_1 X_k^p \tilde{a}_{ij}^k = c_2 (X_{ij}^p + b_{qr}^p X_i^q X_j^r + X_k^p \tilde{a}_{ij}^k). \tag{5}$$

This implies $c_1 = c_2 = 0$ or $c_1 = c_2 = 1$. Assume by induction that our assertion holds in the order $r-1$. Consider the equivariance with respect to the kernel of the jet projection $G^r_m \times G^r_n \to G^{r-1}_m \times G^{r-1}_n$. The action of an element $((\delta^i_j, 0, \dots, 0, \tilde{a}^i_{j_1 \dots j_r}), (\delta^p_q, 0, \dots, 0, b^p_{q_1 \dots q_r}))$ leaves $X_1, \dots, X_{r-1}$ unchanged and it holds

$$\bar{X}^p_{i_1 \dots i_r} = X^p_{i_1 \dots i_r} + b^p_{q_1 \dots q_r} X^{q_1}_{i_1} \dots X^{q_r}_{i_r} + X^p_j \tilde{a}^j_{i_1 \dots i_r}. \tag{6}$$

Then the equivariance condition for $f^p_{i_1 \dots i_r}$ requires

$$\begin{aligned} c_r X^p_{i_1 \dots i_r} &+ (c_1)^r b^p_{q_1 \dots q_r} X^{q_1}_{i_1} \dots X^{q_r}_{i_r} + c_1 X^p_j \tilde{a}^j_{i_1 \dots i_r} \\ &= c_r (X^p_{i_1 \dots i_r} + b^p_{q_1 \dots q_r} X^{q_1}_{i_1} \dots X^{q_r}_{i_r} + X^p_j \tilde{a}^j_{i_1 \dots i_r}). \end{aligned} \tag{7}$$

This implies $c_r = c_1 = 0$ or 1.

For $r = 1$ we have a homothety $f^n \colon X \mapsto k_n X$, $k_n \in \mathbb{R}$, on each subcategory $\mathcal{M}f_m \times \mathcal{M}f_n \subset \mathcal{M}f_m \times \mathcal{M}f$. If we take the value of the transformation $(f^1, \dots, f^n, \dots)$ on the product of $\mathrm{id}_{\mathbb{R}^m}$ with the injection $i_{a,b} \colon \mathbb{R}^a \to \mathbb{R}^b$, $(x^1, \dots, x^a) \mapsto (x^1, \dots, x^a, 0, \dots, 0)$, $a < b$, and apply it to 1-jet at 0 of the map $x^1 = t^1$, $x^2 = 0, \dots, x^a = 0$, $(t^1, \dots, t^m) \in \mathbb{R}^m$, we find $k_a = k_b$. For $r \geq 2$ we have on each subcategory either the identity or the contraction. Applying the latter idea once again, we deduce that the same alternative must take place in all cases. □

32.2. The construction of the r-th jet prolongation $J^r Y$ of a fibered manifold $Y \to X$ can be considered as a bundle functor on the category $\mathcal{FM}_m$. This functor is also denoted by J^r. However, in order to distinguish from 32.1, we shall use J^r_{fib} for J^r in the fibered case here.

Proposition. *The only natural transformation $J^r_{\text{fib}} \to J^r_{\text{fib}}$ is the identity.*

Proof. The construction of product fibered manifolds defines an injection $\mathcal{M}f_m \times \mathcal{M}f \to \mathcal{FM}_m$ and the restriction of J^r_{fib} to $\mathcal{M}f_m \times \mathcal{M}f$ is J^r. For $r = 1$, proposition 31.1 gives a one-parameter family

$$(y^p_i) \mapsto (c y^p_i) \tag{1}$$

of possible candidates for the natural transformation $J^1_{\text{fib}} \to J^1_{\text{fib}}$. But the transformation law of y^p_i with respect to the kernel of the standard homomorphism $G^1_{m,n} \to G^1_m \times G^1_n$ is $\bar{y}^p_i = y^p_i + a^p_i$. The equivariance condition for (1) reads $a^p_i = c a^p_i$, which implies $c = 1$.

For $r \geq 2$, proposition 32.1 offers the contraction and the identity. But the contraction is clearly not natural on the whole category $\mathcal{FM}_m$, so that only the identity remains. □

32.3. Natural transformations $J^1 J^1 \to J^1 J^1$. It is well known that the canonical involution of the second tangent bundle plays a significant role in applications. A remarkable feature of the canonical involution on TTM is that it exchanges both the projections $p_{TM} \colon TTM \to TM$ and $Tp_M \colon TTM \to TM$.

Nowadays, in several problems of the field theory the role of the tangent bundle of a smooth manifold is replaced by the first jet prolongation J^1Y of a fibered manifold $p\colon Y \to M$. On the second iterated jet prolongation $J^1J^1Y = J^1(J^1Y \to M)$ there are two analogous projections to J^1Y, namely the target jet projection $\beta_1\colon J^1J^1Y \to J^1Y$ and the prolongation $J^1\beta\colon J^1J^1Y \to J^1Y$ of the target jet projection $\beta\colon J^1Y \to Y$. Hence one can ask whether there exists a natural transformation of J^1J^1Y into itself exchanging the projections β_1 and $J^1\beta$, provided J^1J^1 is considered as a functor on $\mathcal{FM}_{m,n}$. But the answer is negative.

Proposition. *The only natural transformation $J^1J^1 \to J^1J^1$ is the identity.*

This assertion follows directly from proposition 32.6 below, so that we shall not prove it separately. It is remarkable that we have a different situation on the subspace $\bar{J}^2Y = \{X \in J^1J^1Y,\ \beta_1X = J^1\beta(X)\}$, which is called the *second semiholonomic prolongation of* Y. There is a one-parametric family of natural transformations $\bar{J}^2 \to \bar{J}^2$, see 32.5.

32.4. An exchange map. However, one can construct a suitable exchange map $e_\Lambda\colon J^1J^1Y \to J^1J^1Y$ by means of a linear connection Λ on the base manifold M as follows. Interpreting Λ as a principal connection on the first order frame bundle P^1M of M, we first explain how Λ induces a map $h_\Lambda\colon J^1J^1Y \oplus QP^1M \to T^1_m(T^1_mY)$. Every $X \in J^1J^1Y$ is of the form $X = j^1_x\rho(z)$, where ρ is a local section of $J^1Y \to M$, and for every $u \in P^1_xM$ we have $\Lambda(u) = j^1_x\sigma(z)$, where σ is a local section of $P^1M \subset J^1_0(\mathbb{R}^m, M)$. Taking into account the canonical inclusion $J^1Y \subset J^1(M,Y)$, the jet composition $\rho(z)\circ\sigma(z)$ defines a local map $M \to J^1_0(\mathbb{R}^m, Y) = T^1_mY$, the 1-jet of which $j^1_x(\rho(z)\circ\sigma(z)) \in J^1_x(M, T^1_mY)$ depends on X and $\Lambda(u)$ only. Since $u \in J^1_0(\mathbb{R}^m, M)$, we have $h_\Lambda(X,u) = (j^1_x(\rho(z)\circ\sigma(z)))\circ u \in T^1_mT^1_mY$. Furthermore, there is a canonical exchange map $\kappa\colon T^1_mT^1_mY \to T^1_mT^1_mY$, the definition of which will be presented in the framework of the theory of Weil bundles in 35.18. Using κ and h_Λ, we construct a map $e_\Lambda\colon J^1J^1Y \to J^1J^1Y$.

Lemma. *For every $X \in (J^1J^1Y)_y$ there exists a unique element $e_\Lambda(X) \in J^1J^1Y$ satisfying*

$$\kappa(h_\Lambda(X,u)) = h_{\tilde{\Lambda}}(e_\Lambda(X),u) \tag{1}$$

for any frame $u \in P^1_xM$, $x = p(y)$, provided $\tilde{\Lambda}$ means the conjugate connection of Λ.

Proof consists in direct evaluation, for which the reader is referred to [Kolář, Modugno, 91]. The coordinate form of e_Λ is

$$y^p_i = Y^p_i, \quad Y^p_i = y^p_i, \quad y^p_{ij} = y^p_{ji} + (y^p_k - Y^p_k)\Lambda^k_{ji} \tag{2}$$

where $Y^p_i = \partial y^p/\partial x^i$, $y^p_{ij} = \partial y^p_i/\partial x^j$ are the additional coordinates on $J^1(J^1Y \to M)$.

32.5. Remark. The subbundle $\bar{J}^2Y \subset J^1J^1Y$ is characterized by $y_i^p = Y_i^p$. Formula 32.4.(2) shows that the restriction of e_Λ to $\bar{J}^2Y$ does not depend on Λ, so that we have a natural map $e\colon \bar{J}^2Y \to \bar{J}^2Y$. Since $\bar{J}^2Y \to J^1Y$ is an affine bundle, e generates a one-parameter family of natural transformations $\bar{J}^2 \to \bar{J}^2$

$$X \mapsto kX + (1-k)e(X), \qquad k \in \mathbb{R}.$$

One proves easily that this family represents all natural transformations $\bar{J}^2 \to \bar{J}^2$, see [Kolář, Modugno, 91].

32.6. The map e_Λ was introduced by M. Modugno by another construction, in which the naturality ideas were partially used. Hence it is interesting to study the whole problem purely from the naturality point of view.

Our goal is to find all natural transformations $J^1J^1Y \oplus QP^1M \to J^1J^1Y$. Since $J^1Y \to Y$ is an affine bundle with associated vector bundle $VY \otimes T^*M$, we can define a map

$$\delta\colon J^1J^1Y \to VY \otimes T^*M, \qquad A \mapsto \beta_1(A) - J^1\beta(A). \tag{1}$$

On the other hand, proposition 25.2 implies directly that all natural operators $N\colon QP^1M \rightsquigarrow TM \otimes T^*M \otimes T^*M$ form the 3-parameter family

$$N\colon \Lambda \mapsto k_1 S + k_2 I \otimes \hat{S} + k_3 \hat{S} \otimes I \tag{2}$$

where S is the torsion tensor of Λ, $\hat{S}$ is the contracted torsion tensor and I is the identity of TM. Using the contraction with respect to TM, we construct a 3-parameter family of maps

$$\langle \delta, N(\Lambda) \rangle\colon J^1J^1Y \to VY \otimes T^*M \otimes T^*M. \tag{3}$$

The well known exact sequence of vector bundles over J^1Y

$$0 \to VY \otimes T^*M \to VJ^1Y \xrightarrow{V\beta} VY \to 0 \tag{4}$$

shows that $VY \otimes T^*M \otimes T^*M$ can be considered as a subbundle in $VJ^1Y \otimes T^*M$, which is the vector bundle associated with the affine bundle $\beta_1\colon J^1J^1Y \to J^1Y$.

Proposition. *All natural transformations $f\colon J^1J^1Y \to J^1J^1Y$ depending on a linear connection Λ on the base manifold form the two 3-parameter families*

$$I. \quad f = id + \langle \delta, N(\Lambda) \rangle, \qquad II. \quad f = e_\Lambda + \langle \delta, N(\Lambda) \rangle. \tag{5}$$

Proof. The standard fibers $V = (y_i^p, Y_i^p, y_{ij}^p)$ and $Z = (\Lambda_{jk}^i)$ are $G_{m,n}^2$-spaces and we have to find all $G_{m,n}^2$-equivariant maps $f\colon V \times Z \to V$. The action of $G_{m,n}^2$ on V is

$$\begin{aligned} \bar{y}_i^p &= a_q^p y_j^q \tilde{a}_i^j + a_j^p \tilde{a}_i^j, \qquad \bar{Y}_i^p = a_q^p Y_j^q \tilde{a}_i^j + a_j^p \tilde{a}_i^j \\ \bar{y}_{ij}^p &= a_q^p y_{kl}^q \tilde{a}_i^k \tilde{a}_j^l + a_{qr}^p y_k^q Y_l^r \tilde{a}_i^k \tilde{a}_j^l + a_{qk}^p Y_l^q \tilde{a}_i^k \tilde{a}_j^l + \\ &\quad + a_{ql}^p y_k^q \tilde{a}_i^k \tilde{a}_j^l + a_q^p y_k^q \tilde{a}_{ij}^k + a_{kl}^p \tilde{a}_i^k \tilde{a}_j^l + a_k^p \tilde{a}_{ij}^k \end{aligned} \tag{6}$$

while the action of $G^2_{m,n}$ on Z is given by 25.2.(3).

The coordinate form of an arbitrary map $f\colon V \times Z \to V$ is

$$\begin{aligned} y &= F(y, Y, y_2, \Lambda) \\ Y &= G(y, Y, y_2, \Lambda) \\ y_2 &= H(y, Y, y_2, \Lambda) \end{aligned} \tag{7}$$

where $y = (y^p_i)$, $Y = (Y^p_i)$, $y_2 = (y^p_{ij})$, $\Lambda = (\Lambda^i_{jk})$. Considering equivariance of (7) with respect to the base homotheties we find

$$\begin{aligned} kF(y, Y, y_2, \Lambda) &= F(ky, kY, k^2 y_2, k\Lambda) \\ kG(y, Y, y_2, \Lambda) &= G(ky, kY, k^2 y_2, k\Lambda) \\ k^2 H(y, Y, y_2, \Lambda) &= H(ky, kY, k^2 y_2, k\Lambda). \end{aligned} \tag{8}$$

By the homogeneous function theorem, F and G are linear in y, Y, Λ and independent of y_2, while H is linear in y_2 and bilinear in y, Y, Λ. The fiber homotheties then yield

$$\begin{aligned} kF(y, Y, \Lambda) &= F(ky, kY, \Lambda) \\ kG(y, Y, \Lambda) &= G(ky, kY, \Lambda) \\ kH(y, Y, \Lambda) &= H(ky, kY, ky_2, \Lambda). \end{aligned} \tag{9}$$

Comparing (9) with (8) we find that F and G are independent of Λ and H is linear in y_2 and bilinear in (y, Λ) and in (Y, Λ).

Since f is $GL(m) \times GL(n)$-equivariant, we can apply the generalized invariant tensor theorem. This yields

$$\begin{aligned} F^p_i &= a y^p_i + b Y^p_i \\ G^p_i &= c y^p_i + d Y^p_i \\ H^p_{ij} &= e y^p_{ij} + f y^p_{ji} + \\ &\quad g y^p_i \Lambda^k_{jk} + h y^p_i \Lambda^k_{kj} + i y^p_j \Lambda^k_{ik} + j y^p_j \Lambda^k_{ki} + k y^p_k \Lambda^k_{ij} + l y^p_k \Lambda^k_{ji} + \\ &\quad m Y^p_i \Lambda^k_{jk} + n Y^p_i \Lambda^k_{kj} + p Y^p_j \Lambda^k_{ik} + q Y^p_j \Lambda^k_{ki} + r Y^p_k \Lambda^k_{ij} + s Y^p_k \Lambda^k_{ji}. \end{aligned} \tag{10}$$

The last step consists in expressing the equivariance of (10) with respect to the subgroup of $G^2_{m,n}$ characterized by $a^i_j = \delta^i_j$, $a^p_q = \delta^p_q$. This leads to certain simple algebraic identities, which are equivalent to (5). □

32.7. Remark. The only map in 32.6.(5) independent of Λ is the identity. This proves proposition 32.3.

If we consider a linear symmetric connection Λ, then the whole family $N(\Lambda)$ vanishes identically. This implies

Corollary. *The only two natural transformations $J^1 J^1 Y \to J^1 J^1 Y$ depending on a linear symmetric connection Λ on the base manifold are the identity and* e_Λ.

32.8. Remark. The functors $\bar{J}^2$ and J^1J^1 restricted to the category $\mathcal{M}f_m \times \mathcal{M}f \subset \mathcal{FM}_m$ define the so called semiholonomic and non-holonomic 2-jets in the sense of [Ehresmann, 54]. We remark that all natural transformations of each of those restricted functors into itself are determined in [Kolář, Vosmanská, 87].

Further we remark that [Kurek, to appear b] described all natural transformations $T^{r*} \to T^{s*}$ between any two one-dimensional covelocities functors from 12.8. He also determined all natural tensors of type $\binom{1}{1}$ on $T^{r*}M$, [Kurek, to appear c].

33. Topics from Riemannian geometry

33.1. Our aim is to outline the application of our general procedures to the study of geometric operations on Riemannian manifolds. Since the Riemannian metrics are sections of a natural bundle (a subbundle in S^2T^*), we can always add the metrics to the arguments of the operation in question instead of specializing our general approach to categories over manifolds for the category of Riemannian manifolds and local isometries. In this way, we reduce the problem to the study of some equivariant maps between the standard fibers, in spite of the fact that the Riemannian manifolds are not locally homogeneous in the sense of 18.4. However, at some stage we mostly have to fix the values of the metric entry by restricting ourselves to the invariance with respect to the isometries and so we need description of all tensors invariant under the action of the orthogonal group.

Let us write $S^2_+T^*$ for the natural bundle of elements of Riemannian metrics.

33.2. $O(m)$-invariant tensors. An $O(m)$-invariant tensor is a tensor $B \in \otimes^p\mathbb{R}^m \otimes \otimes^q\mathbb{R}^{m*}$ satisfying $aB = B$ for all $a \in O(m)$. The canonical scalar product on $\mathbb{R}^m$ defines an $O(m)$-equivariant isomorphism $\mathbb{R}^m \cong \mathbb{R}^{m*}$. This identifies B with an element from $\otimes^{p+q}\mathbb{R}^{m*}$, i.e. with an $O(m)$-invariant linear map $\otimes^{p+q}\mathbb{R}^m \to \mathbb{R}$. Let us define a linear map $\varphi_\sigma\colon \otimes^{2s}\mathbb{R}^m \to \mathbb{R}$, by

$$\varphi_\sigma(v_1 \otimes \cdots \otimes v_{2s}) = (v_{\sigma(1)}, v_{\sigma(2)}).(v_{\sigma(3)}, v_{\sigma(4)}) \cdots (v_{\sigma(2k-1)}, v_{\sigma(2s)}),$$

where $(\ ,\ )$ means the canonical scalar product defined on $\mathbb{R}^m$ and $\sigma \in \Sigma_{2s}$ is a permutation. The maps φ_σ are called the *elementary invariants*. The fundamental result due to [Weyl, 46] is

Theorem. *The linear space of all $O(m)$-invariant linear maps $\otimes^k\mathbb{R}^m \to \mathbb{R}$ is spanned by the elementary invariants for $k = 2s$ and is the zero space if k is odd.*

Proof. We present a proof based on the Invariant tensor theorem (see 24.4), following the lines of [Atiyah, Bott, Patodi, 73]. The idea is to involve explicitly all metrics $g_{ij} \in S^2_+\mathbb{R}^{m*}$ and then to look for $GL(m)$-invariant maps. So together with an $O(m)$-invariant map $\varphi\colon \otimes^k\mathbb{R}^m \to \mathbb{R}$ we consider the map $\bar{\varphi}\colon S^2_+\mathbb{R}^{m*} \times \otimes^k\mathbb{R}^m \to \mathbb{R}$, defined by $\bar{\varphi}(\mathbb{I}_m, x) = \varphi(x)$ and $\bar{\varphi}(G, x) = \bar{\varphi}((A^{-1})^TGA^{-1}, Ax)$

for all $A \in GL(m)$, $G \in S^2_+\mathbb{R}^{m*}$. By definition, $\bar{\varphi}$ is $GL(m)$-invariant. With the help of the next lemma, we shall be able to extend the map $\bar{\varphi}$ to the whole $S^2\mathbb{R}^{m*} \times \otimes^k\mathbb{R}^m$.

Let us write $V = \otimes^k\mathbb{R}^m$. The map $\bar{\varphi}$ induces a map $GL(m) \times V \to \mathbb{R}$, $(A, x) \to \bar{\varphi}(A^T A, x) = \varphi(Ax)$ and this map is extended by the same formula to a polynomial map $f\colon \mathfrak{gl}(m) \times V \to \mathbb{R}$, linear in V. So $f_x(A) = f(A, x) = \varphi(Ax)$ is polynomial and $O(m)$-invariant for all $x \in V$, and $f(A, x) = \bar{\varphi}(A^T A, x)$ if A invertible.

Lemma. *Let $h\colon \mathfrak{gl}(m) \to \mathbb{R}$ be a polynomial map such that $h(BA) = h(A)$ for all $B \in O(m)$. Then there is a polynomial F on the space of all symmetric matrices such that $h(A) = F(A^T A)$.*

Proof. In dimension one, we deal with the well known assertion that each even polynomial, i.e. $h(x) = h(-x)$, is a polynomial in x^2. However in higher dimensions, the proof is quite non trivial. We present only the main ideas and refer the reader to our source, [Atiyah, Bott, Patodi, 73, p. 323], for more details.

First notice that it suffices to prove the lemma for non singular matrices, for then the assertion follows by continuity. Next, if $A^T A = P$ with P non singular and if there is a symmetric Q, $Q^2 = P$, then A lies in the $O(m)$-orbit of Q. Indeed, Q is also non singular and $B = AQ^{-1}$ satisfies $B^T B = Q^{-1} A^T A Q^{-1} = \mathbb{I}_m$. So it suffices to restrict ourselves to symmetric matrices.

Hence we want to find a polynomial map g satisfying $h(Q) = g(Q^2)$ for all symmetric matrices. For every symmetric matrix P, there is the square root $\sqrt{P} = Q$ if we extend the field of scalars to its algebraic closure. This can be computed easily if we express $P = B^T D B$ with an orthogonal matrix B and diagonal matrix D, since then $\sqrt{P} = B^T \sqrt{D} B$ and $\sqrt{D}$ is the diagonal matrix with the square roots of the eigen values of P on its diagonal. But we should express Q as a universal polynomial in the elements p_{ij} of the matrix P. Let us assume that all eigenvalues λ_i of P are different. Then we can write

$$Q = \sum_{i=1}^{m} \sqrt{\lambda_i} \prod_{j \neq i} \frac{P - \lambda_j}{\lambda_i - \lambda_j}.$$

Notice that the eigen values λ_i are given by rational functions of the elements p_{ij} of P. Thus, in order to make this to a polynomial expression, we have first to extend the field of complex numbers to the field K of rational functions (i.e. the elements are ratios of polynomials in p_{ij}'s). So for matrices with entries from K, all eigen values depend polynomially on p_{ij}'s. We also need their square roots to express Q, but next we shall prove that after inserting $Q = \sqrt{P}$ into $h(Q)$ all square roots will factor out. For any fixed P, let us consider the splitting field L over K with respect to the roots of the equation $\det(P - \lambda^2) = 0$. So $\sqrt{P}$ is polynomial over L. As a polynomial map, h extends to $\mathfrak{gl}(m, L)$ and the next sublemma shows that it is in fact $O(m, L)$-invariant.

Sublemma. *Let L be any algebraic extension of $\mathbb{R}$ and let $f\colon O(m, L) \to L$ be a rational function. If f vanishes on $O(m, \mathbb{R})$ then f is zero.*

Proof. The Cayley map $C\colon \mathfrak{o}(m,\mathbb{R}) \to O(m,\mathbb{R})$ is a birational isomorphism of the orthogonal group with an affine space. Hence there are 'enough real points' to make zero all coefficients of the rational map. For more details see [Atiyah, Bott, Patodi, 73] □

Now the basic fact is, that for any automorphism $\sigma\colon L \to L$ of the Galois group of L over K we have $(\sigma Q)^2 = \sigma P = P$ and since both Q and σQ are symmetric, $B = \sigma Q Q^{-1}$ is orthogonal. Hence we get $\sigma h(Q) = h(\sigma Q) = h(BQ) = h(Q)$. Since this holds for all σ, $h(Q)$ lies in K and so $h(Q) = g(Q^2)$ for a rational function g.

The latter equality remains true if P is a real symmetric matrix such that all its eigen values are distinct and the denominator of $g(P)$ is non zero. If $g = F/G$ for two polynomials F and G, we get $F(A^TA) = h(A)G(A^TA)$. If we choose A so that $G(A^TA) = 0$, we get $F(A^TA) = 0$. Hence g is a globally defined rational function without poles and so a polynomial.

Thus, we have found a polynomial F on the space of symmetric matrices such that $h(A) = F(A^TA)$ holds for a Zariski open set in $\mathfrak{gl}(m)$. This proves our lemma. □

Let us continue in the proof of the Weyl's theorem. By the lemma, every f_x satisfies $f_x(A) = g_x(A^TA)$ for certain polynomial g_x and so we get a polynomial mapping $g\colon S^2\mathbb{R}^{m*} \times V \to \mathbb{R}$ linear in V. For all $B, A \in GL(m,\mathbb{C})$ we have $g((B^{-1})^TA^TAB^{-1}, Bx) = f(AB^{-1}, Bx) = f(A,x) = g(A^TA, x)$ and so $g\colon S^2\mathbb{R}^{m*} \times V \to \mathbb{R}$ is $GL(m)$-invariant. Then the composition of g with the symmetrization yields a polynomial $GL(m)$-invariant map $\otimes^2\mathbb{R}^{m*} \times \otimes^k\mathbb{R}^m \to \mathbb{R}$, linear in the second entry. Each multi homogeneous component of degree $s+1$ in the sense of 24.11 is also $GL(m)$-invariant and so its total polarization is a linear $GL(m)$-invariant map $H\colon \otimes^{2s}\mathbb{R}^{m*} \otimes \otimes^k\mathbb{R}^m \to \mathbb{R}$. Hence, by the Invariant tensor theorem, $k = 2s$ and H is a sum of complete contractions over possible permutations of indices. Since the original mapping φ is given by $\varphi(x) = g(\mathbb{I}_m, x)$, Weyl's theorem follows. □

33.3. To explain the coordinate form of 33.2, it is useful to consider an arbitrary metric $G = (g_{ij}) \in S^2_+\mathbb{R}^{m*}$. Let $O(G) \subset GL(m)$ be the subgroup of all linear isomorphisms preserving G, so that $O(m) = O(\mathbb{I}_m)$. Clearly, theorem 33.2 holds for $O(G)$-invariant tensors as well. Every $O(G)$-invariant tensor $B = (B^{i_1\dots i_p}_{j_1\dots j_p}) \in \otimes^p\mathbb{R}^m \otimes \otimes^q\mathbb{R}^{m*}$ induces an $O(G)$-invariant tensor $g_{i_1k_1}\dots g_{i_pk_p}B^{k_1\dots k_p}_{j_1\dots j_q} \in \otimes^{p+q}\mathbb{R}^{m*}$. Hence theorem 33.2 implies that all $O(G)$-invariant tensors in $\otimes^p\mathbb{R}^m \otimes \otimes^q\mathbb{R}^{m*}$ with $p+q$ even are linearly generated by

$$g^{i_1k_1}\dots g^{i_pk_p}g_{\sigma(k_1)\sigma(k_2)}\cdots g_{\sigma(j_{q-1})\sigma(j_q)}$$

where $g^{ik}g_{jk} = \delta^i_j$, $g^{ij} = g^{ji}$, for all permutations σ of $p+q$ letters.

Consequently, all $O(G)$-equivariant tensor operations are generated by: tensorizing by the metric tensor $G\colon \mathbb{R}^m \to \mathbb{R}^{m*}$ or by its inverse $\tilde{G}\colon \mathbb{R}^{m*} \to \mathbb{R}^m$, applying contractions and permutations of indices, and taking linear combinations.

33.4. Our next main goal is to prove the famous Gilkey theorem on natural exterior forms on Riemannian metrics, i.e. to determine all natural operators $S^2_+T^* \rightsquigarrow \Lambda^p T^*$. This will be based on 33.2 and on the reduction theorems from section 28. But since the resulting forms come from the Levi-Cività connection via the Chern-Weil construction, we first determine all natural operators transforming linear symmetric connections into exterior forms. This will help us to describe easily the metric operators later on.

Let us start with a description of natural tensors depending on symmetric linear connections, i.e. natural operators $Q_\tau P^1 \rightsquigarrow T^{(p,q)}$, where $T^{(p,q)}\mathbb{R}^m = \mathbb{R}^m \times \otimes^p\mathbb{R}^m \otimes \otimes^q\mathbb{R}^{m*}$. Each covariant derivative of the curvature $R(\Gamma) \in C^\infty(TM \otimes T^*M \otimes \Lambda^2 T^*M)$ of the connection Γ on M is natural. Further every tensor multiplication of two natural tensors and every contraction on one covariant and one contravariant entry of a natural tensor give new natural tensors. Finally we can tensorize any natural tensor with a $GL(m)$-invariant tensor, we can permute any number of entries in the tensor products and we can repeat each of these steps and take linear combinations.

Lemma. *All natural operators $Q_\tau P^1 \rightsquigarrow T^{(p,q)}$ are obtained by this procedure. In particular, there are no non zero operators if $q-p=1$ or $q-p<0$.*

Proof. By 23.5, every such operator has some finite order r and so it is determined by a smooth G^{r+2}_m-equivariant map $f\colon T^r_m Q \to V$, where Q is the standard fiber of the connection bundle and $V = \otimes^p\mathbb{R}^m \otimes \otimes^q\mathbb{R}^{m*}$. By the proof of the theorem 28.6, there is a G^1_m-equivariant map $g\colon W^{r-1} \to V$ such that $f = g \circ C^{r-1}$. Here $W^{r-1} = W \times \ldots \times W_{r-1}$, $W = \mathbb{R}^m \otimes \mathbb{R}^{m*} \otimes \Lambda^2\mathbb{R}^{m*}$, $W_i = W \otimes \otimes^i\mathbb{R}^{m*}$, $i = 1, \ldots, r-1$. Therefore the coordinate expression of a natural tensor is given by smooth maps

$$\omega^{i_1\ldots i_p}_{j_1\ldots j_q}(W^i_{jkl}, \ldots, W^i_{jklm_1\ldots m_{r-1}}).$$

Hence we can apply the Homogeneous function theorem (see 24.1). The action of the homotheties $c^{-1}\delta^i_j \in G^1_m$ gives

$$c^{q-p}\omega^{i_1\ldots i_p}_{j_1\ldots j_q}(W^i_{jkl}, \ldots, W^i_{jklm_1\ldots m_{r-1}}) = \omega^{i_1\ldots i_p}_{j_1\ldots j_q}(c^2W^i_{jkl}, \ldots, c^{r+1}W^i_{jklm_1\ldots m_{r-1}}).$$

Hence the ω's must be sums of homogeneous polynomials of degrees d_s in the variables $W^i_{jklm_1\ldots m_s}$ satisfying

$$2d_0 + \cdots + (r+1)d_{r-1} = q - p. \tag{1}$$

Now we can consider the total polarization of each multi homogeneous component and we obtain linear mappings

$$S^{d_0}W \otimes \cdots \otimes S^{d_{r-1}}W_{r-1} \to V.$$

According to the Invariant tensor theorem, all the polynomials in question are linearly generated by monomials obtained by multiplying an appropriate number of variables $W^i_{jkl\alpha}$ and applying some of the $GL(m)$-equivariant operations.

If $q = p$, then the polynomials would be of degree zero, and so only the $GL(m)$-invariant tensors can appear. If $q-p=1$ or $q-p<0$, there are no non negative integers solving (1). □

33.5. Natural forms depending on linear connections. To determine the natural operators $Q_\tau P^1 \rightsquigarrow \Lambda^q T^*$, we have to consider the case $p = 0$ and apply the alternation to the subscripts. It is well known that the Chern-Weil construction associates a natural form to every polynomial P which is defined on $\mathbb{R}^m \otimes \mathbb{R}^{m*}$ and invariant under the action of $GL(m)$. This natural form is obtained by substitution of the entries of the matrix valued curvature 2-form R for the variables and taking the wedge product for multiplication. So if P is homogeneous of degree j, then $P(R)$ is a natural $2j$-form. Let us denote by ω_q the form obtained from the tensor product of q copies of the curvature R by taking its trace and alternating over the remaining entries. In coordinates, $\omega_q = (R^{k_q}_{k_1ab} R^{k_1}_{k_2cd} \dots R^{k_{q-1}}_{k_qef})$, where we alternate over all indices $a, \dots, f$. One finds easily that the polynomials P_q depending on the entries of the matrix 2-form R correspond to the homogeneous components of degree q in $\det(\mathbb{I}_m + R)$ and so the forms ω_q equal the *Chern forms* c_q up to the constant factor $(i/(2\pi))^q$.

The wedge product on the linear space of all natural forms depending on connections defines the structure of a graded algebra.

Theorem. *The algebra of all natural operators $Q_\tau P^1 \rightsquigarrow \oplus_{p=0}^m \Lambda^p T^*$ is generated by the Chern forms c_q.*

In particular, there are no natural forms with odd degrees and consequently all natural forms are closed.

Proof. We have to continue our discussion from the proof of the lemma 33.4. However, we need some relations on the absolute derivatives $R^i_{jklm_1\dots m_s}$ of the curvature tensor. First recall the antisymmetry, the first and the second Bianchi identity, cf. 28.5

$$R^i_{jkl} = -R^i_{jlk} \tag{1}$$

$$R^i_{jkl} + R^i_{klj} + R^i_{ljk} = 0 \tag{2}$$

$$R^i_{jklm} + R^i_{jlmk} + R^i_{jmkl} = 0 \tag{3}$$

Lemma. *The alternation of $R^i_{jklm_1\dots m_s}$ over any 3 indices among the first four subscripts is zero.*

Proof. Since the covariant derivative commutes with the tensor operations like the permutation of indices, it suffices to discuss the variables R^i_{jkl} and R^i_{jklm}. By (2), the alternation over the subscripts in R^i_{jkl} is zero and (3) yields the same for the alternation over k, l, m in R^i_{jklm}. In view of (1), it remains to discuss the alternation of R^i_{jklm} over j, l, m. (1) implies $R^i_{jkml} = -R^i_{jmkl}$ and so we can rewrite this alternation as follows

$$\begin{aligned} &R^i_{jklm} + R^i_{jmkl} + R^i_{jlmk} - R^i_{jlmk} \\ &+R^i_{mkjl} + R^i_{mlkj} + R^i_{mjlk} - R^i_{mjlk} \\ &+R^i_{lkmj} + R^i_{ljkm} + R^i_{lmjk} - R^i_{lmjk}. \end{aligned}$$

The first three entries on each row form a cyclic permutation and hence give zero. The same applies to the last column. □

Now it is easy to complete the proof of the theorem. Consider first a monomial containing at least one quantity $R^i_{jklm_1\dots m_s}$ with $s > 0$. Then there exists one term of the product with three free subscripts among the first four ones or one term R^i_{jkl} with all free subscripts, so that the monomial vanishes after alternation. Further, (1) and (2) imply $R^i_{jkl} - R^i_{lkj} = -R^i_{klj}$. Hence we can restrict ourselves to contractions with the first subscripts and so all the possible natural forms are generated by the expressions $R^{k_q}_{k_1ab}R^{k_1}_{k_2cd}\dots R^{k_{q-1}}_{k_qef}$ where the indices $a,\dots,f$ remain free for alternation. But these are coordinate expressions of the forms ω_q. □

33.6. Characteristic classes. The dimension of the homogeneous component of the algebra of natural forms of degree $2s$ equals the number $\pi(s)$ of the partitions of s into sums of positive integers. Since all natural forms are closed, they determine cohomology classes in the De Rham cohomologies of the underlying manifolds. It is well known from the Chern-Weil theory that these classes do not depend on the connection. This can be deduced as follows.

Consider two linear connections Γ, $\bar\Gamma$ expressed locally by Γ^i_j, $\bar\Gamma^i_j \in (T^*M \otimes TM)\otimes T^*M$, and their curvatures R^i_j, $\bar R^i_j \in (T^*M\otimes TM)\otimes\Lambda^2T^*M$. Write $\Gamma_t = t\bar\Gamma + (1-t)\Gamma$ and analogously R_t for the curvatures. Let P_q be the polynomial defining the form ω_q and Q be its total polarization. We define $\tau_q(\Gamma,\bar\Gamma) = q\int_0^1 Q(\bar\Gamma - \Gamma, R_t,\dots,R_t)dt$. The structure equation yields $\frac{d}{dt}R_t = \frac{d}{dt}(d\Gamma_t) - \frac{d}{dt}\Gamma_t\wedge\Gamma_t - \Gamma_t\wedge\frac{d}{dt}\Gamma_t = d(\bar\Gamma-\Gamma)$ and we calculate easily in normal coordinates

$$\begin{aligned}\omega_q(\bar\Gamma) - \omega_q(\Gamma) &= \int_0^1 \frac{d}{dt}Q(R_t,\dots,R_t)dt = q\int_0^1 Q(\frac{d}{dt}R_t, R_t,\dots,R_t)dt\\ &= q\int_0^1 dQ(\bar\Gamma-\Gamma,R_t,\dots,R_t)dt = d\tau_q(\Gamma,\bar\Gamma).\end{aligned}$$

In fact, τ_q is one of many natural operators $Q_\tau P^1\times Q_\tau P^1 \rightsquigarrow \Lambda^{2q-1}T^*$ and the integration helps us to find the proper linear combination of more elementary operators which are obtained by a procedure similar to that from 33.4–33.5. The form $\tau_q(\Gamma,\bar\Gamma)$ is called the *transgression.*

33.7. Natural forms on Riemannian manifolds. Since there is the natural Levi-Cività connection, we can evaluate the natural forms from 33.5 using the curvature of this connection. In this case 28.14.(3) holds, i.e.

$$g_{in}W^n_{jklm_1\dots m_r} = -g_{jn}W^n_{iklm_1\dots m_r}. \tag{1}$$

For $g_{ij} = \delta_{ij}$, $r = 0$, this implies

$$R^i_{jkl} = -R^j_{ikl} \tag{2}$$

and so the contractions in a monomial $R^{k_q}_{k_1ab}R^{k_1}_{k_2cd}\dots R^{k_{q-1}}_{k_qef}$ yield zero if q is odd. The natural forms $p_j = (2\pi)^{-2j}\omega_{2j}$ are called the *Pontryagin forms.* The dimension of the homogeneous component of degree $4s$ of the algebra of forms generated by the Pontryagin forms is $\pi(s)$, cf. 33.6.

If we assume the dependence of the natural operators on the metric, then every two indices of any tensor can be contracted. In particular, the complete contractions of covariant derivatives of the curvature of the Levi-Cività connection give rise to natural functions of all even orders grater then one. Composing k natural functions with any fixed smooth function $\mathbb{R}^k \to \mathbb{R}$, we get a new natural function. Since every natural form can be multiplied by any natural function without loosing naturality, we see that there is no hope to describe all natural forms in a way similar to 33.5. However, in Riemannian geometry we often meet operations with a sort of homogeneity with respect to the change of the scale of the metric and these can be described in more details.

Our operators will have several arguments as a rule and we shall use the following brief notation in this section: Given several natural bundles $F_a, \dots, F_b$, we write $F_a \times \dots \times F_b$ for the natural bundle associating to each m-manifold M the fibered product $F_aM \times_M \dots \times_M F_bM$ and similarly on morphisms. (Actually, this is the product in the category of functors, cf. 14.11.) Hence $D\colon F_1 \times F_2 \rightsquigarrow G$ means a natural operator transforming couples of sections from $C^\infty(F_1M)$ and $C^\infty(F_2M)$ to sections from $C^\infty(GM)$ (which is also denoted by $D\colon F_1 \oplus F_2 \rightsquigarrow G$ in this book). Analogously, given natural operators $D_1\colon F_1 \rightsquigarrow G_1$ and $D_2\colon F_2 \rightsquigarrow G_2$, we use the symbol $D_1 \times D_2\colon F_1 \times F_2 \rightsquigarrow G_1 \times G_2$.

Definition. Let E and F be natural bundles over m-manifolds. We say that a natural operator $D\colon S^2_+T^* \times E \rightsquigarrow F$ is *conformal*, if $D(c^2g, s) = D(g, s)$ for all metrics g, sections s, and all positive $c \in \mathbb{R}$. If F is a natural vector bundle and D satisfies $D(c^2g) = c^\lambda D(g)$, then λ is called the *weight* of D.

Let us notice that the weight of the metric g_{ij} is 2 (we consider the inclusion $g\colon S^2_+T^* \rightsquigarrow S^2T^*$), that of its inverse g^{ij} is -2, while the curvature and all its covariant derivatives are conformal.

33.8. Gilkey theorem. *There are no non zero natural forms on Riemannian manifolds with a positive weight. The algebra of all conformal natural forms on Riemannian manifolds is generated by the Pontryagin forms.*

33.9. Let us start the proof with a discussion on the reduction procedure developed in section 28. Even if we have no estimate on the order, we can get an analogous result. Consider an arbitrary natural operator $Q_\tau P^1 \times E \rightsquigarrow F$. By the non-linear Peetre theorem, D is of order infinity and so it is determined by the restriction $\mathcal{D}$ of its associated mapping $J^\infty((Q_\tau P^1 \times E)\mathbb{R}^m) \to F\mathbb{R}^m$ to the fiber over the origin. Moreover, we obtain an open filtration of the whole fiber $J^\infty_0((Q_\tau P^1 \times E)\mathbb{R}^m)$ consisting of maximal G^∞_m-invariant open subsets U_k where the associated mapping $\mathcal{D}$ factorizes through $\mathcal{D}_k\colon \pi^\infty_k(U_k) \subset J^k_0((Q_\tau P^1 \times E)\mathbb{R}^m) \to F_0\mathbb{R}^m$. Now, we can apply the same procedure as in the section 28 to this invariant open submanifolds $\pi^\infty_k(U_k)$.

Let F be a first order bundle functor on $\mathcal{M}f_m$, E be an open natural sub bundle of a vector bundle functor $\bar{E}$ on $\mathcal{M}f_m$. The curvature and its covariant derivatives are natural operators $\rho_k\colon Q_\tau P^1 \rightsquigarrow R_k$, with values in tensor bundles

R_k, $R_k\mathbb{R}^m = \mathbb{R}^m \times W_k$, $W_0 = \mathbb{R}^m \otimes \mathbb{R}^{m*} \otimes \Lambda^2\mathbb{R}^{m*}$, $W_{k+1} = W_k \otimes \mathbb{R}^{m*}$. Similarly, the covariant differentiation of sections of E forms natural operators $d_k\colon Q_\tau P^1 \times E \rightsquigarrow E_k$, where $E_0 = \bar{E}$, $E_0\mathbb{R}^m =: \mathbb{R}^m \times V_0$, d_0 is the inclusion, $E_k\mathbb{R}^m = \mathbb{R}^m \times V_k$, $V_{k+1} = V_k \otimes \mathbb{R}^{m*}$. Let us write $D^k = (\rho_0, \dots, \rho_{k-2}, d_0, \dots, d_k)\colon Q_\tau P^1 \times E \rightsquigarrow R^{k-2} \times E^k$, where $R^l = R_0 \times \dots \times R_l$, $E^l = E_0 \times \dots \times E_l$. All D^k are natural operators. In 28.8 we defined the Ricci sub bundles $Z^k \subset R^{k-2} \times E^k$ and we know $D^k\colon Q_\tau P^1 \times E \rightsquigarrow Z^k$.

Let us further define the functor Z^∞ as the inverse limit of Z^k, $k \in \mathbb{N}$, with respect to the obvious natural transformations (projections) $p^k_\ell\colon Z^k \to Z^\ell$, $k > \ell$, and similarly $D^\infty\colon Q_\tau P^1 \times E \rightsquigarrow Z^\infty$. As a corollary of 28.11 and the non linear Peetre theorem we get

Proposition. *For every natural operator $D\colon Q_\tau P^1 \times E \rightsquigarrow F$ there is a unique natural transformation $\tilde{D}\colon Z^\infty \to F$ such that $D = \tilde{D} \circ D^\infty$. Furthermore, for every m-dimensional compact manifold M and every section $s \in C^\infty(Q_\tau P^1 M \times_M EM)$, there is a finite order k and a neighborhood V of s in the C^k-topology such that $\tilde{D}_M|(D^\infty)_M(V) = (\pi^\infty_k)^*(\tilde{D}_k)_M$, for some $(\tilde{D}_k)_M\colon (D^k)_M(V) \to C^\infty(Z^k M)$, and $D_M|V = (\tilde{D}_k)_M \circ (D^k)_M|V$.*

In words, a natural operator $D\colon Q_\tau \times E \rightsquigarrow F$ is determined in all coordinate charts of an arbitrary m-dimensional manifold M by a universal smooth mapping defined on the curvatures and all their covariant derivatives and on the sections of EM and all their covariant derivatives, which depends 'locally' only on finite number of these arguments.

33.10. The Riemannian case. In section 28, we also applied the reduction procedure to operators depending on Riemannian metrics and general vector fields. In fact we have viewed the operators $D\colon S^2_+T^* \times E \rightsquigarrow F$ as operators $\bar{D}\colon Q_\tau P^1 \times (S^2_+T^* \times E) \rightsquigarrow F$ independent of the first argument and we have used the Levi-Cività connection $\Gamma\colon S^2_+T^* \rightsquigarrow Q_\tau P^1$ to write D as a composition $D = \bar{D} \circ (\Gamma, \mathrm{id})$. Since the covariant derivatives of the metric with respect to the metric connection are zero, we can restrict ourselves to sub bundles in the Ricci subspaces corresponding to the bundle $S^2_+T^* \times E$, which are of the form $S^2_+T^* \times Z^k$ with $Z^k \subset R^{k-2} \times E^k$, cf. 28.14. Let us notice that the bundles $Z^k M$ involve the curvature of the Riemannian connection on M, its covariant derivatives, and the covariant derivatives of the sections of EM. Similarly as above, we define the inverse limits Z^∞ and D^∞ and as a corollary of the non linear Peetre theorem and 28.15 we get

Corollary. *For every natural operator $D\colon S^2_+T^* \times E \rightsquigarrow F$ there is a natural transformation $\tilde{D}\colon S^2_+T^* \times Z^\infty \to F$ such that $D = \tilde{D} \circ D^\infty \circ (\Gamma, \mathrm{id})$. Furthermore, for every m-dimensional compact manifold M and every section $s \in C^\infty(S^2_+T^*M \times_M EM)$, there is a finite order k and a neighborhood V of s in the C^k-topology such that $\tilde{D}_M|(D^\infty \circ (\Gamma, \mathrm{id}))_M(V) = (\pi^\infty_k)^*(\tilde{D}_k)_M$, where $(\tilde{D}_k)_M\colon (D^k \circ (\Gamma, \mathrm{id}))_M(V) \to C^\infty(Z^k M)$, and $D_M|V = (\tilde{D}_k)_M \circ (D^k)_M \circ (\Gamma, \mathrm{id})_M|V$.*

33.11. Polynomiality. Since the standard fiber V_0 of E_0 is embedded identically into $Z_0^k\mathbb{R}^m$ by the associated map to the operator D^k, we can use 28.16 and add the following proposition to the statements of 33.9, or 33.10, respectively.

Corollary. *The operator D is polynomial if and only if the operators $\tilde{D}^k$ are polynomial. Further D is polynomial with smooth real functions on the values of E_0, or $S^2_+T^*$, as coefficients if and only if the operators $\tilde{D}^k$ are polynomial with smooth real functions on the values of E_0, or $S^2_+T^*$, as coefficients, respectively.*

33.12. Natural operators $D\colon S^2_+T^* \rightsquigarrow T^{(p,q)}$. According to 33.9 we find G^∞_m-invariant open subsets U_k in $J^\infty_0(S^2_+T^*\mathbb{R}^m)$ forming a filtration of the whole jet space, such that on these subsets $\mathcal{D}$ factorizes through smooth G^{k+1}_m-equivariant mappings

$$f^{i_1\dots i_p}_{j_1\dots j_q} = f^{i_1\dots i_p}_{j_1\dots j_q}(g_{ij},\dots,g_{ij\ell_1\dots\ell_k})$$

defined on $\pi^\infty_k U_k$. For large k's, the action of the homotheties $c^{-1}\delta^i_j$ on g's is well defined and we get

$$c^{q-p}f^{i_1\dots i_p}_{j_1\dots j_q}(g_{ij},\dots,g_{ij\ell_1\dots\ell_k}) = f^{i_1\dots i_p}_{j_1\dots j_q}(c^2g_{ij},\dots,c^{2+k}g_{ij\ell_1\dots\ell_k}). \tag{1}$$

Now, let us add the assumption that D is homogeneous with weight λ, choose the change $g \mapsto c^{-2}g$ of the scale of the metric and insert this new metric into (1). We get

$$c^{q-p-\lambda}f^{i_1\dots i_p}_{j_1\dots j_q}(g_{ij},\dots,g_{ij\ell_1\dots\ell_k}) = f^{i_1\dots i_p}_{j_1\dots j_q}(g_{ij},c^1g_{ij,\ell_1},\dots,c^kg_{ij\ell_1\dots\ell_k}).$$

This formula shows that the mappings $f^{i_1\dots i_p}_{j_1\dots j_q}$ are polynomials in all variables except g_{ij} with functions in g_{ij} as coefficients.

According to 33.11 and 28.16, the map $\mathcal{D}$ is on U_k determined by a polynomial mapping

$$\omega = (\omega^{i_1\dots i_p}_{j_1\dots j_q}(g_{ij},W^i_{jkl},\dots,W^i_{jklm_1\dots m_{k-2}}))$$

which is G^1_m-equivariant on the values of the covariant derivatives of the curvatures and the sections. If we apply once more the equivariance with respect to the homothety $x \mapsto c^{-1}x$ and at the same time the change of the scale of the metric $g \mapsto c^{-2}g$, we get

$$\begin{aligned} c^{q-p-\lambda}\omega^{i_1\dots i_p}_{j_1\dots j_q}(g_{ij},R^i_{jkl},\dots,R^i_{jklm_1\dots m_{k-2}}) &= \\ &= \omega^{i_1\dots i_p}_{j_1\dots j_q}(g_{ij},c^2R^i_{jkl},\dots,c^kR^i_{jklm_1\dots m_{k-2}}). \end{aligned}$$

This homogeneity shows that the polynomial functions $\omega^{i_1\dots i_p}_{j_1\dots j_q}$ must be sums of homogeneous polynomials with degrees a_ℓ in the variables $R^i_{jklm_1\dots m_\ell}$ satisfying

$$2a_0 + \dots + ka_{k-2} = q - p - \lambda \tag{2}$$

and their coefficients are functions depending on g_{ij}'s.

Now, we shall fix $g_{ij} = \delta_{ij}$ and use the $O(m)$-equivariance of the homogeneous components of the polynomial mapping ω. For this reason, we shall switch to the variables $R_{ijklm_1\dots m_s} = g_{ia}R^a_{jklm_1\dots m_s}$. Using the standard polarization technique and H. Weyl's theorem, we get that each multi homogeneous component in question results from multiplication of variables $R_{ijklm_1,\dots,m_s}$, $s = 0, 1, \dots, r$, and application of some $O(m)$-equivariant tensor operations on the target space. Hence our operators result from a finite number of the following steps.

(a) take tensor product of arbitrary covariant derivatives of the curvature tensor
(b) tensorize by the metric or by its inverse
(c) apply arbitrary $GL(m)$-equivariant operation
(d) take linear combinations.

33.13. Remark. If $q-p = \lambda+1$, then there is no non negative integer solution of 33.12.(2) and so all natural tensors in question are zero. The case $q = 2$, $p = 1$, $\lambda = 0$ implies that *the Levi-Cività connection is the only conformal natural connection on Riemannian manifolds.*

Indeed, the difference of two such connections is a natural tensor twice covariant and once contravariant, and so zero.

33.14. Consider now $\Lambda^p\mathbb{R}^{m*}$ as the target tensor space. So in the above procedure, all indices which were not contracted must be alternated at the end. Since the metric is a symmetric tensor, we get zero whenever using the above step (b) and alternating over both indices. But contracting over any of them has no proper effect, for $\delta_{ij}R_{jklnm_1,\dots,m_s} = R_{iklnm_1,\dots,m_s}$. So we can omit the step (b) at all.

The first Bianchi identity and 33.7.(1) imply $R_{ijkl} = R_{klij}$. Then the lemma in 33.5 and 33.7.(1) yield

Lemma. *The alternation of $R_{ijklm_1\dots m_s}$, $0 \le s$, over arbitrary 3 indices among the first four or five ones is zero.*

Consider a monomial P in the variables $R_{ijkl\alpha}$ with degrees a_s in $R_{ijklm_1\dots m_s}$. In view of the above lemma, if P remains non zero after all alternations, then we must contract over at least two indices in each $R_{ijkl\alpha}$ and so we can alternate over at most $2a_0+\dots+ka_{k-2}$ indices. This means $p \le 2a_0+\dots+ka_{k-2} = p-\lambda$. Consequently $\lambda \le 0$ if there is a non zero natural form with weight λ. This proves the first assertion of theorem 33.8.

Let $\lambda = 0$. Since the weight of g^{ij} is -2, any contraction on two indices in the monomial decreases the weight of the operator by 2. Every covariant derivative $R_{ijklm_1\dots m_s}$ of the curvature has weight 2. So we must contract on exactly two indices in each $R_{ijklm_1\dots m_s}$ which implies there are $s+2$ of them under alternation. But then there must appear three alternated indices among the first five if $s \ne 0$. This proves $a_1 = \dots = a_{k-2} = 0$, so that $p = 2a_0$. Hence all the natural forms have even degrees and they are generated by the forms ω_q, cf. 33.5. As we deduced in 33.7, these forms are zero if their degree is not divisible by four.

This completes the proof of the theorem 33.8. □

33.15. Remark. The original proof of the Gilkey theorem assumes a polynomial dependence of the natural forms on a finite number of the derivatives $g_{ij,\alpha}$ of the metric and on the entries of the inverse matrix g^{ij}, but also the homogeneity in the weight, [Gilkey, 73]. Under such polynomiality assumption, our methods apply to all natural tensors. In particular, it follows easily that the Levi-Cività connection is the only second order polynomial connection on Riemannian manifolds. Of course, the latter is not true in higher orders, for we can contract appropriate covariant derivatives of the curvature and so we get natural tensors in $T \otimes T^* \otimes T^*$ of orders higher than two.

33.16. Operations on exterior forms. The approach from 33.4–33.5 can be easily extended to the study of all natural operators $D\colon Q_rP^1 \times T^{(s,r)} \rightsquigarrow T^{(q,p)}$ with $s < r$ or $s = r = 0$. This was done in [Slovák, 92a], we shall present only the final results. If we omit the assumption on s and r, we have to assume the polynomiality.

Theorem. *All natural operators $D\colon Q_rP^1 \times T^{(s,r)} \rightsquigarrow T^{(q,p)}$, $s < r$, are obtained by a finite iteration of the following steps: take tensor product of arbitrary covariant derivatives of the curvature tensor or the covariant derivatives of the tensor fields from the domain, apply arbitrary $GL(m)$-equivariant operation, take linear combinations. In the case $s = r = 0$ we have to add one more step, the compositions of the functions from the domain with arbitrary smooth functions of one real variable.*

The algebra of all natural operators $D\colon Q_rP^1 \times T^{(0,r)} \rightsquigarrow \Lambda T^*$, $r > 0$, is generated by the alternation, the exterior derivative d and the Chern forms c_q.

The algebra of all natural operators $D\colon Q_rP^1 \times T^{(0,0)} \rightsquigarrow \Lambda T^*$ is generated by the compositions with arbitrary smooth functions of one real variable, the exterior derivative d and the Chern forms c_q.

The proof of these results follows the lines of 33.4–33.5 using two more lemmas: First, the alternation on all indices of the second covariant derivative $\nabla^2 t$ of an arbitrary tensor $t \in C^\infty(\otimes^s\mathbb{R}^{m*})$ is zero (which is proved easily using the Bianchi and Ricci identities) and , second, the alternation of the first covariant derivative of an arbitrary tensor $t \in C^\infty(\otimes^s\mathbb{R}^{m*})$ coincides with the exterior differential of the alternation of t (this well known fact is proved easily in normal coordinates).

33.17. Operations on exterior forms on Riemannian manifolds. A modification of our proof of the Gilkey theorem for operations on exterior forms on Riemannian manifolds, which is also based on the two lemmas mentioned above, appeared in [Slovák, 92a]. The equality 33.7.(2) on the Riemannian curvatures can be expressed as $R_{ijkl} = R_{jikl}$, and this holds for curvatures of metrics with arbitrary signatures. This observation extends our considerations to pseudoriemannian manifolds, see [Slovák, 92b]. In particular, our proof of the Gilkey theorem extends to the classification of natural forms on pseudoriemannian manifolds. Let us write $S^2_{\text{reg}}T^*$ for the bundle functor of all non degenerate symmetric two-forms. The definition of the weight of the operators depending on metrics and the definition of the Pontryagin forms extend obviously to the pseudoriemannian case. All the considerations go also through for metrics with any fixed signature.

Theorem. *All natural operators $D\colon S^2_{\mathrm{reg}}T^* \times T^{(s,r)} \rightsquigarrow T^{(q,p)}$, $s < r$, homogeneous in weight are obtained by a finite iteration of the following steps: take tensor product of arbitrary covariant derivatives of the curvature tensor or the covariant derivatives of the tensor fields from the domain, tensorize by the metric or its inverse, apply arbitrary $GL(m)$-equivariant operation, take linear combinations. In the case $s = r = 0$ we have to add one more step, the compositions of the functions from the domain with arbitrary smooth functions of one real variable.*

There are no non zero operators $D\colon S^2_{\mathrm{reg}}T^* \times T^{(0,r)} \rightsquigarrow \Lambda T^*$, $r \geq 0$, with a positive weight. The algebra of all conformal natural operators $S^2_{\mathrm{reg}}T^* \times T^{(0,r)} \rightsquigarrow \Lambda T^*$, $r > 0$, is generated by the alternation, the exterior derivative d and the Pontryagin forms p_q.

The algebra of all conformal natural operators $D\colon S^2_{\mathrm{reg}}T^* \times T^{(0,0)} \rightsquigarrow \Lambda T^*$ is generated by the compositions with arbitrary smooth functions of one real variable, the exterior derivative d and the Pontryagin forms p_q.

The discussion from the proof of these results can be continued for every fixed negative weight. In particular, the situation is interesting for $\lambda = -2$ and linear operators $D\colon \Lambda^pT^* \rightsquigarrow \Lambda^pT^*$ depending on the metric. Beside the compositions $d \circ \delta$ and $\delta \circ d$ of the exterior differential d and the well known codifferential $\delta\colon \Lambda^p \rightsquigarrow \Lambda^{p-1}$ (the Laplace-Beltrami operator is $\Delta = \delta \circ d + d \circ \delta$), there are only three other generators: the multiplication by the scalar curvature, the contraction with the Ricci curvature and the contraction with the full Riemmanian curvature. This classification was derived under some additional assumptions in [Stredder, 75], see also [Slovák, 92b].

33.18. Oriented pseudoriemannian manifolds. It is also quite important in Riemannian geometry to know what are the operators natural with respect to the orientation preserving local isometries. We shall not go into details here since this would require to extend the description from 33.2 to all $SO(m)$-invariant linear maps and then to repeat some steps of the proof of the Gilkey theorem. This was done in [Stredder, 75] (for the polynomial forms and Riemannian manifolds), and in [Slovák, 92b]. Let us only remark that on oriented pseudoriemannian manifolds we have a natural volume form $\omega\colon S^2_{\mathrm{reg}}T^* \rightsquigarrow \Lambda^mT^*$ and natural transformations $*\colon \Lambda^pT^* \to \Lambda^{m-p}T^*$. All natural operators on oriented pseudoriemannian manifolds homogeneous in the weight are generated by those described above, the volume form ω, and the natural transformations $*$.

As an example, let us draw a diagram which involves all linear natural conformal operators on exterior forms on oriented pseudoriemannian manifolds of even dimensions which do not vanish on flat pseudoriemannian manifolds, up to the possible omitting of the d's on the sides in the operators indicated by the long arrows. (More explicitely, we do not consider any contribution from the curvatures.) The symbols Ω^p refer, as usual, to the p-forms, the plus and minus subscripts indicate the splitting into the selfdual and anti-selfdual forms in the degree $\frac{1}{2}m$.

In the even dimensional case, there are no natural conformal operators between exterior forms beside the exterior derivatives. For the proofs see [Slovák, 92b].

We should also remark that the name 'conformal' is rather misleading in the context of the natural operators on conformal (pseudo-) Riemannian manifolds since we require the invariance only with respect to constant rescaling of the metric (cf. the end of the next section). On the other hand, each natural operator on the conformal manifolds must be conformal in our sense.

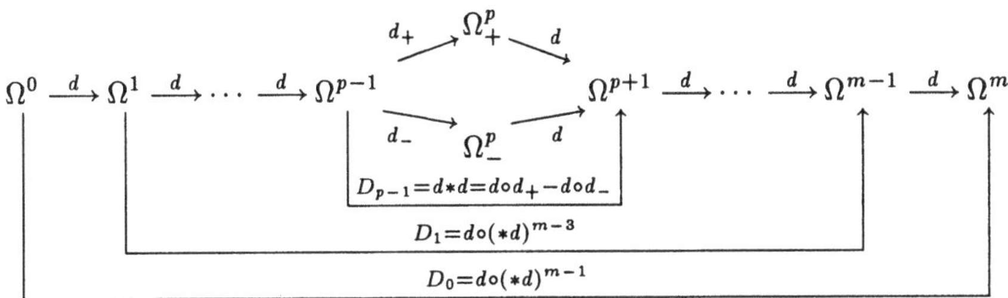

33.19. First order operators. The whole situation becomes much easier if we look for first order natural operators $D\colon S^2_+T^* \rightsquigarrow (F,G)$, where F and G are arbitrary natural bundles, say of order r. Namely, every metric g on a manifold M satisfies $g_{ij} = \delta_{ij}$ and $\frac{\partial g_{ij}}{\partial x^k} = 0$ at the center of any normal coordinate chart. Therefore, if D, $\bar{D}$ are two such operators and if their values $D_{\mathbb{R}^m}(g)$, $\bar{D}_{\mathbb{R}^m}(g)$ on the canonical Euclidean metric g on $\mathbb{R}^m$ coincide on the fiber over the origin, then $D = \bar{D}$. Hence the whole classification problem reduces to finding maps between the standard fibers which are equivariant with respect to the action of the subgroup $O(m) \rtimes B^r_1 \subset G^1_m \rtimes B^r_1 = G^r_m$. In fact we used this procedure in section 29.

Let us notice that the natural operators on oriented Riemannian manifolds are classified on replacing $O(m) \rtimes B^r_1$ by $SO(m) \rtimes B^r_1$. If we modify 29.7 in such a way, we obtain (cf. [Slovák, 89])

Proposition. *All first order natural connections on oriented Riemannian manifolds are*

(1) *The Levi-Cività connection Γ, if $m > 3$ or $m = 2$*
(2) *The one parametric family $\Gamma + kD_1$ where D_1 means the scalar product and $k \in \mathbb{R}$, if $m = 1$*
(3) *The one parametric family $\Gamma + kD_3$ where D_3 means the vector product and $k \in \mathbb{R}$, if $m = 3$.*

33.20. Natural metrics on the tangent spaces of Riemannian manifolds. At the end of this section, we shall describe all first order natural operators transforming metrics into metrics on the tangent bundles. The results were proved in [Kowalski, Sekizava, 88] by the method of differential equations. Let us start with some notation.

We write $\pi_M\colon TM \to M$ for the natural projection and F for the natural bundle with $FM = \pi_M^*(T^*\otimes T^*)M \to M$, $Ff(X_x, g_x) = (Tf.X_x, (T^*\otimes T^*)f.g_x)$ for all manifolds M, local diffeomorphisms f, $X_x \in T_xM$, $g_x \in (T^*\otimes T^*)_xM$. The sections of the canonical projection $FM \to TM$ are called the *F-metrics* in literature. So the F-metrics are mappings $TM \oplus TM \oplus TM \to \mathbb{R}$ which are linear in the second and the third summand. We first show that it suffices to describe all natural F-metrics, i.e. natural operators $S^2_+T^* \rightsquigarrow (T, F)$.

There is the natural Levi-Cività connection $\Gamma\colon TM \oplus TM \to TTM$ and the natural equivalence $\nu\colon TM\oplus TM \to VTM$. There are three F-metrics, naturally derived from sections $G\colon TM \to (S^2T^*)TM$. Given such G on TM, we define

$$
\begin{aligned}
\gamma_1(G)(u,X,Y) &= G(\Gamma(u,X),\Gamma(u,Y))\\
\gamma_2(G)(u,X,Y) &= G(\Gamma(u,X),\nu(u,Y))\\
\gamma_3(G)(u,X,Y) &= G(\nu(u,X),\nu(u,Y)).
\end{aligned}
\tag{1}
$$

Since G is symmetric, we know also $G(\nu(u,X),\Gamma(u,Y)) = \gamma_2(G)(u,Y,X)$. Notice also that γ_1 and γ_3 are symmetric.

The connection Γ defines the splitting of the second tangent space into the vertical and horizontal subspaces. We shall write $X_{x,u} = X^h_u + X^v_u$ for each $X_{x,u} \in T_uTM$, $\pi(u) = x$. Since for every $X_{x,u}$ there are unique vectors $X^h \in T_xM$, $X^v \in T_xM$ such that $\Gamma(u,X^h) = X^h_u$ and $\nu(u,X^v) = X^v_u$, we can recover the values of G from the three F-metrics γ_i,

$$
\begin{aligned}
G(X_{x,u},Y_{x,u}) = \gamma_1(G)(u,X^h,Y^h) &+ \gamma_2(G)(u,X^h,Y^v)\\
&+ \gamma_2(G)(u,Y^h,X^v) + \gamma_3(G)(u,X^v,Y^v).
\end{aligned}
\tag{2}
$$

Lemma. *The formulas (1) and (2) define a bijection between triples of natural F-metrics where the first and the third ones are symmetric, and the natural operators $S^2_+T^* \rightsquigarrow (S^2T^*)T$.* □

33.21. Let us call every section $G\colon TM \to (S^2T^*)TM$ a (possibly degenerated) metric. If we fix an F-metric δ, then there are three distinguished constructions of a metric G.

(1) If δ symmetric, we choose $\gamma_1 = \gamma_3 = \delta$, $\gamma_2 = 0$. So we require that G coincides with δ on both vertical and horizontal vectors. This is called the *Sasaki lift* and we write $G = \delta^s$. If δ is non degenerate and positive definite, the same holds for δ^s.
(2) We require that G coincides with δ on the horizontal vectors, i.e. we put $\gamma_1 = \delta$, $\gamma_2 = \gamma_3 = 0$. This is called the *vertical lift* and G is a degenerate metric which does not depend on the underlying Riemannian metric. We write $G = \delta^v$.
(3) The *horizontal lift* is defined by $\gamma_2 = \delta$, $\gamma_1 = \gamma_3 = 0$ and is denoted by $G = \delta^h$. If δ positive definite, then the signature of G is (m,m).

We can reformulate the lemma 33.20 as

Proposition. *There is a bijective correspondence between the triples of natural F-metrics (α, β, γ), where α and γ are symmetric, and natural (possibly degenerated) metrics G on the tangent bundles given by*

$$G = \alpha^s + \beta^h + \gamma^v. \qquad \square$$

33.22. Proposition. *All first order natural F-metrics α in dimensions $m > 1$ form a family parameterized by two arbitrary smooth functions $\mu, \nu\colon (0,\infty) \to \mathbb{R}$ in the following way. For every Riemannian manifold (M, g) and tangent vectors $u, X, Y \in T_xM$*

$$\text{(1)} \qquad \alpha_{(M,g)}(u)(X,Y) = \mu(g(u,u))g(X,Y) + \nu(g(u,u))g(u,X)g(u,Y).$$

If $m = 1$, then the same assertion holds, but we can always choose $\nu = 0$.

In particular, all first order natural F-metrics are symmetric.

Proof. We have to discuss all $O(m)$-equivariant maps $\alpha\colon \mathbb{R}^m \to \mathbb{R}^{m*} \otimes \mathbb{R}^{m*}$. Denote by $g^0 = \sum_i dx^i \otimes dx^i$ the canonical Euclidean metric and by $|\ |$ the induced norm. Each vector $v \in \mathbb{R}^m$ can be transformed into $|v|\ \frac{\partial}{\partial x^1}\big|_0$. Hence α is determined by its values on the one-dimensional subspace spanned by $\frac{\partial}{\partial x^1}\big|_0$. Moreover, we can also change the orientation on the first axis, i.e. we have to define α only on $t\ \frac{\partial}{\partial x^1}\big|_0$ with positive reals t.

Let us consider the group G of all linear orthogonal transformations keeping $\frac{\partial}{\partial x^1}\big|_0$ fixed. So for every $t \in \mathbb{R}$ the tensor $\beta(t) = \alpha(t\frac{\partial}{\partial x^1}) \in \mathbb{R}^{m*} \otimes \mathbb{R}^{m*}$ is G-invariant. On the other hand, every such smooth map β determines a natural F-metric.

So let us assume $s_{ij} dx^i \otimes dx^j$ is G-invariant. Since we can change the orientation of any coordinate axis except the first one, all s_{ij} with different indices must be zero. Further we can exchange any couple of coordinate axis different from the first one and so all coefficients at $dx^i \otimes dx^i$, $i \neq 1$, must coincide. Hence all G-invariant tensors are of the form

$$\text{(2)} \qquad \nu dx^1 \otimes dx^1 + \mu g^0.$$

The reals μ and ν are independent, if $m > 1$. In dimension one, G is the trivial group and so the whole one dimensional tensor space consists of G-invariant tensors.

Thus, our mapping β is defined by (2) with two arbitrary smooth functions μ and ν (and they can be reduced to one if $m = 1$). Given $v = t\ \frac{\partial}{\partial x^1}\big|_0$, we can write

$$\alpha_{(\mathbb{R}^m, g^0)}(v)(X,Y) = \beta(|v|)(X,Y) = \mu(|v|)g^0(X,Y) + \nu(|v|)|v|^{-2}g^0(v,X)g^0(v,Y)$$

In order to prove that all natural F-metrics are of the form (1), we only have to express $\mu(t)$, $\nu(t)$ as $\bar\nu(t^2) = t^{-2}\nu(t)$ and $\bar\mu(t^2) = \mu(t)$ for all positive reals, see 33.19. Obviously, every such operator is natural and the proposition is proved. $\square$

33.23. If we use the invariance with respect to $SO(m)$ in the proof of the above proposition, we get

Proposition. *All first order natural F-metrics α on oriented Riemannian manifolds of dimensions m form a family parameterized by arbitrary smooth functions μ, ν, κ, λ: $(0,\infty) \to \mathbb{R}$ in the following way. For every Riemannian manifold (M,g) of dimension $m > 3$ and tangent vectors u, X, $Y \in T_xM$*

$$\alpha_{(M,g)}(u)(X,Y) = \mu(g(u,u))g(X,Y) + \nu(g(u,u))g(u,X)g(u,Y).$$

If $m = 3$ then

$$\begin{aligned}\alpha_{(M,g)}(u)(X,Y) = \mu(g(u,u))g(X,Y) + \nu(g(u,u))g(u,X)g(u,Y) \\ + \kappa(g(u,u))g(u,X\times Y)\end{aligned}$$

where $\times$ means the vector product. If $m = 2$, then

$$\begin{aligned}\alpha_{(M,g)}(u)(X,Y) = \mu(g(u,u))g(X,Y) + \nu(g(u,u))g(u,X)g(u,Y) \\ + \kappa(g(u,u))\big(g(J^g(u),X)g(u,Y) + g(J^g(u),Y)g(u,X)\big) \\ + \lambda(g(u,u))\big(g(J^g(u),X)g(u,Y) - g(J^g(u),Y)g(u,X)\big)\end{aligned}$$

where J^g is the canonical almost complex structure on (M,g). In the dimension $m = 1$ we get

$$\alpha_{(M,g)}(u)(X,Y) = \mu(g(u,u))g(X,Y).$$

33.24. If we combine the results from 33.20–33.23 we deduce that all natural metrics on tangent bundles of Riemannian manifolds depend on six arbitrary smooth functions on positive real numbers if $m > 1$, and on three functions in dimension one.

The same result remains true for oriented Riemannian manifolds if $m > 3$ or $m = 1$, but the metrics depend on 7 real functions if $m = 3$ and on 10 real functions in dimension two.

34. Multilinear natural operators

We have already discussed several ways how to find natural operators and all of them involve some results from representation theory. Our general procedures work without any linearity assumption and we also used them in section 30 devoted to the bilinear operators of the type of Frölicher-Nijenhuis bracket. However, there are very effective methods involving much more linear representation theory of the jet groups in question which enable us to solve more general classes of problems concerning linear geometric operations.

In fact, the representation theory of the Lie algebras of the infinite jet groups, i.e. the formal vector fields with vanishing absolute terms, plays an important role. Thus, the methods differ essentially if these Lie algebras have finite dimension in the geometric category in question. The best known example beside the Riemannian manifolds is the category of manifolds with conformal Riemannian structure.

Although we feel that this theory lies beyond the scope of our book, we would like to give at least a survey and a sort of interface between the topics and the terminology of this book and some related results and methods available in the literature. For a detailed survey on the subject we recommend [Kirillov, 80, pp. 3–29]. The linear natural operators in the category of conformal pseudo-Riemannian manifolds are treated in the survey [Baston, Eastwood, 90].

Some basic concepts and results from representation theory were treated in section 13.

34.1. Recall that every natural vector bundles $E_1,\dots,E_m,E\colon \mathcal{M}f_n \to \mathcal{FM}$ of order r correspond to G^r_n-modules $V_1,\dots,V_m,V$. Further, m-linear natural operators $D\colon C^\infty(E_1\oplus\dots\oplus E_m) = C^\infty(E_1)\times\dots\times C^\infty(E_m)\to C^\infty(E)$ are of some finite order k (depending on D), cf. 19.9, and so they correspond to m-linear G^{k+r}_n-equivariant mappings $\mathcal{D}$ defined on the product of the standard fibers $T^k_nV_i$ of the k-th prolongations J^kE_i, $\mathcal{D}\colon T^k_nV_1\times\dots\times T^k_nV_m\to V$, see 14.18 or 18.20. Equivalently, we can consider linear G^{k+r}_n-equivariant maps $\mathcal{D}\colon T^k_nV_1\otimes\dots\otimes T^k_nV_m\to V$. We can pose the problem at three levels.

First, we may fix all bundles $E_1,\dots,E_m,E$ and ask for all m-linear operators $D\colon E_1\oplus\dots\oplus E_m \rightsquigarrow E$. This is what we always have done.

Second, we fix only the source $E_1\oplus\dots\oplus E_m$, so that we search for all m-linear geometric operations with the given source. The methods described below are efficient especially in this case.

Third, both the source and the target are not prescribed.

We shall first proceed in the latter setting, but we derive concrete results only in the special case of first order natural vector bundles and $m=1$. Of course, the results will appear in a somewhat implicit way, since we have to assume that the bundles in question correspond to irreducible representations of $G^1_n = GL(n)$. We do not lose much generality, for all representations of $GL(n)$ are completely reducible, except the exceptional indecomposable ones (cf. [Boerner, 67, chapter V]). But although all tensorial representations are decomposable, it might be a serious problem to find the decompositions explicitly in concrete examples. This also concerns our later discussion on bilinear operations. In particular, we do not know how to deduce explicitly (in some short elementary way) the results from section 30 from the more general results due P. Grozman, see below.

34.2. Given linear representations π, ρ of a connected Lie group G on vector spaces V, W, we know that a linear mapping $\varphi\colon V\to W$ is a G-module homomorphism if and only if it is a $\mathfrak{g}$-module homomorphism with respect to the induced representations $T\pi$, $T\rho$ of the Lie algebra $\mathfrak{g}$ on V, W, see 5.15. So if we find all $\mathfrak{g}^{k+r}_n$-module homomorphisms $\mathcal{D}\colon T^k_nV_1\otimes\dots\otimes T^k_nV_m\to V$, we describe all $(G^{k+r}_n)^+$-equivariant maps and so all operators natural with respect to orientation preserving diffeomorphisms. Hence we shall be able to analyze the problem on the Lie algebra level. But we first continue with some observations concerning the G_n^{r+k}-modules.

Recall that for every G_n^r-module V with homogeneous degree d (as a G_n^1-module) the induced G_n^{r+k}-module $T_n^k V$ decomposes as $GL(n)$-module into the sum $T_n^k V = V_0 \oplus \cdots \oplus V_k$ of $GL(n)$-modules V_i with homogeneous degrees $d-i$. Hence given an irreducible G_n^1-module W and a G_n^{k+r}-module homomorphism $\varphi\colon T_n^k V \to W$ such that $\ker\varphi$ does not include V_k, W must have homogeneous degree $d-k$ and $T_n^k V$ is a decomposable G_n^{r+k}-module by virtue of 13.14. Hence in order to find all G_n^{k+r}-module homomorphisms with source $T_n^k V$ we have to discuss the decomposability of this module. Note $T_n^k V$ is always reducible if $k > 0$, cf. 13.14. A corollary in [Terng, 78, p. 812] reads

If V is an irreducible G_n^1-module, then $T_n^k V$ is indecomposable except $V = \Lambda^p \mathbb{R}^{n*}$, $k = 1$.

So an explicit decomposition of $T_n^1(\Lambda^p \mathbb{R}^{n*})$ leads to

Theorem. *All non zero linear natural operators $D\colon E_1 \rightsquigarrow E$ between two natural vector bundles corresponding to irreducible G_n^1-modules are*

(1) $E_1 = \Lambda^p T^*$, $E = \Lambda^{p+1} T^*$, $D = k.d$, where $k \in \mathbb{R}$, $n > p \geq 0$

(2) $E_1 = E$, $D = k.\mathrm{id}$, $k \in \mathbb{R}$.

This theorem was originally formulated by J. A. Schouten, partially proved by [Palais, 59] and proved independently by [Kirillov, 77] and [Terng, 78]. Terng proved this result by direct (rather technical) considerations and she formulated the indecomposability mentioned above as a consequence. Her methods are not suitable for generalizations to m-linear operations or to more general categories over manifolds.

34.3. If we pass to the Lie algebra level, we can include more information extending the action of $\mathfrak{g}_n^{k+r}$ to an action of the whole algebra $\mathfrak{g} = \mathfrak{g}_{-1} \oplus \mathfrak{g}_0 \oplus \dots$ of formal vector fields $X = \sum_{|\alpha|=0}^{\infty} a_\alpha^j x^\alpha \frac{\partial}{\partial x^j}$ on $\mathbb{R}^n$. In particular, the action of the (abelian) subalgebra of constant vector fields $\mathfrak{g}_{-1}$ will exclude the general reducibility of $T_n^k V$.

Lemma. *The induced action of $\mathfrak{g}_n^{k+r}$ on $T_n^k V = (J^k E)_0 \mathbb{R}^n$ is given by the Lie differentiation $j_0^{r+k} X . j_0^k s = j_0^k(\mathcal{L}_{-X} s)$ and this formula extends the action to the Lie algebra $\mathfrak{g}$ of formal vector fields. Every $\mathfrak{g}_n^{k+r}$-module homomorphism $\varphi\colon T_n^k V \to W$ is a $\mathfrak{g}$-module homomorphism.*

Proof. We have

$$\begin{aligned} j_0^{r+k} X . j_0^k s &= \tfrac{\partial}{\partial t}\big|_0 \ell_{\exp t . j_0^{r+k} X}(j_0^k s) = \quad \text{(by 13.2)} \\ &= \tfrac{\partial}{\partial t}\big|_0 \ell_{j_0^{r+k} \mathrm{Fl}_t^X}(j_0^k s) = \quad \text{(by 14.18)} \\ &= \tfrac{\partial}{\partial t}\big|_0 j_0^k(E(\mathrm{Fl}_t^X) \circ s \circ \mathrm{Fl}_{-t}^X) = \quad \text{(by 6.15)} \\ &= j_0^k \mathcal{L}_{-X} s \end{aligned}$$

Each $\mathfrak{g}_n^{k+r}$-module homomorphism $\varphi\colon T_n^k V \to W$ defines an operator D natural with respect to orientation preserving local diffeomorphisms. It follows from 6.15 that every natural linear operator commutes with the Lie differentiation (this can be seen easily also along the lines of the above computation and we shall discuss even the converse implication in chapter XI). Hence for all $j_0^\infty X \in \mathfrak{g}$, $j_0^k s \in T_n^k V$

$$j_0^\infty X.\varphi(j_0^k s) = \mathcal{L}_{-X} Ds(0) = D(\mathcal{L}_{-X} s)(0) = \varphi(j_0^k(\mathcal{L}_{-X} s)) = \varphi(j_0^\infty X.j_0^k s)$$

and so φ is a $\mathfrak{g}$-module homomorphism. □

34.4. Consider a G_n^r-module V, a $\mathfrak{g}$-module homomorphism $\varphi\colon T_n^k V \to W$ and its dual $\varphi^*\colon W^* \to (T_n^k V)^*$. If W is a G_n^q-module, then the subalgebra $\mathfrak{b}_q = \mathfrak{g}_q \oplus \mathfrak{g}_{q+1} \oplus \dots$ in $\mathfrak{g}$ acts trivially on the image $\operatorname{Im}\varphi^* \subset (T_n^k V)^*$.

We say that a $\mathfrak{g}$-module V is of height p if $\mathfrak{g}_q.V = 0$ for all $q > p$ and $\mathfrak{g}_p.V \neq 0$.

Definition. The vectors $v \in T_n^k V$ with trivial action of all homogeneous components of degrees greater then the height of V are called *singular vectors*.

An analogous definition applies to subalgebras $\mathfrak{a} \subset \mathfrak{g}$ with grading and $\mathfrak{a}$-modules.

So the linear natural operations between irreducible first order natural vector bundles are described by $\mathfrak{g}_n^{k+1}$-submodules of singular vectors in $(T_n^k V)^*$. Similarly we can treat m-linear operators on replacing $(T_n^k V)^*$ by $(T_n^k V_1)^* \otimes \dots \otimes (T_n^k V_m)^*$. Since all modules in question are finite dimensional, it suffices to discuss the highest weight vectors (see 34.8) in these submodules which can also lead to the possible weights of irreducible modules V. For this purpose, one can use the methods developed (for another aim) by Rudakov. Remark that the Kirillov's proof of theorem 34.2 also analyzes the possible weights of the modules V, but by discussing the possible eigen values of the Laplace-Casimir operator.

First we have to derive some suitable formula for the action of $\mathfrak{g}$ on $(T_n^k V)^*$. In what follows, V and W will be G_n^1-modules and we shall write $\partial_i = \frac{\partial}{\partial x^i} \in \mathfrak{g}_{-1}$.

34.5. Lemma. $(T_n^k V)^* = \sum_{i=0}^k S^i(\mathfrak{g}_{-1}) \otimes V^*$.

Proof. Every multi index $\alpha = i_1 \dots i_{|\alpha|}$, $i_1 \le \dots \le i_{|\alpha|}$, yields the linear map

$$\ell_\alpha\colon T_n^k V \to V, \qquad \ell_\alpha(j_0^k s) = (\mathcal{L}_{-\partial_{i_1}} \circ \dots \circ \mathcal{L}_{-\partial_{i_{|\alpha|}}} s)(0).$$

Since the elements in $\mathfrak{g}_{-1}$ commute, we can view the elements in $S^{|\alpha|}(\mathfrak{g}_{-1})$ as linear combinations of maps ℓ_α. Now the contraction with V^* yields a linear map $\sum_{i=0}^k S^i(\mathfrak{g}_{-1}) \otimes V^* \to (T_n^k V)^*$. This map is bijective, since $(T_n^k V)^*$ has a basis induced by the iterated partial derivatives which correspond to the maps ℓ_α. □

This identification is important for our computations. Let us denote $\ell_i = \mathcal{L}_{-\partial_i} \in \mathfrak{g}_{-1}^* = S^1(\mathfrak{g}_{-1})$, so the elements ℓ_α can be viewed as $\ell_\alpha = \ell_{i_1} \circ \dots \circ \ell_{i_{|\alpha|}} \in S^{|\alpha|}(\mathfrak{g}_{-1})$ and we have $\ell_\alpha = 0$ if $|\alpha| > k$. Further, for every $\ell \in \mathfrak{g}$ we shall denote $\operatorname{ad}\ell_\alpha.\ell = (-1)^{|\alpha|}[\partial_{i_1}, [\dots [\partial_{i_{|\alpha|}}, \ell] \dots]]$.

34.6. Lemma. *The action of* $\ell \in \mathfrak{g}_q$ *on* $\ell_\alpha \otimes v \in S^p \otimes V^*$ *is*

$$\ell.\ell_\alpha \otimes v = \sum_{\substack{\beta+\gamma=\alpha \\ |\gamma|=q}} \ell_\beta \otimes (\mathrm{ad}\ell_\gamma.\ell).v + \sum_{\substack{\beta+\gamma=\alpha \\ |\gamma|=q+1}} (\ell_\beta \circ (\mathrm{ad}\ell_\gamma.\ell)) \otimes v.$$

Proof. We compute with $\ell = j_0^k X \in \mathfrak{g}_q$

$$\ell.(\ell_\alpha \otimes v)(j_0^k s) = -(\ell_\alpha \otimes v)(\ell.j_0^k s) = (\ell_\alpha \otimes v)(j_0^k(\mathcal{L}_X s)) = \langle (\ell_\alpha \circ \mathcal{L}_X s)(0), v\rangle.$$

Since $\ell_j \circ \mathcal{L}_Y = \mathcal{L}_Y \circ \ell_j + \mathcal{L}_{[-\partial_j, Y]}$ for all $Y \in \mathfrak{g}$, $1 \le j \le n$, and $[\partial_j, \mathfrak{g}_l] \subset \mathfrak{g}_{l-1}$, we get

$$\ell.(\ell_\alpha \otimes v)(j_0^k s) = \langle \ell_{i_1} \dots \ell_{i_{p-1}} \mathcal{L}_X \ell_{i_p} s(0), v\rangle + \langle \ell_{i_1} \dots \ell_{i_{p-1}} \mathcal{L}_{[-\partial_{i_p}, X]} s(0), v\rangle$$

and the same procedure can be applied p times in order to get the Lie derivative terms just at the left hand sides of the corresponding expressions. Each choice of indices among $i_1, \dots, i_p$ determines just one summand of the outcome. Hence we obtain (the sum is taken also over repeating indices)

$$\ell.(\ell_\alpha \otimes v)(j_0^k s) = \sum_{\beta+\gamma=\alpha} \langle (\mathrm{ad}\ell_\gamma.\ell).\ell_\beta s(0), v\rangle.$$

Further $\mathrm{ad}\ell_\gamma.\ell = 0$ whenever $|\gamma| > q+1$ and for all vector fields $Y \in \mathfrak{g}_0 \oplus \mathfrak{g}_1 \oplus \dots$ we have

$$\langle (\mathcal{L}_Y \circ \ell_\beta s)(0), v\rangle = -\langle (\ell_\beta s)(0), \mathcal{L}_Y v\rangle$$

so that only the terms with $|\gamma| = q$ or $|\gamma| = q+1$ can survive in the sum (notice $Y \in \mathfrak{g}_p$, $p \ge 1$, implies $\mathcal{L}_Y v = 0$). Since $\ell = j_0^k Y \in \mathfrak{g}_0$ acts on (the jet of constant section) v by $\ell.v = \mathcal{L}_{-Y} v(0)$, we get the result. □

34.7. Example. In order to demonstrate the computations with this formula, let us now discuss the linear operations in dimension one.

We say that V is a $\mathfrak{g}_n^k$-module *homogeneous in the order* if there is k_0 such that $\mathfrak{g}_{k_0}.v = 0$ implies $v = 0$ and $\mathfrak{g}_l.v = 0$ for all v and $l > k_0$. Each $\mathfrak{g}_n^k$-module includes a submodule homogeneous in order. Indeed, the isotropy algebra of each vector v contains some kernel $\mathfrak{b}_l$, $l \le k$, denote l_v the minimal one. Let p be the minimum of these l's. Then the set of vectors with $l_v = p$ is a submodule homogeneous in order. In particular, every irreducible module is homogeneous in order.

Consider a $\mathfrak{g}_1^1$ module V homogeneous in order. For every non zero vector $a = \ell_1^p \otimes v \in S^p(\mathfrak{g}_{-1}) \otimes V^* \subset (T_1^k V)^*$ and $\ell \in \mathfrak{g}_1$ we get

$$\ell.a = -p\ell_1^{p-1} \otimes [\partial_1, \ell]v + \binom{p}{2}\ell_1^{p-2} \circ [\partial_1, [\partial_1, \ell]] \otimes v.$$

Take $\ell = x^2 \frac{d}{dx}$ so that $[\partial_1, \ell] = 2x\frac{d}{dx} =: 2h$ and $[\partial_1, [\partial_1, \ell]] = 2\partial_1$.

Assume now $\mathfrak{b}_1.a = 0$. Then

$$0 = \ell.a = \ell_1^{p-1} \otimes (-2ph.v + p(p-1)v)$$

so that $2h.v = (p-1)v$ or $p = 0$.

Further, set $\ell = x^3 \frac{d}{dx}$. We get

$$\begin{aligned} 0 = \ell.a &= \binom{p}{2}\ell_1^{p-2} \otimes [\partial_1, [\partial_1, \ell]].v - \binom{p}{3}\ell_1^{p-3}.[\partial_1, [\partial_1, [\partial_1, \ell]]] \otimes v \\ &= \ell_1^{p-2} \otimes (3p(p-1)h.v - p(p-1)(p-2)v). \end{aligned}$$

Hence either $p = 0$ or $p = 1$ or $3h.v = \frac{3}{2}(p-1)v = (p-2)v$. The latter is not possible, for it says $p = -1$. The case $p = 0$ is not interesting since the action of $\mathfrak{b}_1$ on all vectors in $V^* = S^0(\mathfrak{g}_{-1}) \otimes V^*$ is trivial. But if $p = 1$ we get $h.v = 0$ and so the homogeneity in order implies the action of $\mathfrak{g}_1^1$ on V is trivial. Hence $V = \mathbb{R}$ if irreducible. Moreover, the submodule generated by a in $(T_1^1\mathbb{R})^*$ is $\ell_1 \otimes \mathbb{R}$ with the action $h.t\ell_1 = 0 + t\ell_1$. Hence if $\varphi\colon T_1^1 V \to W$ is a $\mathfrak{g}$-module homomorphism and if both V and W are irreducible, then either φ factorizes through $\varphi\colon V \to W$ which means $V = W$, $\varphi = k.\mathrm{id}_V$, or $V = \mathbb{R}$, $W = \mathbb{R}^*$ with the minus identical action of $\mathfrak{g}_1^1$. In this way we have proved theorem 34.2 in the dimension one.

34.8. The situation in higher dimensions is much more difficult. Let us mention some concepts and results from representation theory. Our source is [Zhelobenko, Shtern, 83] and [Naymark, 76].

Consider a Lie algebra $\mathfrak{g}$ and its representation ρ in a vector space V. An element $\lambda \in \mathfrak{g}^*$ is called a *weight* if there is a non zero vector $v \in V$ such that $\rho(x)v = \lambda(x)v$ for all $x \in \mathfrak{g}$. Then v is called a weight vector (corresponding to λ). If $\mathfrak{h} \subset \mathfrak{g}$ is a subalgebra, then the weights of the adjoint representation of $\mathfrak{h}$ in $\mathfrak{g}$ are called *roots* of the algebra $\mathfrak{g}$ with respect to $\mathfrak{h}$. The corresponding weight vectors are called the root vectors (with respect to $\mathfrak{h}$).

A maximal solvable subalgebra $\mathfrak{b}$ in a Lie algebra $\mathfrak{g}$ is called a *Borel subalgebra.* A maximal commutative subalgebra $\mathfrak{h} \subset \mathfrak{g}$ with the property that all operators $\mathrm{ad}x$, $x \in \mathfrak{h}$, are diagonal in $\mathfrak{g}$ is called a *Cartan subalgebra.*

In our case $\mathfrak{g} = \mathfrak{gl}(n)$, the upper triangular matrices form the Borel subalgebra $\mathfrak{b}_+$ while the diagonal matrices form the Cartan subalgebra $\mathfrak{h}$. Let us denote $\mathfrak{n}_+$ the derived algebra $[\mathfrak{b}_+, \mathfrak{b}_+]$, i.e. the subalgebra of triangular matrices with zeros on the diagonals. Consider a $\mathfrak{gl}(n)$-module V. A vector $v \in V$ is called the *highest weight vector* (with respect to $\mathfrak{b}_+$) if there is a root $\lambda \in \mathfrak{h}^*$ such that $x.v - \lambda(x)v = 0$ for all $x \in \mathfrak{h}$ and $x.v = 0$ for all $x \in \mathfrak{n}_+$. The root λ is called the highest weight. In our case we identify $\mathfrak{h}^*$ with $\mathbb{R}^n$.

The highest weight vectors always exist for complex representations of complex algebras and are uniquely determined for the irreducible ones. The procedure of complexification allows to use this for the real case as well. So each finite dimensional irreducible representation of $\mathfrak{gl}(n)$ is determined by a highest weight $(\lambda_1, \dots, \lambda_n) \in \mathbb{C}$ such that all $\lambda_i - \lambda_{i+1}$ are non negative integers, $i = 1, \dots, n-1$.

34.9. Examples. Let us start with the weight of the canonical representation on $\mathbb{R}^n$ corresponding to the tangent bundle T. The action of $a = (a_l^k)$, $a_l^k = \delta_j^k \delta_l^i$ for some $j < i$, (corresponding to the action of $X = x^i \frac{\partial}{\partial x^j}$ given by the negative of the Lie derivative) on a highest weight vector v must be zero, so that only its first coordinate can be nonzero. Hence the weight is $(1, 0, \dots, 0)$.

Now we compute the weights of the irreducible modules $\Lambda^p \mathbb{R}^{n*}$. The action of $X = x^i \frac{\partial}{\partial x^j}$ on a (constant) form ω is $\mathcal{L}_{-X}\omega$. Since $\mathcal{L}_X dx^l = \delta_j^l dx^i$ we get (cf. 7.6) that if $X.\omega = 0$ for all $j < i$ then ω is a constant multiple of $dx^{n-p+1} \wedge \dots \wedge dx^n$. Further, the action of $\mathcal{L}_{-x^i/\partial x^i}$ on $dx^{i_1} \wedge \dots \wedge dx^{i_p}$ is minus identity if i appears among the indices i_j and zero if not. Hence the highest weight is $(0, \dots, 0, -1, \dots, -1)$ with $n - p$ zeros. Similarly we compute the highest weight of the dual $\Lambda^p \mathbb{R}^n$, $(1, \dots, 1, 0, \dots, 0)$ with $n - p$ zeros.

Analogously one finds that the highest weight vector of $S^p \mathbb{R}^{m*}$ is the symmetric tensor product of p copies of dx^n and the weight is $(0, \dots, 0, -p)$.

34.10. Let us come back to our discussion on singular vectors in $(T_n^k V)^*$ for an irreducible $\mathfrak{gl}(n)$-module V. In our preceding considerations we can take suitable subalgebras with grading instead of the whole algebra $\mathfrak{g}$ of formal vector fields. It turns out that one can describe in detail the singular vectors in dimension two and for the subalgebra of divergence free formal vector fields. We shall denote this algebra by $\mathfrak{s}(2)$ and we shall write $\mathfrak{s}_2^r$ for the Lie algebras of the corresponding jet groups. We shall not go into details here, they can be found in [Rudakov, 74, pp. 853–859]. But let us indicate why this description is useful. A subalgebra $\mathfrak{a} \subset \mathfrak{g}$ is called a testing subalgebra if there is an isomorphism $\mathfrak{s}(2) \to \mathfrak{a} \subset \mathfrak{g}$ of algebras with gradings and a distinguished subspace $w(\mathfrak{a}) \subset \mathfrak{g}_{-1}$ such that $\mathfrak{g}_{-1} = \mathfrak{a}_{-1} \oplus w(\mathfrak{a})$, $[\mathfrak{a}, w(\mathfrak{a})] = 0$.

Lemma. *Let V be a $\mathfrak{g}_n^1$-module, $(T_n^k V)^* = \sum_{i=0}^k S^i(\mathfrak{g}_{-1}) \otimes V^*$ and $\mathfrak{a} \subset \mathfrak{g}$ be a testing subalgebra. Then $\bar{V} = \sum_{i=0}^k S^i(w(\mathfrak{a})) \otimes V^* \subset (T_n^k V)^*$ is an $\mathfrak{a}_0$-module and there is an $\mathfrak{a}$-module isomorphism $(T_n^k V)^* \to \sum_{i=0}^k S^i(\mathfrak{a}_{-1}) \otimes \bar{V}$ onto the image.*

Proof. $\bar{V} = \sum_{i=0}^\infty S^i(w(\mathfrak{a})) \otimes V^*$ is an $\mathfrak{a}_0$-module, for $[\mathfrak{a}, w(\mathfrak{a})] = 0$. We have

$$\sum_{i=0}^\infty S^i(\mathfrak{a}_{-1}) \otimes \bar{V} = \sum_{i=0}^\infty S^i(\mathfrak{a}_{-1}) \otimes \sum_{j=0}^\infty S^j(w(\mathfrak{a})) \otimes V^*$$
$$= \sum_{i=0}^\infty S^i(\mathfrak{a}_{-1} \oplus w(\mathfrak{a})) \otimes V^*. \quad \square$$

34.11. It turns out that there are enough testing subalgebras in the algebra of formal vector fields. Using the results on $\mathfrak{s}(2)$, Rudakov proves that for every $\mathfrak{g}_n^1$-module V the homogeneous singular vectors can appear only in $V^* \oplus S^1(\mathfrak{g}_{-1}) \otimes V^* \subset (T_n^k V)^*$. This is equivalent to the assertion that all linear natural operators are of order one.

Let us remark that this was also proved by [Terng, 78] in a very interesting way. She proved that every tensor field is locally a sum of fields with polynomial coefficients of degree one in suitable coordinates (different for each summand) and so the naturality and linearity imply that the orders must be one.

34.12. Now, we know that if there is a homogeneous singular vector x which does not lie in $V^* \subset (T^1_n V)^*$ then there must be a highest weight singular vector $x \in \mathfrak{g}^*_{-1} \otimes V^*$, for all linear representations in question are finite dimensional. Let us write $x = \sum_{i=1}^k l_i \otimes u_i$, where $k \leq n$ and all ℓ_i are assumed linearly independent.

Proposition. *Let $x = \sum_{i=1}^k l_i \otimes u_i$ be a singular vector of highest weight with respect to the Borel algebra $\mathfrak{b}_+ \subset \mathfrak{g}^1_n$. If $u_i \neq 0$, $i = 1, \dots, p$, and $u_{p+1} = 0$, then $u_i = 0$, $i = p+1, \dots, k$, and u_p is a highest weight vector with weight $\lambda = (1, \dots, 1, 0, \dots, 0)$ with $n - p + 1$ zeros. Then the weight of x is $\mu = (1, \dots, 1, 0, \dots, 0)$ with $n - p$ zeros.*

Proof. Since x is singular, we have for all k, j, l (we do not use summation conventions now)
(1)
$$0 = -x^k x^l \tfrac{\partial}{\partial x^j} . \textstyle\sum_p \ell_p \otimes u_p = \sum_p 1 \otimes [\tfrac{\partial}{\partial x^p}, x^k x^l \tfrac{\partial}{\partial x^j}].u_p = x^l \tfrac{\partial}{\partial x^j}.u_k + x^k \tfrac{\partial}{\partial x^j}.u_l.$$

In particular, for all k, j

$$x^k \tfrac{\partial}{\partial x^j}.u_k = 0 \tag{2}$$
$$x^j \tfrac{\partial}{\partial x^j}.u_k = -x^k \tfrac{\partial}{\partial x^j}.u_j. \tag{3}$$

Further, x is a highest weight vector with weight $\mu = (\mu_1, \dots, \mu_n)$ and for all i, j we have

$$\begin{aligned} x^i \tfrac{\partial}{\partial x^j}.x &= \textstyle\sum_p \ell_p \otimes x^i \tfrac{\partial}{\partial x^j}.u_p + \sum_p [-\tfrac{\partial}{\partial x^p}, x^i \tfrac{\partial}{\partial x^j}] \otimes u_p \\ &= \textstyle\sum_p \ell_p \otimes x^i \tfrac{\partial}{\partial x^j}.u_p + \ell_j \otimes u_i. \end{aligned} \tag{4}$$

If $i > j$, we have $x^i \frac{\partial}{\partial x^j}.x = 0$ and so

$$x^i \tfrac{\partial}{\partial x^j}.u_p = 0 \qquad p \neq j \tag{5}$$
$$x^i \tfrac{\partial}{\partial x^j}.u_j = -u_i. \tag{6}$$

Further, $x^i \frac{\partial}{\partial x^i}.x = \mu_i x$ and so (4) implies for all p, i

$$x^i \tfrac{\partial}{\partial x^i}.u_p = (\mu_i - \delta^p_i) u_p. \tag{7}$$

The latter formula shows that the vectors u_p are either zero or root vectors of V^* with respect to the Cartan algebra $\mathfrak{h}$ with weights $\lambda(p) = (\lambda_1, \dots, \lambda_n)$, $\lambda_i = \mu_i - \delta^p_i$. Formula (2) implies that either $u_p = 0$ or $\mu_p = 1$. If $u_p = 0$, then all $u_l = 0$, $l \geq p$, by (6). Assume $u_p \neq 0$ and $u_{p+1} = 0$, i.e. $\mu_i = 1$, $i \leq p$. Then (5) and (6) show that u_p is a highest weight vector. By (3), $x^j \frac{\partial}{\partial x^j}.u_k = \lambda(k)_j u_k = -x^k \frac{\partial}{\partial x^j}.u_j$, so that for $k = p$, $j > p$, (7) implies $\lambda(p)_j u_p = \mu_j.u_p = -x^p \frac{\partial}{\partial x^j}.u_j = 0$. Hence $\mu_i = 1$, $i = 1, \dots, p$, and $\mu_i = 0$, $i = p+1, \dots, n$. $\square$

34.13. Now it is easy to prove theorem 34.2. If $D\colon E_1 \rightsquigarrow E$ is a linear natural operator between bundles corresponding to irreducible G^1_n-modules V, W, then either $V^* = \Lambda^p\mathbb{R}^n$, $p = 0, \dots, n-1$, and $W^* = \Lambda^{p+1}\mathbb{R}^n$, or D is a zero order operator. The dual of the inclusion $W^* \to (T^1_n V)^*$ corresponds to the exterior differential up to a constant multiple.

Let us remark, that the only part of the proof we have not presented in detail is the estimate of the order, but we mentioned a purely geometric way how to prove this, cf. 34.11. It might be useful in concrete situations to combine some general methods with final computations in the above style.

34.14. The method of testing subalgebras is heavily used in [Rudakov, 75] dealing with subalgebras of divergence free formal vector fields and Hamiltonian vector fields. The aim of all the mentioned papers by Rudakov is the study of infinite dimensional representations of infinite dimensional Lie algebras of formal vector fields. His considerations are based on the study of the space $\sum_{i=0}^{\infty} S^i(\mathfrak{g}_{-1}) \otimes V^*$ and so the results are relevant for our purposes as well. We should remark that in the cited papers the action slightly differs in notation and the vector fields $x^i \frac{\partial}{\partial x^j}$ are identified with the transposed matrix (a^i_j) to our (a^j_i) and so the weights correspond to the Borel subalgebra of lower triangular matrices. Due to Rudakov's results, a description of all linear operations natural with respect to unimodular or symplectic diffeomorphisms is also available. In the unimodular case we get the following result. We write $\mathcal{S}\ell_n$ for the category of n-dimensional manifolds with fixed volume forms and local diffeomorphisms preserving the distinguished forms.

Theorem. *All non zero linear natural operators $D\colon E_1 \rightsquigarrow E$ between two first order natural bundles on category $\mathcal{S}\ell_n$ corresponding to irreducible representations of the first order jet group are*

(1) $E_1 = E$, $D = k.\mathrm{id}$, $k \in \mathbb{R}$

(2) $E_1 = \Lambda^p T^*$, $E = \Lambda^{p+1}T^*$, $D = k.d$, $k \in \mathbb{R}$, $n > p \geq 0$

(3) $E_1 = \Lambda^{n-1}T^*$, $E = \Lambda^1 T^*$, $D = k.(d \circ i \circ d)\colon \Lambda^{n-1}T^* \to \Lambda^n T^* \xrightarrow[\cong]{i} \Lambda^0 T^* \to \Lambda^1 T^*$, $k \in \mathbb{R}$.

Let us point out that this theorem describes all linear natural operations not only up to decompositions into irreducible components but also up to natural equivalences. For example, to find linear natural operations with vector fields we have to notice $\mathbb{R}^n \cong \Lambda^{n-1}\mathbb{R}^{n*}$, $\frac{\partial}{\partial x^p} \mapsto i(\frac{\partial}{\partial x^p})(dx^1 \wedge \dots \wedge dx^n)$. Hence the Lie differentiation of the distinguished volume forms corresponds to the exterior differential on $(n-1)$-forms, the identification of n-forms with functions yields the divergence of vector fields and the exterior differential of the divergence represents the 'composition' of exterior derivatives from point (3). Beside the constant multiples of identity, there are no other linear operations (with irreducible target).

We shall not describe the Hamiltonian case. We remark only that then not even the differential forms correspond to irreducible representations and that the interesting operations live on irreducible components of them.

34.15. Next we shall shortly comment some results concerning m-linear operations. We follow mainly [Kirillov, 80]. So ρ^* will denote a representation dual to a representation ρ of G_n^{1+} and we write Δ for the one-dimensional representation given by $a \mapsto \det a^{-1}$. Further $\Gamma_\rho(M)$ is the space of all smooth sections of the vector bundle E_ρ over M corresponding to ρ. In particular $\Gamma_\Delta(M)$ coincides with $\Omega^n M$. To every representation ρ we associate the representation $\tilde{\rho} := \rho^* \otimes \Delta$. The pointwise pairing on $\Gamma_\rho(M) \times \Gamma_{\rho^*}(M)$ gives rise to a bilinear mapping $\Gamma_\rho(M) \times \Gamma_{\tilde{\rho}}(M) \to \Omega^n(M)$, a natural bilinear operation of order zero. Given two sections $s \in \Gamma_\rho(M)$, $\tilde{s} \in \Gamma_{\tilde{\rho}}(M)$ with compact supports we can integrate the resulting n-form, let us write $\langle s, \tilde{s} \rangle$ for the outcome. We have got a bilinear functional invariant with respect to the diffeomorphism group $\mathrm{Diff}M$.

For every m-linear natural operator D of type $(\rho_1, \dots, \rho_m; \rho)$ we define an $(m+1)$-linear functional

$$F_D(s_1, \dots, s_m, s_{m+1}) = \langle D(s_1, \dots, s_m), s_{m+1} \rangle,$$

defined on sections $s_i \in \Gamma_{\rho_i}(M)$, $i = 1, \dots, m$, $s_{m+1} \in \Gamma_{\tilde{\rho}}(M)$ with compact supports. The functional F_D satisfies

(1) F_D is continuous with respect to the C^∞-topology on Γ_{ρ_i} and $\Gamma_{\tilde{\rho}}$

(2) F_D is invariant with respect to $\mathrm{Diff}M$

(3) $F_D = 0$ whenever $\cap_{i=1}^{m+1} \mathrm{supp} s_i = \emptyset$.

We shall call the functionals with properties (1)–(3) the invariant local functionals of the type $(\rho_1, \dots, \rho_m; \tilde{\rho})$.

Theorem. *The correspondence $D \mapsto F_D$ is a bijection between the m-linear natural operators of type $(\rho_1, \dots, \rho_m; \rho)$ and local linear functionals of type $(\rho_1, \dots, \rho_m; \tilde{\rho})$.*

The proof is sketched in [Kirillov, 80] and consists in showing that each such functional is given by an integral operator the kernel of which recovers the natural m-linear operator.

34.16. The above theorem simplifies essentially the discussion on m-linear natural operations. Namely, there is the action of the permutation group Σ_{m+1} on these operations defined by $(\sigma F_D)(s_1, \dots, s_{m+1}) = F_D(s_{\sigma 1}, \dots, s_{\sigma(m+1)})$, $\sigma \in \Sigma_{m+1}$. Hence a functional of type $(\rho_1, \dots, \rho_m; \rho)$ is transformed into a functional of type $(\rho_{\sigma^{-1}(1)}, \dots, \rho_{\sigma^{-1}(m+1)})$ and so for every operation D of the type $(\rho_1, \dots, \rho_m; \rho)$ there is another operation σD. If $\sigma(m+1) = m+1$, then this new operation differs only by a permutation of the entries but, for example, if σ transposes only m and $m+1$, then σD is of type $(\rho_1, \dots, \rho_{m-1}, \tilde{\rho}; \tilde{\rho}_m)$.

In the simplest case $m = 1$, the exterior derivative $d\colon \Omega^p M \to \Omega^{p+1} M$ is transformed by the only non trivial element in Σ_2 into $d\colon \Omega^{n-p-1} M \to \Omega^{n-p} M$.

If $m = 2$, the action of Σ_3 becomes significant. We shall now describe all operations in this case. Those of order zero are determined by projections of $\rho_1 \otimes \rho_2$ onto irreducible components.

34.17. First order bilinear natural operators. We shall divide these operations into five classes, each corresponding to some intrinsic construction and the action of Σ_3.

1. Write τ for the canonical representation of $G_n^1{}^+$ on $\mathbb{R}^n$, i.e. $\Gamma_\tau(M)$ are the smooth vector fields on M. For every representation ρ we have the Lie derivative $\mathcal{L}\colon \Gamma_\tau(M) \times \Gamma_\rho(M) \to \Gamma_\rho(M)$, a natural operation of type $(\tau, \rho; \rho)$. The action of Σ_3 yields an operation of type $(\rho, \tilde{\rho}; \tilde{\tau})$ allowing to construct invariantly a covector density from any two tensor fields which admit a pointwise pairing into a volume form. This operation appears often in the lagrangian formalism and Nijenhuis called it the lagrangian Schouten concomitant.

2. This class contains the operations of the types $(\Lambda^k\tau\otimes\Delta^\kappa, \Lambda^l\tau\otimes\Delta^\lambda; \Lambda^m\tau\otimes\Delta^\mu)$, where k, l, m are certain integers between zero and n while κ, λ, μ are certain complex numbers.

Assume first $k+l > n+1$. Then an operation exists if $m = k+l-n-1$, $\mu = \kappa+\lambda-1$. Let us choose an auxiliary volume form $v \in \Gamma_\Delta(M)$ and use the identification $\Lambda^k\tau\otimes\Delta^\kappa \cong \Lambda^{n-k}\tau^*\otimes\Delta^{\kappa-1}$, i.e. we shall construct an operation of the type $(\Lambda^{k'}\tau^*\otimes\Delta^{\kappa'}, \Lambda^{l'}\tau^*\otimes\Delta^{\lambda'}; \Lambda^{m'}\tau^*\otimes\Delta^{\mu'})$ with $k'+l' \le n-1$, $m' = k'+l'+1$ and $\mu' = \kappa'+\lambda'$. Then we can write a field of type $\Lambda^k\tau\otimes\Delta^\kappa$ in the form $\omega.v^{\kappa-1}$, $\omega \in \Lambda^{n-k}T^*M$. We define

$$D(\omega_1.v^{\kappa-1}, \omega_2.v^{\lambda-1}) = (c_1 d\omega_1 \wedge \omega_2 + c_2\omega_1 \wedge d\omega_2).v^{\mu-1} \tag{1}$$

where ω_1 is a $(n-k)$-form, ω_2 is a $(n-l)$-form, and c_1, c_2 are constants. The right hand side in (1) should not depend on the choice of v. So let us write $v = \varphi.\tilde{v}$ where φ is a positive function. Then $\omega_1.v^{\kappa-1} = \tilde{\omega}_1.\tilde{v}^{\kappa-1}$, $\omega_2.v^{\lambda-1} = \tilde{\omega}_2.\tilde{v}^{\lambda-1}$, with $\tilde{\omega}_1 = \varphi^{\kappa-1}.\omega_1$, $\tilde{\omega}_2 = \varphi^{\lambda-1}.\omega_2$. After the substitution into (1), there appears the extra summand

$$\begin{aligned}(c_1 d\varphi^{\kappa-1} \wedge \omega_1 \wedge \varphi^{\lambda-1}\omega_2 + c_2\varphi_{\kappa-1}\omega_1 \wedge d\varphi^{\lambda-1} \wedge \omega_2)\tilde{v}^{mu-1}& \\ = \left((\kappa-1)c_1 + (-1)^k(\lambda-1)c_2\right).d(\ln\varphi) \wedge \omega_1 \wedge \omega_2.v^{\mu-1}&.\end{aligned}$$

Thus (1) is a correct definition of an invariant operation if and only if

$$(\kappa-1)c_1 + (-1)^k(\lambda-1)c_2 = 0. \tag{2}$$

Now take $k+l \le n+1$. We find an operation if and only if $m = k+l-1$ and $\mu = \kappa+\lambda$. As before, we fix an auxiliary volume form v and we write the fields of type $\Lambda^k\tau\otimes\Delta^\kappa$ as $a.v^\kappa$ where a is a k-vector field. The usual divergence of vector fields extends to a linear operation δ_v on k-vector fields, $\delta_v(X_1\wedge\dots\wedge X_k) = \sum_{i=1}^k(-1)^{i+1}\operatorname{div}X_i.X_1\wedge\dots\hat{}^i\dots\wedge X_k$, where $\hat{}^i$ means that the entry is missing. Of course, this divergence depends on the choice of v. We have

(3)

$$\delta_{\varphi v}(X_1\wedge\dots\wedge X_k) = \varphi.\delta_v(X_1\wedge\dots\wedge X_k) + \sum_{i=1}^k(-1)^{i+1}X_i(\varphi).X_1\wedge\dots\hat{}^i\dots\wedge X_k.$$

Let us look for a natural operator D of the form

$$D(a.v^\kappa, b.v^\lambda) = \left(c_1\delta_v(a)\wedge b + c_2 a\wedge\delta_v(b) + c_3\delta_v(a\wedge b)\right).v^\mu.$$

Formula (3) implies that D is natural if and only if

$$(4) \qquad (\kappa-1)c_1+(\kappa+\lambda-1)c_3=0, \quad (\lambda-1)c_2+(\kappa+\lambda-1)c_3=0.$$

The formulas (2) and (4) define the constants uniquely except the case $\kappa = \lambda = 1$ when we get two independent operations, see also the fifth class. Let us point out that the second class involves also the Schouten-Nijenhuis bracket $\Lambda^p T \oplus \Lambda^q T \rightsquigarrow \Lambda^{p+q-1}T$ (the case $\kappa=\lambda=0$, $k+l\le n+1$), cf. 30.10, sometimes also caled the antisymmetric Schouten concomitant, which defines the structure of a graded Lie algebra on the fields in question. This bracket is given by

$$\begin{aligned}&[X_1\wedge\cdots\wedge X_k, Y_1\wedge\cdots\wedge Y_l]\\ &\quad=\textstyle\sum_{i,j}(-1)^{i+j}[X_i,Y_j]\wedge X_1\wedge\cdots{}^{\wedge_i}\cdots\wedge X_k\wedge Y_1\wedge\cdots{}^{\wedge_j}\cdots\wedge Y_l.\end{aligned}$$

The second class is invariant under the action of Σ_3.

3. The third class is represented by the so called symmetric Schouten concomitant. This is an operation of type $(S^k\tau, S^l\tau; S^{k+l-1}\tau)$ with a nice geometric definition. The elements in S^kTM can be identified with functions on T^*M fiberwise polynomial of degree k. Since there is a canonical symplectic structure on T^*M, there is the Poisson bracket on $C^\infty(T^*M)$. The bracket of two fiberwise polynomial functions is also fiberwise polynomial and so the bracket gives rise to our operation.

The action of Σ_3 yields an operation of the type $(S^k\tau, S^l\tau^*\otimes\Delta; S^{l-k+1}\tau^*\otimes\Delta)$. If $k=1$, this is the Lie derivative and if $k=l$, we get the lagrangian Schouten concomitant.

4. This class involves the Frölicher-Nijenhuis bracket, an operation of the type $(\tau\otimes\Lambda^k\tau^*, \tau\otimes\Lambda^l\tau^*; \tau\otimes\Lambda^{k+l}\tau^*)$, $k+l\le n$. The tensor spaces in question are not irreducible, $\tau\otimes\Lambda^k\tau^*$ is a sum of $\Lambda^{k-1}\tau^*$ and an irreducible representation ρ_k of highest weight $(1,\dots,1,0,\dots,0,-1)$ where 1 appears k-times (the trace-free vector valued forms). The Frölicher-Nijenhuis bracket is a sum of an operation of type $(\rho_k,\rho_l;\rho_{k+l})$ and several other simpler operations.

If we apply the action of Σ_3 to the Frölicher-Nijenhuis bracket, we get an operation of the type $(\tau\otimes\Lambda^m\tau^*, \tau^*\otimes\Lambda^k\tau^*; \tau^*\otimes\Lambda^{k+m}\tau^*)$ which is expressed through contractions and the exterior derivative.

5. Finally, there are the natural operations which reduce to compositions of wedge products and exterior differentiation. Such operations are always defined if at least one of the representations ρ_1, ρ_2, or one of the irreducible components of $\rho_1\otimes\rho_2$ coincides with $\Lambda^k\tau^*$. Since $\widetilde{\Lambda^k\tau^*}=\Lambda^{n-k}\tau^*$, this class is also invariant under the action of Σ_3.

In [Grozman, 80b] we find the next theorem. Unfortunately its proof based on the Rudakov's algebraic methods is not available in the literature. In an earlier paper, [Grozman, 80a], he classified the bilinear operations in dimension two, including the unimodular case.

34.18. Theorem. *All natural bilinear operators between natural bundles corresponding to irreducible representations of $GL(n)$ are exhausted by the zero* *order operators, the five classes of first order operators described in 34.17, the operators of second and third order obtained by the composition of the first and zero order operators and one exceptional operation in dimension $n = 1$, see the example below.*

In particular, there are no bilinear natural operations of order greater then three.

34.19. Example. A tensor density on the real line is determined by one complex number λ, we write $f(x)(dx)^{-\lambda} \in C^\infty(E_\lambda \mathbb{R})$ for the corresponding fields of geometric objects. There is a natural bilinear operator $D\colon E_{2/3} \oplus E_{2/3} \rightsquigarrow E_{-5/3}$

$$D(f(dx)^{-2/3}, g(dx)^{-2/3}) = \left(2 \begin{vmatrix} f & g \\ d^3f/dx^3 & d^3g/dx^3 \end{vmatrix} + 3 \begin{vmatrix} df/dx & dg/dx \\ d^2f/dx^2 & d^2g/dx^2 \end{vmatrix}\right) . (dx)^{5/3}$$

This is a third order operation which is not a composition of lower order ones.

34.20. The multilinear natural operators are also related to the cohomology theory of Lie algebras of formal vector fields. In fact these operators express zero dimensional cohomologies with coefficients in tensor products of the spaces of the fields in question, see [Fuks, 84]. The situation is much further analyzed in dimension $n = 1$ in [Feigin, Fuks, 82]. In particular, they have described all skew symmetric operations $E_\lambda \oplus \cdots \oplus E_\lambda \rightsquigarrow E_\mu$. They have deduced

Theorem. *For every $\lambda \in \mathbb{C}$, $m > 0$, $k \in \mathbb{Z}$, there is at most one skew symmetric operation $D\colon \Lambda^m C^\infty E_\lambda \to C^\infty E_\mu$ with $\mu = m\lambda - \frac{1}{2}m(m-1) - k$, up to a constant multiple. A necessary and sufficient condition for its existence is the following: either k=0, or $0 < k \leq m$ and λ satisfies the quadratic equation*

$$\left((\lambda + \tfrac{1}{2})(k_1 + 1) - m\right)\left((\lambda + \tfrac{1}{2})(k_2 + 1) - m\right) = \tfrac{1}{2}(k_2 - k_1)^2$$

with arbitrary positive $k_1 \in \mathbb{Z}$, $k_2 \in \mathbb{Z}$, $k_1.k_2 = k$.

The operator corresponding to the first possibility $k = 0$, $D\colon \Lambda^m C^\infty(E_\lambda \mathbb{R}) \to C^\infty(E_{m\lambda - \frac{1}{2}m(m-1)}\mathbb{R})$, admits a simple expression

$$f_1(dx)^{-\lambda} \wedge \cdots \wedge f_m(dx)^{-\lambda} \mapsto \begin{vmatrix} f_1 & f_1' & \dots & f_1^{(m-1)} \\ f_2 & f_2' & \dots & f_2^{(m-1)} \\ \dots & \dots & \dots & \dots \\ f_m & f_m' & \dots & f_m^{(m-1)} \end{vmatrix} (dx)^{-m\lambda + \frac{1}{2}m(m-1)}$$

Grozman's operator from 34.19 corresponds to the choice $m = 2$, $k = 2$, $\lambda = 2/3$, $k_1 = 2$, $k_2 = 1$. The proof of this theorem is rather involved. It is based on the structure of projective representations of the algebra of formal vector fields on the one-dimensional sphere.

34.21. The problem of finding all natural m-linear operations has been also formulated for super manifolds. As far as we know, only the linear operations were classified, see [Bernstein, Leites, 77], [Leites, 80], [Shmelev, 83], but their results include also the unimodular, and Hamiltonian cases. Some more information is also available in [Kirillov, 80].

34.22. The linear natural operations on conformal manifolds. As we have seen, the description of the linear natural operators is heavily based on the structure of the subalgebra in the algebra of formal vector fields which corresponds to the jet groups in the category in question. If the category involves very few morphisms, these algebras become small. In particular, they might have finite dimensions like in the case of Riemannian manifolds or conformal Riemannian manifolds. The former example is not so interesting for the following reasons: Since all irreducible representations of the orthogonal groups are $O(m,\mathbb{R})$-invariant irreducible subspaces in tensor spaces, we can work in the whole category of manifolds in the way demonstrated in section 33. On the other hand, if we include the so called spinor representations of the orthogonal group, we get serious problems with the whole setting. However, the second example is of highest interest for many reasons coming both from mathematics and physics and it is treated extensively nowadays. Let us conclude this section with a very short overview of the known results, for more information see the survey [Baston, Eastwood, 90] or the papers [Baston, 90], [Branson, 85].

Let us write $\mathcal{C}$ for the category of manifolds with a conformal Riemannian structure, i.e. with a distinguished line bundle in $S^2_+T^*M$, and the morphisms keeping this structure. More explicitely, two metrics g, $\hat{g}$ on M are called conformal if there is a positive smooth function f on M such that $\hat{g} = f^2 g$. A conformal structure is an equivalence class with respect to this equivalence relation. The conformal structure on M can also be described as a reduction of the first order frame bundle P^1M to the conformal group $CO(m,\mathbb{R}) = \mathbb{R} \rtimes O(m,\mathbb{R})$, and the conformal morphisms φ are just those local diffeomorphisms which preserve this reduction under the $P^1\varphi$-action. Thus, each linear representation of $CO(m,\mathbb{R})$ on a vector space V defines a bundle functor on $\mathcal{C}$. The category $\mathcal{C}$ is not locally homogeneous, but it is local.

The main difference from the situations typical for this book is that there are new natural bundles in the category $\mathcal{C}$. In fact, we can take any linear representation of $O(m,\mathbb{R})$ and a representation of the center $\mathbb{R} \subset GL(m,\mathbb{R})$ and combine them together. The representations of the center are of the form $(t.\text{id})(v) = t^{-w}.v$ with an arbitrary real number w, which is called the *conformal weight* of the representation or of the corresponding bundle functor. Each tensorial representation of $GL(m,\mathbb{R})$ induces a representation of $CO(m,\mathbb{R})$ with the conformal weight equal to the difference of the number of covariant and contravariant indices. In particular, the convention for the weight is chosen in such a way that the bundle of metrics has conformal weight two. If we restrict our considerations to the tensorial representations, we exclude nearly all natural linear operators.

Each isometry of a conformal manifold with respect to an arbitrary metric from the distinguished class is a conformal morphism. Thus, the Riemannian natural operators described in section 33 can be taken for candidates in the classification. But the remaining problems are still so difficult that a general solution has not been found yet.

Let us mention at least two possibilities how to treat the problem. The first one is to restrict ourselves to locally conformally flat manifolds, i.e. we consider only a subcategory in $\mathcal{C}$ which is admissible in our sense. Thus, the classification problem for linear operators reduces to a (difficult) problem from the representation theory. But what remains then is to distinguish those natural operators on the conformally flat manifolds which are restrictions of natural operators on the whole category, and to find explicite formulas for them. For general reasons, there must be a universal formula in the terms of the covariant derivatives, curvatures and their covariant derivatives. The best known example is the conformal Laplace operator on functions in dimension 4

$$D = \nabla^a \nabla_a + \frac{1}{6} R$$

where $\nabla^a \nabla_a$ means the operator of the covariant differentiation applied twice and followed by taking trace, and R is the scalar curvature. The proper conformal weights ensuring the invariance are -1 on the source and -3 on the target. The first summand $D_0 = \nabla^a \nabla_a$ of D is an operator which is natural on the functions with the specified weights on conformally flat manifolds and the second summand is a correction for the general case. In view of this example, the question is how far we can modify the natural operators (homogeneous in the order and acting between bundles corresponding to irreducible representations of $CO(m, \mathbb{R})$) found on the flat manifolds by adding some corrections. The answer is rather nice: with some few exceptions this is always possible and the order of the correction term is less by two (or more) than that of D_0. Moreover, the correction involves only the Ricci curvature and its covariant derivatives. This was deduced in [Eastwood, Rice, 87] in dimension four, and in [Baston, 90] for dimensions greater than two (the complex representations are treated explicitely and the authors assert that the real analogy is available with mild changes). In particular, there are no corrections necessary for the first order operators, which where completely classified by [Fegan, 76]. Nevertheless, the concrete formulas for the operators (first of all for the curvature terms) are rarely available. Another disadvantage of this approach is that we have no information on the operators which vanish on the conformally flat manifolds, even we do not know how far the extension of a given operator to the whole category is determined.

The description of all linear natural operators on the conformally flat manifolds is based on the general ideas as presented at the begining of this section. This means we have to find the morphisms of $\mathfrak{g}$-modules $W^* \to (T_n^\infty V)^*$, where $\mathfrak{g}$ is the algebra of formal vector fields on $\mathbb{R}^n$ with flows consisting of conformal morphisms. One can show that $\mathfrak{g} = \mathfrak{o}(n+1, 1)$, the pseudo-orthogonal algebra, with grading $\mathfrak{g} = \mathfrak{g}_{-1} \oplus \mathfrak{g}_0 \oplus \mathfrak{g}_1 = \mathbb{R}^n \oplus \mathfrak{co}(n, \mathbb{R}) \oplus \mathbb{R}^{n*}$. The lemmas 34.5 and 34.6 remain true and we see that $(T_n^\infty V)^*$ is the so called generalized Verma module corresponding to the representation of $CO(n, \mathbb{R})$ on V. Each homomorphism $W^* \to (T_n^\infty V)^*$ extends to a homomorphism of the generalized Verma modules $(T_n^\infty W)^* \to (T_n^\infty V)^*$ and so we have to classify all morphisms of generalized Verma modules. These were described in [Boe, Collingwood, 85a, 85b]. In particular, if we start with usual functions (i.e. with conformal weight zero), then all conformally invariant operators which form a 'connected pattern' involving the functions are drawn in 33.18. (The latter means that there are no more operators having one of the bundles indicated on the diagram as the source or target.) A very interesting point is a general principal coming from the representation theory (the so called Jantzen-Zuzkermann functors) which asserts that once we have got such a 'connected pattern' all other ones are obtained by a general procedure. Unfortunately this 'translation procedure' is not of a clear geometric character and so we cannot get the formulas for the corresponding operators in this way, cf. [Baston, 90]. The general theory mentioned above implies that all the operators from the diagram in 33.18 admit the extension to the whole category of conformal manifolds, except the longest arrow $\Omega^0 \to \Omega^m$. By the 'translation procedure', the same is ensured for all such patterns, but the question whether there is an extension for the exceptional 'long arrows' is not solved in general. Some of them do extend, but there are counter examples of operators which do not admit any extension, see [Branson, 89], [Graham, to appear].

Another more direct approach is used by [Branson, 85, 89] and others. They write down a concrete general formula in terms of the Riemannian invariants and they study the action of the conformal rescaling of the metric. Since it is sufficient to study the infinitesimal condition on the invariance with respect to the rescaling of the metric, they are able to find series of conformally invariant operators. But a classification is available for the first and second order operators only.

Remarks

Proposition 30.4 was proved by [Kolář, Michor, 87]. Proposition 31.1 was deduced in [Kolář, 87a]. The natural transformations $J^r \to J^r$ were determined in [Kolář, Vosmanská, 89]. The exchange map e_Λ from 32.4 was introduced by [Modugno, 89a].

The original proof of the Gilkey theorem on the uniqueness of the Pontryagin forms, [Gilkey, 73], was much more combinatorial and had not used H. Weyl's theorem. Our approach is similar to [Atiyah, Bott, Patodi, 73], but we do not need their polynomiality assumption. The Gilkey theorem was generalized in several directions. For the case of Hermitian bundles and connections see [Atiyah, Bott, Patodi, 73], for oriented Riemannian manifolds see [Stredder, 75], the metrics with a general signature are treated in [Gilkey, 75]. The uniqueness of the Levi-Cività connection among the polynomial conformal natural connections on Riemannian manifolds was deduced by [Epstein, 75]. The classification of the first order liftings of Riemannian metrics to the tangent bundles covers the results due to [Kowalski, Sekizawa, 88], who used the so called method of differential equations in their much longer proof. Our methods originate in [Slovák, 89] and an unpublished paper by W. M. Mikulski.

CHAPTER VIII. PRODUCT PRESERVING FUNCTORS

We first present the theory of those bundle functors which are determined by local algebras in the sense of A. Weil, [Weil, 51]. Then we explain that the Weil functors are closely related to arbitrary product preserving functors $\mathcal{M}f \to \mathcal{M}f$. In particular, every product preserving bundle functor on $\mathcal{M}f$ is a Weil functor and the natural transformations between two such functors are in bijection with the homomorphisms of the local algebras in question.

In order to motivate the development in this chapter we will tell first a mathematical short story. For a smooth manifold M, one can prove that the space of algebra homomorphisms $\operatorname{Hom}(C^\infty(M,\mathbb{R}),\mathbb{R})$ equals M as follows. The kernel of a homomorphism $\varphi : C^\infty(M,\mathbb{R}) \to \mathbb{R}$ is an ideal of codimension 1 in $C^\infty(M,\mathbb{R})$. The zero sets $Z_f := f^{-1}(0)$ for $f \in \ker\varphi$ form a filter of closed sets, since $Z_f \cap Z_g = Z_{f^2+g^2}$, which contains a compact set Z_f for a function f which is unbounded on each non compact closed subset. Thus $\bigcap_{f\in\ker\varphi} Z_f$ is not empty, it contains at least one point x_0. But then for any $f \in C^\infty(M,\mathbb{R})$ the function $f-\varphi(f)1$ belongs to the kernel of φ, so vanishes on x_0 and we have $f(x_0)=\varphi(f)$.

An easy consequence is that $\operatorname{Hom}(C^\infty(M,\mathbb{R}),C^\infty(N,\mathbb{R})) = C^\infty(N,M)$. So the category of algebras $C^\infty(M,\mathbb{R})$ and their algebra homomorphisms is dual to the category $\mathcal{M}f$ of manifolds and smooth mappings.

But now let $\mathbb{D}$ be the algebra generated by 1 and ε with $\varepsilon^2 = 0$ (sometimes called the algebra of dual numbers or Study numbers, it is also the truncated polynomial algebra of degree 1). Then it turns out that $\operatorname{Hom}(C^\infty(M,\mathbb{R}),\mathbb{D}) = TM$, the tangent bundle of M. For if φ is a homomorphism $C^\infty(M,\mathbb{R}) \to \mathbb{D}$, then $\pi\circ\varphi : C^\infty(M,\mathbb{R}) \to \mathbb{D} \to \mathbb{R}$ equals ev_x for some $x \in M$ and $\varphi(f) - f(x).1 = X(f).\varepsilon$, where X is a derivation over x since φ is a homomorphism. So X is a tangent vector of M with foot point x. Similarly we may show that $\operatorname{Hom}(C^\infty(M,\mathbb{R}),\mathbb{D}\otimes\mathbb{D}) = TTM$.

Now let A be an arbitrary commutative real finite dimensional algebra with unit. Let $W(A)$ be the subalgebra of A generated by the idempotent and nilpotent elements of A. We will show in this chapter, that $\operatorname{Hom}(C^\infty(M,\mathbb{R}),A) = \operatorname{Hom}(C^\infty(M,\mathbb{R}),W(A))$ is a manifold, functorial in M, and that in this way we have defined a product preserving functor $\mathcal{M}f \to \mathcal{M}f$ for any such algebra. A will be called a *Weil algebra* if $W(A) = A$, since in [Weil, 51] this construction appeared for the first time. We are aware of the fact, that Weil algebras denote completely different objects in the Chern-Weil construction of characteristic classes. This will not cause troubles, and a serious group of mathematicians has already adopted the name Weil algebra for our objects in synthetic differential geometry, so we decided to stick to this name. The functors constructed in this way will be called Weil functors, and we will also present a covariant approach to them which mimics the construction of the bundles of velocities, due to [Morimoto, 69], cf. [Kolář, 86].

We will discuss thoroughly natural transformations between Weil functors and study sections of them, a sort of generalized vector fields. It turns out that the addition of vector fields generalizes to a group structure on the set of all sections, which has a Lie algebra and an exponential mapping; it is infinite dimensional but nilpotent.

Conversely under very mild conditions we will show, that up to some covering phenomenon each product preserving functor is of this form, and that natural transformations between them correspond to algebra homomorphisms. This has been proved by [Kainz-Michor, 87] and independently by [Eck, 86] and [Luciano, 88].

Weil functors will play an important role in the rest of the book, and we will frequently compare results for other functors with them. They can be much further analyzed than other types of functors.

35. Weil algebras and Weil functors

35.1. A real commutative algebra A with unit 1 is called *formally real* if for any $a_1, \dots, a_n \in A$ the element $1 + a_1^2 + \dots + a_n^2$ is invertible in A. Let $E = \{e \in A : e^2 = e, e \neq 0\} \subset A$ be the set of all nonzero idempotent elements in A. It is not empty since $1 \in E$. An idempotent $e \in E$ is said to be *minimal* if for any $e' \in E$ we have $ee' = e$ or $ee' = 0$.

Lemma. *Let A be a real commutative algebra with unit which is formally real and finite dimensional as a real vector space.*

Then there is a decomposition $1 = e_1 + \dots + e_k$ into all minimal idempotents. Furthermore $A = A_1 \oplus \dots \oplus A_k$, where $A_i = e_i A = \mathbb{R} \cdot e_i \oplus N_i$, and N_i is a nilpotent ideal.

Proof. First we remark that every system of nonzero idempotents $e_1, \dots, e_r$ satisfying $e_i e_j = 0$ for $i \neq j$ is linearly independent over $\mathbb{R}$. Indeed, if we multiply a linear combination $k_1 e_1 + \dots + k_r e_r = 0$ by e_i we obtain $k_i = 0$. Consider a non minimal idempotent $e \neq 0$. Then there exists $e' \in E$ with $e \neq ee' =: \bar{e} \neq 0$. Then both $\bar{e}$ and $e - \bar{e}$ are nonzero idempotents and $\bar{e}(e - \bar{e}) = 0$. To deduce the required decomposition of 1 we proceed by recurrence. Assume that we have a decomposition $1 = e_1 + \dots + e_r$ into nonzero idempotents satisfying $e_i e_j = 0$ for $i \neq j$. If e_i is not minimal, we decompose it as $e_i = \bar{e}_i + (e_i - \bar{e}_i)$ as above. The new decomposition of 1 into $r + 1$ idempotents is of the same type as the original one. Since A is finite dimensional this proceedure stabilizes. This yields $1 = e_1 + \dots + e_k$ with minimal idempotents. Multiplying this relation by a minimal idempotent e, we find that e appears exactly once in the right hand side. Then we may decompose A as $A = A_1 \oplus \dots \oplus A_k$, where $A_i := e_i A$.

Now each A_i has only one nonzero idempotent, namely e_i, and it suffices to investigate each A_i separately. To simplify the notation we suppose that $A = A_i$, so that now 1 is the only nonzero idempotent of A. Let $N := \{n \in A : n^k = 0 \text{ for some } k\}$ be the ideal of all nilpotent elements in A.

We claim that any $x \in A \setminus N$ is invertible. If not then $xA \subset A$ is a proper ideal, and since A is finite dimensional the decreasing sequence

$$A \supset xA \supset x^2 A \supset \cdots$$

of ideals must become stationary. If $x^k A = 0$ then $x \in N$, thus there is a k such that $x^{k+\ell}A = x^k A \neq 0$ for all $\ell > 0$. Then $x^{2k}A = x^k A$ and there is some $y \in A$ with $x^k = x^{2k}y$. So we have $(x^k y)^2 = x^k y \neq 0$, and since 1 is the only nontrivial idempotent of A we have $x^k y = 1$. So $x^{k-1}y$ is an inverse of x as required.

So the quotient algebra A/N is a finite dimensional field, so A/N equals $\mathbb{R}$ or $\mathbb{C}$. If $A/N = \mathbb{C}$, let $x \in A$ be such that $x + N = \sqrt{-1} \in \mathbb{C} = A/N$. Then $1 + x^2 + N = N = 0$ in $\mathbb{C}$, so $1 + x^2$ is nilpotent and A cannot be formally real. Thus $A/N = \mathbb{R}$ and $A = \mathbb{R} \cdot 1 \oplus N$ as required. □

35.2. Definition. A *Weil algebra* A is a real commutative algebra with unit which is of the form $A = \mathbb{R} \cdot 1 \oplus N$, where N is a finite dimensional ideal of nilpotent elements.

So by lemma 35.1 a formally real and finite dimensional unital commutative algebra is the direct sum of finitely many Weil algebras.

35.3. Some algebraic preliminaries. Let A be a commutative algebra with unit and let M be a module over A. The *semidirect product* $A[M]$ of A and M or the *idealisator* of M is the algebra $(A \times M, +, \cdot)$, where $(a_1, m_1) \cdot (a_2, m_2) = (a_1 a_2, a_1 m_2 + a_2 m_1)$. Then M is a (nilpotent) ideal of $A[M]$.

Let $M^{m \times n} = \{(t_{ij}) : t_{ij} \in M, 1 \leq i \leq m, 1 \leq j \leq n\}$ be the space of all $(m \times n)$-matrices with entries in the module M. If $S \in A^{r \times m}$ and $T \in M^{m \times n}$ then the product of matrices $ST \in M^{r \times n}$ is defined by the usual formula.

For a matrix $U = (u_{ij}) \in A^{n \times n}$ the determinant is given by the usual formula $\det(U) = \sum_{\sigma \in S_n} \operatorname{sign} \sigma \prod_{i=1}^n u_{i,\sigma(i)}$. It is n-linear and alternating in the columns of U.

Lemma. *If $m = (m_i) \in M^{n \times 1}$ is a column vector of elements in the A-module M and if $U = (u_{ij}) \in A^{n \times n}$ is a matrix with $Um = 0 \in M^{n \times 1}$ then we have $\det(U) m_i = 0$ for each i.*

Proof. We may compute in the idealisator $A[M]$, or assume without loss of generality that all $m_i \in A$. Let u_{*j} denote the j-th column of U. Then $\sum u_{ij} m_j = 0$ for all i means that $m_1 u_{*1} = -\sum_{j>1} m_j u_{*j}$, thus

$$\begin{aligned} \det(U) m_1 &= \det(m_1 u_{*1}, u_{*2}, \ldots, u_{*n}) \\ &= \det(-\textstyle\sum_{j>1} m_j u_{*j}, u_{*2}, \ldots, u_{*n}) = 0 \quad \square \end{aligned}$$

Lemma. *Let I be an ideal in an algebra A and let M be a finitely generated A-module. If $IM = M$ then there is an element $a \in I$ with $(1-a)M = 0$.*

Proof. Let $M = \sum_{i=1}^n A m_i$ for generators $m_i \in M$. Since $IM = M$ we have $m_i = \sum_{j=1}^n t_{ij} m_j$ for some $T = (t_{ij}) \in I^{n \times n}$. This means $(1_n - T)m = 0$ for $m = (m_j) \in M^{n \times 1}$. By the first lemma we get $\det(1_n - T) m_j = 0$ for all j. But $\det(1_n - T) = 1 - a$ for some $a \in I$. □

Lemma of Nakayama. *Let (A, I) be a local algebra (i.e. an algebra with a unique maximal ideal I) and let M be an A-module. Let $N_1, N_2 \subset M$ be submodules with N_1 finitely generated. If $N_1 \subseteq N_2 + IN_1$ then we have $N_1 \subseteq N_2$.*

In particular $IN_1 = N_1$ implies $N_1 = 0$.

Proof. Let $IN_1 = N_1$. By the lemma above there is some $a \in I$ with $(1-a)N_1 = 0$. Since I is a maximal ideal (so A/I is a field), $1 - a$ is invertible. Thus $N_1 = 0$. If $N_1 \subseteq N_2 + IN_1$ we have $I((N_1 + N_2)/N_2) = (N_1 + N_2)/N_2$ thus $(N_1 + N_2)/N_2 = 0$ or $N_1 \subseteq N_2$. □

35.4. Lemma. *Any ideal I of finite codimension in the algebra of germs $\mathcal{E}_n := C_0^\infty(\mathbb{R}^n, \mathbb{R})$ contains some power $\mathcal{M}_n^k$ of the maximal ideal $\mathcal{M}_n$ of germs vanishing at 0.*

Proof. Consider the chain of ideals $\mathcal{E}_n \supseteq I + \mathcal{M}_n \supseteq I + \mathcal{M}_n^2 \supseteq \cdots$. Since I has finite codimension we have $I + \mathcal{M}_n^k = I + \mathcal{M}_n^{k+1}$ for some k. So $\mathcal{M}_n^k \subseteq I + \mathcal{M}_n \mathcal{M}_n^k$ which implies $\mathcal{M}_n^k \subseteq I$ by the lemma of Nakayama 35.3 since $\mathcal{M}_n^k$ is finitely generated by the monomials of order k in n variables. □

35.5. Theorem. *Let A be a unital real commutative algebra. Then the following assertions are equivalent.*

(1) *A is a Weil algebra.*

(2) *A is a finite dimensional quotient of an algebra of germs $\mathcal{E}_n = C_0^\infty(\mathbb{R}^n, \mathbb{R})$ for some n.*

(3) *A is a finite dimensional quotient of an algebra $\mathbb{R}[X_1, \dots, X_n]$ of polynomials.*

(4) *A is a finite dimensional quotient of an algebra $\mathbb{R}[[X_1, \dots, X_n]]$ of formal power series.*

(5) *A is a quotient of an algebra $J_0^k(\mathbb{R}^n, \mathbb{R})$ of jets.*

Proof. Let $A = \mathbb{R} \cdot 1 \oplus N$, where N is the maximal ideal of nilpotent elements, which is generated by finitely many elements, say $X_1, \dots, X_n$. Since $\mathbb{R}[X_1, \dots, X_n]$ is the free real unital commutative algebra generated by these elements, A is a quotient of this polynomial algebra. There is some k such that $x^{k+1} = 0$ for all $x \in N$, so A is even a quotient of the jet algebra $J_0^k(\mathbb{R}^n, \mathbb{R})$. Since the jet algebra is itself a quotient of the algebra of germs and the algebra of formal power series, the same is true for A. That all these finite dimensional quotients are Weil algebras is clear, since they all are formally real and have only one nonzero idempotent. □

If A is a quotient of the jet algebra $J_0^r(\mathbb{R}^n, \mathbb{R})$, we say that the *order* of A is at most r.

35.6. The width of a Weil algebra. Consider the square N^2 of the nilpotent ideal N of a Weil algebra A. The dimension of the real vector space N/N^2 is called the *width* of A.

Let $\mathcal{M} \subset \mathbb{R}[x_1, \dots, x_n]$ denote the ideal of all polynomials without constant term and let $I \subset \mathbb{R}[x_1, \dots, x_n]$ be an ideal of finite codimension which is contained in $\mathcal{M}^2$. Then *the width of the factor algebra $A = \mathbb{R}[x_1, \dots, x_n]/I$*

is n. Indeed the nilpotent ideal of A is $\mathcal{M}/I$ and $(\mathcal{M}/I)^2 = \mathcal{M}^2/I$, hence $(\mathcal{M}/I)/(\mathcal{M}/I)^2 \cong \mathcal{M}/\mathcal{M}^2$ is of dimension n.

35.7. Proposition. *If M is a smooth manifold and I is an ideal of finite codimension in the algebra $C^\infty(M,\mathbb{R})$, then $C^\infty(M,\mathbb{R})/I$ is a direct sum of finitely many Weil algebras.*

If A is a finite dimensional commutative real algebra with unit, then we have $\operatorname{Hom}(C^\infty(M,\mathbb{R}),A) = \operatorname{Hom}(C^\infty(M,\mathbb{R}),W(A))$, where $W(A)$ is the subalgebra of A generated by all idempotent and nilpotent elements of A (the so-called Weil part of A). In particular $W(A)$ is formally real.

Proof. The algebra $C^\infty(M,\mathbb{R})$ is formally real, so the first assertion follows from lemma 35.1. If $\varphi : C^\infty(M,\mathbb{R}) \to A$ is an algebra homomorphism, then the kernel of φ is an ideal of finite codimension in $C^\infty(M,\mathbb{R})$, so the image of φ is a direct sum of Weil algebras and is thus generated by its idempotent and nilpotent elements. □

35.8. Lemma. *Let M be a smooth manifold and let $\varphi : C^\infty(M,\mathbb{R}) \to A$ be an algebra homomorphism into a Weil algebra A.*

Then there is a point $x \in M$ and some $k \geq 0$ such that $\ker\varphi$ contains the ideal of all functions which vanish at x up to order k.

Proof. Since $\varphi(1) = 1$ the kernel of φ is a nontrivial ideal in $C^\infty(M,\mathbb{R})$ of finite codimension.

If Λ is a closed subset of M we let $C^\infty(\Lambda,\mathbb{R})$ denote the algebra of all real valued functions on Λ which are restrictions of smooth functions on M. For a smooth function f let $Z_f := f^{-1}(0)$ be its zero set. For a subset $S \subset C^\infty(\Lambda,\mathbb{R})$ we put $Z_S := \bigcap\{Z_f : f \in S\}$.

Claim 1. Let I be an ideal of finite codimension in $C^\infty(\Lambda,\mathbb{R})$. Then Z_I is a finite subset of Λ and $Z_I = \emptyset$ if and only if $I = C^\infty(\Lambda,\mathbb{R})$.

Z_I is finite since $C^\infty(\Lambda,\mathbb{R})/I$ is finite dimensional. $Z_f = \emptyset$ implies that f is invertible. So if $I \neq C^\infty(\Lambda,\mathbb{R})$ then $\{Z_f : f \in I\}$ is a filter of nonempty closed sets, since $Z_f \cap Z_g = Z_{f^2+g^2}$. Let $h \in C^\infty(M,\mathbb{R})$ be a positive proper function, i.e. inverse images under h of compact sets are compact. The square of the geodesic distance with respect to a complete Riemannian metric on a connected manifold M is such a function. Then we put $f = h|\Lambda \in C^\infty(\Lambda,\mathbb{R})$. The sequence $f, f^2, f^3, \dots$ is linearly dependent $\mod I$, since I has finite codimension, so $g = \sum_{i=1}^n \lambda_i f^i \in I$ for some $(\lambda_i) \neq 0$ in $\mathbb{R}^n$. Then clearly Z_g is compact. So this filter of closed nonempty sets contains a compact set and has therefore nonempty intersection $Z_I = \bigcap_{f\in I} Z_f$.

Claim 2. If I is an ideal of finite codimension in $C^\infty(M,\mathbb{R})$ and if a function $f \in C^\infty(M,\mathbb{R})$ vanishes near Z_I, then $f \in I$.

Let $Z_I \subset U_1 \subset \overline{U}_1 \subset U_2$ where U_1 and U_2 are open in M such that $f|U_2 = 0$. The restriction mapping $C^\infty(M,\mathbb{R}) \to C^\infty(M \setminus U_1,\mathbb{R})$ is a surjective algebra homomorphism, so the image I' of I is again an ideal of finite codimension in $C^\infty(M \setminus U_1,\mathbb{R})$. But clearly $Z_{I'} = \emptyset$, so by claim 1 we have $I' = C^\infty(M \setminus U_1,\mathbb{R})$.

Thus there is some $g \in I$ such that $g|(M \setminus U_1) = f|(M \setminus U_1)$. Now choose $h \in C^\infty(M, \mathbb{R})$ such that $h = 0$ on U_1 and $h = 1$ off U_2. Then $f = fh = gh \in I$.

Claim 3. For the ideal $\ker\varphi$ in $C^\infty(M, \mathbb{R})$ the zero set Z_I consists of one point x only.

Since $\ker\varphi$ is a nontrivial ideal of finite codimension, $Z_{\ker\varphi}$ is not empty and finite by claim 1. For any function $f \in C^\infty(M, \mathbb{R})$ which is 1 or 0 near the points in $Z_{\ker\varphi}$ the element $\varphi(f)$ is an idempotent of the Weil algebra A. Since 1 is the only nonzero idempotent of A, the zero set Z_I consists of one point.

Now by claims 2 and 3 the ideal $\ker\varphi$ contains the ideal of all functions which vanish near x. So φ factors to the algebra $C_x^\infty(M, \mathbb{R})$ of germs at x, compare 35.5.(2). Now $\ker\varphi \subset C_x^\infty(M, \mathbb{R})$ is an ideal of finite codimension, so by lemma 35.4 the result follows. □

35.9. Corollary. *The evaluation mapping $ev : M \to \operatorname{Hom}(C^\infty(M, \mathbb{R}), \mathbb{R})$, given by $ev(x)(f) := f(x)$, is bijective.*

This result is sometimes called the exercise of Milnor, see [Milnor-Stasheff, 74, p. 11]. Another (similar) proof of it can be found in the mathematical short story in the introduction to chapter VIII.

Proof. By lemma 35.8, for every $\varphi \in \operatorname{Hom}(C^\infty(M, \mathbb{R}), \mathbb{R})$ there is an $x \in M$ and a $k \geq 0$ such that $\ker\varphi$ contains the ideal of all functions vanishing at x up to order k. Since the codimension of $\ker\varphi$ is 1, we have $\ker\varphi = \{f \in C^\infty(M, \mathbb{R}) : f(x) = 0\}$. Then for any $f \in C^\infty(M, \mathbb{R})$ we have $f - f(x)1 \in \ker\varphi$, so $\varphi(f) = f(x)$. □

35.10. Corollary. *For two manifolds M_1 and M_2 the mapping*

$$\begin{aligned} C^\infty(M_1, M_2) &\to \operatorname{Hom}(C^\infty(M_2, \mathbb{R}), C^\infty(M_1, \mathbb{R})) \\ f &\mapsto (f^* : g \mapsto g \circ f) \end{aligned}$$

is bijective.

Proof. Let $x_1 \in M_1$ and $\varphi \in \operatorname{Hom}(C^\infty(M_2, \mathbb{R}), C^\infty(M_1, \mathbb{R}))$. Then $ev_{x_1} \circ \varphi$ is in $\operatorname{Hom}(C^\infty(M_2, \mathbb{R}), \mathbb{R})$, so by 35.9 there is a unique $x_2 \in M_2$ such that $ev_{x_1} \circ \varphi = ev_{x_2}$. If we write $x_2 = f(x_1)$, then $f : M_1 \to M_2$ and $\varphi(g) = g \circ f$ for all $g \in C^\infty(M_2, \mathbb{R})$. This also implies that f is smooth. □

35.11. Chart description of Weil functors. Let $A = \mathbb{R} \cdot 1 \oplus N$ be a Weil algebra. We want to associate to it a functor $T_A : \mathcal{M}f \to \mathcal{M}f$ from the category $\mathcal{M}f$ of all finite dimensional second countable manifolds into itself. We will give several descriptions of this functor, and we begin with the most elementary and basic construction, the idea of which goes back to [Weil, 53].

Step 1. If $p(t)$ is a real polynomial, then for any $a \in A$ the element $p(a) \in A$ is uniquely defined; so we have a (polynomial) mapping $T_A(p) : A \to A$.

Step 2. If $f \in C^\infty(\mathbb{R},\mathbb{R})$ and $\lambda 1 + n \in \mathbb{R}\cdot 1 \oplus N = A$, we consider the Taylor expansion $j^\infty f(\lambda)(t) = \sum_{j=0}^\infty \frac{f^{(j)}(\lambda)}{j!}t^j$ of f at λ and we put

$$T_A(f)(\lambda 1 + n) := f(\lambda)1 + \sum_{j=1}^\infty \frac{f^{(j)}(\lambda)}{j!}n^j,$$

which is finite sum, since n is nilpotent. Then $T_A(f) : A \to A$ is smooth and we get $T_A(f \circ g) = T_A(f) \circ T_A(g)$ and $T_A(\mathrm{Id}_\mathbb{R}) = \mathrm{Id}_A$.

Step 3. For $f \in C^\infty(\mathbb{R}^m,\mathbb{R})$ we want to define the value of $T_A(f)$ at the vector $(\lambda_1 1 + n_1, \dots, \lambda_m 1 + n_m) \in A^m = A \times \dots \times A$. Let again $j^\infty f(\lambda)(t) = \sum_{\alpha\in\mathbb{N}^m} \frac{1}{\alpha!}d^\alpha f(\lambda)t^\alpha$ be the Taylor expansion of f at $\lambda \in \mathbb{R}^m$ for $t \in \mathbb{R}^m$. Then we put

$$T_A(f)(\lambda_1 1 + n_1, \dots, \lambda_m 1 + n_m) := f(\lambda)1 + \sum_{|\alpha|\geq 1} \frac{1}{\alpha!}d^\alpha f(\lambda) n_1^{\alpha_1}\dots n_m^{\alpha_m},$$

which is again a finite sum.

Step 4. For $f \in C^\infty(\mathbb{R}^m,\mathbb{R}^k)$ we apply the construction of step 3 to each component $f_j : \mathbb{R}^m \to \mathbb{R}$ of f to define $T_A(f) : A^m \to A^k$.

Since the Taylor expansion of a composition is the composition of the Taylor expansions we have $T_A(f \circ g) = T_A(f) \circ T_A(g)$ and $T_A(\mathrm{Id}_{\mathbb{R}^m}) = \mathrm{Id}_{A^m}$.

If $\varphi : A \to B$ is a homomorphism between two Weil algebras we have $\varphi^k \circ T_A f = T_B f \circ \varphi^m$ for $f \in C^\infty(\mathbb{R}^m,\mathbb{R}^k)$.

Step 5. Let $\pi = \pi_A : A \to A/N = \mathbb{R}$ be the projection onto the quotient field of the Weil algebra A. This is a surjective algebra homomorphism, so by step 4 the following diagram commutes for $f \in C^\infty(\mathbb{R}^m,\mathbb{R}^k)$:

$$\begin{array}{ccc} A^m & \xrightarrow{T_A f} & A^k \\ {\scriptstyle \pi_A^m}\downarrow & & \downarrow{\scriptstyle \pi_A^k} \\ \mathbb{R}^m & \xrightarrow{f} & \mathbb{R}^k \end{array}$$

If $U \subset \mathbb{R}^m$ is an open subset we put $T_A(U) := (\pi_A^m)^{-1}(U) = U \times N^m$, which is an open subset in $T_A(\mathbb{R}^m) := A^m$. If $f : U \to V$ is a smooth mapping between open subsets U and V of $\mathbb{R}^m$ and $\mathbb{R}^k$, respectively, then the construction of steps 3 and 4, applied to the Taylor expansion of f at points in U, produces a smooth mapping $T_A f : T_A U \to T_A V$, which fits into the following commutative diagram:

$$\begin{array}{ccccccc} U \times N^m & = & T_A U & \xrightarrow{T_A f} & T_A V & = & V \times N^k \\ & {\scriptstyle pr_1}\searrow & \downarrow{\scriptstyle \pi_A^m} & & \downarrow{\scriptstyle \pi_A^k} & \swarrow{\scriptstyle pr_1} & \\ & & U & \xrightarrow{f} & V & & \end{array}$$

We have $T_A(f \circ g) = T_Af \circ T_Ag$ and $T_A(\mathrm{Id}_U) = \mathrm{Id}_{T_AU}$, so T_A is now a covariant functor on the category of open subsets of $\mathbb{R}^m$'s and smooth mappings between them.

Step 6. In 1.14 we have proved that the separable connected smooth manifolds are exactly the smooth retracts of open subsets in $\mathbb{R}^m$'s. If M is a smooth manifold, let $i : M \to \mathbb{R}^m$ be an embedding, let $i(M) \subset U \subset \mathbb{R}^m$ be a tubular neighborhood and let $q : U \to U$ be the projection of U with image $i(M)$. Then q is smooth and $q \circ q = q$. We define now $T_A(M)$ to be the image of the smooth retraction $T_Aq : T_AU \to T_AU$, which by 1.13 is a smooth submanifold.

If $f : M \to M'$ is a smooth mapping between manifolds, we define $T_Af : T_AM \to T_AM'$ as

$$T_AM \subset T_AU \xrightarrow{T_A(i' \circ f \circ q)} T_AU' \xrightarrow{T_Aq'} T_AU',$$

which takes values in T_AM'.

It remains to show, that another choice of the data $(i, U, q, \mathbb{R}^m)$ for the manifold M leads to a diffeomorphic submanifold T_AM, and that T_Af is uniquely defined up to conjugation with these diffeomorphisms for M and M'. Since this is a purely formal manipulation with arrows we leave it to the reader and give instead the following:

Step 6'. Direct construction of T_AM for a manifold M using atlases.

Let M be a smooth manifold of dimension m, let (U_α, u_α) be a smooth atlas of M with chart changings $u_{\alpha\beta} := u_\alpha \circ u_\beta^{-1} : u_\beta(U_{\alpha\beta}) \to u_\alpha(U_{\alpha\beta})$. Then the smooth mappings

$$\begin{array}{ccc} T_A(u_\beta(U_{\alpha\beta})) & \xrightarrow{T_A(u_{\alpha\beta})} & T_A(u_\alpha(U_{\alpha\beta})) \\ \downarrow{\scriptstyle \pi_A^m} & & \downarrow{\scriptstyle \pi_A^m} \\ u_\beta(U_{\alpha\beta}) & \xrightarrow{u_{\alpha\beta}} & u_\alpha(U_{\alpha\beta}) \end{array}$$

form again a cocycle of chart changings and we may use them to glue the open sets $T_A(u_\alpha(U_\alpha)) = u_\alpha(U_\alpha) \times N^m \subset T_A(\mathbb{R}^m) = A^m$ in order to obtain a smooth manifold which we denote by T_AM. By the diagram above we see that T_AM will be the total space of a fiber bundle $T(\pi_A, M) = \pi_{A,M} : T_AM \to M$, since the atlas $(T_A(U_\alpha), T_A(u_\alpha))$ constructed just now is already a fiber bundle atlas. Thus T_AM is Hausdorff, since two points x_i can be separated in one chart if they are in the same fiber, or they can be separated by inverse images under $\pi_{A,M}$ of open sets in M separating their projections.

This construction does not depend on the choice of the atlas. For two atlases have a common refinement and one may pass to this.

If $f \in C^\infty(M, M')$ for two manifolds M, M', we apply the functor T_A to the local representatives of f with respect to suitable atlases. This gives local representatives which fit together to form a smooth mapping $T_Af : T_AM \to T_AM'$. Clearly we again have $T_A(f \circ g) = T_Af \circ T_Ag$ and $T_A(\mathrm{Id}_M) = \mathrm{Id}_{T_AM}$, so that $T_A : \mathcal{M}f \to \mathcal{M}f$ is a covariant functor.

35.12. Remark. If we apply the construction of 35.11, step 6' to the algebra $A = 0$, which we did not allow ($1 \neq 0 \in A$), then T_0M depends on the choice of the atlas. If each chart is connected, then $T_0M = \pi_0(M)$, computing the connected components of M. If each chart meets each connected component of M, then T_0M is one point.

35.13. Theorem. Main properties of Weil functors. *Let $A = \mathbb{R} \cdot 1 \oplus N$ be a Weil algebra, where N is the maximal ideal of nilpotents. Then we have:*

1. The construction of 35.11 defines a covariant functor $T_A : \mathcal{M}f \to \mathcal{M}f$ such that $(T_AM, \pi_{A,M}, M, N^{\dim M})$ is a smooth fiber bundle with standard fiber $N^{\dim M}$. For any $f \in C^\infty(M, M')$ we have a commutative diagram

$$\begin{array}{ccc} T_AM & \xrightarrow{T_Af} & T_AM' \\ \Big\downarrow{\scriptstyle \pi_{A,M}} & & \Big\downarrow{\scriptstyle \pi_{A,M'}} \\ M & \xrightarrow{f} & M'. \end{array}$$

So (T_A, π_A) is a bundle functor on $\mathcal{M}f$, which gives a vector bundle on $\mathcal{M}f$ if and only if N is nilpotent of order 2.

2. The functor $T_A : \mathcal{M}f \to \mathcal{M}f$ is multiplicative: it respects products. It maps the following classes of mappings into itself: immersions, initial immersions, embeddings, closed embeddings, submersions, surjective submersions, fiber bundle projections. It also respects transversal pullbacks, see 2.19. For fixed manifolds M and M' the mapping $T_A : C^\infty(M, M') \to C^\infty(T_AM, T_AM')$ is smooth, i.e. it maps smoothly parametrized families into smoothly parametrized families.

3. If (U_α) is an open cover of M then $T_A(U_\alpha)$ is also an open cover of T_AM.

4. Any algebra homomorphism $\varphi : A \to B$ between Weil algebras induces a natural transformation $T(\varphi, \quad) = T_\varphi : T_A \to T_B$. If φ is injective, then $T(\varphi, M) : T_AM \to T_BM$ is a closed embedding for each manifold M. If φ is surjective, then $T(\varphi, M)$ is a fiber bundle projection for each M. So we may view T as a co-covariant bifunctor from the category of Weil algebras times $\mathcal{M}f$ to $\mathcal{M}f$.

Proof. 1. The main assertion is clear from 35.11. The fiber bundle $\pi_{A,M} : T_AM \to M$ is a vector bundle if and only if the transition functions $T_A(u_{\alpha\beta})$ are fiber linear $N^{\dim M} \to N^{\dim M}$. So only the first derivatives of $u_{\alpha\beta}$ should act on N, so any product of two elements in N must be 0, thus N has to be nilpotent of order 2.

2. The functor T_A respects products in the category of open subsets of $\mathbb{R}^m$'s by 35.11, step 4 and 5. All the other assertions follow by looking again at the chart structure of T_AM and by taking into account that f is part of T_Af (as the base mapping).

3. This is obvious from the chart structure.

4. We define $T(\varphi, \mathbb{R}^m) := \varphi^m : A^m \to B^m$. By 35.11, step 4, this restricts to a natural transformation $T_A \to T_B$ on the category of open subsets of $\mathbb{R}^m$'s and by gluing also on the category $\mathcal{M}f$. Obviously T is a co-covariant bifunctor on

the indicated categories. Since $\pi_B \circ \varphi = \pi_A$ (φ respects the identity), we have $T(\pi_B, M) \circ T(\varphi, M) = T(\pi_A, M)$, so $T(\varphi, M) : T_A M \to T_B M$ is fiber respecting for each manifold M. In each fiber chart it is a linear mapping on the typical fiber $N_A^{\dim M} \to N_B^{\dim M}$.

So if φ is injective, $T(\varphi, M)$ is fiberwise injective and linear in each canonical fiber chart, so it is a closed embedding.

If φ is surjective, let $N_1 := \ker\varphi \subseteq N_A$, and let $V \subset N_A$ be a linear complement to N_1. Then for $m = \dim M$ and for the canonical charts we have the commutative diagram:

$$\begin{array}{ccc}
T_A M & \xrightarrow{\quad T(\varphi, M) \quad} & T_B M \\
\uparrow & & \uparrow \\
T_A(U_\alpha) & \xrightarrow{\quad T(\varphi, U_\alpha) \quad} & T_B(U_\alpha) \\
{\scriptstyle T_A(u_\alpha)} \downarrow & & \downarrow {\scriptstyle T_B(u_\alpha)} \\
u_\alpha(U_\alpha) \times N_A^m & \xrightarrow{\quad \operatorname{Id} \times (\varphi|N_A)^m \quad} & u_\alpha(U_\alpha) \times N_B^m \\
\| & & \| \\
u_\alpha(U_\alpha) \times N_1^m \times V^m & \xrightarrow{\quad \operatorname{Id} \times 0 \times Iso \quad} & u_\alpha(U_\alpha) \times 0 \times N_B^m
\end{array}$$

So $T(\varphi, M)$ is a fiber bundle projection with standard fiber $(\ker\varphi)^m$. □

35.14. Theorem. Algebraic description of Weil functors. *There are bijective mappings $\eta_{M,A} : \operatorname{Hom}(C^\infty(M, \mathbb{R}), A) \to T_A(M)$ for all smooth manifolds M and all Weil algebras A, which are natural in M and A. Via η the set $\operatorname{Hom}(C^\infty(M, \mathbb{R}), A)$ becomes a smooth manifold and $\operatorname{Hom}(C^\infty(\ \ , \mathbb{R}), A)$ is a global expression for the functor T_A.*

Proof. Step 1. Let (x^i) be coordinate functions on $\mathbb{R}^n$. By lemma 35.8 for $\varphi \in \operatorname{Hom}(C^\infty(\mathbb{R}^n, \mathbb{R}), A)$ there is a point $x(\varphi) = (x^1(\varphi), \dots, x^n(\varphi)) \in \mathbb{R}^n$ such that $\ker\varphi$ contains the ideal of all $f \in C^\infty(\mathbb{R}^n, \mathbb{R})$ vanishing at $x(\varphi)$ up to some order k, so that $\varphi(x^i) = x^i(\varphi) \cdot 1 + \varphi(x^i - x^i(\varphi))$, the latter summand being nilpotent in A of order $\le k$. Applying φ to the Taylor expansion of f at $x(\varphi)$ up to order k with remainder gives

$$\begin{aligned}
\varphi(f) &= \sum_{|\alpha| \le k} \tfrac{1}{\alpha!} \frac{\partial^{|\alpha|} f}{\partial x^\alpha}(x(\varphi))\, \varphi(x^1 - x^1(\varphi))^{\alpha_1} \dots \varphi(x^n - x^n(\varphi))^{\alpha_n} \\
&= T_A(f)(\varphi(x^1), \dots, \varphi(x^n)).
\end{aligned}$$

So φ is uniquely determined by the elements $\varphi(x^i)$ in A and the mapping

$$\begin{aligned}
&\eta_{\mathbb{R}^n, A} : \operatorname{Hom}(C^\infty(\mathbb{R}^n, \mathbb{R}), A) \to A^n, \\
&\eta(\varphi) := (\varphi(x^1), \dots, \varphi(x^n))
\end{aligned}$$

is injective. Furthermore for $g = (g^1, \dots, g^m) \in C^\infty(\mathbb{R}^n, \mathbb{R}^m)$ and coordinate functions $(y^1, \dots, y^m)$ on $\mathbb{R}^m$ we have

$$\begin{aligned}(\eta_{\mathbb{R}^m,A} \circ (g^*)^*)(\varphi) &= (\varphi(y^1 \circ g), \dots, \varphi(y^m \circ g))\\ &= (\varphi(g^1), \dots, \varphi(g^m))\\ &= \big(T_A(g^1)(\varphi(x^1), \dots, \varphi(x^n)), \dots, T_A(g^m)(\varphi(x^1), \dots, \varphi(x^n))\big),\end{aligned}$$

so $\eta_{\mathbb{R}^n,A}$ is natural in $\mathbb{R}^n$. It is also bijective since any $(a_1, \dots, a_n) \in A^n$ defines a homomorphism $\varphi : C^\infty(\mathbb{R}^n, \mathbb{R}) \to A$ by the prescription $\varphi(f) := T_A f(a_1, \dots, a_n)$.

Step 2. Let $i : U \to \mathbb{R}^n$ be the embedding of an open subset. Then the image of the mapping

$$\operatorname{Hom}(C^\infty(U, \mathbb{R}), A) \xrightarrow{(i^*)^*} \operatorname{Hom}(C^\infty(\mathbb{R}^n, \mathbb{R}), A) \xrightarrow{\eta_{\mathbb{R}^n,A}} A^n$$

is the set $\pi_{A,\mathbb{R}^n}^{-1}(U) = T_A(U) \subset A^n$, and $(i^*)^*$ is injective.

To see this let $\varphi \in \operatorname{Hom}(C^\infty(U, \mathbb{R}), A)$. By lemma 35.8 $\ker \varphi$ contains the ideal of all f vanishing up to some order k at a point $x(\varphi) \in U \subseteq \mathbb{R}^n$, and since $\varphi(x^i) = x^i(\varphi) \cdot 1 + \varphi(x^i - x^i(\varphi))$ we have

$$\pi_{A,\mathbb{R}^n}(\eta_{\mathbb{R}^n,A}(\varphi \circ i^*)) = \pi_A^n(\varphi(x^1), \dots, \varphi(x^n)) = x(\varphi) \in U.$$

As in step 2 we see that the mapping

$$\pi_{A,\mathbb{R}^n}^{-1}(U) \ni (a_1, \dots, a_n) \mapsto (C^\infty(U, \mathbb{R}) \ni f \mapsto T_A(f)(a_1, \dots, a_n))$$

is the inverse to $\eta_{\mathbb{R}^n,A} \circ (i^*)^*$.

Step 3. The two functors $\operatorname{Hom}(C^\infty(\quad, \mathbb{R}), A)$ and $T_A : \mathcal{M}f \to \mathcal{S}et$ coincide on all open subsets of $\mathbb{R}^n$'s, so they have to coincide on all manifolds, since smooth manifolds are exactly the retracts of open subsets of $\mathbb{R}^n$'s by 1.14.1. Alternatively one may check that the gluing process described in 35.11, step 6, works also for the functor $\operatorname{Hom}(C^\infty(\quad, \mathbb{R}), A)$ and gives a unique manifold structure on it which is compatible to $T_A M$. □

35.15. Covariant description of Weil functors. Let A be a Weil algebra, which by 35.5.(2) can be viewed as $\mathcal{E}_n/I$, a finite dimensional quotient of the algebra $\mathcal{E}_n = C_0^\infty(\mathbb{R}^n, \mathbb{R})$ of germs at 0 of smooth functions on $\mathbb{R}^n$.

Definition. Let M be a manifold. Two mappings $f, g : \mathbb{R}^n \to M$ with $f(0) = g(0) = x$ are said to be *I-equivalent*, if for all germs $h \in C_x^\infty(M, \mathbb{R})$ we have $h \circ f - h \circ g \in I$.

The equivalence class of a mapping $f : \mathbb{R}^n \to M$ will be denoted by $j_A(f)$ and will be called the *A-velocity* at 0 of f. Let us denote by $J_A(M)$ the set of all A-velocities on M.

There is a natural way to extend J_A to a functor $\mathcal{M}f \to Set$. For every smooth mapping $f : M \to N$ between manifolds we put $J_A(f)(j_A(g)) := j_A(f \circ g)$ for $g \in C^\infty(\mathbb{R}^n, M)$.

Now one can repeat the development of the theory of (n, r)-velocities for the more general space $J_A(M)$ instead of $J_0^k(\mathbb{R}^n, M)$ and show that $J_A(M)$ is a smooth fiber bundle over M, associated to a higher order frame bundle. This development is very similar to the computations done in 35.11 and we will in fact reduce the whole situation to 35.11 and 35.14 by the following

35.16. Lemma. *There is a canonical equivalence*

$$J_A(M) \to \operatorname{Hom}(C^\infty(M, \mathbb{R}), A),$$
$$j_A(f) \mapsto (C^\infty(M, \mathbb{R}) \ni g \mapsto j_A(g \circ f) \in A),$$

which is natural in A and M and a diffeomorphism, so the functor $J_A : \mathcal{M}f \to \mathcal{FM}$ is equivalent to T_A.

Proof. We just have to note that $J_A(\mathbb{R}) = \mathcal{E}_n/I = A$. □

Let us state explicitly that a trivial consequence of this lemma is that the Weil functor determined by the Weil algebra $\mathcal{E}_n/\mathcal{M}_n^{k+1} = J_0^k(\mathbb{R}^n, \mathbb{R})$ is the functor T_n^r of (n, r)-velocities from 12.8.

35.17. Theorem. *Let A and B be Weil algebras. Then we have:*

(1) *We get the algebra A back from the Weil functor T_A by $T_A(\mathbb{R}) = A$ with addition $+_A = T_A(+_\mathbb{R})$, multiplication $m_A = T_A(m_\mathbb{R})$ and scalar multiplication $m_t = T_A(m_t) : A \to A$.*
(2) *The natural transformations $T_A \to T_B$ correspond exactly to the algebra homomorphisms $A \to B$*

Proof. (1) This is obvious. (2) For a natural transformation $\varphi : T_A \to T_B$ its value $\varphi_\mathbb{R} : T_A(\mathbb{R}) = A \to T_B(\mathbb{R}) = B$ is an algebra homomorphisms. The inverse of this mapping is already described in theorem 35.13.4. □

35.18. The basic facts from the theory of Weil functors are completed by the following assertion, which will be proved in more general context in 36.13.

Proposition. *Given two Weil algebras A and B, the composed functor $T_A \circ T_B$ is a Weil functor generated by the tensor product $A \otimes B$.*

Corollary. *(See also 37.3.) There is a canonical natural equivalence $T_A \circ T_B \cong T_B \circ T_A$ generated by the exchange algebra isomorphism $A \otimes B \cong B \otimes A$.*

36. Product preserving functors

36.1. A covariant functor $F : \mathcal{M}f \to \mathcal{M}f$ is said to be *product preserving*, if the diagram

$$F(M_1) \xleftarrow{F(pr_1)} F(M_1 \times M_2) \xrightarrow{F(pr_2)} F(M_2)$$

is always a product diagram. Then $F(\text{point}) = \text{point}$, by the following argument:

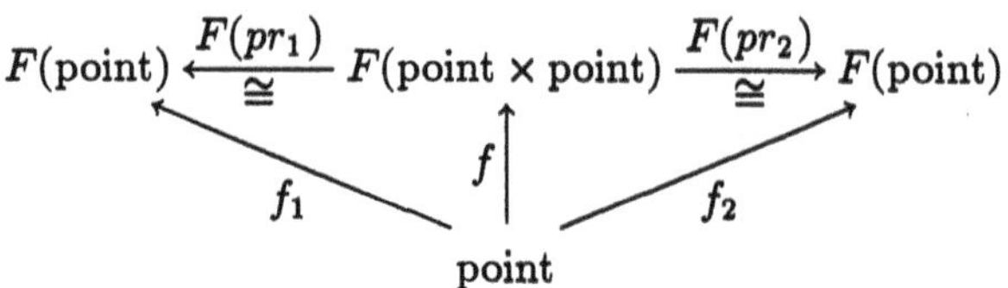

Each of f_1, f, and f_2 determines each other uniquely, thus there is only one mapping $f_1 : \text{point} \to F(\text{point})$, so the space $F(\text{point})$ is single pointed.

The basic purpose of this section is to prove the following

Theorem. *Let F be a product preserving functor together with a natural transformation $\pi_F : F \to \text{Id}$ such that (F, π_F) satisfies the locality condition 18.3.(i).*

Then $F = T_A$ for some Weil algebra A.

This will be a special case of much more general results below. The final proof will be given in 36.12. We will first extract uniquely a sum of Weil algebras from a product preserving functor, then we will reconstruct the functor from this algebra under mild conditions.

36.2. We denote the addition and the multiplication on the reals by $+, m : \mathbb{R}^2 \to \mathbb{R}$, and for $\lambda \in \mathbb{R}$ we let $m_\lambda : \mathbb{R} \to \mathbb{R}$ be the scalar multiplication by λ and we also consider the mapping $\lambda : \text{point} \to \mathbb{R}$ onto the value λ.

Theorem. *Let $F : \mathcal{M}f \to \mathcal{M}f$ be a product preserving functor. Then either $F(\mathbb{R})$ is a point or $F(\mathbb{R})$ is a finite dimensional real commutative and formally real algebra with operations $F(+)$, $F(m)$, scalar multiplication $F(m_\lambda)$, zero $F(0)$, and unit $F(1)$, which is called $\text{Al}(F)$. If $\varphi : F_1 \to F_2$ is a natural transformation between two such functors, then $\text{Al}(\varphi) := \varphi_{\mathbb{R}} : \text{Al}(F_1) \to \text{Al}(F_2)$ is an algebra homomorphism.*

Proof. Since F is product preserving, we have $F(\text{point}) = \text{point}$. All the laws for a commutative ring with unit can be formulated by commutative diagrams of mappings between products of the ring and the point. We do this for the ring $\mathbb{R}$ and apply the product preserving functor F to all these diagrams, so we get the laws for the commutative ring $F(\mathbb{R})$ with unit $F(1)$ with the exception of $F(0) \neq F(1)$ which we will check later for the case $F(\mathbb{R}) \neq \text{point}$. Addition $F(+)$ and multiplication $F(m)$ are morphisms in $\mathcal{M}f$, thus smooth and continuous.

For $\lambda \in \mathbb{R}$ the mapping $F(m_\lambda) : F(\mathbb{R}) \to F(\mathbb{R})$ equals multiplication with the element $F(\lambda) \in F(\mathbb{R})$, since the following diagram commutes:

$$\begin{array}{ccccc}
F(\mathbb{R}) & \xrightarrow{\quad F(m_\lambda) \quad} & F(\mathbb{R}) \times F(\mathbb{R}) & & \\
\cong \downarrow & & & & \\
F(\mathbb{R}) \times \text{point} & \xrightarrow{\text{Id} \times F(\lambda)} & F(\mathbb{R}) \times F(\mathbb{R}) & \longrightarrow & F(\mathbb{R}) \\
\cong \downarrow & & \| & \nearrow F(m) & \\
F(\mathbb{R} \times \text{point}) & \xrightarrow{F(\text{Id} \times \lambda)} & F(\mathbb{R} \times \mathbb{R}) & &
\end{array}$$

We may investigate now the difference between $F(\mathbb{R}) = \text{point}$ and $F(\mathbb{R}) \neq \text{point}$. In the latter case for $\lambda \neq 0$ we have $F(\lambda) \neq F(0)$ since multiplication by $F(\lambda)$ equals $F(m_\lambda)$ which is a diffeomorphism for $\lambda \neq 0$ and factors over a one pointed space for $\lambda = 0$. So for $F(\mathbb{R}) \neq \text{point}$ which we assume from now on, the group homomorphism $\lambda \mapsto F(\lambda)$ from $\mathbb{R}$ into $F(\mathbb{R})$ is actually injective.

In order to show that the scalar multiplication $\lambda \mapsto F(m_\lambda)$ induces a continuous mapping $\mathbb{R} \times F(\mathbb{R}) \to F(\mathbb{R})$ it suffices to show that $\mathbb{R} \to F(\mathbb{R})$, $\lambda \mapsto F(\lambda)$, is continuous.

$(F(\mathbb{R}), F(+), F(m_{-1}), F(0))$ is a commutative Lie group and is second countable as a manifold since $F(\mathbb{R}) \in \mathcal{M}f$. We consider the exponential mapping $\exp : \mathcal{L} \to F(\mathbb{R})$ from the Lie algebra $\mathcal{L}$ into this group. Then $\exp(\mathcal{L})$ is an open subgroup of $F(\mathbb{R})$, the connected component of the identity. Since $\{F(\lambda) : \lambda \in \mathbb{R}\}$ is a subgroup of $F(\mathbb{R})$, if $F(\lambda) \notin \exp(\mathcal{L})$ for all $\lambda \neq 0$, then $F(\mathbb{R})/\exp(\mathcal{L})$ is a discrete uncountable subgroup, so $F(\mathbb{R})$ has uncountably many connected components, in contradiction to $F(\mathbb{R}) \in \mathcal{M}f$. So there is $\lambda_0 \neq 0$ in $\mathbb{R}$ and $v_0 \neq 0$ in $\mathcal{L}$ such that $F(\lambda_0) = \exp(v_0)$. For each $v \in \mathcal{L}$ and $r \in \mathbb{N}$, hence $r \in \mathbb{Q}$, we have $F(m_r)\exp(v) = \exp(rv)$. Now we claim that for any sequence $\lambda_n \to \lambda$ in $\mathbb{R}$ we have $F(\lambda_n) \to F(\lambda)$ in $F(\mathbb{R})$. If not then there is a sequence $\lambda_n \to \lambda$ in $\mathbb{R}$ such that $F(\lambda_n) \in F(\mathbb{R}) \setminus U$ for some neighborhood U of $F(\lambda)$ in $F(\mathbb{R})$, and by considering a suitable subsequence we may also assume that $2^{n^2}(\lambda_{n+1} - \lambda)$ is bounded. By lemma 36.3 below there is a C^∞-function $f : \mathbb{R} \to \mathbb{R}$ with $f(\frac{\lambda_0}{2^n}) = \lambda_n$ and $f(0) = \lambda$. Then we have

$$\begin{aligned}
F(\lambda_n) &= F(f)F(m_{2^{-n}})F(\lambda_0) = F(f)F(m_{2^{-n}})\exp(v_0) = \\
&= F(f)\exp(2^{-n}v_0) \to F(f)\exp(0) = F(f(0)) = F(\lambda),
\end{aligned}$$

contrary to the assumption that $F(\lambda_n) \notin U$ for all n. So $\lambda \mapsto F(\lambda)$ is a continuous mapping $\mathbb{R} \to F(\mathbb{R})$, and $F(\mathbb{R})$ with its manifold topology is a real finite dimensional commutative algebra, which we will denote by $\text{Al}(F)$ from now on.

The evaluation mapping $ev_{\text{Id}_\mathbb{R}} : \text{Hom}(C^\infty(\mathbb{R}, \mathbb{R}), \text{Al}(F)) \to \text{Al}(F)$ is bijective since it has the right inverse $x \mapsto (C^\infty(\mathbb{R}, \mathbb{R}) \ni f \mapsto F(f)_x)$. But by 35.7 the evaluation map has values in the Weil part $W(\text{Al}(F))$ of $\text{Al}(F)$, so the algebra $\text{Al}(F)$ is generated by its idempotent and nilpotent elements and has to be formally real, a direct sum of Weil algebras by 35.1. $\square$

Remark. In the case of product preserving bundle functors the smoothness of $\lambda \mapsto F(\lambda)$ is a special case of the regularity proved in 20.7. In fact one may also conclude that $F(\mathbb{R})$ is a smooth algebra by the results from [Montgomery-Zippin, 55], cited in 5.10.

36.3. Lemma. [Kriegl, 82] *Let $\lambda_n \to \lambda$ in $\mathbb{R}$, let $t_n \in \mathbb{R}$, $t_n > 0$, $t_n \to 0$ strictly monotone, such that*

$$\left\{ \frac{\lambda_n - \lambda_{n+1}}{(t_n - t_{n+1})^k}, n \in \mathbb{N} \right\}$$

is bounded for all k. Then there is a C^∞-function $f : \mathbb{R} \to \mathbb{R}$ with $f(t_n) = \lambda_n$ and $f(0) = \lambda$ such that f is flat at each t_n.

Proof. Let $\varphi \in C^\infty(\mathbb{R}, \mathbb{R})$, $\varphi = 0$ near 0, $\varphi = 1$ near 1, and $0 \le \varphi \le 1$ elsewhere. Then we put

$$f(t) = \begin{cases} \lambda & \text{for } t \le 0, \\ \varphi\left(\dfrac{t - t_{n+1}}{t_n - t_{n+1}}\right)(\lambda_n - \lambda_{n+1}) + \lambda_{n+1} & \text{for } t_{n+1} \le t \le t_n, \\ \lambda_1 & \text{for } t_1 \le t, \end{cases}$$

and one may check by estimating the left and right derivatives at all t_n that f is smooth. □

36.4. Product preserving functors without Weil algebras. Let $F : \mathcal{M}f \to \mathcal{M}f$ be a functor with preserves products and assume that it has the property that $F(\mathbb{R}) =$ point. Then clearly $F(\mathbb{R}^n) = F(\mathbb{R})^n =$ point and $F(M) =$ point for each smoothly contractible manifold M. Moreover we have:

Lemma. *Let $f_0, f_1 : M \to N$ be homotopic smooth mappings, let F be as above. Then $F(f_0) = F(f_1) : F(M) \to F(N)$.*

Proof. A continuous homotopy $h : M \times [0,1] \to N$ between f_0 and f_1 may first be reparameterized in such a way that $h(x,t) = f_0(x)$ for $t < \varepsilon$ and $h(x,t) = f_1(x)$ for $1 - \varepsilon < t$, for some $\varepsilon > 0$. Then we may approximate h by a smooth mapping without changing the endpoints f_0 and f_1. So finally we may assume that there is a smooth $h : M \times \mathbb{R} \to N$ such that $h \circ \text{ins}_i = f_i$ for $i = 0, 1$ where $\text{ins}_t : M \to M \times \mathbb{R}$ is given by $\text{ins}_t(x) = (x, t)$. Since

$$\begin{array}{ccccc} F(M) & \xleftarrow{F(pr_1)} & F(M \times \mathbb{R}) & \xrightarrow{F(pr_2)} & F(\mathbb{R}) \\ & & \| & & \| \\ & & F(M) \times \text{point} & & \text{point} \end{array}$$

is a product diagram we see that $F(pr_1) = \text{Id}_{F(M)}$. Since $pr_1 \circ \text{ins}_t = \text{Id}_M$ we get also $F(\text{ins}_t) = \text{Id}_{F(M)}$ and thus $F(f_0) = F(h) \circ F(\text{ins}_0) = F(h) \circ F(\text{ins}_1) = F(f_1)$. □

Examples. For a manifold M let $M = \bigcup M_\alpha$ be the disjoint union of its connected components and put $\tilde{H}_1(M) := \bigcup_\alpha H_1(M_\alpha; \mathbb{R})$, using singular homology with real coefficients, for example. If M is compact, $\tilde{H}_1(M) \in \mathcal{M}f$ and $\tilde{H}_1$ becomes a product preserving functor from the category of all compact manifolds into $\mathcal{M}f$ without a Weil algebra.

For a connected manifold M the singular homology group $H_1(M, \mathbb{Z})$ with integer coefficients is a countable discrete set, since it is the abelization of the fundamental group $\pi_1(M)$, which is a countable group for a separable connected manifold. Then again by the Künneth theorem $H_1(\quad; \mathbb{Z})$ is a product preserving functor from the category of connected manifolds into $\mathcal{M}f$ without a Weil algebra.

More generally let K be a finite CW-complex and let $[K, M]$ denote the discrete set of all (free) homotopy classes of continuous mappings $K \to M$, where M is a manifold. Algebraic topology tells us that this is a countable set. Clearly $[K, \quad]$ then defines a product preserving functor without a Weil algebra. Since we may take the product of such functors with other product preserving functors we see, that the Weil algebra does not determine the functor at all. For conditions which exclude such behaviour see theorem 36.8 below.

36.5. Convention. Let $A = A_1 \oplus \cdots \oplus A_k$ be a formally real finite dimensional commutative algebra with its decomposition into Weil algebras. In this section we will need the product preserving functor $T_A := T_{A_1} \times \ldots \times T_{A_k} : \mathcal{M}f \to \mathcal{M}f$ which is given by $T_A(M) := T_{A_1}(M) \times \ldots \times T_{A_k}(M)$. Then 35.13.1 for T_A has to be modified as follows: $\pi_{A,M} \colon T_A M \to M^k$ is a fiber bundle. All other conclusions of theorem 35.13 remain valid for this functor, since they are preserved by the product, with exception of 35.13.3, which holds for connected manifolds only now. Theorem 35.14 remains true, but the covariant description (we will not use it in this section) 35.15 and 35.16 needs some modification.

36.6. Lemma. *Let $F : \mathcal{M}f \to \mathcal{M}f$ be a product preserving functor. Then the mapping*

$$\chi_{F,M} : F(M) \to \operatorname{Hom}(C^\infty(M, \mathbb{R}), \operatorname{Al}(F)) = T_{\operatorname{Al}(F)} M$$
$$\chi_{F,M}(x)(f) := F(f)(x),$$

is smooth and natural in F and M.

Proof. Naturality in F and M is obvious. To show that χ is smooth is more difficult. To simplify the notation we let $\operatorname{Al}(F) =: A = A_1 \oplus \cdots \oplus A_k$ be the decomposition of the formally real algebra $\operatorname{Al}(F)$ into Weil algebras.

Let $h = (h^1, \ldots, h^n) : M \to \mathbb{R}^n$ be a closed embedding into some high dimensional $\mathbb{R}^n$. By theorem 35.13.2 the mapping $T_A(h) : T_A M \to T_A \mathbb{R}^n$ is also a closed embedding. By theorem 35.14, step 1 of the proof (and by reordering the product), the mapping $\eta_{\mathbb{R}^n, A} : \operatorname{Hom}(C^\infty(\mathbb{R}^n, \mathbb{R}), A) \to A^n$ is given by $\eta_{\mathbb{R}^n, A}(\varphi) = (\varphi(x^i))_{i=1}^n$, where (x^i) are the standard coordinate functions on $\mathbb{R}^n$. We have

$F(\mathbb{R}^n) \cong F(\mathbb{R})^n \cong A^n \cong T_A(\mathbb{R}^n)$. Now we consider the commuting diagram

$$\begin{array}{ccccc} F(M) & & & & \\ \downarrow{\scriptstyle \chi_{F,M}} & & & & \\ \operatorname{Hom}(C^\infty(M,\mathbb{R}),A) & \xrightarrow{\eta_{M,A}} & T_A(M) & & \\ \downarrow{\scriptstyle (h^*)^*} & & \downarrow{\scriptstyle T_A(h)} & & \\ \operatorname{Hom}(C^\infty(\mathbb{R}^n,\mathbb{R}),A) & \xrightarrow{\eta_{\mathbb{R}^n,A}} & T_A(\mathbb{R}^n) & = & F(\mathbb{R}^n) \end{array}$$

For $z \in F(M)$ we have

$$\begin{aligned} (\eta_{\mathbb{R}^n,A} \circ (h^*)^* \circ \chi_{F,M})(z) &= \eta_{\mathbb{R}^n,A}(\chi_{F,M}(z) \circ h^*) \\ &= \big(\chi_{F,M}(z)(x^1 \circ h), \dots, \chi_{F,M}(z)(x^n \circ h)\big) \\ &= \big(\chi_{F,M}(z)(h^1), \dots, \chi_{F,M}(z)(h^n)\big) \\ &= \big(F(h^1)(z), \dots, F(h^n)(z)\big) = F(h)(z). \end{aligned}$$

This is smooth in $z \in F(M)$. Since $\eta_{M,A}$ is a diffeomorphism and $T_A(h)$ is a closed embedding, $\chi_{F,M}$ is smooth as required. □

36.7. The universal covering of a product preserving functor. Let $F : \mathcal{M}f \to \mathcal{M}f$ be a product preserving functor. We will construct another product preserving functor as follows. For any manifold M we choose a universal cover $q_M : \tilde{M} \to M$ (over each connected component of M separately), and we let $\pi_1(M)$ denote the group of deck transformations of $\tilde{M} \to M$, which is isomorphic to the product of all fundamental groups of the connected components of M. It is easy to see that $\pi_1(M)$ acts strictly discontinuously on $T_A(\tilde{M})$, and by lemma 36.6 therefore also on $F(\tilde{M})$. So the orbit space

$$\tilde{F}(M) := F(\tilde{M})/\pi_1(M)$$

is a smooth manifold. For $f : M_1 \to M_2$ we choose any smooth lift $\tilde{f} : \tilde{M}_1 \to \tilde{M}_2$, which is unique up to composition with elements of $\pi_1(M_i)$. Then $F\tilde{f}$ factors as follows:

$$\begin{array}{ccc} F(\tilde{M}_1) & \xrightarrow{F(\tilde{f})} & F(\tilde{M}_2) \\ \downarrow & & \downarrow \\ \tilde{F}(M) & \xrightarrow{\tilde{F}(f)} & \tilde{F}(M_2). \end{array}$$

The resulting smooth mapping $\tilde{F}(f)$ does not depend on the choice of the lift $\tilde{f}$. So we get a functor $\tilde{F} : \mathcal{M}f \to \mathcal{M}f$ and a natural transformation $q = q_F : \tilde{F} \to F$, induced by $F(q_M) : F(\tilde{M}) \to F(M)$, which is a covering mapping. This functor $\tilde{F}$ is again product preserving, because we may choose $(M_1 \times M_2)^\sim = \tilde{M}_1 \times \tilde{M}_2$ and $\pi_1(M_1 \times M_2) = \pi_1(M_1) \times \pi_1(M_2)$, thus

$$\begin{aligned}\tilde{F}(M_1 \times M_2) &= F((M_1 \times M_2)^\sim)/\pi_1(M_1 \times M_2) = \\ &= F(\tilde{M}_1)/\pi_1(M_1) \times F(\tilde{M}_2)/\pi_1(M_2) = \tilde{F}(M_1) \times \tilde{F}(M_2).\end{aligned}$$

Note finally that $\tilde{T}_A \neq T_A$ if A is sum of at least two Weil algebras. As an example consider $A = \mathbb{R} \oplus \mathbb{R}$, then $T_A(M) = M \times M$, but $\tilde{T}_A(S^1) = \mathbb{R}^2/\mathbb{Z}(2\pi, 2\pi) \cong S^1 \times \mathbb{R}$.

36.8. Theorem. *Let F be a product preserving functor.*

1. *If M is connected, then there exists a unique smooth mapping $\psi_{F,M} : \widetilde{T_{\mathrm{Al}(F)}}(M) \to F(M)$ which is natural in F and M and satisfies $\chi_{F,M} \circ \psi_{F,M} = q_{T_{\mathrm{Al}(F)},M}$:*

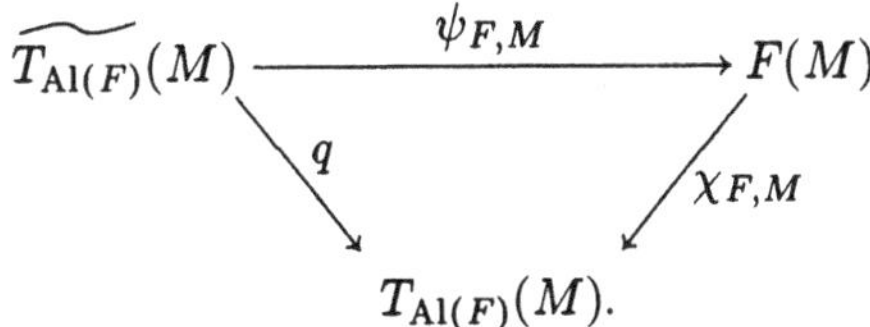

2. *If F maps embeddings to injective mappings, then $\chi_{F,M} : F(M) \to T_{\mathrm{Al}(F)}(M)$ is injective for all manifolds M, and it is a diffeomorphism for connected M.*
3. *If M is connected and $\psi_{F,M}$ is surjective, then $\chi_{F,M}$ and $\psi_{F,M}$ are covering mappings.*

Remarks. Condition (2) singles out the functors of the form T_A among all product preserving functors. Condition (3) singles the coverings of the T_A's. A product preserving functor satisfying condition (3) will be called *weakly local* .

Proof. We let $\mathrm{Al}(F) =: A = A_1 \oplus \dots \oplus A_k$ be the decomposition of the formally real algebra $\mathrm{Al}(F)$ into Weil algebras. We start with a

Sublemma. *If M is connected then $\chi_{F,M}$ is surjective and near each $\varphi \in \mathrm{Hom}(C^\infty(M,\mathbb{R}), A) = T_A(M)$ there is a smooth local section of $\chi_{F,M}$.*

Let $\varphi = \varphi_1 + \dots + \varphi_k$ for $\varphi_i \in \mathrm{Hom}(C^\infty(M,\mathbb{R}), A_i)$. Then by lemma 35.8 for each i there is exactly one point $x_i \in M$ such that $\varphi_i(f)$ depends only on a finite jet of f at x_i. Since M is connected there is a smoothly contractible open set U in M containing all x_i. Let $g : \mathbb{R}^m \to M$ be a diffeomorphism onto U. Then $(g^*)^* : \mathrm{Hom}(C^\infty(\mathbb{R}^m,\mathbb{R}), A) \to \mathrm{Hom}(C^\infty(M,\mathbb{R}), A)$ is an embedding of an open neighborhood of φ, so there is $\bar{\varphi} \in \mathrm{Hom}(C^\infty(\mathbb{R}^m,\mathbb{R}), A)$ depending smoothly on φ such that $(g^*)^*(\bar{\varphi}) = \varphi$. Now we consider the mapping

$$\begin{aligned}\mathrm{Hom}(C^\infty(\mathbb{R}^m,\mathbb{R}), A) &\xrightarrow{\eta_{\mathbb{R}^m}} T_A(\mathbb{R}^m) \cong F(\mathbb{R}^m) \xrightarrow{F(g)} \\ &\xrightarrow{F(g)} F(M) \xrightarrow{\chi_M} \mathrm{Hom}(C^\infty(M,\mathbb{R}), A).\end{aligned}$$

We have $(\chi_M \circ F(g) \circ \eta_{\mathbb{R}^m})(\bar{\varphi}) = ((g^*)^* \circ \chi_{\mathbb{R}^m} \circ \eta_{\mathbb{R}^m})(\bar{\varphi}) = (g^*)^*(\bar{\varphi}) = \varphi$, since it follows from lemma 36.6 that $\chi_{\mathbb{R}^m} \circ \eta_{\mathbb{R}^m} = \mathrm{Id}$. So the mapping $s_U := F(g) \circ \eta_{\mathbb{R}^m} \circ (g^{**})^{-1} : T_A U \to F(M)$ is a smooth local section of χ_M defined near φ. We may also write $s_U = F(i_U) \circ (\chi_{F,U})^{-1} : T_A U \to F(M)$, since for contractible U the mapping $\chi_{F,U}$ is clearly a diffeomorphism. So the sublemma is proved.

(1) Now we start with the construction of $\psi_{F,M}$. We note first that it suffices to construct $\psi_{F,M}$ for simply connected M because then we may induce it for not simply connected M using the following diagram and naturality.

$$\begin{array}{ccc} \widetilde{T_A}(\tilde{M}) = T_A\tilde{M} & \xrightarrow{\psi_{F,\tilde{M}}} & F(\tilde{M}) \\ \downarrow & & \downarrow \\ \widetilde{T_A}(M) & \xrightarrow{\psi_{F,M}} & F(M). \end{array}$$

Furthermore it suffices to construct $\psi_{F,M}$ for high dimensional M since then we have

$$\begin{array}{ccc} \widetilde{T_A}(M \times \mathbb{R}) & \xrightarrow{\psi_{F,M\times\mathbb{R}}} & F(M \times \mathbb{R}) \\ \downarrow & & \downarrow \\ \widetilde{T_A}(M) \times F(\mathbb{R}) & \xrightarrow{\psi_{F,M} \times \mathrm{Id}_{F(\mathbb{R})}} & F(M) \times F(\mathbb{R}). \end{array}$$

So we may assume that M is connected, simply connected and of high dimension. For any contractible subset U of M we consider the local section s_U of $\chi_{F,M}$ constructed in the sublemma and we just put $\psi_{F,M}(\varphi) := s_U(\varphi)$ for $\varphi \in T_A U \subset T_A M$. We have to show that $\psi_{F,M}$ is well defined. So we consider contractible U and U' in M with $\varphi \in T_A(U \cap U')$. If $\pi(\varphi) = (x_1, \dots, x_k) \in M^k$ as in the sublemma, this means that $x_1, \dots, x_k \in U \cap U'$. We claim that there are contractible open subsets V, V', and W of M such that $x_1, \dots, x_k \in V \cap V' \cap W$ and that $V \subset U \cap W$ and $V' \subset U' \cap W$. Then by the naturality of χ we have $s_U(\varphi) = s_V(\varphi) = s_W(\varphi) = s_{V'}(\varphi) = s_{U'}(\varphi)$ as required. For the existence of these sets we choose an embedding $H : \mathbb{R}^2 \to M$ such that $c(t) = H(t, \sin t) \in U$, $c'(t) = H(t, -\sin t) \in U'$ and $H(2\pi j, 0) = x_j$ for $j = 1, \dots, k$. This embedding exists by the following argument. We connect the points by a smooth curve in U and a smooth curve in U', then we choose a homotopy between these two curves fixing the x_j's, and we approximate the homotopy by an embedding, using transversality, again fixing the x_j's. For this approximation we need $\dim M \geq 5$, see [Hirsch, 76, chapter 3]. Then V, V', and W are just small tubular neighborhoods of c, c', and H.

(2) Since a manifold M has at most countably many connected components, there is an embedding $I : M \to \mathbb{R}^n$ for some n. Then from

$$\begin{array}{ccc} F(M) & \xrightarrow{F(i)} & F(\mathbb{R}^n) \\ {\scriptstyle \chi_{F,M}}\downarrow & & \cong\downarrow{\scriptstyle \chi_{F,\mathbb{R}^n}} \\ T_A(M) & \xrightarrow[T_A(i)]{} & T_A(\mathbb{R}^n), \end{array}$$

lemma 36.6, and the assumption it follows that $\chi_{F,M}$ is injective. If M is furthermore connected then the sublemma implies furthermore that $\chi_{F,M}$ is a diffeomorphism.

(3) Since $\chi \circ \psi = q$, and since q is a covering map and ψ is surjective, it follows that both χ and ψ are covering maps. □

In the example $F = T_{\mathbb{R}\oplus\mathbb{R}}$ considered at the end of 36.7 we get that ψ_{F,S^1} : $\tilde{F}(S^1) = \mathbb{R}^2/\mathbb{Z}(2\pi, 2\pi) \to F(S^1) = S^1 \times S^1 = \mathbb{R}^2/(\mathbb{Z}(2\pi, 0) \times \mathbb{Z}(0, 2\pi))$ is the covering mapping induced from the injection $\mathbb{Z}(2\pi, 2\pi) \to \mathbb{Z}(2\pi, 0) \times \mathbb{Z}(0, 2\pi)$.

36.9. Now we will determine all weakly local product preserving functors F on the category $con\mathcal{M}f$ of all connected manifolds with $\mathrm{Al}(F)$ equal to some given formally real finite dimensional algebra A with k Weil components. Let F be such a functor.

For a connected manifold M we define $C(M)$ by the following transversal pullback:

$$\begin{array}{ccc} C(M) & \longrightarrow & F(M) \\ \downarrow & & \downarrow \\ T_{\mathbb{R}^k}(M) = M^k & \xrightarrow{\ 0\ } & T_A M, \end{array}$$

where 0 is the natural transformation induced by the inclusion of the subalgebra $\mathbb{R}^k$ generated by all idempotents into A.

Now we consider the following diagram: In it every square is a pullback, and each vertical mapping is a covering mapping, if F is weakly local, by theorem 36.8.

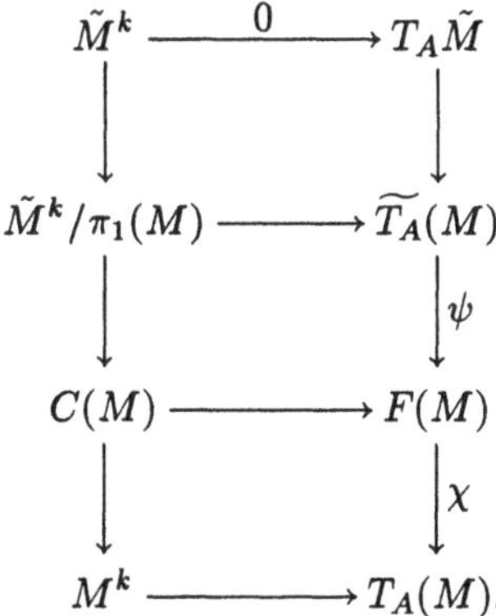

Thus $F(M) = T_A(\tilde{M})/G$, where G is the group of deck transformations of the covering $C(M) \to \tilde{M}^k$, a subgroup of $\pi_1(M)^k$ containing $\pi_1(M)$ (with its diagonal action on $\tilde{M}^k$). Here $g = (g_1, \dots, g_k) \in \pi_1(M)^k$ acts on $T_A(\tilde{M}) = T_{A_1}(\tilde{M}) \times \dots \times T_{A_k}(\tilde{M})$ via $T_{A_1}(g_1) \times \dots \times T_{A_k}(g_k)$. So we have proved

36.10. Theorem. *A weakly local product preserving functor F on the category $con\mathcal{M}f$ of all connected manifolds is uniquely determined by specifying a formally real finite dimensional algebra $A = \mathrm{Al}(F)$ and a product preserving*

functor $G : con\mathcal{M}f \to$ Groups satisfying $\pi_1 \subseteq G \subseteq \pi_1^k$, where π_1 is the fundamental group functor, sitting as diagonal in π_1^k, and where k is the number of Weil components of A.

The statement of this theorem is not completely rigorous, since π_1 depends on the choice of a base point.

36.11. Corollary. *On the category of simply connected manifolds a weakly local product preserving functor is completely determined by its algebra $A = \mathrm{Al}(F)$ and coincides with T_A.*

If the algebra $\mathrm{Al}(F) = A$ of a weakly local functor F is a Weil algebra (the unit is the only idempotent), then $F = T_A$ on the category $con\mathcal{M}f$ of connected manifolds. In particular F is a bundle functor and is local in the sense of 18.3.(i).

36.12. Proof of theorem 36.1. Using the assumptions we may conclude that $\pi_{F,M} : F(M) \to M$ is a fiber bundle for each $M \in \mathcal{M}f$, using 20.3, 20.7, and 20.8. Moreover for an embedding $i_U : U \to M$ of an open subset $F(i_U) : F(U) \to F(M)$ is the embedding onto $F(M)|U = \pi_{F,M}^{-1}(U)$. Let $A = \mathrm{Al}(F)$. Then A can have only one idempotent, for even the bundle functor $pr_1 : M \times M \to M$ is not local. So A is a Weil algebra.

By corollary 36.11 we have $F = T_A$ on connected manifolds. Since F is local, it is fully determined by its values on smoothly contractible manifolds, i.e. all $\mathbb{R}^m$'s. □

36.13. Lemma. *For product preserving functors F_1 and F_2 on $\mathcal{M}f$ we have $\mathrm{Al}(F_2 \circ F_1) = \mathrm{Al}(F_1) \otimes \mathrm{Al}(F_2)$ naturally in F_1 and F_2.*

Proof. Let B be a real basis for $\mathrm{Al}(F_1)$. Then

$$\mathrm{Al}(F_2 \circ F_1) = F_2(F_1(\mathbb{R})) = F_2(\prod_{b \in B} \mathbb{R} \cdot b) \cong \prod_{b \in B} F_2(\mathbb{R}) \cdot b,$$

so the formula holds for the underlying vector spaces. Now we express the multiplication $F_1(m) : \mathrm{Al}(F_1) \times \mathrm{Al}(F_1) \to \mathrm{Al}(F_1)$ in terms of the basis: $b_i b_j = \sum_k c_{ij}^k b_k$, and we use

$$F_2(F_1(m)) = (F_1(m)^*)^* : \mathrm{Hom}(C^\infty(\mathrm{Al}(F_1) \times \mathrm{Al}(F_1), \mathbb{R}), \mathrm{Al}(F_2)) \to \\ \to \mathrm{Hom}(C^\infty(\mathrm{Al}(F_1), \mathbb{R}), \mathrm{Al}(F_2))$$

to see that the formula holds also for the multiplication. □

Remark. We chose the order $\mathrm{Al}(F_1) \otimes \mathrm{Al}(F_2)$ so that the elements of $\mathrm{Al}(F_2)$ stand on the right hand side. This coincides with the usual convention for writing an atlas for the second tangent bundle and will be essential for the formalism developed in section 37 below.

36.14. Product preserving functors on not connected manifolds. Let F be a product preserving functor $\mathcal{M}f \to \mathcal{M}f$. For simplicity's sake we assume that F maps embeddings to injective mappings, so that on connected manifolds it coincides with T_A where $A = \mathrm{Al}(F)$. For a general manifold we have $T_A(M) \cong$

$\operatorname{Hom}(C^\infty(M,\mathbb{R}),A)$, but this is not the unique extension of $F|con\mathcal{M}f$ to $\mathcal{M}f$, as the following example shows: Consider $P_k(M) = M\times\ldots\times M$ (k times), given by the product of Weil algebras $\mathbb{R}^k$. Now let $P_k^c(M) = \bigsqcup_\alpha P_k(M_\alpha)$ be the disjoint union of all $P_k(M_\alpha)$ where M_α runs though all connected components of M. Then P_k^c is a different extension of $P_k|con\mathcal{M}f$ to $\mathcal{M}f$.

Let us assume now that $A = \operatorname{Al}(F)$ is a direct sum on k Weil algebras, $A = A_1\oplus\cdots\oplus A_k$ and let $\pi : T_A \to P_k$ be the natural transformation induced by the projection on the subalgebra $\mathbb{R}^k$ generated by all idempotents. Then also $F^c(M) = \pi^{-1}(P_k^c(M)) \subset T_A(M)$ is an extension of $F|con\mathcal{M}f$ to $\mathcal{M}f$ which differs from T_A. Clearly we have $F^c(M) = \bigsqcup_\alpha F(M_\alpha)$ where the disjoint union runs again over all connected components of M.

Proposition. *Any product preserving functor $F : \mathcal{M}f \to \mathcal{M}f$ which maps embeddings to injective mappings is of the form $F = G_1^c\times\ldots\times G_n^c$ for product preserving functors G_i which also map embeddings to injective mappings.*

Proof. Let again $\operatorname{Al}(F) = A = A_1\oplus\cdots\oplus A_k$ be the decomposition into Weil algebras. We conclude from 36.8.2 that $\chi_{F,M} : F(M)\to T_A(M)$ is injective for each manifold M. We have to show that the set $\{1,\ldots,k\}$ can be divided into equivalence classes $I_1,\ldots,I_n$ such that $F(M)\subseteq T_A(M)$ is the inverse image under $\pi : T_A(M)\to P_k(M)$ of the union of all $N_1\times\ldots\times N_k$ where the N_i run through all connected components of M in such a way that $i,j\in I_r$ for some r implies that $N_i = N_j$. Then each I_r gives rise to $G_r^c = T^c_{\bigoplus_{i\in I_r} A_i}$.

To find the equivalence classes we consider $X = \{1,\ldots,k\}$ as a discrete manifold and consider $F(X)\subseteq T_A(X) = X^k$. Choose an element $i = (i_1,\ldots,i_k)\in F(X)$ with maximal number of distinct members. The classes I_r will then be the non-empty sets of the form $\{s : i_s = j\}$ for $1\le j\le k$. Let n be the number of different classes.

Now let D be a discrete manifold. Then the claim says that

$$F(D) = \{(d_1,\ldots,d_k)\in D^k : s,t\in I_r \text{ implies } d_s = d_t \text{ for all } r\}.$$

Suppose not, then there exist $d = (d_1,\ldots,d_k)\in F(D)$ and r,s,t with $s,t\in I_r$ and $d_s\neq d_t$. So among the pairs $(i_1,d_1),\ldots,(i_k,d_k)$ there are at least $n+1$ distinct ones. Let $f : X\times D\to X$ be any function mapping those pairs to $1,\ldots,n+1$. Then $F(f)(i,d) = (f(i_1,d_1),\ldots,f(i_k,d_k))\in F(X)$ has at least $n+1$ distinct members, contradicting the maximality of n. This proves the claim for D and also $F(\mathbb{R}^m\times D) = A^m\times F(D)$ is of the right form since the connected components of $\mathbb{R}^m\times D$ correspond to the points of D.

Now let M be any manifold, let $p : M\to\pi_0(M)$ be the projection of M onto the (discrete) set of its connected components. For $a\in F(M)$ the value $F(p)(a)\in F(\pi_0(M))$ just classifies the connected component of $P_k(M)$ over which a lies, and this component of $P_k(M)$ must be of the right form. Let $x_1,\ldots,x_k\in M$ such $s,t\in I_r$ implies that x_s and x_t are in the same connected component M_r, say, for all r. The proof will be finished if we can show that the fiber $\pi^{-1}(x_1,\ldots,x_k)\subset T_A(M)$ is contained in $F(M)\subset T_A(M)$. Let $m = \dim M$ (or the maximum of $\dim M_i$ for $1\le i\le n$ if M is not a pure manifold) and let

$N = \mathbb{R}^m \times \{1,\dots,n\}$. We choose $y_1,\dots,y_k \in N$ and a smooth mapping $g : N \to M$ with $g(y_i) = x_i$ which is a diffeomorphism onto an open neighborhood of the x_i (a submersion for non pure M). Then clearly $T_A(g)(\pi^{-1}(y_1,\dots,y_k)) = \pi^{-1}(x_1,\dots,x_k)$, and from the last step of the proof we know that $F(N)$ contains $\pi^{-1}(y_1,\dots,y_k)$. So the result follows. □

By theorem 36.10 we know the minimal data to reconstruct the action of F on connected manifolds. For a not connected manifold M we first consider the surjective mapping $M \to \pi_0(M)$ onto the space of connected components of M. Since $\pi_0(M) \in \mathcal{M}f$, the functor F acts on this discrete set. Since F is weakly local and maps points to points, $F(\pi_0(M))$ is again discrete. This gives us a product preserving functor F_0 on the category of countable discrete sets.

If conversely we are given a product preserving functor F_0 on the category of countable discrete sets, a formally real finite dimensional algebra A consisting of k Weil parts, and a product preserving functor $G : con\mathcal{M}f \to$ groups with $\pi_1 \subseteq G \subseteq \pi_1^k$, then clearly one can construct a unique product preserving weakly local functor $F : \mathcal{M}f \to \mathcal{M}f$ fitting these data.

37. Examples and applications

37.1. The tangent bundle functor. The tangent mappings of the algebra structural mappings of $\mathbb{R}$ are given by

$$\begin{aligned}
T\mathbb{R} &= \mathbb{R}^2, \\
T(+)(a,a')(b,b') &= (a+b, a'+b'), \\
T(m)(a,a')(b,b') &= (ab, ab' + a'b), \\
T(m_\lambda)(a,a') &= (\lambda a, \lambda a').
\end{aligned}$$

So the Weil algebra $T\mathbb{R} = \mathrm{Al}(T) =: \mathbb{D}$ is the algebra generated by 1 and δ with $\delta^2 = 0$. It is sometimes called the algebra of *dual numbers* or also of *Study numbers*. It is also the truncated polynomial algebra of order 1 on $\mathbb{R}$. We will write $(a + a'\delta)(b + b'\delta) = ab + (ab' + a'b)\delta$ for the multiplication in $T\mathbb{R}$.

By 35.17 we can now determine all natural transformations over the category $\mathcal{M}f$ between the following functors.

(1) The natural transformations $T \to T$ consist of all fiber scalar multiplications m_λ for $\lambda \in \mathbb{R}$, which act on $T\mathbb{R}$ by $m_\lambda(1) = 1$ and $m_\lambda(\delta) = \lambda.\delta$.
(2) The projection $\pi : T \to \mathrm{Id}_{\mathcal{M}f}$ is the only natural transformation.

37.2. Lemma. *Let $F : \mathcal{M}f \to \mathcal{M}f$ be a multiplicative functor, which is also a natural vector bundle over $\mathrm{Id}_{\mathcal{M}f}$ in the sense of 6.14, then $F(M) = V \otimes TM$ for a finite dimensional vector space V with fiberwise tensor product. Moreover for the space of natural transformations between two such functors we have $Nat(V \otimes T, W \otimes T) = L(V,W)$.*

Proof. A natural vector bundle is local, so by theorem 36.1 it coincides with T_A, where A is its Weil algebra. But by theorem 35.13.(1) T_A is a natural vector bundle if and only if the nilideal of $A = F(\mathbb{R})$ is nilpotent of order 2, so $A = F(\mathbb{R}) = \mathbb{R} \cdot 1 \oplus V$, where the multiplication on V is 0. Then by construction 35.11 we have $F(M) = V \otimes TM$. Finally by 35.17.(2) we have $Nat(V \otimes T, W \otimes T) = \operatorname{Hom}(\mathbb{R} \cdot 1 \oplus V, \mathbb{R} \cdot 1 \oplus W) \cong L(V, W)$. □

37.3. The most important natural transformations. Let F, F_1, and F_2 be multiplicative bundle functors (Weil functors by theorem 36.1) with Weil algebras $A = \mathbb{R} \oplus N$, $A_1 = \mathbb{R} \oplus N_1$, and $A_2 = \mathbb{R} \oplus N_2$ where the N's denote the maximal nilpotent ideals. We will denote by $N(F)$ the nilpotent ideal in the Weil algebra of a general functor F. By 36.13 we have $\operatorname{Al}(F_2 \circ F_1) = A_1 \otimes A_2$. Using this and 35.17 we define the following natural transformations:

(1) The projections $\pi_1 : F_1 \to \operatorname{Id}$, $\pi_2 : F_2 \to \operatorname{Id}$ induced by $(\lambda.1 + n) \mapsto \lambda \in \mathbb{R}$. In general we will write $\pi_F : F \to \operatorname{Id}$. Thus we have also $F_2\pi_1 : F_2 \circ F_1 \to F_2$ and $\pi_2 F_1 : F_2 \circ F_1 \to F_1$.
(2) The zero sections $0_1 : \operatorname{Id} \to F_1$ and $0_2 : \operatorname{Id} \to F_2$ induced by $\mathbb{R} \to A_1$, $\lambda \mapsto \lambda.1$. Then we have $F_2 0_1 : F_2 \to F_2 \circ F_1$ and $0_2 F_1 : F_1 \to F_2 \circ F_1$.
(3) The isomorphism $A_1 \otimes A_2 \cong A_2 \otimes A_1$, given by $a_1 \otimes a_2 \mapsto a_2 \otimes a_1$ induces the *canonical flip mapping* $\kappa_{F_1,F_2} = \kappa : F_2 \circ F_1 \to F_1 \circ F_2$. We have $\kappa_{F_1,F_2} = \kappa^{-1}_{F_2,F_1}$.
(4) The multiplication m in A is a homomorphism $A \otimes A \to A$ which induces a natural transformation $\mu = \mu_F : F \circ F \to F$.
(5) Clearly the Weil algebra of the product $F_1 \times_{\operatorname{Id}} F_2$ in the category of bundle functors is given by $\mathbb{R}.1 \oplus N_1 \oplus N_2$. We consider the two natural transformations

$$(\pi_2 F_1, F_2 \pi_1), 0_{F_1 \times_{\operatorname{Id}} F_2} \circ \pi_{F_2 \circ F_1} : F_2 \circ F_1 \to (F_1 \times_{\operatorname{Id}} F_2).$$

The equalizer of these two transformations will be denoted by $vl : F_2 * F_1 \to F_2 \circ F_1$ and will be called the *vertical lift*. At the level of Weil algebras one checks that the Weil algebra of $F_2 * F_1$ is given by $\mathbb{R}.1 \oplus (N_1 \otimes N_2)$.
(6) The canonical flip κ factors to a natural transformation $\kappa_{F_2 * F_1} : F_2 * F_1 \to F_1 * F_2$ with $vl \circ \kappa_{F_2 * F_1} = \kappa_{F_2,F_1} \circ vl$.
(7) The multiplication μ induces a natural transformation $\mu \circ vl : F * F \to F$.

It is clear that κ expresses the symmetry of higher derivatives. We will see that the vertical lift vl expresses linearity of differentiation.

The reader is advised to work out the Weil algebra side of all these natural transformations.

37.4. The second tangent bundle. In the setting of 35.5 we let $F_1 = F_2 = T$ be the tangent bundle functor, and we let $T^2 = T \circ T$ be the second tangent bundle. Its Weil algebra is $\mathbb{D}_2 := \operatorname{Al}(T^2) = \mathbb{D} \otimes \mathbb{D} = \mathbb{R}^4$ with generators 1, δ_1, and δ_2 and with relations $\delta_1^2 = \delta_2^2 = 0$. Then $(1, \delta_1; \delta_2, \delta_1\delta_2)$ is the standard basis of $\mathbb{R}^4 = T^2\mathbb{R}$ in the usual description, which we also used in 6.12. From the list of natural transformations in 37.1 we get $\pi T : (\delta_1, \delta_2) \mapsto (\delta, 0)$, $T\pi : (\delta_1, \delta_2) \mapsto (0, \delta)$, and $\mu = + \circ (\pi T, T\pi) : T^2 \to T, (\delta_1, \delta_2) \mapsto (\delta, \delta)$. Then we have $T * T = T$, since $N(T) \otimes N(T) = N(T)$, and the natural transformations from 37.3 have the following form:

$$\kappa : T^2 \to T^2,$$
$$\kappa(a1 + x_1\delta_1 + x_2\delta_2 + x_3\delta_1\delta_2) = a1 + x_2\delta_1 + x_1\delta_2 + x_3\delta_1\delta_2.$$
$$vl : T \to T^2, \quad vl(a1 + x\delta) = a1 + x\delta_1\delta_2.$$
$$m_\lambda T : T^2 \to T^2,$$
$$m_\lambda T(a1 + x_1\delta_1 + x_2\delta_2 + x_3\delta_1\delta_2) = a1 + x_1\delta_1 + \lambda x_2\delta_2 + \lambda x_3\delta_1\delta_2.$$
$$Tm_\lambda : T^2 \to T^2,$$
$$Tm_\lambda(a1 + x_1\delta_1 + x_2\delta_2 + x_3\delta_1\delta_2) = a1 + \lambda x_1\delta_1 + x_2\delta_2 + \lambda x_3\delta_1\delta_2.$$
$$(+T) : T^2 \times_T T^2 \to T^2,$$
$$(+T)((a1 + x_1\delta_1 + x_2\delta_2 + x_3\delta_1\delta_2), (a1 + x_1\delta_1 + y_2\delta_2 + y_3\delta_1\delta_2)) =$$
$$= a1 + x_1\delta_1 + (x_2 + y_2)\delta_2 + (x_3 + y_3)\delta_1\delta_2.$$
$$(T+)((a1 + x_1\delta_1 + x_2\delta_2 + x_3\delta_1\delta_2), (a1 + y_1\delta_1 + x_2\delta_2 + y_3\delta_1\delta_2)) =$$
$$= a1 + (x_1 + y_1)\delta_1 + x_2\delta_2 + (x_3 + y_3)\delta_1\delta_2.$$

The space of all natural transformations $\operatorname{Nat}(T, T^2) \cong \operatorname{Hom}(\mathbb{D}, \mathbb{D}_2)$ turns out to be the real algebraic variety $\mathbb{R}^2 \cup_{\mathbb{R}} \mathbb{R}^2$ consisting of all homomorphisms $\delta \mapsto x_1\delta_1 + x_2\delta_2 + x_3\delta_1\delta_2$ with $x_1x_2 = 0$, since $\delta^2 = 0$. The homomorphism $\delta \mapsto x\delta_1 + y\delta_1\delta_2$ corresponds to the natural transformation $(+T) \circ (vl \circ m_y, 0T \circ m_x)$, and the homomorphism $\delta \mapsto x\delta_2 + y\delta_1\delta_2$ corresponds to $(T+) \circ (vl \circ m_y, T0 \circ m_x)$. So any element in $\operatorname{Nat}(T, T^2)$ can be expressed in terms of the natural transformations $\{0T, T0, (T+), (+T), T\pi, \pi T, vl, m_\lambda \text{ for } \lambda \in \mathbb{R}\}$.

Similarly $\operatorname{Nat}(T^2, T^2) \cong \operatorname{Hom}(\mathbb{D}_2, \mathbb{D}_2)$ turns out to be the real algebraic variety $(\mathbb{R}^2 \cup_{\mathbb{R}} \mathbb{R}^2) \times (\mathbb{R}^2 \cup_{\mathbb{R}} \mathbb{R}^2)$ consisting of all

$$\begin{pmatrix} \delta_1 \\ \delta_2 \end{pmatrix} \mapsto \begin{pmatrix} x_1\delta_1 + x_2\delta_2 + x_3\delta_1\delta_2 \\ y_1\delta_1 + y_2\delta_2 + y_3\delta_1\delta_2 \end{pmatrix}$$

with $x_1x_2 = y_1y_2 = 0$. Again any element of $\operatorname{Nat}(T^2, T^2)$ can be written in terms of $\{0T, T0, (T+), (+T), T\pi, \pi T, \kappa, m_\lambda T, Tm_\lambda \text{ for } \lambda \in \mathbb{R}\}$. If for example $x_2 = y_1 = 0$ then the corresponding transformation is

$$(+T) \circ (m_{y_2}T \circ Tm_{x_1}, (T+) \circ (vl \circ + \circ (m_{x_3} \circ \pi T, m_{y_3} \circ T\pi), 0T \circ m_{x_1} \circ \pi T)).$$

Note also the relations $T\pi \circ \kappa = \pi T$, $\kappa \circ (T+) = (+T) \circ (\kappa \times \kappa)$, $\kappa \circ vl = vl$, $\kappa \circ Tm_\lambda = m_\lambda T$; so κ interchanges the two vector bundle structures on $T^2 \to T$, namely $((+T), m_\lambda T, \pi T)$ and $((T+), Tm_\lambda, T\pi)$, and $vl : T \to T^2$ is linear for both of them. The reader is advised now to have again a look at 6.12.

37.5. In the situation of 37.3 we let now $F_1 = F$ be a general Weil functor and $F_2 = T$. So we consider $T \circ F$ which is isomorphic to $F \circ T$ via $\kappa_{F,T}$. In general we have $(F_1 \times_{\mathrm{Id}} F_2) * F = F_1 * F \times_{\mathrm{Id}} F_2 * F$, so $+ : T \times_{\mathrm{Id}} T \to T$ induces a fiber addition $(+ * F) : T * F \times_{\mathrm{Id}} T * F \to T * F$, and $m_\lambda * F : T * F \to T * F$ is a fiber scalar multiplication. So $T * F$ is a vector bundle functor on the category $\mathcal{M}f$ which can be described in terms of lemma 37.2 as follows.

Lemma. *In the notation of lemma 37.2 we have $T * F \cong \bar{N} \otimes T$, where $\bar{N}$ is the underlying vector spaces of the nilradical $N(F)$ of F.*

Proof. The Weil algebra of $T * F$ is $\mathbb{R}.1 \oplus (N(F) \otimes N(T))$ by 37.3.(5). We have $N(F) \otimes N(T) = N(F) \otimes \mathbb{R}.\delta = \bar{N}$ as vector space, and the multiplication on $N(F) \otimes N(T)$ is zero. □

37.6. Sections and expansions. For a Weil functor F with Weil algebra $A = \mathbb{R}.1 \oplus N$ and for a manifold M we denote by $\mathfrak{X}_F(M)$ the space of all smooth sections of $\pi_{F,M} : F(M) \to M$. Note that this space is infinite dimensional in general. Recall from theorem 35.14 that

$$F(M) = T_A(M) \xleftarrow{\eta_{M,A}} \operatorname{Hom}(C^\infty(M,\mathbb{R}), A)$$

is an isomorphism. For $f \in C^\infty(M,\mathbb{R})$ we can decompose $F(f) = T_A(f) : F(M) \to F(\mathbb{R}) = A = \mathbb{R}.1 \oplus N$ into

$$F(f) = T_A(f) = (f \circ \pi) \oplus N(f),$$
$$N(f) : F(M) \to N.$$

Lemma.

(1) *Each $X_x \in F(M)_x = \pi^{-1}(x)$ for $x \in M$ defines an $\mathbb{R}$-linear mapping*

$$D_{X_x} : C^\infty(M,\mathbb{R}) \to N,$$
$$D_{X_x}(f) := N(f)(X_x) = F(f)(X_x) - f(x).1,$$

which satisfies

$$D_{X_x}(f.g) = D_{X_x}(f).g(x) + f(x).D_{X_x}(g) + D_{X_x}(f).D_{X_x}(g).$$

We call this the expansion property at $x \in M$.

(2) *Each $\mathbb{R}$-linear mapping $\xi : C^\infty(M,\mathbb{R}) \to N$ which satisfies the expansion property at $x \in M$ is of the form $\xi = D_{X_x}$ for a unique $X_x \in F(M)_x$.*

(3) *The $\mathbb{R}$-linear mappings $\xi : C^\infty(M,\mathbb{R}) \to C^\infty(M,N) = N \otimes C^\infty(M,\mathbb{R})$ which have the expansion property*

$$\text{(a)} \qquad \xi(f.g) = \xi(f).g + f.\xi(g) + \xi(f).\xi(g), \quad f,g \in C^\infty(M,\mathbb{R}),$$

are exactly those induced (via 1 and 2) by the smooth sections of $\pi : F(M) \to M$.

Linear mappings satisfying the expansion property 1 will be called *expansions*: if N is generated by δ with $\delta^{k+1} = 0$, so that $F(M) = J_0^k(\mathbb{R}, M)$, then these are parametrized Taylor expansions of f to order k (applied to a k-jet of a curve through each point). For $X \in \mathfrak{X}_F(M)$ we will write $D_X : C^\infty(M,\mathbb{R}) \to$

$C^\infty(M,N) = N \otimes C^\infty(M,\mathbb{R})$ for the expansion induced by X. Note the defining equation

$$\text{(b)} \qquad F(f) \circ X = f.1 + D_X(f) = (f.1, D_X(f)) \text{ or}$$
$$f(x).1 + D_X(f)(x) = F(f)(X(x)) = \eta^{-1}_{M,A}(X(x))(f).$$

Proof. (1) and (2). For $\varphi \in \operatorname{Hom}(C^\infty(M,\mathbb{R}), A) = F(M)$ we consider the foot point $\pi(\eta_{M,A}(\varphi)) = \eta_{M,\mathbb{R}}(\pi(\varphi)) = x \in M$ and $\eta_{M,A}(\varphi) = X_x \in F(M)_x$. Then we have $\varphi(f) = T_A(f)(X_x)$ and the expansion property for D_{X_x} is equivalent to $\varphi(f.g) = \varphi(f).\varphi(g)$.

(3) For each $x \in M$ the mapping $f \mapsto \xi(f)(x) \in N$ is of the form $D_{X(x)}$ for a unique $X(x) \in F(M)_x$ by 1 and 2, and clearly $X : M \to F(M)$ is smooth. □

37.7. Theorem. *Let F be a Weil functor with Weil algebra $A = \mathbb{R}.1 \oplus N$. Using the natural transformations from 37.3 we have:*

1. *$\mathfrak{X}_F(M)$ is a group with multiplication $X \diamond Y = \mu_F \circ F(Y) \circ X$ and identity 0_F.*
2. *$\mathfrak{X}_{T*F}(M)$ is a Lie algebra with bracket induced from the usual Lie bracket on $\mathfrak{X}_T(M)$ and the multiplication $m : N \times N \to N$ by $[a\otimes X, b\otimes Y]_{T*F} = a.b \otimes [X,Y]$.*
3. *There is a bijective mapping $\exp : \mathfrak{X}_{T*F}(M) \to \mathfrak{X}_F(M)$ which expresses the multiplication $\diamond$ by the Baker-Campbell-Hausdorff formula.*
4. *The multiplication $\diamond$, the Lie bracket $[\ ,\ ]_{T*F}$, and exp are natural in F (with respect to natural operators) and M (with respect to local diffeomorphisms).*

Remark. If $F = T$, then $\mathfrak{X}_T(M)$ is the space of all vector fields on M, the multiplication is $X \diamond Y = X + Y$, and the bracket is $[X,Y]_{T*T} = 0$, and exp is the identity. So the multiplication in (1), which is commutative only if F is a natural vector bundle, generalizes the linear structure on $\mathfrak{X}(M)$.

37.8. For the proof of theorem 37.7 we need some preparation. If $a \in N$ and $X \in \mathfrak{X}(M)$ is a smooth vector field on M, then by lemma 37.5 we have $a \otimes X \in \mathfrak{X}_{T*F}(M)$ and for $f \in C^\infty(M,\mathbb{R})$ we use $Tf(X) = f.1 + df(X)$ to get

$$\begin{aligned}(T * F)(f)(a \otimes X) &= (\mathrm{Id}_N \otimes Tf)(a \otimes X)\\ &= f.1 + a.df(X) = f.1 + a.X(f)\\ &= f.1 + D_{a\otimes X}(f) \quad \text{by 37.6.(b). Thus}\end{aligned}$$

$$\text{(a)} \qquad D^{T*F}_{a\otimes X}(f) = D_{a\otimes X}(f) = a.X(f) = a.df(X).$$

So again by 37.5 we see that $\mathfrak{X}_{T*F}(M)$ is isomorphic to the space of all $\mathbb{R}$-linear mappings $\xi : C^\infty(M,\mathbb{R}) \to N \otimes C^\infty(M,\mathbb{R})$ satisfying

$$\xi(f.g) = \xi(f).g + f.\xi(g).$$

These mappings are called ***derivations***.

Now we denote $\mathcal{L} := L_{\mathbb{R}}(C^\infty(M,\mathbb{R}), N \otimes C^\infty(M,\mathbb{R}))$ for short, and for ξ, $\eta \in \mathcal{L}$ we define

$$\text{(b)}\quad \begin{aligned} \xi \bullet \eta := (m \otimes \operatorname{Id}_{C^\infty(M,\mathbb{R})}) \circ (\operatorname{Id}_N \otimes \xi) \circ \eta : C^\infty(M,\mathbb{R}) \to \\ \to N \otimes C^\infty(M,\mathbb{R}) \to N \otimes N \otimes C^\infty(M,\mathbb{R}) \to N \otimes C^\infty(M,\mathbb{R}), \end{aligned}$$

where $m : N \otimes N \to N$ is the (nilpotent) multiplication on N. Note that $D^F : \mathfrak{X}_F(M) \to \mathcal{L}$ and $D^{T*F} : \mathfrak{X}_{T*F}(M) \to \mathcal{L}$ are injective linear mappings.

37.9. Lemma. *1. $\mathcal{L}$ is a real associative nilpotent algebra without unit under the multiplication $\bullet$, and it is commutative if and only if $m = 0 : N \times N \to N$.*

(1) For X, $Y \in \mathfrak{X}_F(M)$ we have $D^F_{X \diamond Y} = D^F_X \bullet D^F_Y + D^F_X + D^F_Y$.
(2) For X, $Y \in \mathfrak{X}_{T*F}(M)$ we have $D^F_{[X,Y]_{T*F}} = D^F_X \bullet D^F_Y - D^F_Y \bullet D^F_X$.
(3) For $\xi \in \mathcal{L}$ define

$$\overline{\exp}(\xi) := \sum_{i=1}^{\infty} \frac{1}{i!} \xi^{\bullet i}$$
$$\overline{\log}(\xi) := \sum_{i=1}^{\infty} \frac{(-1)^{i-1}}{i} \xi^{\bullet i}.$$

Then $\overline{\exp}, \overline{\log} : \mathcal{L} \to \mathcal{L}$ are bijective and inverse to each other. $\overline{\exp}(\xi)$ is an expansion if and only if ξ is a derivation.

Note that $i = 0$ lacks in the definitions of $\overline{\exp}$ and $\overline{\log}$, since $\mathcal{L}$ has no unit.

Proof. (1) We use that m is associative in the following computation.

$$\begin{aligned} \xi \bullet (\eta \bullet \zeta) &= (m \otimes \operatorname{Id}_{C^\infty(M,\mathbb{R})}) \circ (\operatorname{Id}_N \otimes \xi) \circ (\eta \bullet \zeta) \\ &= (m \otimes \operatorname{Id}) \circ (\operatorname{Id}_N \otimes \xi) \circ (m \otimes \operatorname{Id}) \circ (\operatorname{Id}_N \otimes \eta) \circ \zeta \\ &= (m \otimes \operatorname{Id}) \circ (m \otimes \operatorname{Id}_N \otimes \operatorname{Id}) \circ (\operatorname{Id}_{N\otimes N} \otimes \xi) \circ (\operatorname{Id}_N \otimes \eta) \circ \zeta \\ &= (m \otimes \operatorname{Id}) \circ (\operatorname{Id}_N \otimes m \otimes \operatorname{Id}) \circ (\operatorname{Id}_{N\otimes N} \otimes \xi) \circ (\operatorname{Id}_N \otimes \eta) \circ \zeta \\ &= (m \otimes \operatorname{Id}) \circ (\operatorname{Id}_N \otimes ((m \otimes \operatorname{Id}) \circ (\operatorname{Id}_N \otimes \xi) \circ \eta) \circ \zeta \\ &= (\xi \bullet \eta) \bullet \zeta. \end{aligned}$$

So $\bullet$ is associative, and it is obviously $\mathbb{R}$-bilinear. The order of nilpotence equals that of N.

(2) Recall from 36.13 and 37.3 that

$$\begin{aligned} F(F(\mathbb{R})) = A \otimes A = ((\mathbb{R}.1 \otimes \mathbb{R}.1) \oplus (\mathbb{R}.1 \otimes N)) \oplus ((N \otimes \mathbb{R}.1) \oplus (N \otimes N)) \\ \cong A \oplus F(N) \cong F(\mathbb{R}.1) \times F(N) \cong F(\mathbb{R}.1 \times N) = F(A). \end{aligned}$$

We will use this decomposition in exactly this order in the following computation.

$$\begin{aligned} f.1 + D^F_{X\diamond Y}(f) &= F(f)\circ(X\diamond Y) \qquad \text{by 37.6.(b)}\\ &= F(f)\circ\mu_{F,M}\circ F(Y)\circ X \qquad \text{by 37.7(1)}\\ &= \mu_{F,\mathbb{R}}\circ F(F(f))\circ F(Y)\circ X \qquad \text{since } \mu \text{ is natural}\\ &= m\circ F(F(f)\circ Y)\circ X\\ &= m\circ F(f.1, D^F_Y(f))\circ X \qquad \text{by 37.6.(b)}\\ &= m\circ((1\otimes F(f)\circ X).\oplus(F(D^F_Y(f))\circ X))\\ &= m\circ\Big(1\otimes(f.1+D^F_X(f)) + \big(D^F_Y(f)\otimes 1 + (\mathrm{Id}_N\otimes D^F_X)(D^F_Y(f))\big)\Big)\\ &= f.1 + D^F_X(f) + D^F_Y(f) + (D^F_X\bullet D^F_Y)(f). \end{aligned}$$

(3) For vector fields $X, Y\in\mathfrak{X}(M)$ on M and $a, b\in N$ we have

$$\begin{aligned} D^{T*F}_{[a\otimes X, b\otimes Y]_{T*F}}(f) &= D_{a.b\otimes[X,Y]}(f)\\ &= a.b.[X,Y](f) \qquad \text{by 37.8.(a)}\\ &= a.b.(X(Y(f)) - Y(X(f)))\\ &= (m\otimes\mathrm{Id}_{C^\infty(M,\mathbb{R})})\circ(\mathrm{Id}_N\otimes D^{T*F}_{a\otimes X})\circ D^{T*F}_{b\otimes Y}(f) - \dots\\ &= (D^{T*F}_{a\otimes X}\bullet D^{T*F}_{b\otimes Y} - D^{T*F}_{b\otimes Y}\bullet D^{T*F}_{a\otimes X})(f). \end{aligned}$$

(4) After adjoining a unit to $\mathcal{L}$ we see that $\overline{\exp}(\xi) = e^\xi - 1$ and $\overline{\log}(\xi) = \log(1+\xi)$. So $\overline{\exp}$ and $\overline{\log}$ are inverse to each other in the ring of formal power series of one variable. The elements 1 and ξ generate a quotient of the power series ring in $\mathbb{R}.1\oplus\mathcal{L}$, and the formal expressions of $\overline{\exp}$ and $\overline{\log}$ commute with taking quotients. So $\overline{\exp} = \overline{\log}^{-1}$. The second assertion follows from a direct formal computation, or also from 37.10 below. □

37.10. We consider now the $\mathbb{R}$-linear mapping C of $\mathcal{L}$ in the ring of all $\mathbb{R}$-linear endomorphisms of the algebra $A\otimes C^\infty(M,\mathbb{R})$, given by

$$\begin{aligned} C_\xi &:= m\circ(\mathrm{Id}_A\otimes\xi) : A\otimes C^\infty(M,\mathbb{R})\to\\ &\to A\otimes N\otimes C^\infty(M,\mathbb{R})\subset A\otimes A\otimes C^\infty(M,\mathbb{R})\to A\otimes C^\infty(M,\mathbb{R}), \end{aligned}$$

where $m : A\otimes A\to A$ is the multiplication. We have $C_\xi(a\otimes f) = a.\xi(f)$.

Lemma.

(1) *$C_{\xi\bullet\eta} = C_\xi\circ C_\eta$, so C is an algebra homomorphism.*
(2) *$\xi\in\mathcal{L}$ is an expansion if and only if $\mathrm{Id}+C_\xi$ is an automorphism of the commutative algebra $A\otimes C^\infty(M,\mathbb{R})$.*
(3) *$\xi\in\mathcal{L}$ is a derivation if and only if C_ξ is a derivation of the algebra $A\otimes C^\infty(M,\mathbb{R})$.*

37.11. Proof of theorem 37.7. 1. It is easily checked that $\mathcal{L}$ is a group with multiplication $\xi \diamond \eta = \xi \bullet \eta + \xi + \eta$, with unit 0, and with inverse $\xi^{-1} = \sum_{i=1}^{\infty}(-\xi)^{\bullet i}$ (recall that $\bullet$ is a nilpotent multiplication). As noted already at the of 37.8 the mapping $D^F : \mathfrak{X}_F(M) \to \mathcal{L}$ is an isomorphism onto the subgroup of expansions, because $\mathrm{Id} + C \circ D^F : \mathfrak{X}_F(M) \to \mathcal{L} \to \mathrm{End}(A \otimes C^\infty(M, \mathbb{R}))$ is an isomorphism onto the subgroup of automorphisms.

2. $C \circ D^{T*F} : \mathfrak{X}_{T*F}(M) \to \mathrm{End}(A \otimes C^\infty(M, \mathbb{R}))$ is a Lie algebra isomorphism onto the sub Lie algebra of $\mathrm{End}(A \otimes C^\infty(M, \mathbb{R}))$ of derivations.

3. Define $\exp : \mathfrak{X}_{T*F}(M) \to \mathfrak{X}_F(M)$ by $D^F_{\exp(X)} = \overline{\exp}(D^F_X)$. The Baker-Campbell-Hausdorff formula holds for

$$\exp : \mathrm{Der}(A \otimes C^\infty(M, \mathbb{R})) \to \mathrm{Aut}(A \otimes C^\infty(M, \mathbb{R})),$$

since the Lie algebra of derivations is nilpotent.

4. This is obvious since we used only natural constructions. □

37.12. The Lie bracket. We come back to the tangent bundle functor T and its iterates. For T the structures described in theorem 37.7 give just the addition of vector fields. In fact we have $X \diamond Y = X + Y$, and $[X, Y]_{T*T} = 0$.

But we may consider other structures here. We have by 37.1 $\mathrm{Al}(T) = \mathbb{D} = \mathbb{R}.1 \oplus \mathbb{R}.\delta$ for $\delta^2 = 0$. So $N \cong \mathbb{R}$ with the nilpotent multiplication 0, but we still have the usual multiplication, now called m, on $\mathbb{R}$.

For X, $Y \in \mathfrak{X}_T(M)$ we have $D_X \in \mathcal{L} = L_{\mathbb{R}}(C^\infty(M, \mathbb{R}), C^\infty(M, \mathbb{R}))$, a derivation given by $f.1 + D_X(f).\delta = Tf \circ X$, see 37.6.(b) — we changed slightly the notation. So $D_X(f) = X(f) = df(X)$ in the usual sense. The space $\mathcal{L}$ has one more structure now, *composition*, which is determined by specifying a generator δ of the nilpotent ideal of $\mathrm{Al}(T)$. The usual *Lie bracket of vector fields* is now given by $D_{[X,Y]} := D_X \circ D_Y - D_Y \circ D_X$.

37.13. Lemma. *In the setting of 37.12 we have*

$$(-T) \circ (TY \circ X, \kappa_T \circ TX \circ Y) = (T+) \circ (vl \circ [X, Y], 0T \circ Y)$$

in terms of the natural transformations descibed in 37.4

This is a variant of lemma 6.13 and 6.19.(4). The following proof appears to be more complicated then the earlier ones, but it demonstrates the use of natural transformations, and we write out carefully the unusual notation.

Proof. For $f \in C^\infty(M, \mathbb{R})$ and X, $Y \in \mathfrak{X}_T(M)$ we compute as follows using repeatedly the defining equation for D_X from 37.12:

$$\begin{aligned} T^2 f \circ TY \circ X &= T(Tf \circ Y) \circ X = T(f.1 \oplus D_Y(f).\delta_1) \circ X \\ &= (Tf \circ X).1 \oplus (T(D_Y(f)) \circ X).\delta_1, \text{ since } T \text{ preserves products,} \\ &= f.1 + D_X(f).\delta_2 + (D_Y(f).1 + D_X D_Y(f).\delta_2).\delta_1 \\ &= f.1 + D_Y(f).\delta_1 + D_X(f).\delta_2 + D_X D_Y(f).\delta_1\delta_2. \end{aligned}$$

Now we use the natural transformation and their commutation rules from 37.4 to compute:

$$\begin{aligned}
T^2 f \circ (-T) \circ (TY \circ X, \kappa_T \circ TX \circ Y) = \\
&= (-T) \circ (T^2 f \circ TY \circ X, \kappa_T \circ T^2 f \circ TX \circ Y) \\
&= (-T) \circ \big(f.1 + D_Y(f).\delta_1 + D_X(f).\delta_2 + D_X D_Y(f).\delta_1\delta_2, \\
&\qquad \kappa_T(f.1 + D_X(f).\delta_1 + D_Y(f).\delta_2 + D_Y D_X(f).\delta_1\delta_2)\big) \\
&= (-T) \circ \big(f.1 + D_Y(f).\delta_1 + D_X(f).\delta_2 + D_X D_Y(f).\delta_1\delta_2, \\
&\qquad f.1 + D_Y(f).\delta_1 + D_X(f).\delta_2 + D_Y D_X(f).\delta_1\delta_2)\big) \\
&= f.1 + D_Y(f).\delta_1 + (D_X D_Y - D_Y D_X)(f).\delta_1\delta_2 \\
&= (T+) \circ (0T \circ (f.1 + D_Y(f).\delta), vl \circ (f.1 + D_{[X,Y]}(f).\delta)) \\
&= (T+) \circ (0T \circ Tf \circ Y, vl \circ Tf \circ [X,Y]) \\
&= (T+) \circ (T^2 f \circ 0T \circ Y, T^2 f \circ vl \circ [X,Y]) \\
&= T^2 f \circ (T+) \circ (0T \circ Y, vl \circ [X,Y]). \quad \square
\end{aligned}$$

37.14. Linear connections and their curvatures. Our next application will be to derive a global formula for the curvature of a linear connection on a vector bundle which involves the second tangent bundle of the vector bundle. So let (E, p, M) be a vector bundle. Recall from 11.10 and 11.12 that a linear connection on the vector bundle E can be described by specifying its *connector* $K : TE \to E$. By lemma 11.10 and by 11.11 any smooth mapping $K : TE \to E$ which is a (fiber linear) homomorphism for both vector bundle structure on TE, and which is a left inverse to the vertical lift, $K \circ vl_E = pr_2 : E \times_M E \to TE \to E$, specifies a linear connection.

For any manifold N, smooth mapping $s : N \to E$, and vector field $X \in \mathfrak{X}(N)$ we have then the *covariant derivative* of s along X which is given by $\nabla_X s := K \circ Ts \circ X : N \to TN \to TE \to E$, see 11.12.

For vector fields $X, Y \in \mathfrak{X}(M)$ and a section $s \in C^\infty(E)$ the curvature R^E of the connection is given by $R^E(X,Y)s = ([\nabla_X, \nabla_Y] - \nabla_{[X,Y]})s$, see 11.12.

37.15. Theorem.

(1) *Let $K : TE \to E$ be the connector of a linear connection on a vector bundle (E, p, M). Then the curvature is given by*

$$R^E(X,Y)s = (K \circ TK \circ \kappa_E - K \circ TK) \circ T^2 s \circ TX \circ Y$$

for $X, Y \in \mathfrak{X}(M)$ and a section $s \in C^\infty(E)$.

(2) *If $s : N \to E$ is a section along $f := p \circ s : N \to M$ then we have for vector fields $X, Y \in \mathfrak{X}(N)$*

$$\begin{aligned}
\nabla_X \nabla_Y s - \nabla_Y \nabla_X s - \nabla_{[X,Y]} s = \\
&= (K \circ TK \circ \kappa_E - K \circ TK) \circ T^2 s \circ TX \circ Y = \\
&= R^E(Tf \circ X, Tf \circ Y)s.
\end{aligned}$$

(3) *Let $K : T^2M \to M$ be a linear connection on the tangent bundle. Then its torsion is given by*

$$\operatorname{Tor}(X,Y) = (K \circ \kappa_M - K) \circ TX \circ Y.$$

Proof. (1) Let first $m_t^E : E \to E$ denote the scalar multiplication. Then we have $\frac{\partial}{\partial t}\big|_0 m_t^E = vl_E$ where $vl_E : E \to TE$ is the vertical lift. We use then lemma 37.13 and the commutation relations from 37.4 and we get in turn:

$$\begin{aligned}
vl_E \circ K &= \tfrac{\partial}{\partial t}\big|_0 m_t^E \circ K = \tfrac{\partial}{\partial t}\big|_0 K \circ m_t^{TE} \\
&= TK \circ \tfrac{\partial}{\partial t}\big|_0 m_t^{TE} = TK \circ vl_{(TE,Tp,TM)}. \\
R(X,Y)s &= \nabla_X\nabla_Y s - \nabla_Y\nabla_X s - \nabla_{[X,Y]}s \\
&= K \circ T(K \circ Ts \circ Y) \circ X - K \circ T(K \circ Ts \circ X) \circ Y - K \circ Ts \circ [X,Y] \\
K \circ Ts \circ [X,Y] &= K \circ vl_E \circ K \circ Ts \circ [X,Y] \\
&= K \circ TK \circ vl_{TE} \circ Ts \circ [X,Y] \\
&= K \circ TK \circ T^2s \circ vl_{TM} \circ [X,Y] \\
&= K \circ TK \circ T^2s \circ ((TY \circ X - \kappa_M \circ TX \circ Y)\,(T-)\,0_{TM} \circ Y) \\
&= K \circ TK \circ T^2s \circ TY \circ X - K \circ TK \circ T^2s \circ \kappa_M \circ TX \circ Y - 0.
\end{aligned}$$

Now we sum up and use $T^2s \circ \kappa_M = \kappa_E \circ T^2s$ to get the result.

(2) The same proof as for (1) applies for the first equality, with some obvious changes. To see that it coincides with $R^E(Tf \circ X, Tf \circ Y)s$ it suffices to write out (1) and $(T^2s \circ TX \circ Y)(x) \in T^2E$ in canonical charts induced from vector bundle charts of E.

(3) We have in turn

$$\begin{aligned}
\operatorname{Tor}(X,Y) &= \nabla_X Y - \nabla_Y X - [X,Y] \\
&= K \circ TY \circ X - K \circ TX \circ Y - K \circ vl_{TM} \circ [X,Y] \\
K \circ vl_{TM} \circ [X,Y] &= K \circ ((TY \circ X - \kappa_M \circ TX \circ Y)\,(T-)\,0_{TM} \circ Y) \\
&= K \circ TY \circ X - K \circ \kappa_M \circ TX \circ Y - 0. \quad \square
\end{aligned}$$

37.16. Weil functors and Lie groups. We have seen in 10.17 that the tangent bundle TG of a Lie group G is again a Lie group, the semidirect product $\mathfrak{g} \ltimes G$ of G with its Lie algebra $\mathfrak{g}$.

Now let A be a Weil algebra and let T_A be its Weil functor. In the notation of 4.1 the manifold $T_A(G)$ is again a Lie group with multiplication $T_A(\mu)$ and inversion $T_A(\nu)$. By the properties 35.13 of the Weil functor T_A we have a surjective homomorphism $\pi_A : T_AG \to G$ of Lie groups. Following the analogy with the tangent bundle, for $a \in G$ we will denote its fiber over a by $(T_A)_aG \subset T_AG$, likewise for mappings. With this notation we have the following commutative diagram:

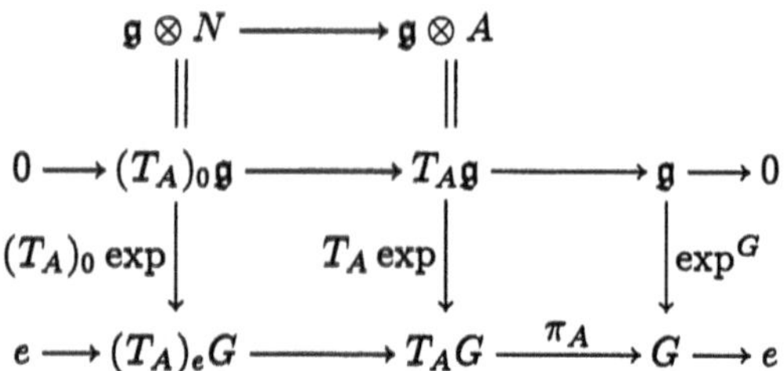

For a Lie group the structural mappings (multiplication, inversion, identity element, Lie bracket, exponential mapping, Baker-Campbell-Hausdorff formula, adjoint action) determine each other mutually. Thus their images under the Weil functor T_A are again the same structural mappings. But note that the canonical flip mappings have to be inserted like follows. So for example

$$\mathfrak{g} \otimes A \cong T_A\mathfrak{g} = T_A(T_eG) \xrightarrow{\kappa} T_e(T_AG)$$

is the Lie algebra of T_AG and the Lie bracket is just $T_A([\ \ ,\ \ ])$. Since the bracket is bilinear, the description of 35.11 implies that $[X \otimes a, Y \otimes b]_{T_A\mathfrak{g}} = [X,Y]_\mathfrak{g} \otimes ab$. Also $T_A \exp^G = \exp^{T_AG}$. Since $\exp^G$ is a diffeomorphism near 0 and since $(T_A)_0(\exp^G)$ depends only on the (invertible) jet of $\exp^G$ at 0, the mapping $(T_A)_0(\exp^G) : (T_A)_0\mathfrak{g} \to (T_A)_eG$ is a diffeomorphism. Since $(T_A)_0\mathfrak{g}$ is a nilpotent Lie algebra, the multiplication on $(T_A)_eG$ is globally given by the Baker-Campbell-Hausdorff formula. The natural transformation $0_G : G \to T_AG$ is a homomorphism which splits the bottom row of the diagram, so T_AG is the semidirect product $(T_A)_0\mathfrak{g} \ltimes G$ via the mapping $T_A\rho : (u,g) \mapsto T_A(\rho_g)(u)$.

Since we will need it later, let us add the following final remark: If $\omega^G : TG \to T_eG$ is the Maurer Cartan form of G (i.e. the left logarithmic derivative of Id_G) then

$$\kappa_0 \circ T_A\omega^G \circ \kappa : TT_AG \cong T_ATG \to T_AT_eG \cong T_eT_AG$$

is the Maurer Cartan form of T_AG.

Remarks

The material in section 35 is due to [Eck,86], [Luciano, 88] and [Kainz-Michor, 87], the original ideas are from [Weil, 51]. Section 36 is due to [Eck, 86] and [Kainz-Michor, 87], 36.7 and 36.8 are from [Kainz-Michor, 87], under stronger locality conditions also to [Eck, 86]. 36.14 is due to [Eck, 86]. The material in section 37 is from [Kainz-Michor, 87].

CHAPTER IX. BUNDLE FUNCTORS ON MANIFOLDS

The description of the product preserving bundle functors on $\mathcal{M}f$ in terms of Weil algebras reflects their general properties in a rather complete way. In the present chapter we use some other procedures to deduce the basic geometric properties of arbitrary bundle functors on $\mathcal{M}f$. Hence the basic subject of this theory is a bundle functor on $\mathcal{M}f$ that does not preserve products. Sometimes we also contrast certain properties of the product-preserving and non-product-preserving bundle functors on $\mathcal{M}f$. First we study the bundle functors with the so-called point property, i.e. the image of a one-point set is a one-point set. In particular, we deduce that their fibers are numerical spaces and that they preserve products if and only if the dimensions of their values behave well. Then we show that an arbitrary bundle functor on manifolds is, in a certain sense, a 'bundle' of functors with the point property. For an arbitrary vector bundle functor F on $\mathcal{M}f$ with the point property we also derive a canonical Lie group structure on the prolongation FG of a Lie group G.

Next we introduce the concept of a flow-natural transformation of a bundle functor F on manifolds. This is a natural transformation $FT \to TF$ with the property that for every vector field $X\colon M \to TM$ its functorial prolongation $FX\colon FM \to FTM$ is transformed into the flow prolongation $\mathcal{F}X\colon FM \to TFM$. We deduce that every bundle functor F on manifolds has a canonical flow-natural transformation, which is a natural equivalence if and only if F preserves products. Then we point out some special features of natural transformations from a Weil functor into an arbitrary bundle functor on $\mathcal{M}f$. This gives a rather effective method for their description. We also deduce that the homotheties are the only natural transformations of the r-th order tangent bundle $T^{(r)}$ into itself. This demonstrates that some properties of $T^{(r)}$ are quite different from those of Weil bundles, where such natural transformations are in bijection with a usually much larger set of all endomorphisms of the corresponding Weil algebras. In the last section we describe basic properties of the so-called star bundle functors, which reflect some constructions of contravariant character on $\mathcal{M}f$.

38. The point property

38.1. Examples. First we mention some examples of vector bundle functors which do not preserve products. In 37.2 we deduced that every product preserving vector bundle functor on $\mathcal{M}f$ is the fibered product of a finite number of copies of the tangent bundle T. In particular, every such functor is of order one. Hence all tensor powers $\otimes^p T$, $p > 1$, their sub bundles like $S^p T$, $\Lambda^p T$ and any combinations of them do not preserve products. This is also easily verified by counting dimensions. An important example of an r-th order vector bundle functor is the r-th tangent functor $T^{(r)}$ described in 12.14 and 41.8. Let us mention that another interesting example of an r-th order vector bundle functor, the bundle of sector r-forms, will be discussed in 48.4.

38.2. Proposition. *Every bundle functor $F\colon \mathcal{M}f \to \mathcal{M}f$ transforms embeddings into embeddings and immersions into immersions.*

Proof. According to 1.14, a smooth mapping $f\colon M \to N$ is an embedding if and only if there is an open neighborhood U of $f(M)$ in N and a smooth map $g\colon U \to M$ such that $g \circ f = \mathrm{id}_M$. Hence if f is an embedding, then $FU \subset FN$ is an open neighborhood of $Ff(FM)$ and $Fg \circ Ff = \mathrm{id}_{FM}$.

The locality of bundle functors now implies the assertion on immersions. However this can be also proved easily considering the canonical local form $i\colon \mathbb{R}^m \to \mathbb{R}^{m+n}$, $x \mapsto (x, 0)$, of immersions, cf. 2.6, and applying F to the composition of i and the projections $\mathrm{pr}_1 : \mathbb{R}^{m+n} \to \mathbb{R}^m$. □

38.3. The point property. Let us write pt for a one-point manifold. A bundle functor F on $\mathcal{M}f$ is said to have the *point property* if $F(\mathrm{pt}) = \mathrm{pt}$. Given such functor F let us consider the maps $i_x\colon \mathrm{pt} \to M$, $i_x(\mathrm{pt}) = x$, for all manifolds M and points $x \in M$. The regularity of bundle functors on $\mathcal{M}f$ proved in 20.7 implies that the maps $c_M\colon M \to FM$, $c_M(x) = Fi_x(\mathrm{pt})$ are smooth sections of $p_M\colon FM \to M$. By definition, $c_N \circ f = Ff \circ c_M$ for all smooth maps $f\colon M \to N$, so that we have found a natural transformation $c\colon \mathrm{Id}_{\mathcal{M}f} \to F$.

If $F = T_A$ for a Weil algebra A, this natural transformation corresponds to the algebra homomorphism $\mathrm{id}_{\mathbb{R}} \oplus 0\colon \mathbb{R} \to \mathbb{R} \oplus N = A$. The r-th order tangent functor has the point property, i.e. we have found a bundle functor which does not preserve the products in any dimension except dimension zero. The technique from example 22.2 yields easily bundle functors on $\mathcal{M}f$ which preserve products just in all dimensions less then any fixed $n \in \mathbb{N}$.

38.4. Lemma. *Let S be an m-dimensional manifold and $s \in S$ be a point. If there is a smoothly parameterized system h_t of maps, $t \in \mathbb{R}$, such that all h_t are diffeomorphisms except for $t = 0$, $h_0(S) = \{s\}$ and $h_1 = id_S$, then S is diffeomorphic to $\mathbb{R}^{\dim S}$.*

Proof. Let us recall that if $S = \cup_{k=0}^{\infty} S_k$ where S_k are open submanifolds diffeomorphic to $\mathbb{R}^m$ and $S_k \subset S_{k+1}$ for all k, then S is diffeomorphic to $\mathbb{R}^m$, see [Hirsch, 76, Chapter 1, Section 2]. So let us choose an increasing sequence of relatively compact open submanifolds $K_n \subset K_{n+1} \subset S$ with $S = \cup_{k=1}^{\infty} K_n$ and a relatively compact neighborhood U of s diffeomorphic to $\mathbb{R}^m$. Put $S_0 = U$. Since S_0 is relatively compact, there is an integer n_1 with $K_{n_1} \supset S_0$ and a $t_1 > 0$ with $h_{t_1}(K_{n_1}) \subset U$. Then we define $S_1 = (h_{t_1})^{-1}(U)$ so that we have $S_1 \supset K_{n_1} \supset S_0$ and S_1 is relatively compact and diffeomorphic to $\mathbb{R}^m$. Iterating this procedure, we construct sequences S_k and n_k satisfying $S_k \supset K_{n_k} \supset S_{k-1}$, $n_k > n_{k-1}$. □

Let us denote by k_m the dimensions of standard fibers $S_m = F_0\mathbb{R}^m$.

38.5. Proposition. *The standard fibers S_m of every bundle functor F on $\mathcal{M}f$ with the point property are diffeomorphic to $\mathbb{R}^{k_m}$.*

Proof. Let us write $s = c_{\mathbb{R}^m}(0)$, $0 \in \mathbb{R}^m$, and let $g_t \colon \mathbb{R}^m \to \mathbb{R}^m$ be the homotheties $g_t(x) = tx$, $t \in \mathbb{R}$. Since $g_0(\mathbb{R}^m) = \{0\}$, the smoothly parameterized family $h_t = Fg_t|S_m \colon S_m \to S_m$ satisfies all assumptions of the previous lemma. □

For a product $M \xleftarrow{p} M \times N \xrightarrow{q} N$ the values $FM \xleftarrow{Fp} F(M \times N) \xrightarrow{Fq} FN$ determine a canonical map $\pi \colon F(M \times N) \to FM \times FN$.

38.6. Lemma. *For every bundle functor F on $\mathcal{M}f$ with the point property all the maps $\pi \colon F(M \times N) \to FM \times FN$ are surjective submersions.*

Proof. By locality of F it suffices to discuss the case $M = \mathbb{R}^m$, $N = \mathbb{R}^n$. Write $0_k = c_{\mathbb{R}^k}(0) \in F\mathbb{R}^k$, $k = 0, 1, \dots$, and denote $i \colon \mathbb{R}^m \to \mathbb{R}^{m+n}$, $i(x) = (x, 0)$, and $j \colon \mathbb{R}^n \to \mathbb{R}^{m+n}$, $j(y) = (0, y)$. In the tangent space $T_{0_{m+n}} F\mathbb{R}^{m+n}$, there are subspaces $V = TFi(T_{0_m} F\mathbb{R}^m)$ and $W = TFj(T_{0_n} F\mathbb{R}^n)$. We claim $V \cap W = 0$. Indeed, if $A \in V \cap W$, i.e. $A = TFi(B) = TFj(C)$ with $B \in T_{0_m} F\mathbb{R}^m$ and $C \in T_{0_n} F\mathbb{R}^n$, then $TFp(A) = TFp(TFi(B)) = B$, but at the same time $TFp(A) = TFp \circ TFj(C) = 0_m$, for $p \circ j$ is the constant map of $\mathbb{R}^n$ into $0 \in \mathbb{R}^m$, and $A = TFi(B) = 0$ follows.

Hence $T\pi|(V \oplus W) \colon V \oplus W \to T_{0_m} F\mathbb{R}^m \times T_{0_n} F\mathbb{R}^n$ is invertible and so π is a submersion at 0_{m+n} and consequently on a neighborhood $U \subset F\mathbb{R}^{m+n}$ of 0_{m+n}. Since the actions of $\mathbb{R}$ defined by the homotheties g_t on $\mathbb{R}^m$, $\mathbb{R}^n$ and $\mathbb{R}^{m+n}$ commute with the product projections p and q, the induced actions on $F\mathbb{R}^m$, $F\mathbb{R}^n$, $F\mathbb{R}^{m+n}$ commute with π as well (draw a diagram if necessary). The family Fg_t is smoothly parameterized and $Fg_0(F\mathbb{R}^{m+n}) = \{0_{m+n}\}$, so that every point of $F\mathbb{R}^{m+n}$ is mapped into U by a suitable Fg_t, $t > 0$. Further all Fg_t with $t > 0$ are diffeomorphisms and so π is a submersion globally. Therefore the image $\pi(F\mathbb{R}^{m+n})$ is an open neighborhood of $(0_m, 0_n) \in F\mathbb{R}^m \times F\mathbb{R}^n$. But similarly as above, every point of $F\mathbb{R}^m \times F\mathbb{R}^n$ can be mapped into this neighborhood by a suitable Fg_t, $t > 0$. This implies that π is surjective. □

It should be an easy exercise for the reader to extend the lemma to arbitrary finite products of manifolds.

38.7. Corollary. *Every bundle functor F on $\mathcal{M}f$ with the point property transforms submersions into submersions.*

Proof. The local canonical form of any submersion is $p \colon \mathbb{R}^n \times \mathbb{R}^k \to \mathbb{R}^n$, $p(x, y) = x$, cf. 2.2. Then $Fp = \mathrm{pr}_1 \circ \pi$ is a composition of two submersions $\pi \colon F(\mathbb{R}^n \times \mathbb{R}^k) \to F\mathbb{R}^n \times F\mathbb{R}^k$ and $\mathrm{pr}_1 \colon F\mathbb{R}^n \times F\mathbb{R}^k \to F\mathbb{R}^n$. Since every bundle functor is local, this concludes the proof. □

38.8. Proposition. *If a bundle functor F on $\mathcal{M}f$ has the point property, then the dimensions of its standard fibers satisfy $k_{m+n} \geq k_m + k_n$ for all $0 \leq m+n < \infty$. Equality holds if and only if F preserves products in dimensions m and n.*

Proof. By lemma 38.6, we have the submersions $\pi \colon F(\mathbb{R}^m \times \mathbb{R}^n) \to F\mathbb{R}^m \times F\mathbb{R}^n$ which implies $k_{m+n} \geq k_m + k_n$. If the equality holds, then π is a local diffeomorphism at each point. Since π commutes with the action of the homotheties, it must be bijective on each fiber over $\mathbb{R}^{m+n}$, and therefore π must be a global diffeomorphism. Given arbitrary manifolds M and N of the proper dimensions, the locality of bundle functors and a standard diagram chasing lead to the conclusion that

$$FM \xleftarrow{Fp} F(M \times N) \xrightarrow{Fq} FN$$

is a product. □

In view of the results of the previous chapter we get

38.9. Corollary. *For every bundle functor F on $\mathcal{M}f$ with the point property the dimensions of its values satisfy* $\dim F\mathbb{R}^m = m \dim F\mathbb{R}$ *if and only if there is a Weil algebra A such that F is naturally equivalent to the Weil bundle T_A.*

38.10. For every Weil algebra A and every Lie group G there is a canonical Lie group structure on $T_A G$ obtained by the application of the Weil bundle T_A to all operations on G, cf. 37.16. If we replace T_A by an arbitrary bundle functor on $\mathcal{M}f$, we are not able to repeat this construction. However, in the special case of a vector bundle functor F on $\mathcal{M}f$ with the point property we can perform another procedure.

For all manifolds M, N the inclusions $i_y\colon M \to M \times N$, $i_y(x) = (x,y)$, $j_x\colon N \to M \times N$, $j_x(y) = (x,y)$, $(x,y) \in M \times N$, form smoothly parameterized families of morphisms and so we can define a morphism $\tau_{M,N}\colon FM \times FN \to F(M \times N)$ by $\tau_{M,N}(z,w) = Fi_{p_N(w)}(z) + Fj_{p_M(z)}(w)$, where $p_M\colon FM \to M$ are the canonical projections. One verifies easily that the diagram

$$\begin{array}{ccc} FM \times FN & \xrightarrow{\tau_{M,N}} & F(M \times N) \\ \Big\downarrow Ff \times Fg & & \Big\downarrow F(f \times g) \\ F\bar{M} \times F\bar{N} & \xrightarrow{\tau_{\bar{M},\bar{N}}} & F(\bar{M} \times \bar{N}) \end{array}$$

commutes for all maps $f\colon M \to \bar{M}$, $g\colon N \to \bar{N}$. So we have constructed a natural transformation $\tau\colon \mathrm{Prod} \circ (F,F) \to F \circ \mathrm{Prod}$, where Prod is the bifunctor corresponding to the products of manifolds and maps. The projections $p\colon M \times N \to M$, $q\colon M \times N \to N$ determine the map $(Fp, Fq)\colon F(M \times N) \to FM \times FN$ and by the definition of $\tau_{M,N}$, we get $(Fp,Fq) \circ \tau_{M,N} = \mathrm{id}_{FM \times FN}$. Now, given a Lie group G with the operations $\mu\colon G \times G \to G$, $\nu\colon G \to G$ and $e\colon \mathrm{pt} \to G$, we define $\mu_{FG} = F\mu \circ \tau_{G,G}$, $\nu_{FG} = F\nu$ and $e_{FG} = Fe = c_G(e)$ where $c_G\colon G \to FG$ is the canonical section. By the definition of τ, we get for every element $(z,w) \in FG \times FG$ over $(x,y) \in G \times G$

$$\mu_{FG}(z,w) = F(\mu(\ ,y))(z) + F(\mu(x,\ ))(w)$$

and it is easy to check all axioms of Lie groups for the operations μ_{FG}, ν_{FG} and e_{FG} on FG. In particular, we have a canonical Lie group structure on the r-th order tangent bundles $T^{(r)}G$ over any Lie group G and on all tensor bundles over G.

Since τ is the identity if F equals to the tangent bundle T, we have generalized the canonical Lie group structure on tangent bundles over Lie groups to all vector bundle functors with the point property, cf. 37.2.

38.11. Remark. Given a bundle functor F on $\mathcal{M}f$ and a principal fiber bundle (P, p, M, G) we might be interested in a natural principal bundle structure on $Fp: FP \to FM$ with structure group FG. If F is a Weil bundle, this structure can be defined by application of F to all maps in question, cf. 37.16. Though we have found a natural Lie group structure on FG for vector bundle functors with the point property which do not preserve products, there is still no structure of principal fiber bundle (FP, Fp, FM, FG) for dimension reasons, see 38.8.

38.12. Let us now consider a general bundle functor F on $\mathcal{M}f$ and write $Q = F(\text{pt})$. For every manifold M the unique map $q_M: M \to \text{pt}$ induces $Fq_M: FM \to Q$ and similarly to 38.3, every point $a \in Q$ determines a canonical natural section $c(a)_M(x) = Fi_x(a)$. Let G be the bundle functor on $\mathcal{M}f$ defined by $GM = M \times Q$ on manifolds and $Gf = f \times \text{id}_Q$ on maps.

Lemma. *The maps $\sigma_M(x,a) = c(a)_M(x)$, $(x,a) \in M \times Q$, and $\rho_M(z) = (p_M(z), Fq_M(z))$, $z \in FM$, define natural transformations $\sigma: G \to F$ and $\rho: F \to G$ satisfying $\rho \circ \sigma = \text{id}$. Moreover the σ_M are embeddings and the ρ_M are submersions for all manifolds M. In particular, for every $a \in Q$ the rule $F_aM = (Fq_M)^{-1}(a)$, $F_af = Ff|F_aM$ determines a bundle functor on $\mathcal{M}f$ with the point property.*

Proof. It is easy to verify that σ and ρ are natural transformations satisfying $\rho \circ \sigma = \text{id}$. This equality implies that σ_M is an embedding and also that ρ_M is a surjective map which has maximal rank on a neighborhood U of the image $\sigma_M(M \times Q)$. It suffices to prove that every $\rho_{\mathbb{R}^m}$ is a submersion. Consider the homotheties $g_t(x) = tx$ on $\mathbb{R}^m$. Then Fg_t is a smoothly parameterized family with $Fg_1 = \text{id}_{\mathbb{R}^m}$ and $Fg_0(F\mathbb{R}^m) = Fi_0 \circ Fq_{\mathbb{R}^m}(F\mathbb{R}^m) \subset \sigma_M(\mathbb{R}^m \times Q)$. Hence every point of $F\mathbb{R}^m$ is mapped into U by some Fg_t with $t > 0$ and so $\rho_{\mathbb{R}^m}$ has maximal rank everywhere.

Since Fq_M is the second component of the surjective submersion ρ_M, all the subsets $F_aM \subset FM$ are submanifolds and one easily checks all the axioms of bundle functors. □

38.13. Proposition. *Every bundle functor on $\mathcal{M}f$ transforms submersions into submersions.*

Proof. By the previous lemma, every value $Ff: FM \to FN$ is a fibered morphism of $Fq_M: FM \to Q$ into $Fq_N: FN \to Q$ over the identity on Q. If f is a submersion, then every $F_af: F_aM \to F_aN$ is a submersion according to 38.7. □

38.14. Proposition. *The dimensions of the standard fibers of every bundle functor F on $\mathcal{M}f$ satisfy $k_{m+n} \geq k_m + k_n - \dim F(\text{pt})$. Equality holds if and only if all bundle functors F_a preserve products in dimensions m and n.* □

38.15. Remarks. If the standard fibers of a bundle functor F on $\mathcal{M}f$ are compact, then all the functors F_a must coincide with the identity functor on $\mathcal{M}f$ according to 38.5. But then the natural transformations σ and ρ from 38.12 are natural equivalences.

38.16. Example. Taking any bundle functor G on $\mathcal{M}f$ with the point property and any manifold Q, we can define $FM = GM \times Q$ and $Ff = Gf \times \mathrm{id}_Q$ to get a bundle functor with $F(\mathrm{pt}) = Q$. We present an example showing that not all bundle functors on $\mathcal{M}f$ are of this type. The basic idea is that some of the individual 'fiber components' F_a of F coincide with the functor T_1^2 of 1-dimensional velocities of the second order while some other ones are the Whitney sums $T \oplus T$ in dependence on the zero values of a smooth function on Q. According to the general theory developed in section 14, it suffices to construct a functor on the second order skeleton of $\mathcal{M}f$. So we take the system of standard fibers $S_n = Q \times \mathbb{R}^n \times \mathbb{R}^n$, $n \in \mathbb{N}_0$, and we have to define the action of all jets from $J_0^2(\mathbb{R}^m, \mathbb{R}^n)_0$ on S_m. Let us write a_i^p, a_{ij}^p for the coefficients of canonical polynomial representatives of the jets in question. Given any smooth function $f\colon Q \to \mathbb{R}$ we define a map $J_0^2(\mathbb{R}^m, \mathbb{R}^n)_0 \times S_m \to S_n$ by

$$(a_i^p, a_{jk}^r)(q, y^\ell, z^m) = (q, a_i^p y^i, f(q) a_{ij}^r y^i y^j + a_i^r z^i).$$

One verifies easily that this is an action of the second order skeleton on the system S_n. Obviously, the corresponding bundle functor F satisfies $F(\mathrm{pt}) = Q$ and the bundle functors F_q coincide with $T \oplus T$ for all $q \in Q$ with $f(q) = 0$. If $f(q) \neq 0$, then F_q is naturally equivalent to the functor T_1^2. Indeed, the maps $\mathbb{R}^{2n} \to \mathbb{R}^{2n}$, $y^i \mapsto y^i$, and $z^i \mapsto f(q) z^i$ are invertible and define a natural equivalence of T_1^2 into F_q, see 18.15 for a help in a more detailed verification.

38.17. Consider a submersion $f\colon Y \to M$ and denote by $\mu\colon FY \to FM \times_M Y$ the induced pullback map, cf. 2.19.

Proposition. *The pullback map $\mu\colon FY \to FM \times_M Y$ of every submersion $f\colon Y \to M$ is a submersion as well.*

We remark that this property represents a special case of the so-called prolongation axiom which was introduced in [Pradines, 74b] for a more general situation.

Proof. In view of 38.12 we may restrict ourselves to bundle functors with point property (in general $Fq_M\colon FM \to F(\mathrm{pt})$ and $Fq_Y\colon FY \to F(\mathrm{pt})$ are fibered manifolds and μ is a fibered morphism so that we can verify our assertion fiberwise). Further we may consider the submersion f in its local form, i.e. $f\colon \mathbb{R}^{m+n} \to \mathbb{R}^m$, $(x,y) \mapsto x$, for then the claim follows from the locality of the functors. Now we can easily choose a smoothly parametrized family of local sections $s\colon Y \times M \to Y$ with $s(y, f(y)) = y$, $s_y \in C^\infty(Y)$, e.g. $s_{(x,y)}(\bar{x}) = (\bar{x}, y)$. Then we define a mapping $\alpha\colon FM \times_M Y \to FY$, $\alpha(z,y) := Fs_y(z)$. Since locally $Ff \circ Fs_y = \mathrm{id}_M$ and $p_Y^F \circ Fs_y = s_y \circ p_M^F$, we have constructed a section of μ. Since the canonical sections $c_M\colon M \to FM$ are natural, we get $\alpha(c_M(x), y) = c_Y(y)$. Hence the section goes through the values of the canonical section c_Y and μ has the maximal rank on a neighborhood of this section. Now the action of homotheties on $Y = \mathbb{R}^{m+n}$ and $M = \mathbb{R}^m$ commute with the canonical local form of f and therefore the rank of μ is maximal globally. □

In particular, given two bundle functors F, G on $\mathcal{M}f$, the natural transformation $\mu\colon FG \to F \times G$ defined as the product of the natural transformations

$F(p^G)\colon FG \to F$ and $p^F \circ G\colon FG \to G$ is formed by surjective submersions $\mu_M\colon F(GM) \to FM \times_M GM$.

38.18. At the end of this section, we shall indicate how the above results can be extended to bundle functors on $\mathcal{FM}_m$. The point property still plays an important role. Since any manifold M can be viewed as the fibered manifold $\mathrm{id}_M\colon M \to M$, we can say that a bundle functor $F\colon \mathcal{FM}_m \to \mathcal{FM}$ has the *point property* if $FM = M$ for all m-dimensional manifolds. Bundle functors on $\mathcal{FM}_m$ with the point property do not admit canonical sections in general, but for every fibered manifold $q_Y: Y \to M$ in $\mathcal{FM}_m$ we have the fibration $Fq_Y: FY \to M$ and $Fq_Y = q_Y \circ p_Y$, where $p_Y: FY \to Y$ is the bundle projection of FY. Moreover, the mapping $C^\infty(q_Y: Y \to M) \to C^\infty(Fq_Y\colon FY \to M)$, $s \mapsto Fs$ is natural with respect to fibered isomorphisms. This enables us to generalize easily the proof of proposition 38.5 to our more general situation, for we can use the image of the section $i\colon \mathbb{R}^m \to \mathbb{R}^{m+n}$, $x \mapsto (x, 0)$ instead of the canonical sections c_M from 38.5. So the standard fibers $S_n = F\mathbb{R}^{m+n}$ of a bundle functor with the point property are diffeomorphic to $\mathbb{R}^{k_n}$.

Proposition. *The dimensions k_n of standard fibers of every bundle functor $F\colon \mathcal{FM}_m \to \mathcal{FM}$ with the point property satisfy $k_{n+p} \geq k_n + k_p$ and for every $\mathcal{FM}_m$-objects $q_Y: Y \to M$, $q_{\bar{Y}}: \bar{Y} \to M$ the canonical map $\pi\colon F(Y \times_M \bar{Y}) \to FY \times_M F\bar{Y}$ is a surjective submersion. Equality holds if and only if F preserves fibered products in dimensions n and p of the fibers. So F preserves fibered products if and only if $k(n) = n.k(1)$ for all $n \in \mathbb{N}_0$.*

Proof. Consider the diagram

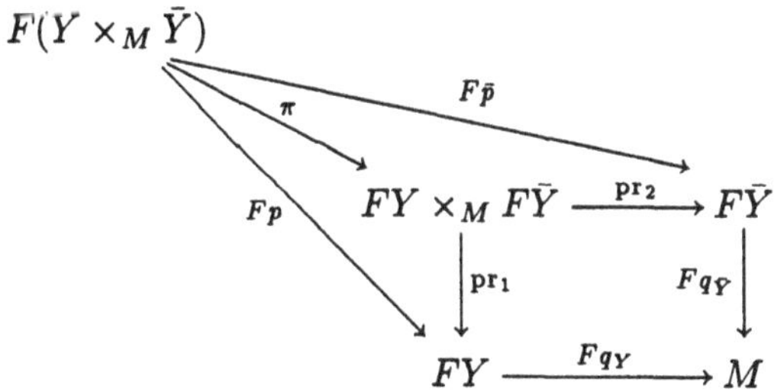

where p and $\bar{p}$ are the projections on $Y \times_M \bar{Y}$.

By locality of bundle functors it suffices to restrict ourselves to objects from a local pointed skeleton. In particular, we shall deal with the values of F on trivial bundles $Y = M \times S$. In the special case $m = 0$, the proposition was proved above.

For every point $x \in M$ we write $(FY)_x := (Fq_Y)^{-1}(x)$ and we define a functor $G = G_x\colon \mathcal{M}f \to \mathcal{FM}$ as follows. We set $G(Y_x) := (FY)_x$ and for every map $f = \mathrm{id}_M \times f_1\colon Y \to \bar{Y}$, $f_1\colon Y_x \to \bar{Y}_x$ we define $Gf_1 := Ff|(FY)_x\colon GY_x \to G\bar{Y}_x$. If we restrict all the maps in the diagram to the appropriate preimages, we get the product $(FY)_x \xleftarrow{\mathrm{pr}_1} (FY)_x \times (F\bar{Y})_x \xrightarrow{\mathrm{pr}_2} (F\bar{Y})_x$ and $\pi_x\colon G(Y_x \times \bar{Y}_x) \to GY_x \times G\bar{Y}_x$. Since G has the point property, π_x is a surjective submersion.

Hence π is a fibered morphism over the identity on M which is fiber wise a surjective submersion. Consequently π is a surjective submersion and the inequality $k_{n+p} \geq k_n + k_p$ follows.

Now similarly to 38.8, if the equality holds, then π is a global isomorphism. □

38.19. Vertical Weil bundles. Let A be a Weil algebra. We define a functor $V_A \colon \mathcal{FM}_m \to \mathcal{FM}$ as follows. For every $q_Y \colon Y \to M$, we put $V_A Y \colon = \cup_{x \in M} T_A Y_x$ and given $f \in \mathcal{FM}_m(Y, \bar{Y})$ we write $f_x = f | Y_x$, $x \in M$, and we set $V_A f | (V_A Y)_x := T_A f_x$. Since $V_A(\mathbb{R}^{m+n} \to \mathbb{R}^m) = \mathbb{R}^m \times T_A \mathbb{R}^n$ carries a canonical smooth structure, every fibered atlas on $Y \to M$ induces a fibered atlas on $V_A Y \to Y$. It is easy to verify that V_A is a bundle functor which preserves fibered products. In the special case of the algebra $\mathbb{D}$ of dual numbers we get the vertical tangent bundle V.

Consider a bundle functor $F \colon \mathcal{FM}_m \to \mathcal{FM}$ with the point property which preserves fibered products, and a trivial bundle $Y = M \times S$. If we repeat the construction of the product preserving functors $G = G_x$, $x \in M$, from the proof of proposition 38.18 we have $G_x = T_{A_x}$ for certain Weil algebras $A = A_x$. So we conclude that $F(\mathrm{id}_M \times f_1)|(FY)_x = G_x(f_1) = V_{A_x}(\mathrm{id}_M \times f_1)|(FY)_x$. At the same time the general theory of bundle functors implies (we take $A = A_0$) $F\mathbb{R}^{m+n} = \mathbb{R}^m \times \mathbb{R}^n \times S_n = \mathbb{R}^m \times A^n = V_A \mathbb{R}^{m+n}$ for all $n \in \mathbb{N}$ (including the actions of jets of maps of the form $\mathrm{id}_{\mathbb{R}^m} \times f_1$). So all the algebras A_x coincide and since the bundles in question are trivial, we can always find an atlas $(U_\alpha, \varphi_\alpha)$ on Y such that the chart changings are over the identity on M. But a cocycle defining the topological structure of FY is obtained if we apply F to these chart changings and therefore the resulting cocycle coincides with that obtained from the functor V_A.

Hence we have deduced the following characterization (which is not a complete description as in 36.1) of the fibered product preserving bundle functors on $\mathcal{FM}_m$.

Proposition. *Let $F \colon \mathcal{FM}_m \to \mathcal{FM}$ be a bundle functor with the point property. The following conditions are equivalent.*

(i) *F preserves fibered products*
(ii) *For all $n \in \mathbb{N}$ it holds $\mathrm{dim} S_n = n(\mathrm{dim} S_1)$*
(iii) *There is a Weil algebra A such that $FY = V_A Y$ for every trivial bundle $Y = M \times S$ and for every mapping $f_1 \colon S \to \bar{S}$ we have $F(id_M \times f_1) = V_A(id_M \times f_1) \colon F(M \times S) \to F(M \times \bar{S})$.*

39. The flow-natural transformation

39.1. Definition. Consider a bundle functor $F \colon \mathcal{M}f \to \mathcal{FM}$ and the tangent functor $T \colon \mathcal{M}f \to \mathcal{FM}$. A natural transformation $\iota \colon FT \to TF$ is called a *flow-natural transformation* if the following diagram commutes for all m-dimensional manifolds M and all vector fields $X \in \mathfrak{X}(M)$ on M.

$$\begin{array}{ccc} FTM & \xrightarrow{F\pi_M} & FM \\ {\scriptstyle FX}\uparrow & \searrow^{\iota_M} & \uparrow{\scriptstyle \pi_{FM}} \\ FM & \xrightarrow{\mathcal{F}X} & TFM \end{array} \tag{1}$$

39.2. Given a map $f\colon Q \times M \to N$, we have denoted by $\tilde{F}f\colon Q \times FM \to FN$ the 'collection' of $F(f(q, \ ))$ for all $q \in Q$, see 14.1. Write $(x, X) = Y \in T\mathbb{R}^m = \mathbb{R}^m \times \mathbb{R}^m$ and define $\mu_{\mathbb{R}^m}\colon \mathbb{R} \times T\mathbb{R}^m \to \mathbb{R}^m$, $\mu_{\mathbb{R}^m}(t, Y) = (x + tX)$ for $t \in \mathbb{R}$,.

Theorem. *Every bundle functor $F\colon \mathcal{M}f \to \mathcal{FM}$ admits a canonical flow-natural transformation $\iota\colon FT \to TF$ determined by*

$$\iota_{\mathbb{R}^m}(z) = j_0^1 \tilde{F}\mu_{\mathbb{R}^m}(\ , z).$$

If F has the point property, then ι is a natural equivalence if and only if F is a Weil functor T_A. In this case ι coincides with the canonical natural equivalence $T_A T \to T T_A$ corresponding to the exchange homomorphism $A \otimes \mathbb{D} \to \mathbb{D} \otimes A$ between the tensor products of Weil algebras.

39.3. The proof requires several steps. We start with a general lemma.

Lemma. *Let M, N, Q be smooth manifolds and let f, $g\colon Q \times M \to N$ be smooth maps. If $j_q^k f(\ , y) = j_q^k g(\ , y)$ for some $q \in Q$ and all $y \in M$, then for every bundle functor F on $\mathcal{M}f$ the maps $\tilde{F}f$, $\tilde{F}g\colon Q \times FM \to FN$ satisfy $j_q^k \tilde{F}f(\ , z) = j_q^k \tilde{F}g(\ , z)$ for all $z \in FM$.*

Proof. It suffices to restrict ourselves to objects from the local skeleton $(\mathbb{R}^m)$, $m = 0, 1, \dots$, of $\mathcal{M}f$. Let r be the order of F valid for maps with source $\mathbb{R}^m$, cf. 22.3, and write p for the bundle projection $p_{\mathbb{R}^m}$. By the general theory of bundle functors the values of F on morphisms $f\colon \mathbb{R}^m \to \mathbb{R}^n$ are determined by the smooth associated map $F_{\mathbb{R}^m,\mathbb{R}^n}\colon J^r(\mathbb{R}^m, \mathbb{R}^n) \times_{\mathbb{R}^m} F\mathbb{R}^m \to F\mathbb{R}^n$, see section 14. Hence the map $\tilde{F}f\colon Q \times F\mathbb{R}^m \to F\mathbb{R}^n$ is defined by the composition of $F_{\mathbb{R}^m,\mathbb{R}^n}$ with the smooth map $f^r\colon Q \times F\mathbb{R}^m \to J^r(\mathbb{R}^m, \mathbb{R}^n) \times_{\mathbb{R}^m} F\mathbb{R}^m$, $(q, z) \mapsto (j^r_{p(z)} f(q, \ ), z)$. Our assumption implies that $f^r(\ , z)$ and $g^r(\ , z)$ have the same k-jet at q, which proves the lemma. □

39.4. Now we deduce that the maps $\iota_{\mathbb{R}^m}$ determine a natural transformation $\iota\colon FT \to TF$ such that the upper triangle in 39.1.(1) commutes. These maps define a natural transformation between the bundle functors in question if they obey the necessary commutativity with respect to the actions of morphisms between the objects of the local skeleton $\mathbb{R}^m$, $m = 0, 1, \dots$. Given such a morphism $f\colon \mathbb{R}^m \to \mathbb{R}^n$ we have

$$\begin{aligned} \iota_{\mathbb{R}^n}(FTf(z)) &= j_0^1 \tilde{F}\mu_{\mathbb{R}^n}(\ , FTf(z)) = j_0^1 F((\mu_{\mathbb{R}^n})_t \circ Tf)(z) \\ TFf(\iota_{\mathbb{R}^m}(z)) &= TFf(j_0^1 \tilde{F}\mu_{\mathbb{R}^m}(\ , z)) = j_0^1 (Ff \circ F(\mu_{\mathbb{R}^m})_t(z)) = \\ &= j_0^1 F(f \circ (\mu_{\mathbb{R}^m})_t)(z). \end{aligned}$$

So in view of lemma 39.3 it is sufficient to prove for all $Y \in T\mathbb{R}^m$, $f\colon \mathbb{R}^m \to \mathbb{R}^n$

$$j_0^1((f \circ (\mu_{\mathbb{R}^m})_t)(Y)) = j_0^1((\mu_{\mathbb{R}^n})_t \circ Tf)(Y).$$

By the definition of μ, the values of both sides are $Tf(Y)$.

Since $(\mu_{\mathbb{R}^m})_0 = \pi_{\mathbb{R}^m} : T\mathbb{R}^m \to \mathbb{R}^m$, we have $\pi_{F\mathbb{R}^m} \circ \iota_{\mathbb{R}^m} = F(\mu_{\mathbb{R}^m})_0 = F\pi_{\mathbb{R}^m}$.

39.5. Let us now discuss the bottom triangle in 39.1.(1). Given a bundle functor F on $\mathcal{M}f$, both the arrows $\mathcal{F}X$ and FX are values of natural operators and ι is a natural transformation. If we fix dimension of the manifold M then these operators are of finite order. Therefore it suffices to restrict ourselves to the fibers over the distinguished points from the objects of a local pointed skeleton. Moreover, if we verify $\iota_{\mathbb{R}^m} \circ FX = \mathcal{F}X$ on the fiber $(FT)_0\mathbb{R}^m$ for a jet of a suitable order of a field X at $0 \in \mathbb{R}^m$, then this equality holds on the whole orbit of this jet under the action of the corresponding jet group. Further, the operators in question are regular and so the equality follows for the closure of the orbit.

Lemma. *The vector field $X = \frac{\partial}{\partial x^1}$ on $(\mathbb{R}^m, 0)$ has the following two properties.*

(1) Its flow satisfies $Fl^X = \mu_{\mathbb{R}^m} \circ (id_{\mathbb{R}} \times X)\colon \mathbb{R} \times \mathbb{R}^m \to \mathbb{R}^m$.

(2) The orbit of the jet $j_0^r X$ under the action of the jet group G_m^{r+1} is dense in the space of r-jets of vector fields at $0 \in \mathbb{R}^m$.

Proof. We have $\mathrm{Fl}_t^X(x) = x + t(1, 0, \dots, 0) = \mu_{\mathbb{R}^m}(t, X(x))$. The second assertion is proved in section 42 below. □

By the lemma, the mappings $\iota_{\mathbb{R}^m}$ determine a flow-natural transformation $\iota\colon FT \to TF$.

Assume further that F has the point property and write k_n for the dimension of the standard fiber of $F\mathbb{R}^n$. If ι is a natural equivalence, then $k_{2n} = 2k_n$ for all n. Hence proposition 38.8 implies that F preserves products and so it must be naturally equivalent to a Weil bundle. On the other hand, assume $F = T_A$ for some Weil algebra A and denote 1 and e the generators of the algebra $\mathbb{D}$ of dual numbers. For every $j_A f \in T_A T\mathbb{R}$, with $f\colon \mathbb{R}^k \to T\mathbb{R} = \mathbb{D}$, $f(x) = g(x) + h(x).e$, take $q\colon \mathbb{R} \times \mathbb{R}^k \to \mathbb{R}$, $q(t, x) = g(x) + th(x)$, i.e. $f(x) = j_0^1 q(\ , x)$. Then we get

$$\iota_{\mathbb{R}}(j_A f) = j_0^1 T_A(\mu_{\mathbb{R}})_t(j_A f) = j_0^1 j_A(g(\ ) + th(\ )) = j_0^1 j_A q(t,\ ).$$

Hence $\iota_{\mathbb{R}}$ coincides with the canonical exchange homomorphism $A \otimes \mathbb{D} \to \mathbb{D} \otimes A$ and so ι is the canonical natural equivalence $T_A T \to TT_A$. □

39.6. Let us now modify the idea from 39.1 to bundle functors on $\mathcal{FM}_m$.

Definition. Consider a bundle functor $F\colon \mathcal{FM}_m \to \mathcal{FM}$ and the vertical tangent functor $V\colon \mathcal{FM}_m \to \mathcal{FM}$. A natural transformation $\iota\colon FV \to VF$ is called a *flow-natural transformation* if the diagram

$$\begin{array}{ccc} FVY & \xrightarrow{F\pi_Y} & FY \\ {\scriptstyle FX}\uparrow & \searrow{\scriptstyle \iota_Y} & \uparrow{\scriptstyle \pi_{FY}} \\ FY & \xrightarrow{\mathcal{F}X} & VFY \end{array} \tag{1}$$

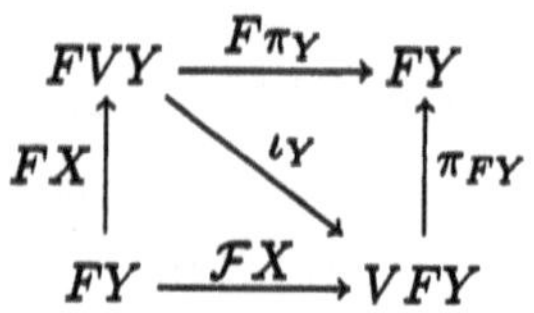

commutes for all fibered manifolds Y with m-dimensional basis and for all vertical vector fields X on Y.

For every fibered manifold $q\colon Y \to M$ in $\mathrm{Ob}\mathcal{FM}_m$, the fibration $q \circ \pi_Y\colon VY \to M$ is an $\mathcal{FM}_m$-morphism. Further, consider the local skeleton $(\mathbb{R}^{m+n} \to \mathbb{R}^m)$ of $\mathcal{FM}_m$ and define

$$\mu_{\mathbb{R}^{m+n}}\colon \mathbb{R} \times V\mathbb{R}^{m+n} = \mathbb{R}^{1+m+n+n} \to \mathbb{R}^{m+n}, \quad (t, x, y, X) \mapsto (x, y + tX).$$

Then every $\mu_{\mathbb{R}^{m+n}}(t, \ )$ is a globally defined $\mathcal{FM}_m$ morphism and we have

$$j_0^1 \mu_{\mathbb{R}^{m+n}}(\ , x, y, X) = (x, y, X).$$

39.7. The proof of 39.3 applies to general categories over manifolds. A bundle functor on an admissible category $\mathcal{C}$ is said to be of a locally finite order if for every $\mathcal{C}$-object A there is an order r such that for all $\mathcal{C}$-morphisms $f\colon A \to B$ the values $Ff(z)$, $z \in FA$, depend on the jets $j^r_{p_A(z)} f$ only. Let us recall that all bundle functors on $\mathcal{FM}_m$ have locally finite order, cf. 22.3.

Lemma. *Let $f, g\colon Q \times mA \to mB$ be smoothly parameterized families of $\mathcal{C}$-morphisms with $j^k_q f(\ , y) = j^k_q g(\ , y)$ for some $q \in Q$ and all $y \in mA$. Then for every regular bundle functor F on $\mathcal{C}$ with locally finite order, the maps $\tilde{F}f$, $\tilde{F}g\colon Q \times FA \to FB$ satisfy $j^k_q \tilde{F}f(\ , z) = j^k_q \tilde{F}g(\ , z)$ for all $z \in FA$.* □

39.8. Let us define $\iota_{\mathbb{R}^{m+n}}(z) = j_0^1 \tilde{F}\mu_{\mathbb{R}^{m+n}}(\ , z)$. If we repeat the considerations from 39.4 we deduce that our maps $\iota_{\mathbb{R}^{m+n}}$ determine a natural transformation $\iota\colon FV \to TF$. But its values satisfy $Tp_Y \circ \iota_Y(z) = j_0^1 p_Y \circ F(\mu_Y)_t(z) = p_Y(z) \in VY$ and so $\iota_Y(z) \in V(FY \to BY)$. So $\iota\colon FV \to VF$ and similarly to 39.4 we show that the upper triangle in 39.6.(1) commutes.

Every non-zero vertical vector field on $\mathbb{R}^{m+n} \to \mathbb{R}^m$ can be locally transformed (by means of an $\mathcal{FM}_m$-morphism) into a constant one and for all constant vertical vector fields X on $\mathbb{R}^{m+n}$ we have $\mathrm{Fl}^X = (\mu_{\mathbb{R}^{m+n}} \circ (\mathrm{id}_{\mathbb{R}^{m+n}} \times X))$. Hence we also have an analogue of lemma 39.5.

Theorem. *For every bundle functor $F\colon \mathcal{FM}_m \to \mathcal{FM}$ there is the canonical flow-natural transformation $\iota\colon FV \to VF$. If F has the point property, then ι is a natural equivalence if and only if F preserves fibered products.*

We have to point out that we consider the fibered manifold structure $FY \to BY$ for every object $Y \to BY \in \mathrm{Ob}\mathcal{FM}_m$, i.e. $\iota_Y\colon F(VY \to BY) \to V(FY \to BY)$.

Proof. We have proved that ι is flow-natural. Assume F has the point property. If ι is a natural equivalence, then proposition 38.18 implies that F preserves fibered products. On the other hand, F preserves fibered products if and only if $F\mathbb{R}^{m+n} = V_A\mathbb{R}^{m+n}$ for a Weil algebra A and then also Ff coincides with $V_A f$ for morphisms of the form $\mathrm{id}_{\mathbb{R}^m} \times g\colon \mathbb{R}^{m+n} \to \mathbb{R}^{m+k}$, see 38.19. But each $\mu_{\mathbb{R}^{m+n}}(t, \ )$ is of this form and any restriction of $\iota_{\mathbb{R}^{m+n}}$ to a fiber $(V_A V\mathbb{R}^{m+n})_x \cong T_A T\mathbb{R}^m$ coincides with the canonical flow natural equivalence $T_A T \to TT_A$, cf. 39.2. Hence ι is a natural equivalence. □

Let us remark that for $F = J^r$ we obtain the well known canonical natural equivalence $J^rV \to VJ^r$, cf. [Goldschmidt, Sternberg, 73], [Mangiarotti, Modugno, 83].

39.9. The action of some bundle functors $F\colon \mathcal{FM}_m \to \mathcal{FM}$ on morphisms can be extended in such a way that the proof of theorem 39.8 might go through for the whole tangent bundle. We shall show that this happens with the functors $J^r\colon \mathcal{FM}_m \to \mathcal{FM}$.

Since $J^r(\mathbb{R}^{m+n} \to \mathbb{R}^m)$ is a sub bundle in the bundle $K^r_m\mathbb{R}^{m+n}$ of contact elements of order r formed by the elements transversal to the fibration, the action of J^rf on a jet $j^r_x s$ extends to all local diffeomorphisms transforming $j^r_x s$ into a jet of a section. Of course, we are not able to recover the whole theory of bundle functors for this extended action of J^r, but one verifies easily that lemma 39.3 remains still valid.

So let us define $\mu_t\colon T\mathbb{R}^{m+n} \to \mathbb{R}^{m+n}$ by $\mu_t(x,z,X,Z) = (x+tX, z+tZ)$. For every section $(x, z(x), X(x), Z(x))$ of $T\mathbb{R}^{m+n} \to \mathbb{R}^m$, its composition with μ_t and the first projection gives the map $x \mapsto x + tX(x)$. If we proceed in a similar way as above, we deduce

Proposition. *There is a canonical flow-natural transformation $\iota\colon J^rT \to TJ^r$ and its restriction $J^rV \to VJ^r$ is the canonical flow-natural equivalence.*

39.10. Remark. Let us notice that $\iota\colon TJ^r \to J^rT$ cannot be an equivalence for dimension reasons if $m > 0$. The flow-natural transformations on jet bundles were presented as a useful tool in [Mangiarotti, Modugno, 83].

It is instructive to derive the coordinate description of $\iota_{\mathbb{R}^{m+n}}$ at least in the case $r = 1$. Let us write a map $f\colon (\mathbb{R}^{m+n} \xrightarrow{\alpha} \mathbb{R}^m) \to (\mathbb{R}^{m+n} \xrightarrow{\beta} \mathbb{R}^m)$ in the form $z^k = f^k(x^i, y^p)$, $w^q = f^q(x^i, y^p)$. In order to get the action of J^1f in the extended sense on $j^1_0 s = (y^p, y^p_i)$ we have to consider the map $(\beta \circ f \circ s)^{-1} = \tilde{f}$, $x^i = \tilde{f}^i(z)$. So $z^k = f^k(\tilde{f}(z), y^p(\tilde{f}(z)))$ and we evaluate that the matrix $\partial \tilde{f}^i/\partial z^k$ is the inverse matrix to $\partial f^k/\partial x^i + (\partial f^k/\partial y^p)y^p_i$ (the invertibility of this matrix is exactly the condition on $j^1_0 s$ to lie in the domain of J^1f). Now the coordinates w^q_k of $J^1f(j^1_0 s)$ are

$$w^q_k = \frac{\partial f^q}{\partial x^j}\frac{\partial \tilde{f}^j}{\partial z^k} + \frac{\partial f^q}{\partial y^p} y^p_j \frac{\partial \tilde{f}^j}{\partial z^q}.$$

Consider the canonical coordinates x^i, y^p on $Y = \mathbb{R}^{m+n}$ and the additional coordinates y^p_i or X^i, Y^p or y^p_i, X^i_j, Y^p_i or y^p_i, ξ^i, η^p, η^p_i on J^1Y or TY or J^1TY or TJ^1Y, respectively. If $j^1_x s = (x^i, y^p, X^j, Y^q, y^r_k, X^\ell_m, Y^s_n)$, then

$$J^1(\mu_Y)_t(j^1_0 s) = (x^i + tX^i, y^p + tY^p, \bar{y}^q_j(t))$$
$$\bar{y}^p_i(t)(\delta^i_j + tX^i_j) = (y^p_i + tY^p_i)\delta^i_j.$$

Differentiating by t at 0 we get

$$\iota_{\mathbb{R}^{m+n}}(x^i, y^p, X^j, Y^q, y^r_k, X^\ell_m, Y^s_n) = (x^i, y^p, y^r_k, X^j, Y^q, Y^s_\ell - y^s_m X^m_\ell).$$

This formula corresponds to the definition in [Mangiarotti, Modugno, 83].

40. Natural transformations

40.1. The first part of this section is concerned with natural transformations with a Weil bundle as the source. In this case we get a result similar to the Yoneda lemma well known from general category theory. Namely, each point in a Weil bundle $T_A M$ is an equivalence class of mappings in $C^\infty(\mathbb{R}^n, M)$ where n is the width of the Weil algebra A, see 35.15, and the canonical projections yield a natural transformation $\alpha\colon C^\infty(\mathbb{R}^n, \;) \to T_A$. Hence given any bundle functor F on $\mathcal{M}f$, every natural transformation $\chi\colon T_A \to F$ gives rise to the natural transformation $\chi \circ \alpha\colon C^\infty(\mathbb{R}^n, \;) \to F$ and this is determined by the value of $(\chi \circ \alpha)_{\mathbb{R}^n}(\mathrm{id}_{\mathbb{R}^n})$. So in order to classify all natural transformations $\chi\colon T_A \to F$ we have to distinguish the possible values $v := \chi_{\mathbb{R}^n} \circ \alpha_{\mathbb{R}^n}(\mathrm{id}_{\mathbb{R}^n}) \in F\mathbb{R}^n$. Let us recall that for every natural transformation χ between bundle functors on $\mathcal{M}f$ all maps χ_M are fibered maps over id_M, see 14.11. Hence $v \in F_0\mathbb{R}^n$ and another obvious condition is $Ff(v) = Fg(v)$ for all maps $f, g\colon \mathbb{R}^n \to M$ with $j_A f = j_A g$. On the other hand, having chosen such $v \in F_0\mathbb{R}^n$, we can define $\chi^v_M(j_A f) = Ff(v)$ and if all these maps are smooth, then they form a natural transformation $\chi^v\colon T_A \to F$.

So from the technical point of view, our next considerations consist in a better description of the points v with the above properties. In particular, we deduce that it suffices to verify $Ff(v) = Fi(v)$ for all maps $f\colon \mathbb{R}^n \to \mathbb{R}^{n+1}$ with $j_A f = j_A i$ where $i\colon \mathbb{R}^n \to \mathbb{R}^{n+1}$, $x \mapsto (x, 0)$.

40.2. Definition. For every Weil algebra A of width n and for every bundle functor F on $\mathcal{M}f$, an element $v \in F_0\mathbb{R}^n$ is called *A-admissible* if $j_A f = j_A i$ implies $Ff(v) = Fi(v)$ for all $f \in C^\infty(\mathbb{R}^n, \mathbb{R}^{n+1})$. We denote by $S_A(F) \subset S = F_0\mathbb{R}^n$ the set of all A-admissible elements.

40.3. Proposition. *For every Weil algebra A of width n and every bundle functor F on $\mathcal{M}f$, the map*

$$\chi \mapsto \chi_{\mathbb{R}^n}(j_A \mathrm{id}_{\mathbb{R}^n})$$

is a bijection between the natural transformations $\chi\colon T_A \to F$ and the subset of A-admissible elements $S_A(F) \subset F_0\mathbb{R}^n$.

The proof consists in two steps. First we have to prove that each $v \in S_A(F)$ defines the transformation $\chi^v\colon T_A \to F$ at the level of sets, cf. 40.1, and then we have to verify that all maps χ^v_M are smooth.

40.4. Lemma. *Let $F\colon \mathcal{M}f \to \mathcal{FM}$ be a bundle functor and A be a Weil algebra of width n. For each point $v \in S_A(F)$ and for all mappings $f, g\colon \mathbb{R}^n \to M$ the equality $j_A f = j_A g$ implies $Ff(v) = Fg(v)$.*

Proof. The proof is a straightforward generalization of the proof of theorem 22.3 with $m = 0$. Therefore we shall present it in a rather condensed form.

During the whole proof, we may restrict ourselves to mappings $f, g\colon \mathbb{R}^n \to \mathbb{R}^k$ of maximal rank. The reason lies in the regularity of all bundle functors on $\mathcal{M}f$, cf. 22.3 and 20.7.

The canonical local form of a map $f\colon \mathbb{R}^n \to \mathbb{R}^{n+1}$ of maximal rank is i and therefore the assertion is trivial for the dimension $k = n+1$.

Since the equivalence on the spaces $C^\infty(\mathbb{R}^n, \mathbb{R}^k)$ determined by A is compatible with the products of maps, we can complete the proof as in 22.3.(b) and 22.3.(e) with $m = 0$, j_0^r replaced by j_A and S_n replaced by $S_A(F)$. □

Let us remark that for $m = 0$ theorem 22.3 follows easily from this lemma. Indeed, we can take the Weil algebra A corresponding to the bundle $T_n^{r_{n+1}}$ of n-dimensional velocities of order r_{n+1}. Then $j_A f = j_A g$ if and only if $j_0^{r_{n+1}} f = j_0^{r_{n+1}} g$ and according to the assumptions in 22.3, $S_A(F) = S_n$. By the general theory, the order r_{n+1} extends from the standard fiber S_n to all objects of dimension n.

40.5. Lemma. *For every Weil algebra A of width n and every smooth curve $c\colon \mathbb{R} \to T_A\mathbb{R}^k$ there is a smoothly parameterized family of maps $\gamma\colon \mathbb{R}\times\mathbb{R}^n \to \mathbb{R}^k$ such that $j_A\gamma_t = c(t)$.*

Proof. There is an ideal $\mathcal{A}$ in the algebra of germs $\mathcal{E}_n = C_0^\infty(\mathbb{R}^n, \mathbb{R})$, cf. 35.5, such that $A = \mathcal{E}_n/\mathcal{A}$. Write $\mathcal{D}_n^r = \mathcal{M}^{r+1}$ where $\mathcal{M}$ is the maximal ideal in $\mathcal{E}_n$, and $\mathbb{D}_n^r = \mathcal{E}_n/\mathcal{D}_n^r$, i.e. $T_{\mathbb{D}_n^r} = T_n^r$. Then $\mathcal{A} \supset \mathcal{D}_n^r$ for suitable r and so we get the linear projection $\mathbb{D}_n^r \to A$, $j_0^r f \mapsto j_A f$. Let us choose a smooth section $s\colon A \to \mathbb{D}_n^r$ of this projection. Now, given a curve $c(t) = j_A f_t$ in $T_A\mathbb{R}^k$ there are the canonical polynomial representatives g_t of the jets $s(j_A f_t)$. If $c(t)$ is smooth, then g_t is a smoothly parameterized family of polynomials and so $j_A g_t$ is a smooth curve with $j_A g_t = c(t)$. □

Proof of proposition 40.3. Given a natural transformation $\chi\colon T_A \to F$, the value $\chi_{\mathbb{R}^n}(j_A \mathrm{id}_{\mathbb{R}^n})$ is an A-admissible element in $F_0\mathbb{R}^n$. On the other hand, every A-admissible element $v \in S_A(F)$ determines the maps $\chi^v_{\mathbb{R}^k}\colon T_A\mathbb{R}^k \to F\mathbb{R}^k$, $\chi^v_{\mathbb{R}^k}(j_A f) = Ff(v)$ and all these maps are smooth. By the definition, $\chi^v_{\mathbb{R}^n}$ obey the necessary commutativity relations and so they determine the unique natural transformation $\chi^v\colon T_A \to F$ with $\chi^v_{\mathbb{R}^n}(v) = v$. □

40.6. Let us apply proposition 40.3 to the case $F = T^{(r)}$, the r-th order tangent functor. The elements in the standard fiber of $T^{(r)}\mathbb{R}^n$ are the linear forms on the vector space $J_0^r(\mathbb{R}^n, \mathbb{R})_0$ and for every Weil algebra A of width n one verifies easily that such a form ω lies in $S_A(T^{(r)})$ if and only if $\omega(j_0^r g) = 0$ for all g with $j_A g = j_A 0$.

As a simple illustration, we find all natural transformations $T_1^q \to T^{(r)}$. Every element $j_0^r f \in T_0^{r*}\mathbb{R} = J_0^r(\mathbb{R}, \mathbb{R})_0$ has the canonical representative $f(x) = a_1 x + a_2 x^2 + \cdots + a_r x^r$. Let us define 1-forms $v_i \in T_0^{(r)}\mathbb{R}$ by $v_i(j_0^r f) = a_i$, $i = 1, 2, \ldots, r$. Since $j_{\mathbb{D}_1^q} f = j_0^q f$, the forms v_i are $\mathbb{D}_1^q$-admissible if and only if $i \le q$. So the linear space of all natural transformations $T_1^q \to T^{(r)}$ is generated by the linearly independent transformations χ^{v_i}, $i = 1, \ldots, \min\{q, r\}$. The maps $\chi_M^{v_i}$ can be described as follows. Every $j_0^q g \in T_1^q M$ determines a curve $g\colon \mathbb{R} \to M$ through $x = g(0)$ up to the order q and given any $j_x^r f \in J_x^r(M, \mathbb{R})_0$ the value $\chi_M^{v_i}(j_0^q g)(j_x^r f)$ is obtained by the evaluation of the i-th order term in $f\circ g\colon \mathbb{R} \to \mathbb{R}$

at $0 \in \mathbb{R}$. So $\chi_M^{v_i}(j_0^q g)$ might be viewed as the i-th derivative on $J_x^r(M,\mathbb{R})_0$ in the direction $j_0^q g$.

In general, given any vector bundle functor F on $\mathcal{M}f$, the natural transformations $T_A \to F$ carry a vector space structure and the corresponding set $S_A(F)$ is a linear subspace in $F_0\mathbb{R}^n$. In particular, the space of all natural transformations $T_A \to F$ is a finite dimensional vector space with dimension bounded by the dimension of the standard fiber $F_0\mathbb{R}^n$.

As an example let us consider the two natural vector bundle structures given by $\pi_{TM}\colon TTM \to TM$ and $T\pi_M\colon TTM \to TM$ which form linearly independent natural transformations $TT \to T$. For dimension reasons these must form a basis of the linear space of all natural transformations $TT \to T$. Analogously the products $T\pi_M \wedge \pi_{TM}\colon TTM \to \Lambda^2 TM$ generate the one-dimensional space of all natural transformations $TT \to \Lambda^2 T$ and there are no non-zero natural transformations $TT \to \Lambda^p T$ for $p > 2$.

40.7. Remark. [Mikulski, to appear a] also determined the natural operators transforming functions on a manifold M of dimension at least two into functions on FM for every bundle functor $F\colon \mathcal{M}f \to \mathcal{FM}$. All of them have the form $f \mapsto h \circ Ff$, $f \in C^\infty(M,\mathbb{R})$, where h is any smooth function $h\colon F\mathbb{R} \to \mathbb{R}$.

40.8. Natural transformations $T^{(r)} \to T^{(r)}$. Now we are going to show that there are no other natural transformations $T^{(r)} \to T^{(r)}$ beside the real multiples of the identity. Thus, in this direction the properties of $T^{(r)}$ are quite different from the higher order product preserving functors where the corresponding Weil algebras have many endomorphisms as a rule. Let us remark that from the technical point of view we shall prove the proposition in all dimensions separately and only then we 'join' all these partial results together.

Proposition. *All natural transformations $T^{(r)} \to T^{(r)}$ form the one-parameter family*

$$X \mapsto kX, \qquad k \in \mathbb{R}.$$

Proof. If x^i are local coordinates on a manifold M, then the induced fiber coordinates $u_i, u_{i_1 i_2}, \dots, u_{i_1\dots i_r}$ (symmetric in all indices) on $T_1^{r*}M$ correspond to the polynomial representant $u_i x^i + u_{i_1 i_2} x^{i_1} x^{i_2} + \dots + u_{i_1\dots i_r} x^{i_1} \dots x^{i_r}$ of a jet from $T_1^{r*}M$. A linear functional on $(T_1^{r*}M)_x$ with the fiber coordinates X^i, $X^{i_1 i_2}, \dots, X^{i_1\dots i_r}$ (symmetric in all indices) has the form

$$X^i u_i + X^{i_1 i_2} u_{i_1 i_2} + \dots + X^{i_1\dots i_r} u_{i_1\dots i_r}. \tag{1}$$

Let y^p be some local coordinates on N, let $Y^p, Y^{p_1 p_2}, \dots, Y^{p_1\dots p_r}$ be the induced fiber coordinates on $T^{(r)}N$ and $y^p = f^p(x^i)$ be the coordinate expression of a map $f\colon M \to N$. If we evaluate the jet composition from the definition of the action of the higher order tangent bundles on morphisms, we deduce by (1) the coordinate expression of $T^{(r)}f$

$$
\begin{aligned}
Y^p &= \frac{\partial f^p}{\partial x^i}X^i + \frac{1}{2!}\frac{\partial^2 f^p}{\partial x^{i_1}\partial x^{i_2}}X^{i_1 i_2} + \dots + \frac{1}{r!}\frac{\partial^r f^p}{\partial x^{i_1}\dots\partial x^{i_r}}X^{i_1\dots i_r}\\
&\vdots\\
Y^{p_1\dots p_s} &= \frac{\partial f^{p_1}}{\partial x^{i_1}}\dots\frac{\partial f^{p_s}}{\partial x^{i_s}}X^{i_1\dots i_s} + \dots \qquad (2)\\
&\vdots\\
Y^{p_1\dots p_r} &= \frac{\partial f^{p_1}}{\partial x^{i_1}}\dots\frac{\partial f^{p_r}}{\partial x^{i_r}}X^{i_1\dots i_r}
\end{aligned}
$$

where the dots in the middle row denote a polynomial expression, each term of which contains at least one partial derivative of f^p of order at least two.

Consider first $T^{(r)}$ as a bundle functor on the subcategory $\mathcal{M}f_m \subset \mathcal{M}f$. According to (2), its standard fiber $S = T_0^{(r)}\mathbb{R}^m$ is a G_m^r-space with the following action

$$
\begin{aligned}
\bar X^i &= a^i_j X^j + a^i_{j_1 j_2}X^{j_1 j_2} + \dots + a^i_{j_1\dots j_r}X^{j_1\dots j_r}\\
&\vdots\\
\bar X^{i_1\dots i_s} &= a^{i_1}_{j_1}\dots a^{i_s}_{j_s}X^{j_1\dots j_s} + \dots \qquad (3)\\
&\vdots\\
\bar X^{i_1\dots i_r} &= a^{i_1}_{j_1}\dots a^{i_r}_{j_r}X^{j_1\dots j_r}
\end{aligned}
$$

where the dots in the middle row denote a polynomial expression, each term of which contains at least one of the quantities $a^i_{j_1 j_2}, \dots, a^i_{j_1\dots j_r}$. Write

$$(X^i, X^{i_1 i_2}, \dots, X^{i_1\dots i_r}) = (X_1, X_2, \dots, X_r).$$

By the general theory, the natural transformations $T^{(r)} \to T^{(r)}$ correspond to G_m^r-equivariant maps $f = (f_1, f_2, \dots, f_r)\colon S \to S$. Consider first the equivariance with respect to the homotheties in $GL(m) \subset G_m^r$. Using (3) we obtain

$$
\begin{aligned}
kf_1(X_1,\dots,X_s,\dots,X_r) &= f_1(kX_1,\dots,k^sX_s,\dots,k^rX_r)\\
&\vdots\\
k^sf_s(X_1,\dots,X_s,\dots,X_r) &= f_s(kX_1,\dots,k^sX_s,\dots,k^rX_r) \qquad (4)\\
&\vdots\\
k^rf_r(X_1,\dots,X_s,\dots,X_r) &= f_r(kX_1,\dots,k^sX_s,\dots,k^rX_r).
\end{aligned}
$$

By the homogeneous function theorem (see 24.1), f_1 is linear in X_1 and independent of $X_2,\dots,X_r$, while $f_s = g_s(X_s) + h_s(X_1,\dots,X_{s-1})$, where g_s is linear in X_s and h_s is a polynomial in $X_1, \dots, X_{s-1}$, $2 \le s \le r$. Further, the equivariancy of f with respect to the whole subgroup $GL(m)$ implies that g_s is a $GL(m)$-equivariant map of the s-th symmetric tensor power $S^s\mathbb{R}^m$ into itself. By the invariant tensor theorem (see 24.4), $g_s = c_s X_s$ (or explicitly, $g^{i_1 \dots i_s} = c_s X^{i_1 \dots i_s}$) with $c_s \in \mathbb{R}$.

Now let us use the equivariance with respect to the kernel B_1^r of the jet projection $G_m^r \to GL(m)$, i.e. $a_j^i = \delta_j^i$. The first line of (3) implies

$$(5) \quad c_1 X^i + a^i_{j_1 j_2}(c_2 X^{j_1 j_2} + h^{j_1 j_2}(X_1)) + \\ + \dots + a^i_{j_1 \dots j_r}(c_r X^{j_1 \dots j_r} + h^{j_1 \dots j_r}(X_1, \dots, X_{r-1})) = \\ = c_1(X^i + a^i_{j_1 j_2} X^{j_1 j_2} + \dots + a^i_{j_1 \dots j_r} X^{j_1 \dots j_r}).$$

Setting $a^i_{j_1 \dots j_s} = 0$ for all $s > 2$, we find $c_2 = c_1$ and $h^{j_1 j_2}(X_1) = 0$. By a recurrence procedure of similar type we further deduce

$$c_s = c_1, \qquad h^{j_1 \dots j_s}(X_1, \dots, X_{s-1}) = 0$$

for all $s = 3, \dots, r$.

This implies that the restriction of every natural transformation $T^{(r)} \to T^{(r)}$ to each subcategory $\mathcal{M}f_m$ is a homothety with a coefficient k_m. Taking into account the injection $\mathbb{R} \to \mathbb{R}^m$, $x \mapsto (x, 0, \dots, 0)$ we find $k_m = k_1$. □

40.9. Remark. We remark that all natural tensors of type $\binom{1}{1}$ on both $T^{(r)}M$ and the so-called extended r-th order tangent bundle $(J^r(M, \mathbb{R}))^*$ are determined in [Gancarzewicz, Kolář, to appear].

41. Star bundle functors

The tangent functor T is a covariant functor on the category $\mathcal{M}f$, but its dual T^* can be interpreted as a covariant functor on the subcategory $\mathcal{M}f_m$ of local diffeomorphisms of m-manifolds only. In this section we explain how to treat functors with a similar kind of contravariant character like T^* on the whole category $\mathcal{M}f$.

41.1. The category of star bundles. Consider a fibered manifold $Y \to M$ and a smooth map $f\colon N \to M$. Let us recall that the induced fibered manifold $f^*Y \to N$ is given by the pullback

$$\begin{array}{ccc} f^*Y & \xrightarrow{f_Y} & Y \\ \downarrow & & \downarrow \\ N & \xrightarrow{f} & M \end{array}$$

The restrictions of the fibered morphism f_Y to individual fibers are diffeomorphisms and we can write

$$f^*Y = \{(x, y);\ x \in N,\ y \in Y_{f(x)}\}, \qquad f_Y(x, y) = y.$$

Clearly $(f \circ g)^*Y \cong g^*(f^*Y)$. Let us consider another fibered manifold $Y' \to M$ over the same base, and a base-preserving fibered morphism $\varphi: Y \to Y'$. Given a smooth map $f: N \to M$, by the pullback property there is a unique fibered morphism $f^*\varphi: f^*Y \to f^*Y'$ such that

$$f_{Y'} \circ f^*\varphi = \varphi \circ f_Y. \tag{1}$$

The pullbacks appear in many well known constructions in differential geometry. For example, given manifolds M, N and a smooth map $f: M \to N$, the cotangent mapping T^*f transforms every form $\omega \in T^*_{f(x)}N$ into $T^*f\omega \in T^*_xM$. Hence the mapping $f^*(T^*N) \to T^*M$ is a morphism over the identity on M. We know that the restriction of T^* to manifolds of any fixed dimension and local diffeomorphisms is a bundle functor on $\mathcal{M}f_m$, see 14.9, and it seems that the construction could be functorial on the whole category $\mathcal{M}f$ as well. However the codomain of T^* cannot be the category $\mathcal{FM}$.

Definition. The category $\mathcal{FM}^*$ of star bundles is defined as follows. The objects coincide with those of $\mathcal{FM}$, but morphisms $\varphi: (Y \to M) \to (Y' \to M')$ are couples (φ_0, φ_1) where $\varphi_0: M \to M'$ is a smooth map and $\varphi_1: (\varphi_0)^*Y' \to Y$ is a fibered morphism over id_M. The composition of morphisms is given by

$$(\psi_0, \psi_1) \circ (\varphi_0, \varphi_1) = (\psi_0 \circ \varphi_0, \varphi_1 \circ ((\varphi_0)^*\psi_1)). \tag{2}$$

Using the formulas (1) and (2) one verifies easily that this is a correct definition of a category. The base functor $B: \mathcal{FM}^* \to \mathcal{M}f$ is defined by $B(Y \to M) = M$, $B(\varphi_0, \varphi_1) = \varphi_0$.

41.2. Star bundle functors. A star bundle functor on $\mathcal{M}f$ is a covariant functor $F: \mathcal{M}f \to \mathcal{FM}^*$ satisfying

(i) $B \circ F = \mathrm{Id}_{\mathcal{M}f}$, so that the bundle projections determine a natural transformation $p: F \to \mathrm{Id}_{\mathcal{M}f}$.

(ii) If $i: U \to M$ is an inclusion of an open submanifold, then $FU = p_M^{-1}(U)$ and $Fi = (i, \varphi_1)$ where $\varphi_1: i^*(FM) \to FU$ is the canonical identification $i^*(FM) \cong p_M^{-1}(U) \subset FM$.

(iii) Every smoothly parameterized family of mappings is transformed into a smoothly parameterized one.

Given a smooth map $f: M \to N$ we shall often use the same notation Ff for the second component φ in $Ff = (f, \varphi)$. We can also view the star bundle functors as rules transforming any manifold M into a fiber bundle $p_M: FM \to M$ and any smooth map $f: M \to N$ into a base-preserving morphism $Ff: f^*(FN) \to FM$ with $F(\mathrm{id}_M) = \mathrm{id}_{FM}$ and $F(g \circ f) = Ff \circ f^*(Fg)$.

41.3. The associated maps. A star bundle functor F is said to be of order r if for every maps $f, g: M \to N$ and every point $x \in M$, the equality $j^r_x f = j^r_x g$ implies $Ff|(f^*(FN))_x = Fg|(g^*(FN))_x$, where we identify the fibers $(f^*(FN))_x$ and $(g^*(FN))_x$.

Let us consider an r-th order star bundle functor $F: \mathcal{M}f \to \mathcal{FM}^*$. For every r-jet $A = j^r_x f \in J^r_x(M, N)_y$ we define a map $FA: F_yN \to F_xM$ by

$$FA = Ff \circ (f_{FN}|(f^*(FN))_x)^{-1}, \tag{1}$$

where $f_{FN}\colon f^*(FN)\to FN$ is the canonical map. Given another r-jet $B = j^r_y g\in J^r_y(N,P)_x$, we have

$$F(B\circ A)=Ff\circ(f^*(Fg))\circ(f_{g^*(FP)}|(f^*g^*(FP))_x)^{-1}\circ(g_{FP}|(g^*(FP))_y)^{-1}.$$

Applying 41.1.(1) to individual fibers, we get

$$(f_{FN}|(f^*(FN))_x)^{-1}\circ Fg=f^*(Fg)\circ(f_{g^*(FP)}|(f^*g^*(FP))_x)^{-1}$$

and that is why

$$\begin{aligned}(2)\qquad F(B\circ A)&=Ff\circ(f_{FN}|(f^*FN)_x)^{-1}\circ Fg\circ(g_{FP}|(g^*FP)_y)^{-1}\\&=FA\circ FB.\end{aligned}$$

For any two manifolds M, N we define

$$(3)\qquad F_{M,N}\colon FN\times_N J^r(M,N)\to FM,\qquad (q,A)\mapsto FA(q).$$

These maps are called the associated maps to F.

Proposition. *The associated maps to any finite order star bundle functor are smooth.*

Proof. This follows from the regularity and locality conditions in the way shown in the proof of 14.4. □

41.4. Description of finite order star bundle functors. Let us consider an r-th order star bundle functor F. We denote $(L^r)^{\mathrm{op}}$ the dual category to L^r, $S_m=F_0\mathbb{R}^m$, $m\in\mathbb{N}_0$, and we call the system $\mathcal{S}=\{S_0,S_1,\dots\}$ the system of standard fibers of F, cf. 14.21. The restrictions $\ell_{m,n}\colon S_n\times L^r_{m,n}\to S_m$, $\ell_{m,n}(s,A)=FA(s)$, of the associated maps 41.3.(3) form the *induced action* of $(L^r)^{\mathrm{op}}$ on $\mathcal{S}$. Indeed, given another jet $B\in L^r(n,p)$ equality 41.3.(2) implies

$$\ell_{m,p}(s,B\circ A)=\ell_{m,n}(\ell_{n,p}(s,B),A).$$

On the other hand, let ℓ be an action of $(L^r)^{\mathrm{op}}$ on a system $\mathcal{S}=\{S_0,S_1,\dots\}$ of smooth manifolds and denote ℓ_m the left actions of G^r_m on S_m given by $\ell_m(A,s)=\ell_{m,m}(s,A^{-1})$. We shall construct a star bundle functor L from these data. We put $LM:=P^rM[S_m;\ell_m]$ for all manifolds M and similarly to 14.22 we also get the action on morphisms. Given a map $f\colon M\to N$, $x\in M$, $f(x)=y$, we define a map $FA\colon F_yN\to F_xM$,

$$FA(\{v,s\})=\{u,\ell_{m,n}(s,v^{-1}\circ A\circ u)\},$$

where $m=\dim M$, $n=\dim N$, $A=j^r_xf$, $v\in P^r_yN$, $s\in S_n$, and $u\in P^r_xM$ is chosen arbitrarily. The verification that this is a correct definition of smooth maps satisfying $F(B\circ A)=FA\circ FB$ is quite analogous to the considerations in 14.22 and is left to the reader. Now, we define $Lf|(f^*(FN))_x=FA\circ f_{FN}$ and it follows directly from 41.1.(1) that $L(g\circ f)=Lf\circ f^*(Lg)$.

Theorem. *There is a bijective correspondence between the set of all r-th order star bundle functors on $\mathcal{M}f$ and the set of all smooth actions of the category $(L^r)^{\mathrm{op}}$ on systems $\mathcal{S}$ of smooth manifolds.*

Proof. In the formulation of the theorem we identify naturally equivalent functors. Given an r-th order star bundle functor F, we have the induced action ℓ of $(L^r)^{\mathrm{op}}$ on the system of standard fibers. So we can construct the functor L. Analogously to 14.22, the associated maps define a natural equivalence between F and L. □

41.5. Remark. We clarified in 14.24 that the actions of the category L^r on systems of manifolds are in fact covariant functors $L^r \to \mathcal{M}f$. In the same way, actions of $(L^r)^{\mathrm{op}}$ correspond to covariant functors $(L^r)^{\mathrm{op}} \to \mathcal{M}f$ or, equivalently, to contravariant functors $L^r \to \mathcal{M}f$, which will also be denoted by F_{inf}. Hence we can summarize: *r-th order bundle functors correspond to covariant smooth functors $L^r \to \mathcal{M}f$ while r-th order star bundle functors to the contravariant ones.*

41.6. Example. Consider a manifold Q and a point $q \in Q$. To any manifold M we associate the fibered manifold $FM = J^r(M,Q)_q \xrightarrow{\alpha} M$ and a map $f\colon N \to M$ is transformed into a map $Ff\colon f^*(FM) \to FN$ defined as follows. Given a point $b \in f^*(J^r(M,Q)_q)$, $b = (x, j^r_{f(x)}g)$, we set $Ff(b) = j^r_x(g \circ f) \in J^r(N,Q)_q$. One verifies easily that F is a star bundle functor of order r. Let us mention the corresponding contravariant functor $L^r \to \mathcal{M}f$. We have $F_{\mathrm{inf}}(m) = J^r_0(\mathbb{R}^m, Q)_q$ and for arbitrary jets $j^r_0 f \in L^r_{m,n}$, $j^r_0 g \in F_{\mathrm{inf}}(m)$ it holds $F_{\mathrm{inf}}(j^r_0 f)(j^r_0 g) = j^r_0(g \circ f)$.

41.7. Vector bundle functors and vector star bundle functors. Let F be a bundle functor or a star bundle functor on $\mathcal{M}f$. By the definition of the induced action and by the construction of the (covariant or contravariant) functor $F_{\mathrm{inf}}\colon L^r \to \mathcal{M}f$, the values of the functor F belong to the subcategory of vector bundles if and only if the functor F_{inf} takes values in the category *Vect* of finite dimensional vector spaces and linear mappings. But using the construction of dual objects and morphisms in the category *Vect*, we get a duality between covariant and contravariant functors $F_{\mathrm{inf}}\colon L^r \to$ *Vect*. The corresponding duality between vector bundle functors and vector star bundle functors is a source of interesting geometric objects like r-th order tangent vectors, see 12.14 and below.

41.8. Examples. Let us continue in example 41.6. If the manifold Q happens to be a vector space and the point q its origin, we clearly get a vector star bundle functor. Taking $Q = \mathbb{R}$ we get the r-th order cotangent functor T^{r*}. If we set $Q = \mathbb{R}^k$, then the corresponding star bundle functor is the functor T^{r*}_k of the (k,r)-covelocities, cf. 12.14.

The dual vector bundle functor to T^{r*} is the r-th order tangent functor. The dual functor to the (k,r)-covelocities is the functor $T^{r\square}_k$, see 12.14.

Remarks

Most of the exposition concerning the bundle functors on $\mathcal{M}f$ is based on [Kolář, Slovák, 89], but the prolongation of Lie groups was described in [Kolář, 83]. The generalization to bundle functors on $\mathcal{FM}_m$ follows [Slovák, 91].

The existence of the canonical flow-natural transformation $FT \to TF$ was first deduced by A. Kock in the framework of the so called synthetic differential geometry, see e.g. [Kock, 81]. His unpublished note originated in a discussion with the first author. Then the latter developed, with consent of the former, the proof of that result dealing with classical manifolds only.

The description of all natural transformations with the source in a Weil bundle by means of some special elements in the standard fiber is a generalization of an idea from [Kolář, 86] due to [Mikulski, 89 b]. The natural transformations $T^{(r)} \to T^{(r)}$ were first classified in [Kolář, Vosmanská, 89].

CHAPTER X.
PROLONGATION OF VECTOR FIELDS AND CONNECTIONS

This section is devoted to systematic investigation of the natural operators transforming vector fields into vector fields or general connections into general connections. For the sake of simplicity we also speak on the prolongations of vector fields and connections. We first determine all natural operators transforming vector fields on a manifold M into vector fields on a Weil bundle over M. In the formulation of the result as well as in the proof we use heavily the technique of Weil algebras. Then we study the prolongations of vector fields to the bundle of second order tangent vectors. We like to comment the interesting general differences between a product-preserving functor and a non-product-preserving one in this case. For the prolongations of projectable vector fields to the r-jet prolongation of a fibered manifold, which play an important role in the variational calculus, we prove that the unique natural operator, up to a multiplicative constant, is the flow operator.

Using the flow-natural equivalence we construct a natural operator transforming general connections on $Y \to M$ into general connections on $T_AY \to T_AM$ for every Weil algebra A. In the case of the tangent functor we determine all first-order natural operators transforming connections on $Y \to M$ into connections on $TY \to TM$. This clarifies that the above mentioned operator is not the unique natural operator in general. Another class of problems is to study the prolongations of connections from $Y \to M$ to $FY \to M$, where F is a functor defined on local isomorphisms of fibered manifolds. If we apply the idea of the flow prolongation of vector fields, we see that such a construction depends on an r-th order linear connection on the base manifold, provided r means the horizontal order of F. In the case of the vertical tangent functor we obtain the operator defined in another way in chapter VII. For the functor J^1 of the first jet prolongation of fibered manifolds we deduce that all natural operators transforming a general connection on $Y \to M$ and a linear connection on M into a general connection on $J^1Y \to M$ form a simple 4-parameter family. In conclusion we study the prolongation of general connections from $Y \to M$ to $VY \to Y$. From the general point of view it is interesting that such an operator exists only in the case of affine bundles (with vector bundles as a special sub case). But we can consider arbitrary connections on them (i.e. arbitrary nonlinear connections in the vector bundle case).

42. Prolongations of vector fields to Weil bundles

Let F be an arbitrary natural bundle over m-manifolds. We first deduce some general properties of the natural operators $A\colon T \rightsquigarrow TF$, i.e. of the natural operators transforming every vector field on a manifold M into a vector field on FM. Starting from 42.7 we shall discuss the case that F is a Weil functor.

42.1. One general example of a natural operator $T \rightsquigarrow TF$ is the flow operator $\mathcal{F}$ of a natural bundle F defined by

$$\mathcal{F}_M X = \tfrac{\partial}{\partial t}\big|_0 F(\mathrm{Fl}_t^X)$$

where Fl^X means the flow of a vector field X on M, cf. 6.19.

The composition $TF = T \circ F$ is another bundle functor on $\mathcal{M}f_m$ and the bundle projection of T is a natural transformation $TF \to F$. Assume we have a natural transformation $i\colon TF \to TF$ over the identity of F. Then we can construct further natural operators $T \rightsquigarrow TF$ by using the following lemma, the proof of which consists in a standard diagram chase.

Lemma. *If $A\colon T \rightsquigarrow TF$ is a natural operator and $i\colon TF \to TF$ is a natural transformation over the identity of F, then $i \circ A\colon T \rightsquigarrow TF$ is also a natural operator.* □

42.2. Absolute operators. This is another class of natural operators $T \rightsquigarrow TF$, which is related with the natural transformations $F \to F$. Let 0_M be the zero vector field on M.

Definition. A natural operator $A\colon T \rightsquigarrow TF$ is said to be an *absolute operator*, if $A_M X = A_M 0_M$ for every vector field X on M.

It is easy to check that, for every natural operator $A\colon T \rightsquigarrow TF$, the operator transforming every $X \in C^\infty(TM)$ into $A_M 0_M$ is also natural. Hence this is an absolute operator called associated with A.

Let L_M be the Liouville vector field on TM, i.e. the vector field generated by the one-parameter group of all homotheties of the vector bundle $TM \to M$. The rule transforming every vector field on M into L_M is the simplest example of an absolute operator in the case $F = T$. The naturality of this operator follows from the fact that every homothety is a natural transformation $T \to T$. Such a construction can be generalized. Let $\varphi(t)$ be a smooth one-parameter family of natural transformations $F \to F$ with $\varphi(0) = \mathrm{id}$, where smoothness means that the map $(\varphi(t))_M\colon \mathbb{R} \times FM \to FM$ is smooth for every manifold M. Then

$$\Phi(M) = \tfrac{\partial}{\partial t}\big|_0 (\varphi(t))_M$$

is a vertical vector field on FM. The rule $X \mapsto \Phi(M)$ for every $X \in C^\infty(TM)$ is an absolute operator $T \rightsquigarrow TF$, which is said to be generated by $\varphi(t)$.

42.3. Lemma. *For an absolute operator $A\colon T \rightsquigarrow TF$ every $A_M 0_M$ is a vertical vector field on FM.*

Proof. Let $\mathcal{J}\colon U \to FM$, $U \subset \mathbb{R} \times FM$, be the flow of $A_M 0_M$ and let $\mathcal{J}_t$ be its restriction for a fixed $t \in \mathbb{R}$. Assume there exists $W \in F_x M$ and $t \in \mathbb{R}$ such that $p_M \mathcal{J}_t(W) = y \neq x$, where $p_M : FM \to M$ is the bundle projection. Take $f \in \mathrm{Diff}(M)$ with the identity germ at x and $f(y) \neq y$, so that the restriction of Ff to $F_x M$ is the identity. Since $A_M 0_M$ is a vector field Ff-related with itself, we have $Ff \circ \mathcal{J}_t = \mathcal{J}_t \circ Ff$ whenever both sides are defined. In particular, $p_M(Ff)\mathcal{J}_t(W) = f p_M \mathcal{J}_t(W) = f(y)$ and $p_M \mathcal{J}_t(Ff)(W) = p_M \mathcal{J}_t(W) = y$, which is a contradiction. Hence the value of $A_M 0_M$ at every $W \in FM$ is a vertical vector. □

42.4. Order estimate. It is well known that every vector field X on a manifold M with non-zero value at $x \in M$ can be expressed in a suitable local coordinate system centered at x as the constant vector field

$$X = \frac{\partial}{\partial x^1}. \tag{1}$$

This simple fact has several pleasant consequences for the study of natural operators on vector fields. The first of them can be seen in the proof of the following lemma.

Lemma. *Let X and Y be two vector fields on M with $X(x) \neq 0$ and $j_x^r X = j_x^r Y$. Then there exists a local diffeomorphism f transforming X into Y such that $j_x^{r+1} f = j_x^{r+1} \mathrm{id}_M$.*

Proof. Take a local coordinate system centered at x such that (1) holds. Then the coordinate functions Y^i of Y have the form $Y^i = \delta_1^i + g^i(x)$ with $j_0^r g^i = 0$. Consider the solution $f = (f^i(x))$ of the following system of equations

$$\delta_1^i + g^i(f^1(x), \dots, f^m(x)) = \frac{\partial f^i(x)}{\partial x^1}$$

determined by the initial condition $f = \mathrm{id}$ on the hyperplane $x^1 = 0$. Then f is a local diffeomorphism transforming X into Y. We claim that the k-th order partial derivatives of f at the origin vanish for all $1 < k \leq r+1$. Indeed, if there is no derivative along the first axis, all the derivatives of order higher than one vanish according to the initial condition, and all other cases follow directly from the equations. By the same argument we find that the first order partial derivatives of f at the origin coincide with the partial derivatives of the identity map. □

This lemma enables us to derive a simple estimate of the order of the natural operators $T \rightsquigarrow TF$.

42.5. Proposition. *If F is an r-th order natural bundle, then the order of every natural operator $A : T \rightsquigarrow TF$ is less than or equal to r.*

Proof. Assume first $X(x) \neq 0$ and $j_x^r X = j_x^r Y$, $x \in M$. Taking a local diffeomorphism f of lemma 42.4, we have locally $A_M Y = (TFf) \circ A_M X \circ (Ff)^{-1}$. But TF is an $(r+1)$-st order natural bundle, so that $j_x^{r+1} f = j_x^{r+1} \mathrm{id}_M$ implies that the restriction of TFf to the fiber of $TFM \to M$ over x is the identity. Hence $A_M Y | F_x M = A_M X | F_x M$. In the case $X(x) = 0$ we take any vector field Z with $Z(x) \neq 0$ and consider the one-parameter families of vector fields $X + tZ$ and $Y + tZ$, $t \in \mathbb{R}$. For every $t \neq 0$ we have $A_M(X + tZ)|F_x M = A_M(Y + tZ)|F_x M$ by the first part of the proof. Since A is regular, this relation holds for $t = 0$ as well. □

42.6. Let S be the standard fiber of an r-th order bundle functor F on $\mathcal{M}f_m$, let Z be the standard fiber of TF and let $q\colon Z \to S$ be the canonical projection. Further, let $V^r_m = J^r_0 T\mathbb{R}^m$ be the space of all r-jets at zero of vector fields on $\mathbb{R}^m$ and let $V_0 \subset V^r_m$ be the subspace of r-jets of the constant vector fields on $\mathbb{R}^m$, i.e. of the vector fields invariant with respect to the translations of $\mathbb{R}^m$. By 18.19 and by proposition 42.5, the natural operators $A\colon T \rightsquigarrow TF$ are in bijection with the associated G^{r+1}_m-equivariant maps $\mathcal{A}\colon V^r_m \times S \to Z$ satisfying $q \circ \mathcal{A} = \mathrm{pr}_2$. Consider the associated maps $\mathcal{A}_1$, $\mathcal{A}_2$ of two natural operators $A_1, A_2\colon T \rightsquigarrow TF$.

Lemma. *If two associated maps $\mathcal{A}_1$, $\mathcal{A}_2\colon V^r_m \times S \to Z$ coincide on $V_0 \times S \subset V^r_m \times S$, then $A_1 = A_2$.*

Proof. If X is a vector field on $\mathbb{R}^m$ with $X(0) \neq 0$, then there is a local diffeomorphism transforming X into the constant vector field 42.1.(1). Hence if the G^{r+1}_m-equivariant maps $\mathcal{A}_1$ and $\mathcal{A}_2$ coincide on $V_0 \times S$, they coincide on those pairs in $V^r_m \times S$, the first component of which corresponds to an r-jet of a vector field with non-zero value at the origin. But this is a dense subset in V^r_m, so that $\mathcal{A}_1 = \mathcal{A}_2$. □

42.7. Absolute operators $T \rightsquigarrow TT_B$. Consider a Weil functor T_B. (We denote a Weil algebra by an unusual symbol B here, since A is taken for natural operators.) By 35.17, for any two Weil algebras B_1 and B_2 there is a bijection between the set of all algebra homomorphisms $\mathrm{Hom}(B_1, B_2)$ and the set of all natural transformations $T_{B_1} \to T_{B_2}$ on the whole category $\mathcal{M}f$. To determine all absolute operators $T \rightsquigarrow TT_B$, we shall need the same result for the natural transformations $T_{B_1} \to T_{B_2}$ on $\mathcal{M}f_m$, which requires an independent proof. If $B = \mathbb{R} \times N$ is a Weil algebra of order r, we have a canonical action of G^r_m on $(T_B\mathbb{R}^m)_0 = N^m$ defined by

$$(j^r_0 f)(j_B g) = j_B(f \circ g)$$

Assume both B_1 and B_2 are of order r. In 14.12 we have explained a canonical bijection between the natural transformations $T_{B_1} \to T_{B_2}$ on $\mathcal{M}f_m$ and the G^r_m-maps $N^m_1 \to N^m_2$. Hence it suffices to deduce

Lemma. *All G^r_m-maps $N^m_1 \to N^m_2$ are induced by algebra homomorphisms $B_1 \to B_2$.*

Proof. Let $H\colon N^m_1 \to N^m_2$ be a G^r_m-map. Write $H = (h_i(y_1, \dots, y_m))$ with $y_i \in N_1$. The equivariance of H with respect to the homotheties in $i(G^1_m) \subset G^r_m$ yields $kh_i(y_1, \dots, y_m) = h_i(ky_1, \dots, ky_m)$, $k \in \mathbb{R}$, $k \neq 0$. By the homogeneous function theorem, all h_i are linear maps. Expressing the equivariance of H with respect to the multiplication in the direction of the i-th axis in $\mathbb{R}^m$, we obtain $h_j(0, \dots, y_i, \dots, 0) = h_j(0, \dots, ky_i, \dots, 0)$ for $j \neq i$. This implies that h_j depends on y_j only. Taking into account the exchange of the axis in $\mathbb{R}^m$, we find $h_i = h(y_i)$, where h is a linear map $N_1 \to N_2$. On the first axis in $\mathbb{R}^m$ consider the map $x \mapsto x + x^2$ completed by the identities on the other axes. The equivariance of H with respect to the r-jet at zero of the latter map implies $h(y) + h(y)^2 = h(y + y^2) = h(y) + h(y^2)$. This yields $h(y^2) = (h(y))^2$ and by polarization we obtain $h(y\bar{y}) = h(y)h(\bar{y})$. Hence h is an algebra homomorphism $N_1 \to N_2$, that is uniquely extended to a homomorphism $B_1 \to B_2$ by means of the identity of $\mathbb{R}$. $\square$

42.8. The group $\mathrm{Aut}B$ of all algebra automorphisms of B is a closed subgroup in $GL(B)$, so that it is a Lie subgroup by 5.5. Every element of its Lie algebra $D \in \mathfrak{Aut}\,B$ is tangent to a one-parameter subgroup $d(t)$ and determines a vector field $D(M)$ tangent to $(d(t))_M$ for $t = 0$ on every bundle T_BM. By 42.2, the constant maps $X \mapsto D(M)$ for all $X \in C^\infty(TM)$ form an absolute operator $\mathrm{op}(D)\colon T \rightsquigarrow TT_B$, which will be said to be generated by D.

Proposition. *Every absolute operator $A\colon T \rightsquigarrow TT_B$ is of the form $A = \mathrm{op}(D)$ for a $D \in \mathfrak{Aut}\,B$.*

Proof. By 42.3, $A_M 0_M$ is a vertical vector field. Since $A_M 0_M$ is Ff-related with itself for every $f \in \mathrm{Diff}(M)$, every transformation $\mathcal{J}_t$ of its flow corresponds to a natural transformation of T_B into itself. By lemma 42.7 there is a one-parameter group $d(t)$ in $\mathrm{Aut}B$ such that $\mathcal{J}_t = (d(t))_M$. $\square$

42.9. We recall that a derivation of B is a linear map $D\colon B \to B$ satisfying $D(ab) = D(a)b + aD(b)$ for all a, $b \in B$. The set of all derivations of B is denoted by $\mathrm{Der}\,B$. The Lie algebra of $GL(B)$ is the space $L(B,B)$ of all linear maps $B \to B$. We have $\mathrm{Der}\,B \subset L(B,B)$ and $\mathrm{Aut}\,B \subset GL(B)$.

Lemma. $\mathrm{Der}B$ *coincides with the Lie algebra of* $\mathrm{Aut}B$.

Proof. If h_t is a one-parameter subgroup in $\mathrm{Aut}\,B$, then its tangent vector belongs to $\mathrm{Der}\,B$, since

$$\tfrac{\partial}{\partial t}\big|_0 h_t(ab) = \tfrac{\partial}{\partial t}\big|_0 h_t(a)h_t(b) = \left(\tfrac{\partial}{\partial t}\big|_0 h_t(a)\right) b + a\left(\tfrac{\partial}{\partial t}\big|_0 h_t(b)\right).$$

To prove the converse, let us consider the exponential mapping $L(B,B) \to GL(B)$. For every derivation D the Leibniz formula

$$D^k(ab) = \sum_{i=0}^{k} \binom{k}{i} D^i(a)D^{k-i}(b)$$

holds. Hence the one-parameter group $h_t = \sum_{k=0}^\infty \frac{t^k}{k!}D^k$ satisfies

$$\begin{aligned} h_t(ab) &= \sum_{k=0}^{\infty}\sum_{i=0}^{k} \tfrac{t^k}{k!}\binom{k}{i} D^i(a)D^{k-i}(b) \\ &= \sum_{k=0}^{\infty}\sum_{i=0}^{k} \tfrac{t^i}{i!}D^i(a)\tfrac{t^{k-i}}{(k-i)!}D^{k-i}(b) \\ &= \left(\sum_{k=0}^{\infty} \tfrac{t^k}{k!}D^k(a)\right)\left(\sum_{j=0}^{\infty} \tfrac{t^j}{j!}D^j(b)\right) = h_t(a)h_t(b). \quad \square \end{aligned}$$

42.10. Using the theory of Weil algebras, we determine easily all natural transformations $TT_B \to TT_B$ over the identity of T_B. The functor TT_B corresponds to the tensor product of algebras $B \otimes \mathbb{D}$ of B with the algebra $\mathbb{D}$ of dual numbers, which is identified with $B \times B$ endowed with the following multiplication

$$(1) \qquad (a,b)(c,d) = (ac, ad + bc)$$

the products of the components being in B. The natural transformations of TT_B into itself over the identity of T_B correspond to the endomorphisms of (1) over the identity on the first factor.

Lemma. *All homomorphisms of $B \otimes \mathbb{D} \cong B \times B$ into itself over the identity on the first factor are of the form*

$$(2) \qquad h(a,b) = (a, cb + D(a))$$

with any $c \in B$ and any $D \in \operatorname{Der} B$.

Proof. On one hand, one verifies directly that every map (2) is a homomorphism. On the other hand, consider a map $h\colon B \times B \to B \times B$ of the form $h(a,b) = (a, f(a) + g(b))$, where f, $g\colon B \to B$ are linear maps. Then the homomorphism condition for h requires $af(c) + ag(d) + cf(a) + cg(b) = f(ac) + g(bc + ad))$. Setting $b = d = 0$, we obtain $af(c) + cf(a) = f(ac)$, so that f is a derivation. For $a = d = 0$ we have $g(bc) = cg(b)$. Setting $b = 1$ and $c = b$ we find $g(b) = g(1)b$. □

42.11. There is a canonical action of the elements of B on the tangent vectors of T_BM, [Morimoto, 76]. It can be introduced as follows. The multiplication of the tangent vectors of M by reals is a map $m\colon \mathbb{R} \times TM \to TM$. Applying the functor T_B, we obtain $T_Bm\colon B \times T_BTM \to T_BTM$. By 35.18 we have a natural identification $TT_BM \cong T_BTM$. Then T_Bm can be interpreted as a map $B \times TT_BM \to TT_BM$. Since the algebra multiplication in B is the T_B-prolongation of the multiplication of reals, the action of $c \in B$ on $(a_1, \dots, a_m, b_1, \dots, b_m) \in TT_B\mathbb{R}^m = B^{2m}$ has the form

$$(1) \qquad c(a_1, \dots, a_m, b_1, \dots, b_m) = (a_1, \dots, a_m, cb_1, \dots, cb_m).$$

In particular this implies that for every manifold M the action of $c \in B$ on TT_BM is a natural tensor $\mathrm{af}_M(c)$ of type $\binom{1}{1}$ on M. (The tensors of type $\binom{1}{1}$ are sometimes called affinors, which justifies our notation.)

By lemma 42.1 and 42.10, if we compose the flow operator $\mathcal{T}_B$ of T_B with all natural transformations $TT_B \to TT_B$ over the identity of T_B, we obtain the following system of natural operators $T \rightsquigarrow TT_B$

$$(2) \qquad \mathrm{af}(c) \circ \mathcal{T}_B + \mathrm{op}(D) \qquad \text{for all } c \in B \text{ and all } D \in \operatorname{Der} B.$$

42.12. Theorem. *All natural operators $T \rightsquigarrow TT_B$ are of the form 42.11.(2).*

Proof. The standard fibers in the sense of 42.6 are $S = N^m$ and $Z = N^m \times B^m$. Let $\mathcal{A}\colon V_m^r \times N^m \to N^m \times B^m$ be the associated map of a natural operator $A\colon T \rightsquigarrow TT_B$ and let $\mathcal{A}_0 = \mathcal{A}|V_0 \times N^m$. Write $y \in N$, $(X, Y) \in B = \mathbb{R} \times N$ and $(v_i) \in V_0$, so that $v_i \in \mathbb{R}$. Then the coordinate expression of $\mathcal{A}_0$ has the form $y_i = y_i$ and

$$X_i = f_i(v_i, y_i), \qquad Y_i = g_i(v_i, y_i)$$

Taking into account the inclusion $i(G_m^1) \subset G_m^{r+1}$, one verifies directly that V_0 is a G_m^1-invariant subspace in V_m^r. If we study the equivariance of (f_i, g_i) with respect to G_m^1, we deduce in the same way as in the proof of lemma 42.7

$$X_i = f(y_i) + kv_i, \qquad Y_i = g(y_i) + h(v_i) \tag{1}$$

where $f\colon N \to \mathbb{R}$, $g\colon N \to N$, $h\colon \mathbb{R} \to N$ are linear maps and $k \in \mathbb{R}$.

Setting $v_i = 0$ in (1), we obtain the coordinate expression of the absolute operator associated with A in the sense of 42.2. By proposition 42.8 and lemmas 42.3 and 42.9, $f = 0$ and g is a derivation in N, which is uniquely extended into a derivation D_A in B by requiring $D_A(1) = 0$. On the other hand, $h(1) \in N$, so that $c_A = k + h(1)$ is an element of B.

Consider the natural transformation $H_A\colon TT_B \to TT_B$ determined by c_A and D_A in the sense of lemma 42.10. Since the flow of every constant vector field on $\mathbb{R}^m$ is formed by the translations, its T_B-prolongation on $T_B\mathbb{R}^m = \mathbb{R}^m \times N^m$ is formed by the products of the translations on $\mathbb{R}^m$ and the identity map on N^m. This implies that $\mathcal{A}$ and the associated map of $H_A \circ \mathcal{T}_B$ coincide on $V_0 \times N^m$. Applying lemma 42.6, we prove our assertion. □

42.13. Example. In the special case of the functor T_1^r of 1-dimensional velocities of arbitrary order r, which is used in the geometric approach to higher order mechanics, we interpret our result in a direct geometric way. Given some local coordinates x^i on M, the r-th order Taylor expansion of a curve $x^i(t)$ determines the induced coordinates $y_1^i, \dots, y_r^i$ on $T_1^r M$. Let $X^i = dx^i$, $Y_1^i = dy_1^i, \dots, Y_r^i = dy_r^i$ be the additional coordinates on $TT_1^r M$. The element $x + \langle x^{r+1} \rangle \in \mathbb{R}[x]/\langle x^{r+1} \rangle$ defines a natural tensor $\mathrm{af}_M(x + \langle x^{r+1} \rangle) =: Q_M$ of type $\binom{1}{1}$ on $T_1^r M$, the coordinate expression of which is $Q_M(X^i, Y_1^i, Y_2^i, \dots, Y_r^i) = (0, X^i, Y_1^i, \dots, Y_{r-1}^i)$. We remark that this tensor was introduced in another way by [de León, Rodriguez, 88]. The reparametrization $x^i(t) \mapsto x^i(kt)$, $0 \neq k \in \mathbb{R}$, induces a one-parameter group of diffeomorphisms of $T_1^r M$ that generates the so called generalized Liouville vector field L_M on $T_1^r M$ with the coordinate expression $X^i = 0$, $Y_s^i = sy_s^i$, $s = 1, \dots, r$. This gives rise to an absolute operator $L\colon T \rightsquigarrow TT_1^r$. If we 'translate' theorem 42.12 from the language of Weil algebras, we deduce that *all natural operators $T \rightsquigarrow TT_1^r$ form a $(2r+1)$-parameter family linearly generated by the following operators*

$$\mathcal{T}_1^r,\ Q \circ \mathcal{T}_1^r, \dots,\ Q^r \circ \mathcal{T}_1^r,\ L,\ Q \circ L, \dots,\ Q^{r-1} \circ L.$$

For $r = 1$, i.e. if we have the classical tangent functor T, we obtain a 3-parameter family generated by the flow operator $\mathcal{T}$, by the so-called vertical lift

$Q \circ \mathcal{T}$ and by the classical Liouville field on TM. (The vertical lift transforms every section $X\colon M \to TM$ into a vertical vector field on TM determined by the translations in the individual fibers of TM.) The latter result was deduced by [Sekizawa, 88a] by the method of differential equations and under an additional assumption on the order of the operators.

42.14. Remark. The natural operators $T \rightsquigarrow TT_k^r$ were studied from a slightly different point of view by [Gancarzewicz, 83a]. He has assumed in addition that all maps $A_M\colon C^\infty(TM) \to C^\infty(TT_k^r M)$ are $\mathbb{R}$-linear and that every $A_M X$, $X \in C^\infty(TM)$ is a projectable vector field on $T_k^r M$. He has determined and described geometrically all such operators. Of course, they are of the form $\text{af}(c) \circ \mathcal{T}_k^r$, for all $c \in \mathbb{D}_k^r$. It is interesting to remark that from the list 42.11.(2) we know that for every natural operator $A\colon T \rightsquigarrow TT_B$ every $A_M X$ is a projectable vector field on $T_B M$. The description of the absolute operators in the case of the functor T_k^r is very simple, since all natural equivalences $T_k^r \to T_k^r$ correspond to the elements of G_k^r acting on the velocities by reparametrization. We also remark that for $r = 1$ Janyška determined all natural operators $T \rightsquigarrow TT_k^1$ by direct evaluation, [Krupka, Janyška, 90].

43. The case of the second order tangent vectors

Theorem 42.12 implies that the natural operators transforming vector fields to product preserving bundle functors have several nice properties. Some of them are caused by the functorial character of the Weil algebras in question. It is useful to clarify that for the non-product-preserving functors on $\mathcal{M}f$ one can meet a quite different situation. As a concrete example we discuss the second order tangent vectors defined in 12.14. We first deduce that all natural operators $T \rightsquigarrow TT^{(2)}$ form a 4-parameter family. Then we comment its most significant properties which differ from the product-preserving case.

43.1. Since $T^{(2)}$ is a functor with values in the category of vector bundles, the multiplication of vectors by real numbers determines the Liouville vector field L_M on every $T^{(2)}M$. Clearly, $X \mapsto L_M$, $X \in C^\infty(TM)$ is an absolute operator $T \rightsquigarrow TT^{(2)}$. Further, we have a canonical inclusion $TM \subset T^{(2)}M$. Using the fiber translations on $T^{(2)}M$, we can extend every section $X\colon M \to TM$ into a vector field $V(X)$ on $T^{(2)}M$. This defines a second natural operator $V\colon T \rightsquigarrow TT^{(2)}$. Moreover, if we iterate the derivative $X(Xf)$ of a function $f\colon M \to \mathbb{R}$ with respect to a vector field X on M, we obtain, at every point $x \in M$, a linear map from $(T_1^{2*}M)_x$ into the reals, i.e. an element of $T_x^{(2)}M$. This determines a first order operator $C^\infty(TM) \to C^\infty(T^{(2)}M)$, the coordinate form of which is

$$X^i \frac{\partial}{\partial x^i} \mapsto X^j \frac{\partial X^i}{\partial x^j} \frac{\partial}{\partial x^i} + X^i X^j \frac{\partial^2}{\partial x^i \partial x^j} \tag{1}$$

Since every section of the vector bundle $T^{(2)}$ can be extended, by means of fiber translations, into a vector field constant on each fiber, we get from (1) another natural operator $D\colon T \rightsquigarrow TT^{(2)}$. Finally, $\mathcal{T}^{(2)}$ means the flow operator as usual.

43.2. Proposition. *All natural operators $T \rightsquigarrow TT^{(2)}$ form the 4-parameter family*

$$k_1 T^{(2)} + k_2 V + k_3 L + k_4 D, \qquad k_1,\, k_2,\, k_3,\, k_4 \in \mathbb{R}. \tag{1}$$

Proof. By proposition 42.5, every natural operator $A\colon T \rightsquigarrow TT^{(2)}$ has order ≤ 2. Let $V^2_m = J^2_0(T\mathbb{R}^m)$, $S = T^{(2)}_0\mathbb{R}^m$, $Z = (TT^{(2)})_0\mathbb{R}^m$ and $q\colon Z \to S$ be the canonical projection. We have to determine all G^3_m-equivariant maps $f\colon V^2_m \times S \to Z$ satisfying $q \circ f = \mathrm{pr}_2$. The action of G^3_m on V^2_m is

$$\bar{X}^i = a^i_j X^j, \qquad \bar{X}^i_j = a^i_{kl}\tilde{a}^k_j X^l + a^i_k X^k_l \tilde{a}^l_j \tag{2}$$

while for X^i_{jk} we shall need the action

$$\bar{X}^i_{jk} = X^i_{jk} + a^i_{jkl}X^l \tag{3}$$

of the kernel K_3 of the jet projection $G^3_m \to G^2_m$ only. The action of G^2_m on S is

$$\bar{u}^i = a^i_j u^j + a^i_{jk}u^{jk}, \qquad \bar{u}^{ij} = a^i_k a^j_l u^{kl}, \tag{4}$$

see 40.8.(2). The induced coordinates on Z are $Y^i = dx^i$, $U^i = du^i$, $U^{ij} = du^{ij}$, and (4) implies

$$\begin{aligned} \bar{Y}^i &= a^i_j Y^j \\ \bar{U}^i &= a^i_{jk}u^j Y^k + a^i_j U^j + a^i_{jkl}Y^l u^{jk} + a^i_{jk}U^{jk} \\ \bar{U}^{ij} &= a^i_{km}a^j_l u^{kl}Y^m + a^i_k a^j_{lm}u^{kl}Y^m + a^i_k a^j_l U^{kl}. \end{aligned} \tag{5}$$

Using (4) we find the following coordinate expression of the flow operator $\mathcal{T}^{(2)}$

$$X^i \tfrac{\partial}{\partial x^i} + (X^i_j u^j + X^i_{jk}u^{jk}) \tfrac{\partial}{\partial u^i} + \left(X^i_k u^{kj} + X^j_k u^{ik}\right) \tfrac{\partial}{\partial u^{ij}}. \tag{6}$$

Consider the first series of components

$$Y^i = f^i(X^j, X^k_l, X^m_{np}, u^q, u^{rs})$$

of the associated map of A. The equivariance of f^i with respect to the kernel K_3 reads

$$f^i(X^j, X^k_l, X^m_{np}, u^q, u^{rs}) = f^i(X^j, X^k_l, X^m_{np} + a^m_{npt}X^t, u^q, u^{rs}).$$

This implies that f^i are independent of X^i_{jk}. Then the equivariance with respect to the subgroup $a^i_j = \delta^i_j$ yields

$$f^i(X^j, X^k_l, u^m, u^{np}) = f^i(X^j, X^k_l + a^k_{lq}X^q, u^m + a^m_{rs}u^{rs}, u^{np}).$$

This gives $f^i = f^i(X^j, u^{kl})$. Using the homotheties in $i(G^1_m) \subset G^3_m$, we obtain $f^i = f^i(X^j)$. Example 24.14 then implies

$$Y^i = kX^i. \tag{7}$$

Consider further the difference $A - kT^{(2)}$ with k taken from (7) and denote by h^i, h^{ij} its components. We evaluate easily

$$a^i_k a^j_l h^{kl}(X^m, X^n_p, X^q_{rs}, u^t, u^{uv}) = h^{ij}(\bar{X}^m, \bar{X}^n_p, \bar{X}^q_{rs}, \bar{u}^t, \bar{u}^{uv}). \tag{8}$$

Quite similarly as in the first step we deduce $h^{ij} = h^{ij}(X^k, u^{lm})$. By homogeneity and the invariant tensor theorem, we then obtain

$$h^{ij} = cu^{ij} + aX^iX^j. \tag{9}$$

For h^i, we find

$$\begin{aligned} a^i_j h^j(X^k, X^l_m, X^n_{pq}, u^r, u^{st}) + ca^i_{jk}u^{jk} + aa^i_{jk}X^jX^k = \\ = h^i(\bar{X}^k, \bar{X}^l_m, \bar{X}^n_{pq}, \bar{u}^r, \bar{u}^{st}). \end{aligned} \tag{10}$$

By (3), h^i is independent of X^i_{jk}. Then the homogeneity condition implies

$$h^i = f^i_j(X^k_l)X^j + g^i_j(X^k_l)u^j. \tag{11}$$

For $X^i = 0$, the equivariance of (11) with respect to the subgroup $a^i_j = \delta^i_j$ reads

$$g^i_j(X^k_l)u^j + ca^i_{jk}u^{jk} = g^i_j(X^k_l)(u^j + a^j_{kl}u^{kl}). \tag{12}$$

Hence $g^i_j(X^k_l) = c\delta^i_j$. The remaining equivariance condition is

$$f^i_j(X^k_l)X^j + aa^i_{jk}X^jX^k = f^i_j(X^k_l + a^k_{lm}X^m)X^j. \tag{13}$$

This implies that all the first order partial derivatives of $f^i_j(X^k_l)$ are constant, so that f^i_j are at most linear in X^k_l. By the invariant tensor theorem, $f^i_j(X^k_l)X^j = eX^jX^i_j + bX^i$. Then (13) yields $e = a$, i.e.

$$h^i = cu^i + bX^i + aX^jX^i_j. \tag{14}$$

This gives the coordinate expression of (1). □

43.3. Remark. For a Weil functor T_B, all natural operators $T \rightsquigarrow TT_B$ are of the form $H \circ T_B$, where H is a natural transformation $TT_B \to TT_B$ over the identity of T_B. For $T^{(2)}$, one evaluates easily that all natural transformations $H\colon TT^{(2)} \to TT^{(2)}$ over the identity of $T^{(2)}$ form the following 3-parameter family

$$\begin{aligned} Y^i &= k_1Y^i, \\ U^i &= k_1U^i + k_2Y^i + k_3u^i, \\ U^{ij} &= k_1U^{ij} + k_3u^{ij}, \end{aligned}$$

see [Doupovec, 90]. Hence the operators of the form $H \circ T^{(2)}$ form a 3-parameter family only, in which the operator D is not included.

43.4. Remark. In the case of Weil bundles, theorem 42.12 implies that the difference between a natural operator $T \rightsquigarrow TT_B$ and its associated absolute operator is a linear operator. This is no more true for the non-product-preserving functors, where the operator D is the simplest counter-example.

43.5. Remark. The operators $\mathcal{T}^{(2)}$, $\mathcal{V}$ and $\mathcal{L}$ transform every vector field on a manifold M into a vector field on $T^{(2)}M$ tangent to the subbundle $TM \subset T^{(2)}M$, but D does not. With a little surprise we can express it by saying that the natural operator $D: T \rightsquigarrow TT^{(2)}$ is not compatible with the natural inclusion $TM \subset T^{(2)}M$.

43.6. Remark. Recently [Mikulski, to appear b], has solved the general problem of determining all natural operators $T \rightsquigarrow TT^{(r)}$, $r \in \mathbb{N}$. All such operators form an $(r+2)$-parameter family linearly generated by the flow operator, by the Liouville vector field of $T^{(r)}$ and by the analogies of the operator D from 43.1 defined by $f \mapsto \underbrace{X \cdots X}_{k\text{-times}} f$, $k = 1, \dots, r$.

44. Induced vector fields on jet bundles

44.1. Let F be a bundle functor on $\mathcal{FM}_{m,n}$. The idea of the flow prolongation of vector fields can be applied to the projectable vector fields on every object $p: Y \to M$ of $\mathcal{FM}_{m,n}$. The flow Fl^η_t of a projectable vector field η on Y is formed by the local isomorphisms of Y and we define the flow operator $\mathcal{F}$ of F by

$$\mathcal{F}_Y\eta = \tfrac{\partial}{\partial t}\big|_0 F(\mathrm{Fl}^\eta_t).$$

The general concept of a natural operator A transforming every projectable vector field on $Y \in \mathrm{Ob}\mathcal{FM}_{m,n}$ into a vector field on FY was introduced in section 18. We shall denote such an operator briefly by $A: T_{\mathrm{proj}} \rightsquigarrow TF$.

44.2. Lemma. *If F is an r-th order bundle functor on $\mathcal{FM}_{m,n}$, then the order of every natural operator $T_{\mathrm{proj}} \rightsquigarrow TF$ is $\le r$.*

Proof. This is quite similar to 42.5, see [Kolář, Slovák, 90] for the details. □

44.3. We shall discuss the case F is the functor J^r of the r-th jet prolongation of fibered manifolds. We remark that a simple evaluation leads to the following coordinate formula for $\mathcal{J}^1\eta$

$$\mathcal{J}^1\eta = \eta^i \tfrac{\partial}{\partial x^i} + \eta^p \tfrac{\partial}{\partial y^p} + \left(\tfrac{\partial \eta^p}{\partial x^i} + \tfrac{\partial \eta^p}{\partial y^q} y^q_i - \tfrac{\partial \eta^j}{\partial x^i} y^p_j \right) \tfrac{\partial}{\partial y^p_i}$$

provided $\eta = \eta^i(x)\frac{\partial}{\partial x^i} + \eta^p(x,y)\frac{\partial}{\partial y^p}$, see [Krupka, 84]. To evaluate $\mathcal{J}^r\eta$, we have to iterate this formula and use the canonical inclusion $J^r(Y \to M) \hookrightarrow J^1(J^{r-1}(Y \to M) \to M)$.

Proposition. *Every natural operator $A\colon T_{\text{proj}} \rightsquigarrow TJ^r$ is a constant multiple of the flow operator $\mathcal{J}^r$.*

Proof. Let V^r be the space of all r-jets of the projectable vector fields on $\mathbb{R}^{n+m} \to \mathbb{R}^m$ with source $0 \in \mathbb{R}^{m+n}$, let $V^0 \subset V^r$ be the space of all r-jets of the constant vector fields and $V_0 \subset V^0$ be the subset of all vector fields with zero component in $\mathbb{R}^n$. Further, let S^r or Z^r be the fiber of $J^r(\mathbb{R}^{m+n} \to \mathbb{R}^m)$ or $TJ^r(\mathbb{R}^{m+n} \to \mathbb{R}^m)$ over $0 \in \mathbb{R}^{m+n}$, respectively. By lemma 44.2 and by the general theory, we have to determine all $G^{r+1}_{m,n}$-maps $\mathcal{A}\colon V^r \times S^r \to Z^r$ over the identity of S^r. Analogously to section 42, every projectable vector field on Y with non-zero projection to the base manifold can locally be transformed into the vector field $\frac{\partial}{\partial x^1}$. Hence $\mathcal{A}$ is determined by its restriction $\mathcal{A}_0$ to $V_0 \times S^r$. However, in the first part of the proof we have to consider the restriction $\mathcal{A}^0$ of $\mathcal{A}$ to $V^0 \times S^r$ for technical reasons.

Having the canonical coordinates x^i and y^p on $\mathbb{R}^{m+n}$, let X^i, Y^p be the induced coordinates on V^0, let y^p_α, $1 \le |\alpha| \le r$, be the induced coordinates on S^r and $Z^i = dx^i$, $Z^p = dy^p$, $Z^p_\alpha = dy^p_\alpha$ be the additional coordinates on Z^r. The restriction $\mathcal{A}^0$ is given by some functions

$$Z^i = f^i(X^j, Y^q, y^s_\beta)$$
$$Z^p = f^p(X^i, Y^q, y^s_\beta)$$
$$Z^p_\alpha = f^p_\alpha(X^i, Y^q, y^s_\beta).$$

Let us denote by g^i, g^p, g^p_α the restrictions of the corresponding f's to $V_0 \times S^r$. The flows of constant vector fields are formed by translations, so that their r-jet prolongations are the induced translations of $J^r(\mathbb{R}^{m+n} \to \mathbb{R}^m)$ identical on the standard fiber. Therefore $\mathcal{J}^r \frac{\partial}{\partial x^1} = \frac{\partial}{\partial x^1}$ and it suffices to prove

$$g^i = kX^i, \qquad g^p = 0, \qquad g^p_\alpha = 0.$$

We shall proceed by induction on the order r. It is easy to see that the action of $i(G^1_m \times G^1_n) \subset G^{r+1}_{m,n}$ on all quantities is tensorial. Consider the case $r = 1$. Using the equivariance with respect to the homotheties in $i(G^1_n)$, we obtain $f^i(X^j, Y^p, y^q_l) = f^i(X^j, kY^p, ky^q_l)$, so that f^i depends on X^i only. Then the equivariance of f^i with respect to $i(G^1_m)$ yields $f^i = kX^i$ by 24.7. The equivariance of f^p with respect to the homotheties in $i(G^1_n)$ gives $kf^p(X^i, Y^q, y^s_j) = f^p(X^i, kY^q, ky^s_j)$. This kind of homogeneity implies $f^p = h^p_q(X^i)Y^q + h^{pj}_q(X^i)y^q_j$ with some smooth functions h^p_q, h^{pj}_q. Using the homotheties in $i(G^1_m)$, we further obtain $h^p_q(kX) = h^p_q(X)$ and $h^{pj}_q(kX) = kh^{pj}_q(X)$. Hence $h^p_q = \text{const}$ and h^{pj}_q is linear in X^i. Then the generalized invariant tensor theorem yields $f^p = aY^p + by^p_i X^i$, a, $b \in \mathbb{R}$. Applying the same procedure to f^p_i, we find $f^p_i = cy^p_i$, $c \in \mathbb{R}$.

Consider the injection $G^2_n \hookrightarrow G^2_{m,n}$ determined by the products with the identities on $\mathbb{R}^m$. The action of an element (a^p_q, a^r_{st}) of the latter subgroup is given by

$$\bar{y}^p_i = a^p_q y^q_i \tag{2}$$

$$\bar{Z}^p_i = a^p_{qt} y^q_i Z^t + a^p_q Z^q_i \tag{3}$$

and V_0 is an invariant subspace. In particular, (3) with $a^p_q = \delta^p_q$ gives an equivariance condition

$$cy^p_i = ba^p_{qt}y^q_i y^t_j X^j + cy^p_i.$$

This yields $b = 0$, so that $g^p = 0$. Further, the subspace V_0 is invariant with the respect to the inclusion of $G^1_{m,n}$ into $G^2_{m,n}$. The equivariance of f^p_i with respect to an element $(\delta^i_j, \delta^p_q, a^p_i) \in G^1_{m,n}$ means $cy^p_i = c(y^p_i + a^p_i)$. Hence $c = 0$, which completes the proof for $r = 1$.

For $r \geq 2$ it suffices to discuss the g's only. Using the homotheties in $i(G^1_n)$, we find that $g^p_{i_1\dots i_s}(X^j, y^q_\beta)$, $1 \leq |\beta| \leq r$, is linear in y^q_β. The homotheties in $i(G^1_m)$ and the generalized invariant tensor theorem then yield

$$g^p_{i_1\dots i_s} = W^p_{i_1\dots i_s} + c_s y^p_{i_1\dots i_s i_{s+1}\dots i_r} X^{i_{s+1}} \dots X^{i_r} \tag{4}$$

where $W^p_{i_1\dots i_s}$ do not depend on $y^p_{i_1\dots i_r}$, $s = 1, \dots, r-1$, and

$$g^p_{i_1\dots i_r} = c_r y^p_{i_1\dots i_r} \tag{5}$$

$$g^p = b_1 y^p_i X^i + \dots + b_r y^p_{i_1\dots i_r} X^{i_1} \dots X^{i_r}. \tag{6}$$

Similarly to the first order case, we have an inclusion $G^{r+1}_n \hookrightarrow G^{r+1}_{m,n}$ determined by the products of diffeomorphisms on $\mathbb{R}^n$ with the identity of $\mathbb{R}^m$. One finds easily the following transformation law

$$\bar{y}^p_{i_1\dots i_s} = a^p_q y^q_{i_1\dots i_s} + F^p_{i_1\dots i_s} + a^p_{q_1\dots q_s} y^{q_1}_{i_1} \dots y^{q_s}_{i_s} \tag{7}$$

where $F^p_{i_1\dots i_s}$ is a polynomial expression linear in a^p_α with $2 \leq |\alpha| \leq s-1$ and independent of $y^p_{i_1\dots i_s}$. This implies

$$\bar{Z}^p_{i_1\dots i_s} = a^p_q Z^q_{i_1\dots i_s} + G^p_{i_1\dots i_s} + a^p_{q_1\dots q_s q_{s+1}} y^{q_1}_{i_1} \dots y^{q_s}_{i_s} Z^{q_{s+1}} \tag{8}$$

where $G^p_{i_1\dots i_s}$ is a polynomial expression linear in a^p_α with $2 \leq |\alpha| \leq s$ and linear in Z^p_α, $0 \leq |\alpha| \leq s-1$.

We deduce that every $g^p_{i_1\dots i_s}$, $0 \leq s \leq r-1$, is independent of $y^p_{i_1\dots i_r}$. On the kernel of the jet projection $G^{r+1}_n \to G^r_n$, (8) for $r = s$ gives

$$0 = a^p_{q_1\dots q_r q_{r+1}} y^{q_1}_{i_1} \dots y^{q_r}_{i_r} g^{q_{r+1}}.$$

Hence $g^p = 0$. On the kernel of the jet projection $G^r_n \to G^{r-1}_n$, (8) with $s = 1, \dots, r-1$, implies

$$0 = c_s a^p_{q_1\dots q_r} y^{q_1}_{i_1} \dots y^{q_r}_{i_r} X^{i_{s+1}} \dots X^{i_r},$$

i.e. $c_s = 0$. By projectability, g^i and g^p_α, $0 \leq |\alpha| \leq r-1$, correspond to a $G^r_{m,n}$-equivariant map $V_0 \times S^{r-1} \to Z^{r-1}$. By the induction hypothesis, $g^p_\alpha = 0$ for all $0 \leq |\alpha| \leq r-1$. Then on the kernel of the jet projection $G^{r+1}_n \to G^{r-1}_n$ (8) gives $0 = c_r a^p_{q_1\dots q_r} y^{q_1}_{i_1} \dots y^{q_r}_{i_r}$, i.e. $g^p_{i_1\dots i_r} = 0$. □

44.4. Bundles of contact elements. Consider the bundle functor K_n^r on $\mathcal{M}f_m$ of the n-dimensional contact elements of order r defined in 12.15.

Proposition. *Every natural operator $A: T \rightsquigarrow TK_n^r$ is a constant multiple of the flow operator $\mathcal{K}_n^r$.*

Proof. It suffices to discuss the case $M = \mathbb{R}^m$. Consider the canonical fibration $\mathbb{R}^m = \mathbb{R}^n \times \mathbb{R}^{m-n} \to \mathbb{R}^n$. As remarked at the end of 12.16, there is an identification of an open dense subset in $K_n^r\mathbb{R}^m$ with $J^r(\mathbb{R}^m \to \mathbb{R}^n)$. By definition, on this subset it holds $\mathcal{J}^r\xi = \mathcal{K}_n^r\xi$ for every projectable vector field ξ on $\mathbb{R}^m \to \mathbb{R}^n$. Since the operator A commutes with the action of all diffeomorphisms preserving fibration $\mathbb{R}^m \to \mathbb{R}^n$, the restriction of A to $\frac{\partial}{\partial x^1}$ is a constant multiple of $\mathcal{K}_n^r(\frac{\partial}{\partial x^1})$ by proposition 44.3. But every vector field on $\mathbb{R}^m$ can be locally transformed into $\frac{\partial}{\partial x^1}$ in a neighborhood of any point where it does not vanish. □

We find it interesting that we have finished our investigation of the basic properties of the natural operators $T \rightsquigarrow TF$ for different bundle functors on $\mathcal{M}f_m$ by an example in which the constant multiples of the flow operator are the only natural operators $T \rightsquigarrow TF$.

44.5. Remark. [Kobak, 91] determined all natural operators $T \rightsquigarrow TT^*$ and $T \rightsquigarrow T(TT^*)$ for manifolds of dimension at least two. Let $\mathcal{T}^*$ be the flow operator of the cotangent bundle, $L_M: T^*M \to TT^*M$ be the vector field generated by the homotheties of the vector bundle T^*M and $\omega_M: TM \times_M T^*M \to \mathbb{R}$ be the evaluation map. Then all natural operators $T \rightsquigarrow TT^*$ are of the form $f(\omega)\mathcal{T}^* + g(\omega)L$, where f, $g \in C^\infty(\mathbb{R},\mathbb{R})$ are any smooth functions of one variable. In the case $F = TT^*$ the result is of similar character, but the complete list is somewhat longer, so that we refer the reader to the above mentioned paper.

45. Prolongations of connections to $FY \to M$

45.1. In 31.1 we deduced that there is exactly one natural operator transforming every general connection on $Y \to M$ into a general connection on $VY \to M$. However, one meets a quite different situation when replacing fibered manifold $VY \to M$ e.g. by the first jet prolongation $J^1Y \to M$ of Y. Pohl has observed in the vector bundle case, [Pohl, 66], that one needs an auxiliary linear connection on the base manifold M to construct an induced connection on $J^1Y \to M$. Our first goal is to clarify this difference from the conceptual point of view.

45.2. Bundle functors of order (r,s). We recall that two maps f, g of a fibered manifold $p: Y \to M$ into another manifold determine the same (r,s)-jet $j_y^{r,s}f = j_y^{r,s}g$ at $y \in Y$, $s \geq r$, if $j_y^rf = j_y^rg$ and the restrictions of f and g to the fiber $Y_{p(y)}$ satisfy $j_y^s(f|Y_{p(y)}) = j_y^s(g|Y_{p(y)})$, see 12.19.

Definition. A bundle functor on a category $\mathcal{C}$ over $\mathcal{FM}$ is said to be of order (r,s), if for any two $\mathcal{C}$-morphisms f, g of Y into $\bar{Y}$

$$j_y^{r,s}f = j_y^{r,s}g \quad \text{implies} \quad (Ff)|(FY)_y = (Fg)|(FY)_y.$$

For example, the order of the vertical functor V is $(0,1)$, while the functor of the first jet prolongation J^1 has order $(1,1)$.

45.3. Denote by $J^{r,s}TY$ the space of all (r,s)-jets of the projectable vector fields on $Y \to M$. This is a vector bundle over Y. Let F be a bundle functor on $\mathcal{FM}_{m,n}$ and $\mathcal{F}$ denote its flow operator $T_{\text{proj}} \rightsquigarrow TF$.

Proposition. *If the order of F is (r,s) and η is a projectable vector field on Y, then the value $(\mathcal{F}\eta)(u)$ at every $u \in (FY)_y$ depends only on $j_y^{r,s}\eta$. The induced map*

$$FY \oplus J^{r,s}TY \to T(FY)$$

is smooth and linear with respect to $J^{r,s}TY$.

Proof. Smoothness can be proved in the same way as in 14.14. Linearity follows directly from the linearity of the flow operator $\mathcal{F}$. □

45.4. Let Γ be a general connection on $p\colon Y \to M$. Considering the Γ-lift $\Gamma\xi$ of a vector field ξ on M, one sees directly that $j_y^{r,s}\Gamma\xi$ depends on $j_{p(y)}^r\xi$ only, $y \in Y$. Let F be a bundle functor on $\mathcal{FM}_{m,n}$ of order (r,s). If we combine the map of proposition 45.3 with the lifting map of Γ, we obtain a map $\widetilde{F\Gamma}\colon FY \oplus J^rTM \to TFY$ linear in J^rTM. Let $\Lambda\colon TM \to J^rTM$ be an r-th order linear connection on M, i.e. a linear splitting of the projection $\pi_0^r\colon J^rTM \to TM$. By linearity, the composition

$$\widetilde{F\Gamma} \circ (\mathrm{id}_{FY} \oplus \Lambda)\colon FY \oplus TM \to TFY \tag{1}$$

is a lifting map of a general connection on $FY \to M$.

Definition. The general connection $\mathcal{F}(\Gamma,\Lambda)$ on $FY \to M$ with lifting map (1) is called the F-prolongation of Γ with respect to Λ.

If the order of F is $(0,s)$, we need no connection Λ on M. In particular, every connection Γ on $Y \to M$ induces in such a way a connection $\mathcal{V}\Gamma$ on $VY \to M$, which was already mentioned in remark 31.4.

45.5. We show that the construction of $\mathcal{F}(\Gamma,\Lambda)$ behaves well with respect to morphisms of connections. Given an $\mathcal{FM}$-morphism $f\colon Y \to \bar{Y}$ over $f_0\colon M \to \bar{M}$ and two general connections Γ on $p\colon Y \to M$ and $\bar{\Gamma}$ on $\bar{p}\colon \bar{Y} \to \bar{M}$, one sees easily that Γ and $\bar{\Gamma}$ are f-related in the sense of 8.15 if and only if the following diagram commutes

$$\begin{array}{ccc} TY & \xrightarrow{Tf} & T\bar{Y} \\ \uparrow{\scriptstyle\Gamma} & & \uparrow{\scriptstyle\bar{\Gamma}} \\ Y \oplus TM & \xrightarrow{f \oplus Tf_0} & \bar{Y} \oplus T\bar{M} \end{array}$$

In such a case f is also called a *connection morphism* of Γ into $\bar{\Gamma}$. Further, two r-th order linear connections $\Lambda\colon TM \to J^rTM$ and $\bar{\Lambda}\colon T\bar{M} \to J^rT\bar{M}$ are called f_0-related, if for every $z \in T_xM$ it holds

$$\bar{\Lambda}(Tf_0(z)) \circ (j_x^r f_0) = (j_x^r Tf_0) \circ \Lambda(z).$$

Let F be as in 45.4.

Proposition. *If Γ and $\bar{\Gamma}$ are f-related and Λ and $\bar{\Lambda}$ are f_0-related, then $\mathcal{F}(\Gamma, \Lambda)$ and $\mathcal{F}(\bar{\Gamma}, \bar{\Lambda})$ are Ff-related.*

Proof. The lifting map of $\mathcal{F}(\Gamma, \Lambda)$ can be determined as follows. For every $X \in T_xM$ we take a vector field ξ on M such that $j^r_x\xi = \Lambda(X)$ and we construct its Γ-lift $\Gamma\xi$. Then $\mathcal{F}(\Gamma, \Lambda)(u)$ is the value of the flow prolongation $\mathcal{F}(\Gamma\xi)$ at $u \in F_xY$. Let $\bar{\Lambda}(Tf_0(X)) = j^r_{\bar{x}}\bar{\xi}$, $\bar{x} = f_0(x)$. If Λ and $\bar{\Lambda}$ are f_0-related, the vector fields ξ and $\bar{\xi}$ are f_0-related up to order r at x. Since Γ and $\bar{\Gamma}$ are f-related, the restriction of $\mathcal{F}(\Gamma\xi)$ over x and the restriction of $\mathcal{F}(\bar{\Gamma}\xi)$ over $\bar{x}$ are Ff-related. □

45.6. In many concrete cases, the connection $\mathcal{F}(\Gamma, \Lambda)$ is of special kind. We are going to deduce a general result of this type.

Let $\mathcal{C}$ be a category over $\mathcal{FM}$, cf. 51.4. Analogously to example 1 from 18.18, a projectable vector field η on $Y \in \mathrm{Ob}\mathcal{C}$ is called a *$\mathcal{C}$-field*, if its flow is formed by local $\mathcal{C}$-morphisms. For example, for the category $\mathcal{PB}(G)$ of smooth principal G-bundles, a projectable vector field η on a principal fiber bundle is a $\mathcal{PB}(G)$-field if and only if η is right-invariant. For the category $\mathcal{VB}$ of smooth vector bundles, one deduces easily that a projectable vector field η on a vector bundle E is a $\mathcal{VB}$-field if and only if η is a linear morphism $E \to TE$, see 6.11. A connection Γ on $(p\colon Y \to M) \in \mathrm{Ob}\mathcal{C}$ is called a *$\mathcal{C}$-connection*, if $\Gamma\xi$ is a $\mathcal{C}$-field for every vector field ξ on M. Obviously, a $\mathcal{PB}(G)$-connection or a $\mathcal{VB}$-connection is a classical principal or linear connection, respectively.

More generally, a projectable family of tangent vectors along a fiber Y_x, i.e. a section $\sigma\colon Y_x \to TY$ such that $Tp \circ \sigma$ is a constant map, is said to be a $\mathcal{C}$-family, if there exists a $\mathcal{C}$-field η on Y such that σ is the restriction of η to Y_x. We shall say that the category $\mathcal{C}$ is *infinitesimally regular*, if any projectable vector field on a $\mathcal{C}$-object the restriction of which to each fiber is a $\mathcal{C}$-family is a $\mathcal{C}$-field.

Proposition. *If F is a bundle functor of a category $\mathcal{C}$ over $\mathcal{FM}$ into an infinitesimally regular category $\mathcal{D}$ over $\mathcal{FM}$ and Γ is a $\mathcal{C}$-connection, then $\mathcal{F}(\Gamma, \Lambda)$ is a $\mathcal{D}$-connection for every Λ.*

Proof. By the construction that we used in the proof of proposition 45.5, the $\mathcal{F}(\Gamma, \Lambda)$-lift of every vector $X \in TM$ is a $\mathcal{D}$-family. Since $\mathcal{D}$ is infinitesimally regular, the $\mathcal{F}(\Gamma, \Lambda)$-lift of every vector field on TM is a $\mathcal{D}$-field. □

45.7. In the special case $F = J^1$ we determine all natural operators transforming a general connection on $Y \to M$ and a first order linear connection Λ on M into a general connection on $J^1Y \to M$. Taking into account the rigidity of the symmetric linear connections on M deduced in 25.3, we first assume Λ to be without torsion. Thus we are interested in the natural operators $J^1 \oplus Q_\tau P^1B \rightsquigarrow J^1(J^1 \to B)$.

On one hand, Γ and Λ induce the J^1-prolongation $\mathcal{J}^1(\Gamma, \Lambda)$ of Γ with respect to Λ. On the other hand, since J^1Y is an affine bundle with associated vector bundle $VY \otimes T^*M$, the section $\Gamma\colon Y \to J^1Y$ determines an identification $I_\Gamma\colon J^1Y \cong VY \otimes T^*M$. The vertical prolongation $\mathcal{V}\Gamma$ of Γ is linear over Y, see 31.1.(3), so that we can construct the tensor product $\mathcal{V}\Gamma \otimes \Lambda^*$ with the dual connection Λ^* on T^*M, see 47.14 and 47.15. The identification I_Γ transforms $\mathcal{V}\Gamma \otimes \Lambda^*$ into another connection $P(\Gamma, \Lambda)$ on $J^1Y \to M$.

45.8. Proposition. *All natural operators $J^1 \oplus Q_\tau P^1 B \rightsquigarrow J^1(J^1 \to B)$ form the one-parameter family*

$$tP + (1-t)\mathcal{J}^1, \qquad t \in \mathbb{R}. \tag{1}$$

Proof. In usual local coordinates, let

$$dy^p = F_i^p(x,y)dx^i \tag{2}$$

be the equations of Γ and

$$d\xi^i = \Lambda^i_{jk}(x)\xi^j dx^k \tag{3}$$

be the equations of Λ. By direct evaluation, one finds the equations of $\mathcal{J}^1(\Gamma, \Lambda)$ in the form (2) and

$$dy_i^p = \left(\frac{\partial F_j^p}{\partial x^i} + \frac{\partial F_j^p}{\partial y^q} y_i^q + \Lambda^k_{ji}(F_k^p - y_k^p)\right) dx^j \tag{4}$$

while the equations of $P(\Gamma, \Lambda)$ have the form (2) and

$$dy_i^p = \left(\frac{\partial F_j^p}{\partial y^q}(y_i^q - F_i^q) + \frac{\partial F_i^p}{\partial x^j} + \frac{\partial F_i^p}{\partial y^q} F_j^q - \Lambda^k_{ij}(y_k^p - F_k^p)\right) dx^j. \tag{5}$$

First we discuss the operators of first order in Γ and of order zero in Λ. Let $S_1 = J_0^1(J^1(\mathbb{R}^{n+m} \to \mathbb{R}^m) \to \mathbb{R}^{n+m})$ be the standard fiber from 27.3, $S_0 = J_0^1(\mathbb{R}^{m+n} \to \mathbb{R}^m)$, $\Lambda = (Q_\tau P^1\mathbb{R}^m)_0$ and $Z = J_0^1(J^1(\mathbb{R}^{m+n} \to \mathbb{R}^m) \to \mathbb{R}^m)$. By using the general theory, the operators in question correspond to $G^2_{m,n}$-maps $S_1 \times \Lambda \times S_0 \to Z$ over the identity of S_0. The canonical coordinates on S_1 are y_i^p, y_{iq}^p, y_{ij}^p and the action of $G^2_{m,n}$ is given by 27.3.(1)-(3). On S_0 we have the well known coordinates Y_i^p and the action

$$\bar{Y}_i^p = a_q^p Y_j^q \tilde{a}_i^j + a_j^p \tilde{a}_i^j. \tag{6}$$

The standard coordinates on Λ are $\Lambda^i_{jk} = \Lambda^i_{kj}$ and the action is

$$\bar{\Lambda}^i_{jk} = a_l^i \Lambda^l_{mn} \tilde{a}_j^m \tilde{a}_k^n + a^i_{lm} \tilde{a}_j^l \tilde{a}_k^m. \tag{7}$$

The induced coordinates on Z are z_i^p, Z_i^p, Z_{ij}^p and one evaluates easily that the action on both z_i^p and Z_i^p has form (6), while

$$\begin{aligned}\bar{Z}_{ij}^p = {} & a_q^p Z_{kl}^q \tilde{a}_i^k \tilde{a}_j^l + a_{qr}^p z_k^q Z_l^r \tilde{a}_i^k \tilde{a}_j^l + a_{qk}^p Z_l^q \tilde{a}_i^k \tilde{a}_j^l \\ & + a_{ql}^p z_k^q \tilde{a}_i^k \tilde{a}_j^l + a_q^p z_k^q \tilde{a}_{ij}^k + a_k^p \tilde{a}_{ij}^k + a_{kl}^p \tilde{a}_i^k \tilde{a}_j^l.\end{aligned} \tag{8}$$

Write $Y = (Y_i^p)$, $y = (y_i^p)$, $y_1 = (y_{iq}^p)$, $y_2 = (y_{ij}^p)$, $\Lambda = (\Lambda_{jk}^i)$. Then the coordinate form of a map $f\colon S_1 \times \Lambda \times S_0 \to Z$ over the identity of S_0 is $z_i^p = Y_i^p$ and

$$(9)\qquad \begin{aligned} Z_i^p &= f_i^p(Y, y, y_1, y_2, \Lambda) \\ Z_{ij}^p &= f_{ij}^p(Y, y, y_1, y_2, \Lambda). \end{aligned}$$

The equivariance of f_i^p with respect to the homotheties in $i(G_m^1)$ yields

$$kf_i^p = f_i^p(kY, ky, ky_1, k^2 y_2, k\Lambda)$$

so that f_i^p is linear in Y, y, y_1, Λ and independent of y_2. The homotheties in $i(G_n^1)$ give that f_i^p is independent of y_1 and Λ. By the generalized invariant tensor theorem 27.1, the equivariance with respect to $i(G_m^1 \times G_n^1)$ implies

$$f_i^p = aY_i^p + by_i^p.$$

Then the equivariance with respect to the subgroup K characterized by $a_j^i = \delta_j^i$, $a_q^p = \delta_q^p$ yields

$$b = 1 - a.$$

For f_{ij}^p the homotheties in $i(G_m^1)$ and $i(G_n^1)$ give

$$\begin{aligned} k^2 f_{ij}^p &= f_{ij}^p(kY, ky, ky_1, k^2 y_2, k\Lambda) \\ kf_{ij}^p &= f_{ij}^p(kY, ky, y_1, ky_2, \Lambda) \end{aligned}$$

so that f_{ij}^p is linear in y_2 and bilinear in the pairs (Y, y_1),(y, y_1), (Y, Λ), (y, Λ). Considering equivariance with respect to $i(G_m^1 \times G_n^1)$, we obtain f_{ij}^p in the form of a 16-parameter family

$$\begin{aligned} f_{ij}^p = {} & k_1 y_{ij}^p + k_2 y_{ji}^p + k_3 Y_i^p y_{qj}^q + k_4 Y_j^p y_{qi}^q + k_5 Y_i^q y_{qj}^p + k_6 Y_j^q y_{qi}^p \\ & + k_7 y_i^p y_{qj}^q + k_8 y_j^p y_{qi}^q + k_9 y_i^q y_{qj}^p + k_{10} y_j^q y_{qi}^p + k_{11} Y_k^p \Lambda_{ij}^k \\ & + k_{12} Y_i^p \Lambda_{kj}^k + k_{13} Y_j^p \Lambda_{ki}^k + k_{14} y_k^p \Lambda_{ij}^k + k_{15} y_i^p \Lambda_{kj}^k + k_{16} y_j^p \Lambda_{ki}^k. \end{aligned}$$

Evaluating the equivariance with respect to K, we find $a = 0$ and such relations among $k_1, \dots, k_{16}$, which correspond to (1).

Furthermore, 23.7 implies that every natural operator of our type has finite order. Having a natural operator of order r in Γ and of order s in Λ, we shall deduce $r = 1$ and $s = 0$, which corresponds to the above case. Let α and γ be multi indices in x^i and β be a multi index in y^p. The associated map of our operator has the form $z_i^p = Y_i^p$ and

$$Z_i^p = f_i^p(Y, y_{\alpha\beta}, \Lambda_\gamma), \quad Z_{ij}^p = f_{ij}^p(Y, y_{\alpha\beta}, \Lambda_\gamma)$$

where $|\alpha| + |\beta| \le r$, $|\gamma| \le s$. Using the homotheties in $i(G_m^1)$, we obtain

$$kf_i^p = f_i^p(kY, k^{1+|\alpha|} y_{\alpha\beta}, k^{1+|\gamma|} \Lambda_\gamma).$$

Hence f_i^p is linear in Y, y_β and Λ, and is independent of the variables with $|\alpha| > 0$ or $|\gamma| > 0$. The homotheties in $i(G_n^1)$ then imply that f_i^p is independent of y_β with $|\beta| > 1$. For f_{ij}^p, the homotheties in $i(G_m^1)$ yield

$$k^2 f_{ij}^p = f_{ij}^p(kY, k^{1+|\alpha|}y_{\alpha\beta}, k^{1+|\gamma|}\Lambda_\gamma) \tag{10}$$

so that f_{ij}^p is a polynomial independent of the variables with $|\alpha| > 1$ or $|\gamma| > 1$. The homotheties in $i(G_n^1)$ imply

$$k f_{ij}^p = f_{ij}^p(kY, k^{1-|\beta|}y_{\alpha\beta}, \Lambda_\gamma) \tag{11}$$

for $|\alpha| \leq 1$, $|\gamma| \leq 1$. Combining (10) with (11) we deduce that f_{ij}^p is independent of $y_{\alpha\beta}$ for $|\alpha| + |\beta| > 1$ and Λ_γ for $|\gamma| > 0$. □

45.9. Using a similar procedure as in 45.8 one can prove that the use of a linear connection on the base manifold for a natural construction of an induced connection on $J^1Y \to M$ is unavoidable. In other words, the following assertion holds, a complete proof of which can be found in [Kolář, 87a].

Proposition. *There is no natural operator* $J^1 \rightsquigarrow J^1(J^1 \to B)$.

45.10. If we admit an arbitrary linear connection Λ on the base manifold in the above problem, the natural operators $QP^1 \rightsquigarrow QP^1$ from proposition 25.2 must appear in the result. By proposition 25.2, all natural operators $QP^1 \rightsquigarrow T \otimes T^* \otimes T^*$ form a 3-parameter family

$$N(\Lambda) = k_1 S + k_2 I \otimes \hat{S} + k_3 \hat{S} \otimes I.$$

By 12.16, $J^1(J^1Y \to M)$ is an affine bundle with associated vector bundle $VJ^1Y \otimes T^*M$. We construct some natural 'difference tensors' for this case. Consider the exact sequence of vector bundles over J^1Y established in 12.16

$$0 \to VY \otimes_{J^1Y} T^*M \to VJ^1Y \xrightarrow{V\beta} VY \to 0$$

where $\otimes_{J^1Y}$ denotes the tensor product of the pullbacks over J^1Y. The connection Γ determines a map $\delta(\Gamma)\colon J^1Y \to VY \otimes T^*M$ transforming every $u \in J^1Y$ into the difference $u - \Gamma(\beta u) \in VY \otimes T^*M$. Hence for every k_1, k_2, k_3 we can extend the evaluation map $TM \oplus T^*M \to \mathbb{R}$ into a contraction $\langle \delta(\Gamma), N(\Lambda)\rangle : J^1Y \to VY \otimes_{J^1Y} T^*M \otimes T^*M \subset VJ^1Y \otimes T^*M$. By the procedure used in 45.8 one can prove the following assertion, see [Kolář, 87a].

Proposition. *All natural operators transforming a connection Γ on Y into a connection on $J^1Y \to M$ by means of a linear connection Λ on the base manifold form the 4-parameter family*

$$tP(\Gamma, \tilde{\Lambda}) + (1-t)\mathcal{J}^1(\Gamma, \Lambda) + \langle \delta(\Gamma), N(\Lambda)\rangle$$

t, k_1, k_2, $k_3 \in \mathbb{R}$, where $\tilde{\Lambda}$ means the conjugate connection of Λ.

46. The cases $FY \to FM$ and $FY \to Y$

46.1. We first describe a geometrical construction transforming every connection Γ on a fibered manifold $p\colon Y \to M$ into a connection $\mathcal{T}_A\Gamma$ on $T_Ap\colon T_AY \to T_AM$ for every Weil functor T_A. Consider Γ in the form of the lifting map

$$\Gamma\colon Y \oplus TM \to TY. \tag{1}$$

Such a lifting map is characterized by the condition

$$(\pi_Y, Tp) \circ \Gamma = \mathrm{id}_{Y \oplus TM} \tag{2}$$

where $\pi\colon T \to \mathrm{Id}$ is the bundle projection of the tangent functor, and by the fact that, if we interpret (1) as the pullback map

$$p^*TM \to TY,$$

this is a vector bundle morphism over Y. Let $\kappa\colon T_AT \to TT_A$ be the flow-natural equivalence corresponding to the exchange homomorphism $A \otimes \mathbb{D} \to \mathbb{D} \otimes A$, see 35.17 and 39.2.

Proposition. *For every general connection $\Gamma\colon Y \oplus TM \to TY$, the map*

$$\mathcal{T}_A\Gamma := \kappa_Y \circ (T_A\Gamma) \circ (\mathrm{id}_{T_AY} \oplus \kappa_M^{-1})\colon T_AY \oplus TT_AM \to TT_AY \tag{3}$$

is a general connection on $T_Ap\colon T_AY \to T_AM$.

Proof. Applying T_A to (2), we obtain

$$(T_A\pi_Y, T_ATp) \circ T_A\Gamma = \mathrm{id}_{T_AY \oplus T_ATM}.$$

Since κ is the flow-natural equivalence, it holds $\kappa_M \circ T_ATp \circ \kappa_Y^{-1} = TT_Ap$ and $T_A\pi_Y \circ \kappa_Y^{-1} = \pi_{T_AY}$. This yields

$$(\pi_{T_AY}, TT_Ap) \circ \mathcal{T}_A\Gamma = \mathrm{id}_{T_A \oplus TT_AM}$$

so that $\mathcal{T}_A\Gamma$ satisfies the analog of (2). Further, one deduces easily that κ_Y : $T_ATY \to TT_AY$ is a vector bundle morphism over T_AY. Even $\kappa_M^{-1}\colon TT_AM \to T_ATM$ is a linear morphism over T_AM, so that the pullback map $(T_Ap)^*\kappa_M^{-1}$: $(T_Ap)^*TT_AM \to (T_Ap)^*T_ATM$ is also linear. But we have a canonical identification $(T_Ap)^*T_ATM \cong T_A(p^*TM)$. Hence the pullback form of $\mathcal{T}_A\Gamma$ on $(T_Ap)^*TT_AM \to TT_AY$ is a composition of three vector bundle morphisms over T_AY, so that it is linear as well. □

46.2. Remark. If we look for a possible generalization of this construction to an arbitrary bundle functor F on $\mathcal{M}f$, we realize that we need a natural equivalence $FT \to TF$ with suitable properties. However, the flow-natural transformation $FT \to TF$ from 39.2 is a natural equivalence if and only if F preserves products, i.e. F is a Weil functor. We remark that we do not know any natural operator transforming general connections on $Y \to M$ into general connections on $FY \to FM$ for any concrete non-product-preserving functor F on $\mathcal{M}f$.

46.3. Remark. Slovák has proved in [Slovák, 87a] that if Γ is a linear connection on a vector bundle $p\colon E \to M$, then $\mathcal{T}_A\Gamma$ is also a linear connection on the induced vector bundle $T_Ap\colon T_AE \to T_AM$. Furthermore, if $p\colon P \to M$ is a principal bundle with structure group G, then $T_Ap\colon T_AP \to T_AM$ is a principal bundle with structure group T_AG. Using the ideas from 37.16 one deduces directly that for every principal connection Γ on $P \to M$ the induced connection $\mathcal{T}_A\Gamma$ is also principal on $T_AP \to T_AM$.

46.4. We deduce one geometric property of the connection $\mathcal{T}_A\Gamma$. If we consider a general connection Γ on $Y \to M$ in the form $\Gamma\colon Y \oplus TM \to TY$, the Γ-lift $\Gamma\xi$ of a vector field $\xi\colon M \to TM$ is given by

$$(1) \qquad (\Gamma\xi)(y) = \Gamma(y, \xi(p(y))), \text{ i.e. } \Gamma\xi = \Gamma \circ (\mathrm{id}_Y, \xi \circ p).$$

On one hand, $\Gamma\xi$ is a vector field on Y and we can construct its flow prolongation $\mathcal{T}_A(\Gamma\xi) = \kappa_Y \circ T_A(\Gamma\xi)$. On the other hand, the flow prolongation $\mathcal{T}_A\xi = \kappa_M \circ T_A\xi$ of ξ is a vector field on T_AM and we construct its $\mathcal{T}_A\Gamma$-lift $(\mathcal{T}_A\Gamma)(\mathcal{T}_A\xi)$. The following assertion is based on the fact that we have used a flow-natural equivalence in the definition of $\mathcal{T}_A\Gamma$.

Proposition. *For every vector field ξ on M, we have $(\mathcal{T}_A\Gamma)(\mathcal{T}_A\xi) = \mathcal{T}_A(\Gamma\xi)$.*

Proof. By (1), we have $\mathcal{T}_A\Gamma(\mathcal{T}_A\xi) = \mathcal{T}_A\Gamma \circ (\mathrm{id}_{T_AY}, \mathcal{T}_A\xi \circ T_Ap) = \kappa_Y \circ T_A\Gamma \circ (\mathrm{id}_{T_AY}, \kappa_M^{-1} \circ \kappa_M \circ T_A\xi \circ T_Ap) = \kappa_Y \circ T_A(\Gamma \circ (\mathrm{id}_Y, \xi \circ p)) = \mathcal{T}_A(\Gamma\xi)$. □

We remark that several further geometric properties of $\mathcal{T}_A\Gamma$ are deduced in [Slovák, 87a].

46.5. Let $\bar\Gamma$ be another connection on another fibered manifold $\bar Y$ and let $f\colon Y \to \bar Y$ be a connection morphism of Γ into $\bar\Gamma$, i.e. the following diagram commutes

$$(1) \qquad \begin{array}{ccc} TY & \xrightarrow{\;Tf\;} & T\bar Y \\ \uparrow{\scriptstyle\Gamma} & & \uparrow{\scriptstyle\bar\Gamma} \\ Y \oplus TBY & \xrightarrow{\;f \oplus TBf\;} & \bar Y \oplus TB\bar Y \end{array}$$

Proposition. *If $f\colon Y \to \bar Y$ is a connection morphism of Γ into $\bar\Gamma$, then $T_Af\colon T_AY \to T_A\bar Y$ is a connection morphism of $\mathcal{T}_A\Gamma$ into $\mathcal{T}_A\bar\Gamma$.*

Proof. Applying T_A to (1), we obtain $T_ATf \circ T_A\Gamma = (T_A\bar\Gamma) \circ (T_Af \oplus T_ATBf)$. From 46.1.(3) we then deduce directly $TT_Af \circ \mathcal{T}_A\Gamma = \mathcal{T}_A\bar\Gamma \circ (T_Af \oplus TT_ABf)$. □

46.6. The problem of finding all natural operators transforming connections on $Y \to M$ into connections on $T_AY \to T_AM$ seems to be much more complicated than e.g. the problem of finding all natural operators $T \rightsquigarrow TT_A$ discussed in section 42. We shall clarify the situation in the case that T_A is the classical tangent functor T and we restrict ourselves to the first order natural operators.

Let $\mathcal{T}$ be the operator from proposition 46.1 in the case $T_A = T$. Hence $\mathcal{T}$ transforms every element of $C^\infty(J^1Y)$ into $C^\infty(J^1(TY \to TBY))$, where

J^1 and $J^1(T \to TB)$ are considered as bundle functors on $\mathcal{FM}_{m,n}$. Further we construct a natural 'difference tensor field' $[C_Y\Gamma]$ for connections on $TY \to TBY$ from the curvature of a connection Γ on Y. Write $BY = M$. In general, the difference of two connections on Y is a section of $VY \otimes T^*M$, which can be interpreted as a map $Y \oplus TM \to VY$. In the case of $TY \to TM$ we have $TY \oplus TTM \to V(TY \to TM)$. To define the operator $[C]$, consider both canonical projections $p_{TM}, Tp_M \colon TTM \to TM$. If we compose $(p_{TM}, Tp_M)\colon TTM \to TM \oplus TM$ with the antisymmetric tensor power and take the fibered product of the result with the bundle projection $TY \to Y$, we obtain a map $\mu_Y \colon TY \oplus TTM \to Y \oplus \Lambda^2 TM$. Since $C_Y\Gamma \colon Y \oplus \Lambda^2TM \to VY$, the values of $C_Y\Gamma \circ \mu_Y$ lie in VY. Every vector $A \in VY$ is identified with a vector $i(A) \in V(VY \to Y)$ tangent to the curve of the scalar multiples of A. Then we construct $[C_Y\Gamma](U, Z)$, $U \in TY$, $Z \in TTM$ by translating $i(C_Y\Gamma(\mu_Y(U, Z)))$ to the point U in the same fiber of $V(TY \to TM)$. This yields a map $[C_Y\Gamma]\colon TY \oplus TTM \to V(TY \to TM)$ of the required type.

46.7. Proposition. *All first order natural operators $J^1 \rightsquigarrow J^1(T \to TB)$ form the following one-parameter family*

$$\mathcal{T} + k[C], \qquad k \in \mathbb{R}.$$

Proof. Let

$$dy^p = F_i^p(x, y)dx^i \tag{1}$$

be the equations of Γ. Evaluating 46.1.(3) in the case $T_A = T$, one finds that the equations of $\mathcal{T}\Gamma$ are (1) and

$$d\eta^p = \left(\tfrac{\partial F_i^p}{\partial x^j}\xi^j + \tfrac{\partial F_i^p}{\partial y^q}\eta^q\right)dx^i + F_i^p(x, y)d\xi^i \tag{2}$$

where $\xi^i = dx^i$, $\eta^p = dy^p$ are the induced coordinates on TY. The equations of $[C_Y\Gamma]$

$$dy^p = 0, \quad d\eta^p = \left(\tfrac{\partial F_i^p}{\partial x^j} + \tfrac{\partial F_i^p}{\partial y^q}F_j^q\right)\xi^j \wedge dx^i \tag{3}$$

follow directly from the definition.

Let $S_1 = J_0^1(J^1(\mathbb{R}^{m+n} \to \mathbb{R}^m) \to \mathbb{R}^{m+n})$, $Q = T_0(\mathbb{R}^{m+n})$, $Z = J_0^1(T\mathbb{R}^{m+n} \to T\mathbb{R}^m)$ be the standard fibers in question and $q\colon Z \to Q$ be the canonical projection. According to 18.19, the first order natural operators $A\colon J^1 \rightsquigarrow J^1(T \to TB)$ are in bijection with the $G^2_{m,n}$-maps $\mathcal{A}\colon S_1 \times Q \to Z$ satisfying $q \circ \mathcal{A} = \mathrm{pr}_2$. The canonical coordinates y_i^p, y_{iq}^p, y_{ij}^p on S_1 and the action of $G^2_{m,n}$ on S_1 are described in 27.3. It will be useful to replace y_{ij}^p by S_{ij}^p and R_{ij}^p in the same way as in 28.2. One sees directly that the action of $G^2_{m,n}$ on Q with coordinates ξ^i, η^p is

$$\bar\xi^i = a_j^i\xi^j, \qquad \bar\eta^p = a_i^p\xi^i + a_q^p\eta^q. \tag{4}$$

The coordinates on Z are ξ^i, η^p and the quantities A^p_i, B^p_i, C^p_i, D^p_i determined by

$$dy^p = A^p_i dx^i + B^p_i d\xi^i, \quad d\eta^p = C^p_i dx^i + D^p_i d\xi^i. \tag{5}$$

A direct calculation yields that the action of $G^2_{m,n}$ on Z is (4) and

$$\begin{aligned} \bar{A}^p_i &= a^p_q \left(A^q_j \tilde{a}^j_i - \tilde{a}^q_i + B^q_j \tilde{a}^j_{ik} a^k_l \xi^l \right) \\ \bar{B}^p_i &= a^p_q B^q_j \tilde{a}^j_i \\ \bar{C}^p_i &= a^p_q (-\tilde{a}^q_{ij} \bar{\xi}^j - \tilde{a}^q_{jr} \bar{\xi}^j \bar{A}^r_i - \tilde{a}^q_{ir} \bar{\eta}^r - \tilde{a}^q_{rs} \bar{\eta}^r \bar{A}^s_i \\ &\quad + C^q_j \tilde{a}^j_i + D^q_j \tilde{a}^j_{ik} \bar{\xi}^k) \\ \bar{D}^p_i &= a^p_q (-\tilde{a}^q_{jr} a^j_k \xi^k a^r_s B^s_l \tilde{a}^l_i - \tilde{a}^q_i - \tilde{a}^q_{rs} a^r_k \xi^k a^s_u B^u_j \tilde{a}^j_i \\ &\quad - \tilde{a}^q_{rs} a^r_t \eta^t a^s_u B^u_j \tilde{a}^j_i + D^q_j \tilde{a}^j_i). \end{aligned} \tag{6}$$

Write $\xi = (\xi^i)$, $\eta = (\eta^p)$, $y = (y^p_i)$, $y_1 = (y^p_{iq})$, $S = (S^p_{ij})$, $R = (R^p_{ij})$.

I. Consider first the coordinate functions $B^p_i(\xi, \eta, y, y_1, S, R)$ of $\mathcal{A}$. The common kernel L of $\pi^2_1 \colon G^2_{m,n} \to G^1_{m,n}$ and of the projection $G^2_{m,n} \to G^2_m \times G^2_n$ described in 28.2 is characterized by $a^i_j = \delta^i_j$, $a^p_q = \delta^p_q$, $a^p_i = 0$, $a^i_{jk} = 0$, $a^p_{qr} = 0$. The equivariance of B^p_i with respect to L implies that B^p_i are independent of y_1 and S. Then the homotheties in $i(G^1_n) \subset G^2_{m,n}$ yield a homogeneity condition

$$kB^p_i = B^p_i(\xi, k\eta, ky, kR).$$

Therefore we have

$$B^p_i = f^p_{iq}(\xi)\eta^q + f^{pj}_{iq}(\xi) y^q_j + f^{pjk}_{iq}(\xi) R^q_{jk}$$

with some smooth functions of ξ. Now the homotheties in $i(G^1_m)$ give

$$k^{-1} B^p_i = f^p_{iq}(k\xi)\eta^q + f^{pj}_{iq}(k\xi) k^{-1} y^q_j + f^{pjk}_{iq}(k\xi) k^{-2} R^q_{jk}.$$

Hence it holds a) $f^p_{iq}(\xi) = k f^p_{iq}(k\xi)$, b) $f^{pj}_{iq}(\xi) = f^{pj}_{iq}(k\xi)$, c) $k f^{pjk}_{iq}(\xi) = f^{pjk}_{iq}(k\xi)$. If we let $k \to 0$ in a) and b), we obtain $f^p_{iq} = 0$ and $f^{pj}_{iq} = \text{const}$. The relation c) yields that f^{pjk}_{iq} is linear in ξ. The equivariance of B^p_i with respect to the whole group $i(G^1_m \times G^1_n)$ implies that f^{pj}_{iq} and f^{pjk}_{iq} correspond to the generalized invariant tensors. By theorem 27.1 we obtain

$$B^p_i = c_1 R^p_{ij} \xi^j + c_2 y^p_i$$

with real parameters c_1, c_2. Consider further the equivariance of B^p_i with respect to the subgroup $K \subset G^2_{m,n}$ characterized by $a^i_j = \delta^i_j$, $a^p_q = \delta^p_q$. This yields

$$c_1 R^p_{ij} \xi^j + c_2 y^p_i = c_1 R^p_{ij} \xi^j + c_2 (y^p_i + a^p_i).$$

This relation implies $c_2 = 0$.

II. For A^p_i we obtain in the same way as in I

$$A^p_i = aR^p_{ij}\xi^j + c_3 y^p_i.$$

The equivariance with respect to subgroup K gives $c_3 = 1$ and $c_1 = 0$.

III. Analogously to I and II we deduce

$$D^p_i = bR^p_{ij}\xi^j + c_4 y^p_i.$$

Taking into account the equivariance of D^p_i with respect to K, we find $c_4 = 1$.

IV. Here it is useful to summarize. Up to now, we have deduced

$$A^p_i = aR^p_{ij}\xi^j + y^p_i, \quad B^p_i = 0, \quad D^p_i = bR^p_{ij}\xi^j + y^p_i. \tag{7}$$

Consider the difference $A - \mathcal{T}$, where $\mathcal{T}$ is the operator (1) and (2). Write

$$E^p_i = C^p_i - y^p_{ij}\xi^j - y^p_{iq}\eta^q. \tag{8}$$

Using a^p_{ij}, we find easily that E^p_i does not depend on S^p_{ij}. By (6) and (8), the action of K on E^p_i is

$$\begin{aligned} -a\tilde{a}^p_{jq}\xi^j R^q_{ik}\xi^k + aa^p_{qr}a^q_j\xi^j R^r_{ik}\xi^k + aa^p_{qr}\eta^q R^r_{ij}\xi^j + E^p_i - bR^p_{jk}\xi^k a^j_{il}\xi^l \\ = E^p_i\left(\xi, \eta^q + a^q_j\xi^j, y^r_j + a^r_j, y^s_{kt} + a^s_{kt} + a^s_{tu}y^u_k, R\right). \end{aligned} \tag{9}$$

If we set $E^p_i = ay^p_{jq}\xi^j R^q_{ik}\xi^k + F^p_i$, then (9) implies that F^p_i is independent of y_1. The action of $i(G^1_m \times G^1_n)$ on $F^p_i(\xi, \eta, y, R)$ is tensorial. Hence we have the same situation as for B^p_i in I. This implies $F^p_i = kR^p_{ij}\xi^j + ey^p_i$. Using once again (9) we obtain $a = b = e = 0$. Hence $E^p_i = kR^p_{ij}\xi^j$ and $C^p_i = y^p_{ij}\xi^j + y^p_{iq}\eta^q + kR^p_{ij}\xi^j$. Thus we have deduced the coordinate form of our statement. □

46.8. Prolongation of connections to $FY \to Y$. Given a bundle functor F on $\mathcal{M}f$ and a fibered manifold $Y \to M$, there are three canonical structures of a fibered manifold on FY, namely $FY \to M$, $FY \to FM$ and $FY \to Y$. Unlike the first two cases, it seems that there should be only poor results on the prolongation of connections to $FY \to Y$. We first present a negative result for the case of the tangent functor T.

Proposition. *There is no first order natural operator transforming connections on $Y \to M$ into connections on $TY \to Y$.*

Proof. We shall use the notation from the proof of proposition 46.7. The equations of a connection on $TY \to Y$ are

$$d\xi^i = M^i_j dx^j + N^i_p dy^p, \quad d\eta^p = P^p_i dx^i + Q^p_q dy^q.$$

One evaluates easily the action formulae $\bar{\xi}^i = a^i_j \xi^j$ and

$$\bar{M}^i_j = a^i_k M^k_l \tilde{a}^l_j + a^i_k N^k_p \tilde{a}^p_j - a^i_l \tilde{a}^l_{jk} a^k_m \xi^m$$
$$\bar{N}^i_p = a^i_j N^j_q \tilde{a}^q_p.$$

The homotheties in $i(G^1_n)$ give

$$N^i_p = kN^i_p(\xi^j, k\eta^q, ky^r_k, y^s_{tl}, ky^u_{mn}).$$

Hence $N^i_p = 0$. For M^i_j, the homotheties in $i(G^1_n)$ imply the independence of M^i_j of η^p, y^p_i, y^p_{ij}. The equivariance of M^i_j with respect to the subgroup K means

$$M^i_j(\xi^j, y^p_{kq}) + a^i_{jk}\xi^k = M^i_j(\xi^j, y^p_{kq} + a^p_{kq}).$$

Since the expressions M^i_j on both sides are independent of a^i_{jk}, the differentiation with respect to a^i_{jk} yields some relations among ξ^i only. □

46.9. Prolongation of connections to $VY \to Y$. We pay special attention to this problem because of its relation to Finslerian geometry. We are going to study the first order natural operators transforming connections on $Y \to M$ into connections on $VY \to Y$, i.e. the natural operators $J^1 \rightsquigarrow J^1(V \to \mathrm{Id})$ where Id means the identity functor. In this case it will be instructive to start from the computational aspect of the problem.

Using the notation from 46.7, the equations of a connection on $VY \to Y$ are

$$d\eta^p = A^p_i(x^j, y^q, \eta^r)dx^i + B^p_q(x^j, y^r, \eta^s)dy^q. \tag{1}$$

The induced coordinates on the standard fiber $Z = J^1_0(V(\mathbb{R}^{m+n} \to \mathbb{R}^m) \to \mathbb{R}^{m+n})$ are η^p, A^p_i, B^p_i and the action of $G^2_{m,n}$ on Z has the form

$$\bar{\eta}^p = a^p_q \eta^q \tag{2}$$
$$\bar{A}^p_i = a^p_{qj}\tilde{a}^j_i \eta^q + a^p_q A^q_j \tilde{a}^j_i - a^p_q B^q_r \tilde{a}^r_s a^s_j \tilde{a}^j_i - a^p_{rs}\tilde{a}^s_q \eta^r a^q_j \tilde{a}^j_i \tag{3}$$
$$\bar{B}^p_q = a^p_r B^r_s \tilde{a}^s_q + a^p_{rs}\tilde{a}^s_q \eta^r. \tag{4}$$

Our problem is to find all $G^2_{m,n}$-maps $S_1 \times \mathbb{R}^n \to Z$ over the identity on $\mathbb{R}^n$. Consider first the component $B^p_q(\eta^r, y^s_i, y^t_{ju}, y^v_{kl})$ of such a map. The homotheties in $i(G^1_n)$ yield

$$B^p_q(\eta^r, y^s_i, y^t_{ju}, y^v_{kl}) = B^p_q(k\eta^r, ky^s_i, y^t_{ju}, ky^v_{kl})$$

so that B^p_q depends on y^r_{is} only. Then the homotheties in $i(G^1_m)$ give $B^p_q(y^r_{is}) = B^p_q(ky^r_{is})$, which implies $B^p_q = \mathrm{const}$. By the invariant tensor theorem, $B^p_q = k\delta^p_q$. The invariance under the subgroup K reads

$$k\delta^p_q + a^p_{qr}\eta^r = k\delta^p_q.$$

This cannot be satisfied for any k. Thus, *there is no first order operator $J^1 \rightsquigarrow J^1(V \to Id)$ natural on the category $\mathcal{FM}_{m,n}$.*

46.10. However, the obstruction is a^p_{qr} and the condition $a^p_{qr} = 0$ characterizes the affine bundles (with vector bundles as a special case). Let us restrict ourselves to the affine bundles and continue in the previous consideration. By 46.9.(3), the action of $i(G^1_m \times G^1_n)$ on $A^p_i(\eta^q, y^r_i, y^s_{jt}, y^u_{kl})$ is tensorial. Using homotheties in $i(G^1_m)$, we find that A^p_i is linear in y^p_i, y^p_{iq}, but the coefficients are smooth functions in η^p. Using homotheties in $i(G^1_n)$, we deduce that the coefficients by y^p_i are constant and the coefficients by y^p_{iq} are linear in η^p. By the generalized invariant tensor theorem, we obtain

$$A^p_i = ay^p_i + by^q_{qi}\eta^p + cy^p_{iq}\eta^q \qquad a,\, b,\, c \in \mathbb{R}. \tag{1}$$

The equivariance of (1) on the subgroup K implies $a = -k$, $b = 0$, $c = 1$. Thus we have proved

Proposition. *All first order operators $J^1 \rightsquigarrow J^1(V \to Id)$ which are natural on the local isomorphisms of affine bundles form the following one-parameter family*

$$d\eta^p = y^p_{iq}\eta^q dx^i + k(dy^p - y^p_i dx^i), \qquad k \in \mathbb{R}.$$

Remarks

Section 42 is based on [Kolář, 88a]. The order estimate in 42.4 follows an idea by [Zajtz, 88b] and the proof of lemma 42.7 was communicated by the second author. The results of section 43 were deduced by [Doupovec, 90]. Section 44 is based on [Kolář, Slovák, 90]. The construction of the connection $\mathcal{F}(\Gamma, \Lambda)$ from 45.4 was first presented in [Kolář, 82b]. Proposition 46.7 was proved by [Doupovec, Kolář, 88]. The relation of proposition 46.10 to Finslerian geometry was pointed out by B. Kis.

CHAPTER XI. GENERAL THEORY OF LIE DERIVATIVES

It has been clarified recently that one can define the generalized Lie derivative $\tilde{\mathcal{L}}_{(\xi,\eta)}f$ of any smooth map $f\colon M \to N$ with respect to a pair of vector fields ξ on M and η on N. Given a section s of a vector bundle $E \to M$ and a projectable vector field η on E over a vector field ξ on M, the second component $\mathcal{L}_\eta s\colon M \to E$ of the generalized Lie derivative $\tilde{\mathcal{L}}_{(\xi,\eta)}s$ is called the Lie derivative of s with respect to η. We first show how this approach generalizes the classical cases of Lie differentiation. We also present a simple, but useful comparison of the generalized Lie derivative with the absolute derivative with respect to a general connection. Then we prove that every linear natural operator commutes with Lie differentiation. We deduce a similar condition in the non linear case as well. An operator satisfying the latter condition is said to be infinitesimally natural. We prove that every infinitesimally natural operator is natural on the category of oriented m-dimensional manifolds and orientation preserving local diffeomorphisms.

A significant advantage of our general theory is that it enables us to study the Lie derivatives of the morphisms of fibered manifolds (our feeling is that the morphisms of fibered manifolds are going to play an important role in differential geometry). To give a deeper example we discuss the Euler operator in the higher order variational calculus on an arbitrary fibered manifold. In the last section we extend the classical formula for the Lie derivative with respect to the bracket of two vector fields to the generalized Lie derivatives.

47. The general geometric approach

47.1. Let M, N be two manifolds and $f\colon M \to N$ be a map. We recall that a *vector field along* f is a map $\varphi\colon M \to TN$ satisfying $p_N \circ \varphi = f$, where $p_N\colon TN \to N$ is the bundle projection.

Consider further a vector field ξ on M and a vector field η on N.

Definition. The generalized Lie derivative $\tilde{\mathcal{L}}_{(\xi,\eta)}f$ of $f\colon M \to N$ with respect to ξ and η is the vector field along f defined by

$$\tilde{\mathcal{L}}_{(\xi,\eta)}f\colon Tf \circ \xi - \eta \circ f. \tag{1}$$

By the very definition, $\tilde{\mathcal{L}}_{(\xi,\eta)}$ is the zero vector field along f if and only if the vector fields η and ξ are f-related.

47.2. Definition 47.1 is closely related with the kinematic approach to Lie differentiation. Using the flows Fl^ξ_t and Fl^η_t of vector fields ξ and η, we construct a curve

$$t \mapsto (\mathrm{Fl}^\eta_{-t} \circ f \circ \mathrm{Fl}^\xi_t)(x) \tag{1}$$

for every $x \in M$. Differentiating it with respect to t for $t = 0$ we obtain the following

Lemma. *$\tilde{\mathcal{L}}_{(\xi,\eta)} f(x)$ is the tangent vector to the curve (1) at $t = 0$, i.e.*

$$\tilde{\mathcal{L}}_{(\xi,\eta)} f(x) = \tfrac{\partial}{\partial t}\big|_0 (\mathrm{Fl}^\eta_{-t} \circ f \circ \mathrm{Fl}^\xi_t)(x).$$

47.3. In the greater part of differential geometry one meets definition 47.1 in certain more specific situations. Consider first an arbitrary fibered manifold $p\colon Y \to M$, a section $s\colon M \to Y$ and a projectable vector field η on Y over a vector field ξ on M. Then it holds $Tp \circ (Ts \circ \xi - \eta \circ s) = 0_M$, where 0_M means the zero vector field on M. Hence $\tilde{L}_{(\xi,\eta)} s$ is a section of the vertical tangent bundle of Y. We shall write

$$\tilde{\mathcal{L}}_{(\xi,\eta)} s =: \tilde{\mathcal{L}}_\eta s\colon M \to VY$$

and say that $\tilde{\mathcal{L}}_\eta$ is the *generalized Lie derivative of s with respect to η*. If we have a vector bundle $E \to M$, then its vertical tangent bundle VE coincides with the fibered product $E \times_M E$, see 6.11. Then the generalized Lie derivative $\tilde{\mathcal{L}}_\eta s$ has the form

$$\tilde{\mathcal{L}}_\eta s = (s, \mathcal{L}_\eta s)$$

where $\mathcal{L}_\eta s$ is a section of E.

47.4. Definition. Given a vector bundle $E \to M$ and a projectable vector field η on E, the second component $\mathcal{L}_\eta s\colon M \to E$ of the generalized Lie derivative $\tilde{\mathcal{L}}_\eta s$ is called the Lie derivative of s with respect to the field η.

If we intend to contrast the Lie derivative $\mathcal{L}_\eta s$ with the generalized Lie derivative $\tilde{\mathcal{L}}_\eta s$, we shall say that $\mathcal{L}_\eta s$ is the restricted Lie derivative. Using the fact that the second component of $\tilde{\mathcal{L}}_\eta s$ is the derivative of $\mathrm{Fl}^\eta_{-t} \circ s \circ \mathrm{Fl}^\xi_t$ for $t = 0$ in the classical sense, we can express the restricted Lie derivative in the form

$$(\mathcal{L}_\eta s)(x) = \lim_{t\to 0} \tfrac{1}{t}(\mathrm{Fl}^\eta_{-t} \circ s \circ \mathrm{Fl}^\xi_t - s)(x). \tag{1}$$

47.5. It is useful to compare the general Lie differentiation with the covariant differentiation with respect to a general connection $\Gamma\colon Y \to J^1Y$ on an arbitrary fibered manifold $p\colon Y \to M$. For every $\xi_0 \in T_xM$, let $\Gamma(y)(\xi_0)$ be its lift to the horizontal subspace of Γ at $p(y) = x$. For a vector field ξ on M, we obtain in this way its Γ-lift $\Gamma\xi$, which is a projectable vector field on Y over ξ. The connection map $\omega_\Gamma\colon TY \to VY$ means the projection in the direction of the horizontal subspaces of Γ. The *generalized covariant differential* $\tilde{\nabla}^\Gamma s$ of a section s of Y is defined as the composition of ω_Γ with Ts. This gives a linear map $T_xM \to V_{s(x)}Y$ for every $x \in M$, so that $\tilde{\nabla}^\Gamma s$ can be viewed as a section $M \to VY \otimes T^*M$, which was introduced in another way in 17.8. The *generalized covariant derivative* $\tilde{\nabla}^\Gamma_\xi s$ *of* s *with respect to a vector field* ξ *on* M is then defined by the evaluation

$$\tilde{\nabla}^\Gamma_\xi s := \langle \xi, \tilde{\nabla}^\Gamma s\rangle : M \to VY.$$

Proposition. *It holds*

$$\tilde{\nabla}^\Gamma_\xi s = \tilde{\mathcal{L}}_{\Gamma\xi} s.$$

Proof. Clearly, the value of ω_Γ at a vector $\eta_0 \in T_yY$ can be expressed as $\omega_\Gamma(\eta_0) = \eta_0 - \Gamma(y)(Tp(\eta_0))$. Hence $\tilde{\mathcal{L}}_{\Gamma\xi}s(x) = Ts(\xi(x)) - \Gamma\xi(s(x))$ coincides with $\omega_\Gamma(Ts(\xi(x)))$. □

In the case of a vector bundle $E \to M$, we have $VE = E \oplus E$ and $\tilde{\nabla}^\Gamma_\xi s = (s, \nabla^\Gamma_\xi s)$. The second component $\nabla^\Gamma_\xi : M \to E$ is called *the covariant derivative of* s *with respect to* ξ, see 11.12. In such a situation the above proposition implies

$$\nabla^\Gamma_\xi s = \mathcal{L}_{\Gamma\xi} s. \tag{1}$$

47.6. Consider further a natural bundle $F\colon \mathcal{M}f_m \to \mathcal{FM}$. For every vector field ξ on M, its flow prolongation $\mathcal{F}\xi$ is a projectable vector field on FM over ξ. If F is a natural vector bundle, we have $VFM = FM \oplus FM$.

Definition. Given a natural bundle F, a vector field ξ on a manifold M and a section s of FM, the generalized Lie derivative

$$\tilde{\mathcal{L}}_{\mathcal{F}\xi} s =: \tilde{\mathcal{L}}_\xi \colon M \to VFM$$

is called the *generalized Lie derivative of* s *with respect to* ξ. In the case of a natural vector bundle F,

$$\mathcal{L}_{\mathcal{F}\xi} s =: \mathcal{L}_\xi s \colon M \to FM$$

is called the *Lie derivative of* s *with respect to* ξ.

47.7. An important feature of our general approach to Lie differentiation is that it enables us to study the Lie derivatives of the morphisms of fibered manifolds. In general, consider two fibered manifolds $p\colon Y \to M$ and $q\colon Z \to M$ over the same base, a base preserving morphism $f\colon Y \to Z$ and a projectable vector field η or ζ on Y or Z over the same vector field ξ on M. Then $Tq \circ (Tf \circ \eta - \zeta \circ f) = 0_M$, so that the values of the generalized Lie derivative $\tilde{\mathcal{L}}_{(\eta,\zeta)} f$ lie in the vertical tangent bundle of Z.

Definition. If Z is a vector bundle, then the second component

$$\mathcal{L}_{(\eta,\zeta)} f \colon Y \to Z$$

of $\tilde{\mathcal{L}}_{(\eta,\zeta)} f \colon Y \to VZ$ is called the Lie derivative of f with respect to η and ζ.

47.8. Having two natural bundles FM, GM and a base-preserving morphism $f\colon FM \to GM$, we can define the Lie derivative of f with respect to a vector field ξ on M. In the case of an arbitrary G, we write

$$\tilde{\mathcal{L}}_{(\mathcal{F}\xi,\mathcal{G}\xi)} f =: \tilde{\mathcal{L}}_\xi f\colon FM \to VGM. \tag{1}$$

If G is a natural vector bundle, we set

$$\mathcal{L}_{(\mathcal{F}\xi,\mathcal{G}\xi)} f =: \mathcal{L}_\xi f\colon FM \to GM. \tag{2}$$

47.9. Linear vector fields on vector bundles. Consider a vector bundle $p: E \to M$. By 6.11, $Tp\colon TE \to TM$ is a vector bundle as well. A projectable vector field η on E over ξ on M is called a *linear vector field*, if $\eta\colon E \to TE$ is a linear morphism of $E \to M$ into $TE \to TM$ over the base map $\xi\colon M \to TM$.

Proposition. *η is a linear vector field on E if and only if its flow is formed by local linear isomorphisms of E.*

Proof. Let x^i, y^p be some fiber coordinates on E such that y^p are linear coordinates in each fiber. By definition, the coordinate expression of a linear vector field η is

$$\xi^i(x)\tfrac{\partial}{\partial x^i} + \eta^p_q(x)y^q\tfrac{\partial}{\partial y^p}. \tag{1}$$

Hence the differential equations of the flow of η are

$$\tfrac{dx^i}{dt} = \xi^i(x), \qquad \tfrac{dy^p}{dt} = \eta^p_q(x)y^q.$$

Their solution represents the linear local isomorphisms of E by virtue of the linearity in y^p. On the other hand, if the flow of η is linear and we differentiate it with respect to t, then η must be of the form (1). □

47.10. Let $\bar\eta$ be another linear vector field on another vector bundle $\bar E \to M$ over the same vector field ξ on the base manifold M. Using flows, we define a vector field $\eta\otimes\bar\eta$ on the tensor product $E\otimes\bar E$ by

$$\eta\otimes\bar\eta = \tfrac{\partial}{\partial t}\big|_0 (\mathrm{Fl}^\eta_t)\otimes(\mathrm{Fl}^{\bar\eta}_t).$$

Proposition. *$\eta\otimes\bar\eta$ is the unique linear vector field on $E\otimes\bar E$ over ξ satisfying*

$$\mathcal{L}_{\eta\otimes\bar\eta}(s\otimes\bar s) = (\mathcal{L}_\eta s)\otimes\bar s + s\otimes(\mathcal{L}_{\bar\eta}\bar s) \tag{1}$$

for all sections s of E and $\bar s$ of $\bar E$.

Proof. If 47.9.(1) is the coordinate expression of η and $y^p = s^p(x)$ is the coordinate expression of s, then the coordinate expression of $\mathcal{L}_\eta s$ is

$$\tfrac{\partial s^p(x)}{\partial x^i}\xi^i(x) - \eta^p_q(x)s^q(x). \tag{2}$$

Further, let

$$\xi^i(x)\tfrac{\partial}{\partial x^i}+\bar\eta^a_b(x)z^b\tfrac{\partial}{\partial z^a}$$

be the coordinate expression of $\bar\eta$ in some linear fiber coordinates x^i, z^a on $\bar E$. If w^{pa} are the induced coordinates on the fibers of $E\otimes\bar E$ and $\bar x^i=\varphi^i(x,t)$, $\bar y^p=\varphi^p_q(x,t)y^q$ or $\bar z^a=\bar\varphi^a_b(x,t)z^b$ is the flow of η or $\bar\eta$, respectively, then $\mathrm{Fl}^\eta_t\otimes\mathrm{Fl}^{\bar\eta}_t$ is

$$\bar x^i=\varphi^i(x,t),\quad \bar w^{pa}=\varphi^p_q(x,t)\bar\varphi^a_b(x,t)w^{qb}.$$

By differentiating at $t=0$, we obtain

$$\eta\otimes\bar\eta=\xi^i(x)\tfrac{\partial}{\partial x^i}+(\eta^p_q(x)\delta^a_b+\delta^p_q\bar\eta^a_b(x))w^{qb}\tfrac{\partial}{\partial w^{pa}}.$$

Thus, if $z^a=\bar s^a(x)$ is the coordinate expression of $\bar s$, we have

$$\mathcal{L}_{\eta\otimes\bar\eta}(s\otimes\bar s)=\left(\tfrac{\partial s^p}{\partial x^i}\bar s^a+s^p\tfrac{\partial\bar s^a}{\partial x^i}\right)\xi^i-\eta^p_q s^q\bar s^a-\bar\eta^a_b s^p\bar s^b.$$

This corresponds to the right hand side of (1). □

47.11. On the dual vector bundle $E^*\to M$ of E, we define the *vector field η^* dual to a linear vector field η on E* by

$$\eta^*=\tfrac{\partial}{\partial t}\big|_0(\mathrm{Fl}^\eta_{-t})^*.$$

Having a vector field ζ on M and a function $f\colon M\to\mathbb{R}$, we can take the zero vector field $0_{\mathbb{R}}$ on $\mathbb{R}$ and construct the generalized Lie derivative

$$\tilde{\mathcal{L}}_{(\zeta,0_{\mathbb{R}})}f=Tf\circ\zeta\colon M\to T\mathbb{R}=\mathbb{R}\times\mathbb{R}.$$

Its second component is the usual Lie derivative $\mathcal{L}_\zeta f=\zeta f$, i.e. the derivative of f in the direction ζ.

Proposition. *η^* is the unique linear vector field on E^* over ξ satisfying*

$$\mathcal{L}_\xi\langle s,\sigma\rangle=\langle\mathcal{L}_\eta s,\sigma\rangle+\langle s,\mathcal{L}_{\eta^*}\sigma\rangle$$

for all sections s of E and σ of E^*.

Proof. Let v_p be the coordinates on E^* dual to y^p. By definition, the coordinate expression of η^* is

$$\xi^i(x)\tfrac{\partial}{\partial x^i}-\eta^q_p(x)v_q\tfrac{\partial}{\partial v_p}.$$

Then we prove the above proposition by a direct evaluation quite similar to the proof of proposition 47.10. □

47.12. A vector field η on a manifold M is a section of the tangent bundle TM, so that we have defined its Lie derivative $\mathcal{L}_\xi\eta$ with respect to another vector field ξ on M as the second component of $T\eta\circ\xi-T\xi\circ\eta$. In 3.13 it is deduced that $\mathcal{L}_\xi\eta=[\xi,\eta]$. Then 47.10 and 47.11 imply, that for the classical tensor fields the geometrical approach to the Lie differentiation coincides with the algebraic one.

47.13. In the end of this section we remark that the operations with linear vector fields discussed here can be used to define, in a short way, the corresponding operation with linear connections on vector bundles. We recall that a linear connection Γ on a vector bundle $E \to M$ is a section $\Gamma\colon E \to J^1E$ which is a linear morphism from vector bundle $E \to M$ into vector bundle $J^1E \to M$. Using local trivializations of E we find easily that this condition is equivalent to the fact that the Γ-lift $\Gamma\xi$ of every vector field ξ on M is a linear vector field on E. By 47.9, the coordinate expression of a linear connection Γ on E is

$$dy^p = \Gamma^p_{qi}(x)y^q dx^i.$$

Let $\bar\Gamma$ be another linear connection on a vector bundle $\bar E \to M$ over the same base with the equations

$$dz^a = \bar\Gamma^a_{bi}(x)z^b dx^i.$$

Using 47.10 and 47.11, we obtain immediately the following two assertions.

47.14. Proposition. *There is a unique linear connection $\Gamma \otimes \bar\Gamma$ on $E \otimes \bar E$ satisfying*

$$(\Gamma \otimes \bar\Gamma)(\xi) = (\Gamma\xi) \otimes (\bar\Gamma\xi)$$

for every vector field ξ on M.

47.15. Proposition. *There is a unique linear connection Γ^* on E^* satisfying $\Gamma^*(\xi) = (\Gamma\xi)^*$ for every vector field ξ on M.*

Obviously, the equations of $\Gamma \otimes \bar\Gamma$ are

$$dw^{pa} = (\Gamma^p_{qi}(x)\delta^a_b + \delta^p_q\bar\Gamma^a_{bi}(x))w^{qb}dx^i$$

and the coordinate expression of Γ^* is

$$dv_p = -\Gamma^q_{pi}(x)v_q dx^i.$$

48. Commuting with natural operators

48.1. The Lie derivative commutes with the exterior differential, i.e. $d(\mathcal{L}_X\omega) = \mathcal{L}_X(d\omega)$ for every exterior form ω and every vector field X, see 7.9.(5). Our geometrical analysis of the concept of the Lie derivative leads to a general result, which clarifies that the specific property of the exterior differential used in the above formula is its linearity.

Proposition. *Let F and G be two natural vector bundles and $A\colon F \rightsquigarrow G$ be a natural linear operator. Then*

$$A_M(\mathcal{L}_X s) = \mathcal{L}_X(A_M s) \tag{1}$$

for every section s of FM and every vector field X on M.

In the special case of a linear natural transformation this is lemma 6.17.

Proof. The explicit meaning of (1) is $A_M(\mathcal{L}_{\mathcal{F}X}s) = \mathcal{L}_{\mathcal{G}X}(A_Ms)$. By the Peetre theorem, A_M is locally a differential operator, so that A_M commutes with limits. Hence

$$\begin{aligned} A_M(\mathcal{L}_{\mathcal{F}X}s) &= \lim_{t\to 0}\frac{1}{t}[A_M(F(\mathrm{Fl}^X_{-t})\circ s\circ \mathrm{Fl}^X_t) - A_Ms] \\ &= \lim_{t\to 0}\frac{1}{t}[G(\mathrm{Fl}^X_{-t}\circ A_Ms\circ \mathrm{Fl}^X_t - A_Ms] = \mathcal{L}_{\mathcal{G}X}(A_Ms) \end{aligned}$$

by linearity and naturality. □

48.2. A reasonable result of this type can be deduced even in the non linear case. Let F and G be arbitrary natural bundles on $\mathcal{M}f_m$, $D\colon C^\infty(FM)\to C^\infty(GM)$ be a local regular operator and $s\colon M\to FM$ be a section. The generalized Lie derivative $\tilde{\mathcal{L}}_Xs$ is a section of VFM, so that we cannot apply D to $\tilde{\mathcal{L}}_Xs$. However, we can consider the so called *vertical prolongation* $VD\colon C^\infty(VFM)\to C^\infty(VGM)$ *of the operator* D. This can be defined as follows.

In general, let $N\to M$ and $N'\to M$ be arbitrary fibered manifolds over the same base and let $D\colon C^\infty(N)\to C^\infty(N')$ be a local regular operator. Every local section q of $VN\to M$ is of the form $\frac{\partial}{\partial t}|_0 s_t$, $s_t\in C^\infty(N)$ and we set

$$VDq = VD(\tfrac{\partial}{\partial t}|_0 s_t) = \tfrac{\partial}{\partial t}|_0(Ds_t)\in C^\infty(VN'). \tag{1}$$

We have to verify that this is a correct definition. By the nonlinear Peetre theorem the operator D is induced by a map $D : J^\infty N\to N'$. Moreover each infinite jet has a neighborhood in the inverse limit topology on $J^\infty N$ on which D depends only on r-jets for some finite r. Thus, there is neighborhood U of x in M and a locally defined smooth map $D^r : J^rN\to N'$ such that $Ds_t(y) = D^r(j^r_ys_t)$ for $y\in U$ and for t sufficiently small. So we get

$$(VD)q(x) = \tfrac{\partial}{\partial t}|_0(D^r(j^r_xs_t)) = TD^r(\tfrac{\partial}{\partial t}|_0 j^r_xs_t) = (TD^r\circ\kappa)(j^r_xq)$$

where κ is the canonical exchange map, and thus the definition does not depend on the choice of the family s_t.

48.3. A local regular operator $D\colon C^\infty(FM)\to C^\infty(GM)$ is called *infinitesimally natural* if it holds

$$\tilde{\mathcal{L}}_X(Ds) = VD(\tilde{\mathcal{L}}_Xs)$$

for all $X\in\mathfrak{X}(M)$, $s\in C^\infty(FM)$.

Proposition. *If $A : F\rightsquigarrow G$ is a natural operator, then all operators A_M are infinitesimally natural.*

Proof. By lemma 47.2, 48.2.(1) and naturality we have

$$\begin{aligned} VA_M(\tilde{\mathcal{L}}_{\mathcal{F}X}s) &= VA_M\left(\tfrac{\partial}{\partial t}|_0(F(\mathrm{Fl}^X_{-t})\circ s\circ\mathrm{Fl}^X_t\right) \\ &= \tfrac{\partial}{\partial t}|_0 A_M(F(\mathrm{Fl}^X_{-t})\circ s\circ\mathrm{Fl}^X_t) = \tfrac{\partial}{\partial t}|_0(G(\mathrm{Fl}^X_{-t})\circ A_Ms\circ\mathrm{Fl}^X_t) \\ &= \tilde{\mathcal{L}}_{\mathcal{G}X}A_Ms. \end{aligned}$$

□

48.4. Let $\mathcal{M}f_m^+$ be the category of oriented m-dimensional manifolds and orientation preserving local diffeomorphisms.

Theorem. *Let F and G be two bundle functors on $\mathcal{M}f_m^+$, M be an oriented m-dimensional manifold and let $A_M : C^\infty(FM) \to C^\infty(GM)$ be an infinitesimally natural operator. Then A_M is the value of a unique natural operator $A: F \rightsquigarrow G$ on M.*

We shall prove this theorem in several steps.

48.5. Let us fix an infinitesimally natural operator $D: C^\infty(F\mathbb{R}^m) \to C^\infty(G\mathbb{R}^m)$ and let us write S and Q for the standard fibers $F_0\mathbb{R}^m$ and $G_0\mathbb{R}^m$. Since each local operator is locally of finite order by the nonlinear Peetre theorem, there is the induced map $\mathcal{D}: T_m^\infty S \to Q$. Moreover, at each $j_0^\infty s \in T_m^\infty S$ the application of the Peetre theorem (with $K = \{0\}$) yields a smallest possible order $r = \chi(j_0^\infty s)$ such that for every section q with $j_0^r q = j_0^r s$ we have $Ds(0) = Dq(0)$, see 23.1. Let us define $\tilde{V}_r \subset T_m^\infty S$ as the subset of all jets with $\chi(j_0^\infty s) \le r$. Let V_r be the interior of $\tilde{V}_r$ in the inverse limit topology and put $U_r := \pi_r^\infty(V_r) \subset T_m^r S$.

The Peetre theorem implies $T_m^\infty S = \cup_r V_r$ and so the sets V_r form an open filtration of $T_m^\infty S$. On each V_r, the map $\mathcal{D}$ factors to a map $\mathcal{D}_r : U_r \to Q$.

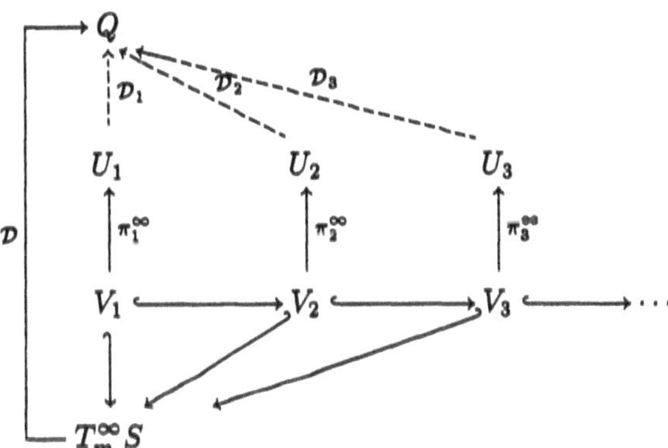

Since there are the induced actions of the jet groups G_m^{r+k} on $T_m^r S$ (here k is the order of F), we have the fundamental field mapping $\zeta^{(r)}: \mathfrak{g}_m^{r+k} \to \mathfrak{X}(T_m^r S)$ and we write ζ^Q for the fundamental field mapping on Q. There is an analogy to 34.3.

Lemma. *For all $X \in \mathfrak{g}_m^{r+k}$, $j_0^r s \in T_m^r S$ it holds*

$$\zeta_X^{(r)}(j_0^r s) = \kappa(j_0^r(\tilde{\mathcal{L}}_{-X} s)).$$

Proof. Write λ for the action of the jet group on $T_m^r S$. We have

$$\begin{aligned} \zeta_X^{(r)}(j_0^r s) &= \tfrac{\partial}{\partial t}\big|_0 \lambda(\exp tX)(j_0^r s) = \tfrac{\partial}{\partial t}\big|_0 j_0^r(F(\mathrm{Fl}_t^X) \circ s \circ \mathrm{Fl}_{-t}^X) \\ &= \kappa(j_0^r(\tfrac{\partial}{\partial t}\big|_0 (F(\mathrm{Fl}_t^X) \circ s \circ \mathrm{Fl}_{-t}^X))) = \kappa(j_0^r(\tilde{\mathcal{L}}_{-X} s)). \quad \square \end{aligned}$$

48.6. Lemma. *For all $r \in \mathbb{N}$ and $X \in \mathfrak{g}_m^{r+k}$ we have $T\mathcal{D}^r \circ \zeta_X^{(r)} = \zeta_X^Q \circ \mathcal{D}^r$ on U_r.*

Proof. Recall that $(VD)q(0) = (T\mathcal{D}^r \circ \kappa)(j_0^r q)$ for all $j_0^r q \in \kappa^{-1}(TU_r)$. Using the above lemma and the infinitesimal naturality of D we compute

$$(T\mathcal{D}_r \circ \zeta_X^{(r)})(j_0^r s) = T\mathcal{D}_r(\kappa(j_0^r(\tilde{\mathcal{L}}_{-X}s))) = VD(\tilde{\mathcal{L}}_{-X}s)(0) =$$
$$= \tilde{\mathcal{L}}_{-X}(Ds)(0) = \zeta_X^Q(Ds(0)) = \zeta_X^Q(\mathcal{D}_r(j_0^r s)). \quad \square$$

48.7. Lemma. *The map $\mathcal{D}\colon T_m^\infty S \to Q$ is $G_m^{\infty+}$-equivariant.*

Proof. Given $a = j_0^\infty f \in G_m^{\infty+}$ and $y = j_0^\infty s \in T_m^\infty S$ we have to show $\mathcal{D}(a.y) = a.\mathcal{D}(y)$. Each a is a composition of a jet of a linear map f and of a jet from the kernel B_1^∞ of the jet projection π_1^∞. If f is linear, then there are linear maps g_i, $i = 1, 2, \dots, l$, lying in the image of the exponential map of G_m^1 such that $f = g_1 \circ \dots \circ g_l$. Since $T_m^\infty S = \cup_r V_r$ there must be an $r \in \mathbb{N}$ such that y and all elements $(j_0^\infty g_p \circ \dots \circ j_0^\infty g_l) \cdot y$ are in V_r for all $p \le l$. Thus, $\mathcal{D}(a.y) = \mathcal{D}_r(j_0^{r+k} f.j_0^r s) = j_0^{r+k} f.\mathcal{D}_r(j_0^r s) = a.\mathcal{D}(y)$, for $\mathcal{D}_r$ preserves all the fundamental fields.

Since the whole kernel B_1^r lies in the image of the exponential map for each $r < \infty$, an analogous consideration for $j_0^\infty f \in B_1^\infty$ completes the proof of the lemma. $\square$

48.8. Lemma. *The natural operator A on $\mathcal{M}f_m^+$ which is determined by the $G_m^{\infty+}$-equivariant map $\mathcal{D}$ coincides on $\mathbb{R}^m$ with the operator D.*

Proof. There is the associated map $\mathcal{A}\colon J^\infty F\mathbb{R}^m \to G\mathbb{R}^m$ to the operator $A_{\mathbb{R}^m}$. Let us write $\mathcal{A}_0$ for its restriction $(J^\infty F)_0\mathbb{R}^m \to G_0\mathbb{R}^m$ and similarly for the map $\mathcal{D}$ corresponding to the original operator D. Now let $t_x : \mathbb{R}^m \to \mathbb{R}^m$ be the translation by x. Then the map $\mathcal{A}$ (and thus the operator A) is uniquely determined by $\mathcal{A}_0$ since by naturality of A we have $(t_{-x})^* \circ A_{\mathbb{R}^m} \circ (t_x)^* = A_{\mathbb{R}^m}$. But t_x is the flow at time 1 of the constant vector field X. For every vector field X and section s we have

$$\tilde{\mathcal{L}}_X((\mathrm{Fl}_t^X)^* s) = \tilde{\mathcal{L}}_X(F(\mathrm{Fl}_{-t}^X) \circ s \circ \mathrm{Fl}_t^X) = \tfrac{\partial}{\partial t}(F(\mathrm{Fl}_{-t}^X) \circ s \circ \mathrm{Fl}_t^X)$$
$$= T(F(\mathrm{Fl}_{-t}^X)) \circ \tilde{\mathcal{L}}_X s \circ \mathrm{Fl}_t^X = (\mathrm{Fl}_t^X)^*(\tilde{\mathcal{L}}_X s)$$

and so using infinitesimal naturality, for every complete vector field X we compute

$$\tfrac{\partial}{\partial t}((\mathrm{Fl}_{-t}^X)^*(D(\mathrm{Fl}_t^X)^* s)) =$$
$$= -(\mathrm{Fl}_{-t}^X)^* \tilde{\mathcal{L}}_X(D(\mathrm{Fl}_t^X)^* s) + (\mathrm{Fl}_{-t}^X)^*((VD)((\mathrm{Fl}_t^X)^* \tilde{\mathcal{L}}_X s)) =$$
$$= (\mathrm{Fl}_{-t}^X)^*(-\tilde{\mathcal{L}}_X(D(\mathrm{Fl}_t^X)^* s) + (VD)(\tilde{\mathcal{L}}_X((\mathrm{Fl}_t^X)^* s))) = 0.$$

Thus $(t_{-x})^* \circ D \circ (t_x)^* = D$ and since $\mathcal{A}_0 = \mathcal{D}_0$ this completes the proof. $\square$

Lemmas 48.7 and 48.8 imply the assertion of theorem 48.4. Indeed, if $M = \mathbb{R}^m$ we get the result immediately and it follows for general M by locality of the operators in question.

48.9. As we have seen, if F is a natural vector bundle, then VF is naturally equivalent to $F \oplus F$ and the second component of our general Lie derivative is just the usual Lie derivative. Thus, the condition of the infinitesimal naturality becomes the usual form $D \circ \mathcal{L}_X = \mathcal{L}_X \circ D$ if $D\colon C^\infty(FM) \to C^\infty(GM)$ is linear.

More generally, if F is a sum $F = E_1 \oplus \cdots \oplus E_k$ of k natural vector bundles, G is a natural vector bundle and D is k-linear, then we have

$$\begin{aligned}\mathrm{pr}_2 \circ VD(\tilde{\mathcal{L}}_X(s_1,\dots,s_k)) &= \tfrac{\partial}{\partial t}\big|_0 D\big(F(\mathrm{Fl}^X_{-t}) \circ (s_1,\dots,s_k) \circ \mathrm{Fl}^X_t\big)\\ &= \sum_{i=1}^k D(s_1,\dots,\mathcal{L}_X s_i,\dots,s_k).\end{aligned}$$

Hence for the k-linear operators we have

Corollary. *Every natural k-linear operator $A : E_1 \oplus \cdots \oplus E_k \rightsquigarrow F$ satisfies*

$$\mathcal{L}_X A_M(s_1,\dots,s_k) = \textstyle\sum_{i=1}^k A_M(s_1,\dots,\mathcal{L}_X s_i,\dots,s_k) \tag{1}$$

for all $s_1 \in C^\infty E_1 M,\dots,s_k \in C^\infty E_k M$, $X \in C^\infty TM$.

Formula (1) covers, among others, the cases of the Frölicher-Nijenhuis bracket and the Schouten bracket discussed in 30.10 and 8.5.

48.10. The converse implication follows immediately for vector bundle functors on $\mathcal{M}f_m^+$. But we can prove more.

Let $E_1,\dots,E_k$ be r-th order natural vector bundles corresponding to actions λ_i of the jet group G^r_m on standard fibers S_i, and assume that with the restricted actions $\lambda_i|G^1_m$ the spaces S_i are invariant subspaces in spaces of the form $\oplus_j(\otimes^{p_j}\mathbb{R}^m \otimes \otimes^{q_j}\mathbb{R}^{m*})$. In particular this applies to all natural vector bundles which are constructed from the tangent bundle. Given any natural vector bundle F we have

Theorem. *Every local regular k-linear operator*

$$A_M\colon C^\infty(E_1M) \oplus \cdots \oplus C^\infty(E_kM) \to C^\infty(FM)$$

over an m-dimensional manifold M which satisfies 48.9.1 is a value of a unique natural operator A on $\mathcal{M}f_m$.

The theorem follows from the theorem 48.4 and the next lemma

Lemma. *Every k-linear natural operator $A : E_1 \oplus \cdots \oplus E_k \rightsquigarrow F$ on $\mathcal{M}f_m^+$ extends to a unique natural operator on $\mathcal{M}f_m$.*

Let us remark, the proper sense of this lemma is that every operator in question obeys the necessary commutativity properties with respect to all local diffeomorhpisms between oriented m-manifolds and hence determines a unique natural operator over the whole $\mathcal{M}f_m$.

Proof. By the multilinear Peetre theorem A is of some finite order ℓ. Thus A is determined by the associated k-linear $(G^{r+\ell}_m)^+$-equivariant map $\mathcal{A}\colon T^\ell_m S_1 \times \ldots \times$

$T^\ell_m S_k \to Q$. Recall that the jet group $G^{r+\ell}_m$ is the semidirect product of $GL(m)$ and the kernel $B^{r+\ell}_1$, while $(G^{r+\ell}_m)^+$ is the semidirect product of the connected component $GL^+(m)$ of the unit and the same kernel $B^{r+\ell}_1$. Thus, in particular the map $\mathcal{A}\colon T^\ell_m S_1 \times \ldots \times T^\ell_m S_k \to Q$ is k-linear and $GL^+(m)$-equivariant. By the descriptions of $(G^{r+\ell}_m)^+$ and $G^{r+\ell}_m$ above we only have to show that any such map is $GL(m)$ equivariant, too. Using the standard polarization technique we can express the map $\mathcal{A}$ by means of a $GL^+(m)$ invariant tensor. But looking at the proof of the Invariant tensor theorem one concludes that the spaces of $GL^+(m)$ invariant and of $GL(m)$ invariant tensors coincide, so the map $\mathcal{A}$ is $GL(m)$ equivariant. □

48.11. Lie derivatives of sector forms. At the end of this section we present an original application of proposition 48.1. This is related with the differentiation of a certain kind of r-th order forms on a manifold M. The simplest case is the 'ordinary' differential of a classical 1-form on M. Such a 1-form ω can be considered as a map $\omega : TM \to \mathbb{R}$ linear on each fiber. Beside its exterior differential $d\omega : \wedge^2 TM \to \mathbb{R}$, E. Cartan and some other classical geometers used another kind of differentiating ω in certain concrete geometric problems. This was called the ordinary differential of ω to be contrasted from the exterior one. We can define it by constructing the tangent map $T\omega : TTM \to T\mathbb{R} = \mathbb{R} \times \mathbb{R}$, which is of the form $T\omega = (\omega, \delta\omega)$. The second component $\delta\omega : TTM \to \mathbb{R}$ is said to be the *(ordinary) differential of* ω . In an arbitrary order r we consider the r-th iterated tangent bundle $T^r M = T(\cdots T(TM)\cdots)$ (r times) of M. The elements of $T^r M$ are called the r-sectors on M. Analogously to the case $r = 2$, in which we have two well-known vector bundle structures p_{TM} and Tp_M on TTM over TM, on $T^r M$ there are r vector bundle structures $p_{T^{r-1}M}$, $Tp_{T^{r-2}M}, \ldots, T\cdots Tp_M$ ($r-1$ times) over $T^{r-1}M$.

Definition. A *sector r-form on M* is a map $\sigma : T^r M \to \mathbb{R}$ linear with respect to all r vector bundle structures on $T^r M$ over $T^{r-1}M$.

A *sector r-form on M at a point x* is the restriction of a sector r-form an M to the fiber $(T^r M)_x$. Denote by $T^r_* M \to M$ the fiber bundle of all sector r-forms at the individual points on M, so that a sector r-form on M is a section of $T^r_* M$. Obviously, $T^r_* M \to M$ has a vector bundle structure induced by the linear combinations of $\mathbb{R}$-valued maps. If $f : M \to N$ is a local diffeomorphism and $A\colon (T^r M)_x \to \mathbb{R}$ is an element of $(T^r_* M)_x$, we define $(T^r_* f)(A) = A \circ (T^r f^{-1})_{f(x)}\colon (T^r N)_{f(x)} \to \mathbb{R}$, where f^{-1} is constructed locally. Since $T^r f$ is a linear morphism for all r vector bundle structures, $(T^r_* f)(A)$ is an element of $(T^r_* N)_{f(x)}$. Hence T^r_* is a natural bundle. In particular, for every vector field X on M and every sector r-form σ on M we have defined the Lie derivative

$$\mathcal{L}_X \sigma = \mathcal{L}_{T^r_* X}\sigma\colon M \to T^r_* M.$$

For every sector r-form $\sigma\colon T^r \to \mathbb{R}$ we can construct its tangent map $T\sigma\colon TT^r M \to T\mathbb{R} = \mathbb{R}\times\mathbb{R}$, which is of the form $(\sigma, \delta\sigma)$. Since the tangent functor preserves vector bundle structures,

$$\delta\sigma : T^{r+1}M \to \mathbb{R}$$

is linear with respect to all $r+1$ vector bundle structures on $T^{r+1}M$ over T^rM, so that this is a sector $(r+1)$-form on M.

48.12. Definition. The operator $\delta : C^\infty T^r_* M \to C^\infty T^{r+1}_* M$ will be called the differential of sector forms.

By definition, δ is a natural operator. Obviously, δ is a linear operator as well. Applying proposition 48.1, we obtain

48.13. Corollary. *δ commutes with the Lie differentiation, i.e.*

$$\delta(\mathcal{L}_X \sigma) = \mathcal{L}_X(\delta\sigma)$$

for every sector r-form σ and every vector field X.

49. Lie derivatives of morphisms of fibered manifolds

We are going to show a deeper application of the geometrical approach to Lie differentiation in the higher order variational calculus in fibered manifolds. For the sake of simplicity we restrict ourselves to the geometrical aspects of the problem.

49.1. By an r-th order Lagrangian on a fibered manifold $p\colon Y \to M$ we mean a base-preserving morphism

$$\lambda\colon J^rY \to \Lambda^m T^*M, \qquad m = \dim M.$$

For every section $s\colon M \to Y$, we obtain the induced m-form $\lambda \circ j^r s$ on M. We underline that from the geometrical point of view the Lagrangian is not a function on J^rY, since m-forms (and not functions) are the proper geometric objects for integration on X. If x^i, y^p are local fiber coordinates on Y, the induced coordinates on J^rY are x^i, y^p_α for all multi indices $|\alpha| \le r$. The coordinate expression of λ is

$$L(x^i, y^p_\alpha) dx^i \wedge \dots \wedge dx^m$$

but such a decomposition of λ into a function on J^rY and a volume element on M has no geometric meaning.

If η is a projectable vector field on Y over ξ on M, we can construct, similarly to 47.8.(2), the Lie derivative $\mathcal{L}_\eta\lambda$ of λ with respect to η

$$\mathcal{L}_\eta\lambda := \mathcal{L}_{(\mathcal{J}^r\eta, \Lambda^m T^*\xi)}\lambda\colon J^rY \to \Lambda^m T^*M$$

which coincides with the classical variation of λ with respect to η.

49.2. The geometrical form of the Euler equations for the extremals of λ is the so-called *Euler morphism* $E(\lambda)\colon J^{2r}Y \to V^*Y \otimes \Lambda^m T^*M$. Its geometric definition is based on a suitable decomposition of $\mathcal{L}_\eta\lambda$. Here it is useful to introduce an appropriate geometric operation.

Definition. Given a base-preserving morphism $\varphi\colon J^qY \to \Lambda^k T^*M$, its formal exterior differential $D\varphi\colon J^{q+1}Y \to \Lambda^{k+1}T^*M$ is defined by

$$D\varphi(j^{q+1}_x s) = d(\varphi \circ j^q s)(x)$$

for every local section s of Y, where d means the exterior differential at $x \in M$ of the local exterior k-form $\varphi \circ j^q s$ on M.

If $f\colon J^qY \to \mathbb{R}$ is a function, we have a coordinate decomposition

$$Df = (D_i f)dx^i$$

where $D_i f = \frac{\partial f}{\partial x^i} + \sum_{|\alpha|\le q} \frac{\partial f}{\partial y^p_\alpha} y^p_{\alpha+i}\colon J^{q+1}Y \to \mathbb{R}$ is the so called formal (or total) derivative of f, provided $\alpha+i$ means the multi index arising from α by increasing its i-th component by 1. If the coordinate expression of φ is $a_{i_1\dots i_k}dx^{i_1}\wedge\dots\wedge dx^{i_k}$, then

$$D\varphi = D_i a_{i_1\dots i_k} dx^i \wedge dx^{i_1} \wedge \dots \wedge dx^{i_k}.$$

To determine the Euler morphism, it suffices to discuss the variation $\mathcal{L}_\eta\lambda$ with respect to the vertical vector fields. If $\eta^p(x,y)\frac{\partial}{\partial y^p}$ is the coordinate expression of such a vector field, then the coordinate expression of $\mathcal{J}^r\eta$ is

$$\sum_{|\alpha|\le r} (D_\alpha \eta^p)\frac{\partial}{\partial y^p_\alpha}$$

where D_α means the iterated formal derivative with respect to the multi index α. In the following assertion we do not indicate explicitly the pullback of $\mathcal{L}_\eta\lambda$ to $J^{2r}Y$.

49.3. Proposition. *For every r-th order Lagrangian $\lambda\colon J^rY \to \Lambda^m T^*M$, there exists a morphism $K(\lambda)\colon J^{2r-1}Y \to V^*J^{r-1}Y \otimes \Lambda^{m-1}T^*M$ and a unique morphism $E(\lambda)\colon J^{2r}Y \to V^*Y \otimes \Lambda^m T^*M$ satisfying*

$$\mathcal{L}_\eta\lambda = D\left(\langle \mathcal{J}^{r-1}\eta, K(\lambda)\rangle\right) + \langle \eta, E(\lambda)\rangle \tag{1}$$

for every vertical vector field η on Y.

Proof. Write $\omega = dx^1 \wedge \dots \wedge dx^m$, $\omega_i = i_{\frac{\partial}{\partial x^i}}\omega$, $K(\lambda) = \sum_{|\alpha|\le r-1} k^{\alpha i}_p dy^p_\alpha \otimes \omega_i$, $E(\lambda) = E_p dy^p \otimes \omega$. Since $\mathcal{L}_\eta\lambda = T\lambda \circ \mathcal{J}^r\eta$, the coordinate expression of $\mathcal{L}_\eta\lambda$ is

$$\sum_{|\alpha|\le r} \frac{\partial L}{\partial y^p_\alpha} D_\alpha \eta^p. \tag{2}$$

Comparing the coefficients of the individual expressions $D_\alpha\eta^p$ in (1), we find the following relations

$$\begin{aligned}
L^{j_1\dots j_r}_p &= K^{(j_1\dots j_r)}_p \\
&\ \vdots \\
L^{j_1\dots j_q}_p &= D_i K^{j_1\dots j_q i}_p + K^{(j_1\dots j_q)}_p \\
&\ \vdots \\
L^j_p &= D_i K^{ji}_p + K^i_p \\
L_p &= D_i K^i_p + E_p
\end{aligned} \tag{3}$$

where $L_p^{j_1\dots j_q} = \frac{\alpha!}{q!}\frac{\partial L}{\partial y_\alpha^p}$ and $K_p^{j_1\dots j_q i} = \frac{\alpha!}{q!}k_p^{\alpha i}$, provided α is the multi index corresponding to $j_1 \dots j_q$, $|\alpha| = q$. Evaluating E_p by a backward procedure, we find

$$E_p = \sum_{|\alpha|\le r}(-1)^{|\alpha|}D_\alpha \frac{\partial L}{\partial y_\alpha^p} \tag{4}$$

for any K's, so that the Euler morphism is uniquely determined. The quantities $K_p^{j_1\dots j_q i}$, which are not symmetric in the last two superscripts, are not uniquely determined by virtue of the symmetrizations in (3). Nevertheless, the global existence of a $K(\lambda)$ can be deduced by a recurrent construction of some sections of certain affine bundles. This procedure is straightforward, but rather technical. The reader is referred to [Kolář, 84b] □

We remark that one can prove easily by proposition 49.3 that a section s of Y is an extremal of λ if and only if $E(\lambda) \circ j^{2r}s = 0$.

49.4. The construction of the Euler morphism can be viewed as an operator transforming every base-preserving morphism $\lambda\colon J^rY \to \Lambda^m T^*M$ into a base-preserving morphism $E(\lambda)\colon J^{2r}Y \to V^*Y \otimes \Lambda^m T^*M$. Analogously to $\mathcal{L}_\eta\lambda$, the Lie derivative of $E(\lambda)$ with respect to a projectable vector field η on Y over ξ on M is defined by

$$\mathcal{L}_\eta E(\lambda) := \mathcal{L}_{(\mathcal{J}^{2r}\eta, V^*\eta\otimes\Lambda^m T^*\xi)}E(\lambda).$$

An important question is whether the Euler operator commutes with Lie differentiation. From the uniqueness assertion in proposition 49.3 it follows that E is a natural operator and from 49.3.(4) we see that E is a linear operator.

49.5. We first deduce a general result of such a type. Consider two natural bundles over m-manifolds F and H, a natural surjective submersion $q\colon H \to F$ and two natural vector bundles over m-manifolds G and K.

Proposition. *Every linear natural operator $A\colon (F,G) \rightsquigarrow (H,K)$ satisfies*

$$\mathcal{L}_\xi(Af) = A(\mathcal{L}_\xi f)$$

for every base-preserving morphism $f\colon FM \to GM$ and every vector field ξ on M.

Proof. By 47.8.(2) and an analogy of 47.4.(1), we have

$$\mathcal{L}_\xi f = \lim_{t\to 0}\tfrac{1}{t}\left(G(\mathrm{Fl}_{-t}^\xi)\circ f \circ F(\mathrm{Fl}_t^\xi) - f\right).$$

Since A commutes with limits by 19.9, we obtain by linearity and naturality

$$A\mathcal{L}_\xi(f) = \lim_{t\to 0}\tfrac{1}{t}\left(K(\mathrm{Fl}_{-t}^\xi)\circ Af \circ H(\mathrm{Fl}_t^\xi) - Af\right) = \mathcal{L}_\xi(Af) \quad \square$$

49.6. Our original problem on the Euler morphism can be discussed in the same way as in the proof of proposition 49.5, but the functors in question are defined on the local isomorphisms of fibered manifolds. Hence the answer to our problem is affirmative.

Proposition. *It holds*

$$\mathcal{L}_\eta E(\lambda) = E(\mathcal{L}_\eta \lambda)$$

for every r-th order Lagrangian λ and every projectable vector field η on Y.

49.7. A projectable vector field η on Y is said to be a ***generalized infinitesimal automorphism of an r-th order Lagrangian*** λ, if $\mathcal{L}_\eta E(\lambda) = 0$. By proposition 49.6 we obtain immediately the following interesting assertion.

Corollary. Higher order Noether-Bessel-Hagen theorem. *A projectable vector field η is a generalized infinitesimal automorphism of an r-th order Lagrangian λ if and only if $E(\mathcal{L}_\eta \lambda) = 0$.*

49.8. An infinitesimal automorphism of λ means a projectable vector field η satisfying $\mathcal{L}_\eta \lambda = 0$. In particular, corollary 49.7 and 49.3.(4) imply that every infinitesimal automorphism is a generalized infinitesimal automorphism.

50. The general bracket formula

50.1. The generalized Lie derivative of a section s of an arbitrary fibered manifold $Y \to M$ with respect to a projectable vector field η on Y over ξ on M is a section $\tilde{\mathcal{L}}_\eta s \colon M \to VY$. If $\bar{\eta}$ is another projectable vector field on Y over $\bar{\xi}$ on M, a general problem is whether there exists a reasonable formula for the generalized Lie derivative $\tilde{\mathcal{L}}_{[\eta,\bar{\eta}]} s$ of s with respect to the bracket $[\eta, \bar{\eta}]$. Since $\mathcal{L}_\eta s$ is not a section of Y, we cannot construct the generalized Lie derivative of $\tilde{\mathcal{L}}_\eta s$ with respect to $\bar{\eta}$. However, in the case of a vector bundle $E \to M$ we have defined $\mathcal{L}_{\bar{\eta}} \mathcal{L}_\eta s \colon M \to E$.

Proposition. *If η and $\bar{\eta}$ are two linear vector fields on a vector bundle $E \to M$, then*

$$\mathcal{L}_{[\eta,\bar{\eta}]} s = \mathcal{L}_\eta \mathcal{L}_{\bar{\eta}} s - \mathcal{L}_{\bar{\eta}} \mathcal{L}_\eta s \tag{1}$$

for every section s of E.

At this moment, the reader can prove this by direct evaluation using 47.10.(2). But we shall give a conceptual proof resulting from more general considerations in 50.5. By direct evaluation, the reader can also verify that the above proposition *does not hold* for arbitrary projectable vector fields η and $\bar{\eta}$ on E. However, if FM is a natural vector bundle, then $\mathcal{F}\xi$ is a linear vector field on FM for every vector field ξ on M, so that we have

Corollary. *If FM is a natural vector bundle, then*

$$\mathcal{L}_{[\xi,\bar{\xi}]}s = \mathcal{L}_{\xi}\mathcal{L}_{\bar{\xi}}s - \mathcal{L}_{\bar{\xi}}\mathcal{L}_{\xi}s$$

for every section s of FM and every vector fields ξ, $\bar{\xi}$ on M.

This result covers the classical cases of Lie differentiation.

50.2. We are going to discuss the most general situation. Let M, N be two manifolds, $f\colon M \to N$ be a map, ξ, $\bar{\xi}$ be two vector fields on M and η, $\bar{\eta}$ be two vector fields on N. Our problem is to find a reasonable expression for

$$\tilde{\mathcal{L}}_{([\xi,\bar{\xi}],[\eta,\bar{\eta}])}f\colon M \to TN. \tag{1}$$

Since $\tilde{\mathcal{L}}_{(\xi,\eta)}f$ is a map of M into TN, we cannot construct its Lie derivative with respect to the pair $(\bar{\xi},\bar{\eta})$, since $\bar{\eta}$ is a vector field on N and not on TN. However, if we replace $\bar{\eta}$ by its flow prolongation $T\bar{\eta}$, we have defined

$$\tilde{\mathcal{L}}_{(\bar{\xi},T\bar{\eta})}\tilde{\mathcal{L}}_{(\xi,\eta)}f\colon M \to TTN. \tag{2}$$

On the other hand, we can construct

$$\tilde{\mathcal{L}}_{(\xi,T\eta)}\tilde{\mathcal{L}}_{(\bar{\xi},\bar{\eta})}f\colon M \to TTN. \tag{3}$$

Now we need an operation transforming certain special pairs of the elements of the second tangent bundle TTQ of any manifold Q into the elements of TQ. Consider $A,B \in TT_zQ$ satisfying

$$\pi_{TQ}(A) = T\pi_Q(B) \quad \text{and} \quad T\pi_Q(A) = \pi_{TQ}(B). \tag{4}$$

Since the canonical involution $\kappa\colon TTQ \to TTQ$ exchanges both projections, we have $\pi_{TQ}(A) = \pi_{TQ}(\kappa B)$, $T\pi_Q(A) = T\pi_Q(\kappa B)$. Hence A and κB are in the same fiber of TTQ with respect to projection π_{TQ} and their difference $A - \kappa B$ satisfies $T\pi_Q(A - \kappa B) = 0$. This implies that $A - \kappa B$ is a tangent vector to the fiber T_zQ of TQ and such a vector can be identified with an element of T_zQ, which will be denoted by $A \div B$.

50.3. Definition. $A \div B \in TQ$ is called the strong difference of A, $B \in TTQ$ satisfying 50.2.(4).

In the case $Q = \mathbb{R}^m$ we have $TT\mathbb{R}^m = \mathbb{R}^m \times \mathbb{R}^m \times \mathbb{R}^m \times \mathbb{R}^m$. If $A = (x,a,b,c) \in TT\mathbb{R}^m$, then B satisfying 50.2.(4) is of the form $B = (x,b,a,d)$ and one finds easily

$$A \div B = (x, c-d) \tag{1}$$

From the geometrical definition of the strong difference it follows directly

Lemma. *If A, $B \in TTQ$ satisfy 50.2.(4) and $f\colon Q \to P$ is any map, then $TTf(A)$, $TTf(B) \in TTP$ satisfy the condition of the same type and it holds*

$$TTf(A) \dotplus TTf(B) = Tf(A \dotplus B) \in TP.$$

50.4. We are going to deduce the bracket formula for generalized Lie derivatives. First we recall that lemma 6.13 reads

$$[\zeta, \bar\zeta] = T\bar\zeta \circ \zeta \dotplus T\zeta \circ \bar\zeta \tag{1}$$

for every two vector fields on the same manifold.

The maps 50.2.(2) and 50.2.(3) satisfy the condition for the existence of the strong difference. Indeed, we have $\pi_{TN} \circ \tilde{\mathcal{L}}_{(\bar\xi,\mathcal{T}\bar\eta)}\tilde{\mathcal{L}}_{(\xi,\eta)}f = \tilde{\mathcal{L}}_{(\xi,\eta)}f$ since any generalized Lie derivative of $\tilde{\mathcal{L}}_{(\xi,\eta)}f$ is a vector field along $\tilde{\mathcal{L}}_{(\xi,\eta)}f$. On the other hand, $T\pi_N \circ (\tilde{\mathcal{L}}_{(\bar\xi,\mathcal{T}\bar\eta)}\tilde{\mathcal{L}}_{(\xi,\eta)}f) = T\pi_N\big(\frac{\partial}{\partial t}\big|_0 T(\mathrm{Fl}^{\bar\eta}_{-t}) \circ \tilde{\mathcal{L}}_{(\xi,\eta)}f \circ \mathrm{Fl}^{\bar\xi}_t\big) = \frac{\partial}{\partial t}\big|_0 (\mathrm{Fl}^{\bar\eta}_{-t} \circ f \circ \mathrm{Fl}^{\bar\xi}_t) = \tilde{\mathcal{L}}(\bar\xi, \bar\eta)f$.

Proposition. *It holds*

$$\tilde{\mathcal{L}}_{([\xi,\bar\xi],[\eta,\bar\eta])}f = \tilde{\mathcal{L}}_{(\xi,\mathcal{T}\eta)}\tilde{\mathcal{L}}_{(\bar\xi,\bar\eta)}f \dotplus \tilde{\mathcal{L}}_{(\bar\xi,\mathcal{T}\bar\eta)}\tilde{\mathcal{L}}_{(\xi,\eta)}f \tag{2}$$

Proof. We first recall that the flow prolongation of η satisfies $\mathcal{T}\eta = \kappa \circ T\eta$. By 47.1.(1) we obtain $\tilde{\mathcal{L}}_{(\bar\xi,\mathcal{T}\bar\eta)}\tilde{\mathcal{L}}_{(\xi,\eta)}f = T(Tf \circ \xi - \eta \circ f) \circ \bar\xi - \mathcal{T}\bar\eta \circ (Tf \circ \xi - \eta \circ f) = TTf \circ T\xi \circ \bar\xi - T\eta \circ Tf \circ \bar\xi - \kappa \circ T\bar\eta \circ Tf \circ \xi + \kappa \circ T\bar\eta \circ \eta \circ f$ as well as a similar expression for $\tilde{\mathcal{L}}_{(\xi,\mathcal{T}\eta)}\tilde{\mathcal{L}}_{(\bar\xi,\bar\eta)}f$. Using (1) we deduce that the right hand side of (2) is equal to $Tf \circ (T\xi \circ \bar\xi \dotplus T\bar\xi \circ \xi) - (T\eta \circ \bar\eta \dotplus T\bar\eta \circ \eta)f = Tf \circ [\xi, \bar\xi] - [\eta, \bar\eta] \circ f$. □

50.5. In the special case of a section $s\colon M \to Y$ of a fibered manifold $Y \to M$ and of two projectable vector fields η and $\bar\eta$ on Y, 50.4.(2) is specialized to

$$\tilde{\mathcal{L}}_{[\eta,\bar\eta]}s = \tilde{\mathcal{L}}_{\mathcal{V}\eta}\tilde{\mathcal{L}}_{\bar\eta}s \dotplus \tilde{\mathcal{L}}_{\mathcal{V}\bar\eta}\tilde{\mathcal{L}}_{\eta}s \tag{1}$$

where $\mathcal{V}\eta$ or $\mathcal{V}\bar\eta$ is the restriction of $\mathcal{T}\eta$ or $\mathcal{T}\bar\eta$ to the vertical tangent bundle $VY \subset TY$. Furthermore, if we have a vector bundle $E \to M$ and a linear vector field on E, then $\mathcal{V}\eta$ is of the form $\mathcal{V}\eta = \eta \oplus \eta$, since the tangent map of a linear map coincides with the original map itself. Thus, if we separate the restricted Lie derivatives in (1) in the case η and $\bar\eta$ are linear, we find $\mathcal{L}_{[\eta,\bar\eta]}s = \mathcal{L}_\eta\mathcal{L}_{\bar\eta}s - \mathcal{L}_{\bar\eta}\mathcal{L}_\eta s$. This proves proposition 50.1.

Remarks

The general concept of Lie derivative of a map $f\colon M \to N$ with respect to a pair of vector fields on M and N was introduced by [Trautman, 72]. The operations with linear vector fields from the second half of section 47 were described in [Janyška, Kolář, 82]. In the theory of multilinear natural operators, the commutativity with the Lie differentiation is also used as the starting point, see [Kirillov, 77, 80]. Proposition 48.4 was proved by [Cap, Slovák, 92]. According to [Janyška, Modugno, to appear], there is a link between the infinitesimally natural operators and certain systems in the sense of [Modugno, 87a]. The concept of a sector r-form was introduced in [White, 82].

The Lie derivatives of morphisms of fibered manifolds were studied in [Kolář, 82a] in connection with the higher order variational calculus in fibered manifolds. We remark that a further analysis of formula 49.3.(3) leads to an interesting fact that a Lagrangian of order at least three with at least two independent variables does not determine a unique Poincaré-Cartan form, but a family of such forms only, see e.g. [Kolář, 84b], [Saunders, 89]. The general bracket formula from section 50 was deduced in [Kolář, 82c].

CHAPTER XII. GAUGE NATURAL BUNDLES AND OPERATORS

In chapters IV and V we have explained that the natural bundles coincide with the associated fiber bundles to higher order frame bundles on manifolds. However, in both differential geometry and mathematical physics one can meet fiber bundles associated to an 'abstract' principal bundle with an arbitrary structure group G. If we modify the idea of bundle functor to such a situation, we obtain the concept of gauge natural bundle. This is a functor on principal fiber bundles with structure group G and their local isomorphisms with values in fiber bundles, but with fibration over the original base manifold. The most important examples of gauge natural bundles and of natural operators between them are related with principal connections. In this chapter we first develop a description of all gauge natural bundles analogous to that in chapter V. In particular, we prove that the regularity condition is a consequence of functoriality and locality and that any gauge natural bundle is of finite order. We also present sharp estimates of the order depending on the dimensions of the standard fibers. So the r-th order gauge natural bundles coincide with the fiber bundles associated to r-th principal prolongations of principal G-bundles (see 15.3), which are in bijection with the actions of the group $W^r_m G$ on manifolds.

Then we discuss a few concrete problems on finding gauge natural operators. The geometrical results of section 52 are based on a generalization of the Utiyama theorem on gauge natural Lagrangians. First we determine all gauge natural operators of the curvature type. In contradistinction to the essential uniqueness of the curvature operator on general connections, this result depends on the structure group in a simple way. Then we study the differential forms of Chern-Weil type with values in an arbitrary associated vector bundle. We find it interesting that the full list of all gauge natural operators leads to a new geometric result in this case. Next we determine all first order gauge natural operators transforming principal connections to the tangent bundle. In the last section we find all gauge natural operators transforming a linear connection on a vector bundle and a classical linear connection on the base manifold into a classical linear connection on the total space.

51. Gauge natural bundles

We are going to generalize the description of all natural bundles $F: \mathcal{M}f_m \to \mathcal{FM}$ derived in sections 14 and 22 to the gauge natural case. Since the concepts and considerations are very similar to some previous ones, we shall proceed in a rather brief style.

51.1. Let $B\colon \mathcal{FM} \to \mathcal{M}f$ be the base functor. Fix a Lie group G and recall the category $\mathcal{PB}_m(G)$, whose objects are principal G-bundles over m-manifolds and whose morphisms are the morphisms of principal G-bundles $f\colon P \to \bar P$ with the base map $Bf\colon BP \to B\bar P$ lying in $\mathcal{M}f_m$.

Definition. A *gauge natural bundle* over m-dimensional manifolds is a functor $F\colon \mathcal{PB}_m(G) \to \mathcal{FM}$ such that

(a) every $\mathcal{PB}_m(G)$-object $\pi\colon P \to BP$ is transformed into a fibered manifold $q_P\colon FP \to BP$ over BP,

(b) every $\mathcal{PB}_m(G)$-morphism $f\colon P \to \bar P$ is transformed into a fibered morphism $Ff\colon FP \to F\bar P$ over Bf,

(c) for every open subset $U \subset BP$, the inclusion $i\colon \pi^{-1}(U) \to P$ is transformed into the inclusion $Fi\colon q_P^{-1}(U) \to FP$.

If we intend to point out the structure group G, we say that F is a G-natural bundle.

51.2. If two $\mathcal{PB}_m(G)$-morphisms f, $g\colon P \to \bar P$ satisfy $j^r_y f = j^r_y g$ at a point $y \in P_x$ of the fiber of P over $x \in BP$, then the fact that the right translations of principal bundles are diffeomorphisms implies $j^r_z f = j^r_z g$ for every $z \in P_x$. In this case we write $\mathbf{j}^r_x f = \mathbf{j}^r_x g$.

Definition. A gauge natural bundle F is said to be of order r, if $\mathbf{j}^r_x f = \mathbf{j}^r_x g$ implies $Ff|F_xP = Fg|F_xP$.

51.3. Definition. A G-natural bundle F is said to be regular if every smoothly parameterized family of $\mathcal{PB}_m(G)$-morphisms is transformed into a smoothly parameterized family of fibered maps.

51.4. Remark. By definition, a G-natural bundle $F\colon \mathcal{PB}_m(G) \to \mathcal{FM}$ satisfies $B \circ F = B$ and the projections $q_P\colon FP \to BP$ form a natural transformation $q\colon F \to B$.

In general, we can consider a category $\mathcal{C}$ over fibered manifolds, i.e. $\mathcal{C}$ is endowed with a faithful functor $m\colon \mathcal{C} \to \mathcal{FM}$. If $\mathcal{C}$ admits localization of objects and morphisms with respect to the preimages of open subsets on the bases with analogous properties to 18.2, we can define the gauge natural bundles on $\mathcal{C}$ as functors $F\colon \mathcal{C} \to \mathcal{FM}$ satisfying $B \circ F = B \circ m$ and the locality condition 51.1.(c). Let us mention the categories of vector bundles as examples. The different way of localization is the source of a crucial difference between the bundle functors on categories over manifolds and the (general) gauge natural bundles. For any two fibered maps f, $g\colon Y \to \bar Y$ we write $\mathbf{j}^r_x f = \mathbf{j}^r_x g$, $x \in BY$, if $j^r_y f = j^r_y g$ for all $y \in Y_x$. Then we say that f and g have the same *fiber r-jet at x*. The space of fiber r-jets between $\mathcal{C}$-objects Y and $\bar Y$ is denoted by $\mathbf{J}^r(Y, \bar Y)$. For a general category $\mathcal{C}$ over fibered manifolds the finiteness of the order of gauge natural bundles is expressed with the help of the fiber jets. The description of finite order bundle functors as explained in section 18 could be generalized now, but there appear difficulties connected with the (generally) infinite dimension of the corresponding jet groups. Since we will need only the gauge natural bundles on $\mathcal{PB}_m(G)$ in the sequel, we will restrict ourselves to this category. Then the description will be quite analogous to that of classical natural bundles. Some basic steps towards the description in the general case were done in [Slovák, 86] where the infinite dimensional constructions are performed with the help of the smooth spaces in the sense of [Frölicher, 81].

51.5. Examples.
(1) The choice $G = \{e\}$ reproduces the natural bundles on $\mathcal{M}f_m$
(2) The functors $Q^r\colon \mathcal{PB}_m(G) \to \mathcal{FM}$ of r-th order principal connections mentioned in 17.4 are examples of r-th order regular gauge natural bundles.
(3) The gauge natural bundles $W^r\colon \mathcal{PB}_m(G) \to \mathcal{PB}_m(W^r_m G)$ of r-th principal prolongation defined in 15.3 play the same role as the frame bundles $P^r\colon \mathcal{M}f_m \to \mathcal{FM}$ did in the description of natural bundles.
(4) For every manifold S with a smooth left action ℓ of $W^r_m G$, the construction of associated bundles to the principal bundles $W^r P$ yields a regular gauge natural bundle $L\colon \mathcal{PB}_m(G) \to \mathcal{FM}$. We shall see that all gauge natural bundles are of this type.

51.6. Proposition. *Every r-th order regular gauge natural bundle is a fiber bundle associated to W^r.*

Proof. Analogously to the case of natural bundles, an r-th order regular gauge natural bundle F is determined by the system of smooth associated maps

$$F_{P,\bar{P}}\colon \mathbf{J}^r(P,\bar{P}) \times_{BP} FP \to F\bar{P}$$

and the restriction of $F_{\mathbb{R}^m \times G, \mathbb{R}^m \times G}$ to the fiber jets at $0 \in \mathbb{R}^m$ yields an action of $W^r_m G = \mathbf{J}^r_0(\mathbb{R}^m \times G, \mathbb{R}^m \times G)_0$ on the fiber $S = F_0(\mathbb{R}^m \times G)$. The same considerations as in 14.6 complete now the proof. □

51.7. Theorem. *Let $F\colon \mathcal{PB}_m(G) \to \mathcal{M}f$ be a functor endowed with a natural transformation $q\colon F \to B$ such that the locality condition 51.1.(c) holds. Then $S := (q_{\mathbb{R}^m \times G})^{-1}(0)$ is a manifold of dimension $s \geq 0$ and for every $P \in Ob\mathcal{PB}_m(G)$, the mapping $q_P\colon FP \to BP$ is a locally trivial fiber bundle with standard fiber S, i.e. $F\colon \mathcal{PB}_m(G) \to \mathcal{FM}$. The functor F is a regular gauge natural bundle of a finite order $r \leq 2s+1$. If moreover $m > 1$, then*

$$r \leq \max\{\frac{s}{m-1}, \frac{s}{m} + 1\}. \tag{1}$$

All these estimates are sharp.

Briefly, every gauge natural bundle on $\mathcal{PB}_m(G)$ with s-dimensional fibers is one of the functors defined in example 51.5.(4) with r bounded by the estimates from the theorem depending on m and s but not on G. The proof is based on the considerations from chapter V and it will require several steps.

51.8. Let us point out that the restriction of any gauge natural bundle F to trivial principal bundles $M \times G$ and to morphisms of the form $f \times \mathrm{id}\colon M \times G \to N \times G$ can be viewed as a natural bundle $\mathcal{M}f_m \to \mathcal{FM}$. Hence the action τ of the abelian group of fiber translations $t_x\colon \mathbb{R}^m\times G\to\mathbb{R}^m\times G$, $(y,a)\mapsto(x+y,a)$, i.e. $\tau_x = Ft_x$, is a smooth action by 20.3. This implies immediately the assertion on fiber bundle structure in 51.7, cf. 20.3. Further, analogously to 20.5.(1) we find that the regularity of F follows if we verify the smoothness of the induced action of the morphisms keeping the fiber over $0\in\mathbb{R}^m$ on the standard fiber $S = F_0(\mathbb{R}^m\times G)$.

51.9. Lemma. *Let $U\subset S$ be a relatively compact open set and write*

$$Q_U = \bigcup_{\varphi} F\varphi(U)\subset S$$

where the union goes through all $\varphi\in\mathcal{PB}_m(G)(\mathbb{R}^m\times G,\mathbb{R}^m\times G)$ with $\varphi_0(0) = (0)$. Then there is $r\in\mathbb{N}$ such that for all $z\in Q_U$ and all $\mathcal{PB}_m(G)$-morphisms $\varphi,\psi\colon\mathbb{R}^m\times G\to\mathbb{R}^m\times G$, $\varphi_0(0)=\psi_0(0)=0$, the condition $\mathrm{j}^r_0\varphi=\mathrm{j}^r_0\psi$ implies $F\varphi(z)=F\psi(z)$.

Proof. Every morphism $\varphi\colon\mathbb{R}^m\times G\to\mathbb{R}^m\times G$ is identified with the couple $\varphi_0\in C^\infty(\mathbb{R}^m,\mathbb{R}^m)$, $\bar\varphi\in C^\infty(\mathbb{R}^m,G)$. So F induces an operator $\tilde F\colon C^\infty(\mathbb{R}^m,\mathbb{R}^m\times G)\to C^\infty(F(\mathbb{R}^m\times G),F(\mathbb{R}^m\times G))$ which is $q_{\mathbb{R}^m\times G}$-local and the map $q_{\mathbb{R}^m\times G}$ is locally non-constant. Consider the constant map $\hat e\colon\mathbb{R}^m\to G$, $x\mapsto e$, and the map $\mathrm{id}_{\mathbb{R}^m}\times\hat e\colon\mathbb{R}^m\to\mathbb{R}^m\times G$ corresponding to $\mathrm{id}_{\mathbb{R}^m\times G}$. By corollary 19.8, there is $r\in\mathbb{N}$ such that $j^r_0f=j^r_0(\mathrm{id}_{\mathbb{R}^m}\times\hat e)$ implies $\tilde Ff(z)=z$ for all $z\in U$. Hence if $\mathrm{j}^r_0\varphi=\mathrm{j}^r_0\mathrm{id}_{\mathbb{R}^m\times G}$, then $F\varphi(z)=z$ for all $z\in U$ and the easy rest of the proof is quite analogous to 20.4. □

51.10. Proposition. *Every gauge natural bundle is regular.*

Proof. The whole proof of 20.5 goes through for gauge natural bundles if we choose local coordinates near to the unit in G and replace the elements $j^\infty_0 f_n\in G^\infty_m$ by the couples $(j^\infty_0 f_n, j^\infty_0\bar\varphi_n)\in G^\infty_m\rtimes T^\infty_m G$ and $\mathrm{id}_{\mathbb{R}^m}$ by $\mathrm{id}_{\mathbb{R}^m}\times\hat e$. Let us remark that also τ_x gets the new meaning of $F(t_x)$. □

51.11. Since every natural bundle $F\colon\mathcal{M}f\to\mathcal{FM}$ can be viewed as the gauge natural bundle $\bar F = F\circ B\colon\mathcal{PB}_m(G)\to\mathcal{FM}$, the estimates from theorem 51.7 must be sharp if they are correct, see 22.1. Further, the considerations from 22.1 applied to our situation show that we complete the proof of 51.7 if we deduce that every smooth action of W^r_mG on a smooth manifold S factorizes to an action of W^k_mG, $k\le r$, with k satisfying the estimates from 51.7.

So let us consider a continuous action $\rho\colon W^r_mG\to\mathrm{Diff}(S)$ and write H for its kernel. Hence H is a closed normal Lie subgroup and the kernel $H_0\subset G^r_m$ of the restriction $\rho_0=\rho|G^r_m$ always contains the normal Lie subgroup $B^r_k\subset G^r_m$ with $k=2\mathrm{dim}S+1$ if $m=1$ and $k=\max\{\frac{\mathrm{dim}S}{m-1},\frac{\mathrm{dim}S}{m}+1\}$ if $m>1$. Let us denote K^r_k the kernel of the jet projection $W^r_mG\to W^k_mG$.

Lemma. *For every Lie group G and all r, $k\in\mathbb{N}$, $r>k\ge1$, the normal closed Lie subgroup in W^r_mG generated by $B^r_k\rtimes\{e\}$ equals to K^r_k.*

Proof. The Lie group W^r_mG can be viewed as the space of fiber jets $\mathrm{J}^r_0(\mathbb{R}^m\times G,\mathbb{R}^m\times G)_0$ and so its Lie algebra $\mathfrak{w}^r_m\mathfrak{g}$ coincides with the space of fiber jets at

$0 \in \mathbb{R}^m$ of (projectable) right invariant vector fields with projections vanishing at the origin. If we repeat the consideration from the proof of 13.2 with jets replaced by fiber jets, we get the formula for Lie bracket in $\mathfrak{w}_m^r\mathfrak{g}$, $[\mathfrak{j}_0^r X, \mathfrak{j}_0^r Y] = -\mathfrak{j}_0^r[X,Y]$. Since every polynomial vector field in $\mathfrak{w}_m^r\mathfrak{g}$ decomposes into a sum of $X_1 \in \mathfrak{g}_m^r$ and a vertical vector field X_2 from the Lie algebra $T_m^r\mathfrak{g}$ of $T_m^r G$, we get immediately the action of $\mathfrak{g}_m^r$ on $T_m^r\mathfrak{g}$, $[j_0^r X_1 + 0, 0 + \mathfrak{j}_0^r X_2] = -\mathfrak{j}_0^r \mathcal{L}_{X_1} X_2$.

Now let us fix a base e_i of $\mathfrak{g}$ and elements $Y_i \in T_m^r\mathfrak{g}$, $Y_i = \mathfrak{j}_0^r x^1 e_i$. Taking any functions f_i on $\mathbb{R}^m$ with $j_0^k f_i = 0$, the r-jets of the fields $X_i = f_i \partial/\partial x^1$ lie in the kernel $\mathfrak{b}_k^r \subset \mathfrak{g}_m^r$ and we get

$$\sum_i [j_0^r X_i, \mathfrak{j}_0^r Y_i] = -\mathfrak{j}_0^r f_i e_i \in T_m^r \mathfrak{g}.$$

Hence $[\mathfrak{b}_k^r, T_m^r\mathfrak{g}]$ contains the whole Lie algebra of the kernel K_k^r and so the latter algebra must coincide with the ideal in $\mathfrak{w}_m^r\mathfrak{g}$ generated by $\mathfrak{b}_k^r \rtimes \{0\}$. Since the kernel K_1^r is connected this completes the proof. □

51.12. Corollary. *Let G be a Lie group and S be a manifold with a continuous left action of $W_m^r G$, $\dim S = s \geq 0$. Then the action factorizes to an action of $W_m^k G$ with $k \leq 2s + 1$. If $m > 1$, then $k \leq \max\{\frac{s}{m-1}, \frac{s}{m} + 1\}$. These estimates are sharp.*

The corollary concludes the proof of theorem 51.7.

51.13. Given two G-natural bundles F, E: $\mathcal{PB}_m(G) \to \mathcal{FM}$, every natural transformation T: $F \to E$ is formed be a system of base preserving $\mathcal{FM}$-morphisms, cf. 14.11 and 51.8. In the same way as in 14.12 one deduces

Proposition. *Natural transformations $F \to E$ between two r-th order G-natural bundles over m-dimensional manifolds are in a canonical bijection with the $W_m^r G$-equivariant maps $F_0 \to E_0$ between the standard fibers $F_0 = F_0(\mathbb{R}^m \times G)$, $E_0 = E_0(\mathbb{R}^m \times G)$.*

51.14. Definition. Let F and E be two G-natural bundles over m-dimensional manifolds. A *gauge natural operator* D: $F \rightsquigarrow E$ is a system of regular operators D_P: $C^\infty FP \to C^\infty EP$ for all $\mathcal{PB}_m(G)$-objects π: $P \to BP$ such that

(a) $D_{\bar{P}}(Ff \circ s \circ Bf^{-1}) = Ff \circ D_P s \circ Bf^{-1}$ for every $s \in C^\infty FP$ and every $\mathcal{PB}_m(G)$-isomorphism f: $P \to \bar{P}$,

(b) $D_{\pi^{-1}(U)}(s|U) = (D_P s)|U$ for every $s \in C^\infty FP$ and every open subset $U \subset BP$.

51.15. For every $k \in \mathbb{N}$ and every gauge natural bundle F of order r its composition $J^k \circ F$ with the k-th jet prolongation defines a gauge natural bundle functor of order $k + r$, cf. 14.16. In the same way as in 14.17 one deduces

Proposition. *The k-th order gauge natural operators $F \rightsquigarrow E$ are in a canonical bijection with the natural transformations $J^k F \to E$.*

In particular, this proposition implies that the k-th order G-natural operators $F \rightsquigarrow E$ are in a canonical bijection with the $W_m^s G$-equivariant maps $J_0^k F \to E_0$, where s is the maximum of the orders of $J^k F$ and E and $J_0^k F = J_0^k F(\mathbb{R}^m \times G)$.

51.16. Consider the G-natural connection bundle Q and an arbitrary G-natural bundle E.

Proposition. *Every gauge natural operator $A\colon Q \rightsquigarrow E$ has finite order.*

Proof. By 51.8, every G-natural bundle F determines a classical natural bundle NF by $NF(M) = F(M \times G)$, $NF(f) = F(f \times \mathrm{id}_G)$. Given a G-natural operator $D\colon F \rightsquigarrow E$, we denote by ND its restriction to NF, i.e. $ND_M = D_{M\times G}$. Clearly, ND is a classical natural operator $NF \to NE$.

Since our operator A is determined locally, we may restrict ourselves to the product bundle $M \times G$. Then we have a classical natural operator NA. In this situation the standard fiber $\mathfrak{g} \otimes \mathbb{R}^{m*}$ of Q coincides with the direct product of $\dim G$ copies of $\mathbb{R}^{m*}$. Hence we can apply proposition 23.5. □

52. The Utiyama theorem

52.1. The connection bundle. First we write the equations of a connection Γ on $\mathbb{R}^m \times G$ in a suitable form. Let e_p be a basis of $\mathfrak{g}$ and let ω^p be the corresponding (left) Maurer-Cartan forms given by $\sum_p \omega^p(X_g)e_p = T(\lambda_{g^{-1}})(X)$. Let

$$(\omega^p)_e = \Gamma^p_i(x)dx^i \tag{1}$$

be the equations of $\Gamma(x,e)$, $x \in \mathbb{R}^m$, $e =$ the unit of G. Since Γ is right-invariant, its equations on the whole space $\mathbb{R}^m \times G$ are

$$\omega^p = \Gamma^p_i(x)dx^i. \tag{2}$$

The connection bundle $QP = J^1P/G$ is a first order gauge natural bundle with standard fiber $\mathfrak{g} \otimes \mathbb{R}^{m*}$. Having a $\mathcal{PB}_m(G)$-isomorphism Φ of $\mathbb{R}^m \times G$ into itself

$$\bar{x} = f(x), \quad \bar{y} = \varphi(x)\cdot y, \qquad f(0) = 0 \tag{3}$$

with $\varphi\colon \mathbb{R}^m \to G$, its 1-jet $\mathbf{j}^1_0\Phi \in W^1_mG$ is characterized by

$$a = \varphi(0) \in G, \quad (a^p_i) = j^1_0(a^{-1}\cdot\varphi(x)) \in \mathfrak{g}\otimes\mathbb{R}^{m*}, \quad (a^i_j) = j^1_0 f \in G^1_m. \tag{4}$$

Let $A^p_q(a)$ be the coordinate expression of the adjoint representation of G. In 15.6 we deduced the following equations of the action of W^1_mG on $\mathfrak{g}\otimes\mathbb{R}^{m*}$

$$\bar{\Gamma}^p_i = A^p_q(a)(\Gamma^q_j + a^q_j)\tilde{a}^j_i. \tag{5}$$

The first jet prolongation J^1QP of the connection bundle is a second order gauge natural bundle, so that its standard fiber $S_1 = J^1_0Q(\mathbb{R}^m \times G)$, with the coordinates Γ^p_i, $\Gamma^p_{ij} = \partial\Gamma^p_i/\partial x^j$, is a W^2_mG-space. The second order partial derivatives a^p_{ij} of the map $a^{-1}\cdot\varphi(x)$ together with $a^i_{jk} = \partial^2_{jk}f^i(0)$ are the additional coordinates on W^2_mG. Using 15.5, we deduce from (5) that the action of W^2_mG on S_1 has the form (5) and

$$\begin{aligned}\bar{\Gamma}^p_{ij} = {} & A^p_q(a)\Gamma^q_{kl}\tilde{a}^k_i\tilde{a}^l_j + A^p_q(a)a^q_{kl}\tilde{a}^k_i\tilde{a}^l_j + \\ & + D^p_{qr}(a)\Gamma^q_k a^r_l\tilde{a}^k_i\tilde{a}^l_j + E^p_{qr}(a)a^q_k a^r_l\tilde{a}^k_i\tilde{a}^l_j + A^p_q(a)(\Gamma^q_k + a^q_k)\tilde{a}^k_{ij}\end{aligned} \tag{6}$$

where the D's and E's are some functions on G, which we shall not need.

52.2. The curvature. To deduce the coordinate expression of the curvature tensor, we shall use the structure equations of Γ. By 52.1.(1), the components φ^p of the connection form of Γ are

$$\varphi^p = \omega^p - \Gamma^p_i(x)dx^i. \tag{1}$$

The structure equations of Γ reads

$$d\varphi^p = c^p_{qr}\varphi^q \wedge \varphi^r + R^p_{ij}dx^i \wedge dx^j \tag{2}$$

where c^p_{qr} are the structure constants of G and R^p_{ij} is the curvature tensor. Since ω^p are the Maurer-Cartan forms of G, we have $d\omega^p = c^p_{qr}\omega^q \wedge \omega^r$. Hence the exterior differentiation of (1) yields

$$d\varphi^p = c^p_{qr}(\varphi^q + \Gamma^q_i dx^i) \wedge (\varphi^r + \Gamma^r_j dx^j) + \Gamma^p_{ij}(x)dx^i \wedge dx^j. \tag{3}$$

Comparing (2) with (3), we obtain

$$R^p_{ij} = \Gamma^p_{[ij]} + c^p_{qr}\Gamma^q_i\Gamma^r_j. \tag{4}$$

52.3. Generalization of the Utiyama theorem. The curvature of a connection Γ on P can be considered as a section $C_P\Gamma\colon BP \to LP \otimes \Lambda^2 T^*BP$, where $LP = P[\mathfrak{g}, \mathrm{Ad}]$ is the so-called adjoint bundle of P, see 17.6. Using the language of the theory of gauge natural bundles, D. J. Eck reformulated a classical result by Utiyama in the following form: *All first order gauge natural Lagrangians on the connection bundle are of the form $A \circ C$, where A is a zero order gauge natural Lagrangian on the curvature bundle and C is the curvature operator,* [Eck, 81]. By 49.1, a first order Lagrangian on a connection bundle QP is a morphism $J^1QP \to \Lambda^m T^*BP$, so that the Utiyama theorem deals with first order gauge natural operators $Q \rightsquigarrow \Lambda^m T^*B$. We are going to generalize this result. Since the proof will be based on the orbit reduction, we shall directly discuss the standard fibers in question.

Denote by $\gamma\colon S_1 \to \mathfrak{g} \otimes \Lambda^2\mathbb{R}^{m*}$ the formal curvature map 52.2.(4). One sees easily that γ is a surjective submersion. The semi-direct decomposition $W^2_mG = G^2_m \rtimes T^2_mG$ together with the target jet projection $T^2_mG \to G$ defines a group homomorphism $p\colon W^2_mG \to G^2_m \times G$. Let Z be a $G^2_m \times G$-space, which can be considered as a W^2_mG-space by means of p. The standard fiber $\mathfrak{g} \otimes \Lambda^2\mathbb{R}^{m*}$ of the curvature bundle is a $G^1_m \times G$-space, which can be interpreted as $G^2_m \times G$-space by means of the jet homomorphism $\pi^2_1\colon G^2_m \to G^1_m$.

Proposition. *For every W^2_mG-map $f\colon S_1 \to Z$ there exists a unique $G^2_m \times G$-map $g\colon \mathfrak{g} \otimes \Lambda^2\mathbb{R}^{m*} \to Z$ satisfying $f = g \circ \gamma$.*

Proof. On the kernel K of $p\colon W^2_mG \to G^2_m \times G$ we have the coordinates a^p_i, $a^p_{ij} = a^p_{ji}$ introduced in 52.1. Let us replace the coordinates Γ^p_{ij} on S_1 by

$$R^p_{ij} = \Gamma^p_{[ij]} + c^p_{qr}\Gamma^q_i\Gamma^r_j, \qquad S^p_{ij} = \Gamma^p_{(ij)}, \tag{1}$$

while Γ^p_i remain unchanged. Hence the coordinate form of γ is $(\Gamma^p_i, R^p_{ij}, S^p_{ij}) \mapsto (R^p_{ij})$. From 52.1.(5) and 52.1.(6) we can evaluate a^p_i and a^p_{ij} in such a way that $\bar\Gamma^p_i = 0$ and $\bar S^p_{ij} = 0$. This implies that each fiber of γ is a K-orbit. Then we apply 28.1. □

52.4. To interpret the proposition 52.3 in terms of operators, it is useful to introduce a more subtle notion of principal prolongation $W^{s,r}P$ of order (s,r), $s \geq r$, of a principal fiber bundle $P(M,G)$. Formally we can construct the fiber product over M

$$W^{s,r}P = P^sM \times_M J^rP \tag{1}$$

and the semi-direct product of Lie groups

$$W^{s,r}_m G = G^s_m \rtimes T^r_m G \tag{2}$$

with respect to the right action $(A,B) \mapsto B \circ \pi^s_r(A)$ of G^s_m on $T^r_m G$. The right action of $W^{s,r}_m G$ on $W^{s,r}P$ is given by a formula analogous to 15.4

$$(u,v)(A,B) = (u \circ A, v.(B \circ \pi^s_r(A^{-1} \circ u^{-1}))),$$

$u \in P^sM$, $v \in J^rP$, $A \in G^s_m$, $B \in T^r_m G$. In the case $r = 0$ we have a direct product of Lie groups $W^{s,0}_m G = G^s_m \times G$ and the usual fibered product $W^{s,0}P = P^sM \times_M P$ of principal fiber bundles.

To clarify the geometric substance of the previous construction, we have to use the concept of (r,s,q)-jet of a fibered manifold morphism introduced in 12.19. Then $W^{s,r}P$ can be defined as the space of all (r,r,s)-jets at $(0,e)$ of the local principal bundle isomorphisms $\mathbb{R}^m \times G \to P$ and the group $W^{s,r}_m G$ is the fiber of $W^{s,r}(\mathbb{R}^m \times G)$ over $0 \in \mathbb{R}^m$ endowed with the jet composition. The proof is left to the reader as an easy exercise. Furthermore, in the same way as in 51.2 we deduce that if two $\mathcal{PB}_m(G)$-morphisms $f,g\colon P \to \bar{P}$ satisfy $j^{r,r,s}_y f = j^{r,r,s}_y g$ at a point $y \in P_x$, $x \in BP$, then this equality holds at every point of the fiber P_x. In this case we write $\mathrm{j}^{r,r,s}_x f = \mathrm{j}^{r,r,s}_x g$.

Now we can say that *natural bundle F is of order (s,r), $s \geq r$,* if $\mathrm{j}^{r,r,s}_x f = \mathrm{j}^{r,r,s}_x g$ implies $Ff|F_xP = Fg|F_xP$. Using the proposition 51.10 we deduce quite similarly to 51.6 that *every gauge natural bundle of order (s,r) is a fiber bundle associated to $W^{s,r}$.*

Then the proposition 52.3 is equivalent to the following assertion.

General Utiyama theorem. *Let F be a gauge natural bundle of order $(2,0)$. Then for every first order gauge natural operator $A\colon Q \rightsquigarrow F$ there exists a unique natural transformation $\bar{A}\colon L \otimes \Lambda^2T^*B \to F$ satisfying $A = \bar{A} \circ C$, where $C\colon Q \rightsquigarrow L \otimes \Lambda^2T^*B$ is the curvature operator.*

In all concrete problems in this chapter the result will be applied to gauge natural bundles of order (1,0). By definition, every such a bundle has the order (2,0) as well.

52.5. Curvature-like operators. The curvature operator $C\colon Q \rightsquigarrow L\otimes\Lambda^2T^*B$ is a gauge natural operator because of the geometric definition of the curvature. We are going to determine all gauge natural operators $Q \rightsquigarrow L \otimes \otimes^2T^*B$. (We shall see that the values of all of them lie in $L \otimes \Lambda^2T^*B$. But this is an interesting geometric result that the antisymmetry of such operators is a consequence of their gauge naturality.) Let $Z \subset L(\mathfrak{g},\mathfrak{g})$ be the subspace of all linear maps commuting with the adjoint action of G. Since every $z \in Z$ is an equivariant linear map between the standard fibers, it induces a vector bundle morphism $\bar{z}_P \colon LP \to LP$. Hence we can construct a *modified curvature operator* $C(z)_P \colon (\bar{z}_P \otimes \Lambda^2 T^* \mathrm{id}_{BP}) \circ C_P$.

Proposition. *All gauge natural operators $Q \rightsquigarrow L \otimes \otimes^2 T^* B$ are the modified curvature operators $C(z)$ for all $z \in Z$.*

Proof. By 51.16, every gauge natural operator A on the connection bundle has finite order. The r-th order gauge natural operators correspond to the $W_m^{r+1}G$-equivariant maps $J_0^r Q \to \mathfrak{g} \otimes \otimes^2 \mathbb{R}^{m*}$. Let $\Gamma^p_{i\alpha}$ be the induced coordinates on $J_0^r Q$, where α is a multi index of range m with $|\alpha| \le r$. On $\mathfrak{g} \otimes \otimes^2 \mathbb{R}^{m*}$ we have the canonical coordinates R^p_{ij} and the action

$$\bar{R}^p_{ij} = A^p_q(a) R^q_{kl} \tilde{a}^k_i \tilde{a}^l_j. \tag{1}$$

Hence the coordinate components of the map associated to A are some functions $f^p_{ij}(\Gamma^q_{k\alpha})$. If we consider the canonical inclusion of G^1_m into $W_m^{r+1}G$, then analogously to 14.20 the transformation laws of all quantities $\Gamma^p_{i\alpha}$ are tensorial. The equivariance with respect to the homotheties in G^1_m gives a homogeneity condition

$$c^2 f^p_{ij}(\Gamma^q_{k\alpha}) = f^p_{ij}(c^{1+|\alpha|}\Gamma^q_{k\alpha}) \qquad 0 \ne c \in \mathbb{R}. \tag{2}$$

By the homogeneous function theorem, f^p_{ij} is independent of $\Gamma^p_{i\alpha}$ with $|\alpha| \ge 2$. Hence A is a first order operator and we can apply the general Utiyama theorem.

The associated map

$$g \colon \mathfrak{g} \otimes \Lambda^2 \mathbb{R}^{m*} \to \mathfrak{g} \otimes \otimes^2 \mathbb{R}^{m*} \tag{3}$$

of the induced natural transformation $L \otimes \Lambda^2 T^* B \to L \otimes \otimes^2 T^* B$ is of the form $g^p_{ij}(R^q_{kl})$. Using the homotheties in G^1_m we find that g is linear. If we fix one coordinate in $\mathfrak{g}$ on the right-hand side of the arrow (3), we obtain a linear G^1_m-map $\times^n \Lambda^2 \mathbb{R}^{m*} \to \otimes^2 \mathbb{R}^{m*}$. By 24.8.(5), this map is a linear combination of the individual inclusions $\Lambda^2 \mathbb{R}^{m*} \hookrightarrow \otimes^2 \mathbb{R}^{m*}$, i.e.

$$g^p_{ij} = z^p_q R^q_{ij}. \tag{4}$$

Using the equivariance with respect to the canonical inclusion of G into $W^2_m G$, we find that the linear map $(z^p_q) \colon \mathfrak{g} \to \mathfrak{g}$ commutes with the adjoint action. □

52.6. Remark. In the case that the structure group is the general linear group $GL(n)$ of an arbitrary dimension n, the invariant tensor theorem implies directly that the Ad-invariant linear maps $\mathfrak{gl}(n) \to \mathfrak{gl}(n)$ are generated by the identity and the map $X \mapsto (\mathrm{trace} X)\mathrm{id}$. Then the proposition 52.5 gives a two-parameter family of all $GL(n)$-natural operators $Q \rightsquigarrow L \otimes \otimes^2 T^* B$, which the first author deduced by direct evaluation in [Kolář, 87b]. In general it is remarkable that the study of the case of the special structure group $GL(n)$, to which we can apply the generalized invariant tensor theorem, plays a useful heuristic role in the theory of gauge natural operators.

Further we remark that all gauge natural operators $Q \oplus Q \rightsquigarrow L \otimes \otimes^2 T^* B$ transforming pairs of connections on an arbitrary principal fiber bundle P into sections of $LP \otimes \otimes^2 T^* BP$ are determined in [Kurek, to appear a].

52.7. Generalized Chern-Weil forms. We recall that for every vector bundle $E \to M$, a section of $E \otimes \Lambda^r T^*M$ is called an *E-valued r-form*, see 7.11. For $E = M \times \mathbb{R}$ we obtain the usual exterior forms on M. Consider a linear action ρ of a Lie group G on a vector space V and denote by $\tilde{V}$ the G-natural bundle over m-manifolds determined by this action of $G = W^0_m G$. We are going to construct some gauge natural operators transforming every connection Γ on a principal bundle $P(M, G)$ into a $\tilde{V}(P)$-valued exterior form. In the special case of the identity action of G on $\mathbb{R}$, i.e. $\rho(g) = \mathrm{id}_{\mathbb{R}}$ for all $g \in G$, we obtain the classical Chern-Weil forms of Γ, [Kobayashi, Nomizu, 69].

Let $h\colon S^r\mathfrak{g} \to V$ be a linear G-map. We have $S^r(\mathfrak{g} \otimes \Lambda^2\mathbb{R}^{m*}) = S^r\mathfrak{g} \otimes S^r\Lambda^2\mathbb{R}^{m*}$, so that we can define $\bar{h}\colon \mathfrak{g} \otimes \Lambda^2\mathbb{R}^{m*} \to V \otimes \Lambda^{2r}\mathbb{R}^{m*}$ by

$$\bar{h}(A) = (h \otimes \mathrm{Alt})(A \otimes \cdots \otimes A), \quad A \in \mathfrak{g} \otimes \Lambda^2\mathbb{R}^{m*}, \tag{1}$$

where $\mathrm{Alt}\colon S^r\Lambda^2\mathbb{R}^{m*} \to \Lambda^{2r}\mathbb{R}^{m*}$ is the tensor alternation. Since $\mathfrak{g} \otimes \Lambda^2\mathbb{R}^{m*}$ or $V \otimes \Lambda^{2r}\mathbb{R}^{m*}$ is the standard fiber of the curvature bundle or of $\tilde{V}(P) \otimes \Lambda^{2r}T^*M$, respectively, $\bar{h}$ induces a bundle morphism $\bar{h}_P\colon L(P) \otimes \Lambda^2T^*M \to \tilde{V}(P) \otimes \Lambda^{2r}T^*M$. For every connection $\Gamma\colon M \to QP$, we first construct its curvature $C_P\Gamma$ and then a $\tilde{V}(P)$-valued $2r$-form

$$\tilde{h}_P(\Gamma) = \bar{h}_P(C_P\Gamma). \tag{2}$$

Such forms will be called *generalized Chern-Weil forms.*

Let $I(\mathfrak{g}, V)$ denote the space of all polynomial G-maps of $\mathfrak{g}$ into V. Every $H \in I(\mathfrak{g}, V)$ is determined by a finite sequence of linear G-maps $h^{r_i}\colon S^{r_i}\mathfrak{g} \to V$, $i = 1, \dots, n$. Then

$$\tilde{H}_P(\Gamma) = \tilde{h}^{r_1}_P(\Gamma) + \cdots + \tilde{h}^{r_n}_P(\Gamma)$$

is a section of $\tilde{V}(P) \otimes \Lambda T^*M$ for every connection Γ on P. By definition, $\tilde{H}$ is a gauge natural operator $Q \rightsquigarrow \tilde{V} \otimes \Lambda T^*B$.

52.8. Theorem. *All G-natural operators $Q \rightsquigarrow \tilde{V} \otimes \Lambda T^*B$ are of the form $\tilde{H}$ for all $H \in I(\mathfrak{g}, V)$.*

Proof. Consider some linear coordinates y^p on $\mathfrak{g}$ and z^a on V and the induced coordinates y^p_{ij} on $\mathfrak{g} \otimes \Lambda^2\mathbb{R}^{m*}$ and $z^a_{i_1 \dots i_s}$ on $V \otimes \Lambda^s\mathbb{R}^{m*}$.

By 51.16 every G-natural operator $A\colon Q \rightsquigarrow \tilde{V} \otimes \Lambda^sT^*B$ has a finite order k. Hence its associated map $f\colon J^k_0Q \to V \otimes \Lambda^s\mathbb{R}^{m*}$ is of the form

$$z^a_{i_1 \dots i_s} = f^a_{i_1 \dots i_s}(\Gamma^p_{i\alpha}), \quad 0 \le |\alpha| \le k.$$

The homotheties in G^1_m give a homogeneity condition

$$k^s f^a_{i_1 \dots i_s}(\Gamma^p_{i\alpha}) = f^a_{i_1 \dots i_s}(k^{1+|\alpha|}\Gamma^p_{i\alpha}).$$

This implies that f is a polynomial map in $\Gamma^p_{i\alpha}$. Fix a, p_1, $|\alpha_1|$, ..., p_r, $|\alpha_r|$ with $|\alpha_1| \ge 2$ and consider the subpolynomial of the a-th component of f which is formed by the linear combinations of $\Gamma^{p_1}_{i_1\alpha_1} \dots \Gamma^{p_r}_{i_r\alpha_r}$. It represents a $GL(m)$-map $\mathbb{R}^{m*} \otimes S^{|\alpha_1|}\mathbb{R}^{m*} \times \dots \times \mathbb{R}^{m*} \otimes S^{|\alpha_r|}\mathbb{R}^{m*} \to \Lambda^p\mathbb{R}^{m*}$. Analogously to 24.8 we deduce that this is the zero map because of the symmetric component $S^{|\alpha_1|}\mathbb{R}^{m*}$. Hence A is a first order operator.

Applying the general Utiyama theorem, we obtain $f = g \circ \gamma$, where g is a $G^1_m \times G$-map $\mathfrak{g} \otimes \Lambda^2\mathbb{R}^{m*} \to V \otimes \Lambda^s\mathbb{R}^{m*}$. The coordinate form of g is

$$z^a_{i_1 \dots i_s} = g^a_{i_1 \dots i_s}(y^p_{ij}).$$

Using the homotheties in G^1_m we find that $s = 2r$ and g is a polynomial of degree r in y^p_{ij}. Its total polarization is a linear map $S^r(\mathfrak{g} \otimes \Lambda^2\mathbb{R}^{m*}) \to V \otimes \Lambda^{2r}\mathbb{R}^{m*}$. If we fix one coordinate in V and any r-tuple of coordinates in $\mathfrak{g}$, we obtain an underlying problem of finding all linear G^1_m-maps $\otimes^r\Lambda^2\mathbb{R}^{m*} \to \Lambda^{2r}\mathbb{R}^{m*}$. By 24.8.(5) each this map is a constant multiple of $y^{p_1}_{[i_1 i_2} \dots y^{p_r}_{i_{2r-1} i_{2r}]}$. Hence g is of the form

$$c^a_{p_1 \dots p_r} y^{p_1}_{[i_1 i_2} \dots y^{p_r}_{i_{2r-1} i_{2r}]}.$$

The equivariance with respect to the canonical inclusion of G into $W^2_m G$ implies that $(c^a_{p_1 \dots p_r})\colon S^r\mathfrak{g} \to V$ is a G-map. □

52.9. Consider the special case of the identity action of G on $\mathbb{R}$. Then every linear G-map $S^r\mathfrak{g} \to \mathbb{R}$ is identified with a G-invariant element of $S^r\mathfrak{g}^*$ and the $(M \times \mathbb{R})$-valued forms are the classical differential forms on M. Hence 52.7.(2) gives the classical Chern-Weil forms of a connection. In this case the theorem 52.8 reads that *all gauge natural differential forms on connections are the classical Chern-Weil forms.* All of them are of even degree. The exterior differential of a Chern-Weil form is a gauge natural form of odd degree. By the theorem 52.8 it must be a zero form. This gives an interesting application of gauge naturality for proving the following classical result.

Corollary. *All classical Chern-Weil forms are closed.*

52.10. In general, if one has a vector bundle $E \to M$, an E-valued r-form $\omega\colon \Lambda^r TM \to E$ and a linear connection Δ on E, one introduces the covariant exterior derivative $d_\Delta\omega\colon \Lambda^{r+1}TM \to E$, see 11.14. Consider the situation from 52.7. For every $H \in I(\mathfrak{g}, V)$ and every connection Γ on P we have constructed a $\tilde{V}(P)$-valued form $\tilde{H}_P(\Gamma)$, which is of even degree. According to 11.11, Γ induces a linear connection Γ_V on $\tilde{V}(P)$. Then $d_{\Gamma_V}\tilde{H}_P(\Gamma)$ is a gauge natural $\tilde{V}(P)$-valued form of odd degree. By the theorem 52.8 it is a zero form. Thus, we have proved the following interesting geometric result.

Proposition. *For every $H \in I(\mathfrak{g}, V)$ and every connection Γ on P, it holds* $d_{\Gamma_V}\tilde{H}_P(\Gamma) = 0$.

52.11. Remark. We remark that another generalization of Chern-Weil forms is studied in [Lecomte, 85].

52.12. Gauge natural approach to the Bianchi identity. It is remarkable that the Bianchi identity for a principal connection $\Gamma\colon BP \to QP$ can be deduced in a similar way. Using the notation from 52.5, we first prove an auxiliary result.

Lemma. *The only gauge natural operator $Q \rightsquigarrow L \otimes \Lambda^3 T^* B$ is the zero operator.*

Proof. By 51.16, every such operator A has finite order. Let

$$f^p_{ijk}(\Gamma^q_{l\alpha}), \qquad 0 \le |\alpha| \le r$$

be its associated map. The homotheties in G^1_m yield a homogeneity condition

$$c^3 f^p_{ijk}(\Gamma^q_{l\alpha}) = f^p_{ijk}(c^{1+|\alpha|}\Gamma^q_{l\alpha}), \quad c \in \mathbb{R} \setminus \{0\}. \tag{1}$$

Hence f is polynomial in Γ^p_i, Γ^p_{ij} and Γ^p_{ijk} of degrees d_0, d_1 and d_2 satisfying

$$3 = d_0 + 2d_1 + 3d_2.$$

This implies f is linear in Γ^p_{ijk}. But Γ^p_{ijk} represent a linear $GL(m)$-map $\mathbb{R}^{m*} \otimes S^2\mathbb{R}^{m*} \to \Lambda^3\mathbb{R}^{m*}$ for each $p = 1, \dots, n$. By 24.8 the only possibility is the zero map. Hence A is a first order operator. By the general Utiyama theorem, f factorizes through a map $g\colon \mathfrak{g} \otimes \Lambda^2\mathbb{R}^{m*} \to \mathfrak{g} \otimes \Lambda^3\mathbb{R}^{m*}$. The equivariance of g with respect to the homotheties in G^1_m yields a homogeneity condition $c^3 g(y) = g(c^2 y)$, $y \in \mathfrak{g} \otimes \Lambda^2\mathbb{R}^{m*}$. Since there is no integer satisfying $3 = 2d$, g is the zero map. □

The curvature of Γ is a section $C_P\Gamma\colon BP \to LP \otimes \Lambda^2 T^* BP$. According to the general theory, Γ induces a linear connection $\tilde{\Gamma}$ on the adjoint bundle LP. Hence we can construct the covariant exterior differential

$$\nabla_{\tilde{\Gamma}} C_P\Gamma\colon BP \to LP \otimes \Lambda^3 T^* BP. \tag{2}$$

By the geometric character of this construction, (2) determines a gauge natural operator. Then our lemma implies

$$\nabla_{\tilde{\Gamma}} C_P(\Gamma) = 0. \tag{3}$$

By 11.15, this is the Bianchi identity for Γ.

53. Base extending gauge natural operators

53.1. Analogously to 18.17, we now formulate the concept of gauge natural operators in more general situation. Let F, E and H be three G-natural bundles over m-manifolds.

Definition. A gauge natural operator $D\colon F \rightsquigarrow (E, H)$ is a system of regular operators $D_P\colon C^\infty FP \to C^\infty_{BP}(EP, HP)$ for every $\mathcal{PB}_m(G)$-object P satisfying $D_{\bar{P}}(Ff \circ s \circ Bf^{-1}) = Hf \circ D_P(s) \circ Ef^{-1}$ for every $s \in C^\infty FP$ and every $\mathcal{PB}_m(G)$-isomorphism $f\colon P \to \bar{P}$, as well as a localization condition analogous to 51.14.(b).

53.2. Quite similarly to 18.19, one deduces

Proposition. *k-th order gauge natural operators $F \rightsquigarrow (E, H)$ are in a canonical bijection with the natural transformations $J^r F \oplus E \to H$.*

If we have a natural transformation $q\colon H \to E$ such that every $q_P\colon HP \to EP$ is a surjective submersion and we require every $D_P(s)$ to be a section of q_P, we write $D\colon F \rightsquigarrow (H \to E)$. Then we find in the same way as in 51.15 that the G-natural operators $F \rightsquigarrow (H \to E)$ are in bijection with the W^s_m-equivariant maps $f\colon J^k_0 F \times E_0 \to H_0$, satisfying $q_0 \circ f = \mathrm{pr}_2$, where $q_0\colon H_0 \to E_0$ is the restriction of $q_{\mathbb{R}^m \times G}$ and s is the maximum of the orders in question.

53.3. Gauge natural operators $Q \rightsquigarrow (QT \to TB)$. In 46.3 we deduced that every connection Γ on principal bundle $P \to M$ with structure group G induces a connection $\mathcal{T}\Gamma$ on the principal bundle $TP \to TM$ with structure group TG. Hence $\mathcal{T}$ is a (first-order) G-natural operator $Q \rightsquigarrow (QT \to TB)$. Now we are going to determine all first-order G-natural operators $Q \rightsquigarrow (QT \to TB)$. Since the difference of two connections on $TP \to TBP$ is a section of $L(TP) \otimes T^*TBP$, it suffices to determine all first-order G-natural operators $Q \rightsquigarrow (LT \otimes T^*TB \to TB)$. The fiber of the total projection $L(T(\mathbb{R}^m \times G)) \otimes T^*T\mathbb{R}^m \to T\mathbb{R}^m \to \mathbb{R}^m$ over $0 \in \mathbb{R}^m$ is the product of $\mathbb{R}^m$ with $\mathfrak{tg} \otimes T^*_0 T\mathbb{R}^m$, $0 \in T\mathbb{R}^m = \mathbb{R}^{2m}$. By 53.2 our operators are in bijection with the $W^2_m G$-equivariant maps $J^1_0 Q(\mathbb{R}^m \times G) \times \mathbb{R}^m \to \mathbb{R}^m \times \mathfrak{tg} \otimes T^*_0 T\mathbb{R}^m$ over the identity of $\mathbb{R}^m$.

We know from 10.17 that TG coincides with the semidirect product $G \rtimes \mathfrak{g}$ with the following multiplication

$$(g_1, X_1)(g_2, X_2) = (g_1 g_2, \mathrm{Ad}(g_2^{-1})(X_1) + X_2) \tag{1}$$

where Ad means the adjoint action of G. This identifies the Lie algebra $\mathfrak{tg}$ of TG with $\mathfrak{g} \times \mathfrak{g}$ and a direct calculation yields the following formula for the adjoint action Ad_{TG} of TG

$$\mathrm{Ad}_{TG}(g, X)(Y, V) = (\mathrm{Ad}(g)(Y), \mathrm{Ad}(g)([X, Y] + V)). \tag{2}$$

Hence the subspace $0 \times \mathfrak{g} \subset \mathfrak{tg}$ is Ad_{TG}-invariant, so that it defines a subbundle $K(TP) \subset L(TP)$. The injection $V \mapsto (0, V)$ induces a map $I_P\colon LP \to K(TP)$.

Every modified curvature $C(z)_P(\Gamma)$ of a connection Γ on P, see 52.5, can be interpreted as a linear morphism $\Lambda^2 TBP \to LP$. Then we can define a linear map $\mu(C(z)_P(\Gamma))\colon TTBP \to L(TP)$ by

$$\mu(C(z)_P(\Gamma))(A) = I_P(C(z)_P(\Gamma)(\pi_1 A \wedge \pi_2 A)), \quad A \in TTBP \tag{3}$$

where $\pi_1\colon TTBP \to TBP$ is the bundle projection and $\pi_2\colon TTBP \to TBP$ is the tangent map of the bundle projection $TBP \to BP$. This determines one series $\mu(C(z))$, $z \in Z$, of G-natural operators $Q \rightsquigarrow (LT \otimes T^*TB \to TB)$.

Moreover, if we consider a modified curvature $C(z)_P(\Gamma)$ as a map $C(z)\colon P \oplus \Lambda^2 TBP \to \mathfrak{g}$, we can construct its vertical prolongation with respect to the first factor

$$V_1 C(z)_P(\Gamma)\colon VP \oplus \Lambda^2 TBP \to T\mathfrak{g} = \mathfrak{tg}.$$

Then we add the vertical projection $\nu\colon TP\to VP$ of the connection Γ and we use the projections π_1 and π_2 from (3). This yields a map

$$\tau(C(z)_P(\Gamma))\colon TP\oplus TTBP\to \mathfrak{tg} \tag{4}$$
$$\tau(C(z)_P(\Gamma))(U,A)=V_1C(z)_P(\Gamma)(\nu U,\pi_1A\wedge\pi_2A),\quad U\in TP,\quad A\in TTBP.$$

The latter map can be interpreted as a section of $L(TP)\otimes T^*TBP$, which gives another series $\tau(C(z))$, $z\in Z$, of G-natural operators $Q\rightsquigarrow(LT\otimes T^*TB\to TB)$.

Proposition. *All first-order gauge natural operators $Q\rightsquigarrow(QT\to TB)$ form the following* $2\dim Z$*-parameter family*

$$T+\mu(C(z))+\tau(C(\bar z)),\qquad z,\bar z\in Z. \tag{5}$$

The proof will occupy the rest of this section.

53.4. Let Γ be a connection on $\mathbb{R}^m\times G$ with equations

$$\omega^p=\Gamma^p_i(x)dx^i. \tag{1}$$

Let (ε^p) be the second component of the Maurer-Cartan form of TG (the first one is (ω^p)) and let X^i be the induced coordinates on $T_0\mathbb{R}^m$. Applying the description of the Maurer-Cartan form of TG from 37.16 to (1), we find the equation of $T\Gamma$ is of the form (1) and

$$\varepsilon^p=\frac{\partial\Gamma^p_i}{\partial x^j}X^jdx^i+\Gamma^p_idX^i. \tag{2}$$

53.5. Remark first that every $\frac{d}{dt}\big|_0 x(t)\in T_0\mathbb{R}^m$ defines a map

$$T^1_mG\to TG,\quad j^1_0\varphi\mapsto \tfrac{d}{dt}\big|_0(\varphi\circ x)(t). \tag{1}$$

Consider an isomorphism $\bar x=f(x)$, $\bar y=\varphi(x)\cdot y$ of $\mathbb{R}^m\times G$ and an element of $V(T(\mathbb{R}^m\times G)\to T\mathbb{R}^m)$. Clearly, such an element can be generated by a map $(x(t),y(t,u))\colon\mathbb{R}^2\to\mathbb{R}^m\times G$, t, $u\in\mathbb{R}$. This map is transformed into

$$\bar x=f(x(t)),\quad \bar y=\varphi(x(t))\cdot y(t,u). \tag{2}$$

Differentiating with respect to t, we find

$$\frac{d\bar y}{dt}=T\mu\Big(\frac{d\varphi(x(0))}{dt},\frac{dy(0,u)}{dt}\Big) \tag{3}$$

where $\mu\colon G\times G\to G$ is the group composition. This implies that the next differentiation with respect to u yields the adjoint action of TG with respect to (1). Thus, if (Y^p,V^p) are the coordinates in $\mathfrak{tg}$ given by our basis in $\mathfrak{g}$, then we deduce by the latter observation that the action of W^2_mG on $\mathbb{R}^m\times\mathfrak{tg}$ is $\bar X^i=a^i_jX^j$ and

$$\bar Y^p=A^p_q(a)Y^q,\quad \bar V^p=A^p_q(a)(c^q_{rs}a^r_jX^jY^s+V^q). \tag{4}$$

On the other hand, the action of W^2_mG on $T_0T\mathbb{R}^m$ goes through the projection into G^2_m and has the standard form

$$d\bar x^i=a^i_jdx^j,\quad d\bar X^i=a^i_{jk}X^jdx^k+a^i_jdX^j. \tag{5}$$

53.6. Our problem is to find all $W^2_m G$-equivariant maps $f\colon \mathbb{R}^m \times J^1_0 Q \to \mathbb{R}^m \times \mathfrak{g} \otimes T^*_0 T\mathbb{R}^m$ over $\mathrm{id}_{\mathbb{R}^m}$. On $J^1_0 Q$, we replace Γ^p_{ij} by R^p_{ij} and S^p_{ij} as in 52.3. The coordinates on $\mathfrak{g} \otimes T^*_0 T\mathbb{R}^m$ are given by

$$Y^p = B^p_i dx^i + C^p_i dX^i \tag{1}$$

$$V^p = D^p_i dx^i + E^p_i dX^i. \tag{2}$$

Hence all components of f are smooth functions of $X = (X^i)$, $\Gamma = (\Gamma^p_i)$, $R = (R^p_{ij})$, $S = (S^p_{ij})$. Using 53.5.(4)–(5), we deduce from (1) the transformation laws

$$\bar{C}^p_i = A^p_q(a) C^q_j \tilde{a}^j_i \tag{3}$$

$$\bar{B}^p_i = A^p_q(a) B^q_j \tilde{a}^j_i - A^p_q(a) C^q_k \tilde{a}^k_j a^j_{li} X^l. \tag{4}$$

Let us start with the component $C^p_i(X, \Gamma, R, S)$ of f. Using a^p_{ij} and a^p_i, we deduce that C's are independent of Γ and S. Then we have the situation of the following lemma.

Lemma. *All* $\mathrm{Ad}_G \times GL(m,\mathbb{R})$*-equivariant maps* $\mathbb{R}^m \times \mathfrak{g} \otimes \Lambda^2 \mathbb{R}^{m*} \to \mathfrak{g} \otimes \mathbb{R}^{m*}$ *have the form* $\mu^p_q R^q_{ij} X^j$ *with* $(\mu^p_q) \in Z$.

Proof. First we determine all $GL(m,\mathbb{R})$-maps $h\colon \mathbb{R}^m \times \times^n \otimes^2 \mathbb{R}^{m*} \to \times^n \mathbb{R}^{m*}$, $h = (h^p_i(b^q_{jk}, X^l))$. If we consider the contraction $\langle h, v\rangle$ of h with $v = (v^i) \in \mathbb{R}^m$, we can apply the tensor evaluation theorem to each component of $\langle h, v\rangle$. This yields

$$h^p_i v^i = \varphi^p(b^q_{ij} X^i X^j, b^r_{ij} v^i X^j, b^s_{ij} X^i v^j, b^t_{ij} v^i v^j).$$

Differentiating with respect to v^i and setting $v^i = 0$, we obtain

$$h^p_i = \varphi^p_q(b^r_{kl} X^k X^l) b^q_{ij} X^j + \psi^p_q(b^r_{kl} X^k X^l) b^q_{ji} X^j \tag{5}$$

with arbitrary smooth functions φ^p_q, ψ^p_q of n variables. If $b^p_{ij} = R^p_{ij}$ are antisymmetric, we have $R^p_{ij} X^i X^j = 0$ and $R^p_{ij} X^j = -R^p_{ji} X^j$, so that

$$h^p_i = \mu^p_q R^q_{ij} X^j, \qquad \mu^p_q \in \mathbb{R}. \tag{6}$$

The equivariance with respect to G then yields $A^p_q(a)\mu^q_r = \mu^p_q A^q_r(a)$, i.e. $(\mu^p_q) \in Z$. □

Thus our lemma implies $C^p_i = \mu^p_q R^q_{ij} X^j$, $(\mu^p_q) \in Z$. For the components B^p_i of f, the use of a^p_i and a^p_{ij} gives that B's are independent of Γ and S. Then the equivariance with respect to a^i_{jk} yields

$$C^p_i = 0. \tag{7}$$

Using our lemma again, we obtain

$$B^p_i = \gamma^p_q R^q_{ij} X^j, \qquad (\gamma^p_q) \in Z. \tag{8}$$

53.7. From 53.6.(2) we deduce the transformation laws

$$\bar{E}^p_i = A^p_q(a)E^q_j\tilde{a}^j_i + A^p_q(a)c^q_{rs}a^r_jX^jC^s_i \tag{1}$$

$$\bar{D}^p_i = A^p_q(a)D^q_j\tilde{a}^j_i + A^p_q(a)c^q_{rs}a^r_jX^jB^s_i - \bar{E}^p_ja^j_{ki}X^k. \tag{2}$$

By 53.6.(7), the first equation implies $E^p_i = \mu^p_qR^q_{ij}X^j$, $(\mu^p_q) \in Z$, in the same way as in 53.6. Using a^p_{ij} in the second equation, we find that the D's are independent of S. Then the use of a^i_{jk} implies

$$E^p_i = 0. \tag{3}$$

The equivariance of D's with $a = e$, $a^i_j = \delta^i_j$ now reads

$$D^p_i(X, \Gamma^q_j + a^q_j, R) = D^p_i(X, \Gamma, R) + c^p_{qr}a^q_jX^j\gamma^r_sR^s_{ik}X^k.$$

Differentiating with respect to a^q_j and setting $a^q_j = 0$, we find that the D's are of the form

$$D^p_i = c^p_{qr}\Gamma^q_jX^j\gamma^r_sR^s_{ik}X^k + F^p_i(X^j, R^q_{kl}X^l).$$

The 'absolute terms' F^p_i can be determined by lemma 53.6. This yields

$$D^p_i = c^p_{qr}\Gamma^q_jX^j\gamma^r_sR^s_{ik}X^k + k^p_qR^q_{ij}X^j, \quad (k^p_q) \in Z. \tag{4}$$

One verifies easily that (3), (4) together with 53.6.(7)–(8) and 53.4.(1)–(2) is the coordinate form of proposition 53.3. □

54. Induced linear connections on the total space of vector and principal bundles

54.1. Gauge natural operators $Q \oplus QTB \rightsquigarrow QT$. Given a vector bundle $\pi\colon E \to BE$ of fiber dimension n, we denote by $GL(\mathbb{R}^n, E) \to BE$ the bundle of all linear frames in the individual fibers of E, see 10.11. This is a principal bundle with structure group $GL(n)$, $n =$ the fiber dimension of E. Clearly E is identified with the fiber bundle associated to $GL(\mathbb{R}^n, E)$ with standard fiber $\mathbb{R}^n$. The construction of associated bundles establishes a natural equivalence between the category $\mathcal{PB}_m(GL(n))$ and the category $\mathcal{VB}_{m,n} := \mathcal{VB} \cap \mathcal{FM}_{m,n}$.

A linear connection D on a vector bundle E is usually defined as a linear morphism $D\colon E \to J^1E$ splitting the target jet projection $J^1E \to E$, see section 17. One finds easily that there is a canonical bijection between the linear connections on E and the principal connections on $GL(\mathbb{R}^n, E)$, see 11.11. That is why we can say that $Q(GL(\mathbb{R}^n, E)) =: QE$ is the bundle of linear connections on E. In the special case $E = TBE$ this gives a well-known fact from the theory of classical linear connections on a manifold.

An interesting geometrical problem is how a linear connection D on a vector bundle E and a classical linear connection Λ on the base manifold BE can induce a classical linear connection on the total space E. More precisely, we are looking for operators which are natural on the category $\mathcal{VB}_{m,n}$. Taking into account the natural equivalence between $\mathcal{VB}_{m,n}$ and $\mathcal{PB}_m(GL(n))$, we see that this is a problem on base-extending $GL(n)$-natural operators. But we find it more instructive to apply the direct approach in this section. Thus, our problem is to find all operators $Q \oplus QTB \rightsquigarrow QT$ which are natural on $\mathcal{VB}_{m,n}$.

54.2. First we describe a concrete construction of such an operator. Let us denote the covariant differentiation with respect to a connection by the symbol of the connection itself. Thus, if X is a vector field on BE and s is a section of E, then $D_X s$ is a section of E. Further, let X^D denote the horizontal lift of vector field X with respect to D. Moreover, using the translations in the individual fibers of E, we derive from every section $s\colon BE \to E$ a vertical vector field s^V on E called the vertical lift of s.

Proposition. *For every linear connection D on a vector bundle E and every classical linear connection Λ on BE there exists a unique classical linear connection $\Gamma = \Gamma(D,\Lambda)$ on the total space E with the following properties*

$$\begin{aligned} \Gamma_{X^D} Y^D &= (\Lambda_X Y)^D, &\qquad \Gamma_{X^D} s^V &= (D_X s)^V, \\ \Gamma_{s^V} X^D &= 0, &\qquad \Gamma_{s^V} \sigma^V &= 0, \end{aligned} \tag{1}$$

for all vector fields X, Y on BE and all section s, σ of E.

Proof. We use direct evaluation, because we shall need the coordinate expressions in the sequel. Let x^i, y^p be some local linear coordinates on E and $X^i = dx^i$, $Y^p = dy^p$ be the induced coordinates on TE. If

$$dy^p = D^p_{qi}(x) y^q dx^i \tag{2}$$

are the equations of D and $\xi^i(x)\frac{\partial}{\partial x^i}$ or $s^p(x)$ is the coordinate form of X or s, respectively, then $D_X s$ is expressed by

$$\frac{\partial s^p}{\partial x^i}\xi^i - D^p_{qi} s^q \xi^i. \tag{3}$$

The coordinate expression of X^D is

$$\xi^i \frac{\partial}{\partial x^i} + D^p_{qi} y^q \xi^i \frac{\partial}{\partial y^p} \tag{4}$$

and s^V is given by

$$s^p(x)\frac{\partial}{\partial y^p}. \tag{5}$$

Let

$$dX^i = \Lambda^i_{jk} X^j dx^k \tag{6}$$

be the coordinate expression of Λ and let

$$\begin{aligned} dX^i &= (\Gamma^i_{jk} X^j + \Gamma^i_{pk} Y^p) dx^k + (\Gamma^i_{jq} X^j + \Gamma^i_{pq} Y^p) dy^q, \\ dY^p &= (\Gamma^p_{ij} X^i + \Gamma^p_{qj} Y^q) dx^j + (\Gamma^p_{ir} X^i + \Gamma^p_{qr} Y^q) dy^r \end{aligned} \tag{7}$$

be the coordinate expression of Γ. Evaluating (1), we obtain

$$\begin{aligned} &\Gamma^i_{jk} = \Lambda^i_{jk}, \quad \Gamma^p_{ij} = \left(\frac{\partial}{\partial x^j} D^p_{qi} - D^p_{rj} D^r_{qi} + D^p_{qk} \Lambda^k_{ij} \right) y^q, \\ &\Gamma^j_{ip} = \Gamma^j_{pi} = 0, \quad \Gamma^p_{iq} = \Gamma^p_{qi} = D^p_{qi}, \quad \Gamma^i_{pq} = 0, \quad \Gamma^p_{qr} = 0. \end{aligned} \tag{8}$$

This proves the existence and the uniqueness of Γ. □

54.3. Since the difference of two classical linear connections on E is a tensor field of $TE \otimes T^*E \otimes T^*E$, we shall heavily use the gauge natural difference tensors in characterizing all gauge natural operators $Q \oplus QTB \rightsquigarrow QT$.

The projection $T\pi\colon TE \to TBE$ defines the dual inclusion $E \oplus T^*BE \hookrightarrow T^*E$. The contracted curvature $\kappa(D)$ of D is a tensor field of $T^*BE \otimes T^*BE$. On the other hand, the Liouville vector field L of E is a section of TE. Hence $L \otimes \kappa(D)$ is one of the difference tensors we need.

Let Δ be the horizontal form of D in the sense of 31.5, so that Δ is a tensor of $TE \otimes T^*E$. The contracted torsion tensor $\hat{S}$ of Λ is a section of T^*BE and we construct two kinds of tensor product $\Delta \otimes \hat{S}$ and $\hat{S} \otimes \Delta$.

According to 28.13, all natural operators transforming Λ into a section of $T^*BE \otimes T^*BE$ form an 8-parameter family, which we denote by $G(\Lambda)$. Hence $L \otimes G(\Lambda)$ is an 8-parameter family of gauge natural difference tensors. Finally, let $N(\Lambda)$ be the 3-parameter family defined in 45.10.

Proposition. *All gauge natural operators $Q \oplus QTB \rightsquigarrow QT$ form the following 15-parameter family*

$$\begin{aligned} &(1 - k_1)\Gamma(D, N(\Lambda)) + k_1 \bar{\Gamma}(D, \overline{N(\Lambda)}) + k_2 L \otimes \kappa(D) + \\ &\qquad k_3 \Delta \otimes \hat{S} + k_4 \hat{S} \otimes \Delta + L \otimes G(\Lambda) \end{aligned} \tag{1}$$

where bar denotes the conjugate connection.

We remark that the list (1) is essentially simplified if we assume Λ to be without torsion. Then $\hat{S}$ vanishes, $N(\Lambda)$ is reduced to Λ and the 8-parameter family $G(\Lambda)$ is reduced to a two-parameter family generated by the two different contractions R_1 and R_2 of the curvature tensor of Λ. This yields the following

Corollary. *All gauge natural operators transforming a linear connection D on E and a linear symmetric connection Λ on TBE into a linear connection on TE form the following 4-parameter family*

$$(1 - k_1)\Gamma(D, \Lambda) + k_1 \bar{\Gamma}(D, \Lambda) + L \otimes (k_2 \kappa(D) + k_3 R_1 + k_4 R_2). \tag{2}$$

54.4. To prove proposition 54.3, first we take into account that, analogously to 51.16 and 23.7, every gauge natural operator $A\colon Q \oplus QTB \rightsquigarrow QT$ has a finite order. Let $S^r = J_0^r Q(\mathbb{R}^m \times \mathbb{R}^n \to \mathbb{R}^m)$ be the fiber over $0 \in \mathbb{R}^m$ of the r-th jet prolongation of the connection bundle of the vector bundle $\mathbb{R}^m \times \mathbb{R}^n \to \mathbb{R}^m$, let $Z^r = J_0^r T\mathbb{R}^m$ be the fiber over $0 \in \mathbb{R}^m$ of the r-th jet prolongation of the connection bundle of $T\mathbb{R}^m$ and V be the fiber over $0 \in \mathbb{R}^m$ of the connection bundle of $T(\mathbb{R}^m \times \mathbb{R}^n)$ with respect to the total projection $QT(\mathbb{R}^m \times \mathbb{R}^n) \to (\mathbb{R}^m \times \mathbb{R}^n) \to \mathbb{R}^m$. Then all S^r, Z^r, $\mathbb{R}^n$ and V are $W_m^{r+1}(GL(n)) =: W_{m,n}^{r+1}$-spaces. In fact, $W_{m,n}^{r+1}$ acts on Z^r by means of the base homomorphisms into G_m^{r+1}, on $\mathbb{R}^n$ by means of the canonical projection into $GL(n)$ and on V by means of the jet homomorphism into $W_{m,n}^1$. The r-th order gauge natural operators $A\colon Q \oplus QTB \rightsquigarrow QT$ are in bijection with $W_{m,n}^{r+1}$-equivariant maps (denoted by the same symbol) $A\colon S^r \times Z^r \times \mathbb{R}^n \to V$ satisfying $q \circ A = \mathrm{pr}_3$, where $q\colon V \to \mathbb{R}^n$ is the canonical projection.

Formula 54.2.(2) induces on S^r the jet coordinates

$$D_\alpha = (D^p_{qi\alpha}), \qquad 0 \le |\alpha| \le r \tag{1}$$

where α is a multi index of range m. On Z^r, 54.2.(6) induces analogously the coordinates

$$\Lambda_\beta = (\Lambda^i_{jk\beta}), \qquad 0 \le |\beta| \le r. \tag{2}$$

On V, we consider the coordinates $y = (y^p)$ and

$$\Gamma^A_{BC}, \qquad A,\ B,\ C = 1, \dots, m+n \tag{3}$$

given by 54.2.(7). Hence the coordinate expression of any smooth map $f\colon S^r \times Z^r \times \mathbb{R}^n \to V$ satisfying $q \circ f = \mathrm{pr}_3$ is $y^p = y^p$ and

$$\Gamma^A_{BC} = f^A_{BC}(D_\alpha, \Lambda_\beta, y). \tag{4}$$

The coordinate form of a linear isomorphism of vector bundle $\mathbb{R}^m \times \mathbb{R}^n \to \mathbb{R}^m$ is

$$\bar{x}^i = f^i(x), \quad \bar{y}^p = f^p_q(x) y^q. \tag{5}$$

The induced coordinates on $W_{m,n}^{r+1}$ are

$$a^i_\alpha = \partial_\alpha f^i(0), \quad a^p_{q\beta} = \partial_\beta f^p_q(0), \qquad 0 < |\alpha| \le r+1,\ 0 \le |\beta| \le r+1.$$

The above-mentioned homomorphism $W_{m,n}^{r+1} \to G_m^{r+1}$ consists in suppressing the coordinates $a^p_{q\beta}$.

The standard action of G_m^2 on Z^0 is given by 25.2.(3). The action of $W_{m,n}^1$ on S^0 is a special case of 52.1.(5) for the group $G = GL(n)$. This yields

$$\bar{D}^p_{qi} = a^p_r \Gamma^r_{sj} \tilde{a}^s_q \tilde{a}^j_i + a^p_{rj} \tilde{a}^r_q \tilde{a}^j_i. \tag{6}$$

The canonical action of $GL(n)$ on $\mathbb{R}^n$ is

$$\bar{y}^p = a^p_q y^q. \tag{7}$$

Using standard evaluation, we deduce from 54.2.(7) that the action of $W^1_{m,n}$ on V is (7) and

$$a^i_{jk} + a^i_l \Gamma^l_{jk} = \bar{\Gamma}^i_{lm} a^l_j a^m_k + \bar{\Gamma}^i_{pl} a^p_{qj} y^q a^l_k + \bar{\Gamma}^i_{lp} a^l_j a^p_{qk} y^q + \bar{\Gamma}^i_{pq} a^p_{rj} y^r a^q_{sk} y^s \tag{8}$$

$$a^i_l \Gamma^l_{pj} = \bar{\Gamma}^i_{qk} a^q_p a^k_j + \bar{\Gamma}^i_{qr} a^q_p a^r_{sj} y^s \tag{9}$$

$$a^i_l \Gamma^l_{jp} = \bar{\Gamma}^i_{kq} a^k_j a^q_p + \bar{\Gamma}^i_{qr} a^q_{sj} y^s a^r_p \tag{10}$$

$$a^i_l \Gamma^l_{pq} = \bar{\Gamma}^i_{rs} a^r_p a^s_q \tag{11}$$

$$\begin{aligned} a^p_{qij} y^q + a^p_{qk} y^q \Gamma^k_{ij} + a^p_q \Gamma^q_{ij} = {} & \bar{\Gamma}^p_{kl} a^k_i a^l_j + \bar{\Gamma}^p_{ql} a^q_{ri} y^r a^l_j + \\ & \bar{\Gamma}^p_{kq} a^k_i a^q_{rj} y^r + \bar{\Gamma}^p_{qr} a^q_{si} y^s a^r_{tj} y^t \end{aligned} \tag{12}$$

$$a^p_{rk} y^r \Gamma^k_{qi} + a^p_{qi} + a^p_r \Gamma^r_{qi} = \bar{\Gamma}^p_{rk} a^r_q a^k_i + \bar{\Gamma}^p_{rs} a^r_q a^s_{ti} y^t \tag{13}$$

$$a^p_{rk} y^r \Gamma^k_{iq} + a^p_{qi} + a^p_r \Gamma^r_{iq} = \bar{\Gamma}^p_{jr} a^j_i a^r_q + \bar{\Gamma}^p_{rs} a^r_{ti} y^t a^s_q \tag{14}$$

$$a^p_{sj} y^s \Gamma^j_{qr} + a^p_s \Gamma^s_{qr} = \bar{\Gamma}^p_{st} a^s_q a^t_r \tag{15}$$

54.5. Let $H \subset W^{r+1}_{m,n}$ be the subgroup determined by the $(r+1)$-th jets of the products of linear isomorphisms on both $\mathbb{R}^m$ and $\mathbb{R}^n$, which is canonically isomorphic to $GL(m) \times GL(n)$. The standard prolongation procedure and 54.5.(8)–(15) imply that the actions of H on $D^p_{qi\alpha}$, $\Lambda^i_{jk\beta}$ and Γ^A_{BC} are tensorial.

Consider the equivariance of f^i_{pq} with respect to the fiber homotheties. This yields

$$k^{-2} f^i_{pq} = f^i_{pq}(D_\alpha, \Lambda_\beta, ky).$$

Multiplying by k^2 and letting $k \to 0$, we obtain

$$\Gamma^i_{pq} = f^i_{pq} = 0. \tag{1}$$

The equivariance of f^i_{jp} with respect to the fiber homotheties gives

$$k^{-1} f^i_{jp} = f^i_{jp}(D_\alpha, \Lambda_\beta, ky).$$

This implies in the same way

$$\Gamma^i_{jp} = 0. \tag{2}$$

For f^i_{pj} and f^p_{qr} we find quite similarly

$$\Gamma^i_{pj} = 0, \quad \Gamma^p_{qr} = 0. \tag{3}$$

For f^p_{qi} the fiber homotheties give

$$f^p_{qi} = f^p_{qi}(D_\alpha, \Lambda_\beta, ky).$$

Letting $k \to 0$ we find f^p_{qi} are independent of y^p. Then the base homotheties yield

$$kf^p_{qi} = f^p_{qi}(k^{1+|\alpha|}D_\alpha, k^{1+|\beta|}\Lambda_\beta).$$

By the homogeneous function theorem, f^p_{qi} are linear in D^p_{qi}, Λ^i_{jk} and independent of D_α, Λ_β with $|\alpha| > 0$, $|\beta| > 0$. By the generalized invariant tensor theorem, we obtain

$$f^p_{qi} = aD^p_{qi} + b\delta^p_q D^r_{ri} + c\delta^p_q \Lambda^j_{ji} + d\delta^p_q \Lambda^j_{ij}. \tag{4}$$

Let $K \subset W^{r+1}_{m,n}$ be the subgroup characterized by $a^i_j = \delta^i_j$, $a^p_q = \delta^p_q$. By 25.2.(3), 54.4.(6) and 54.4.(13), the equivariance of (4) on K reads

$$a^p_{qi} = aa^p_{qi} + b\delta^p_q a^r_{ri} + c\delta^p_q a^j_{ji} + d\delta^p_q a^i_{ij}.$$

This implies $a = 1$, $b = 0$, $c + d = 0$, i.e.

$$\Gamma^p_{qi} = D^p_{qi} + c_1\delta^p_q(\Lambda^j_{ji} - \Lambda^j_{ij}), \qquad c_1 \in \mathbb{R}. \tag{5}$$

For f^p_{iq} we deduce in the same way

$$\Gamma^p_{iq} = D^p_{qi} + c_2\delta^p_q(\Lambda^j_{ji} - \Lambda^j_{ij}), \qquad c_2 \in \mathbb{R}. \tag{6}$$

The fiber homotheties yield that f^i_{jk} is independent of y^p. Then the base homotheties imply that f^i_{jk} is linear in D^p_{qi}, Λ^i_{jk} and independent of $D^p_{qi\alpha}$, $\Lambda^i_{jk\beta}$ with $|\alpha| > 0$, $|\beta| > 0$. By the generalized invariant tensor theorem, we obtain

$$\begin{aligned} f^i_{jk} = {} & a\Lambda^i_{jk} + b\Lambda^i_{kj} + c\delta^i_j\Lambda^l_{lk} + d\delta^i_j\Lambda^l_{kl} + \\ & e\delta^i_k\Lambda^l_{lj} + f\delta^i_k\Lambda^l_{jl} + g\delta^i_j D^p_{pk} + h\delta^i_k D^p_{pj}. \end{aligned} \tag{7}$$

By 25.2.(3), 54.4.(6) and 54.4.(8), the equivariance of (7) on K reads

$$a^i_{jk} = (a+b)a^i_{jk} + (c+d)\delta^i_j a^l_{lk} + (e+f)\delta^i_k a^l_{lj} + g\delta^i_j a^p_{pk} + h\delta^i_k a^p_{pj}.$$

This implies $a + b = 1$, $c + d = e + f = g = h = 0$, i.e.

$$\Gamma^i_{jk} = (1 - c_3)\Lambda^i_{jk} + c_3\Lambda^i_{kj} + c_4\delta^i_j(\Lambda^l_{lk} - \Lambda^l_{kl}) + c_5\delta^i_k(\Lambda^l_{lj} - \Lambda^l_{jl}). \tag{8}$$

54.6. The study of f^p_{ij} is quite analogous to 54.5, but it leads to more extended evaluations. That is why we do not perform all of them in detail here. The fiber homotheties yield

$$kf^p_{ij} = f^p_{ij}(D_\alpha, \Lambda_\beta, ky).$$

By the homogeneous function theorem, f^p_{ij} is linear in y^p, i.e.

$$f^p_{ij} = F^p_{ijq}(D_\alpha, \Lambda_\beta)y^q.$$

The base homotheties then imply

$$k^2 F^p_{ijq} = F^p_{ijq}(k^{1+|\alpha|} D_\alpha, k^{1+|\beta|}\Lambda_\beta).$$

By the homogeneous function theorem, F^p_{ijq} is linear in D^p_{qij}, Λ^i_{jkl}, bilinear in D^p_{qi}, Λ^i_{jk} and independent of $D^p_{qi\alpha}$, $\Lambda^i_{jk\beta}$ with $|\alpha| > 1$, $|\beta| > 1$. Using the generalized invariant tensor theorem, we obtain F^p_{ijq} in the form of a 40-parameter family. The equivariance of f^p_{ij} with respect to K then yields

$$\begin{aligned}(1)\qquad \Gamma^p_{ij} = \Big(&(1-c_6)D^p_{qij} + c_6 D^p_{qji} + c_7\delta^p_q(D^r_{rij} - D^r_{rji}) - \\ &c_6 D^p_{ri} D^r_{qj} + (c_6 - 1)D^p_{rj} D^r_{qi} + (1-c_3)D^p_{qk}\Lambda^k_{ij} + c_3 D^p_{qk}\Lambda^k_{ji} + \\ &(c_4 - c_1)D^p_{qi}(\Lambda^l_{lj} - \Lambda^l_{jl}) + (c_5 - c_2)D^p_{qj}(\Lambda^l_{li} - \Lambda^l_{il}) + \\ &\delta^p_q G_{ij}(\Lambda)\Big) y^q\end{aligned}$$

where $G_{ij}(\Lambda)$ is the coordinate form of $G(\Lambda)$.

One verifies directly that (1) and 54.5.(1)–(3), (5), (6), (8) is the coordinate expression of 54.3.(1). □

54.7. The case of principal bundles. An analogous problem is to study the gauge natural operators transforming a connection D on a principal G-bundle $\pi\colon P \to BP$ and a classical linear connection Λ on the base manifold BP into a classical linear connection on the total space P. First we present a geometrical construction of such an operator.

Let vA be the vertical component of a vector $A \in T_yP$ and bA be its projection to the base manifold. Consider a vector field X on BP such that $j^1_x X = \Lambda(bA)$, $x = \pi(y)$. Construct the lift X^D of X and the fundamental vector field $\varphi(vA)$ determined by vA. An easy calculation shows that the rule

$$(1)\qquad A \mapsto j^1_y(X^D + \varphi(vA))$$

determines a classical linear connection $N_P(D,\Lambda)\colon TP \to J^1(TP \to P)$ on P.

54.8. We are going to determine all gauge natural operators of the above type. The result of 54.3 suggests us that the case Λ is without torsion is much simpler than the general case. That is why we restrict ourselves to a symmetric Λ. Since the difference of two classical linear connections on P is a tensor field of type $TP \otimes T^*P \otimes T^*P$, we characterize all gauge natural operators in question as a sum of the operator N from 54.7 and of the gauge natural difference tensor fields. We construct geometrically the following 3 systems of difference tensor fields.

I. The connection form of D is a linear map $\omega\colon TP \to \mathfrak{g}$. Take any bilinear map $f_1 : \mathfrak{g} \times \mathfrak{g} \to \mathfrak{g}$ and compose $\omega \oplus \omega$ with f_1. This defines an n^3-parameter system of difference tensor fields $TP \otimes TP \to VP$, $n = \dim G$.

II. The curvature form $D\omega$ of ω is a bilinear map $TP \oplus TP \to \mathfrak{g}$. Take any linear map $f_2 \colon \mathfrak{g} \to \mathfrak{g}$ and compose $D\omega$ with f_2. This yields an n^2-parameter system of difference tensor fields.

III. By 28.7, all natural operators transforming a linear symmetric connection Λ on BP into a tensor field of $T^*BP \otimes T^*BP$ form a 2-parameter family linearly generated by both different contractions R_1 and R_2 of the curvature tensor of Λ. The tangent map of the bundle projection $P \to BP$ defines the dual injection $P \oplus T^*BP \to T^*P$. Taking any fundamental vector field $\bar{Y}$ determined by a vector $Y \in \mathfrak{g}$, we obtain a $2n$-parameter system of difference tensor fields linearly generated by $\bar{Y} \otimes R_1$ and $\bar{Y} \otimes R_2$.

54.9. Proposition. *All gauge natural operators transforming a connection on P and a classical linear symmetric connection of the base manifold BP into a classical linear connection on P form the $(n^3 + n^2 + 2n)$-parameter family generated by operator N and by the above families I, II, and III of the difference tensor fields.*

The proof consists in straightforward application of our techniques, but it is too long to be performed here. We refer the reader to [Kolář, to appear a].

Remarks

Our approach to gauge natural bundles and operators generalizes directly the theory of natural bundles. So we also prove the regularity originally assumed in [Eck, 81]. Let us mention that, analogously to chapter XI, we can define the Lie derivative of sections of gauge natural bundles with respect to the right invariant vector fields on the corresponding principal fiber bundles and then the infinitesimally gauge natural operators. The relation between the gauge naturality and infinitesimal gauge naturality is similar to the case of natural bundles if the gauge group is connected; more information can be found in [Cap, Slovák, 92].

The first application of our methods for finding gauge natural operators was presented in [Kolář, 87b]. The considerations in that paper are restricted to the case the structure group is the general linear group $GL(n)$ in an arbitrary dimension (independent of the dimension of the base manifold), for in such a case one can apply directly the results from chapter VI. [Kolář, 87b] has also determined all $GL(n)$-natural operators transforming a principal connection on a principal bundle P and a classical linear connection on the base manifold into a principal connection on W^1P. The curvature-like operators were found in the special case $G = GL(n)$ in [Kolář, 87b] and the general problem was solved in [Kolář, to appear a]. The greater part of the results from section 52 was deduced in [Kolář, to appear b]. Proposition 53.3 was proved for the special case $G = GL(n)$ in [Kolář, 91], the general result is first presented in this book. Section 54 is based on [Gancarzewicz, Kolář, 91].

References

Albert, C.; Molino, P., *Pseudogroupes de Lie transitifs I. Structures differentiables*, Collection Travaux en cours, Hermann, Paris, 1984.

Atiyah, M.; Bott, R.; Patodi, V.K., *On the heat equation and the index theorem*, Inventiones Math. **19** (1973), 279–330.

Baston, R. J., *Verma modules and differential conformal invariants*, J. Differential Geometry **32** (1990), 851–898.

Baston, R. J., *Almost Hermitian symmetric manifolds, I: Local twistor theory; II: Differential invariants*, Preprints (1990).

Baston, R.J.; Eastwood, M.G., *Invariant operators*, Twistors in mathematics and physics, Lecture Notes in Mathematics 156, Cambridge University Press, 1990.

Boe, B. D.; Collingwood, D. H., *A comparison theory for the structure of induced representations I.*, J. of Algebra **94** (1985), 511-545.

Boe, B. D.; Collingwood, D. H., *A comparison theory for the structure of induced representations II.*, Math. Z. **190** (1985), 1-11.

Boerner, H., *Darstellungen von Gruppen*, 2nd ed., Springer, Grundlehren der math. Wissenschaften 74, Berlin Heidelberg New York, 1967.

Boman, J., *Differentiability of a function and of its composition with a function of one variable*, Math. Scand. **20** (1967), 249–268.

Branson, T. P., *Differential operators canonically associated to a conformal structure*, Math. Scand. **57** (1985), 293–345.

Branson, T. P., *Conformal transformations, conformal change, and conformal covariants*, Proceedings of the Winter School on Geometry and Physics, Srní 1988, Suppl. Rendiconti Circolo Mat. Palermo, Serie II **21** (1989), 115–134.

Branson T. P., *Second-order conformal covariants I., II.*, Københavns universitet matematisk institut, Preprint Series, No. 2, 3, (1989).

Bröcker, T.; Jänich, K., *Einführung in die Differentialtopologie*, Springer-Verlag, Heidelberg, 1973.

Cabras, A.; Canarutto, D.; Kolář, I.; Modugno, M., *Structured bundles*, Pitagora Editrice, Bologna, 1991.

Cahen, M.; de Wilde, M.; Gutt, S., *Local cohomology of the algebra of C^∞-functions on a connected manifold*, Lett. Math. Phys. **4** (1980), 157–167.

Cantrijn, F.; Crampin, M.; Sarlet, W; Saunders, D., *The canonical isomorphism between T^kT^*M and T^*T^kM*, CRAS Paris, Série II **309** (1989), 1509–1514.

Cap, A., *All linear and bilinear natural concomitants of vector valued differential forms*, Comment. Math. Univ. Carolinae **31** (1990), 567–587.

Cap, A., *Natural operators between vector valued differential forms*, Proccedings of the Winter School on Geometry and Physics (Srní 1990), Rend. Circ. Mat. Palermo (2) Suppl. **26** (1991), 113–121.

Cap, A.; Slovák, J., *Infinitesimally natural operators are natural*, Diff. Geo. Appl. **2** (1992), 45–55.

Cartan, É., *Leçons sur la théorie des espaces à connexion projective*, Paris, 1937.

Chrastina, J., *On a theorem of J. Peetre*, Arch. Math. (Brno) **23** (1987), 71–76.

Cohen, R., *A proof of the immersion conjecture*, Proc. Nat. Acad. Sci. USA **79** (1982), 3390–3392.

Crittenden, R., *Covariant differentiation*, Quart. J. Math. Oxford Ser. **13** (1962), 285–298.

Dekrét, A., *On canonical forms on non-holonomic and semi-holonomic prolongations of principal fibre bundles*, Czech. Math. J. **22** (1972), 653–662.

Dekrét, A., *Mechanical structures and connections*, Proceedings, Dubrovnik, Yugoslavia, 1988, pp. 121–132.

Dekrét, A., *Vector fields and connections on fibred manifolds*, Suppl. Rendiconti del Circolo Matematico di Palermo, Serie II **22** (1990), 25–34.

Dekrét, A., *On mechanical systems of second order on manifolds*, to appear in Proc. Conference Eger.

Dekrét, A., *Vector fields and connections on TM*, to appear in Časopis pěst. mat.

Dekrét, A.; Vosmanská, G., *Natural transformations of 2-quasijets*, to appear.

de León, M.; Rodrigues, P. R., *Generalized Classical Mechanics and Field Theory*, North-Holland Mathematical Studies, 112, Amsterdam, 1985.

de León, M.; Rodrigues, P. R., *Dynamical connections and non-autonomous Lagrangian systems*, Ann. Fac. Sci. Toulouse Math. **IX** (1988), 171–181.

de Wilde, M.; Lecomte, P., *Algebraic characterizations of the algebra of functions and of the Lie algebra of vector fields on a manifold*, Compositio Math. **45** (1982), 199–205.

Dieudonné, J. A., *Foundations of modern analysis, I*, Academic Press, New York – London, 1960.

Dieudonné, J. A.; Carrell, J. B., *Invariant Theory, Old and New*, Academic Press, New York – London, 1971.

Donaldson, S., *An application of gauge theory to the topology of 4-manifolds*, J. Diff. Geo. **18** (1983), 269–316.

Doupovec, M., *Natural operators transforming vector fields to the second order tangent bundle*, Čas. pěst. mat. **115** (1990), 64–72.

Doupovec, M., *Natural transformations between $T_1^2T^*M$ and $T^*T_1^2M$*, Ann. Polon. Math. **LVI** (1991), 67-77.

Doupovec, M., *Natural transformations of third tangent and cotangent bundles*, to appear in Czechoslovak Math. J.

Doupovec, M.; Kolář, I., *On the natural operators transforming connections to the tangent bundle of a fibred manifold*, Knižnice VUT Brno **B-119** (1988), 47–56.

Doupovec, M.; Kolář, I., *Natural affinors on time-dependent Weil bundles*, to appear in Arch. Math. (Brno).

Eastwood, M. G.; Graham, C. R., *Invariants of CR Densities*, Proccedings of Symposia in Pure Mathematics, Part 2 **52** (1991), 117–133.

Eastwood, M. G.; Graham, C. R., *Invariants of conformal densities*, Duke Math. J. **63** (1991), 633–671.

Eastwood, M. G.; Rice, J. W., *Conformally invariant differential operators on Minkowski space and their curved analogues*, Commun. Math. Phys. **109** (1987), 207–228.

Eastwood, M. G.; Tod, P., *Edth – a differential operator on the sphere*, Math. Proc. Cambr. Phil. Soc. **92** (1982), 317–330.

Eck, D. J., *Gauge-natural bundles and generalized gauge theories*, Mem. Amer. Math. Soc. **247** (1981).

Eck, D. J., *Invariants of k-jet actions*, Houston J. Math. **10** (1984), 159–168.

Eck, D. J., *Product preserving functors on smooth manifolds*, J. Pure and Applied Algebra **42** (1986), 133–140.

Eck, D. J., *Natural sheaves*, Illinois J. Math. **31** (1987), 200–207.

Ehresmann, C., *Les connections infinitésimales dans un espace fibré différentiable*, Coll. de Topo., Bruxelles, C.B.R.M., 1950, pp. 29–55.

Ehresmann, C., *Les prolongements d'une variété différentiable, I: Calcul des jets, prolongement principal*, C. R. A. S. Paris **233** (1951), 598–600.

Ehresmann, C., *Les prolongements d'une variété différentiable, II: L'espace des jets d'ordre r de V_n dans V_m*, C. R. A. S. Paris **233** (1951), 777–779.

Ehresmann, C., *Structures locales et structures infinitésimales*, C. R. A. S. Paris **234** (1952), 1081–1083.

Ehresmann, C., *Les prolongements d'une variété différentiable, IV: Eléments de contact et éléments d'enveloppe*, C. R. A. S. Paris **234** (1952), 1028–1030.

Ehresmann, C., *Extension du calcul des jets aux jets non holonomes*, C. R. A. S. Paris **239** (1954), 1762–1764.

Ehresmann, C., *Les prolongements d'un espace fibré différentiable*, C. R. A. S. Paris (1955), 1755–1757.

Ehresmann, C., *Sur les connexions d'ordre supériour*, Atti V Cong. Un. Mat. Italiana, Pavia–Torino, 1956, pp. 326–328.

Ehresmann, C., *Gattungen von Lokalen Strukturen*, Jahresber. d. Deutschen Math. **60-2** (1957), 49–77.

Eilenberg, S., *Foundations of fiber bundles*, University of Chicago, Summer 1957.

Epstein, D. B. A., *Natural vector bundles*, Category Theory, Homology Theory and their Applications III,, Lecture Notes in Mathematics 99, 1969, pp. 171–195.

Epstein, D.B.A., *Natural tensors on Riemannian manifolds*, J. Diff. Geom. **10** (1975), 631–645.

Epstein, D. B. A.; Thurston W. P., *Transformation groups and natural bundles*, Proc. London Math. Soc. **38** (1979), 219–236.

Fefferman, C., *Parabolic invariant theory in complex analysis*, Adv. Math. **31** (1979), 131–262.

Fefferman, C.; Graham, R., *Conformal invariants*, Élie Cartan et les Mathématiques d'Aujourd'hui, Astérisque, hors serie, 1985, pp. 95–116.

Fegan, H. D., *Conformally invariant first order differential operators*, Quart. J. Math. **27** (1976), 371–378.

Feigin, B. L.; Fuks, D. B., *On invariant differential operators on the line*, (Russian), Funct. Anal. and its Appl. **13 No 4** (1979), 91–92.

Ferraris, M.; Francaviglia, M.; Reina, C., *Sur les fibrés d'objets géometriques et leurs applications physiques*, Ann. Inst. Henri Poincaré, Section A **38** (1983), 371–383.

Freedman, M. H., *The topology of four dimensional manifolds*, J. Diff. Geo. **17** (1982), 357–454.

Freedman, M. H.; Luo, Feng, *Selected Applications of Geometry to Low-Dimensional Topology*, University Lecture Series 1, Amer. Math. Soc., 1989.

Frölicher, A., *Smooth structures*, Category theory 1981, LN 962, Springer-Verlag, pp. 69–82.

Frölicher, A.; Kriegl, A., *Linear spaces and differentiation theory*, Pure and Applied Mathematics, J. Wiley, Chichester, 1988.

Frölicher, A.; Nijenhuis, A., *Theory of vector valued differential forms. Part I.*, Indagationes Math **18** (1956), 338–359.

Frölicher, A.; Nijenhuis, A., *Invariance of vector form operations under mappings*, Comm. Math. Helv. **34** (1960), 227–248.

Fuks, D. B., *Cohomology of infinite dimensional Lie algebras*, (Russian), Nauka, Moscow, 1984.

Gancarzewicz, J., *Local triviality of a bundle of geometric objects*, Zeszyty Nauk. Uniw. Jagiellon. Prace Mat. **23** (1982), 39–42.

Gancarzewicz, J., *Des relévements des fonctions aux fibrés naturels*, CRAS Paris, Série I **295** (1982), 743–746.

Gancarzewicz, J., *Liftings of functions and vector fields to natural bundles*, Warszawa 1983, Dissertationes Mathematicae **CCXII**.

Gancarzewicz, J., *Relévements des champs de vecteurs aux fibrés naturels*, CRAS Paris, Série I **296** (1983), 59–61.

Gancarzewicz, J., *Differential Geometry (Polish)*, PWN, Warszawa, 1987.

Gancarzewicz, J., *Horizontal lifts of linear connections to the natural vector bundles*, Research Notes in Math. 121, Pitman, 1985, pp. 318-341.

Gancarzewicz, J.; Mahi, S., *Lifts of 1-forms to the tangent bundle of higher order*, to appear.

Gancarzewicz, J.; Kolář, I., *Some gauge-natural operators on linear connections*, Monatsh. Math. **111** (1991), 23–33.

Gancarzewicz, J.; Kolář, I., *Natural affinors on the extended r-th order tangent bundles*, to appear in Rendiconti del Circolo Matematico di Palermo.

Gheorghiev, G., *Sur les prolongements réguliers des espaces fibrés et les groupes de Lie associés*, C. R. Acad. Sci. Paris **266, A** (1968), 65–68.

Gilkey, P. B., *Curvature and the eigenvalues of the Laplacian for elliptic complexes*, Advances in Math. **10** (1973), 344–382.

Gilkey, P. B., *The spectral geometry of a Riemannian manifold*, J. Diff. Geom. **10** (1975), 601–618.

Gilkey, P.B., *Local invariants of a pseudo-Riemannian manifold*, Math. Scand. **36** (1975), 109–130.

Gilkey, P. B., *Invariance Theory, The Heat Equation, And the Atiyah-Singer Index Theorem*, Mathematics Lecture Series 11, Publish or Perish Inc., Wilmington, Delaware, 1984.

Goldschmidt, H; Sternberg, S., *The Hamilton formalism in the calculus of variations*, Ann. Inst. Fourier (Grenoble) **23** (1973), 203–267.

Gollek, H., *Anwendungen der Jet-Theorie auf Faserbündel und Liesche Transformationgruppen*, Math. Nachr. **53** (1972), 161–180.

Gompf, R., *Three exotic* $\mathbb{R}^4$*'s and other anomalies*, J. Diff. Geo. **18** (1983), 317–328.

Gover, A. R., *Conformally invariant operators of standard type*, Quart. J. Math. **40** (1989), 197–208.

Graham, C. R., *Non-existence of curved conformally invariant operators*, Preprint (1990).

Greub, W.; Halperin, S.; Vanstone, R., *Connections, Curvature, and Cohomology I, II, III*, Academic Press, New York and London, 1972, 1973, 1976.

Grozman, P. Ya., *Local invariant bilinear operators on tensor fields on the plane*, (Russian), Vestnik MGU, Ser. Mat., Mech. **No 6** (1980), 3–6.

Grozman, P. Ya., *Classification of bilinear invariant operators on tensor fields*, (Russian), Funct. Anal. and its Appl. **14 No 2** (1980), 58–59.

Gurevich, G. B., *Foundations of the theory of algebraic invariants*, (Russian), OGIZ, Moscow-Leningrad, 1948.

Hilgert, J.; Neeb, K.-H., *Lie-Gruppen und Lie-Algebren*, Vieweg, Braunschweig, 1991.

Hirsch, M. W., *Differential topology*, GTM 33, Springer-Verlag, New York, 1976.

Hochschild, G. P., *Introduction to affine algebraic groups*, Holden Day, San Francisco, 1971.

Jacobson, N., *Lie algebras*, J. Wiley-Interscience, 1962.

Jakobsen, H. P., *Conformal invariants*, Publ. RIMS, Kyoto Univ **22** (1986), 345–364.

Janyška, J., *Geometric properties of prolongation functors*, Časopis pěst. mat. **108** (1983).

Janyška J., *On natural operations with linear connections*, Czechoslovak Math. J. **35** (1985), 106–115.

Janyška J., *Natural and gauge-natural operators on the space of linear connections on a vector bundle*, Differential Geometry and Its Applications, Proc. Conf. Brno 1989, World Scientific Publ. Co., Singapore, 1990, pp. 58–68.

Janyška, J., *Natural operations with projectable tangent valued forms on a fibred manifold*, Annali di Mat. **CLIX** (1991), 171–187.

Janyška J., *Remarks on almost complex connections*, to appear.

Janyška J.; Kolář I., *Lie derivatives on vector bundles*, Proceedings of the conference on differential geometry and its applications, Universita Karlova Praha, 1982, pp. 111–116.

Janyška J.; Kolář I., *On the connections naturally induced on the second order frame bundle*, Arch. Math. (Brno) **22** (1986), 21–28.

Janyška J.; Modugno, M., *Infinitesimal natural and gauge-natural lifts*, to appear.

Joris, H, *Une* C^∞*-application non-immersive qui possède la propriété universelle des immersions.*, Archiv Math. **39** (1982), 269–277.

Kainz, G.; Kriegl, A.; Michor, P. W., C^∞*-algebras from the functional analytic viewpoint*, J. pure appl. Algebra **46** (1987), 89-107.

Kainz, G.; Michor, P. W., *Natural transformations in differential geometry*, Czechoslovak Math. J. **37** (1987), 584-607.

Kirillov, A. A., *On invariant differential operations over geometric quantities*, (Russian), Func. Anal. and its Appl. **11 No 2** (1977), 39–44.

Kirillov, A. A., *Invariant operators over geometric quantities*, Current problems in mathematics, Vol 16, (Russian), VINITI, 1980, pp. 3–29.

Klein, T., *Connections on higher order tangent bundles*, Časopis pěst. mat. **106** (1981), 414–421.

Kobak, P., *Natural liftings of vector fields to tangent bundles of 1-forms*, Mathematica Bohemica **116** (1991), 319–326.

Kobayashi, S., *Theory of connections*, Annali di matematica pura et applicata **43** (1957), 119–185.

Kobayashi, S., *Canonical forms on frame bundles of higher order contact*, Proc. of symposia in pure math., vol III, A.M.S., 1961, pp. 186–193.

Kobayashi, S.; Nomizu, K., *Foundations of Differential Geometry. Vol I*, J. Wiley-Interscience, 1963; *Vol. II*, 1969.
Kock, A., *Synthetic Differential Geometry*, London Mathematical Society Lecture Notes Series 51, Cambridge, 1981.
Kolář, I., *Complex velocities on real manifolds*, Czechoslovak Math. J. **21 (96)** (1971), 118–123.
Kolář, I., *Canonical forms on the prolongations of principal fibre bundles*, Rev. Roumaine Math. Pures Appl. **16** (1971), 1091–1106.
Kolář, I., *On the prolongations of geometric object fields*, An. Sti. Univ. "Al. I. Cuza" Iasi **17** (1971), 437–446.
Kolář, I., *On manifolds with connection*, Czechoslovak Math. J. **23 (98)** (1973), 34–44.
Kolář, I., *On the absolute differentiation of geometric object fields*, Annales Polonici Mathematici **27** (1973), 293–304.
Kolář, I., *On some operations with connections*, Math. Nachr. **69** (1975), 297–306.
Kolář, I., *Generalized G-structures and G-structures of higher order*, Boll. Un. Mat. Ital., Supl fasc. 3 **12** (1975), 245–256.
Kolář, I., *On the invariant method in differential geometry of submanifolds*, Czechoslovak Math. J. **27 (102)** (1977), 96–113.
Kolář, I., *On the automorphisms of principal fibre bundles*, CMUC **21** (1980), 309–312.
Kolář, I., *Structure morphisms of prolongation functors*, Mathematica Slovaca **30** (1980), 83–93.
Kolář, I., *Connections in 2-fibred manifolds*, Arch. Math. (Brno) **XVII** (1981), 23–30.
Kolář, I., *Fiber parallelism and connections*, Proceedings of Colloquium Global Differential Geometry / Global Analysis, West Berlin, Lecture Notes in Mathematics 838, 1981, pp. 165–173.
Kolář, I., *On generalized connections*, Beiträge zur Algebra und Geometrie **11** (1981), 29–34.
Kolář, I., *Lie derivatives and higher order Lagrangians*, Proceedings of the conference on differential geometry and its applications, Universita Karlova Praha, 1982, pp. 117–123.
Kolář, I., *Prolongations of generalized connections*, Colloq. Math. Societatis János Bolyai, 31. Differential Geometry, Budapest (Hungary) 1979, North Holland, 1982, pp. 317–325.
Kolář, I., *On the second tangent bundle and generalized Lie derivatives*, Tensor, N. S. **38** (1982), 98–102.
Kolář, I., *Functorial prolongations of Lie groups and their actions*, Čas. pěst. mat. **108** (1983), 289–293.
Kolář, I., *Natural transformations of the second tangent functor into itself*, Arch. Math. (Brno) **XX** (1984), 169–172.
Kolář, I., *A geometrical version of the higher order Hamilton formalism in fibred manifolds*, Journal of Geometry and Physics **1** (1984), 127–137.
Kolář, I., *Higher order absolute differentiation with respect to generalized connections*, Differential Geometry, Banach Center Publications, Volume 12, Warsaw, 1984, pp. 153–162.
Kolář, I., *Natural operations with connections on second order frame bundles*, Colloquia Mathematica Societatis János Bolyai, 46. Topics in Differential Geometry, Debrecen (Hungary), 1984, pp. 715–732.
Kolář, I., *Covariant approach to natural transformations of Weil functors*, Comment. Math. Univ. Carolin. **27** (1986), 723–729.
Kolář, I., *Some natural operations with connections*, Journal of National Academy of Mathematics, India **5** (1987), 127–141.
Kolář, I., *Some natural operators in differential geometry*, Proc. "Conf. Diff. Geom. and Its Applications", Brno 1986, Dordrecht, 1987, pp. 91–110.
Kolář, I., *On the natural operators on vector fields*, Ann. Global Anal. Geom. **6** (1988), 109–117.
Kolář, I., *Lie derivatives of sectorform fields*, Colloq. math. **LV** (1988), 71–78.
Kolář, I., *Functorial prolongations of r-connections*, Lecture Notes, Universität Leipzig, 1988, 10p.

Kolář, I., *On some gauge-natural operators*, Seminari dell' Instituto di Matematica Applicata "Giovanni Sansone", Firenze 1989, 98 p.

Kolář, I., *General natural bundles and operators*, Proceedings of Conference on Differential Geometry and Applications, Brno 1989, World-Scientific, Singapure, 1990, pp. 69–78.

Kolář, I., *Natural operators transforming vector fields and exterior forms into exterior forms*, Scientific Periodical of Salesian Technical University **64** (1990), 135–139.

Kolář, I., *Gauge-natural operators transforming connections to the tangent bundle*, The Mathematical Heritage of C. F. Gauss, World Scientific Publ. Co., Singapore, 1991, pp. 416–426.

Kolář, I., *Some gauge-natural operators on connections*, Colloquia Mathematica Societatis János Bolyai, 56. Differential Geometry, Eger (Hungary), 1989 (J. Szenthe, L. Tamássy, eds.), János Bolyai Math. Soc. and Elsevier, Budapest and Amsterdam, 1992, pp. 435–446.

Kolář, I., *Gauge natural forms of Chern Weil type*, to appear in Ann. Global Anal. Geom..

Kolář, I., *An abstract characterization of the jet spaces*, to appear.

Kolář, I.; Lešovský, V., *Structure equations of generalized connections*, Čas. pěst. mat. **107** (1982), 253–256.

Kolář, I.; Michor, P. W., *All natural concomitants of vector valued differential forms*, Proceedings of the Winter School on Geometry and Physics, Srní 1987, Suppl. Rendiconti Circolo Mat. Palermo, Serie II **16** (1987), 101-108.

Kolář, I.; Modugno, M., *Natural maps on the iterated jet prolongation of a fibred manifold*, Annali di Matematica **CLVIII** (1991), 151–165.

Kolář, I.; Modugno, M., *Torsions of connections on some natural bundles*, Differential Geometry and Its Applications **2** (1992), 1–16.

Kolář, I.; Radziszewski, Z., *Natural transformations of second tangent and cotangent functors*, Czechoslovak Math. J. **38 (113)** (1988), 274–279.

Kolář, I.; Slovák, J., *On the geometric functors on manifolds*, Proceedings of the Winter School on Geometry and Physics, Srní 1988, Suppl. Rendiconti Circolo Mat. Palermo, Serie II **21** (1989), 223–233.

Kolář, I.; Slovák, J., *Prolongation of vector fields to jet bundles*, Proceedings of the Winter School on Geometry and Physics, Srní 1989,, Suppl. Rendiconti Circolo Mat. Palermo, Serie II **22** (1990), 103–111.

Kolář, I.; Vosmanská, G., *Natural operations with second order jets*, Supplemento ai Rendiconti del Circolo Matematico di Palermo, Serie II **14** (1987), 179–186.

Kolář, I.; Vosmanská, G., *Natural transformations of higher order tangent bundles and jet spaces*, Čas. pěst. mat. **114** (1989), 181–186.

Kosmann-Schwarzbach, Y., *Vector fields and generalized vector fields on fibred manifolds*, Lecture Notes in Mathematics 792, 1980, pp. 307–355.

Koszul, J. L., *Fibre bundles and differential geometry*, Tata Institute of Fundamental Research, Bombay, 1960.

Kowalski, O.; Sekizawa, M., *Natural transformations of Riemannian metrics on manifolds to metrics on linear frame bundles — a classification*, Differential Geometry and Its Applications, Proceedings, D. Reidel Publishing Company, 1987, pp. 149–178.

Kowalski, O.; Sekizawa, M., *Natural transformations of Riemannian metrics on manifolds to metrics on tangent bundles — a classification*, Bulletin of Tokyo Gakugei University, Sect. IV **40** (1988), 1–29.

Kriegl, A., *Die richtigen Räume für Analysis im Unendlich - Dimensionalen*, Monatshefte Math. **94** (1982), 109–124.

Kriegl, A.; Michor, P. W., *Aspects of the theory of infinite dimensional manifolds*, Differential Geometry and Applications **1** (1991), 159–176.

Krupka, D., *Elementary theory of differential invariants*, Arch. Math. (Brno) **XIV** (1978), 207–214.

Krupka, D., *Fundamental vector fields on type fibres of jet prolongations of tensor bundles*, Math. Slovaca **29** (1979), 159–167.

Krupka, D., *On the Lie algebras of higher differential groups*, Bull. Acad. Polon. Sci., Sér. math. **27** (1979), 235–239.

Krupka, D., *Reducibility theorems for differentiable liftings in fiber bundles*, Arch. Math. (Brno) **XV** (1979), 93–106.

Krupka, D., *A remark on algebraic identities for the covariant derivative of the curvature tensor*, Arch. Math. (Brno) **XVI** (1980), 205–212.

Krupka, D., *Finite order liftings in principal fibre bundles*, Beiträge zur Algebra und Geometrie **11** (1981), 21–27.

Krupka, D., *Local invariants of a linear connection*, Colloq. Math. Societatis János Bolyai, 31. Differential Geometry, Budapest (Hungary) 1979, North Holland, 1982, pp. 349–369.

Krupka, D., *Natural Lagrangian structures*, Differential Geometry, Banach Center Publications, volume 12, Warsaw, 1984, pp. 185–210.

Krupka D.; Janyška J., *Lectures on differential invariants*, Univerzita J. E. Purkyně, Brno, 1990.

Krupka, D.; Mikolášová V., *On the uniqueness of some differential invariants: d, [,], ∇*, Czechoslovak Math. J. **34** (1984), 588–597.

Krupka, D.; Trautman, A., *General Invariance of Lagrangian Structures*, Bull. Acad. Polon. Sci., Sér. math. **22** (1974), 207–211.

Kurek, J., *On natural operators on sectorform fields*, Čas. pěst. mat. **115** (1990), 232–239.

Kurek, J., *On the natural operators of vertical prolongation type on 1-connection bundle*, Demonstratio Mathematica **XXIII** (1990), 161–173.

Kurek, J., *On the natural operators of Bianchi type*, Arch. Math (Brno) **27a** (1991), 25–29.

Kurek, J., *On gauge-natural operators of curvature type on pairs of connections*, to appear in Matematica Slovaca.

Kurek, J., *Natural transformations of higher order cotangent functors*, to appear in Annales Polonici Matematici.

Kurek, J., *Natural affinors on higher order cotangent bundles*, to appear in Annales UMCS Lublin.

Laptěv, G. F., *Structure equations of principal fibre bundles*, Russian, Trudy Geometr. Sem., II, Moscow, 1969, pp. 161-178.

Lecomte, P. B. A., *Sur la suite exacte canonique associée à un fibré principal*, Bull. Soc. Math. France **113** (1985), 259–271.

Lecomte, P. B. A., *Applications of the cohomology of graded Lie algebras to formal deformations of Lie algebras*, Letters in Math. Physics **13** (1987), 157–166.

Lecomte, P., *Accessibilité et gouvernabilité en theorie du controle*, Notes, Université de Liége.

Lecomte, P.; Michor, P. W.; Schicketanz, H., *The multigraded Nijenhuis-Richardson Algebra, its universal property and application*, J. Pure Applied Algebra **77** (1992), 87–102.

Leicher, H., *Natural operations on covariant tensor fields*, J. Differential Geometry **8** (1973), 117–123.

Leites, D. A., *Formulas for characters of irreducible finite dimensional representations of Lie superalgebras*, (Russian), Funct. Anal. and its Appl. **14 No 2** (1980), 35–38.

Libermann, P., *Sur la géométrie des prolongements des espaces fibrés vectoriels*, Ann. Inst. Fourier, Grenoble **14** (1964), 145–172.

Libermann, P., *Sur les prolongements des fibrés principaux et des groupoides différentiables banachiques*, Analyse global, Sém. Math. Supérieures, No. 42 (Été, 1969), Presses Univ. Montreal (1971), 7–108.

Libermann, P., *Parallélismes*, J.Differential Geometry **8** (1973), 511–539.

Lichnerowicz, A., *Global theory of connections and holonomy groups*, Nordhoff International Publshing, 1976.

Lubczonok, G., *On the reduction theorems*, Ann. Polon. Math. **26** (1972), 125–133.

Luciano, O. O., *Categories of multiplicative functors and Weil's infinitely near points*, Nagoya Math. J. **109** (1988), 69–89.

Luna, D., *Fonctions différentiables invariantes sous l'operation d'un groupe réductif*, Ann. Inst. Fourier, Grenoble **26** (1976), 33–49.

Malgrange, B., *Ideals of differentiable functions*, Oxford University Press, 1966.

Mangiarotti, L.; Modugno, M., *Some results on calculus of variations on jet spaces*, Ann. Inst. H. Poincaré **39 (1)** (1983), 29–43.

Mangiarotti, L.; Modugno, M., *New operators on jet spaces*, Ann. Fac. Scien. Toulouse **2** (5) (1983), 171-198.

Mangiarotti, L.; Modugno, M., *Graded Lie algebras and connections on a fibred space*, Journ. Math. Pures et Appl. **83** (1984), 111–120.

Mangiarotti, L.; Modugno, M., *Fibered spaces, Jet spaces and Connections for Field Theories*, Proceedings of the International Meeting on Geometry and Physics, Florence 1982, Pitagora, Bologna, 1983.

Mangiarotti, L.; Modugno, M., *On the geometric structure of gauge theories*, J. Math. Phys. **26** (6) (1985), 1373–1379.

Mangiarotti, L.; Modugno, M., *Connections and differential calculus on fibered manifolds. Applications to field theory.*, preprint 1989.

Marathe, K.B.; Modugno, M., *Polynomial connections on affine bundles*, Preprint 1988.

Mauhart, M., *Iterierte Lie Ableitungen und Integrabilität*, Diplomarbeit, Universität Wien, 1990.

Mauhart, M.; Michor, P. W., *Commutators of flows and fields*, to appear, Archivum Math. (Brno) (1992).

Michor, P. W., *Manifolds of differentiable mappings*, Shiva, Orpington, 1980.

Michor, P. W., *Manifolds of smooth mappings IV: Theorem of De Rham*, Cahiers Top. Geo. Diff. **24** (1983), 57–86.

Michor, P. W., *Remarks on the Frölicher-Nijenhuis bracket*, Proceedings of the Conference on Differential Geometry and its Applications, Brno 1986, D. Reidel, 1987, pp. 197–220.

Michor, P. W., *Remarks on the Schouten-Nijenhuis bracket*, Supplemento ai Rendiconti del Circolo Matematico di Palermo, Serie II **16** (1987), 207–215.

Michor, P. W., *Gauge theory for diffeomorphism groups*, Proceedings of the Conference on Differential Geometric Methods in Theoretical Physics, Como 1987 (K. Bleuler, M. Werner, eds.), Kluwer, Dordrecht, 1988, pp. 345–371.

Michor, P. W., *Graded derivations of the algebra of differential forms associated with a connection*, Differential Geometry, Proceedings, Peniscola 1988,, Springer Lecture Notes 1410, 1989a, pp. 249–261.

Michor, P. W., *Knit products of graded Lie algebras and groups*, Proceedings of the Winter School on Geometry and Physics, Srní 1989, Suppl. Rendiconti Circolo Matematico di Palermo, Ser. II **22** (1989b), 171–175.

Michor, P. W., *The relation between systems and associated bundles*, to appear, Annali di Matematica (Firenze).

Michor, P. W., *Gauge theory for fiber bundles*, Monographs and Textbooks in Physical Sciences, Vol. 19, Bibliopolis, Napoli, 1991.

Mikulski, W. M., *Locally determined associated spaces*, J. London Math. Soc. **32** (1985), 357–364.

Mikulski, W. M., *There exists a prolongation functor of infinite order*, Časopis pěst. mat. **114** (1989), 57–59.

Mikulski, W. M., *Natural transformations of Weil functors into bundle functors*, Supplemento di Rendiconti del Circolo Matematico di Palermo, Serie II **22** (1989), 177–191.

Mikulski, W. M., *Natural transformations transforming functions and vector fields to functions on some natural bundles*, to appear.

Mikulski, W. M., *Some natural operations on vector fields*, to appear.

Milnor, J., *On manifolds homeomorphic to the 7-sphere*, Annals of Math. **64** (1956), 399–405.

Milnor, J. W.; Stasheff J. D., *Characteristic classes*, Annals of Mathematics Studies 76, Princeton, 1974.

Modugno, M., *An introduction to systems of connections*, Sem Ist. Matem. Appl. Firenze **7** (1986), 1–76.

Modugno, M., *Systems of vector valued forms on a fibred manifold and applications to gauge theories*,, Lecture Notes in Math., vol. 1251, Springer-Verlag, 1987, pp. 238–264.

Modugno, M., *New results on the theory of connections: systems, overconnections and prolongations*, Differential Geometry and Its Applications, Proceedings, D. Reidel Publishing Company, 1987, pp. 243–269.

Modugno, M., *Systems of connections and invariant Lagrangians*, Differential geometric methods in theoretical physics, Proc. XV. Conf. Clausthal 1986, World Scientific Publishing, Singapore, 1987.

Modugno, M., *Linear overconnections*, Proceedings Journées Relat. Toulouse, 1988, pp. 155–170.

Modugno, M., *Jet involution and prolongation of connections*, Časopis Pěst. Mat. **114** (1989), 356–365.

Modugno, M., *Non linear connections*, Ann. de Physique, Colloque no 1, supplément au no 6 **14** (1989), 143–147.

Modugno, M., *Torsion and Ricci tensor for non-linear connections*, Diferential Geometry and Its Applications **1** (1991), 177–192.

Modugno, M.; Ragionieri, R; Stefani, G., *Differential pseudoconnections and field theories*, Ann. Inst. H. Poincaré **34** (**4**) (1981), 465–493.

Modugno, M.; Stefani, G., *Some results on second tangent and cotangent spaces*, Quaderni dell' Instituto di Matematica dell' Università di Lecce **Q. 16** (1978).

Moerdijk, I.; Reyes G. E., *Rings of smooth functions and their localizations, I*, J. Algebra **99** (1986), 324-336.

Monterde, J.; Montesinos, A., *Integral curves of derivations*, Ann. Global Anal. Geom. **6** (1988), 177-189.

Montgomery, D.; Zippin, L., *Transformation groups*, J. Wiley-Interscience, New York, 1955.

Morimoto, A., *Prolongations of geometric structures*, Math. Inst. Nagoya University, Chikusa-Ku, Nagoya, Japan, 1969.

Morimoto, A., *Prolongations of connections to bundles of infinitely near points*, J. Diff. Geo. **11** (1976), 479–498.

Morrow, J., *The denseness of complete Riemannian metrics*, J. Diff. Geo. **4** (1970), 225–226.

Nagata, J., *Modern dimension theory*, North Holland, 1965.

Naymark, M. A., *Group representation theory*, (Russian), Nauka, Moscow, 76.

Newlander, A.; Nirenberg, L., *Complex analytic coordinates in almost complex manifolds*, Ann. of Math. **65** (1957), 391–404.

Nijenhuis, A., *Jacobi-type identities for bilinear differential concomitants of certain tensor fields I*, Indagationes Math. **17** (1955), 390–403.

Nijenhuis, A., *Geometric aspects of formal differential operations on tensor fields*, Proc. Internat. Congress Math. (Edinburgh, 1958), Cambridge University Press, Cambridge, 1960, pp. 463–469.

Nijenhuis, A., *Natural bundles and their general properties*, in Differential Geometry in Honor of K. Yano, Kinokuniya, Tokyo, 1972, pp. 317–334.

Nijenhuis, A., *Invariant differentiation techniques*, Geometrodynamics Proceedings, World Scientific Publishing Co., 1985, pp. 249–266.

Nijenhuis, A.; Richardson, R., *Deformation of Lie algebra structures*, J. Math. Mech. **17** (1967), 89–105.

Nomizu, K.; Ozeki, H., *The existence of complete Riemannian metrics*, Proc. AMS **12** (1961), 889–891.

Okassa, E., *Prolongements des champs de vecteurs à des variétés de points proches*, CRAS Paris, série I **300** (1985), 173–176.

Okassa, E., *Prolongement des champs de vecteurs à des variétés de points proches*, Ann. Fac. Sci. Toulouse Math. **8** (1986-1987), 349–366.

Oproiu, V., *Connections on the semiholonomic frame bundle of the second order*, Revue Roumaine Mat. Pures et Appliq. **XIV** (1969), 661–672.

Ørsted, B., *Conformally invariant differential equations and projective geometry*, J. Funct. Anal. **44** (1981), 1–23.

Palais, R. S., *A global formulation of the Lie theory of transformation groups*, Mem. AMS **22** (1957).

Palais, R., *Natural operations on differential forms*, Trans. Amer. Math. Soc. **92** (1959), 125–141.

Palais, R. S.; Terng, C. L., *Natural bundles have finite order*, Topology **16** (1977), 271–277.

Peetre, J., *Rectifications à l'article "Une caractérisation abstraite des opérateurs différentiels"*, Math. Scand. **8** (1960), 116–120.

Pogoda, Z., *The regularity condition for some natural functors*, Supplemento ai Rendiconti del Circolo Matematico di palermo, Serie II **14** (1987), 105–115.

Pohl, F.W., *Differential geometry of higher order*, Topology **1** (1962), 169–211.

Pohl, F. W., *Connections in differential geometry of higher order*, Trans. Amer. Math. Soc. **125** (1966), 310–325.

Pradines, J., *Représentation des jets non holonomes par des morphismes vectoriels doubles sondés*, CRAS, Paris, série A **278** (1974), 1523–1526.

Pradines, J., *Fibrés vectoriels doubles symétriques et jets holonomes d'ordre 2*, CRAS, Paris, série A **278** (1974), 1557–1560.

Que, N., *Du prolongements des espaces fibrés et des structures infinitésimales*, Ann. Inst. Fourier, Grenoble **17** (1967), 157–223.

Que, N., *Sur l'espace de prolongement différentiable*, J. Differential Geometry **2** (1968), 33–40.

Quinn, F., *Ends III*, J. Diff. Geo. **17** (1982), 503–521.

Reinhart, B. L., *Differential Geometry of Foliations (The Fundamental Integrability Problem)*, Ergebnisse d. Mathematik und ihrer Grenzgebiete, Bel. 99, 2. Folge, Springer-Verlag, 1983.

Rudakov, A. N., *Irreducible representations of infinite dimensional Lie algebras of Cartan type*, (Russian), Izv. AN SSSR, Ser. Mat. **38** (1974), 835–866.

Rudakov, A. N., *Irreducible representations of infinite dimensional Lie algebras of types S and H*, (Russian), Izv. AN SSSR, Ser. Mat. **39** (1975), 496–511.

Salvioli, S., *On the theory of geometric objects*, J. Differential Geometry **7** (1972), 257–278.

Sattinger; Weaver, *Lie groups and algebras with applications to physics, geometry, and mechanics*, Applied Mathematical Sciences, Springer-Verlag, 1986.

Saunders, D. J., *The Geometry of Jet Bundles*, London Mathematical Society Lecture Notes Series 142, Cambridge, 1989.

Schicketanz H., *On derivations and cohomology of the Lie algebra of vector valued forms related to a smooth manifold*, Bull. Soc. Roy. Sci. Liége **57,6** (1988), 599-617.

Schouten, J. A., *Über Differentialkonkomitanten zweier kontravarianten Grössen*, Indagationes Math. **2** (1940), 449–452.

Schouten, J. A., *Ricci calculus*, Berlin-Göttingen, 1954.

Sekizawa, M., *Natural transformations of affine connections on manifolds to metrics on cotangent bundles*, Supplemento ai Rendiconti del Circolo Matematico di Palermo, Serie II **16** (1987), 101–108.

Sekizawa, M., *Natural transformations of vector fields on manifolds to vector fields on tangent bundles*, Tsukuba J. Math. **12** (1988), 115–128.

Sekizawa, M., *Natural transformations of symmetric affine connections on manifolds to metrics on linear frame bundle: a classification*, Mh. Math. **105** (1988), 229–243.

Shmelev, G. S., *Differential $H(2n,m)$-invariant operators and indecomposable $osp(2,2n)$-representations*, Russian, Funktsional. Anal. i Prilozh. **17** (1983), 94-95.

Slovák, J., *Smooth structures on fibre jet spaces*, Czech. Math. J. **36** (1986), 358–375.

Slovák, J., *Prolongations of connections and sprays with respect to Weil functors*, Proceedings of the Winter School on Geometry and Physics, Srní 1986, Suppl. Rendiconti Circolo Mat. Palermo, Serie II **14** (1987).

Slovák, J., *On the finite order of some operators*, Proceedings of the Conference on Differential Geometry and its Applications, Communications, Brno 1986, D. Reidel, 1987, pp. 283–294.

Slovák, J., *Peetre theorem for nonlinear operators*, Ann. Global Anal. Geom. **6** (1988), 273–283.

Slovák, J., *On natural connections on Riemannian Manifolds*, Comment. Math. Univ. Carolinae **30** (1989), 389–393.

Slovák, J., *Action of jet groups on manifolds*, Proceeding of Conference on Differential Geometry and Applications, Brno 1989, World Scientific, Singapure, 1990, pp. 178–186.

Slovák, J., *Bundle functors on fibred manifolds*, Ann. Global Anal. Geom. **9** (1991), 129-143.

Slovák, J., *On invariant operations on a manifold with connection or metric*, to appear in J. Diff. Geom. (1992).

Slovák, J., *Invariant operators on pseudo-Riemannian manifolds*, to appear in Comment. Math. Univ. Carolinae **33** (1992).
Shtern, A. I.; Zhelobenko, D. P., *Representations of Lie groups*, Russian, Nauka, Moscow, 83.
Stefan, P., *Accessible sets, orbits and foliations with singularities*, Proc. London Math. Soc. **29** (1974), 699–713.
Stredder, P., *Natural differential operators on Riemannian manifolds and representations of the orthogonal and special orthogonal group*, J. Diff. Geom. **10** (1975), 647–660.
Sussman, H. J., *Orbits of families of vector fields and integrability of distributions*, Trans. AMS **180** (1973), 171-188.
Szybiak, A., *Covariant differentiation of geometric objects*, Dissertationes Mathematicae, LVI, Warszawa (1967).
Terng C. L., *Natural vector bundles and natural differential operators*, American J. of Math. **100** (1978), 775–828.
Tong, Van Duc, *Sur la géométrie différentielle des fibrés vectoriels*, Kódai Math. Sem. Rep. **26** (1975), 349–408.
Tougeron, J. C., *Idéaux de fonctions différentiables*, Ergebnisse 71, Springer-Verlag, 1972.
Trautman, A., *Invariance of Lagrangian systems, General Relativity*, Papers in honour of J. L. Synge, Clarenden Press, Oxford, 1972, pp. 85–99.
van Strien, S., *Unicity of the Lie product*, Comp. Math. **40** (1980), 79–85.
Varadarajan, V.S., *Lie groups, Lie algebras, and their representations*, 2-nd edition, Springer-Verlag, New York, 1984.
Veblen, O., *Differential invariants and geometry*, Atti del Congr. Int. Mat., Bologna, 1928.
Virsik, J., *A generalized point of view to higher order connections on fibre bundles*, Czechoslovak Math. J. **19** (1969), 110–142.
Weil, André, *Théorie des points proches sur les variétés differentielles*, Colloque de topologie et géométrie différentielle, Strasbourg, 1953, pp. 111–117.
Weyl, H., *The classical groups*, Princeton University Press, Princeton, 1946.
White, J. E., *The methods of iterated tangents with applications to local Riemannian geometry*, Pitman Press, 1982.
Whitney, H., *Analytic extensions of differentiable functions defined on closed sets*, Trans. Am. Math. Soc. **36** (1934), 63–89.
Whitney, H., *The selfintersections of a smooth n-manifold in 2n-space*, Annals of Math. **45** (1944), 220–293.
Wolf, J. A., *Spaces of constant curvature*, McGraw-Hill, New York, 1967.
Wünsch, V., *On conformally invariant differential operators*, Math. Nachr. **129** (1986), 269–281.
Yamabe, H., *On an arcwise connected subgroup of a Lie group*, Osaka Math. J. **2** (1950), 13-14.
Yano, K.; Ishihara, S., *Tangent and cotangent bundles*, Marcel Dekker, Inc., New York, 1973.
Yuen, D. C., *Higher order frames and linear connections*, Cahiers Topol. Géom. Diff. **12** (1971), 333–371.
Zajtz, A., *A classification of natural bundles*, Colloquia Matematica Societatis János Bolyai, 46, Topics in Differential Geometry,, Debrecen (Hungary), 1984, pp. 1333–1357.
Zajtz. A., *The sharp upper bound on the order of natural bundles of given dimensions*, Bull. Soc. Math. Belg. **Ser B39** (1987), 347–357.
Zajtz, A., *Geometric objects with finitely determined germs*, Ann. Polon. Math. **49** (1988), 157–168.
Zajtz, A., *On the order of natural operators and liftings*, Ann. Polon. Math. **49** (1988), 169–178.
Zajtz, A., *Foundations of differential geometry of natural bundles*, to appear.

List of symbols

1_j the multi index with j-th component one and all others zero, 13.2
$\alpha: J^r(M,N) \to M$ the source mapping of jets, 12.2
$\beta: J^r(M,N) \to N$ the target mapping of jets, 12.2
$B: \mathcal{FM} \to \mathcal{M}f$ the base functor, 2.20
$C^\infty E$, also $C^\infty(E \to M)$ the space of smooth sections of a fiber bundle
$C^\infty(M,N)$ the space of smooth maps of M into N
$C^\infty_x(M,N)$ the space of germs at $x \in M$, 1.4
$\mathrm{conj}_a : G \to G$ the conjugation in a Lie group G by $a \in G$, 4.24
d usually the exterior derivative, 7.8
$\mathbb{D}$ the algebra of dual numbers, 37.1
$\mathbb{D}^r_n = J^r_0(\mathbb{R}^n, \mathbb{R})$ the algebra of r-jets of functions, 40.5
(E,p,M,S), also simply E usually a fiber bundle with total space E, base M, and standard fiber S, 9.1
$\mathcal{F}$ usually the flow operator of a natural bundle F, 6.19, 42.1
Fl^X_t, also $\mathrm{Fl}(t,X)$ the flow of a vector field X, 3.7
$\mathcal{FM}$ the category of fibered manifolds and fiber respecting mappings, 2.20
$\mathcal{FM}_m$ the category of fibered manifolds with m-dimensional bases and fiber respecting mappings with local diffeomorphisms as base maps, 12.16
$\mathcal{FM}_{m,n}$ the category of fibered manifolds with m-dimensional bases and n-dimensional fibers and locally invertible fiber respecting mappings, 17.1
$\mathcal{FM}^*$ the category of star bundles, 41.1
G usually a general Lie group with multiplication $\mu : G \times G \to G$, left translation λ, and right translation ρ
$\mathfrak{g}$ the Lie algebra of a Lie group G
G^r_m the jet group (differential group) of order r in dimension m, 12.6
$G^r_{m,n}$ the jet group of order r of the category $\mathcal{FM}_{m,n}$, 18.8
$GL(n)$ the general linear group in dimension n with real coefficients, 4.30
$GL(\mathbb{R}^n, E)$ the linear frame bundle of a vector bundle E, 10.11
$\mathbb{I}_k$ short for the $k \times k$-identity matrix $\mathrm{Id}_{\mathbb{R}^k}$
$\mathrm{inv}J^r(M,N)$ the bundle of invertible r-jets of M into N, 12.3
J^rE the bundle of r-jets of local sections of a fiber bundle $E \to M$, 12.16
$J^r(M,N)$ the bundle of r-jets of smooth functions from M to N, 12.2
$j^rf(x)$, also j^r_xf the r-jet of a mapping f at x, 12.2
K^r_n the functor of (n,r)-contact elements, 12.15
$\mathcal{L}$ the Lie derivative, 6.15, 47.4
$\ell: G \times S \to S$ usually a left action of a Lie group
$L(V,W)$ the space of all linear maps of vector space V into a vector space W
$LP = P[\mathfrak{g}, \mathrm{Ad}]$ the adjoint bundle of principal bundle $P(M,G)$, 17.6
L^r the r-th order skeleton of the category $\mathcal{M}f$, 12.6
M usually a (base) manifold
$\mathcal{M}f$ the category of manifolds and smooth mappings, 1.2
$\mathcal{M}f_m$ the category of m-dimensional manifolds and local diffeomorphisms, 6.14

$\mathbb{N}$ natural numbers
$\Omega^k(M)$ the space of k-forms on a manifold M, 7.4
$\Omega^k(M,E)$ the space of E-valued k-forms, 7.11
$P(M,G)$, also (P,p,M,G) a principal fiber bundle with structure group G, 10.6
$P[S,\ell]$, also $P[S]$ the associated bundle to a principal bundle $P(M,G)$ for the action $\ell\colon G\times S\to S$, 10.7
$\mathcal{PB}$ the category of principal fiber bundles, 10.6
$\mathcal{PB}_m$ the category of principal bundles over m-dimensional manifolds and of $\mathcal{PB}$-morphisms covering local diffeomorphisms, 17.4
$\mathcal{PB}(G)$ the category of principal G-bundles, 10.6
$\mathcal{PB}_m(G)$ the category of principal G-bundles over m-dimensional manifolds and local isomorphisms, 15.1
$P^rM = \text{inv}J^r_0(\mathbb{R}^{\dim M}, M)$ the r-th order frame bundle of a manifold M, 12.12
$\pi^r_s : J^r(M,N)\to J^s(M,N)$ projection of r-jets into s-jets, $s\le r$, 12.2
QP the connection bundle of a principal bundle P, 17.4
$Q_\tau P^1M$ the bundle of torsion free linear connections, 25.3
$\mathbb{R}$ real numbers
$r\colon P\times G\to P$ usually a right action, in particular the principal right action of a principal bundle
TM the tangent bundle of a manifold M with projection $\pi_M : TM\to M$ 1.7
$T^{(r)}M = J^r(M,\mathbb{R})^*_0$ the r-order tangent bundle, 12.14
$T^r_k = J^r_0(\mathbb{R}^k,\ )$ the functor of (k,r)-velocities, 12.8
T_A the Weil functor corresponding to the Weil algebra A, 35.11
t_x usually the translation $\mathbb{R}^m\to\mathbb{R}^m$, $y\mapsto y+x$
$\mathcal{VB}$ the category of vector bundles, 6.3
W^r_mG the (m,r)-principal prolongation of a Lie group G, 15.2
W^rP the r-th principal prolongation of a principal bundle P, 15.3
$\mathfrak{X}(M)$ the set of all vector fields on a manifold M, 3.1
$Y\to M$ usually a fibered manifold
$\mathbb{Z}$ integers

Author index

Index

L

M

N

O

P

Q

R

S